T5-AFQ-891

SIMULATION IN
ENGINEERING SCIENCES

organized by
AFCET
and
ENSM

sponsored by
IFAC
(Applications Committee)

with the financial aid of
MATRA
Electricité de France
Centre National de la Recherche Scientifique
Conseil Général de Loire Atlantique
Université de Nantes
Ville de Nantes

NORTH-HOLLAND
AMSTERDAM · NEW YORK · OXFORD

SIMULATION IN ENGINEERING SCIENCES

Applications to the Automatic Control of Mechanical and Energy Systems

---

Proceedings of the IMACS International Symposium
Nantes, France, 9-11 May, 1983

edited by

J. BURGER

and

Y. JARNY

Ecole Nationale Supérieure de Mécanique
Nantes, France

1983

NORTH-HOLLAND
AMSTERDAM · NEW YORK · OXFORD

ISBN: 0 444 86795 3

Published by:
ELSEVIER SCIENCE PUBLISHERS B.V.
P.O. Box 1991
1000 BZ Amsterdam
The Netherlands

Sole distributors for the U.S.A. and Canada:
ELSEVIER SCIENCE PUBLISHERS COMPANY, INC.
52 Vanderbilt Avenue
New York, N.Y. 10017
U.S.A.

Library of Congress Cataloging in Publication Data

Main entry under title:

Simulation in engineering sciences.

1. Automatic control--Mathematical models--Congresses. 2. Automatic control--Data processing--Congresses. 3. Computer simulation--Congresses. I. Burger, J. (Jacques), 1942- . II. Jarny, Y. (Yvon), 1950- . III. International Association for Mathematics and Computers in Simulation.
TJ212.2.S55 1983 629.8'312 83-16557
ISBN 0-444-86795-3

PRINTED IN THE NETHERLANDS

PREFACE

These Proceedings contain the texts of the papers presented at the IMACS Symposium SIMULATION IN ENGINEERING SCIENCES, which was held in Nantes, France, on 9-11 May 1983.

The International Program Committee accepted about 60 papers among 120 projects that had been presented.

The various chapters of this volume take into account the topics used at the sessions of the Symposium.

The 4 invited survey papers are followed by the contributions related to simulation methods and tools, studied from both the software and the hardware points of view. Then a great number of papers are devoted to technical applications, i.e. to the simulation of actual processes. The contributions concerning mainly modelling are not separated from the others because of the strong connection between modelling and simulation.

The Committee has put particular emphasis on papers related to mechanical systems, vehicles and robots, considering the high interest of a great many people in the subject.

Distributing the communications into the various sessions proved to be rather difficult and this arrangement reflects the editors' personal opinion.

J. Burger – Y. Jarny
May '83

INTERNATIONAL PROGRAMME COMMITTEE

Chairman: R. Vichnevetsky (USA)
Vice Chairman: R. Mezencev (F)

Members:

D.P. Atherton (GB
C. Brebbia (GB)
H. Chareton (F)
M. Charpentier (F)
G. Davoust (F)
K. Furuta (F)
G. Gilles (F)
A. Javor (H)
R. Kulikowski (PL)
H. Kusnetsov (USSR)
G.G. Leininger (USA)
L. Le Letty (F)
J.L. Lions (F)
C. Maffezzoni (I)
M. Mansour (CH)
M. Matausek (YU)
A. Munack (D)
J.C. Nihoul (B)
J. O'Shea
M. Roseau (F)
A. Sydow (GDR)
I. Troch (A)
P. van Remoortere (B)
P. Wahlstrom (SF)

Secretary: J. Burger

NATIONAL ORGANIZING COMMITTEE

Chairman: R. Mezencev

Members:

J.P. Babary
G. Davoust
G. Gilles
L. Le Letty

Secretary: Y. Jarny
Secretariat: E. Fayola

LIST OF CONTENTS

INVITED PAPERS

Simulation Software: Today and Tomorrow
F.E. CELLIER 3

Parallel Processors in Engineering Sciences
L. DEKKER 21

Comparaison des Optimums Energétique, Economique et Ecologique d'un Système Industriel
P. LE GOFF 33

Fourier Methods in Computational Fluid and Field Dynamics
R. VICHNEVETSKY 41

I. SIMULATION TOOLS – Software

A Language for Real Time Simulation of Processes with Boolean Inputs and Outputs
E.F. CAMACHO, L.G. FRANQUELO and J. LOZANO 55

Computer Aided Modelling of Complex Processes: A Program Package
L. MARCOCCI and S. SPELTA 61

Flexible Software Package for Railcar Design
R.C. WHITE and A.A. MERABET 67

CATPAC: A Software Package for Computer-Aided Control Engineering
M. BARTHELMES, P. BRESSLER, D. BÜNZ, K. GÜTSCHOW, J. HEEGER and H.-J. LEMKE 73

II. SIMULATION METHODS – NUMERICAL METHODS

Simulation of On-Line State Estimation for Distributed Dynamic Systems
A. MASLOWSKI 81

The Use of Numerical Simulation to Verify the Efficiency of New PWM Strategies
for the Feedback Control of a D.C. Motor Drive
L. FORTUNA, A. GALLO and M. LA CAVA 87

Differential Dynamic Programming – Implementation of Algorithms and Applications
J. LOPEZ-CORONADA and L. LE LETTY 93

Linear Approximation of Nonlinear Systems Based on Least Squares Methods
J.G. DEN HOLLANDER, J.A. HOOGSTRATEN and G.A.J. VAN DE MOESDIJK 103

Simulation of Engineering Problems Using Boundary Elements
C.A. BREBBIA 109

III. SIMULATION TOOLS – Hardware

Graphic Model Building System –GMBS–
Y. YAMAMOTO and M. LENNGREN 121

The Sole Simulation Package in PASCAL
J.E. ROODA and S. JOOSTEN 127

UNISYS – Computer Assisted Modelling and Simulation System
K.A. GRABOWIECKI 133

Dynamic System Simulation in Designing Computer Peripherals
M.H. DOST 139

IV. SIMULATION AND CONTROL METHODS AND TECHNIQUES

Applying a Simulation Tool to the Design of Warped Surfaces: Notion of Interpolating Process
R. HAJ NASSAR, D. MEIZEL and P. BIELEC 147

Simulation – Aid to Process Interaction
D. DE BUYSER, L. DE WAEL and G.C. VANSTEENKISTE 153

Digital Simulation of Two Level Suboptimal Control Systems
A. ŚWIERNIAK 159

Mean Square Stability and Feedback Control of a Discrete Macro Economic Dynamic System with Small Stochastic Perturbations in Parameters
M.E. ELARABY and A.A. EL-KADER 165

V. BIOLOGICAL SYSTEMS

Résolution, Identification d'un Modèle Stochastique de Transit Cérébral
M. SIBONY 173

Neurodynamical Simulation of Peristaltic Movement
Y. UMETANI and N. INOU 179

Simulation sur Micro-Ordinateur de Signaux Electro-Physiologiques
D. GITTON and C. DONCARLI 185

VI. SIMULATION AND CONTROL METHODS AND TECHNIQUES

Controllability and Observability of Linear Time Delay Systems
V.M. MARCHENKO 193

Fractional Order Position Control and Comparison with the Adaptive Control
A. OUSTALOUP 199

NETSY: Network System Simulator (From NETSY to NESSY – A Case-Study)
D. AMBRÓZY 207

Numerical Simulation of the Control of the Space-Time Thermal Profile in an Epitaxial Reactor
J. SOUZA LEAO, J.P. GOUYON and J.P. BABARY 213

Simulation of Systems with Mobile Power Sources
E.P. CHUBAROV and V.A. KUBYSHKIN 219

VII. ENERGY SYSTEMS

Modelling and Control of an Underground Gasification Well
R. GOREZ and C. DUQUÉ 227

Modelling Analysis of Solar Energy Systems
L. FORTUNA, A. GALLO, M. LA CAVA and A. BARBARINO 233

Simulation of the Fuel Elements, for their Optimization
N. DĂNILĂ, N. STAN and D.C. IONESCU 239

Optimization of Operating Policies in Hydro Electric Power Systems
H. HABIBOLLAHZADEH, J. BUBENKO, D. SJELVGREN, N. ANDERSSON and G. LARSSON 249

VIII. MECHANICAL SYSTEMS

Simulation and Control of a Nonlinear Electromagnetic Suspension System
P.K. SINHA and A.J. HULME 257

Hybrid Simulator Assisted Design of a Feed-back Control to Resonance for a Vibrating Soil Compactor
G. GARCIN, M. GUESDON and F. DEGRAEVE 263

Simulation of Flexible Structures with Multiple Modes for a White Noise Input
V.F. POTERASU, M. DI PAOLA and G. MUSCOLINO 271

IX. VEHICLES – TRANSPORTATION (1)

The Modelling and Control of RC Helicopter
K. FURUTA, Y. OHYAMA and O. YAMANO 279

The Design of an On-Board Look-Ahead-Simulation for Approach
J. RIEPE 285

Simulation and Control of a Long Freight Train
M.E. AHMED and M.M. BAYOUMI 291

X. MECHANICAL SYSTEMS

A Dynamic Simulation of Water Supply Systems and its Application to Advanced Control Systems
S. KOBAYASHI, S. SAITO, Y. ANBE and T. ARAKAWA 299

Mathematical Model for the Solidification of Metal Casting in a Mold
A. BERMÚDEZ and J. DURANY 305

Computer-Simulation of an Experimental Method for Notch-Shape-Optimization
E. SCHNACK 311

XI. CHEMICAL SYSTEMS

Numerical Simulation of the Control System of a Nitrogen Gasification Plant
M. METZGER 319

Modelling and Simulation of a Catalytic Reformer Plant
M. PERROUD and G. BORNARD 325

Importance of Simulation in Deep Bed Filtration
S. VIGNESWARAN 331

XII. MANIPULATORS – ROBOTS (1)

Simulation and Dynamical Control of a Buffer Storage System in Chemical Industry
J.M. CHARTRES, J.M. BARBEZ and D. MEIZEL 337

Computer Assisted Synthesis of an Optimal Dynamic Control for a Robot Manipulator
J.Y. GRANDIDIER, D. GIRARDEAU, M. MONSION and C. BOUVET 343

Design of a High Speed Seven-Link-Biped Stabilized by State Feedback Control
T. MITA 351

Simulation sur un Modèle de Robot Rigide à Deux Degrés de Liberté d'une Commande Non Linéaire Découplée
F. BOURNONVILLE and J. DESCUSSE 357

XIII. THERMIC SYSTEMS

Dynamic Analysis of the Room Temperature with Nonstationary Heat Transfer
M. POPOVIĆ and E. HUMO 367

Numerical Solutions of the Thermal Behaviour of an Extruder
Y. JARNY and J. BURGER 373

Optimiser-Regulator for Large Buildings or Homes
V.J. BERWAERTS, J. BROEKX, M. SEGERS, J. STEVERLINCK, G. BERWAERTS and E. CLAESEN 377

XIV. VEHICLES – TRANSPORTATION (2)

Simulation of Train Operation under Automated Control on a Metro Line
J. CASTET 387

Simulating the Road Behaviour of Private Cars
M. DEGONDE 395

Modular Programming Structure Applied to the Simulation of Non-Linear Aircraft Models
J.A. HOOGSTRATEN and G.A.J. VAN DE MOESDIJK 401

XV. MANIPULATORS – ROBOTS (2)

Validation of an Adaptive Robot Control Structure by means of Simulation
R. CHAUDET and J. O'SHEA 411

Simulation of an Electropneumatic Drive for Industrial Robots
K. DESOYER, P. KOPACEK and I. TROCH 417

XVI. ELECTRICAL AND ELECTROMECHANICAL SYSTEMS

Simulation of Piecewise Linear Structured Electric Circuits – Petri-Net Representation of Semi-Conductors Functioning in Commutation
F. BORDRY, H. FOCH and M. METZ 425

Non Linear Identification, Application to an Electromechanical System
M. GAUTIER 431

AUTHOR INDEX 437

ACKNOWLEDGEMENTS

The editors wish to thank the members of the International Program Committee for their help in working out the program, in diffusing the call for papers and in selecting these.

We also thank the AFCET which assumed the scientific secretariat and the financial control of the Symposium. Mrs. E. Fayola has been particularly helpful with her experience in organzing congresses and we do thank her.

The editors are very grateful to the following organizations for their valuable finanical aid:

MATRA
ELECTRICITE DE FRANCE
CENTRE NATIONAL DE LA RECHERCHE SCIENTIFIQUE
CONSEIL GENERAL DE LOIRE ATLANTIQUE
UNIVERSITE DE NANTES
VILLE DE NANTES

AFCET

AFCET is a non-profit scientific organization concerned with Computer Science, Automatic Control and Operations Research. AFCET is aimed at promoting the various sciences and techniques used in managing and controlling activities of complex organizations.

AFCET gathers experts who are attentive to every progress made in those fields. It ensures the dissemination of all related information while looking for a proper insertion of those sciences and techniques in our society.

Naturally maintaining the French scientific community in contact with similar foreign societies, AFCET is also intended to be a priviledged interlocutor with government agencies and other organizations when impartial professional advice is required.

AFCET, 156, Bld. Péreire, F-75017 Paris, France

IMACS

IMACS is an international association of professionals and scientists concerned with computing with special emphasis on the simulation of systems. This includes mathematical modelling, numerical analysis, approximation theory, computer hardware and software, programming languages and compilers. Applications are concerned with the engineering, physical, social, life and environmental sciences.

IMACS, c/o Professor P. van Remoortere (General Secretary), E.R.M. Electricifé, Av. de la Renaissance, 30, B-1040 Brussels, Belgium

INVITED PAPERS

Simulation in Engineering Sciences
J. Burger and Y. Jarny (eds.)
Elsevier Science Publishers B.V. (North-Holland)

SIMULATION SOFTWARE: TODAY AND TOMORROW

Francois E. Cellier

Institute for Automatic Control
The Swiss Federal Institute of Technology Zurich
ETH - Zentrum
CH-8092 Zurich
Switzerland

This paper describes briefly the current situation on the simulation software market. A list of simulation software features is presented which is then graphed in tabular form versus a couple of current simulation languages and packages. In a second part, some of the major shortcomings of current simulation systems are outlined, and some prospectives for development are given.

1. INTRODUCTION

Eight years ago I have been asked already once to survey the numerical techniques used in continuous simulation together with the major software systems which existed at that time for this purpose [8]. When I was now asked once more to repeat this task, I tried to figure out whether our knowledge about simulation techniques has sufficiently advanced over the past eight years to justify a reconsideration. I then came to the conclusion that most of the "prospectives for development" mentioned in that paper had meanwhile become everydays state-of-the-art issues, while most of the software systems considered at that time are meanwhile obsolete. Moreover, I have some new ideas about future development of simulation software which were not present yet in 1975. Therefore, I considered the time come to write another survey now.

In this paper, I shall not review the basic features (such as numerical integration) dealt with in my previous survey [8]. I shall assume that the reader of this article has already acquired a basic knowledge of the functioning of simulation software. Moreover, I shall extend my view to discrete simulation as well, as it was realized in the mean time that the techniques used in these two classes of simulation systems are very much related to each other, and as there exists now a considerable number of software systems capable of performing combined continuous and discrete simulation.

There are meanwhile so many simulation software systems on the market that it has become impossible to review even the major ones within a reasonably limited number of pages. For this reason, the mentioned software systems (which are those that I know best) are meant to be respresentative for many other software systems showing only minor differences.

This paper shall basically consist of three parts. In a first section, I shall try to present a crude clustering of simulation software systems based on a selection of classifying features. In a second part, I shall then list a few simulation systems which are either available on the market now or which are currently under development. The paper shall then be concluded with a list of proposals how some of the more important shortcomings of the available systems might be overcome in the future.

2. SIMULATION SOFTWARE FEATURES

In the following table, a rather large set of characterizing features of nowadays simulation software systems are listed. Altogether 15 software systems have been analysed with respect to the availability of these features. If a feature is not available in a system, this is indicated by (-). If the feature is available, this is marked by (x). In many cases however, the implementation of a feature being present in several software systems differs with respect to the degree of sophistication (e.g. comfort of usage). In such cases, the better implementation is indicated by (xx). Sometimes, a foot note is added to explain the difference. Please note that a (-) does not necessarily indicate that the respective feature is not programmable in that language. It just means that no particular provision is taken by the language for that purpose. To cite an example: it is of course possible to receive a histogram by use of SIMULA (as SIMULA is a flexible general purpose programming language). Still, the respective box in the table is marked by (-) as no particular provision is taken by SIMULA to relieve the user from writing his own program to collect statistics and get a histogram printed.

Features: \ Languages:	ACSL	DARE-P	SIMNON	DYMOLA	SYSMOD	FORSIM-6	SIMULA	PROSIM	SIMSCRIPT	GPSS-F-2	GPSS-F-3	SLAM-2	GASP-5	GASP-6	COSY
EXPRESSIVENESS OF THE LANGUAGE															
Continuous Systems															
Ordinary Differential Equations	xx	x	x	xx	xx	x	-	x	-	-	x	x	x	x	xx
Partial Differential Equations	x1	x2	-	-	-	xx	-	-	-	-	-	-	x3	xx	xx
Difference Equations	x	-	xx4	-	x	-	x	x	x	x	x	x	x	x	x
Discrete Systems															
Event Handling	x	-	-	-	x	-	x	x	xx	xx	xx	xx	xx	xx	xx
Process Interaction	-	-	-	-	-	-	xx	xx	-	xx	xx	xx	-	xx	xx
Activity Scanning	-	-	-	-	-	-	xx	xx	-	xx	xx	xx	-	xx	xx
Symbolic Library (at Source Level)	xx	-	-	xx	xx	-	-	-	-	-	-	-	-	-	xx
Run-Time Library															
for Continuous Systems	xx	x	x	x	xx	x	-	x	-	-	x	x	xx	xx	xx
for Discrete Systems	-	-	-	-	x	-	x	x	x	xx	xx	xx	xx	xx	xx
Data Handling (Data-Base Management)	-	x	-	-	x	-	-	-	-	-	x	xx5	x	x	x
NUMERICAL BEHAVIOR															
Integration															
Library of Integration Methodes	x	xx	x	x	x	x	-	-	-	-	x	-	x	x	x
Own Integration Method	-	xx	-	-	-	-	-	-	-	-	x	-	x	x	x
Automatic Selection of Method	-	-	-	-	-	-	-	-	-	-	-	-	x6	x6	x6
Partitioning of State-Space															
Manual Partitioning	xx	-	x	x	xx	-	-	-	-	-	x	-	-	-	-
Automatic Partitioning	-	-	-	-	-	xx7	-	-	-	-	-	-	-	-	-
Algebraic Loop Solver	xx	-	-	-	xx	xx	-	-	-	-	-	-	x	x	xx
Root Solver (State Events)	x	-	-	-	xx	x	-	-	-	-	x	x	xx	xx	xx
Steady-State Solver	x	-	-	-	-	-	-	-	-	-	-	-	-	-	-
Tracking Problems	x	-	-	-	-	-	-	-	-	-	-	-	-	-	-
Integral-Differential Equations	-	-	-	-	-	-	-	-	-	-	-	-	-	-	-
Sparse Linear System Solver	-	-	-	-	-	x	-	-	-	-	-	-	-	x	x
Stiff Systems	xx	xx	x	x	xx	xx	-	-	-	-	x	-	xx	xx	xx
Highly Oscillatory Systems	-	-	-	-	-	-	-	-	-	-	-	-	-	-	-
Linear Systems (Special Method)	-	-	-	-	-	-	-	-	-	-	-	-	-	-	-
Noisy Systems	x	x	x	-	x	x	-	x	-	-	x	x	x	x	x
Partial Differential Equations (PDE)	x	x	-	-	-	xx	-	-	-	-	-	-	x	xx	xx
Spatial Derivatives Computation															
One-Dimensional Derivatives	-	x	-	-	-	xx	-	-	-	-	-	-	xx	xx	xx
Two-Dimensional Derivatives	-	-	-	-	-	xx	-	-	-	-	-	-	-	xx	xx

1) Formulation through vector-integration and interpretive macros.
2) By use of the REPEAT-operator.
3) As GASP-V is restricted to 100 ODE's, only PDE's in one space dimension can be formulated.
4) Special operator. All other systems treat difference equations by means of time events.
5) By use of a special simulation data language (SDL) which is available through Pritsker & Assoc.
6) Provided to date only in an experimental version which is not yet made available.
7) Not in the public version of FORSIM-VI, but well documented, tested and available from AECL.

Features: \ Languages:	ACSL	DARE-P	SIMNON	DYMOLA	SYSMOD	FORSIM-6	SIMULA	PROSIM	SIMSCRIPT	GPSS-F-2	GPSS-F-3	SLAM-2	GASP-5	GASP-6	COSY
Three-Dimensional Derivatives	-	-	-	-	-	xx	-	-	-	-	-	-	-	xx	xx
Library of Methods	-	-	-	-	-	xx	-	-	-	-	-	-	x	xx	xx
Spline Interpolation	-	-	-	-	-	x	-	-	-	-	-	-	-	x	x
Up-wind Interpolation	-	-	-	-	-	x	-	-	-	-	-	-	-	x	x
Own Interpolation Method	-	-	-	-	-	-	-	-	-	-	-	-	-	-	-
Automatic Selection of Method	-	-	-	-	-	-	-	-	-	-	-	-	-	-	-
Variable Grid Width	-	-	-	-	-	-	-	-	-	-	-	-	-	-	-
Automatic Grid Width Control	-	-	-	-	-	-	-	-	-	-	-	-	-	-	-
Error Estimation	-	-	-	-	-	-	-	-	-	-	-	-	-	-	-
Parabolic PDE's	x	x	-	-	-	xx	-	-	-	-	-	-	x	xx	xx
Hyperbolic PDE's	x	x	-	-	-	x	-	-	-	-	-	-	x	x	x
Ellyptic PDE's	x	-	-	-	-	x	-	-	-	-	-	-	-	x	x
Shock Waves	-	-	-	-	-	-	-	-	-	-	-	-	-	-	-
Mixed PDE's and ODE's	x	x	-	-	-	xx	-	-	-	-	-	-	xx	xx	xx
Statistical Analysis															
Random Number Generation	x	x	-	-	x	-	x	x	xx	x	xx	xx	xx	xx	xx
Distribution Function Library	-	-	-	-	-	-	x	-	xx	x	x	xx	xx	xx	xx
Tabular Distribution Functions	-	-	-	-	-	-	-	-	x	x	x	x	x	x	x
Distr. Funct. Parameter Fitting	-	-	-	-	-	-	-	-	-	-	-	xx1	-	-	-
Statistical Report Generation															
Sampling Statistics	-	-	-	-	-	-	-	x	xx	xx	xx	xx	xx	xx	xx
Time-Persistant Statistics	-	-	-	-	-	-	-	-	xx	xx	xx	xx	xx	xx	xx
Histograms	-	-	-	-	-	-	-	xx	-	xx	xx	xx	xx	xx	xx
Confidence Intervals	-	-	-	-	-	-	-	-	-	x	x	xx1	-	-	-
Variance Estimation															
Variance of the Mean	-	-	-	-	-	-	-	-	-	x	x	x	x	x	x
Run-Length Determination	-	-	-	-	-	-	-	-	-	x	x	-	-	-	-
Transient Period Duration	-	-	-	-	-	-	-	-	-	x	x	-	-	-	-
Significance Tests	-	-	-	-	-	-	-	-	-	x	x	xx1	-	-	-
Variance Reduction															
Replication and Batch	-	-	-	-	-	-	-	-	-	-	-	-	-	-	-
Subinterval Analysis	-	-	-	-	-	-	-	-	-	-	-	-	-	-	-
Sensitivity Analysis															
Linear Approximation (Metamodel)	-	-	-	-	-	-	-	-	-	-	-	-	-	-	-
Replication	-	-	-	-	-	-	-	-	-	-	-	-	-	-	-
Table Look-up															
Two-Dimensional Tables	x	x	-	-	x	x	-	-	-	-	-	x	x	x	x
Three-Dimensional Tables	x	x	-	-	-	x	-	-	-	-	-	-	-	x	x
Linear Interpolation	x	x	-	-	x	x	-	-	-	-	-	x	x	x	x
Non-linear Interpolation	-	-	-	-	x	-	-	-	-	-	-	-	-	x	x
Spline Interpolation	-	-	-	-	x	-	-	-	-	-	-	-	-	x	x
Dynamic Table Load	-	-	-	-	-	-	-	-	-	-	-	x	x	x	x
Sequential Interpolation	-	-	-	-	x	-	-	-	-	-	-	-	-	x	x
Mass-Storage Interpolation	-	-	-	-	x	-	-	-	-	-	-	-	-	x	x

1) By use of a separate program (AID) also provided by Pritsker & Assoc.

Features: \ Languages:	ACSL	DARE-P	SIMNON	DYMOLA	SYSMOD	FORSIM-6	SIMULA	PROSIM	SIMSCRIPT	GPSS-F-2	GPSS-F-3	SLAM-2	GASP-5	GASP-6	COSY
STRUCTURAL FEATURES															
Application Program Development															
Model Structuring Capabilities															
Parallel Structures															
Continuous Systems (Sorting)	x	x	x	xx	x	-	-	-	-	-	-	-	-	-	xx
Discrete Systems (Networks)	-	-	-	-	-	-	-	-	-	x	x	xx	-	-	-
Procedural Structures															
Continuous Systems (Nosort)	xx	x	-	-	xx	x	-	x	-	-	x	x	x	x	xx
Discrete Systems (Algorithmic)	x	-	-	-	x	-	xx	xx	x	x	x	x	x	x	xx
Event Handling															
Time Events	x	-	-	-	xx	-	x	x	xx	xx	xx	xx	xx	xx	xx
State Events	x	-	-	-	xx	-	-	x	-	-	xx	xx	xx	xx	xx
External Events															
Operator Intervention	-	-	xx	xx	xx	-	-	-	-	-	-	-	-	-	-
Real-Time Interrupts	-	-	-	-	-	-	-	-	-	-	-	-	-	-	-
Process Interaction															
Continuous Processes	x	-	xx	xx	xx	-	-	-	-	-	x	-	-	xx	xx
Discrete Processes	-	-	-	-	-	-	xx	xx	-	xx	xx	xx	-	xx	xx
Network Description	-	-	-	-	-	-	-	-	-	x	x	xx	-	-	-
Activity Scanning	-	-	-	-	-	-	xx	xx	-	xx	xx	x	-	xx	xx
Submodel Definition															
Continuous Submodels	x	-	xx	xx	xx	-	-	-	-	-	x	-	-	x	xx
Discrete Submodels	-	-	-	-	-	-	xx	xx	-	xx	xx	-	-	xx	xx
Hierarchical Model Definition															
for Continuous Systems	-	-	-	xx	xx	-	-	-	-	-	-	-	-	-	xx
for Discrete Systems	-	-	-	-	-	-	xx	xx	-	-	-	-	-	-	xx
Initial Computations	x	x	x	-	x	x	-	-	x	x	x	x	x	x	x
Terminal Computations	x	x	x	-	x	x	-	-	x	x	x	x	x	x	x
Controlled Experiments	x	xx	xx	xx	xx	-	-	-	x	-	-	xx	xx	xx	xx
Optimization	-	-	-	-	-	-	-	-	-	-	-	-	-	-	xx1
Parameter Fitting (provided)	-	-	-	-	-	-	-	-	-	-	-	-	-	-	xx1
Modularity															
Macro Feature	xx	-	-	-	xx	-	-	-	-	-	-	-	-	-	xx
Interpretive Macros	xx	-	-	-	xx	-	-	-	-	-	-	-	-	-	xx
Module Feature	-	-	-	xx	-	-	-	-	-	-	-	-	-	-	xx
Graphical Model Specification	-	-	-	-	-	-	-	-	-	-	-	xx2	-	-	-
Program Validation and Verification															
Model Comparison	-	x	x	-	xx	-	-	-	-	-	-	xx3	x	x	x
Sensitivity Analysis	-	-	-	-	-	-	-	-	-	-	-	-	-	-	-
Linearized Model Analysis	-	-	-	-	-	-	-	-	-	-	-	-	-	-	-
Parameter Fitting (possible)	x	xx	xx	xx	xx	-	-	-	-	-	-	xx	xx	xx	xx
Eigenvalue- Eigenvector Analysis	x	-	-	-	-	-	-	-	-	-	-	-	-	-	-
Debugging Aids															
Dimensional Analysis	-	-	-	-	-	-	-	-	-	-	-	-	-	-	-
Declaration of Variables	-	-	x	xx	xx	-	xx	xx	x	-	-	-	-	-	xx
Steady-State Finding	x	-	-	-	-	-	-	-	-	-	-	-	-	-	-
Graphical Model Representation	-	-	-	-	-	-	-	-	-	-	-	xx2	-	-	-

1) By use of a separate run-time library (NLP) also available from ETH Zurich.
2) In an experimental version which is currently under development by Pritsker & Assoc.
3) By use of a special simulation data language (SDL) provided by Pritsker & Assoc.

Features \ Languages	ACSL	DARE-P	SIMNON	DYMOLA	SYSMOD	FORSIM-6	SIMULA	PROSIM	SIMSCRIPT	GPSS-F-2	GPSS-F-3	SLAM-2	GASP-5	GASP-6	COSY
Tracing															
Event Monitoring	-	-	-	-	xx	-	-	xx	x	x	xx	xx	x	xx	xx
Range Surveillance	-	-	-	-	-	-	-	-	-	-	-	-	-	-	xx
Deadlock Detection	-	-	-	-	-	-	-	-	-	-	-	-	-	-	-
Range Analysis	-	-	-	-	-	-	-	-	-	-	-	-	-	-	-
Program Execution															
Batch Processing	x	x	x	x	x	x	x	x	x	x	x	x	x	x	x
Interactive Operation															
Parameter Setting	x	x1	x	x	x	-	-	-	-	-	-	-	-	-	-
Parameter Tuning	-	-	-	-	-	-	-	-	-	-	-	-	-	-	-
Output Selection	x	x1	x	x	x	-	-	-	-	-	-	-	-	-	-
Run-Time Display	-	x1	-	-	-	-	-	-	-	-	-	-	-	-	-
Operator Intervention	-	-	x	x	x	-	-	-	-	-	-	-	-	-	-
Execution of Monitors	-	-	-	-	x	-	-	-	-	-	-	-	-	-	-
Real-Time Synchronization	-	-	-	-	-	-	-	-	-	-	-	-	-	-	-
Data Handling															
Library of Parametric Models	-	-	-	-	x	-	-	-	-	-	-	-	-	-	xx
Library of Parameter Sets	x	x	x	x	x	-	-	-	-	-	-	-	-	-	-
Library of Experimental Frames	x	-	xx	xx	xx	-	-	-	-	-	-	-	-	-	x
Library of Trajectories	-	xx	x	x	x	-	-	-	-	-	-	xx2	xx	xx	xx
Numerical Manipulations	-	-	-	-	-	-	-	-	-	-	-	xx2	-	-	-
Structural Manipulations	-	-	-	-	-	-	-	-	-	-	-	-	-	-	-
Input/Output															
Model Parameters	x	x	x	x	xx	x	-	-	-	x	x	-	-	-	xx
Driving Functions															
Tabular Functions	-	-	x	-	xx	-	-	-	-	-	-	-	-	-	xx
Real-Time Input	-	-	-	-	-	-	-	-	-	-	-	-	-	-	-
Graphical Model Representation	-	-	-	-	-	-	-	-	-	-	-	xx3	-	-	-
Graphical Model Specification	-	-	-	-	-	-	-	-	-	-	-	xx3	-	-	-
Storing Output Data (on Files)	-	xx	x	x	xx	-	-	-	-	x	xx	xx2	xx	xx	xx
Retrieving Input Data (from Files)	-	-	x	-	xx	-	-	-	-	-	-	-	-	xx	xx
Crossplots	x	x	x	x	x	-	-	-	-	-	-	xx4	x	x	x
Overplots	x	xx	x	x	xx	-	-	-	-	-	x	xx4	xx	xx	xx
Three-Dimensional Plots	-	x5	-	-	-	x6	-	-	-	-	-	-	x	x	x
Histograms	-	-	-	-	-	-	-	x	-	-	xx7	xx8	x	x	x
Bar Charts	-	-	-	-	-	-	-	-	-	-	-	xx4	-	-	-
Pie Graphs	-	-	-	-	-	-	-	-	-	-	-	xx4	-	-	-

1) Not in the official version provided by the University of Arizona. There exist however several interactive versions of DARE-P, one VAX version available from ETH Zurich, and one portable (but less interactive) version from the University of Arizona.
2) By use of a special simulation data language (SDL) provided by Pritsker & Assoc.
3) In an experimental vesion which is currently under development by Pritsker & Assoc.
4) By use of a separate program (SIMCHART) also provided by Pritsker & Assoc.
5) Not in the official version of the DARE-P postprocessor as provided by the University of Arizona. An enhanced postprocessor is available from ETH Zurich.
6) Only provided in an internal version from the Atomic Energy of Canada, Ltd. but not in the version handed out to external users (bases upon a local graphics software package).
7) Quality plots base on the Erlanger graphics system. Not provided for external use.
8) Quality histograms only in connection with the separate program AID from Pritsker & Assoc.

Features: \ Languages:	ACSL	DARE-P	SIMNON	DYMOLA	SYSMOD	FORSIM-6	SIMULA	PROSIM	SIMSCRIPT	GPSS-F-2	GPSS-F-3	SLAM-2	GASP-5	GASP-6	COSY
Output Quality															
Lineprinter Plot	x	x	x	x	x	x	-	x	-	x	x	x	x	x	x
Plotter output monochrome	x	x	x	x	x	x1	-	-	-	-	xx2	x3	x	x	x
Color Graphics	-	-	x	x	-	-	-	-	-	-	-	-	-	-	-
STATUS OF IMPLEMENTATION															
Compiler	x	x	x	x	-	-4	x	x	x	-4	-4	x	-4	-4	-
Run-Time System	x	x	x	-5	-	x	x	x	x	x	-6	x	x	-7	-8
Environment	-	x	x	x	-	-	-	-	-	-	-	-	-	-	-
PORTABILITY	x	xx	-	x	xx	-	-	-	-	x	x	xx	xx	xx	xx
DOCUMENTATION	xx	x	x	xx	-	x	xx	x	xx	xx	-6	xx	x	-	-
Machine Readable	-	-	-	x	-	-	-	-	-	-	-	-	-	x	x
On-Line HELP Information	-	-	x	-	-	-	-	-	-	-	-	-	-	-	-

1) Only provided in an internal version of the Atomic Energy of Canada, Ltd. but not in the version which is handed out to others (bases upon a local graphics package).
2) Quality plots base on the Erlanger graphics system. Not provided for external use.
3) By use of a separate program (SIMCHART) provided by Pritsker & Assoc.
4) These are FORTRAN packages. A compiler is therefore not required.
5) The run-time system of DYMOLA is SIMNON. A special run-time system is therefore not required.
6) The run-time system exists already at the University of Erlangen, but has not yet been released for external use. An official release is planned for late 1983 as soon as the complete documentation becomes ready.
7) A preliminary version of GASP-VI exists at ETH. This version is however not yet sufficiently debugged and consolidated to allow for external distribution.
8) The run-time system of COSY is GASP-VI. A special run-time system is therefore not required.

A first set of characteristics describes general features, e.g. whether a system is meant for purely discrete simulation (like SIMULA) or for purely continuous simulation (like DARE-P). This group is headed "expressiveness of the language".

Continuous systems may be described by several formalisms, ordinary differential equations (ODE's), partial differential equations (PDE's), or possibly difference equations. Please, note that systems described by difference equations are called _continuous_ here, although many people (e.g. most control engineers) would call them _discrete_. This is due to the fact, that the simulation methodology required for the solution of difference equation systems is much closer related to the continuous methodology than to the "real" discrete methodology, a term which I reserve for event oriented systems.

Discrete systems may either provide for a (primitive) event description only which may result in rather unreadable models when the systems are large, or for a process interaction mechanism. Activity scanning means that the duration of an activity (e.g. service time) may depend on some other parts in the system, e.g. the completion of another service performed on another transaction at the same time.

Symbolic libraries are libraries at source level, e.g. macro libraries for the description of submodels. Such source libraries provide in some sense for an open ended operator set of the language.

Data handling describes in global terms the way in which the simulation data are managed by the simulation system. I shall not further emphasis on this point now, as this shall become a central aspect of my chapter on future developments.

The next section is headed "numerical behavior", and describes many aspects of how the simulation software treats a model numerically. Four major topics here concern the numerical integration of differential equations, computation of spatial derivatives (for PDE's), statistical analysis, and table look-up.

Almost any system able to cope with ODE's provides nowadays for a library of integration algorithms. This is important as there does not exist any integration algorithm which is equally well suited for all types of ODE problems. In particular, methods for the solution of stiff systems are quite common in current simulation software. SLAM-II (which is otherwise a strong piece of software) is particularly weak in this respect as it offers just one Runge-Kutta algorithm for integration. Many continuous systems can therefore not be successfully treated by use of SLAM. By far the best library is present in DARE-P. This software offers more than a dozen different integration algorithms both in single and double precision. A special conversion routine allows to convert single precision to double precision and vice-versa, such that only one version of each algorithm needs to be stored at any time.

However, many users are demanded too much by a rich selection of routines. It would be highly welcome, if the simulation software could find out automatically which routine is best to apply in each case, and load that routine from the library. First steps towards an achievement of this goal are implemented in an experimental version of GASP-V [10]. A first commercially available integration routine which is able to switch between a stiff and a non-stiff algorithm has been described by Petzold and Hindmarsh [21]. However, this algorithm has to my knowledge not yet been implemented in any ready-to-use simulation software.

For some problems, it may be beneficial to split them up into submodels which are integrated independently by either a different step size or even a different integration algorithm. This feature is called "partitioning of the state-space". Some software systems (such as ACSL) allow for a definition of submodels, whereby each submodel may be integrated by use of a separate integration routine. Communication between submodels takes place after each communication interval only, that is: the variables of each subsystem are considered constant during each communication interval by each other subsystem. This is not necessarily optimal as: (a) the so introduced discontinuities may create serious numerical stability problems; (b) the so introduced artificial sampling may introduce analytical stability problems; and (c) the model splitting which is optimal for model structuring and readability is not necessarily also optimal for numerical integration. Some other systems (such as SIMNON) try to overcome these problems by letting the user separately specify submodels (which exist only at source level but no longer during run-time) from groups of state variables which are to be integrated together (slow subsystem versus fast subsystem). The integration method offered for that purpose in SIMNON is somewhat obscure. It works neatly with some adaptive control systems where it is quite clear that the inner control loop consists of considerably faster modi than the outer adaptation loop, but in other cases the method tends to fail. Another such algorithm has been described by Eitelberg [14]. A third (and rather promising) approach is taken by Carver who developed an automated partitioning scheme which is rather simple to use and even works well on non-linear problems [7].

Algebraic loop solvers are quite common in current simulation software. However, their use tends to be costly, as an iteration has to take place during each function evaluation. For that reason, it may be advantageous to apply an integration routine which solves the algebraic equations at once during integration. Such a code has e.g. been described by Petzold [34]. This approach shall lead away from the currently used state-space description of ODE's (that is: $x'=f(x,t)$) to the more general form: $f(x,x',t)=0$.

Root solvers are required for the location of state events, and are essential algorithms for combined continuous and discrete simulation.

Steady-state solvers allow for the computation of the steady-state at less computational cost than by simply integrating there. Such algorithms are rare. ACSL offers something, but the offered algorithm fails often in non-linear cases.

Tracking problems ask for "freezing" of state variables during simulation. This feature is currently only provided for by ACSL.

Integro-differential equations are numerically harmful, and there is no system which is currently able to deal with them successfully. Typical examples include the flow through a pipe with variable delay. Some systems (like DARE-P) claim to offer a solution which, however, mostly fails when applied.

A sparse linear system solver is very much needed for efficient integration of high dimensional problems (e.g. PDE's). The economization there may be dramatic. Still, most simulation systems do not offer this feature, though it is extremely simple to implement.

A stiff system solver is nowadays very commonly found. However, it is rarely realized that not all "difficult" problems are really "stiff". A very good discussion of this point may be found from Gear [18]. One other class of difficult problems to solve is the class of "highly oscillatory problems". Possible solutions may be by means of stroboscopic integration methods which use a low order integration scheme to integrate over a short time by use of a small step length while storing the maxima of the oscillations away, and then a superimposed higher order integration scheme which uses these maxima as supporting values to extrapolate by use of a much larger step length. One such algorithm has been described by Petzold [33]. Another possible solution to this problem may be to solve the problem x''=f(x,x',t), a representation into which many highly oscillatory systems may be transformed, by means of a Fourier rather than a Taylor series expansion. I am unaware of any such code though.

Linear systems may be more efficiently solved by special integration techniques (implicit integration) which is commonly done in network analysis programs but never in general purpose simulation software, although this feature would be very simple to implement.

Noisy systems are formulatable in all languages, but they are not necessarily also solvable, as the variable step length integration algorithms tend to fail, while fixed step algorithms produce rubbish. A (x) here means simply that there is provision for a random number generator, but nothing more than that.

PDE's are solved in simulation by the method-of-lines approach. They are converted to sets of ODE's by means of discretizing them in the space dimensions. The time dimension is kept "continuous" for integration. This methodology works well on parabolic PDE's. The integration over time should use a stiff system solver, as the discretization in the space domain creates stiff sets of ODE's. A sparse linear system solver is here very advantageous.

Hyperbolic PDE's create some headache. A step forward has been made by Carver and Hinds with their up-wind interpolation routine [5]. The idea behind this algorithm is very simple. A central interpolation scheme does not make much sense when a wave moves towards one direction only (direction of the characteristic). Such a scheme would mean to interpolate by use of the unknown future. It is then better to use a biased scheme which only requires values from one side, that is: from the past. The trick in this algorithm is to detect the direction of the characteristic automatically, a task which is much simpler to accomplish than finding the characteristics as a whole.

The Jacobian of hyperbolic PDE's tends to have its eigenvalues also wide spread. However, the resulting ODE's are not stiff, as some of the eigenvalues are complex and close to the imaginary axis. The Gear algorithm does an awful job on them. More research for adequate integration schemes is still required.

Hyperbolic PDE's moreover tend to develop "shock waves" which move through space with time. Such shock waves require an adaptive spatial grid. A first code which offers this feature has been presented recently by Schiesser [42]. It might meanwhile be implemented in either the LEANS or DSS software which are not described further in this survey, as they differ not sufficiently much from FORSIM-VI to demand for a separate discussion. An excellent survey on simulation software for PDE problems has been very recently worked out by Karplus [24].

Elliptic PDE's may be solved by means of invariant embedding techniques. This approach is mostly less efficient though than using a finite element technique. A powerful steady-state solver might make invariant embedding somewhat more attractive again.

An entirely different topic is the discussion of statistical analysis techniques. In current simulation software, these techniques are much further developed in discrete event software than in continuous software, although they may be used to a large extent also there. Some systems (such as SLAM-II) provide for about a dozen different distribution functions. Some systems allow to describe the distribution function in a tabular form to allow for even more general distributions. This feature is very useful when the distributions result from measured quantities. SLAM-II (in connection with the separate program AID) even allows to fit the statistical parameters of some distribution functions such as the BETA and GAMMA distributions from measured data.

Statistical report generation is today only available for discrete system simulation, although it may be equally important to know from a noisy continuous systems in which range the results are expected to lie rather than getting one single trajectory displayed (which may be rather at random). Statistical analysis of the results is available in few systems only. SLAM-II does it in connection with a simulation data-base system (SDL), while GPSS-FORTRAN offers a few functions for that purpose. The separation of the statistical analysis from the simulation task may on the long run be more fruitful, as I shall discuss later.

Table look-up is another critical issue. Most continuous systems provide for means of handling at least two-dimensional tables. Some others also allow to handle three-dimensional tables. Interpolation formulae are mostly linear, but some systems offer also non-linear interpolation (e.g. CSMP-III,) some others also spline interpolation. However, most systems just provide for static tables (e.g. "FUNCTION" in CSMP). This is a nice toy for school examples. Real problems (e.g. wind tunnel experiments) tend to require very huge tables which cannot be handled in this way. A minimum requirement is a dynamic table load feature (e.g. offered in CSMP-III). Even better is mass storage interpolation which allows to keep the data in the data base (where they belong), rather than to copy them into the central memory. A useful feature is also sequential interpolation which allows to interpolate between data which are cuncurrently produced (e.g. for real-time experiments).

Structural features describe the comfort with which models may be generated by use of a particular simulation software. Subheadings are: application program development, program validation and verification, program execution, data handling and I/O.

A sorting algorithm is available in most continuous simulation systems. It enables the user to specify his set of equations in virtually any sequence. DYMOLA and MODEL [39] (which is not described here as being too similar to DYMOLA to demand for separate discussion) go both a giant step further, in that they allow for the syntax: expression = expression instead of the commonly used syntax: variable = expression. DYMOLA tries to use formula manipulation to solve for the required output variable (which may be context dependent). This procedure shall be only successful if the output variable appears linearly in the equation (though all other variables may appear as non-linear as they like). MODEL keeps the equations as they are, while utilizing an integration algorithm which is able to solve the problem: f(x,x',t)=0. Thus, DYMOLA shall require more compilation but less execution time than MODEL for the solution of the same problem.

Discrete systems may be "parallelized" by means of a PERT network type description or similar. Concurrent processes are another (algorithmical rather than graphical) approach.

All discrete simulation systems offer time events, all combined systems also state events. However, there exist some more event types to be mentioned which are not commonly found in current simulation software. These are the external events. In interactive simulation, the user may wish to suspend a simulation run (e.g. by typing CTRL_C). He may then want to modify a parameter value and resume the simulation thereafter from where it was suspended. Another typical example of an external event would be a real-time interrupt.

Modularity can be guaranteed by quite different approaches. Quite common are macros (partly even rather powerful, e.g. in ACSL). However, macros imply that it is known beforehand, what are inputs and what are outputs of a macro. This is not necessarily always the case. A statement such as: U=I*R may well reappear as: I=U/R in another context. Therefore, macros are modular only in a restricted sense. A much more powerful concept is the concept of "modules" (or "models" as they are called in DYMOLA). The functioning of this mechanism is quite simple. As there exist formula manipulation routines in DYMOLA, it is immaterial, whether the above equation is coded as: U=I*R or I=U/R or even U-I*R=0. The formula solver shall produce the required form automatically depending on the context in which the equation is used. An alternative approach to the previously presented module approach consists of a graphical model specification which is e.g. provided in MODEL. (In fact, this is just another layer of sophistication.)

How are models validated, once they are defined? Obviously, there is no firm and final answer to this question. A first step may be to be able to compare the results from a simulation run to some measured data or to some data produced by another model. Both cases would require the (simulation <u>and</u> measurement) data to be stored away in one and the same data base for later reuse. Sensitivity analysis provides another means for gaining confidence into simulation results. It is possible to automate this procedure even for non-linear models. We have an ALGOL program available which produces the derivative of another ALGOL program with respect to any variable or array of variables [23]. There exists also an experimental version of a PASCAL program for the computation of derivatives of FORTRAN subroutines. Linearized model analysis is very similar. These models may, however, be of lower order. A discussion of so called metamodels can be found from Kleijnen [26]. Also eigenvalue and eigenvector analysis are useful tools to find out whether the modi of the model are within a reasonable domain. This works particularly well for linear models where modal decomposition may be applied.

Debugging of application programs seems not very in vogue today. Some systems at least ask the variables to be declared to allow for some redundancy. Most typing errors can be detected in this way. It is not necessarily true that a simulation system which can represent the Van-der-Pol equation elegantly (in terms of a short user code) is equally well suited for large scale models. Dimensional analysis (by asking the user to provide the dimensions for all of his variables) might be another way of checking correctness. Most systems provide for event monitoring (run-time check). However, other run-time checks would also make sense, like range surveillance (for large scale models) or automated deadlock detection (e.g. timeout).

Most simulation systems today are batch operated. Some systems provide for (still rather moderate) means of interactive operation. Only DARE-ELEVEN, a "dialect" of DARE-P, provides for a run-time display which allows for "on-line" surveillance of some simulation trajectories. There exist some simulation systems for real-time execution (e.g. MICRODARE, another DARE "slang",) which are however not surveyed here as they are supposed to run on very small computers and offer little to no user comfort. These systems hardly offer any of the other features (beside from the run-time synchronization), and make a discussion therefore not very profitable. The problems here are still so special and machine dependent that it is far too early for a survey.

Data handling and I/O shall be described in due course.

3. SURVEY OF EXISTING SOFTWARE

ACSL [28]: is a powerful continuous simulation language. It is also representative for another program: CSSL-IV [30] which is therefore not further discussed. Both supersede the famous and widely spread CSMP-III software [22]. A current extension to ACSL introduces state events and time events, making ACSL also usable for combined problems [29]. Still, the discrete features offered are very limited, and it is suggested to use ACSL only for "mildly" combined problems (that is: continuous problems with some discontinuities).

DARE [27,49]: stands for an entire family of simulation languages. Described in this survey is the version DARE-P which is the most portable, most powerful, and best tested dialect within the DARE family. DARE-P is certainly less powerful than ACSL, but it has also its distinct advantages. In particular, it provides for a (very primitive) data base interface which allows to compare different simulation results with each other on one sheet of paper. By separating out the postprocessor from the simulation run, DARE-P offers much more flexibility with respect to output representations. The offered portable high quality output is a real jewel. DARE-P is also particularly strong with respect to integration techniques. This software is very simple to teach and to learn, and it is available at nominal cost. The jungest "child" in the DARE family is EARLY DESIRE. An interactive version of DARE-P, offering most of the features of the previous DARE-ELEVEN software, but running on VAX under VMS, is meanwhile made available from ETH Zurich. Another DARE dialect is PSCSP [19], a real-time simulation software from ETH Zurich. There exist also multiprocessor implementations of PSCSP and of MICRODARE.

SIMNON [15,16]: is a highly interactive continuous simulation language running on VAX under VMS. Other versions exist for UNIVAC and DEC-10. A special feature of SIMNON is its notation of subsystems described by difference equations. Such subsystems may freely be mixed with other subsystems described by ODE's. In this way, SIMNON is particularly useful for control engineers simulating continuous plants with digital controllers. For model description, ACSL is in almost any respect superior to SIMNON. However, SIMNON offers a much higher degree of interactiveness than ACSL. Controlled experiments can be executed in a much broader sense, as the "MACRO" feature of SIMNON (for repetitive execution of command sequences) is much more powerful than ACSL's "PROCED" feature. SIMNON provides also for something like a primitive data base mechanism which bridges the gap to some other interactive programs, e.g. IDPAC for parameter identification (available from the same source).

DYMOLA [17]: is really a modeling language rather than a simulation language. A PASCAL coded preprocessor (there exists also a SIMULA coded version of the preprocessor) translates DYMOLA programs into either SIMNON models or FORTRAN subsystems which may be loaded and executed together with the SIMNON system. The beauty of DYMOLA lies in its modularity. Large scale models can be coded much easier and less error prone than in other languages. DYMOLA is in so far experimental, as it offers no other simulation features. That is: the user is often forced to access the precompiled SIMNON program to add some other features at that level. DYMOLA is also representative for MODEL [28] which is therefore not further discussed.

SYSMOD [1]: is basically a workable subset of COSY (discussed further down). The main reason to include this software in this review lies in the assumption that SYSMOD shall be sooner available and better (because industrially) maintained than COSY. Moreover, SYSMOD has also some nice extensions. SYSMOD is certainly the simulation language with the nicest features for description of experimental frames. Controlled experiments can therefore be expressed particularly nicely in SYSMOD. Also the run-time system is improved. Submodels can be separately compiled but nevertheless be jointly integrated. SYSMOD is predominantly a continuous simulation software. There exist discrete events and waiting queues, but no process mechanism is foreseen. SYSMOD is meant for large scale models (with declaration of variables, good structuring capabilities), and is also designed to digest large amount of data (e.g. measurements from wind tunnel experiments or similar). SYSMOD is currently under development by a British Company (Systems Designers, Ltd.) under contract from the British Ministry of Defence. Current plans are to release this software in 1984.

FORSIM [6]: is a continuous simulation software primarily designed for the solution of PDE problems (by use of the method-of-lines approach). Parabolic PDE's can be solved efficiently in one, two or three space dimensions. A very good implementation of the Gear stiff system solver together with use of the Reid sparse linear matrix routines makes FORSIM-VI a very powerful tool for that purpose. Both the integration method as the spatial discretization method are parameterized. The user can select among a large variety of different algorithms by simply changing one single parameter. In this way, FORSIM-VI is also a very nice experimentation tool, in that a large variety of alternative algorithms can be tested out by minor program modifications. Hyperbolic PDE's can be solved to a lesser extent. Up-wind interpolation implements a "pseudo-characteristic method", but more research is still required here to find better suited integration algorithms and adaptive spatial grids (for shock wave treatment). Elliptic PDE's may be solved to a still lower extent. Finite element methods are far more efficient here for most applications. FORSIM-VI is also representative for some other systems such as LEANS-III [41] or DSS [50] which are therefore not further discussed here.

SIMULA [2]: is a powerful programming language. Its strength lies in the fact that virtually any problem can be solved in a highly structured way. SIMULA'67 is a very good language to implement compilers. As a simulation language, SIMULA offers far too little simulation specific support though. Looking into our table, SIMULA must leave the impression of being a very poor simulation language. In its basic version, SIMULA may be used for the solution of discrete event problems only, but even for that purpose, the available support is minimal. The powerful CLASS concept of SIMULA provides for a virtually unlimited open-ended operator set. No wonder therefore, that there exist several "extensions" to SIMULA (which are basically collections of precut SIMULA classes) to make SIMULA better usable for simulation purposes. One such extension is DEMOS [3] which adds to SIMULA a transaction flow view (comparable to GPSS), tabular distribution functions, statistical report generation (including histograms), event monitoring, and automated deadlock detection (a feature which is rarely found in today's simulation software). Another extension is DISCO [20] which adds some primitives for ODE solution, making DISCO a combined simulation language (although the continuous aspects of DISCO are by far insufficient for complex continuous applications). The class concept of SIMULA is possibly not optimal for maintaining continuous attributes in a user friendly way. Another extension, COSMOS (which is currently under development by Kettenis from the Agricultural University of Wageningen, The Netherlands), goes therefore another way by implementing a (SIMULA coded) preprocessor which translates COSMOS programs down to an intermediate SIMULA code.

PROSIM [45]: is another SIMULA dialect. Its implementation is such that all SIMULA features have been reimplemented in a PL/I environment. PROSIM offers also some combined features. Sierenberg has implemented ODE's in a rather original way, in that the ODE's in PROSIM are not implemented as code, but rather as a data structure. The specification of an ODE in PROSIM is done by declaring a continuous attribute. This genuine solution looks very interesting. It is, however, rather difficult to use in more complex application, as logical connections between different continuous equations are almost unexpressible. All continuous problems which would require somewhere a NOSORT section are almost uncodeable in PROSIM (and these are, to my experience, almost all realistically large problems). PROSIM is therefore basically a discrete event simulation language which allows a few attributes of entities to change continuously in time rather than discretely.

SIMSCRIPT [25]: is another rather popular simulation language. The version described in this report is SIMSCRIPT-II. As with SIMULA, there exist several extensions to this version. SIMSCRIPT-II.5 [40] adds a process interaction to the software comparable to GPSS. Another extension is C-SIMSCRIPT which adds some means for ODE solution, making C-SIMSCRIPT another combined simulation language. The version SIMSCRIPT-II.5 is well maintained by C.A.C.I.

GPSS FORTRAN [43]: is a GPSS dialect which implements most features of GPSS-V [44] in terms of a FORTRAN program. This implementation makes the coding of small GPSS models somewhat cumbersome (longer code) and more error prone (FORTRAN and common blocks), but on the other hand makes this software much more flexible for larger applications. Logical branching is much more generally possible and easier accomplishable in GPSS_FORTRAN-II than in GPSS-V. I, therefore, have not included GPSS-V in my table, as I consider this software to be really obsolete by now. GPSS_FORTRAN-III shall add means for ODE handling, making also this software a combined simulation language.

SLAM [37]: is a very powerful combined continuous and discrete simulation language. Discrete systems are modelled in terms of PERT networks. However, to enhance the flexibility of the tool, special event nodes have been introduced. Whenever a transaction passes through such an event node, an event subroutine (to be coded in FORTRAN by the user) is executed. In this way, SLAM-II combines the comfortable modelling capabilities of its predecessor Q-GERT [36] with the flexibility of its other predecessor GASP-IV [35]. Continuous models are expressed in terms of a FORTRAN subroutine (thus no sorting capability is provided) precisely as in GASP-IV. SLAM-II is still rather weak with respect to its continuous simulation features. In particular, the integration algorithm (a Runge-Kutta code) has been coded directly into the execution control routine, making SLAM unusable for stiff systems (which most of the higher order systems are). SLAM-II is therefore highly recommended for predominantly discrete problems, whereas predominantly continuous problems are better solved by use of other software.

GASP [9,35,38]: is a library of extremely portable FORTRAN coded subroutines for combined continuous and discrete simulation. GASP-IV (which is meawhile obsolete) was designed by Pritsker & Assoc. GASP-V added to GASP-IV many continuous simulation features, making GASP-V as versatile for continuous as for discrete simulation problems. GASP-V can therefore be considered to be the first fully combined simulation package. GASP-VI finally adds to GASP-V a process interaction view similar to those of GPSS or SIMULA (transactions may pass through work stations as in GPSS, but they may alternatively also suspend processes as in SIMULA). GASP-V is available from ETH Zurich at nominal cost, whereas GASP-VI is still experimental.

COSY [11,12]: was the first simulation language to be formally designed from the beginning with the help of a syntax analysis program [4]. It uses strictly a LL-1 grammar which makes its compiler easily maintainable and upgradable. (Only SYSMOD shares this advantage with COSY). COSY was originally designed as a front end to GASP-V,. to make the use of that software somewhat more comfortable, but COSY has meanwhile advanced much further. COSY is by far the most general and versatile simulation language proposed to date. A price, which we have to pay for this versatility and universality, is the complexity of the software. Although it is possible to write users manuals for subsets of COSY, making the use of that subset extremely simple, it is quite difficult to master COSY in its entirety. Thus, COSY shares some of the disadvantages of PL/I.

We may once try to add up all (x) and (xx) over each of the columns, to get a measurement unit for the universality of these simulation software systems. Doing so, we receive the following table:

ACSL	62
DARE-P	51
SIMNON	45
DYMOLA	47
SYSMOD	84
FORSIM-VI	45
SIMULA'67	27
PROSIM	37
SIMSCRIPT-II	26
GPSS_FORTRAN-II	45
GPSS_FORTRAN-III	64
SLAM-II	92
GASP-V	71
GASP-VI	104
COSY	136

Obviously, such a quality measurement has to be taken with care. SIMULA and SIMSCRIPT-II turn out badly, because they do not offer anything for continuous systems, and because their simulation support is insufficient. This does not necessarily mean that there is no space for these languages. Very powerful simulation systems can be coded on the basis of these general purpose programming languages. A very nice continuous simulation language is ACSL, probably the best among the currently available systems. SYSMOD shall still be superior when it becomes available. SLAM-II is highly recommended for discrete simulation. Truely combined simulation should at the moment probably best be performed by using GASP-V.

4. SIMULATION SOFTWARE IN THE FUTURE

One of the problems of modern simulation software lies in its complexity. The appetite of the users has grown drastically, and with it also the number of features, the software is supposed to offer. This is in deed a problem, as it becomes more and more difficult to (a) implement such software, and (b) learn to master it as a user. Every language designer knows that a "good" and "successful" computer language should offer less than about 100 reserved keywords, otherwise the compiler gets large and clumsy. What happens to languages which do not obey that rule, we know at least since the days of PL/I. The user manual of any language should be completely expressible in less than about 100 pages, otherwise the average user shall never be able to master all features offered by the language. Here we run into a serious problem with the design of simulation software, as the number of required keywords is dictated by the complexity of the task rather than by the wishes of the purist language designer.

What can we do to overcome that problem: As I believe, the key lies in the data handling. The base of a simulation system should be a database management system (DBMS) adapted to the needs of simulation users. A first step into that direction has been reported recently by Standridge [46,47,48]. Several independent programs for different aspects of system analysis and/or synthesis may then be implemented independently of each other, programs which communicate through their database interface. Advantages of this approach are manifold. Let me start with some advantages which concern simulation alone:

1) Representation of one variable trajectory from several runs (overplot). Many current simulation languages offer this feature ("PAGE MERGE" in CSMP-III). However, while current languages require an additional concept to be mastered, implementation of this feature becomes most natural when the data are stashed away into a data-base during simulation, while data retrieval and display are accomplished by a separate postprocessor grouped around the same data-base as the simulation program. The only means of communication between the two programs takes place through the data-base.

2) Representation of one variable trajectory from both simulation and real experiment on one graph. The implementation of this feature requires yet another program for real-time data acquisition grouped around the same data-base. Otherwise, there is nothing new about. This extremely useful

feature (for model validation) does not exist in any of the current simulation languages I am aware of.

3) Dynamic table load. Tabular data need no longer be coded directly into the simulation program (e.g. by a CSMP "FUNCTION" statement), but may be stored in the data-base. These data may be user generated, generated by another (previous) simulation run, or even generated by real measurements. They may then be used as driving functions to another simulation model. One application of this technique could be the solution of the finite-time Matrix-Riccati differential equation where the Riccati equation needs to be computed first backward in time from the final time (T) to initial time (0.), while the system equations must be computed thereafter forward in time from initial time (0.) to final time (T) making use of the previously stored trajectory of the Riccati matrix.

4) Statistical analysis of noisy data. It may often be interesting to analyse stochastic data for their statistical parameters. Again, this is not really a "job" for the simulation language to accomplish. It is much more natural to store the stochastic simulation data away into the data-base, and to analyse them thereafter by an independent statistics program grouped around the data-base.

5) After many replications of a stochastic model, one may wish to display not a particular time history, but a range in which the results are expected to be found. Such a representation is much more useful for a manager, as he may see trends in that curve which are not easily expressible in figures, as not a single digit of the results may be significant. Again, the implementation of such a feature would be a command belonging to one of the postprocessors, and need not really be mixed up with the simulation language features.

6) There may exist several models for the same system, or several submodels to form an entire model. It may be very useful to store also parametric models, sets of parameter values for these models, experimental frames, and possibly some other structures in the data-base for more comfortable model manipulations. These considerations have been discussed in several articles by Oren and Zeigler [31,32].

One may easily find more examples to show that this concept is fruitful. Another advantage of this concept is that independent manuals may be written for the independent program modules. The manager may then e.g. only study the postprocessor manual, as his only direct access to the computer will be to display the data which have been gathered by other people beforehand. In this way, the concept also allows to split portions of the modeling business among several individuals, a separation which is much more natural and much easier accomplishable then to ask several people to write independently different submodels.

One of the beauties of good old closed-shop-batch-processing lay in the fact that the user needed not to learn anything beside the simulation language, where to deliver his cards and from where to get his listings. In the "ideal" case, there existed a special "box" for CSMP input such that the user even was released from those magic control cards he otherwise had to add to the job (and never understood). The introduction of interactive operation made the task somewhat more difficult to the novice user, as suddenly he had to learn something about file manipulation programs (copy, delete, concatenate, etc.) and data manipulation programs (full screen editor). The introduction of our highly recommended new DBMS concept makes his life by no means easier. It is no longer sufficient to specify what results he wants to see, he has to say what data are to be stored, where they are to be stored, where they are to be refound, how they are to be retrieved, what has to happen with these data after the session is over, and so forth. We suddenly realize that simulation no longer consists of a single well defined and well confined task, but that there exists now a SIMULATION ENVIRONMENT similar to the "environment" definition in ADA. A simulation software no longer consists of a simulation run-time system alone possibly preceded by a compilation step (simulation compiler), but -- equally important -- of a simulation environment definition describing the way in which the simulation software is embedded in the operating system of the implementation machine.

Again, there seems to be a problem we should try to do something about. One of the nicest features of modern operating systems is the fact that they allow to describe their own features in their own terms. To state an example: it is possible to implement the operating system UNIX in terms of the operating system VMS running on a VAX computer (which has been done several times). If this program is then automatically executed from within the LOGIN file, the user gets the impression that his VAX machine is running UNIX and not VMS. Obviously, this way of running UNIX is less efficient than running UNIX in native mode, and this technique is therefore not necessarily recommended for a general purpose operating system such as UNIX. However, the nice thing about this technique is that it may very profitably be used for the implementation of a special purpose "SIMULATION OPERATING SYSTEM", a new term which I want to introduce as an alternative to the classical "simulation languages" and "simulation packages". Let me cite a very simple example: to run ACSL on a VAX 11/780 installation, I have implemented a command procedure (roughly 200 lines of code) to compile, link, and execute ACSL. This command procedure may be used in three different modes:

```
@ACSL pilot LIST FORT GIGI
```

would effect the pilot ejection study (on file PILOT.CSL) to be compiled with production of a listing of both the ACSL program itself (option LIST) and its FORTRAN precompilation (option FORT). Thereafter, the program is linked together with the graphics driver for the GIGI terminal (from which the program is operated). Finally, the program is executed. Upon termination, the user is asked whether he wished to receive a hardcopy of both print- and plot-files on a Versatek printer-plotter, and whether he wants to clean up his intermediary files thereafter. Specification of:

```
@ACSL HELP
```

would explain what parameters are at the users disposal. A third possibility then would be to specify:

```
@ACSL
```

alone in which case the command procedure would enter an interrogative mode and ask for all parameters needed.

This command procedure alone is already quite useful. However, it does not solve all problems for use of ACSL in a class environment. Still, the students would have to learn how to call the editor, copy programs, etc.. For that purpose, another command procedure SIM.COM has been coded which resides on my own accounting number (with read permission only) while being strapped to act as LOGIN file on the students account. This second command procedure (roughly 400 lines of code) enables a menu-type interaction with the user. Each user logging in to the students account is asked first for a second password (his name) after which he enters his own directory. Thereafter, he gets a menu of possible commands displayed which reads as follows:

```
Current ACSL Problem :  NONE

Code: (A) Run ACSL problem
      (C) Clean up files
      (D) Delete ACSL problem
      (E) Edit ACSL problem file
      (F) Edit ACSL data file
      (G) Display general HELP information
      (H) Start/stop HELP menu.
      (L) List of existing ACSL problems
      (M) Display the message of the day
      (N) Print Non-ACSL files (after error)
      (O) Make old version current again
      (P) Purge old versions
      (Q) Show disk quota
      (R) Read file from other problem
      (S) Select ACSL problem
      (T) Display status of queues
      (V) Display VAX-specific information
      (W) Write file to other problem
      (Z) Exit from ACSL account

--------------->
```

which we felt to be about the minimum number of commands, a user must have at his disposal. The user may now e.g. press "L <CR>" to obtain a directory of all of his currently defined ACSL problems. Thereafter, the HELP menu is repeated. Now, the user may press "S <CR>" to be prompted for the name of the problem to be executed (e.g. PILOT). "A <CR>" would then result in a call to the previously described command procedure ACSL.COM for compilation, linkage, and execution of the currently selected ACSL problem. "E <CR>" would call the editor with the currently selected ACSL problem file (depending on the terminal type, this automatically results in either a call to EDT or in a call to TECO). The students obviously must get the impression that their computer runs a special purpose ACSL operating system, even though in reality it runs under VMS (just hidden from the students). The students remain within the LOGIN file throughout their entire session. Our experiences with this mode of operation were extremely positive. The students were able to master this simple simulation operating system without any difficulties after a short demonstration of 20 minutes length.

Currently I am implementing a similar simulation operating system for the second course on discrete event simulation. For that purpose, I base on the software by Pritsker and Associates, that is: the SDL data-base management system together with the SLAM-II simulation software. Lateron, we shall add also the other programs AID (for statistical analysis and interactive fitting of distribution function parameters) and SIMCHART (graphical postprocessor). The full beauties of such a combination would be rather difficult to feel if these programs were used independently of each other. It is really the introduction of the simulation operating system which makes such a system easily manageable and useful.

The idea of such a mode of operation is in deed not really new. Already 10 years ago people were talking about Management Information Systems (MIS) as a cure-all to any disease. Unfortunately, these systems (as far as they ever got) turned out to be diseases in themselves, in that:

1) the systems were huge, clumsy, slow, and unflexible,

2) they never were able to fulfil the task they were build for, as managers were unable to understand what was going on, and therefore (and for good reason) had little trust in the results produced.

Now, 10 years later, the idea of using a computer to take decisions has been burried, still the concept survived in a new outlook, nowadays called Decision Support System (DSS). The idea here is that the manager should no longer be replaced by a computer. Instead, the computer should provide the manager with all available data to take a correct decision. Obviously, the heart of a DSS is again a DBMS, possibly enhanced by some additional modules for statistical analysis, econometric modeling, data display, and similar -- as one can see, precisely the concept which I advertised in this chapter. However, I prefer the term "simulation operating system" over DSS, as this new term does not imply that the tool is to be used by managers only which (partly and at times) well may be the case, but certainly need not. There are many technical applications for which this concept (as shown above) is useful. Moreover, modern operating systems shall now allow much more efficient implementations than what was possible 10 years ago, as this concept lends itself readily to programming at the operating system level (by means of a command language) rather than at the language level (which to a good extent would require assembly programming, as most computer systems would not allow to call a system program (e.g. the FORTRAN compiler) from within a (e.g. in FORTRAN coded) user program). Such comfortable command languages were unavailable 10 years ago.

To close up this short discussion, I strongly advertise a solution in which the many demands for system analysis (and synthesis) features are properly separated into independent modules (with independent documentation) which communicate with each other through a data-base interface. By these means, the flexibility of the resulting tool is drastically enhanced (as has been shown at some examples) while keeping each of the program modules sufficiently simple to let them be user manageable on one side, and efficiently implementable on the other hand.

REFERENCES:

[1] Baker, N. J. C. and Smart, P. J., The SYSMOD Language and Run Time Facilities Definition, Techn Note 6.82, Royal Aircraft Establishment, Farnborough, Hampshire, United Kingdom. (March 1982).

[2] Birtwistle, G. M., Dahl, O.-J., Myhrhaug, B. and Nygaard, K., SIMULA BEGIN. (Studentlitteratur Sweden and van Nostrand Reinhold, New York, 1973).

[3] Birtwistle, G.M., DEMOS: Discrete Event Modelling on SIMULA. (Macmillan, London and Basingstoke, 1979).

[4] Bongulielmi, A. P. and Cellier, F. E., On the Usefulness of Deterministic Grammars for Simulation Languages, Proc. of the SWISSL Workshop, St. Agata, Italy. Shall appear also in Simuletter. (September 1979).

[5] Carver, M. B. and Hinds, H. W., The Method of Lines and the Advective Equation, in Proc. of the ACM SIGNUM Meeting, Albuquerque, New Mexico. (November 1977).

[6] Carver, M. B., Stewart, D. G., Blair, J. M. and Selander, W. N., The FORSIM VI Simulation Package for the Automated Solution of Arbitrarily Defined Partial and/or Ordinary Differential Equation Systems, Report AECL-5821, Atomic Energy of Canada Ltd., Chalk River, Ontario. (February 1978).

[7] Carver, M. B. and MacEwen, S. R., Automatic Partitioning in Ordinary Differential Equation Integration, in Cellier, F. E. (ed.), Progress in Modelling and Simulation. (Academic Press, London and New York, 1982).

[8] Cellier, F. E., Continuous-System Simulation by Use of Digital Computers: A State-of-the-Art Survey and Prospectives for Development, in Hamza, M. H. (ed.), Proc. of the International Symposium and Course SIMULATION'75. (Acta Press, Calgary and Zurich, 1975).

[9] Cellier, F. E. and Blitz, A. E., GASP-V: A Universal Simulation Package, in Dekker, L. (ed.), Simulation of Systems, Proc. of the 8th AICA Congress. (North-Holland, Amsterdam, 1976).

[10] Cellier, F. E. and Moebius, P. J., Towards Robust General Purpose Simulation Software, in Skeel, R. D. (ed.), Proc. of the ACM SIGNUM Meeting on Numerical Ordinary Differential Equations, Dept. of Computer Science, University of Illinois at Urbana-Champaign. (March 1979).

[11] Cellier, F. E. and Bongulielmi, A. P., The COSY Simulation Language, in Dekker, L., Savastano, G. and VanSteenkiste G. C. (eds.), Simulation of Systems, Proc. of the 9th IMACS Congress. (North-Holland, Amsterdam, 1979).

[12] Cellier, F. E., Rimvall, M. C. and Bongulielmi, A. P., Discrete Processes in COSY, in Maceri, F. (ed.), Proc. of the European Simulation Meeting held in Cosenza, Italy. (April 1981). Also in Crosbie R. E. and Cellier, F. E. (eds.), TC3-IMACS, Simulation Software, Committee Newsletter, No 11. (July 1982).

[13] Cellier, F. E., (ed.), Progress in Modelling and Simulation. (Academic Press, London and New York, 1982).

[14] Eitelberg, E., Modular Simulation of Large Stiff Systems, in Cellier, F. E. (ed.), Progress in Modelling and Simulation. (Academic Press, London and New York, 1982).

[15] Elmqvist, H., SIMNON - An Interactive Simulation Program for Nonlinear Systems - User's Manual, Report TFRT-3091, Dept. of Automatic Control, Lund Institute of Technology, Lund, Sweden. (April 1975).

[16] Elmqvist, H., SIMNON - An Interactive Simulation Program for Nonlinear Systems, in Hamza, M. H. (ed.), Proc. of the International Symposium SIMULATION'77. (Acta Press, Anaheim, Calgary and Zurich, 1977).

[17] Elmqvist, H., DYMOLA - A Structured Model Language for Large Continuous Systems - User's Manual, in Crosbie, R. E. and Cellier, F. E. (eds.), TC3-IMACS, Simulation Software, Committee Newsletter, No 10. (September 1981).

[18] Gear, C. W., Stiff Software: What Do We Have and What Do We Need?, in Aiken R. C. (ed.), Proc. of the International Conference on Stiff Computation, Dept. of Chemical Engineering, University of Utah, Salt Lake City. (April 1982).

[19] Halin, H. J., et alia, The ETH Multiprocessor Project: Parallel Simulation of Continuous Systems, Simulation, Vol 35, No 4. (October 1980).

[20] Helsgaun, K., DISCO - A SIMULA-based Language for Continuous Combined and Discrete Simulation, Simulation, Vol 35, No 1. (July 1980).

[21] Hindmarsh, A. C., Stiff System Problems and Solutions at LLNL, in Aiken R. C. (ed.), Proc. of the International Conference on Stiff Computation, Dept. of Chemical Engineering, University of Utah, Salt Lake City. (April 1982).

[22] IBM, Continuous System Modeling Program III (CSMP III) Program Reference Manual, Program Number 5734-XS9, Form SH19-7001-2, IBM Canada Ltd., Program Product Centre, 1150 Eglington Ave. East, Don Mills 402, Ontario. (September 1972).

[23] Joss J., Algorithmisches Differentieren, Ph.D. Thesis, ETH Zurich, Diss. ETH 5757. (1976).

[24] Karplus W. J., Software for Distributed System Simulation, in Cellier, F. E. (ed.), Progress in Modelling and Simulation. (Academic Press, London and New York, 1982).

[25] Kiviat P. J., Villanueva, R. and Markowitz, H. M., The SIMSCRIPT II Programming Language. (Prentice-Hall, 1968).

[26] Kleijnen, J. P. C., Experimentation with Models: Statistical Design and Analysis Techniques, in Cellier, F. E. (ed.), Progress in Modelling and Simulation. (Academic Press, London and New York, 1982).

[27] Korn, G. A. and Wait, J. V., Digital Continuous-System Simulation. (Prentice-Hall, 1978).

[28] Mitchell and Gauthier, Assoc., ACSL: Advanced Continuous Simulation Language - User Guide / Reference Manual, P.O.Box 685, Concord, Mass. (1981).

[29] Mitchell, E. E. L., Advanced Continuous Simulation Language (ACSL): An Update, in Ames, W. F. (ed.), System Simulation and Scientific Computation, Proc. of the 10th IMACS Congress, Dept. of Computer Science, Rutgers University, New Brunswick, New Jersey. (August 1982).

[30] Nilsen, R. N., The CSSL-IV Simulation Language, User Manual. Simulation Services, 20926 Germain Street, Chatsworth, California.

[31] Oren, T. I. and Zeigler, B. P., Concepts for Advanced Simulation Methodologies, Simulation, Vol 32, No 3. (March 1979).

[32] Oren, T. I., Computer-Aided Modelling Systems, in Cellier, F. E. (ed.), Progress in Modelling and Simulation. (Academic Press, London and New York, 1982).

[33] Petzold, L. R., An Efficient Numerical Method for Highly Oscillatory Ordinary Differential Equations, Dept. of Computer Science, University of Illinois at Urbana-Champaign, Form UIUCDCS-R-78-933. (August 1978).

[34] Petzold, L. R., A Description of DASSL: A Differential/Algebraic System Solver, in Ames, W. F. (ed.), System Simulation and Scientific Computation, Proc. of the 10th IMACS Congress, Dept. of Computer Science, Rutgers University, New Brunswick, New Jersey. (August 1982).

[35] Pritsker, A. A. B., The GASP IV Simulation Language. (Wiley, New York, 1974).

[36] Pritsker, A. A. B., Modeling and Analysis Using Q-GERT Networks. (Halsted Press, New York, 1977).

[37] Pritsker, A. A. B. and Pegden, C. D., Introduction to Simulation and SLAM. (Halsted Press, New York and Systems Publishing Corp., West Lafayette, 1979).

[38] Rimvall M. C. and Cellier, F. E., The GASP-VI Simulation Package for Process-Oriented Combined Continuous and Discrete System Simulation, in Ames, W. F. (ed.), System Simulation and Scientific Computation, Proc. of the 10th IMACS Congress, Dept. of Computer Science, Rutgers University, New Brunswick, New Jersey. (August 1982).

[39] Roth M. G. and Runge T. F., Simulation of Continuous Networks with MODEL, Dept. of Computer Science, University of Illinois at Urbana-Champaign, Form UIUCDCS-R-78-921. (December 1978).

[40] Russell, E. C., Building Simulation Models with SIMSCRIPT II.5. C.A.C.I., 12011 San Vicente Boulevard, Los Angeles, California.

[41] Schiesser, W. E., LEANS - III Introductory Programming Manual, Computing Center, Lehigh University, Bethlehem, Penna. (September 1971).

[42] Schiesser, W. E., Some Characteristics of ODE Problems Generated by the Numerical Method of Lines, in Aiken R. C. (ed.), Proc. of the International Conference on Stiff Computation, Dept. of Chemical Engineering, University of Utah, Salt Lake City. (April 1982).

[43] Schmidt B., GPSS-FORTRAN Version II Einfuehrung in die Simulation diskreter Systeme mit Hilfe eines FORTRAN-Programmpaketes. (Springer, Berlin, Heidelberg and New York, 1977).

[44] Schriber T. J., Simulation Using GPSS. (Wiley, New York, 1974).

[45] Sierenberg R. and de Gans, O., PROSIM Textbook, Dept. of Applied Mathematics, Delft Technical University, Delft, The Netherlands. (1982).

[46] Standridge, C. R., Using the Simulation Data Language (SDL), Simulation, Vol 37, No 3. (September 1981).

[47] Standridge, C. R., The Simulation Data Language (SDL): Applications and Examples, Simulation, Vol 37, No 4. (October 1981).

[48] Standridge, C. R. and Pritsker, A. A. B., Using Data Base Capabilities in Simulation, in Cellier, F. E. (ed.), Progress in Modelling and Simulation. (Academic Press, London and New York, 1982).

[49] Wait, J. V. and Clarke III, D., DARE P User's Manual, Dept. of Electrical Engineering, University of Arizona, Tucson. (December 1976).

[50] Zellner, M. G., DSS - Distributed System Simulator, Computing Center, Lehigh University, Bethlehem, Penna. (May 1970).

Simulation in Engineering Sciences
J. Burger and Y. Jarny (eds.)
Elsevier Science Publishers B.V. (North-Holland)

PARALLEL PROCESSORS IN ENGINEERING SCIENCES

L. Dekker

Delft University of Technology
Netherlands

The aim of this paper is to give an impression about the power that parallel data processing will have in the near future, in particular in the field of engineering. The applicability of parallel data processing will be illustrated by examples, realized on the Delft Parallel Processor.

1. INTRODUCTION

Very typical for sequential data processing is the necessity to decompose a given task in smaller ones. These smaller tasks have to be executed successively: time-sequential subtasks (sequentialisation of a given task). As a consequence the execution time may be (much) longer than for parallel data processing. In the latter case a given task is decomposed in time-parallel subtasks in such a way that these subtasks can be treated simultaneously (parallelization of a given task).

Not long ago the only tools for parallel data processing were analog/hybrid computers. In the engineering sciences even today these tools are applied successfully, because analog/hybrid computation has several very interesting features, a.o.:

- both the processing and the transfer of data can easily be realized;
- a large computation speed can be achieved because of the parallel data processing;
- real-time graphical display of the time-parallel variables is easily possible;
- the analog/hybrid computers are well-adapted to the field of systems simulation because of:
  - its multi-processing structure, having a truly time-parallel simultaneity
  - its data flow structure.

These characteristics imply that it is not difficult to parallelize data processing tasks. But analog/hybrid computation has also several serious disadvantages. The present hardware technology permits to eliminate these disadvantages, except one:

- the set of implementable parallel algorithms is very limited because of the fact that all mathematical operations are based on electric laws of analogy.

For this reason the applicability has been rather limited.

The production technology of LSI and VLSI hardware components is already well-developed. This technology makes it possible to develop and construct parallel processors with an algorithmic based on laws of logic: discrete-time parallel data processing.

2. PARALLELIZATION OF DATA PROCESSING (Brok et al 83, Dekker et al 82, 83a,b, Haynes 82, IEEE 77, 79, Kuhn et al 81, Sips 82a,b)

In the classical digital computer the data processing is realized merely by applying the principle of "extension in time". The instructions are executed in the CPU one after another; this leads to one operation at a time on one word in the accumulator: "Single Instruction stream, Single Data stream"-architecture, (SISD). This main feature is the limiting factor for the data processing speed of digital computation. Besides extension in time there is also the principle "extension in space", i.e. the execution of data processing tasks simultaneously in different parts of a data processor. In this way both the data processing power and speed can be enlarged. In practice one has to apply both principles, because a multi-processor will consist of a finite number of simultaneously operating components, but also because both principles often results from systems modelling. In the recent past parallel (data) processors have been constructed, able to perform the same stream of instruction on different streams of data: "Single Instruction stream, Multiple Data stream"-architecture, (SIMD). In a SIMD-processor the parallel tasks have to be identical. Hence the applicability is restricted to special purposes. Parallel processors also have become available, able to perform different streams of instructions on different streams of data: "Multiple Instruction stream, Multiple Data stream"-architecture, (MIMD). In a MIMD-processor each subprocessor has its own control unit. In the case that the subprocessors are classical digital computers, there will arise difficulties with respect to a.o. the exchange of data. Moreover, the speed ratio has an upper bound equal to the number of subprocessors. It is better to apply a type of subprocessor, also having parallel data processing power. Or, in other words, the parallel data processing power has to be distributed over all subsystems of a parallel processor (Dekker 83b): "Distributed Multiple Instruction-, and Multiple Data stream"-architecture, (DMIMD). A DMIMD-processor is a spatial hierarchy of subsystems. Thus, a DMIMD-processor consists of a number of simultaneously operating subprocessors, here to be called processor-modules (PM), while again each PM consists of a number of simultaneously operating processing elements (PE), etc.

In this paper we assume that a processing element is the smallest hardware subsystem, containing a number of hardware components (to be called multi-operand operators, (MOP)) with a programmable spatial structure. Again a multi-operand operator has parallel data processing power, but now the spatial structure of its hardware subsystems is not arbitrarily programmable.

The applicability of a DMIMD-processor will depend much on the detail-design of its architecture. The efficiency of a parallel data processor can degrade considerably for several reasons. For example, the architecture of a multi-processor may fail with respect to the power of the interconnection structure between the subsystems of the multi-processor. A poor interconnection structure gives rise to problems of congestion with respect to the exchange, storage and retrieval of data. In this respect it has to be avoided that a parallel processor contains components, which can be commonly accessed by other components during a parallel run. For example, the exchange of data can not appropriately be realized by means of one common data bus. From its stored parallel program each data processing subsystem always knows the variables for which values are needed. Hence, an attractive conceptual approach is the principle used for the distribution of newspapers, etc. For all output variables of a closely connected cluster of subprocessors, the most recent data will be transfered regularly and simultaneously to the input of all these subprocessors, like the data in a newspaper are sent in parallel to all subscribers. As a consequence many parallel data buses are needed. It is also necessary to realize, in each subprocessor of the cluster, a multi-accessible memory of registers in order to be able to store all data in parallel. In bit-parallel processing it is necessary to transfer a digital word before it is possible to utilize this digital word in an operation as the recent value of a variable. A maximal optimal data processing speed would require also bit-parallel data transfer. But then each data bus will consist of a number of electrical wires, equal to the number of bits of the digital words. Note that bit-sequential data transfer, is a must for the production of VLSI chips, but that it is also attractive for the realization of the many data buses between the subsystems of a parallel processor. We mention that there are attempts going on to design standard algorithms, based on time-parallel but bit-sequential data processing of all input variables, in such a way that the pass-through time of a multi-operand operation is only slighty dependent on the complexity of an algorithm (Sips, 82b).

Obviously the realization problems around the design and the construction can only be solved for parallel processors with a finite number of subsystems. As a consequence the power to treat data simultaneously (time-parallel) will be limited too. It implies that computational modelling concerns both "algorithm modelling" and "format modelling". The format modelling takes care for the transformation of an arbitrary task such that it is implementable on a parallel processor of a given size. In both regards the computational modelling for time-parallel data processing in discrete-time sets is still underdeveloped. This fact of a finite time-parallel processing power results into two non-independent design problems:

- the design of the hardware/software architecture,
- the design of a portable "format modelling" software package, able to handle conveniently and user-friendly the implementation problem for an arbitrary finite size of the parallel data processing power.

As a consequence of format modelling "block-iterative" parallel algorithms will become important (Dekker et al. 1976).

It is practically impossible, because of hardware and costs limitations, to realize full interconnectability between all hardware subsystems of a large parallel processor. Because large credible system models will normally have a sparse interaction matrix, a "one-to-one analogy in space" implementation does not require full interconnectability for large parallel processors. The demand of a modular hardware/software architecture implies that the "reduction of interconnectability" has to be designed in a recurrent way. Quite often successive decomposition of a system model finally results in system models of first order. For this reason a processing element PE must be capable to implement first order systems in a close one-to-one analogy in space (and in time). Normally a small system, described as a system of first order systems, has an (almost) a full interaction matrix. As a consequence there must be full interconnectability between all processing elements of a processing module PM. Because in many cases larger systems can be described in a natural way as a spatial hierarchy, this description can be a key to realise the interconnectability reduction. However, there must also be included a neighbour interconnectability for those systems having a neighbour interaction structure. Note that there has to be also full interconnectability between the multi-operand processors MOP's of a processing element PE, in order to be able to program an arbitrary spatial configuration between the MOP's. Also a MOP will have some parallel processing power; it will consist of some hardware/software components ALU's (but with preprogrammed structures in space and in time).

3. PARALLELIZATION OF ALGORITHMS

(Andriessen et al 81, Andriessen 80, Dekker et al 82, Dekker 76, 81, 83b)

It makes only sense to apply parallel data processing, when it yields a considerable profit for reasonable costs. Note that the parallelization of the data processing power of a computer concerns all activities: data handling, storage and retrieval. The design of parallel algorithms depends much on the type of the interaction structure in time between the different tasks. There are two extreme cases:

- the tasks have a time-full interaction structure,
- the tasks have a time-sparse interaction structure.

In the first case there exist interactions between the tasks at all times of one time set, while in the other case there only exist interactions at a few times. In other words the tasks are continuously conflicting (with respect to the exchange of data), or most of the time independent, nonconflicting.

We will illustrate that for tasks of arithmetical nature indeed parallelization can yield much profit. Arithmetical parallel algorithms normally have a rather time-full interaction structure. Generally speaking, more "extension in space" yields a faster algorithm and more "extension in time" requires less hardware. Important figures for the profit and the costs of a parallelization are the "speed ratio" and the "hardware ration" in the extreme cases: pure extension in space, and pure extension in time. The ratio between the speed ratio and the hardware ratio is a measure for the "efficiency of the parallelization". The parallelization efficiency can be maximally equal to *1*, but in practice it can be much smaller because in a hardware configuration often the hardware components do not operate only "parallel in space", but also "sequential in space". Let us assume that all parallel algorithms are derived from a set of "standard algorithms":

- each standard algorithm operates in a discrete time set T with a cycle time τ;
- the standard algorithms will be implemented in MOP's; programmable are only "names" and "parameters".

In practice parallelization can often be performed in a natural way. For example, in matrix calculus for the mapping:

$$y = A * u, \quad u,y \in R^n \tag{3,1}$$

the first stage of parallelization can be realized by means of n inner product processors, because this mapping can be decomposed in:

$$y_j = a_j . u \;,\; j = 1,\dots,n;\; A^T = (a_1^T \dots a_n^T) \tag{3,2}$$

The efficiency of this first parallelization stage is maximal, equal to *1*. The speed ratio is equal to the number of inner processors.

The inner product mapping (3,2) is an important example of a multi-operand operator. For this reason we consider the design of a parallel inner product algorithm of two vectors u and v, $u, v \in R^n$. We assume that there are available

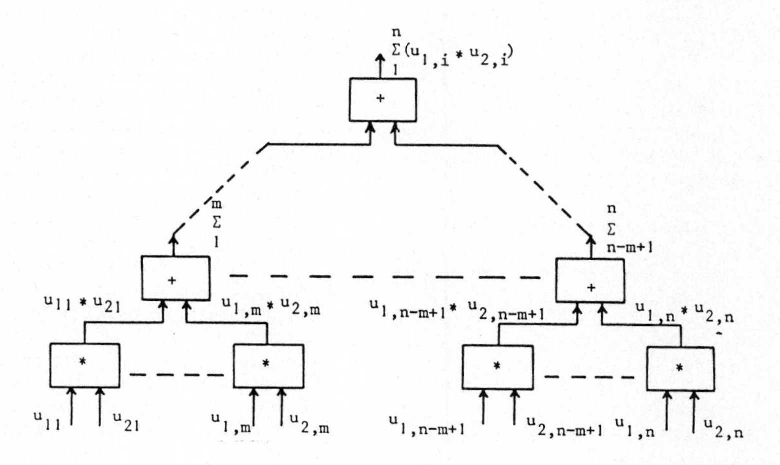

Figure 3,1: Spatial configuration of a parallel algorithm of an inner product mapping realized by means of multipliers and m-input adders.

ALU's, able to perform a.o.: a "2-input multiplier" algorithm and a "m-input adder" algorithm for digital words of n^* bits. Note that for the normal value of m, $m=2$, these algorithms are unbalanced in complexity; for values of m and n^* in the same order of magnitude the algorithms are roughly of equal complexity. Let us first consider the case that the algorithm is designed by applying merely extension in space. Only for convenience we take $n=m^L$. The optimal parallelization proves to be a pipeline of ALU's, programmed partly as 2-input multipliers and partly as m-input adders, see figure 3,1. This inner product algorithm can be used in a parallel program just by calling its name. The ALU's have to execute their operations $L+1$ times; so the organization in time of the calculations can be realized easily. The right exchange of data is automaticly obtained by means of the data flow via the connections between the ALU's. Figure 3,1 shows an ordering of the ALU's both "parallel in space" and "sequential in space". As a consequence of the sequential ordering in space this inner product algorithm belongs to the class of delay algorithms. The delay time is equal to $L+1$ cycle times, being the pass-through time (defined as the time period needed for the current values of the input variables u and v to pass completely through the hardware configuration from the input to the output). In the contrary case that only extension in time is applied in the design of the inner product algorithm by using only one ALU, it is necessary to apply storage/retrieval of intermediate results. The successive operations have to be organized by means of a stored program. The different tasks:

- multiply serially all pairs of input values and store the resulting products in a sequence.
- add all m-tuples of successive elements of the sequence of products and store all intermediate sums at the end of the sequence.

The sum or the last m-tuple is the value of the inner product.

The profit of parallelization depends much on the interval of the input variable u of a function f, for which the output variable y has to be determined. The profit can be quite different for:

- a point mapping,i.e. a mapping $y := f(u)$ for only one point u.
- a nonrecursive multi-point mapping, i.e. a mapping $y := f(u)$ for a series of points $(u_1, y_1), (u_2, y_1)$, ...,
- a recursive multi-point mapping, i.e. a mapping $y := f(u,y)$ has to be executed for a series of points.

First we consider the "inner product" algorithm for a point mapping. Then we find:

- speed ratio = (mn-1)/(m-1)(L+1), L = ^{m}log n,
- hardware ratio = (mn-1)/(m-1),
- parallelization efficiency = 1/(L+1).

The minimal length of the spatial hierarchy of figure 3,1 is equal to two for $m=n$, because the multiplications have to be finished before the addition(s) can start. Thus, in case a n-input adder belongs to the set of standard algorithms, the parallelization efficiency is maximal and equal to one ha. For $n = 32$ and the usual value of m, $m = 2$, the efficiency is only $1/6$; in this case the profit in speed of spending more hardware seems to be small. But one has to be aware that the decrease in costs and the increase in power of hardware components have already become so large that in practice the achievable degree of parallelization will often be quite satisfactory. For a nonrecursive multi-point mapping, the parallelization of the inner product algorithm of figure 3,1 offers much more profit. Because of the pipeline structure the inner product algorithm u is able "to accept input data" and "to produce output data" at all times of the time set T. Hence:

- speed ratio = (mn-1)/(m-1),
- parallelization efficiency = 1

In this way the value of L has no influence. The value of L is virtually equal to zero and the profit of parallelization is maximal. In practice nonrecursive multi-point algorithms occur frequently, not only in special-purpose applications like signal and image processing and computational physics but also in general-purpose matrix calculus. This is the reason why the class of spatially sequential pipelines is becoming so important. In case of a recursive multi-point mapping, the inner product algorithm can only accept new samples of the input variables each time the pass-through time $(L+1)\tau$ has been finished. In other words, for a point mapping and a recursive multi-point mapping the speed ratios are equal.

In matrix calculus it is normally easy to apply decomposition in a way that it results in tasks of both equal complexity and equal parallelization. For this reason in matrix calculus parallelization can yield a large total speed ratio. For example, let us consider the point mapping $y = A * u$ with A is a full $k*k$-matrix in the case that k inner product processors are available, able to implement a parallel inner product algorithm of dimension n, $k = n*\kappa$. first this point mapping will be decomposed in k parallel mappings:

$$y_i = a_i.u, \quad i = 1,\dots,k; \quad A^T = (a_1^T \dots a_k^T) \qquad (3,3)$$

and further these parallel mappings can be transformed in multi-point mappings:

$$z_{i,j+1} = z_{ij}+a_{ij}.u_j, \quad i = 1,\dots,k; \quad j = 1,\dots,\kappa;$$

$$z_{i1} = 0, \quad z_{ik} = y_i; \qquad (3,4)$$

$$a_i^T = (a_{i,1}^T,\dots,a_{i,\kappa}^T);$$

$$u^T = (u_1^T,\dots,u_\kappa^T); \quad a_{ij}, u_j \in R^n.$$

It is not unrealistic that inner product

processors can be made with a speed ratio between *10* and *100*. It also looks feasible that parallel processors will become available with a number of subprocessors between *100* and *1000*. Hence, in matrix calculus in cases of maximal parallelization efficiency a speed ratio can be obtained in the order of 10^3 up to 10^6.

For nonlinear multi-operand operations it is not obvious that parallel algorithms can be designed with such a large speed ratio. For this reason for general-purpose parallel processors "piece-wise linearization methods" will become important in algorithm modelling, because of speed considerations (Dekker,70)

So far in this contribution the parallelization of algorithms has been considered in the case that the choice of the time-parallel algorithms was restricted to the class of delay algorithms. But then, as we have seen, in the decomposition of a given mapping there arise complications with respect to the parallelization of a task, the complexities of the arising subtasks, the synchronization at all levels of decomposition. Now we will review an approach in the design of a type of parallel algorithms, having none of these difficulties. For the inner product algorithms of figure 3,1 there exists a natural synchronization, because the multipliers and adders were assumed to execute standard algorithms. For this reason they are capable to accept input data and to produce output data at all times of *T*. There would also be a natural synchronization, if time-parallel algorithms of arbitrary complexity could be designed able to operate in *T* and able both "to produce useable output data" and "to accept input data" at all times of *T*. It depends on the static and dynamic accuracy and on other properties whether the output data can be considered to be "useable". The spatial configuration, in general form for a mapping $y = f(u)$, of this type of parallel algorithms is shown in figure 3,2.

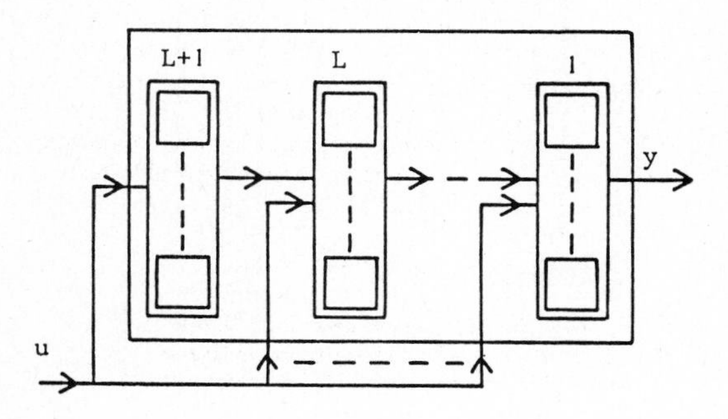

Figure 3,2 Spatial configuration of a hybrid pipeline algorithm for y = f(u).

The blocks of this spatial configuration do not only operate serially in space but also parallel in space with regard to the input variable *u*. This is the reason why this pipeline structure is called a (spatially) "hybrid pipeline". Several hybrid pipeline algorithms of good quality have already been developed (Andriessen, 80), but the practical application will have to wait until the implementation by means of VLSI techniques is feasible.

In practice it must be possible to allow closed-loops in the application of parallel algorithms. It requires that hybrid pipeline algorithms have good stability properties and a good dynamic behaviour (Dekker, 81a). In order to illustrate that for the development of parallel data processing also knowledge is required, quite unusually so far in informatics, we summarize the design of a hybrid pipeline algorithm for the simplest mapping $y = u$. A hybrid pipeline algorithm, of a similar structure as shown in figure 3,2, with a maximal stability range proves to be:

$$y_k = (u_{k-1} + \dots + u_{k-L})/L \qquad (3,5)$$

Replacing u by $u+g*y$, the stability interval of (3,5) is equal to the g-interval for which the algorithm is stable. Stability analysis yields a stability interval, proportional to L:

$$-L < g < 1 \qquad (3,6)$$

Unfortunately it proves that the dynamic behaviour is quite unattractive in a significant part of the stability interval. It can be understood because the algorithm (3,5) is a discretized version of the undamped transport equation. One can expect to improve the dynamic behaviour by means of a modification such that the algorithm has a combined transport-type and diffusion-type behaviour:

$$y_{j,k+1} = (1-a)(y_{j+1,k} + u_k/L) + ay_{j,k} + b(y_{j+1,k} - 2y_{j,k} + y_{j-1,k}) \qquad (3,7)$$

$$y = y_{0,k} = (L+1)y_{1,k}/L; \; y_{L,k} = 0$$

Indeed, this algorithm proves to have a good dynamic behaviour over the whole stability range.

4. INFLUENCE OF SYSTEMS MODELLING AND OF VLSI TECHNOLOGY ON THE ARCHITECTURE AND THE PROGRAMMATURE OF A PARALLEL PROCESSOR (Dekker et al 79,82, Dekker 83b)

The only reason that there is a need of data processing is the fact that in human societies it is important to analyse, to simulate, to control, to modify, etc. systems. In this respect systems modelling will have much influence on the design of parallel processors. In the future because of costs and size considerations, the hardware of a parallel processor will be based on the application of VLSI techniques. As a consequence the architecture of a parallel processor will have to be based on a modular structure for the hardware as well as for the software; this will also be attractive for systems modelling (Dekker, 83b). The components of a parallel processor have to be configurated, in space and in time, in a recurrent way. For the realization of modifications in a system model, caused by discrete events, it is necessary to incorporate special hardware/software components, "model generators", to be able to realize model modifications at all levels of a spatial hierarchy. Because the loading of the parallel program of the successor (sub)system after a discrete event would take too much time, it is necessary to load in advance the parallel (sub)programs of the possible successor (sub)systems in multi-read accessible, multi-buffered storage devices. The state of the hardware technology allows that in the near future all demands from a practical point of view are taken as design criteria for the construction of a parallel processor. The data processing systems are on the way to become hardware/software systems, capable to perform data handling in a time-parallel way by means of a rigorous distribution of hardware/software. In the future a parallel "digital" processor can offer the users for a reasonable price a data processing power, to such an extent that the set of nonsolvable computational problems will diminish a lot.

5. THE DELFT PARALLEL PROCESSOR, AN EXAMPLE OF A MIMD-PROCESSOR (Brok et al 83, Dekker 82,83b, Sips 82a)

Since the middle of *1981* the Delft Parallel Processor (DPP) has become operational at the Delft University of Technology. This first DPP is a small MIMD-processor consisting of only eight time-parallel subprocessors, called processing elements (PE). Moreover a PE is still a mono-processor; it does not yet have parallel data processing power. Figure 5,1 shows the special architecture of the DPP in the case that there is only one processing module PM, consisting of eight PE's. So far the data processing power of a PE is formed by an arithmetical microprocessor AM9511a, capable of performing a large set of arithmetical functions on digital words in the formats "32 bits floating point" or "16 or 32 bits fixed point". Intermediate results can be stored in a data memory. Each PE has one arithmetical output variable and several boolean output variables. The time sets of the PE's can be chosen individually per PE. In this way the PE's can execute their programs in either a synchronized or an unsynchronized mode of operation. Because the time set of the data transfer and those of the PE's can be chosen differently, adaptation is needed, realized by means of input buffers and output buffers. All output variables of the

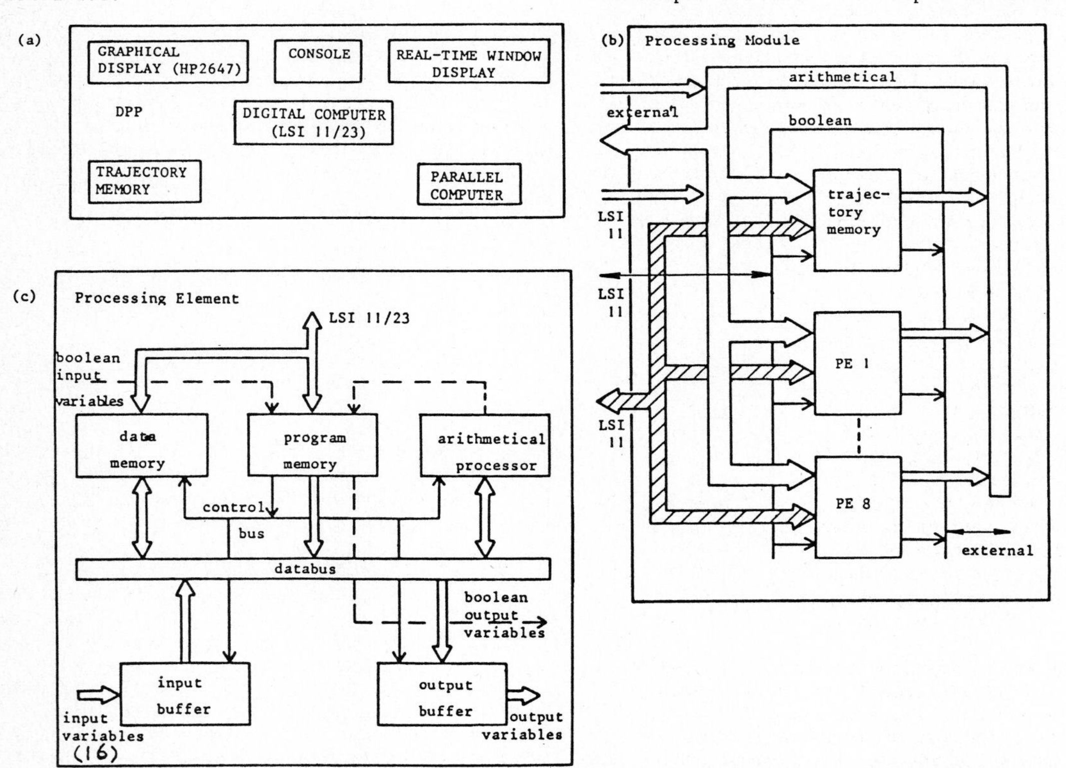

Figure 5,1: The architecture of a Delft Parallel Processor consisting of *8* processing elements: (a) total configuration, (b) parallel computer, (c) a processing element.

PE's are transfered time-parallel via the data transfer system at each data transfer time. In this way at the input each PE can sample 16 arithmetical input variables. The functioning of a PE is defined by a parallel program, stored in a program memory. A parallel program of a PE defines: the internal data transfer of the PE, the operation modes of the input and output registers, and the operations to be executed by the arithmetical microprocessor of the PE. The boolean output variables of the PE's can be used to influence the control of their parallel programs. A sequential host computer LSI-11/23 initializes and supervises the parallel programs via a sequential program written in FORTRAN. In the parallel run stage each PE autonomously executes the prestored parallel program. The software tools of the DPP guarantee that the writing of parallel and sequential programs is simple. It allows a user to implement system models in an easy and highly interactive way. During a parallel run one can observe some of the external input and output variables of the PE's, without or after some signal processing, on a microprocessor-controlled real-time window display. For instance it will soon become possible to realize in real-time a 32 point F(ast) F(ourier) T(ransform). During a parallel run the external input and output variables of the PE's are automatically stored in a trajectory memory. In this way, after a parallel run, a user can decide to have a closer look at the data, the DPP has produced. Because of this input-output trajectory storage these is no need for repetitive mode like in analog computation. Normally the interactive observation of these input-output trajectories takes much time compared to the parallel run itself. In the meanwhile the parallel computer of the DPP can be used for other purposes:

- time-sliced use of the parallel computer by several users,
- robot-experimentation, in order to support a user during a simulation run.

There is no doubt that the architecture of the DPP permits to enlarge gradually the data processing power. For example, in the near future the PE's will be provided with parallel data-processing power by incorporating a MOP. This MOP will be capable to perform an inner product $u.v$, $u,v \in R^n$, $n>>2$, based, by means of extension in time, on the fast execution of the multi-operand operator: $a*b+c*d+e$. Moreover, the parallel data processing power can be improved by the implementation of hybrid pipeline algorithms, but that requires special VLSI components. In this way the DPP will become gradually a DMIMD-processor.

At present full interconnectability between the PE's in a PM is only realizable for roughly up to sixteen PE's. It looks evident to try to realize the reduction of interconnectability without the application of the principle "extension in time" to the data transfer system (Dekker 82; Sips 82b). Because normally the data transfer time τ_t will be much smaller than the data processing time (of the smallest independent subtasks), τ_c, it will be attractive to apply also the principle of extension in time. Inside a DPP of 16 PE's there is full interconnectability for totally 32 variables: sixteen output variables of the PE's and sixteen external input variables. In case of a DPP consisting of two PM's, in order to preserve an adequate interconnectability with the environment, it is necessary to subdivide the 16 input variables of a PM in two groups, for instance 8 variables per group. One can use the variables of one group of a PM as input variables of the DPP and the variables of the other group as output variables of the other PM (see figure 5,2a). But that also requires that the output variables are subdivided in two groups, having as a consequence a real interconnectability reduction of 50 percent. However, still it will often be possible to implement easily both distributed parameter systems and sytems described by a spatial hierarchy in a one-to-one way. For example, in the case that the 16 input variables of a PM are subdivided in 12 and 4 variables, the interconnectability reduction is correct for the simulation of a 2D-diffusion system with 4*8 nodes (4*4 & 4*4, but also: 2*8 & 2*8), when the second space derivative is approximated by the second order difference quotient. In practice an interconnectability reduction, like shown in figure 5,2a, will not often give great troubles

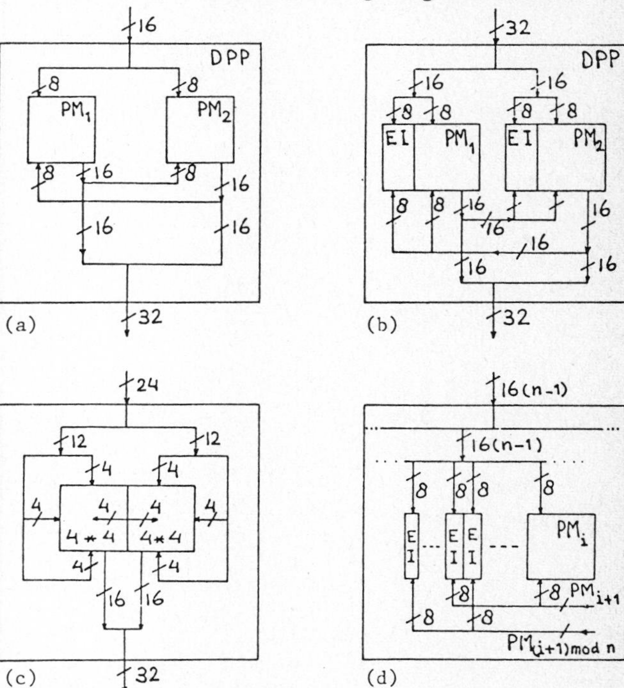

Figure 5,2: Spatial reduction of interconnectability; full interconnectability by extension in time.

with respect to the implementation of system models. But still it would be attractive if the interconnectability reduction could always be removed by applying extension in time of the tranfer of data *ettd*. Indeed that can be realied by means of interconnectability extenders EI, consisting of sixteen registers, in the way shown in figure 5,2b. Normally *ettd* will be equal to *1*, but if one or several of the numbers of variables of an input group or an interconnection group is larger than *8*: *ettd = 2*. For example *ettd* equals *2* for a 2D diffusion system with *4*8* nodes (*4*4 & 4*4* of *2*8 & 2*8*), see figure 5,2c. In this way all systems for which the state variable has a dimension less than *32*, can be simulated in a one-to-one way.

6. PARALLELIZATION OF SOME ENGINEERING SYSTEMS (Kerckhoffs et al 82, Kerckhoffs 82,83, DPP 83)

Finally we want to discuss the parallelization of two systems in the field of engineering, simulated on the DPP. The first example concerns a nonlinear, stiff differential system of the second order: a symmetric, double slope spring, having very different slopes. Such systems are important subsystems for the simulation of the control forces of an aeroplane in a flight simulator ("Fokker Control Loading System"). The Coulomb friction implies a relation between the mass speed $\dot{x}_m$ and the mass force F_m according the following mathematical model:

$$\begin{aligned} F_m &= F-F_c \text{ if } \dot{x}_m>0 \text{ or (if } \dot{x}_m=0 \text{ and if } F>F_c), \\ F_m &= 0 \qquad \text{(if } \dot{x}_m=0 \text{ and if } F<F_c), \\ F_m &= F+F_c \text{ if } \dot{x}_m<0 \text{ or (if } \dot{x}_m=0 \text{ and if } F<-F_c), \end{aligned} \tag{6,1}$$

where F is the input force and F_c is the Coulomb force. The computational precision of a data processor is not sufficient to simulate exactly a precise state of a continuous variable, like $\dot{x}_m=0$. Thus, the system model (6,1) must be modified slightly, before it can be used as a simulation model, for example:

$$\begin{aligned} F_m &= F-F_c, \text{ if } \dot{x}_\varepsilon>\varepsilon \text{ or (if } -\varepsilon\leqslant\dot{x}_\varepsilon\leqslant\varepsilon \text{ and if } F>F) \\ F_m &= 0, \qquad \text{(if } -\varepsilon\leqslant\dot{x}_\varepsilon\leqslant\varepsilon \text{ and if } |F|<F_c) \\ F_m &= F+F_c, \text{ if } \dot{x}_\varepsilon<-\varepsilon \text{ or (if } -\varepsilon\leqslant\dot{x}_\varepsilon\leqslant\varepsilon \text{ and if } F>-F_c) \\ \dot{x}_m &= \dot{x}_\varepsilon-\varepsilon, \text{ if } \dot{x}_\varepsilon>\varepsilon \\ \dot{x}_m &= 0, \text{ if } |\dot{x}_\varepsilon|\leqslant\varepsilon \\ \dot{x}_m &= \dot{x}_\varepsilon+\varepsilon, \text{ if } \dot{x}_\varepsilon<-\varepsilon \end{aligned} \tag{6,2}$$

```
SEQ
    PAR y1:=ẋε-ε; b1:=y1>0,
        y2:=ẋε+ε; b2:=y2<0,
        y3:=F-Fc; b3:=y3>0,
        y4:=F+Fc; b4:=y4<0/
    SWI b1:PAR ẋm:=y1, Fm:=y3,
        b2:PAR ẋm:=y2, Fm:=y4,
      ELSE:PAR ẋm:=0,
          SWI b3: Fm:=y3,
              b4: Fm:=y4,
            ELSE: Fm:=0
          END
    END
END
```

(a)

```
SEQ
    PAR y1:=ẋε-ε; b1:=y1>0,
        y2:=ẋε+ε; b2:=y2<0,
        y3:=F-Fc; b3:=y3>0,
        y4:=F+Fc; b4:=y4<0/
    PAR b5:= NOT(b1ORb2)AND b3,
        b6:= NOT(b1ORb2)AND b4,
        b7:= NOT(b1ORb2ORb3ORb4)/
    SWI b1: PAR ẋm:=y1, Fm:=y3,
        b2: PAR ẋm:=y2, Fm:=y4,
        b5: PAR ẋm:=0,  Fm:=y3,
        b6: PAR ẋm:=0,  Fm:=y4,
        b7: PAR ẋm:=0,  Fm:=0
    END
END
```

(b)

```
SEQ
    PAR y1:=ẋε-ε; b1:=y1>0,
        y2:=ẋε+ε; b2:=y2<0,
        y3:=F-Fc; b3:=y3>0,
        y4:=F+Fc; b4:=y4<0/
    PAR b5:=NOT(b1ORb2),
        b6:=b1OR((NOTb2)ANDb3),
        b7:=b2OR((NOTb1)ANDb4),
        b8:=NOT(b1ORb2ORb3ORb4)/
    SWI b1: ẋm:=y1,
        b2: ẋm:=y2,
        b5: ẋm:=0
    END
    SWI b6: Fm:=y3,
        b7: Fm:=y4,
        b8: Fm:=0
    END
END
```

(c)

Legend of pseudo-instructions:

SEQ	sequential execution of the following tasks until
END	the end of the sequential execution
PAR	simultaneous execution of the following tasks;
...;...	sequential execution of the two subtasks, at the left and right hand side;
...,...	simultaneous execution of the two subtasks, at the left and right hand side;
/	end of a program PAR.
SWI $b_1:t_1,...$	execution of a selection branch; if (b_1 OR...) = 1: execution of that task t_i for which $b_i=1$ for the smallest value i, if not:
ELSE $b_1^*:t_1^*,...$	if (b,OR...) = 0 and (b_1^*OR...) = 1: execution of that task t_j^* for which $b_j^*=1$ for the smallest value of j, if not:
END	end of the execution of the selection branch.
y:=x	y becomes equal to x.
$b:=(y\lesseqgtr 0)$	if $y\lesseqgtr 0$ becomes true, then the boolean b becomes 1.

Figure 6,1: Three parallel algorithms for the realization of the Coulomb friction

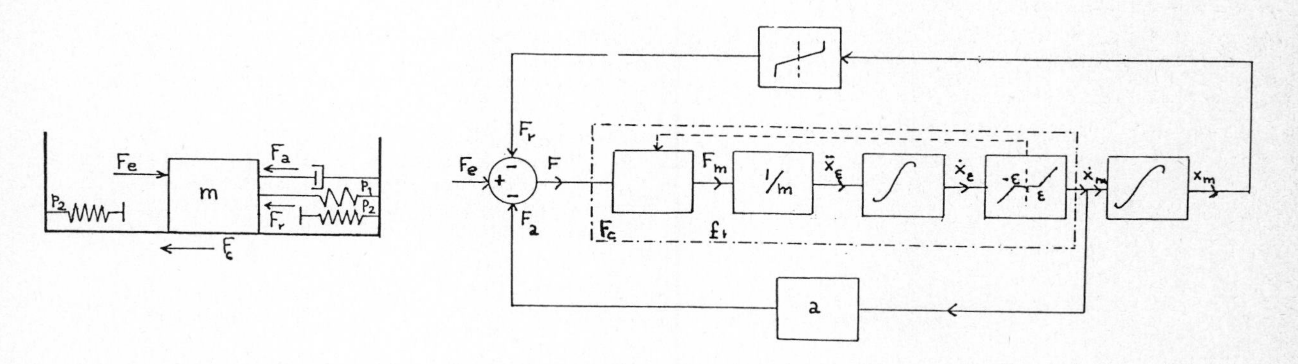

Figure 6,2: Simulation model of a mass/spring system

These are several possibilities to transform this simulation model in a parallel algorithm. We will demonstrate this by designing some parallel algorithms by means of pseudo-instructions. One can see from (6,2) that the simulation model realizes the flow of the data $\dot{x}_m$ and F by means of the following stages: arithmetical comparisons, boolean operations, conditional transfers, arithmetical operations. In this way the simulation model (6,1) has by nature a partly sequential behaviour. The programs of three corresponding parallel algorithms are shown in figure 6,1. The parallel program of figure 6,1a concerns a realization that yields the shortest execution time under all circumstances. Thus, this algorithm is not adequate for a real-time simulation. The other two algorithms have always a constant execution time and so they can be used in real-time applications. (Note that the small variations, caused by the fact that the calculations not always take the same time, can be eliminated by a simple synchronization). The third parallel algorithm has the advantage that for each of the variables $\dot{x}_m$ and F_m the conditional assignment expressions appear all in one selection branch (SWI...END). As a consequence the interaction structure of the PE's of this parallel algorithm will be the most simple. The figure 6,2 shows the block diagram of the simulation model. The three other subprograms of the parallel algorithm of this simulation model, for the simulation of:

- the spring force F_r,
- the input mass force F,
- the mass displacement x_m,

are described in the figure 6,3a,b,d respectively, where p_1 and p_2 are the two spring slopes and $a * \dot{x}_m$ is the damping force. The total program can be compressed in four macro-instructions: "spring force", "input force", "friction force", "displacement".

```
SEQ
 PAR z1:=xm-x1 ; c1:=(z1>0),           (a)
     z2:=xm+x1 ; c2:=(z2<0)/            .
 SWI c1: Fr:=p1*x1+p2*z1,               .
     c2: Fr:=-p1*x1+p2*z2,              .
   ELSE: Fr:=p1*xm                      .
 END                                   (a)
 F:Fe-Fr-a*ẋm                          (b)
 parallel algorithm                    (c)
 for the                                .
 Coulomb friction                      (c)
 PAR ẍm =Fm/m,                         (d)
     ẋ :=x +h*ẍ                         .
     xm:=xm^ε+h*ẋm^ε/                  (d)
```

Figure 6,3 The three other subprograms

The parallel algorithm of a second example, a distributed parameter system, will only be summarized. First we will consider the simulation of a simple heat diffusion system in two dimensions (see figure 6,4), according to the mathematical model:

$$T_t=d_1*T_{xx}+d_2*T_{yy}$$

$$T(x,y,0)=T_0;T_b(x,t)=T_b,T_h(x,t)=T_h, \qquad (6,3)$$

$$T_g(y,t)=T_g,T_d(y,t)=T_d.$$

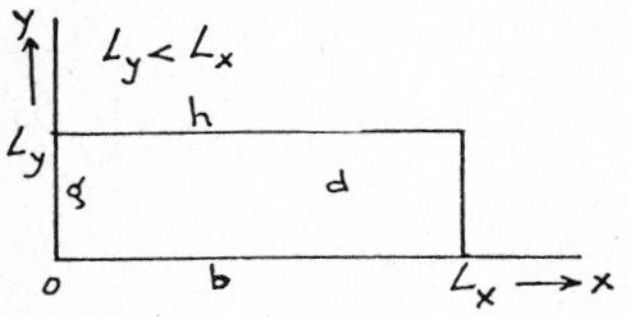

Figure 6,4 The boundaries b,g,h,d of the heat diffusion system.

The simulation model is obtained by means of the approximation of the partial derivatives by sampling the continuous variables t,x,y, for example as follows:

$$u:=u+h(c_b(u_b-u)+c_g(u_g-u)+c_h(u_h-u)+c_d(u_d-u)) \qquad (6,4)$$

where u, u_b, u_g, u_h, u_d are the approximations of the temperatures T in a node and its neighbour nodes. For convience we assume that the value of L_y = (m+1) y is so small that the number of the PE's is sufficient to treat the nodes (x_i, y_j), $j = 1,\dots,m$ simultaneously. Then, in the y-direction one can apply the principle of extension in space to deal with the time-parallel subtasks $t_1,\dots,t_m$. In the x-direction one has to apply extension in time: $t_i = t_{i1};\dots;t_{in}$, $i = 1,\dots,m$, where $L_x = n * x$. One can realize the sequence of subtasks t_{ij}, $j = 1,\dots,n$ by means of: (a) a decentralization management or (b) a centralized management:

```
    SEQ                                  SEQ DO  j=1,...,n
         PAR t11;...;t1n                      PAR tij,...,tmj/
(a)           .                 (b)           END
              .                          END                  (6,5)
            tm1;...;tmn/
```

Note that certain modifications of the diffusion system (6,3) can easily be included in the subtasks t_{ij}, like for the boundaries g and d an irregular geometry as well as a boundary control.

In similar cases a decentralized management will normally result in a better parallelization efficiency. The subtasks of distributed-parameter systems have an identical interaction structure. As a consequence the algorithm modelling of a distributed-parameter system can become simple with the help of interconnection and interaction macro-instructions.

BIBLIOGRAPHY:

Andriessen, J.H.M. ('80), "Discrete-time parallel algorithms with continuous-time response", in Proc. Simulation '80, Interlaken, Acta Press, pp. 98-103.

Andriessen, J.H.M., Dekker, L., Kerckhoffs, E.J.H. (eds.) ('81), in Proc. Algorithmics in parallel processing and simulation, 6th European Simulation Meeting, Delft, 1981.

Brok, S.W., Dekker, L., Kerckhoffs, E.J.H., Ruighaver, A.B., Sips, H.J. ('83), "Architecture and programmature of the MIMD-structured Delft Parallel processor", 1st European Simulation Congress ESC 83, Aachen, 1983, Springer Verlag Berlin.

Dekker, L. ('70), "Applicability of hybrid computation", in Proc. Hybrid Computation, AICA-IFIP, Munich, 1970.

Dekker, L., Gelderen, J.A. van, Kerckhoffs, E.J.H., ('73), "A block-iterative method for the solution of large systems of ordinary differential equations", in Proc. AICA, Prague, 1973, pp. 125-129.

Dekker, L. ('76), "Algorithms for parallel processing", in Proc. Microprocessors and simulation, 3rd European Simulation Meeting, Capri, 1976, pp. 5-39.

Dekker, L., Kerckhoffs, E.J.H., Vansteenkiste, G.C., Zuidervaart, J.C. ('79), "Outline of a future parallel simulator", in Proc. Simulation of systems, IMACS, Sorrento, 1979, North Holland Publishing Co., pp. 837-864.

Dekker, L. ('81), "Stability as a design criterium for true parallel algorithms", in proc. Algorithmics in parallel processsing and simulation, 6th European Simulation Meeting, Delft, 1981, pp. 3.1-3.33.

Dekker, L., Sips, H.J., Zuidervaart, J.C. ('82), "A view on parallel simulation", Computer-aided modelling and simulation, J.A. Spriet and G.C. Van-steenkiste (eds.), Acad. Press, 1981, pp. 429-454.

Dekker, L. ('82), "Delft Parallel Processor: A missing link to the future of simulation", SIMS '82 Annual Meeting, Trondheim, 1982.

Dekker, L. ('83a), "Future of parallel data processing", in Proc. Special computers (bijzondere rekenautomaten), KIVI, The Hague, 1983.

Dekker, L. ('83b), "Concepts for an Advanced Parallel Simulation Architecture", Contribution to "Simulation and modelbased methodologies: an integrative view", T.I. Oren, B.P. Ziegler, M.S. Elzas (Ed.), Springer, Heidelberg, 1983.

DPP ('83), "Delft Parallel Processor: general information and applications", Delft University of Technology, Delft, 1983.

Haynes, L.S., (guest ed.) ('82), Highly Parallel Computing, Computer, 15, 1, pp. 7-96.

IEEE Trans. Comp. ('77), "Parallel processing", Vol. C-26, no. 2, 1977
('79), "Data base systems", Vol. C-12, no. 3, 1979.

Kuhn, R.H., Padua, D.A. ('81), Tutorial on parallel processing, IEEE Computer Society, Los Angeles, U.S.A..

Kerckhoffs, E.J.H. ('82), "The application of an experimental parallel processor for the simulation of systems in biotechnology and medical engineering", in Proc. Modelling and data-analysis in biotechnology and medical engineering, working conference, IFIP, Brussels, 1982.

Kerckhoffs, E.J.H., and Meiboom, W. ('82), "Simulation of data-base functions in an experimental parallel processor/some simple examples", in Proc. Modelling and simulation, AMSE, Paris, 1982.

Kerckhoffs, E.J.H., ('83), "The applicability of the Delft Parallel Processor", in Proc. Special computers (bijzondere rekenautomaten), KIVI, The Hague, 1983.

Sips, H.J. ('82), "Philosophy behind the design and construction of the Delft Parallel Processor as a simulation tool", SIMS '82 Annual Meeting, Trondheim, 1982.

Sips, H.J. ('82b), "A bit-sequential multi-operand inner product processor", in Proc. Parallel Processing K.E. Batcher et al, (eds.), IEEE Comp. Soc., Los Angeles, 1982, pp. 301-303.

Simulation in Engineering Sciences
J. Burger and Y. Jarny (eds.)
Elsevier Science Publishers B.V. (North-Holland)

COMPARAISON DES OPTIMUMS ENERGETIQUE, ECONOMIQUE ET ECOLOGIQUE D'UN SYSTEME INDUSTRIEL

Pierre Le Goff

Laboratoire des Sciences du Génie Chimique CNRS-ENSIC
Institut National Polytechnique de Lorraine - Nancy

Dans un projet industriel, il n'existe généralement pas un objectif unique à atteindre, mais plusieurs fonctions-objectifs à optimiser, selon le choix du décideur : - pour l'ingénieur, responsable du projet, au sein d'une entreprise privée, l'objectif est de minimiser le coût total, en francs, de l'opération ; - pour les Pouvoirs Publics, l'objectif est de minimiser les paiements, nets, en devises étrangères ; - pour un hypothétique décideur mondial, responsable du capital d'énergie fossile de la planète, l'objectif serait de minimiser la consommation totale d'énergie primaire...
L'auteur présente et compare, pour un même projet, les trois comptabilités : en francs, en devises et en énergie primaire. Il montre comment les trois optimums sont parfois en conjonction et parfois en contradiction. Plusieurs exemples concrets portant sur les coûts d'exploitation et les coûts d'investissements sont ainsi analysés.

1. LES RESSOURCES RARES

1.1 La Welmmite

Toute opération industrielle a pour objectif plus ou moins direct de chercher à satisfaire des besoins d'êtres humains, en partant de l'exploitation des ressources naturelles. Comme le montre la figure 1, cette opération peut être considérée comme un "réacteur" ouvert, qui transforme des flux de grandeurs d'entrée en flux de grandeurs de sortie. Les principales grandeurs d'entrée (les "intrants") sont : l'Eau, l'Energie, le Territoire, les Matériaux, la Main-d'oeuvre. Ce sont les grandeurs *"WELMM"*, d'après la nomenclature proposée par l'I.I.A.S.A. (1) Les grandeurs de sortie (les "extrants") sont d'une part tous les *biens et services*, destinés aux consommateurs, et d'autre part les *sous-produits*, dont une partie est ré-utilisée et le reste rejeté dans l'environnement (déchets matériels, énergie dégradée...).

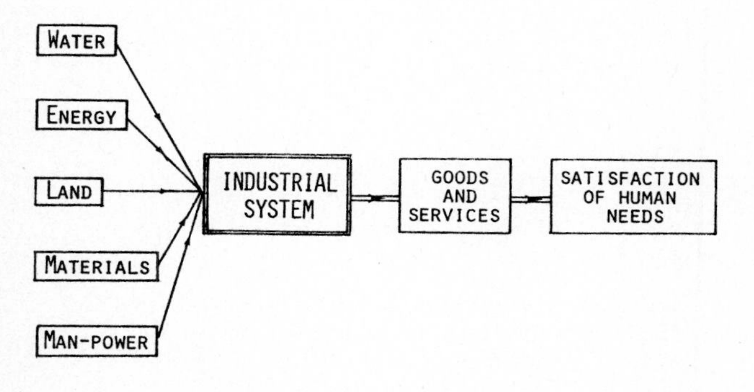

Figure 1 : The WELMM approach

Le décideur technique responsable d'une telle opération a généralement pour mission de satisfaire l'objectif qui lui est assigné (- produire une quantité imposée de biens et de services -) en *consommant le moins possible des grandeurs WELMM*, grandeurs qui sont plus ou moins rares, donc plus ou moins "coûteuses". Mais il est important de remarquer que ces grandeurs ne sont pas indépendantes : par exemple une économie d'énergie sera souvent associée à une sur-consommation d'un matériau et d'autre part elle pourra entraîner soit un sur-emploi, soit un sous-emploi de main-d'oeuvre.

Les auteurs de l'I.I.A.S.A. disent que l'on ne consomme pas séparément de l'énergie, des matériaux... mais une grandeur composite unique, qu'ils ont baptisée la *WELMMITE !* Ce concept convient certainement fort bien pour les études macro-économiques, à l'échelle d'une nation, ou de toute la planète. Des résultats intéressants ont été publiés.(2) Mais ce concept n'est pas suffisant pour les études micro-économiques à l'échelle d'une entreprise industrielle.

1.2 La Meltide

Il nous semble qu'au moins deux grandeurs coûteuses importantes doivent être ajoutées aux 5 grandeurs définies par l'I.I.A.S.A., à savoir :

- *L'INFORMATION* : c'est le contenu des *connaissances* scientifiques et du *savoir-faire* technique des opérateurs humains. C'est précisément l'apport d'information qui permet de faire des économies sur les autres grandeurs coûteuses, grâce à une amélioration du système de production donc de la productivité.
- *LE TEMPS* : il intervient sous diverses formes :
  . c'est d'abord le *délai de réalisation*, c'est-à-dire le temps écoulé entre l'instant où l'on décide de produire et l'instant où le produit est disponible pour l'utilisateur,

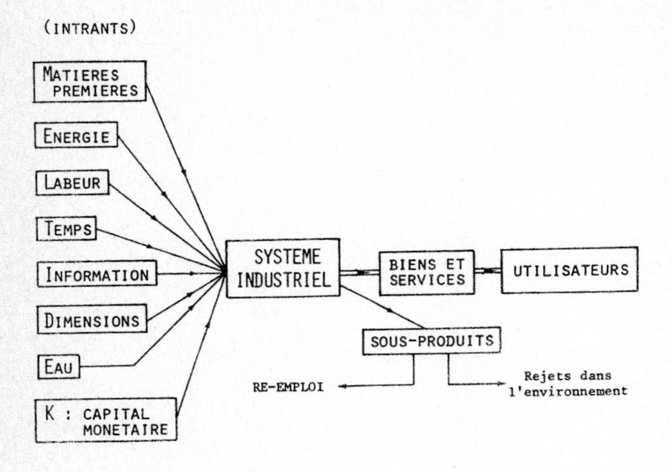

Figure 2 : La Meltide (+K) et ses constituants

. c'est ensuite la *durée d'utilisation* des équipements installés, donc la durée d'amortissement du capital investi,

. c'est enfin, et plus généralement, le *temps "entropique"* qui traduit le phénomène universel d'*usure* des machines, donc de baisse de productivité.

Mais pour compléter cet inventaire du patrimoine des ressources rares, disponibles à un instant donné, il faut encore ajouter le CAPITAL MONETAIRE qui doit être considéré comme une autre "matière première" également soumise aux lois du marché. Les 7 premiers intrants sont partiellement substituables les uns aux autres, comme expliqué dans la suite de ce rapport, tandis que ce 8ième intrant présente l'avantage unique d'être entièrement substituable aux 7 autres, tout au moins en économie de libre échange (voir réf. 3, volume 2, pour plus de détails).

Les intrants seront donc constitués des grandeurs coûteuses énumérées sur la figure 2 et nous parlerons, en français, de consommation de MELTIDE[1].

1.3 La fonction de Cobb-Douglas généralisée

Considérons une opération industrielle qui produit une certaine quantité P_1 d'un bien (ou d'un service) en consommant certaines quantités des intrants coûteux : $M_1, E_1, L_1, T_1, I_1, D_1, E_1, K_1$... L'"Etat" de fonctionnement du système est défini par les valeurs de ces 8 variables d'état. Pour représenter l'influence, sur la production P, de *petites variations* de chacune des 8 variables à partir de cet état de référence, le modèle mathématique le plus simple est la fonction puissance :

$$\frac{P}{P_1} = \left(\frac{M}{M_1}\right)^{\alpha_m} \cdot \left(\frac{E}{E_1}\right)^{\alpha_e} \cdot \left(\frac{L}{L_1}\right)^{\alpha_l} \cdot \left(\frac{T}{T_1}\right)^{\alpha_t} \cdot \left(\frac{I}{I_1}\right)^{\alpha_i} \cdot \left(\frac{D}{D_1}\right)^{\alpha_d} \cdot \left(\frac{Eau}{Eau_1}\right)^{\alpha_o} \cdot \left(\frac{K}{K_1}\right)^{\alpha_k} \quad (1)$$

appelée : fonction de *COBB-DOUGLAS généralisée*. Rappelons en effet que les économistes COBB et DOUGLAS ont proposé, en 1928, de représenter la relation entre la production P, la main-d'oeuvre L et le capital K, par la relation :

$$P = k.L^{\alpha}K^{\beta} \quad \text{avec } \alpha+\beta = 1 \quad (2)$$

La relation (1) apparaît bien comme une généralisation de (2).

Dans le vocabulaire des économistes, les α s'appellent des *élasticités*. Par exemple α_e est l'élasticité de la production par rapport à l'énergie consommée.

2. SUBSTITUTION ENERGIE-MATIERE PREMIERE

2.1 Isoquantes de production

Supposons que l'opération ait pour objectif de produire une quantité imposée P d'un certain bien. Le décideur a le choix entre de nombreuses combinaisons des divers intrants : les intrants sont, au moins en partie, *substituables* les uns aux autres.

Considérons à titre d'exemple l'extraction d'un métal à partir d'un minerai, par les opérations successives de concassage-broyage, lavage par solvant, précipitation, calcination, réduction.

Chacune de ces opérations a un rendement en métal inférieur à l'unité et ce rendement est, toutes choses égales par ailleurs, une fonction *croissante* de l'énergie qui y est consommée. Pour produire P kg de métal, on a donc le choix entre consommer beaucoup de matière première et peu d'énergie, ou bien faire l'inverse.

Cette substituabilité de la matière première par de l'énergie est avantageusement représentée sur un diagramme bidimensionnel (figure 3a). On y trace un réseau de courbes appelées ISOQUANTES de PRODUCTION correspondant à une quantité P de produit donnée et d'équation :

$$\dot{M}.\dot{E}^{\beta} = \text{cste} \quad \text{avec } \beta \equiv \frac{\alpha_e}{\alpha_m} \quad (3)$$

β est l'"élasticité" de la substitution matière/énergie.

La consommation *totale* d'énergie E_{tot} (pour un kg de produit P) est la somme d'un terme *croissant*, l'énergie opératoire $\dot{E}$ et d'un terme *décroissant*, le contenu énergétique CE de la matière première, soit :

$$\dot{E}_{tot} = \dot{E} + \dot{M}.CE \quad (4)$$

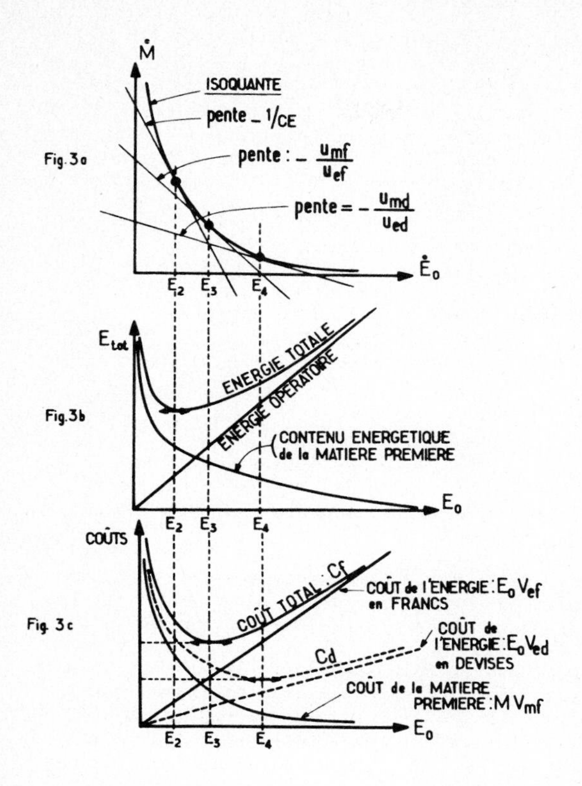

Le "contenu énergétique" CE est la somme de toutes les quantités d'énergie consommées dans les opérations placées en amont de l'opération étudiée, c'est-à-dire tout le long de la filière d'élaboration de l'intrant-matière à partir de la ressource naturelle (= le minerai).
La figure 3b montre que l'énergie totale présente un minimum pour la valeur E_2 de l'énergie opératoire, soit :

$$E_2 = \left[\frac{\beta \dot{M}_1}{E_1} \cdot CE\right]^{1/1+\beta} \qquad (5)$$

Mais l'équation (4) s'écrit aussi :

$$\dot{M} = \frac{E_{tot}}{CE} - \frac{E}{CE} \qquad (6)$$

et se représente par une droite de pente $-1/CE$ sur la figure 3a. Il est facile de vérifier que le point de l'isoquante qui correspond au minimum de consommation d'énergie totale est celui dont la tangente a pour pente $-1/CE$.

2.2 Optimums en francs, en devises et en énergie primaire

Reprenons le même raisonnement que ci-dessus mais en comptabilisant maintenant des francs et non plus de l'énergie. Soit v_{mf} la valeur en francs, sur le marché national, d'un kg de matière première et v_{ef} la valeur, en francs, d'une unité d'énergie. Le coût total, en francs, de fabrication d'un kg de produit est encore ici (fig. 3c) la somme d'un terme croissant $E_o.v_{ef}$ et d'un terme décroissant $M.v_{mf}$:

$$C_f = E_o v_{ef} + M v_{mf} \qquad (7)$$

Le minimum du coût total en francs a pour abscisse $E = E_3$. Il correspond au point de l'isoquante de production dont la tangente a pour pente $-v_{ef}/v_{mf}$ (figure 3a).

Mais supposons que la matière première et l'énergie soient importées, au moins en partie. Il faut alors considérer cette même opération sur le marché international et comptabiliser les valeurs en devises étrangères. Soient v_{ed} et v_{md} les valeurs, en devises, d'une unité d'énergie et d'un kg de matière première. Le coût total en devises est alors :

$$C_d = E_o v_{ed} + M v_{md} \qquad (8)$$

La figure 3c montre, à titre d'exemple, le cas où la matière première serait payée à 100 % en devises, alors que l'énergie entraînerait une sortie de devises correspondant seulement à 40 % de sa valeur en francs. Dans ce cas, le minimum du coût total en devises correspond à un point d'abscisse E_4 nettement différent du point E_3.

2.3 La "CAREC" : CARactéristique Energie-Coûts

Pour chaque opération industrielle, pour chaque projet, qu'il soit à l'échelle micro-économique ou à l'échelle macro-économique, on pourra donc établir parallèlement les 3 bilans comptables de *valeur* des divers intrants :

1/ la *valeur en monnaie nationale* : en économie de marché, tout décideur technique appartenant à une entreprise privée cherchera évidemment à *minimiser le coût de fabrication* du produit et il évaluera ce coût en francs, qui est l'unité de valeur sur le marché national ;
2/ la *valeur en devises étrangères* : pour un responsable du ministère des finances, cette même opération sera jugée du point de vue de la balance des paiements sur le marché international. Son objectif sera de *minimiser les sorties de devises*. La valeur devra être mesurée en devises, pratiquement en US-dollars ;
3/ la *valeur en énergie primaire* : pour un hypothétique décideur de niveau planétaire, qui se donnerait comme seul objectif de *préserver le capital*, limité, *d'énergie fossile* dont dispose la terre, la valeur de chaque bien économique serait donnée par son *contenu en énergie primaire*.

Pour un même projet les 3 solutions optimales qui correspondent respectivement aux minimums de chacun des 3 coûts, en francs, en devises et en énergie primaire seront comparés avantageusement à l'aide d'une représentation tridimensionnelle (figure 4). Un "état" de fonctionnement du système est représenté par un point de coordonnées E_{tot}, C_f, C_d. Pour une production P imposée avec 2 degrés de liberté, ce point décrit une *surface* isoquante dont la forme est généralement celle d'un paraboloïde. Cette surface présente 3 plans tangents remarquables, correspondant respectivement aux minimums :
- de la consommation d'énergie primaire : E_{min}
- du coût en francs : C_{fmin}
- du coût en devises : C_{dmin}.

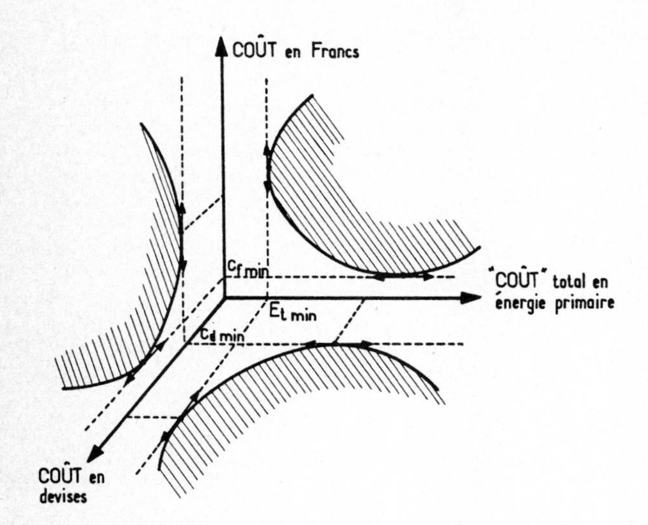

Figure 4 : Surface Caractéristique Energie-Coûts ou "CAREC"

Nous proposons d'appeler *CAREC* (pour CARactéristique Energie-Coûts) une telle surface, ainsi que les courbes constituées par ses projections sur les trois plans principaux. (Il sera en effet souvent plus simple de raisonner dans les 3 espaces à 2 dimensions). La *CAREC* d'une opération industrielle sera l'instrument de base pour procéder à *l'étude comparée des 3 comptabilités*, en francs, en devises et en énergie primaire.

Bien entendu cette méthode ne s'applique pas seulement à l'étude de la substitution globale : énergie/matière première. Elle s'applique aussi à toute autre substitution entre divers intrants coûteux, notamment entre deux matières premières substituables (par exemple métaux/matières plastiques ou encore cuivre-aluminium) ou entre deux sources d'énergie, comme combustible fossile et électricité.

3. OPTIMISATION D'UN SYSTEME A DEUX SOURCES D'ENERGIE

La plupart des procédés industriels utilisent à la fois de l'énergie thermique et de l'énergie mécanique. La première est fournie sous la forme d'un fluide caloporteur (fumées de combustion, vapeur surchauffée) produit dans l'usine même à partir d'un combustible (pétrole, charbon, gaz). La seconde est généralement produite par des moteurs électriques et l'électricité est achetée au réseau national. Or ces deux formes d'énergie peuvent souvent être *substituées l'une à l'autre* au moins en partie.

Dans les industries *agro-alimentaires*, on trouve le *pressage-séchage des pulpes* gorgées d'eau (pulpes de betterave, fourrages, pâtes à papier, drêches de brasserie...). Le surpressage des pulpes, avant séchage, a pour but d'enlever par un moyen mécanique, une partie importante de l'eau qui les accompagne. Le surpressage permet en consommant une faible quantité d'énergie mécanique, d'éviter la consommation d'une quantité considérablement supérieure d'énergie thermique qui serait nécessaire à l'évaporation de cette eau (figure 5).

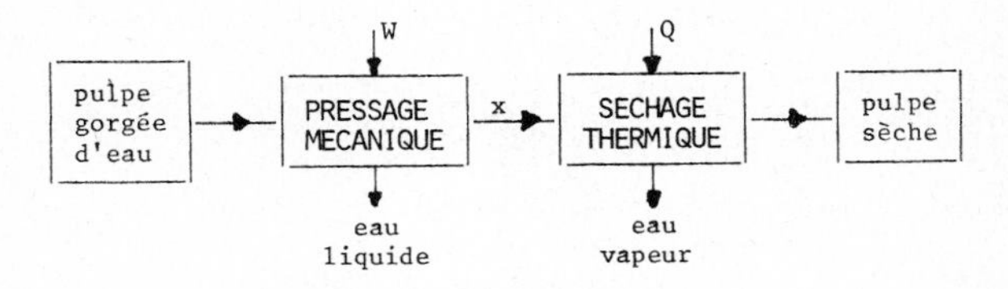

Figure 5 : Le pressage-séchage des pulpes

Soit x, le *taux de siccité* de la pulpe entre le pressage et le séchage (x ≡ pourcentage de matière sèche dans la pulpe). L'énergie mécanique W à fournir à la presse est une fonction *croissante* de x. Tandis que l'énergie thermique Q à fournir au séchoir est une fonction *décroissante* de x. Il existe donc une valeur optimale de x qui correspond au minimum de consommation d'énergie primaire (figure 6).

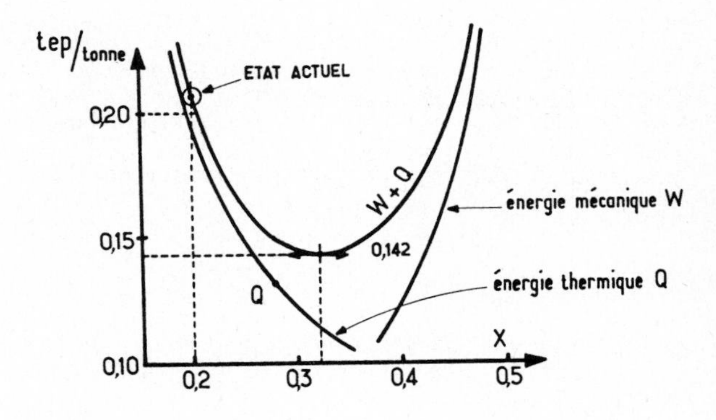

Figure 6 : Consommations d'énergie primaire en fonction du taux de siccité intermédiaire x

Cette courbe a été tracée sur l'exemple d'une sucrerie de taille moyenne, qui traite 4000 tonnes de betteraves par jour (pendant une campagne de 80 jours par an) et produit ainsi 22 000 tonnes par an de pellets de pulpe séchée à 88 %, destinées à l'alimentation du bétail. La consommation totale d'énergie primaire est :

$$E_{tot} = 5{,}25.x^{4,36}+0{,}05.x^{-1}-0{,}051 \text{ tep/tonne} \quad (9)$$

Sur la figure 7 est portée, en abscisse, la consommation totale d'énergie primaire, et en ordonnées, à la fois le coût en francs et le coût en devises. Ainsi nous superposons sur un même plan, les projections sur deux plans de la CAREC (CARactéristique Energie/Coût) qui devrait normalement être présentée dans un espace à 3 dimensions comme sur la figure 4.

On remarque que l'isoquante de coût en francs a une *extrémité très pointue*, du fait que le minimum en énergie (point de tangente verticale) et le minimum en francs (point de tangente horizontale) sont très voisins. Dans l'état actuel de fonctionnement, la siccité $x_1 = 0,20$ est *nettement inférieure* à celles des trois optimums calculés ci-dessus. Elle est même inférieure à celle de l'optimum monétaire en francs (0,30) ce qui semble contraire au principe même de la société à économie de marché. Cette apparente divergence vient de ce que nous ne considérons ici que les coûts opératoires, *hors investissements*. Or la situation actuelle résulte d'une optimisation des investissements et des coûts opératoires.

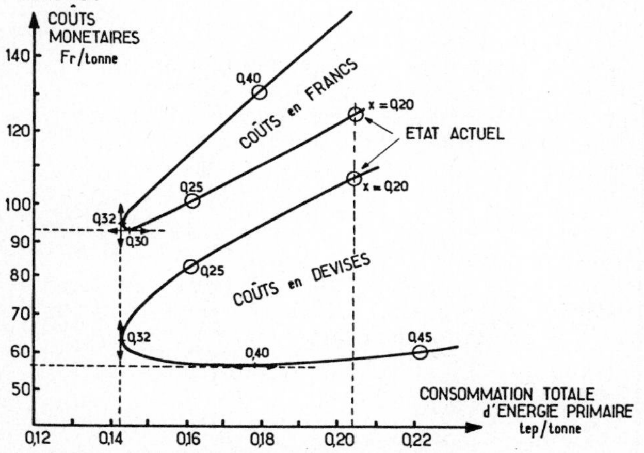

Figure 7 : CARACTERISTIQUE ENERGIE-COUT (CAREC) pour le pressage-séchage des pulpes de betterave

Le tracé de ces CAREC révèle l'importance des économies réalisables si l'on passait de l'état actuel (x = 0,20) à la valeur x = 0,30 à 0,32, correspondant aux minimums en énergie et en francs : l'économie d'énergie serait de 30 %, celle en francs serait de 25 % et celle en devises de plus de 40 %. Il est donc urgent que l'on mette au point des presses permettant ces performances.

4. OPTIMISATION DES INVESTISSEMENTS ECONOMISEURS D'ENERGIE

D'après le raisonnement précédent, tout déplacement vers "la gauche" à partir de l'état actuel semblerait bénéfique du point de vue de l'intérêt public, puisqu'il conduit à une économie d'énergie opératoire. Mais cette économie d'énergie a été obtenue grâce à un sur-équipement.. Or ce sur-équipement a lui-même nécessité de l'énergie pour être fabriqué. Alors ne faut-il pas craindre que l'une compense l'autre... et qu'au total l'économie d'énergie soit négligeable... sinon négative !

La réponse à cette question implique que, dans un projet, on ne s'occupe pas seulement de l'énergie opératoire, mais que l'on fasse le *bilan énergétique complet* en tenant compte du CONTENU ENERGETIQUE de tous les équipements.

Ainsi le diagramme de la figure 7 devra être modifié : on portera en abscisse, non plus seulement l'énergie opératoire $\dot{E}_0 N$ consommée pendant N années mais aussi le contenu énergétique CE des équipements.

Considérons, à titre d'exemple, un des problèmes les plus classiques dans les entreprises d'ingénierie : il s'agit de choisir le diamètre optimal d'une canalisation destinée à transporter un débit de fluide $\dot{M}$ imposé, sur une distance donnée L, par exemple d'un réservoir à un autre réservoir (figure 8).

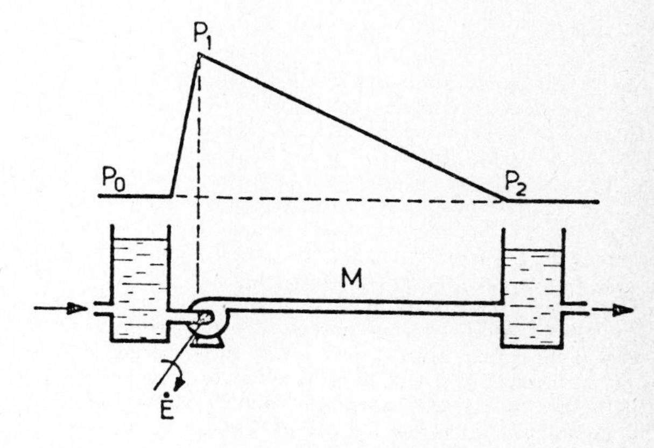

Figure 8

Un tuyau de grand diamètre d représente un investissement élevé, mais la perte de pression y est faible (elle varie comme d^{-5} en régime turbulent) et l'énergie mécanique dégradée par frottement est donc également faible. De plus le prix du groupe moto-pompe nécessaire pour véhiculer le fluide croît avec sa puissance nominale et donc aussi avec l'énergie dégradée. Ainsi l'investissement total minimal sera le compromis entre deux systèmes : une grosse canalisation associée à une petite pompe ou bien une petite canalisation associée à une grosse pompe (point I sur la figure 9).

En ajoutant maintenant à cet investissement le coût opératoire, provenant de l'énergie consommée durant les N années de fonctionnement du système, on obtient un autre optimum (point P) pour une valeur du diamètre de la canalisation nettement différente du précédent. Ensuite, ces calculs pourront être recommencés en tenant compte du contenu énergétique du tuyau et de la moto-pompe, ce qui conduira à deux nouveaux optimums, l'un pour le minimum de l'"investissement énergétique" (c'est-à-dire la somme des contenus énergétiques des équipements) et l'autre pour le minimum de la consommation totale d'énergie primaire.

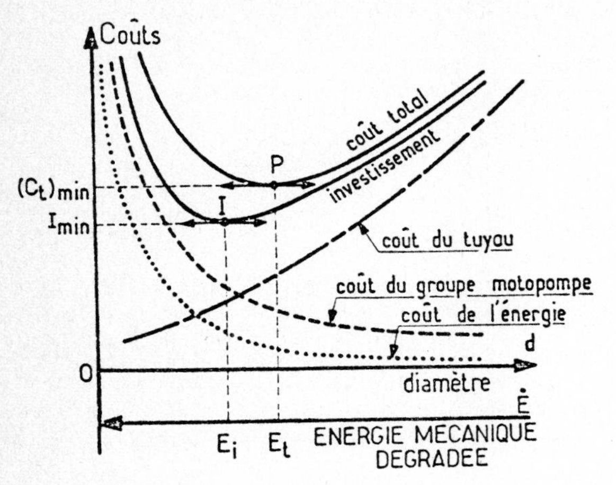

Figure 9

Ces 4 optimums sont présentés simultanément sur le diagramme CAREC de la figure 10, calculé sur l'exemple du transport d'un débit de 120 m^3/heure d'eau sur une distance de un kilomètre. On peut y lire que le diamètre optimal varie de 5,7 cm (pour l'investissement énergétique minimal) à 20,3 cm (pour la consommation totale d'énergie minimale), les deux optimums monétaires étant respectivement de 9,44 cm et de 15,9 cm.

Les coordonnées de chacun de ces 4 points donnent les coûts respectifs en francs et en énergie primaire. Il est donc facile d'en déduire les économies réalisées, ou les surcoûts, quand on passe d'un optimum à un autre.

De nombreux procédés industriels sont des généralisations de l'exemple précédent. Du point de vue de l'énergie dissipée dans le système, on peut classer les équipements en trois parties : les éléments qui *fournissent de l'énergie* au système, les éléments dans lesquels cette *énergie est dégradée*, enfin ceux qui sont *indépendants* de l'énergie.

Nous appellerons "*pro-énergétique*", l'investissement qui correspond aux machines qui fournissent de l'énergie au système (pompes, moteurs, ventilateurs, chaudières...). Leur coût est une fonction *croissante* de leur puissance nominale et sera donc presque toujours une fonction croissante du flux d'énergie dégradée $\dot{E}$ (en kwh/an par exemple).

De la même manière, l'investissement "*anti-énergétique*" sera constitué par les parties de l'installation où se dégradent les différentes énergies mises en cause ; il s'agira par exemple des conduites dans lesquelles l'énergie mécanique de pression est dégradée en chaleur par suite des frottements, des parois plus ou moins efficacement isolées à travers lesquelles les calories s'échappent vers l'extérieur... Le coût de tels systèmes sera presque toujours une fonction décroissante du flux d'énergie qui s'y dégrade.

Le coût total d'une opération est donc la somme de quatre termes :

(COUT TOTAL) = (COUT DES MACHINES FOURNISSANT L'ENERGIE) + (COUT DES EQUIPEMENTS QUI REDUISENT LA CONSOMMATION D'ENERGIE) + (COUT DE L'ENERGIE CONSOMMEE) + (COUTS INDEPENDANTS DE L'ENERGIE)

investissement pro-énergétique — investissement anti-énergétique

$$C_t = A.\dot{E}^a + B.\dot{E}^{-b} + D.\dot{E}^d + K \quad (10)$$

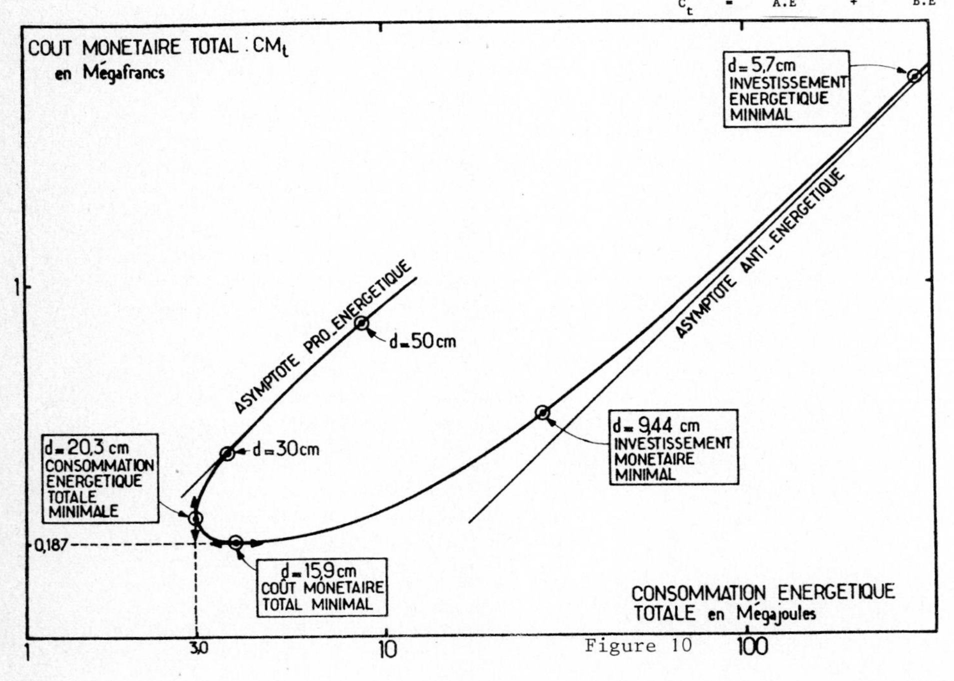

Figure 10

La technique d'optimisation présentée ci-dessus sur des exemples simples a donc pu être généralisée à des systèmes plus complexes, avec des équations du type (10).(3)

5. LA SUBSTITUTION : INFORMATION/ENERGIE + MATIERES PREMIERES

Tout système industriel de production implique des apports d'information : INFORMATION *EXTERNE* constituée par l'ensemble des *connaissances* scientifiques et techniques et aussi du *savoir faire* des opérateurs humains (la "matière grise") ; INFORMATION *INTERNE* issue des capteurs de mesures physiques et chimiques placés sur les grandeurs de sorties (mesures de températures, pressions, concentrations...) (figure 11).

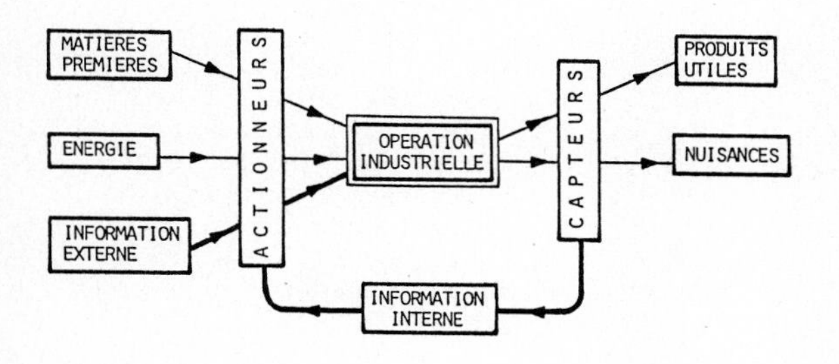

Figure 11

Les raisonnements qui précèdent sur la substitution énergie/matières peuvent être presque exactement transposés à la substitution : information/autres intrants.

Considérons à titre d'exemple un projet d'automatisation d'un procédé industriel existant : il s'agit d'installer sur le procédé, un ensemble de capteurs d'information sur les extrants, et d'actionneurs sur les intrants reliés par un système de régulation automatique.

On espère ainsi réduire la consommation d'énergie et de matières du procédé grâce à un meilleur contrôle continu de son fonctionnement. Mais cette chaîne informatique entraîne de son côté un surcoût qu'il faudra amortir sur toute la production.

Comme le montre la figure 12, il existera toujours un état optimal correspondant au minimum du coût total.

Ce même raisonnement peut être appliqué aux apports d'information externe, tels que achats de brevets et de licences, investissements dans des laboratoires de recherche scientifique.

Une augmentation ΔI des connaissances scientifiques et techniques du système dont le coût d'acquisition est ΔC_i doit normalement entraîner une réduction $\Delta \dot{C}_e$ du coût annuel de l'énergie et des matière premières consommées et des nuisances produites.

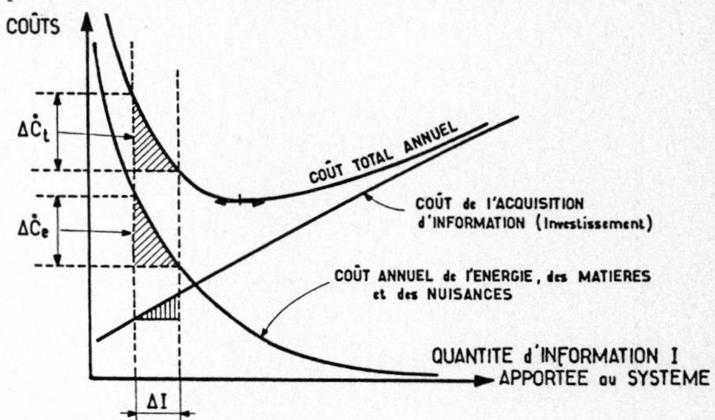

Figure 12 : Bilan de la substitution : information/énergie + matières + nuisances

Le "*temps de récupération*" d'un investissement informationnel T_r est donc donné par le rapport

$$T_r \equiv \frac{\Delta C_i}{\Delta \dot{C}_e} \quad \text{(en années)}$$

Bien entendu, un tel investissement ne sera rentable économiquement que si son temps de récupération T_r est *inférieur* à la durée de fonctionnement de l'installation, soit N années. La réduction annuelle du coût total $\Delta \dot{C}_t$ (figure 12) est alors donnée par :

$$\Delta \dot{C}_t = \Delta \dot{C}_e - \Delta C_i / N$$

Nous avons appliqué un tel calcul aux subventions accordées par les Pouvoirs Publics à divers organismes de recherches industrielles et universitaires dans le cadre de l'Action concertée "*Energie et Génie Chimique*" pendant les années 1976-79. Nous avons pu vérifier que les temps de récupération s'échelonnaient, pour la plupart, entre quelques mois et 3 à 4 années. Il s'agissait donc d'investissements parfaitement rentables du point de vue national.

CONCLUSION

Les modèles mathématiques que nous avons présentés ici sous une forme simple et qui sont développés dans un ouvrage publié récemment ne prétendent absolument pas se substituer aux opérateurs humains.(3) Ces modèles ne sont que des AIDES A LA DECISION. Ils ont, à notre avis, le grand intérêt de faciliter le choix entre plusieurs solutions possibles, en apportant au décideur des informations sur l'influence quantitative des variations éventuelles des divers paramètres.

Mais les décisions finales resteront toujours le privilège des humains.

REFERENCES

(1) I.I.A.S.A. : Institut International d'Analyse des Systèmes Appliqués, Laxenburg, Autriche.

(2) Revue de l'Energie n° 316, juin-juillet 1979, numéro entièrement consacré à la méthode WELMM.

(3) ENERGETIQUE INDUSTRIELLE (P. LE GOFF, coordonnateur), Librairie Lavoisier, Paris
Vol. 1 : Analyse thermodynamique et mécanique des économies d'énergie (1979)
Vol. 2 : Analyse économique et optimisation des procédés (1980)
Vol. 3 : application en Génie Chimique : échangeurs, séparateurs, réacteurs (1982).

NOTE SUR LA FIGURE 2

1 En toute rigueur, l'eau devrait être incluse dans l'une des rubriques précédentes, puisqu'elle intervient, soit comme matière première, soit comme vecteur d'énergie (réfrigération). Mais vu l'importance de ce fluide dans toutes les activités humaines, il est tout à fait justifié de lui réserver une rubrique particulière.

Simulation in Engineering Sciences
J. Burger and Y. Jarny (eds.)
Elsevier Science Publishers B.V. (North-Holland)

FOURIER METHODS IN COMPUTATIONAL FLUID AND FIELD DYNAMICS

Robert VICHNEVETSKY
Rutgers University
New Brunswick, New Jersey, 08903, U.S.A.

ABSTRACT

This paper is a review, with examples, of those areas where the theory of Fourier transforms has played a role in the development of computational methods in fluid and field dynamics. These may be separated into (i) the development of computing algorithms using Fourier transforms and (ii) the analysis of standard finite difference and finite element algorithms by Fourier methods.

Recent results in the analysis of spurious reflection phenomena at computational boundaries of fluid flow simulation, which invoke the theory of wave propagation and the concept of group velocity are given.

1. INTRODUCTION

There are essentially two areas where the theory of Fourier transforms plays a role in the computation of flows and fields.

1. Fourier algorithms which rely on the property of Fourier series to carry out specific calculations.

2. Fourier analysis which rely on the theory as a tool for the analysis of error and stability properties of otherwise standard finite difference or finite element algorithms. Albeit theoretical, this second use of Fourier transforms seems, in a way, to have played a role more important than the first.

The relationship of trigonometric series with the hyperbolic partial differential equations of mathematical physics has been known since the 18th century. The first contributions to the theory of vibrating strings by Brook Taylor (1685–1731), Leonhard Euler (1707–1783), Daniel Bernoulli (1700–1782), and Louis Lagrange (1736–1813) among others, do contain indeed many uses of this analytic tool. The name of Fourier became attached later on to trigonometric series, after he applied them to the solution of the heat equation which he had derived (Fourier, - 1807).

2. THE FAST FOURIER TRANSFORM

An algorithm that is fundamental to the implementation of Fourier calculations is of course the Fast Fourier Transform (FFT). It uses the symmetry of trigonometric functions to regroup the equations so as to minimize the computational effort. Whereas computing the discrete Fourier transform of a function represented by N discrete data points requires on the order of N^2 operations, when implemented in the straightforward way, organizing the same calculation as prescribed by the FFT requires only on the order of $N \log_2 N$ operations (for N=1024, this represents a saving of more than 100 to 1).

The original description of the fast Fourier transform was given in a paper by Cooley and Tukey published in *Mathematics of Computation* in 1965, although the technique was not unknown in the field. For example, the basic idea of the method may be found on page 239 of C. Lanczos' *Applied Analysis*, published in 1956, and earlier results are described in Danielson and Lanczos (1942). Danielson and Lanczos, in turn, refer to Runge (1903, 1905) for the source of their method. Other pre-1965 users of "special techniques" which were in the same vein as the FFT are given in the book *The Fast Fourier Transform* by Brigham (1974). "Computational tricks" were not held in high esteem in the precomputer days, and attempts to give them visibility under a unified heading were few. The name of "numerical analysis" was coined soon after the advent of electronic computers. The number of people involved in doing scientific calculations grew dramatically, and many techniques that had not been well known came to the surface or were reinvented at that time. The "invention" of the fast Fourier transform in the 1960's is an example.

3. FOURIER METHODS IN COMPUTATIONAL FLUID DYNAMICS

It is after fast, large scale computers had become available (in the early 1960's), and that the fast Fourier transform algorithm had been popularized (in the mid to late 1960's) that Fourier methods in fluid flow calculations made their appearance. We shall briefly describe their principal fields of application :

Computational fluid dynamics imply the approximation of both first and second order spatial derivatives. The equations which illustrate the relevant concepts are :

(i) Those of the motion of a non viscous fluid (often called Euler's equations : he first described them in 1752–1755), which may be written as :

$$\frac{DU}{Dt} = \frac{\partial U}{\partial t} + U^T \nabla \cdot U = g - \frac{1}{\rho} \nabla p \tag{1}$$

where $U(x,t)$ is a vector and $\nabla \cdot$ is the gradient operator. Solving these equations numerically requires that the advection term $U^T \nabla \cdot U$ be approximated, and Fourier methods have been used mostly with the intent of improving the accuracy of this computation.

(ii) Those of viscous flow, illustrated by the Navier Stokes equation for an incompressible flow, in primitive form :

$$\frac{DU}{Dt} = g - \frac{1}{\rho} \nabla p + \nu \nabla^2 U \tag{2}$$

where ∇^2 is the Laplacian operator and ν is the dynamic viscosity of the fluid. An alternate formulation is the stream function vorticity form :

$$\frac{D\zeta}{Dt} = \nu \nabla^2 \zeta \qquad U = \begin{bmatrix} \partial\Psi/\partial y \\ -\partial\Psi/\partial x \end{bmatrix}$$
$$\nabla^2 \Psi = \zeta \tag{3}$$

The numerical treatment of these equations requires that the Laplacian ∇^2 also be approximated. In many cases (in particular if the vorticity form (3) is used, or if (2) is integrated by an implicit time marching method), the solution process requires that a form of Poisson's equation

$$\nabla^2 U = \rho(x,y) \tag{4}$$

be solved at each time step. In this case, Fourier methods are often used to reduce the computing time in the solution of (4).

4. APPLICATIONS IN PLASMA PHYSICS

Questions of the stability of nuclear plasma in the context of the development of nuclear fusion energy devices are often investigated by direct simulation. In one type of approach, the "fluid" consisting of the moving charged particles is represented by a finite number of superparticles : on the order of 10^4 superparticles are used in realistic simulation (it is reported that physicists doing these kinds of simulation are sometimes called "particle pushers" by those doing the more conventional – continuous model simulations). The particles are in motion in an exterior magnetic field and a self generated electrical field. The latter is computed by solving Poisson's equation at each time step (or every few time steps). Since many time steps are needed, it is vital, as with Navier Stokes' equations, that the solution of Poisson's equation be obtained as economically as possible.

5. FOURIER METHODS FOR FIRST ORDER HYPERBOLIC EQUATIONS

Consider as a model of (1) the linear, scalar equation :

$$\frac{\partial U}{\partial t} + c\frac{\partial U}{\partial x} = G(x,t) \tag{5}$$

over $x \in [0, \ell)$ with boundary conditions :

$$U(0,t) = U(\ell,t) \tag{6}$$

Periodic boundary conditions of this kind occur for instance in mathematical models of the global atmosphere and Fourier methods have indeed been developed and used to a large extent by geophysical fluid dynamicists and meteorologists.

$U(x,t)$ may be expressed in Fourier series form as :

$$U(x,t) = \frac{1}{\ell} \sum_{k=-\infty}^{\infty} \hat{U}_k e^{2ik\pi x/\ell} \tag{7}$$

where

$$\hat{U}_k(t) = \int_0^\ell U(x,t) e^{-2ik\pi x/\ell} dx \tag{8}$$

The form of $U(x,t)$ which solves ( ) is then easily obtained. Since

$$c\frac{\partial U}{\partial x} = \frac{c}{\ell} \sum \frac{2ik\pi}{\ell} \hat{U}_k e^{2ik\pi x/\ell} \tag{9}$$

it follows that each $\hat{U}_k(t)$ satisfies

$$\frac{d\hat{U}_k}{dt} + c\frac{2ik\pi}{\ell} \hat{U}_k = \hat{G}_k \tag{10}$$

where $\{\hat{G}_k\}$ is the Fourier transform of $G(x,t)$

It is Daniel Bernoulli who, with his 1753 paper, may be credited with paternity of the idea of the separation of the dynamics of a string into sinusoidal modes which do not interact with one another. Once more, history had its word to say, and the method of separation of variables is generally called the method of Fourier, who used it more than half a century later.

Approximations are obtained when k is restricted in (7) to within a finite range of values, say (–N,N), i.e., U(x,t) is approximated by a truncated Fourier series. Using this concept leads to several types of feasible approximations, such as :

(i) truncated Fourier series methods in which time integration is done in Fourier space and all calculations take place in the frequency domain. This is restricted to linear problems.

(ii) collocation methods in which time integration is done by expressing the solution in a finite number of discrete points in physical space, called collocation points, and trigonometric interpolation is used solely with the intent of approximating the spatial partial derivatives in these points.

(iii) collocation methods in which time integration is done in Fourier space, but transformation is done at each time step to and from physical space, where the non-linear terms are being computed.

Of these, only (ii) and (iii) have general applicability to a variety of problems in fluid dynamics, since they can be applied to non-linear equations.

6. FOURIER METHODS FOR SECOND ORDER EQUATIONS

Special Poisson's equation solvers which use Fourier transforms have been developed in the mid to late 1960's. Their first use was in the simulation of nuclear plasmas, and they have been applied to fluid dynamics later on. Fast Poisson Solvers were initially restricted to simple rectangular regions. (Reference to subsequent work toward the application of Poisson solvers to less regular domains may be found for instance in Proskurowski - 1981; see also Vichnevetsky - 1981-a). Consider for instance the solution of (4) in a square with the simple boundary conditions $U = 0$ on ∂D. The simplest method uses a double Fourier series :

$$U(x,y) = \sum_{p,q} \hat{U}_{p,q} \sin\left(\frac{p\pi x}{x_{max}}\right) \sin\left(\frac{q\pi y}{y_{max}}\right) \tag{11}$$

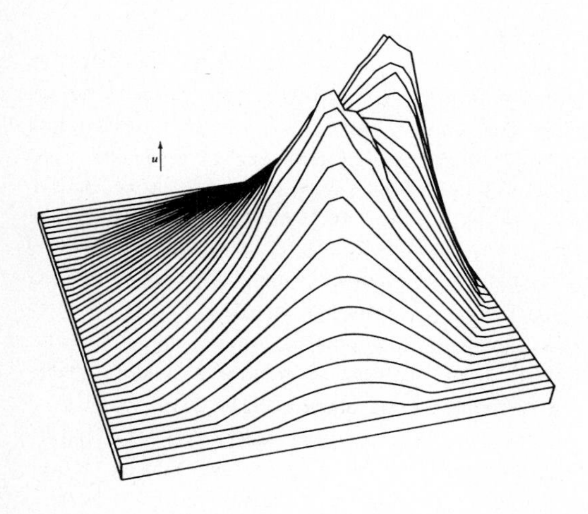

Figure 1 - A solution of Poisson's equation obtained with a solver using fast Fourier transforms.

which is a solution of (4) when the coefficients $\{\hat{U}_{p,q}\}$ are :

$$\hat{U}_{p,q} = \frac{-\hat{\rho}_{p,q}}{(p\pi/x_{max})^2 + (q\pi/y_{max})^2} \tag{12}$$

and $\{\hat{\rho}_{p,q}\}$ is the Fourier transform of $\rho(x,y)$

There is also the popular Hockney method which consist in using Fourier transforms in x and finite differences in y. The resulting system of tridiagonal equations is then solved by recursive cyclic reduction.

Original ideas and analyses of the efficiency of the Fast Poisson solvers with Fourier transforms may be found in Hockney (1965), Gentleman and Sande (1966), Buneman (1969), Colony and Reynolds (1970), and Le Bail (1972).

7. COMPARISONS

Fourier methods achieve one thing that finite difference methods generally do not : they approximate the partial derivatives with the maximum accuracy which is permitted by a representation of the variables in a finite number of discrete points. Essentially, the derivatives are approximated by the analytic derivative of the trigonometric interpolant between all these points.

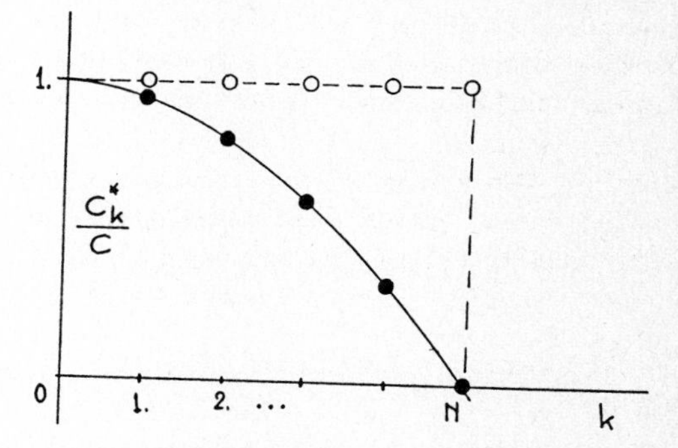

Figure 2 - Numerical phase velocity for Fourier components of wave number k in finite difference ● and Fourier O approximations.

Consider the following error analysis :

A numerical solution of (5) consisting of 2N discrete values in physical space may be expressed in Fourier series forms such as (7) with k=-N,...N. The approximation of the advective term $c\, \partial U/\partial x$ by standard three point finite differences then results in a Fourier series representation of the advective term by :

$$\frac{1}{\ell} \sum c_k^* \frac{2ik\pi}{\ell} \hat{U}_k e^{2ik\pi x/\ell} \tag{13}$$

where c_k^*, the phase velocity of the kth Fourier component is given by (see section 9 below) :

$$c_k^* = c\,\frac{\sin(2k\pi h/\ell)}{(2k\pi h/\ell)} \tag{14}$$

The difference $c^*_k - c$ measures the error in the approximation of $c\,\partial U/\partial x$. By contrast, Fourier methods approximate this term with (9) where the correct velocity c is preserved in all the Fourier components which are included in the computation, and the only source of error is that which results from the elimination or aliasing of the higher wave number components.

Comparisons of the efficiency of Fourier methods with that of standard higher order finite difference approximations have been carried out in some detail in those cases where high accuracy is sought (as in long term atmospheric simulations). These may be found in Ozszag (1971), Kreiss and Oliger (1972) and Fornberg (1975).

There are however, cases in the simulation of engineering systems where a lower order of accuracy is sufficient. The possible advantage of Fourier methods then stems from the fact that a very few number of Fourier components may have the potential of giving a realistic approximation, while a very few number of discrete points in physical space may not have that ability. See for instance, discussions about this in Clymer (1966). Applications of this concept are found in recent work in the simulation and control of what engineers refer to as distributed paramenter systems. In fact, these "modal methods" had been known for a long time by nuclear physicists working in the area of the control of nuclear reactors. Not only do they consider sinusoidal or "Fourier" modes, but also other modes which have spatial shapes given by the eigenfunctions of an associated Sturm-Liousville problem - see for instance Kaplan (1964) and Stacey (1967).

The limitation of Fourier methods lies of course in the fact that they prefer regular domains and simple boundary conditions, and that their efficiency fades away when those requisites are not met.

8. SIMULATION OF ATMOSPHERIC TURBULENCE

A typical application is in the simulation of large scale atmospheric turbulence.

An example is illustrated in Figure 3 . These results give the stream function and temperature distribution obtained as part of a simulation of the atmosphere in the northern hemisphere by Peskin and Kowalski (1983), using a Fourier method that was originally given by Ozszag (1976) : The numerical solution follows (iii) above, namely :

- All the problem's variables are expressed in Fourier space,
- They are transformed (by an FFT) to physical space at each time step for the calculation, point by point, of the non-linear terms,
- Non linear terms are re-transformed to Fourier space by an FFT,
- Marching to the next time step is done in Fourier space - (a combination of leap-frog and Crank-Nicolson methods was used).

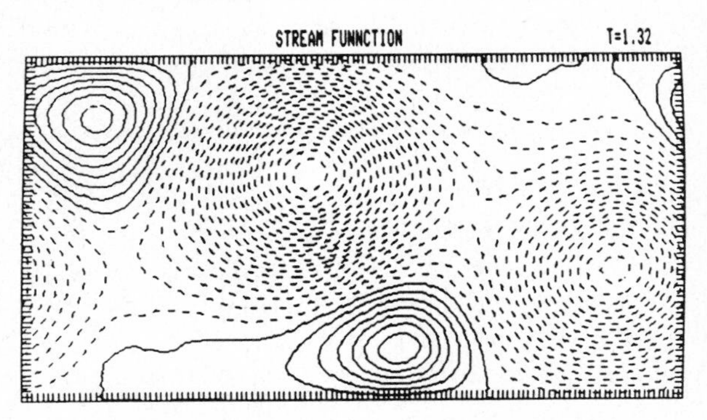

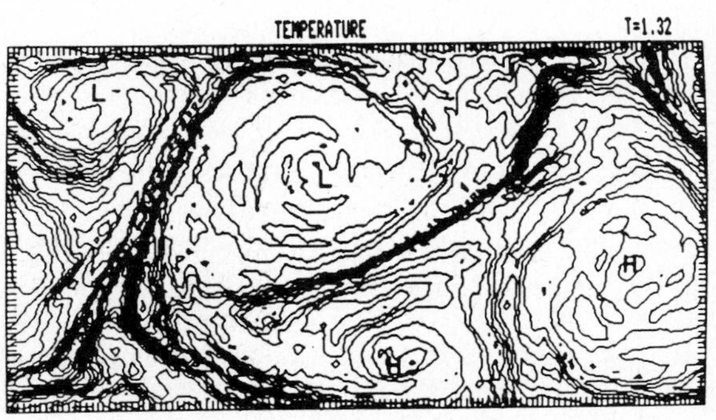

Figure 3 - Numerical simulation of atmospheric turbulence in the Northern hemisphere obtained with a spectral method (Peskin and Kowalski, 1983).

9. FOURIER ANALYSIS

Fourier analysis of the accuracy and stability properties of numerical algorithms may be achieved by observing the time evolution of sinusoidal (sometimes called "trial") solutions. This is the engineers' and physicists' approach. A mathematical refinement consists in observing instead the time evolution of the Fourier transform of numerical solutions. The mathematics are almost identical, but the second viewpoint gives a foundation for more detailed analyses, such as the calculation of energy norms of the error.

In both cases, one assumes that the equations may be locally linearized. The simple scalar equation :

$$\frac{\partial U}{\partial t} + c\frac{\partial U}{\partial x} = 0 \qquad (15)$$

is used as a model where c is to assume the value of the characteristic velocities of the system of equations being integrated. In the case of compressible gas dynamics, these are the particle velocity, and the sound velocities upstream and downstream of the flow, respectively.

The underlying assumption in those analyses is that the models are meant to represent the local behaviour of small perturbations of numerical solutions. It may appear as a paradox (but it is not) that such local analyses are modeled by sinusoidal solutions which must be defined on the entire real axis to allow Fourier transform theory to apply with its full analytic power.

Semi-Discretizations

For illustration we consider typical semi-discretizations such as the explicit (finite-difference)

$$\frac{dU_n}{dt} = -c\left(\frac{U_{n+1} - U_{n-1}}{2h}\right) \tag{16a}$$

and implicit (linear finite element/Galerkin) :

$$\frac{1}{6}\left(\frac{dU_{n-1}}{dt} + 4\frac{dU_n}{dt} + \frac{dU_{n+1}}{dt}\right) = -c\left(\frac{U_{n+1} - U_{n-1}}{2h}\right) \tag{16b}$$

which may both be expressed in operator notations as :

$$\frac{dU_n}{dt} = A \cdot U_n \tag{17}$$

where $A\cdot$ is an appropriately defined discrete (or Toeplitz) operator. Fully discrete algorithms are then obtained by applying to these some numerical time-marching procedure. But a first type of error analysis consists in assuming that time-marching errors may be neglected. The error may then be analyzed by defining the discrete Fourier transform of $\{U_n(t)\}$

$$\overline{U}(\omega,t) \equiv h \sum_n U_n e^{-i\omega nh} \tag{18}$$

Taking the Fourier transform of (17) then results in

$$\frac{d\overline{U}}{dt} = \hat{A}(\omega) \cdot \overline{U} \tag{19}$$

where

$$\hat{A}(\omega) \equiv A \cdot e^{i\omega nh} / e^{i\omega nh} \tag{20}$$

is the spectral functions of the operator A. Comparing this with the exact :

$$\frac{d\hat{U}(\omega,t)}{dt} = -ic\omega\hat{U}(\omega,t) \tag{21}$$

defines

$$c^*(\omega) = -\,\mathrm{I_m}\hat{A}(\omega)/\omega \tag{22}$$

as the phase velocity at which numerical Fourier components of spatial frequency, or wave number, ω propagate, and

$$\mathrm{R_e}\,\hat{A}(\omega) \tag{23}$$

as the measure of an amplitude error. Semi-discretizations for which $\mathrm{R_e}\hat{A}(\omega) \equiv 0$. are called conservative semi-discretizations. Both (16a) and (16b) are conservative, with

finite differences:

$$c^*(\omega) = c\,\frac{\sin(\omega h)}{\omega h} \tag{24a}$$

and

finite elements :

$$c^*(\omega) = c\,\frac{3\sin(\omega h)}{\omega h\,(2 + \cos(\omega h))} \tag{24b}$$

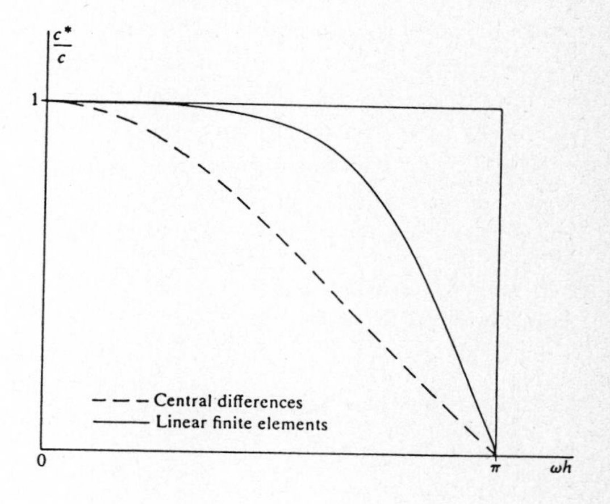

Figure 4 - Phase velocities of semi-discretizations versus wave number.

Fully Discrete Algorithms

Many time-marching methods for a system of equations written in the general form (17) may be expressed in operator notation as :

$$U_n^{j+1} = Z(\Delta t A) \cdot U_n^j \tag{25}$$

where $Z\cdot$ (sometimes called the amplification operator) is a function of the operator $A\cdot$, or more precisely of the dimensionless grouping $\Delta t A\cdot$. We have for example, for some of the well known time-marching methods :

Euler's Method :

$$Z\cdot = 1 + \Delta t A\cdot$$

Crank Nicolson Method :

$$Z\cdot = \frac{1 + \Delta t A\cdot/2}{1 - \Delta t A\cdot/2}$$

Leapfrog Method :

$$Z\cdot = \Delta t A\cdot \pm \sqrt{1 + (\Delta t A)^2}$$

To carry out a Fourier analysis of these fully discrete approximations, one replaces, as in the semi-discrete case, U_n^j by its Fourier transform $\overline{U}^j$

whereupon (25) becomes

$$\overline{U}^{j+1}(\omega) = \hat{Z}(\omega)\cdot\overline{U}^{j}(\omega) \tag{26}$$

obtained by replacing the operator $A\cdot$ by its spectral function $\hat{A}(\omega)$ in the expression of Z . This was first done by von Neumann in his classical analysis of numerical stability (somewhat differently : he used sinusoidal trial solutions and inserted them without preliminaries in the full discretizations), and $\hat{Z}(\omega)$ is traditionally called the amplification factor.

But $\hat{Z}(\omega)$ contains information about accuracy as well : By comparison with (26), we have the ratio of Fourier transforms for an exact solution of (15) :

$$\hat{U}(\omega,t^{j+1}) = e^{-i\omega c\Delta t}\,\hat{U}(\omega,t^{j}) \tag{27}$$

which allows one to define

$$c^*(\omega) = -\angle\hat{Z}(\omega)/\omega\Delta t \tag{28}$$

as the numerical phase velocity, and

$$|\hat{Z}(\omega)| - 1. \tag{29}$$

as an amplitude error : When $|\hat{Z}(\omega)| < 1$, there is a numerical or spurious damping And $|\hat{Z}(\omega)| > 1$. corresponds to numerical instability (von Neumann).

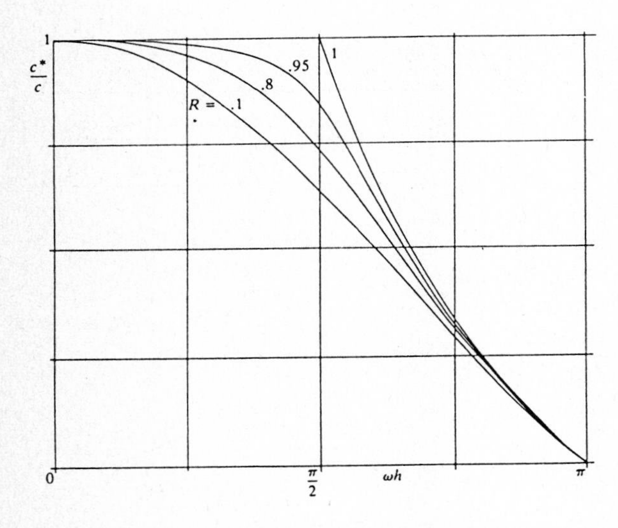

Figure 5 - Phase velocities versus wave number for the finite difference in space, leapfrog in time full discretization of (15). R = Courant number$=c\Delta t/h$

Analysis and discussion of specific expressions of numerical phase velocities may be found interspersed in the computational and fluid dynamics literature, for instance in Vichnevetsky and De Schutter (1975), Gray and Pinder (1976), Chin, Hedstrom and Karlsson (1979), Vichnevetsky and Bowles (1982), and Lapidus and Pinder (1982). Equivalent analyses in the case of the second order wave equation are given in Dougalis and Serbin (1981) and, with specific reference to finite element approximations of the equations describing vibrations in mechanical structures, in Krieg and Key (1973) and in Belytchsko and Mullen (1978). (See also Section 15 below.)

10. TRANSFER FUNCTIONS

Engineers are familiar with the use of transfer function for the analysis of input/output properties of linear systems. Transfer functions are defined as a ratio of Fourier transforms. The same concept may be used to evaluate properties of semi-discrete approximations of hyperbolic equations. Consider the Fourier transforms in time :

$$\hat{U}(x,\Omega) = \int U(x,t)\, e^{-i\Omega t}\, dt \tag{30}$$

If $U(x,t)$ is an exact solution of the advection equation (15), then

$$\hat{U}(x,\Omega) = \hat{U}(0,\Omega)\, e^{-i\Omega x/c} \tag{31}$$

and for two adjacent mesh points :

$$\hat{U}(x_{n+1},\Omega) = e^{-i\Omega h/c}\,\hat{U}(x_n,\Omega) \tag{32}$$

where $e^{-i\Omega h/c}$ is the transfer function between those two points.

Semi-discretizations have the effect of replacing this exact transfer function by approximations. For instance, to the simple upwind approximation

$$\frac{dU_n}{dt} = -c\,\frac{U_n - U_{n-1}}{h} \tag{33}$$

corresponds the transfer function

$$\hat{U}_{n+1}/\hat{U}_n = \frac{1}{1+i\Omega h/c} \tag{34}$$

which we call "cell transfer function" of the approximation. By comparison with (32) we find

$$c^*(\Omega) = \frac{\Omega h}{\arctan(\Omega h/c)} \tag{35}$$

for the numerical phase velocity, and the difference

$$\left(1+(\Omega h/c)^2\right)^{-1/2} - 1.$$

is an amplitude error which may be associated to spurious damping or spurious diffusion.

11. DISPERSION AND GROUP VELOCITY

The concept of group velocity occurs in describing the propagation of wave packets in dispersive media (defined as media where sinusoidal waves of the form $e^{i\omega(x-c^*(\omega)t)}$ may exist with a frequency dependent phase velocity $c^*(\omega)$. A wave packet is then defined as a single wave number sinusoidal function modulated by a smooth envelope $\phi(x,t)$, such as :

$$U = \phi(x,t)\, e^{i\omega(x-c^*t)} \quad (37)$$

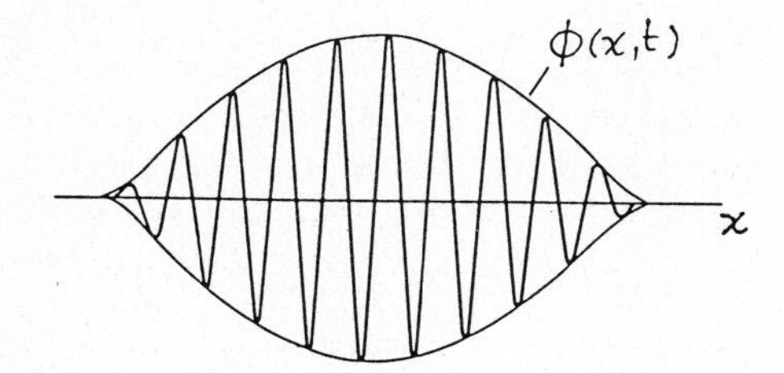

Figure 6 - Wave packet.

It is then found that the envelope of the packet does not propagate in the dispersive medium at the phase velocity c*, but instead at the group velocity given by :

$$\gamma(\omega) = \frac{d}{d\omega}(\omega c^*(\omega)) \quad (38)$$

Or, in terms of time Fourier transforms,

$$1/\gamma(\Omega) = \frac{d}{d\Omega}(\Omega/c^*(\Omega)) \quad (39)$$

Applying this to (24a) (24b), one finds that the group velocity for sinusoidal solutions of wavelength near 2h is negative. This explains the presence of reflected perturbations when a smooth solution passes across a discontinuity or a boundary of the computational domain : since all numerical phase velocities are positive, a cursory observation of the Fourier series form of the numerical solution does not suggest that any part of it might propagate in the negative direction. But the group velocity analysis shows that reflected solutions may indeed exist in the form of wave packets of short wavelength numerical oscillations which propagate upstream of the flow after having been spuriously generated.

The mathematics of propagation in dispersive media and many of the attending developments in Fourier analysis had been well established by the end of the nineteenth century. The concept of group velocity had been used conceptually by Hamilton (1839), was known to Rayleigh (Theory of Sound - 1877) and was fully investigated in the early part of this century with publications of Sommerfeld (1912, 1914) being prominent. An important class of dispersive media is that of crystals, which present periodic structures to the propagation of light. Of interest is to note the fact that numerical discretizations of hyperbolic equations have created new families of similarly periodic structures. It comes therefore as no surprise that applying the same analysis to those discretizations proves to be a very fruitful endeavor.

12. REFLECTION PHENOMENA

The appropriate tool for the analysis of spurious reflection phenomena in discretizations of hyperbolic equations is that given by time Fourier transforms. If we denote by

$$\{\hat{U}_n(\Omega)\} = \int \{U_n(t)\}\, e^{-i\Omega t} dt \quad (40)$$

the set of time Fourier transforms of a solution of (16a) or (16b), then an analytic treatment reveals the following (Vichnevetsky 1981, a, b, 1983 a) :

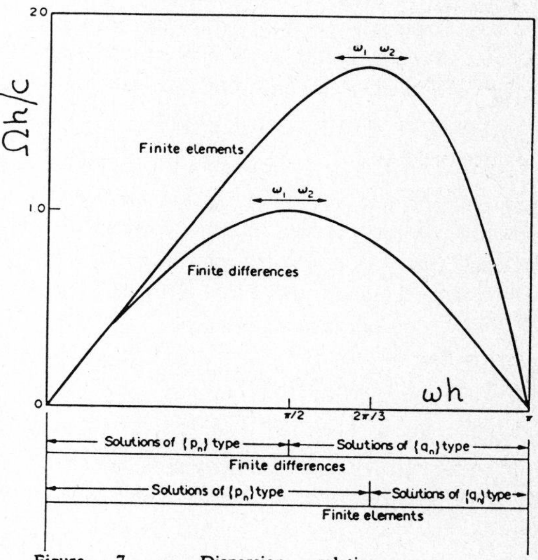

Figure 7 - Dispersion relation (frequency Ω versus wave number ω) To each $\Omega < \Omega_c$ correspond two wave numbers ω_1 and ω_2 which correspond to one another in reflection phenomena. (From reference 63).

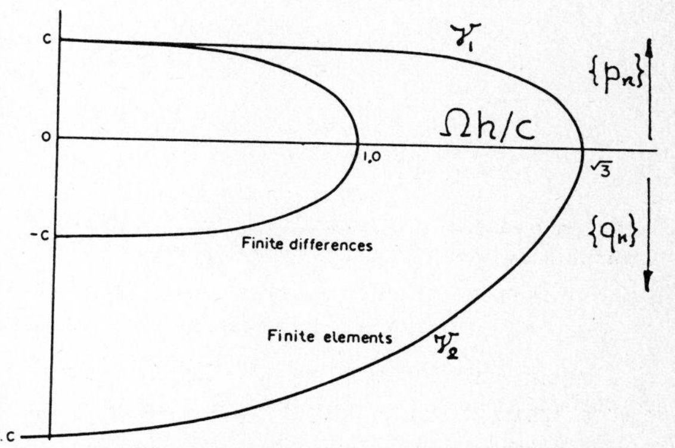

Figure 8 - Group velocities. To each frequency Ω below the cutoff frequency corresponds a positive group velocity (solution of $\{p_n\}$ type) and a negative group velocity (solution of $\{q_n\}$ type).

Any solution may be decomposed into the sum of two fundamental types

$$\{U_n(t)\} = \{p_n(t)\} + \{q_n(t)\} \tag{41}$$

whose Fourier transforms satisfy :

$$\hat{p}_{n+1}/\hat{p}_n = \hat{E}_1(\Omega) \; ; \; \hat{q}_{n+1}/\hat{q}_n = \hat{E}_2(\Omega) \tag{42}$$

where $\hat{E}_1$ and $\hat{E}_2$ are cell transfer functions of the semi-discretization, and are given in reference (63).

While solutions of the $\{p_n\}$ type have positive phase and group velocities, those of $\{q_n\}$ type have positive phase but negative group velocities : solutions of $\{p_n\}$ type are convergent approximations of genuine solutions, but solutions of $\{q_n\}$ type are entirely spurious. They carry energy upstream of the flow and are precisely the kind of extraneous solutions that are generated when spurious reflection occurs at boundaries of computational domains and at interfaces in numerical mesh refinement.

13. REFLECTION AT BOUNDARIES

When the simple advection equation (15) is approximated over a finite domain (say $x\in[0,1]$), the approximation at the downstream boundary is different from the approximation in the interior points. This creates spurious reflection. Consider, for instance, the simple upstream approximation in :

$$\frac{dU_N}{dt} = -c\left(\frac{U_N - U_{N-1}}{h}\right) \; ; \; \left(U_N \simeq U(\ell,t)\right) \tag{43}$$

Observe that :

(i) any incident solution arriving in $x=\ell$ from $x<0$ is of $\{p_n\}$ type, since it carries energy downstream.

(ii) the corresponding reflected solution must be of $\{q_n\}$ type, since it carries energy upstream

(iii) to a component of frequency Ω in the incident $\{p_n\}$ correspond a component at the same frequency Ω in the reflected $\{q_n\}$. To quantify the reflection, we take the t-Fourier transform of (43) :

$$i\Omega\hat{U}_N = -\frac{c}{h}(\hat{U}_N - \hat{U}_{N-1}) \tag{44a}$$

which together with

$$\hat{p}_{N-1} = \hat{E}_1^{-1}\hat{p}_N \; ; \; \hat{q}_{N-1} = \hat{E}_2^{-1}\hat{q}_N \tag{44b}$$

gives the reflection ratio :

$$\rho(\Omega) \equiv \frac{\hat{q}_N}{\hat{p}_N} = -\frac{1+i\Omega h/c - \hat{E}_1^{-1}}{1+i\Omega h/c - \hat{E}_2^{-1}} = O(\Omega h) \tag{45}$$

Better treatments of the boundary may be sought by replacing (43) with improved expressions aimed at making $\rho(\Omega)$ as small as possible. This has been a subject of considerable interest in recent years .

Less obvious is the question of spurious reflection at an upstream or inlet boundary. The usual numerical treatment consists in letting

$$U_0(t) = U(0,t) \tag{46}$$

i.e., letting the numerical value be equal to the imposed value. But this creates reflection when a spurious solution (of $\{p_n\}$ type) arrives at the boundary. An improved treatment should attempt to absorb those spurious solutions instead of reflecting them. Appropriate non-reflecting inlet boundaries for finite difference discretizations are given in Vichnevetsky, Sciubba and Pak (1981).

14. REFLECTIONS IN MESH REFINEMENT

Mesh refinement is commonly used when a greater accuracy is sought in some subdomain of the problem. A typical example is that of grids used in numerical weather prediction. Whereas coarse grids may be used over the oceans, a fine grid is desirable over land where the predictions count. A simple model of mesh refinement is illustrated in figure . The mathematics for the description of reflection at the interface are similar to (44), leading to the reflection ratio for the spurious solution reflected toward $x<0$:

$$\rho(\Omega) = \frac{\sqrt{1-(\Omega h/c)^2} - \sqrt{1-(\Omega k/c)^2}}{\sqrt{1-(\Omega h/c)^2} - \sqrt{1-(\Omega k/c)^2}} = O(\Omega(h-k))$$

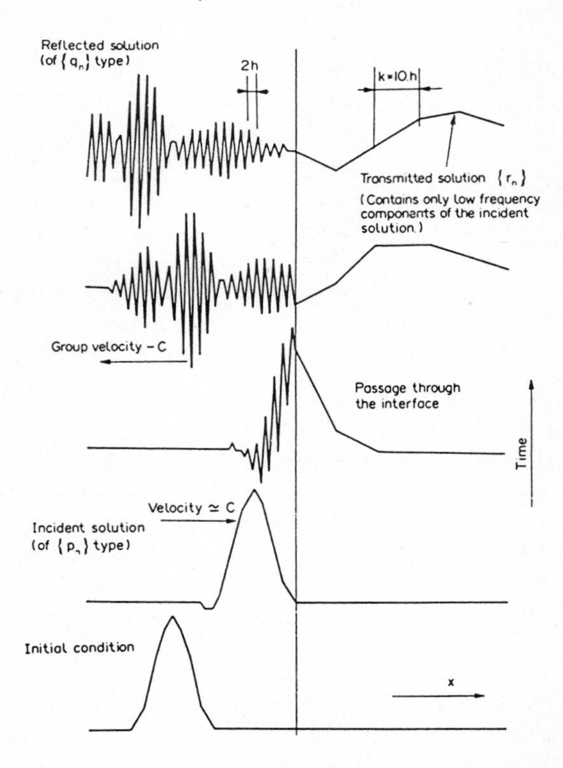

Figure 9 – Reflection of a smooth function at a mesh coarsening interface. The reflected solution is a wave packet of wave length near 2h.

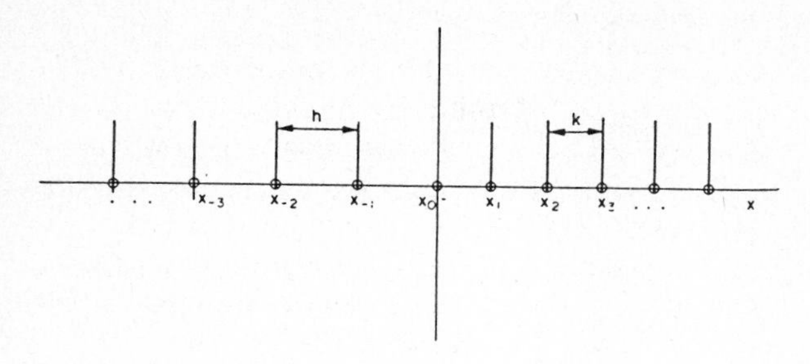

Figure 10 - Mesh refinement.

15. NUMERICAL WAVE PROPAGATION IN 2 DIMENSIONS - ANISOTROPY

The wave equation in two space dimensions

$$\frac{\partial^2 U}{\partial t^2} = c^2 \left(\frac{\partial^2 U}{\partial x^2} + \frac{\partial^2 U}{\partial y^2}\right) \tag{47}$$

has propagation characteristics which are isotropic : a plane sinusoidal wave

$$U = e^{i\omega(x\cos\alpha + y\sin\alpha \pm ct)} \tag{48}$$

propagates in the 1_α direction at the phase velocity $\pm c$ which is independent of α. By contrast, numerical simulations on a discrete grid introduce, in addition to dispersion (c^* depending on ω) also anisotropy (c^* depending on α). Consider, for instance, the simple semi-discretization of (47) on a square grid

$$\frac{d^2 U_{m,n}}{dt^2} = c^2 \left[\left(\frac{U_{m+1} - 2U_m + U_{m-1}}{h^2}\right)_n + \left(\frac{U_{n-1} - 2U_n + U_{n+1}}{h^2}\right)_m\right] \tag{49}$$

If we insert the expression of a plane sinusoidal trial solution such as (48), we find, after integration

$$U_{m,n} = e^{i\omega(x\cos\alpha + y\sin\alpha \pm c^*(\omega,\alpha)t)} \tag{50}$$

where the numerical phase velocity is :

$$c^*(\alpha,\omega) = c\frac{\sqrt{2}}{\omega h}\left(2 - \cos(\omega h\cos\alpha) - \cos(\omega h\sin\alpha)\right) \tag{51}$$

(see also Birkhoff and Dougalis (1975)).

Anisotropy properties of general explicit and implicit semi-discretizations of the wave equation on a square grid are given in Vichnevetsky and Bowles (1982). Similar results for finite element discretizations (bilinear elements on rectangles and linear elements on triangles) are given by Mullen and Belytschko (1982). Anisotropy of phase velocity implies of course anisotropy of group velocity as well. These properties have been analysed by Trefethen (1982) for semi- and full discretizations.- figure below:

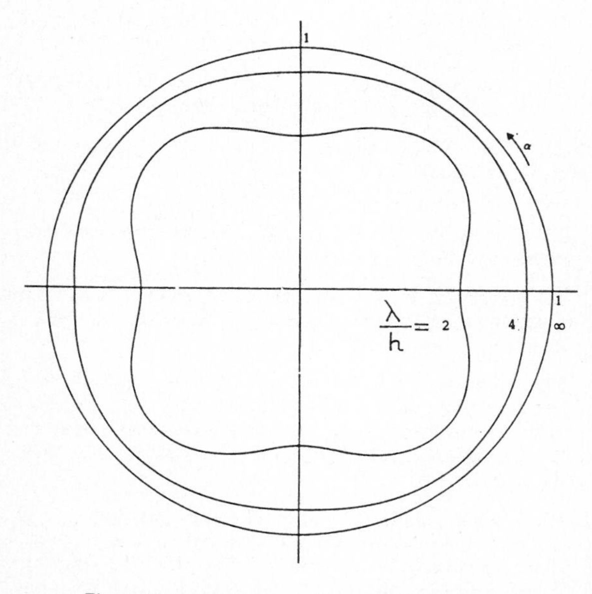

Figure 11 - Anisotropy : phase velocity of the semi-discretization (49) versus direction (α) and wave length (λ).

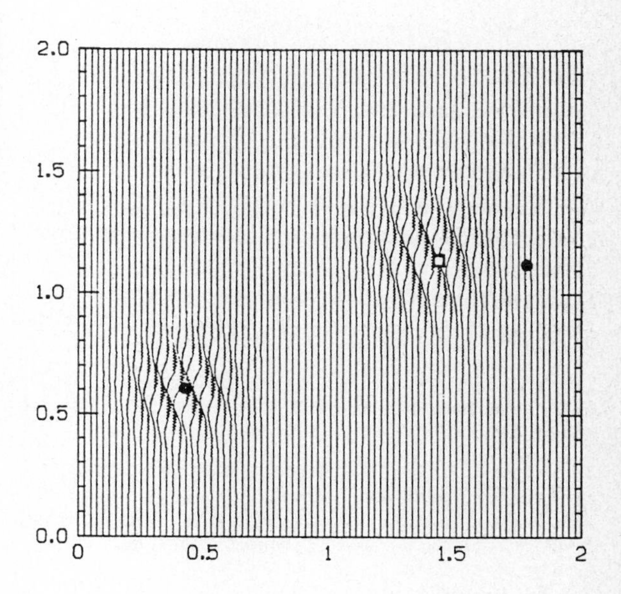

Figure 12 - Propagation of a wave packet in a discretization of the wave equation on a square grid.
● is the ideal position of the packet and □ is the position predicted by the group velocity analysis. (from reference (55) by permission)

16. ACKNOWLEDGMENTS

The assistance of Elissa C. Pariser in the preparation of this paper is gratefully acknowledged.

17. BIBLIOGRAPHY AND REFERENCES

1. Abbot, M. B. "Computational hydraulics, A short pathology." *J. Hydraulic Res. 14* (1976), 271-285.

2. Belytschko, T. and T. Mullen. "On dispersive properties of finite element solutions". In *Modern Problems in Elastic Wave Propagations*. I. Miklowitz and J. D. Achenbach, Eds., J. Wiley and Sons. 1978.

3. Benyon, P. "A review of numerical methods for simulation." *Simulation 11* (1968).

4. Bernoulli, Daniel. "Reflexions et eclaircissemens sur les nouvelles vibrations des cordes." *Mem. Acad. Sci. Berlin 9* (1753a), 147-172. (1753: publ. 1755)

5. Bernoulli, Daniel. "Sur les melange de plusiers especes de vibrations simples isochrones." *Mem. Acad. Sci. Berlin 8* (1753b), 173-195. (1753: publ. 1755)

6. Birkhoff, G. and V. A. Dougalis. "Numerical solution of hydrodynamic problems". In *Advances in Computer Methods for Partial Differential Equations*, R. Vichnevetsky, Ed.,AICA, Rutgers University, New Brunswick, NJ, 1975.

7. Birkhoff, G. "Numerical fluid dynamics." *SIAM Review 25*, 1 (1983), 1-34.

8. Brigham, E. O.. *The Fast Fourier Transform*. Prentice-Hall, Englewood Cliffs, NJ, 1974.

9. Brillouin, L.. *Wave Propagation in Periodic Structures*. McGraw Hill, New York, 1946.

10. Brillouin, L.. *Wave Propagation and Group Velocity*. Academic Press, New York, 1960.

11. Buneman, O. "A compact non-iterative Poisson solver." *SUIPR Report 294* (1974).

12. Carver, M. B. and H. W. Hinds. "The method of lines and the advection equation." *Simulation 31* (1978), 59-69.

13. Chin, R. C. Y., G. W. Hedstrom, and K. E. Karlsson. "A simplified Galerkin method for hyperbolic equations." *Math. Comput. 33* (1979), 647-658.

14. Churchill, R. V.. *Fourier Series and Boundary Value Problems*. McGraw Hill, New York, 1941. (revised edition 1963)

15. Clymer, A. B. "Methods of simulating structural dynamics." *Simulation 7* (1966).

16. Colony, R. and R. R. Reynolds. "An application of Hockney's method for solving Poisson's equation." *Proc. Spring Joint Comp. Conf., AFIPS* (1970).

17. Cooley, J. W. and J. W. Tukey. "An algorithm for the machine calculation of complex Fourier series." *Math. Comput. 19* (1965), 297-301.

18. Dorr, F. W. "The direct solutin of the discrete Poisson equation on a rectangle." *SIAM 12*, 8 (March 1970), 248-263.

19. Dougalis, V. A. and G. Birkhoff. "A comparison of numerical methods for solving wave equations". In *First International Congress on Numerical Ship Hydrodynamics*, J. W. Schot and N. Salvesen, Eds., U.S. Naval R&D Center, (D. W. Taylor Model Basin), 1975.

20. Dougalis, V. A. and S. M. Serbin. "On the efficiency of some fully discrete Galerkin methods for second order hyperbolic equations." *Computers and Math with Appl.* (1981).

21. Engquist, B. and A. Majda. "Absorbing boundary conditions for the numerical simulation of waves." *Math. Comp. 31*, 139 (July 1977), 629-651.

22. Fix, G. and G. Strang. "Fourier analysis of the finite element method in Ritz-Galerkin theory." *Studies in Appl. Math 48* (1969), 265-273.

23. Fornberg, B. "On a Fourier method for the integration of hyperbolic equations." *SIAM J. Num. Anal. 12* (1975), 509-528.

24. Fourier, J. "Theorie de le propagation de le chaleur dans les solides." *Monograph presented to the Institut de France* (1807). reproduced in Grattan-Guinness (1972)

25. Fromm, J. E. "A method for reducing dispersion in convective difference schemes." *J. Comp. Phys. 3* (1968), 176-189.

26. Gentleman, W. M. and G. Sande. "Fast Fourier Transforms - for fun and profit". Proc. Spring Joint Comp. Conf., AFIPS, 1966".

27. Gilliland, M. C. "A spectral stability analysis of finite-difference operations." *IEEE Trans. Electronic Computers EC-15* (1966), 849-854.

28. Gottlieb, D. and S. A. Orszag. "Numerical Analysis of Spectral Methods". In *CBMS Regional Conference Series*, SIAM, Philadelphia, 1977.

29. Grattan-Guinness, I.. *The Development of the Foundations of Mathematical Analysis from Euler to Riemann*. MIT Press, Cambridge, MA, 1970.

30. Grattan-Guinness, I.. *Joseph Fourier 1768-1830 - His Life and Work*. MIT Press, Cambridge, MA, 1972.

31. Gray, W. G., and G. F. Pinder. "An analysis of the numerical solution of the transport equation." *Water Resources Res. 12*, 3 (1976), 547-555.

32. Hamming, R. W.. *Digital Filters*. Prentice Hall, Englewood Cliffs, NJ, 1977.

33. Hockney, R. W. "A fast direct solution of Poisson's equation using Fourier analysis." *JACM 12*, 1 (1965), 95-113.

34. Hyman, J. M. "A method of lines approach to the numerical solution of conservation laws". In *Advances in Computer Methods for Partial Differential Equations*, R. Vichnevetsky, Ed.,IMACS, New Brunswick, NJ, 1979, pp. 313-321.

35. Kaplan, S., O. J. Marlowe, and J. Berwick. "Applications of synthesis techniques to problems with time-dependence." *Nuclear Science and Engineering 18* (1964), 163-176.

36. Kreiss, H. O. and J. Oliger. "Comparison of accurate methods for the integration of hyperbolic equations." *TELLUS XXIV* (1972), 119-215.

37. Kreiss, H. O. and J. Oliger. "Methods for the approximate solution of time dependent problems." *World Meteorological Organization/International Council of Scientific Unions* (1973).

38. Krieg, D. R. and S. W. Key. "Transient shell response by numerical time integration." *Internat. J. for Num. Meth. in Eng. 7* (1973), 273–286.

39. Lagrange, J. L. "Recherches sur la nature de le propagation du son." *Miscell. Taurin 1* (1759). also in Works, Vol. 1, pp. 39–148

40. Lanczos, C.. *Applied Analysis*. Prentice Hall, Englewood Cliffs, NJ, 1956.

41. Lapidus, L. and G. F. Pinder. *Numerical Solutions of Partial Differential Equations in Science and Engineering*. J. Wiley and Sons, 1982.

42. Le Bail, R. C. . "Use of Fast Fourier Transforms for solving partial differential equations in physics." *J. Comp. Phys. 9* (1972), 440–465.

43. Miranker, W. L. "Difference schemes with best possible truncation error." *Num. Math. 17* (1971), 124–142.

44. Mullen, R. and T. Belytschko. "Dispersion analysis of finite element semidiscretizations of the two-dimensional wave equations." *Internat. J. for Num. Meth. in Eng. 18* (1982), 11–29.

45. O'Brien, G., M. A. Hyman, and S. Kaplan. "A study of the numerica solutin of partial differential equations." *J. Math and Physics* (1952), 223 ff..

46. Orszag, S. "Numerical simulation of incompressible flow within simple boundaries: Accuracy." *J. Fluid Mech. 49* (1971), 75 ff..

47. Papoulis, H.. *The Fourier Integral and Its Application*. McGraw Hill, New York, 1962.

48. Peskin, R. L. and A. D. Kowalski. "A Simulation of Large Scale Atmospheric Turbulence and Diffusion". Proceedings Sixth Symposium on Turbulence and Diffusion, 1983.

49. Proskurowski, W. "Capacitance Matrix Methods – Available software and possible extensions". In *Proceedings, 10th IMACS World Congress*, IMACS, New Brunswick, NJ, 1982.

50. Lord Rayleigh, 3rd Baron (J. W. Strutt). *The Theory of Sound*. Macmillan, London, 1894. reprinted by Dover, New York, 1945

51. Roache, P. J.. *Computational Fluid Dynamics*. Hermosa Publisher, Albuquerque, NM, 1972.

52. Roberts, R. V. and N. O. Weiss. "Convective difference schemes." *Math. Comput. 20* (1966), 272–299.

53. Rosenbrock, H. H. "Distinctive probems of process control." *Chem. Eng. Progress 9* (1962), 43–50.

54. Shannon, C. E. and W. Weaver. *The Mathematical Theory of Communication*. University of Illinois Press, Urbana, 1949.

55. Stacey, W. M. Jr.. *Modal Approximations*. MIT Press, Cambridge, MA, 1967.

56. Strang. "Trigonometric polynomials and difference methods of maximum accuracy." *J. Math and Phys. 41* (1962), 147–354.

57. Swartz, B. "Comparing certain classes of difference and finite element methods for a hyperbolic problem". In *Advances in Computer Methods for Partial Differential Equations*, R. Vichnevetsky, Ed.,AICA, Rutgers University, New Brunswick, NJ, 1975.

58. Trefethen, L. F. "Group velocity in finite difference schemes." *SIAM Review 24* (1982), 113–136.

59. Vichnevetsky, R. "Physical criteria in computer methods for partial differential equations". Proc. 7th AICA International Congr., Prague, 1973. reprinted in Proc. AICA, XVI, January, 1974, Brussels.

60. Vichnevetsky, R. "Stability charts in the numerical approximation of partial differential equations: a review." *Math Comput. Simulation 21* (1979), 170–177.

61. Vichnevetsky, R. "Energy and group velocity in semi-discretizations of hyperbolic equations." *Math Comput. Simulation 23* (1981a), 333–343.

62. Vichnevetsky, R. "Propagation through numerical mesh refinement for hyperbolic equations." *Math Comput. Simulation 23* (1981b), 344–353.

63. Vichnevetsky, R.. *Computer Methods for Partial Differential Equations*. Prentice Hall, Englewood Cliffs, NJ, 1981c.

64. Vichnevetsky, R. "Group Velocity and Reflection Phenomena in Numerical Approximations of Hyperbolic Equations." *Journal of the Franklin Institute* (1983).

65. Vichnevetsky, R. and J. B. Bowles. *Fourier Analysis of Numerical Approximations of Hyperbolic Equations*. SIAM, Philadelphia, PA, 1982.

66. Vichnevetsky, R. and F. De Schutter. "A frequency analysis of finite element methods for initial value problems". In *Advances in Computer Methods for Partial Differential Equations*, R. Vichnevetsky, Ed.,AICA, Rutgers University, New Brunswick, NJ, 1975.

67. Vichnevetsky, R. and B. Peiffer. "Error waves in finite element and finite difference methods for hyperbolic equations". In *Advances in Computer Methods for Partial Differential Equations*, R. Vichnevetsky, Ed.,AICA, Rutgers University, New Brunswick, NJ, 1975.

68. Vichnevetsky, R., E.Sciubba and Y Pak. "Non-reflecting upstream boundary conditions". In *Advances in Computer Methods for Partial Differential Equations IV*, R. Vichnevetsky, Ed.,IMACS, Rutgers University, New Brunswick, NJ, 1981.

69. Vichnevetsky, R. and A. W. Tomalesky. "Spurious error waves in numerical approximations of hyperbolic equations". Proc. 5th Princeton Conference on Information Science and Systems, 1971.

70. Vliegenthart, A. C. "Dissipative difference schemes for shallow water equations." *J. Engrg. Math. 3* (1969), 81–94.

71. Von Neumann, J. and R. D. Richtmeyer. "A method for the numerical calculation of hydrodynamic shocks." *J. Appl. Phys. 21* (1950), 232–237.

72. Wesseling, P. "Accuracy of the third-order predictor-corrector difference schemes for hyperbolic problems." *AIAA J. 10* (1972), 948–949.

I. SIMULATION TOOLS – Software

Simulation in Engineering Sciences
J. Burger and Y. Jarny (eds.)
Elsevier Science Publishers B.V. (North-Holland)

A LANGUAGE FOR REAL TIME SIMULATION OF PROCESSES WITH BOOLEAN INPUTS AND OUTPUTS

E.F.Camacho, L.G.Franquelo and J.Lozano

E.T.S. Ing. Industriales Univ. Sevilla

This paper deals with the problem of real time simulation of processes with boolean inputs and outputs. A language for this purpose and the programs that processes it is presented. The language allows the description of processes with simultaneous evolutions as a timed petri net type of description is used. Random failures can also be introduced in the behaviour of the model.
The language allows the control of a semigraphic CRT in order to facilitate the task of following the model behaviour.

1. INTRODUCTION

Simulation is a fundamental tool when developing logical automatas, especially when those automatas are designed to control complex processes or processes where testing is expensive. As an example the starting up and shutting down procedure of a hydroelectric power unit or controlling substation operations, where on-line testing of the automata should be avoided as much as possible due to the risk of damaging expensive equipement while running the experiment.
The simplest of all possible boolean simulators consists of a set of switches, simulating process inputs, and lights, simulating process outputs. The automata is connected to these and a human operator moves the switches as the plant would do according to the sequence of orders received from the automata. The human operator is in this way simulating the behaviour of the plant. This method of testing an automata has tree major drawbacks. The first one is that human operators are very slow; it can take 15 seconds or more for the operator to decide which switches must be changed if the process he is simulating is complex enough. The second disavantage is that due to frequent errors the system is not properly simulated. Finally it is very difficult to carry out systematic tests with this method. Therefore only very slow or simple processes can be simulated with this method in real time.
A hardware model of the plant overrides all these problems but it is normally very costly and unflexible.
This paper presents a software simulator implemented in a PDP 11/23 computer. The simulator consists of a language and a collection of programs for processing it. The language is based on the Petri Nets approach for describing automatas.

The overall system structure is described in the next section. The language is treated in section 3 and some examples are given in section 4.

2. SYSTEM STRUCTURE AND FUNCTIONING

The structure chosen for the simulator can be seen in figure 1. It consists of a computer connected to the automata to be tested and to a semi-graphic CRT that allows an interactive simulation. The automata is connected to the computer via parallel input-output digital ports.

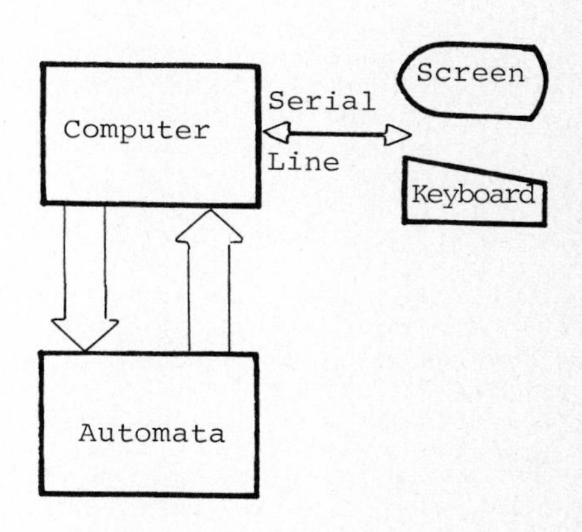

Figure 1. System structure

Once the model is running, manual operations can be introduced easily using the keyboard attached to the CRT. As an example, when simulating an electrical substation a breaker can be manually opened or closed by the operator while the model is running.
A mimic can be related to the model and it will appear on the screen as soon as the simulation program begins. Up to 128

active points can be defined for each -- application. These points are related to 4 boolean variables two of which are associated to the keyboard (inputs for the simulator) and the other two to the screen (outputs for the simulator).
The keyboard signals are activated positioning the screen cursor on the active position and pressing one of the four -- predefined keys that are associated with the values 00, 01, 10, 11 for the two -- keyboard signals mentioned above. As was mentioned before, these keyboard variables will simulate manual operations and will be considered as input signals for the model. In the example of the circuit breaker mentioned above, an active point could be related to it and the open and close would be associated to the two keyboard signals corresponding to that active point with values 10 and 01 respectively. The non logical conditions (11) of these signals can be used within the model to declare a defective breaker -- which can be useful to test the automata under malfunctioning of the process.
The two screen signals mentioned before are output variables for the model. Up to four symbols can be associated to the four possible values of these two signals. The simulator will represent in -- the related active point the symbol corresponding to the values of the variables. These variables are very useful -- for an interactive simulation of the process, as the model behaviour can be easily followed on the screen.
Besides the keyboard and screen variables mentioned above, the simulator allows the use of 512 input-output signals connected to the automata. Internal boolean variables (up to 256) can also be used. These internal signals are useful for connecting Petri Nets.
In order to simulate stochastics failures or evolutions in the model, up to 32 randomly generated boolean signals are provided. The first 16 of these signals are generated by a 1 second clock whilst the other 16 with a 1 minute clock.
The way of operating the simulator can be seen in figure 2. The model is defined in a simulation language that will be described in the next section. The model is compiled and the tables and code necessary for the simulation program are produced as is shown in figure 2.
The simulation program reads the output data of the compiler and the graphic representing the process from a disk.
With this information and the input signals (external, internal, keyboard and random) the simulator moves the output signals (external, internal and screen) according to the process description and its actual state.
A matrix method (2) is used to compute - the marking of the nets.
The amount of memory and the computation required decrease considerably if instead of using a single big net for modelling the system, various small nets are used.

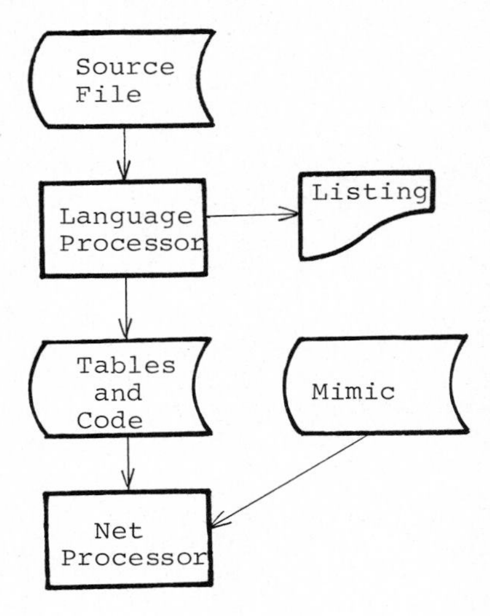

Figure 2. System operation.

3. SIMULATION LANGUAGE.

To facilitate the task of modelling, a simulation language has been defined -- and its processor implemented. The language uses the Petri Nets approach allowing a model to contain several Petri Nets.
Each net defined in the model begins -- and finishes with reserved words. Labels can be associated with the booleans variables described in the previous section. The labels can be then used as outputs associated to the marking of any place or can be used as part of an expression in a transition.
Labels can be global, valid for all --- nets, or local to a net. Labels can be associated with a boolean signal specifying type and number of signal or indicating only type. The processor will associate the next free signal of that - type in this last case. This and the possibility of using local labels allows the effective use of a macro processor, which is implemented within the program, thus facilitating the modelling of systems with repetitive parts (see example of substation given below).
Four labels can be assigned to a screen active point, giving the coordinates. - The first two correspond to the keyboard and the last two to the screen signals. The four graphic simbols associated with the screen labels must be given in the asignation instruction.

Transitions are defined indicating their number (within the net), the places entering and leaving the transition and a boolean infix expression associated with it. The expressions can contain any label previously defined, the boolean operators NOT, AND, OR and parenthesis.
The net structure is defined once all the transitions have been specified. Outputs are, in this model, associated to the marking of the places. Therefore it is necessary to use another type of instruction specifying the initial marking of the places and the outputs related to them if any. It is also possible to introduce timed Petri Nets, this is achieved by specifying the time delay associated with each place if any. The marking of a place is not effective (for outputs or validating transitions) until this delay time has elapsed.

4. EXAMPLES

To illustrate the scope of the language, two of the applications where the simulator has been used are described. The first one is a model for an electrical substation. The second example is a hydroturbine generation unit. Automatas to control some functions of these two systems are being developed and the simulator is being used to test the behaviour of the automatas.

4.1 Electrical substation

This example shows how easily systems with repetitive parts can be described with the language presented.
As was mentioned in the previous paragraph, an automata for controlling certain aspects of the operation of electrical substations is being developed. This automata is based on microprocessors and has three main functions: load shedding, automatic reclosure and faulty ground detection. The model of the substation should therefore reproduce in its behaviour those aspects of the behaviour of the substation which are relevant to the functions mentioned above.
The two main elements in a substation are circuit breakers and line switches - as a substation contains several of these elements, they will be defined as macros.
The program listing for the macro describing the circuit breaker is the following:

```
01   .MACRO BREAK XX,YY
02  NET
03  &I=X+
04  &SA=O+
05  &SC=O+
06  &EA=I+
07 &EC=I+
08 &TA,&TC,&PA,&PC=XX,YY 96,97,0,X
09 TR 1 FROM 1 TO 2 EX &TC*´&TA+&EC*´&EA
10 TR 2 FROM 2 TO 3 EX  ´(&TC*&TA)
11 TR 3 FROM 3 TO 4 EX ´&TC*&TA+´&EC*&EA
12 TR 4 FROM 4 TO 1 EX ´(&TC*&TA)
13 PL 1 M 1 &PA,&SA
14 PL 2 T S 2
15 PL 3 &PC,&SC,&I
16 PL 4 T S 2
17 ENDNET
18 .ENDM
```

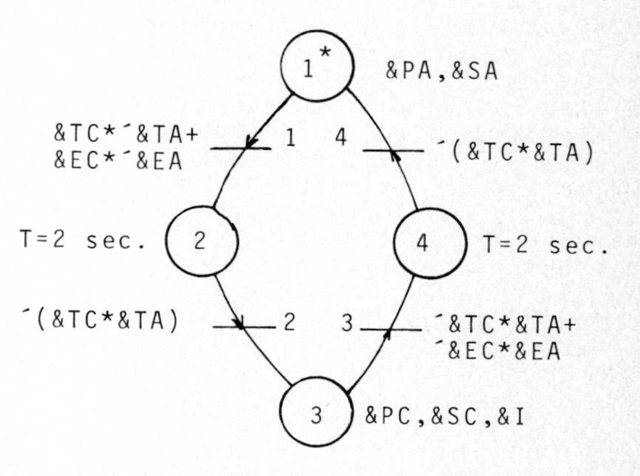

Figure 3 Circuit breaker Petri Net.

Lines 2 and 17 correspond to the instructions specifying the begining and end of a Petri Net.
Lines 3 to 7 are label definitions, the symbol & specifies that these labels are local to the net. The label &I will be associated with the next available internal signal, the labels following, &SA, &SC, &EA and &EC will be associated with the following two available output and input signals respectively.
Signals &EA and &EC are associated to the open and closure orders to the circuit breaker, whilst the local labels &SA and &SC are associated to the open and closed position switches of the circuit breaker. The internal signal &I will be used inother nets where the circuit breaker position needs to be known, as is the case of the faulty ground detection where topological considerations are needed. Line 8 of the listing corresponds to the definition of an active point on the screen. In this case only one active point is associated with the circuit breaker. The signals &TA and &TC correspond to the keyboard signals associated with the breaker, simulating as was mentioned before, manual operation on the circuit breaker. &PA and &PC correspond to the screen variables for the active point and the four parameters at the end to the symbols that will be associated with the four possible values of the signals

&PA and &PC.The parameters XX and yy are the relative x-y position of the active point on the screen.

Lines 9 to 12 describe the transitions - of this particular net. As it can be seen, they are defined by the input and output places and by the associated boolean expression.

Lines 13 to 16 define the initial marking of the places, the associated output and time-lag. Line 14 specifies that a 2 second time delay should be observed for - place number 2.

A line switch can be modelled using an identical net to the one described above except for the external input signals , which are non existent as the automata is not going to alter line switches,and the screen representation which is also different.

Figure 4 shows the screen representations of a coupling cell and a line cell.These types of cells can be defined as follows

```
.MACRO CELLIN OX,OY
;        LINE CELL POS. OX,OY
X=OX
Y=OY
;        LINE BREAKER POS. OX+5,OY+4
BREAK 5,4
;        LINE SWITCH BUS 1 POS. OX+7,OY+2
SWITCH 7,2
;        LINE SWITCH BUS 2 POS. OX+3,OY+2
SWITCH 3,2
;        SWITCH BYPASS POS. OX,OY+2
SWITCH 0,2
;        LINE SWITCH POS. OX+5,OY+6
SWITCH 5,6
;        LINE VOLTAGE  POS. OX+5,OY+9
VL 5,9
.ENDM
```

```
.MACRO COUPL OX,OY
;        COUPLING CELL POS. OX,OY
X=OX
Y=OY
;        COUPLING BREAKER POS OX+2,OY+4
SWITCH 2,4
;        SWITCH BUS 1  POS. OX+4,OY+2
SWITCH 4,2
;        SWITCH BUS 2  POS. OX,OY+2
SWITCH 0,2
.ENDM
```

The program line with VL is a macro call to a Petri Net where the line voltage is simulated. The general program of a model that only takes into account the behaviour of the switches and breakers is given in the following listing.

```
0074  ;
0075  ;          PROGRAM
0076  ;
0077  COUPL 3,12
0078  CELLIN 11,12
0079  CELLIN 22,12
0080  CELLIN 33,12
0081  CELLIN 44,12
0082  CELLIN 55,12
0083  CELLIN 66,12
0084  COUPLU 3,10
0085  CELLIU 11,10
0086  CELLIU 22,10
0087  CELLIU 33,10
0088  CELLIU 44,10
0089  CELLIU 55,10
0090  CELLIU 66,10
0091  X=0
0092  Y=0
0093  ;          SWITCH BUS 1
0094  SWITCH 76,11
0095  ;          SWITCH BUS 2
0096  SWITCH 78,11
0097  END
```

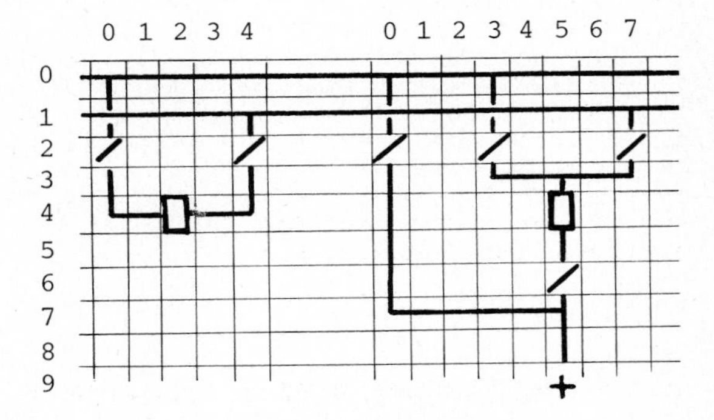

Figure 4. Coupling and line cells screen representations

The mimic representation associated to this program can be seen in figure 5.As it can be observed this model only considers the isolated functioning of the main elements of the substation. More Petri Nets have been developed to simulate other aspects of the behaviour of the process. As an example, let us consider the faulty ground signals generation.

These signals are obtained at transformers and should be computed (by the simulator) taking into account the faulty lines and busbars (declared by the operator via keyboard signals) and topological considerations (depending on the status of the circuit breakers and line switches). For this purpose four internal signals are defined. These signals represent a faulty ground transmited to one of the four busbars, and are set if one or more faulty grounded lines are -

connected to the respective busbar or if a busbar ground signal has been set by the operator. To obtain these signals a trivial Petri Net with four places is associated with each line cell.

The faulty ground signals on the transformers are then easily computed from the busbars faulty ground signals and very simple topological considerations on the transformer cells status.

4.2. Hydroturbine power unit.

A model of a hydroturbine power unit has been developed to test on automata controlling its starting up and shutting down procedures. To show the scope of the model let us sumarize the starting up procedure taken from (3).

After a local or remote start up order has been received and once the security conditions have been checleed the starting pilot valve is operated. After that, the limiter of the hydraulic actuator is reset to the no-load position, the gate apparatus is opened and control is transfered to the automatic synchronization equipment once the generator speed is over a porcentage of its nominal value.

The secuence of operations in the normal shutting down procedure is practically the same but in reverse order. In the case of an emergency shut down the secuence is simpler as most of the elements of the unit are turned off at the same time.

From the automata point of view, the hydroturbine power unit consists of a set of elements that behave as logical systems. For example the oil system with two inputs (activating and disactivating the oil pumps) and one output that is set after a time lag which depends on the oil preassure is adecuate. This behaviour is very easily modelled with a two places Petri net.

The behaviour of the hydroturbine is modelled with a Petri Net composed of a doubled linked chain of timed places. The outputs of these places correspond to the speed relays of 0.5, 20, 90 %, and overspeed and the evolution of the marks depending on the valves positions. Various types of failures can be introduced in the behaviour of the model via keyboard signals. For instance keyboard signal is used in the transition between places with outputs associated with 20 and 90 %. When this keyboard signal is set the transition is not validated and a time out alarm should be produced in the automata. Other aspects of the unit are modelled this way. The

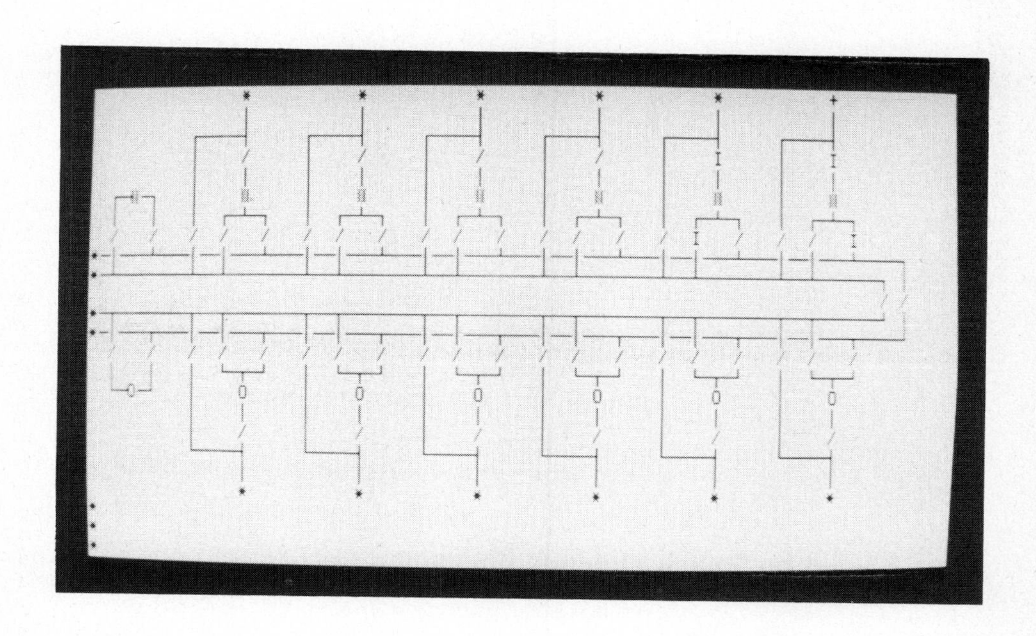

Figure 5. Substation Screen Mimic.

screen mimic associated with the model - can be seen in figure 6.

For these two examples presented, the cy cle time has been smaller then 0.2 se -- conds. Both of them containing over 400 places and about the same number of tran sitions.

CONCLUSIONS.

A software simulator for modelling boolean systems bases on the Petri Nets - description approach has been implemented. The simulator allows an interactive operation via a CRT, which has pro - ved very helpful in the cases simulated to increase the description power of the lenguage, output functions are now being implemented.

ACKNOWLEDGEMENT.

The authors would like to acknowledge - the help of the Cia. Sevillana de Electricidad for supporting this project.Co mments by Mr. J.C. Serrano, J. Colmenero, F. Mateo and J. Montaner were especially helpful.

REFERENCES:

1 Peterson,J.L. Petri Nets, A.C.M.Comp. Surveys, vol. 9, n°3 (1977), 223-251.

2 Daclin, E. and Blanchard,M., Synthese des Systémes Logiques (Cepadues, 1976).

3 Barzan A., Automation in Electrical Power Systems.(Mir, Moscow, 1981).

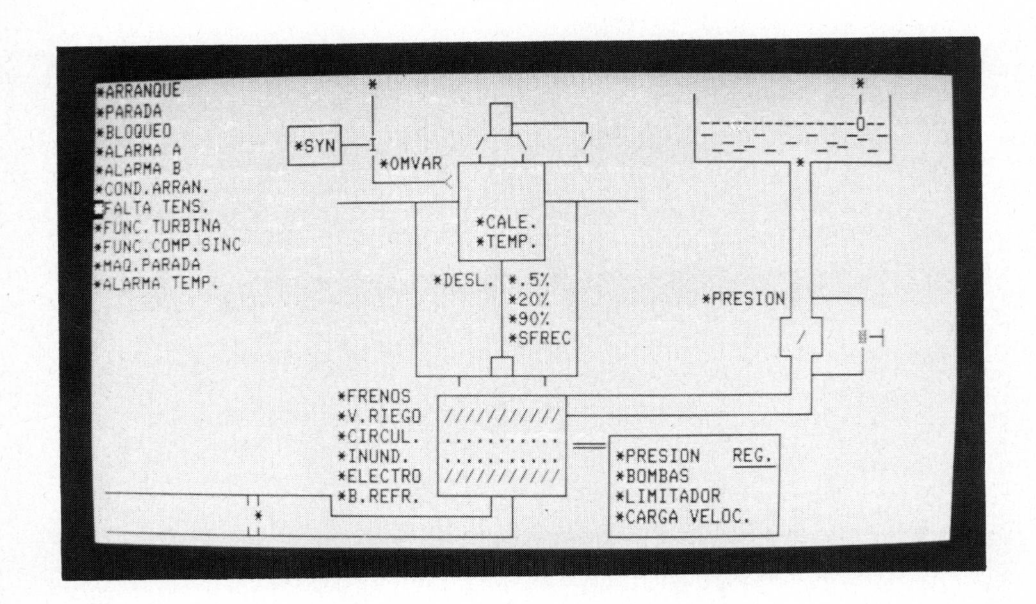

Figure 6. Hydroturbine Screen Mimic.

Simulation in Engineering Sciences
J. Burger and Y. Jarny (eds.)
Elsevier Science Publishers B.V. (North-Holland)

COMPUTER AIDED MODELLING OF COMPLEX PROCESSES: A PROGRAM PACKAGE

L. Marcocci - S. Spelta

ENEL, DSR - Centro Ricerca di Automatica, Via V. Peroni 77 - 20133 MILANO ITALY

A Computer Aided Modelling Package devoloped for modelling and simulation of continuous processes is presented. The package is conceived to carry out the whole pattern of a simulation study; distinctive feature is the ability of modelling a complex process simply by interconnecting off-the-shelf process component models in the required configuration. An application of the package to a large size electrical generating plant is outlined. Comparison with experimental results is also included.

1. INTRODUCTION

Computer simulation languages (see e.g. [1] and [2]) do generally not adequately support the user in the critical phase of the model building, which takes very long time for complex real processes, particularly when the model aims at describing the full range (nonlinear) behaviour of the real plant, as in the case of large thermoelectric power plants [3].
More recently, a positive solution of the above problem has been searched for by developing computer packages, which, while keeping the benefits of any simulation language, are able to support the user in all the phases of modelling and simulation so that they could be better defined as packages for "Computer Aided Modelling (CAM)" (see e.g. [4], [5]). Their effective application to industrial processes is based on the following remarks:
(i) Most industrial processes are very complex but they are generally constituted by many "components" belonging to a limited number of types, that is they have a "modular" structure.
(ii) Modern computer systems are suitable for advanced computer aided work.
(iii) Process complexity and modularity generally entail "sparsity" of the corresponding system of equations describing the model, so that efficient numerical methods for sparse system can be conveniently exploited.
(iv) Process engineers, who are the typical users, are familiar with description of the process in terms of geometrical and topological data and process operating ratings, so the user should be required to supply this kind of information for the model building.
The present paper describes a CAM package developed for modelling and simulation of continuous processes, particularly oriented to large size electrical generating plants. The package is conceived to carry out the whole pattern of a simulation study, starting from the plant data entering till to the graphic presentation of the simulation results. Moreover, it is equipped with special outputs (linearized system matrices) for the interface with control design programs.
Particular care has been taken of the steady states computation as this phase can be considered as a keystone both for the model building and of the dynamic simulation.
Basic features of the package are:

- easy man-computer interaction for data entering, data debugging, selection of operations to be performed, storing and presentation of results;
- modular structure of the process model, realized by organizing the package in two hyerarchical levels: the higher one containing a master program taking care of the global algebraic-differential system constituting the process model, the lower one consisting of a modules-shelf, where each module describes in a fairly general way a typical plant component;
- ability of automatically solving possible algebraic loops hidden into the system equations;
- possibility of interchanging process inputs, outputs and uncertain parameters during the steady state computation in order to exploit the redundant information coming from process data.

The last part of the paper is devoted to the description of the application of the package to a 320 MW once-through boiler power plant, which was the most significant real-size problem dealth with by the authors.

2. BASIC CONCEPTS AND PACKAGE ORGANIZATION

The CAM Package (CAMP) described in the present paper consists of two main parts:
(i) The "library" of the Modules (ML), made up of self-consistent mathematical models each one representing a (fairly general) component of the class of industrial processes we are concerned with.
(ii) The computation master program (CMP), devoted to handling and coordinating all the computations activities and, in particular, to merging the necessary elements of ML and to solving large scale systems of nonlinear algebraic and differential equations.

Distinctive feature of CAMP is that the user is allowed to build up the model of a real process simply describing the connections among process subsystems (i.e. the topology) and the relevant design data of each component. The mathematical model and the simulation of an industrial process are therefore naturally obtained by putting together the CMP, the relevant part of the ML and the user's information specifying actual realization and connection of the process components describable by the selected Modules.

2.1 The Modules Library (ML)

A Module, that is the mathematical model of an fairly general physical subsystem able to be fit to a process component, essentially consists of a system of nonlinear algebraic and differential equations of the form:

$$\dot{x}_M = f_M (x_M, y_M, u_M, q_M) , \qquad n_M \text{ equations,} \qquad (1.a)$$

$$0 = g_M (x_M, y_M, u_M, q_M) , \qquad p_M \text{ equations,} \qquad (1.b)$$

where $u_M \in R^{m_M}$ is the vector of the input variables, $x \in R^{n_M}$ and $y \in R^{p_M}$ are the vectors of the differential (states) and algebraic outputs, respectively, and $q_M \in R^{r_M}$ is the vector of geometrical-phisical parameters (design data).
Input and output variables of a Module are the variables by which the mathematical connection of elementary models describing process subsystems can be specified.
Thus, the Modules are selfconsistent and essentially independent one of each other, so that the ML can be easily extended to deal with a larger class of processes or real cases, provided that the choice of input and output variables of new Modules allows connections with the already existent ones. As for the computer implementation, each Module is constituted by 3 subroutines (SUB) called by CMP while carrying out the different functions relative to the model building and to the simulation. The tasks of the three subroutines are the following.

SUB 1: Interactively communicating to the user the physical meaning of the input and output variables of a Module.

SUB 2: Reading and storing the geometrical and physical data which specify a Module as a given process subsystem.

SUB 3: For any values of the variables x_M, y_M and u_M and of the parameter q_M appearing in (1), computing the values taken on by $f_M(x_M,y_M,u_M,q_M)$ and $g_M(x_M,y_M,u_M,q_M)$ and by their partial derivatives with respect to x_M, y_M and u_M.

The above described organization implies the following properties:

a) Module simplicity and readability. In fact only SUB 3 is devoted to computing. Neither equation (1.b) is necessarily solved with respect to y_M ("algebraic loops" are also allowed) nor the Module is concerned with the computation of initial conditions of (1).

b) Module generality. SUB 1-3 are written in view of a certain type of plant component, yet they are independent of the data process variables specifying a definite real system. Therefore, a single Module can be used to describe both different components of the same process and similar components of different processes.

c) Module reliability. It is a consequence of the above two properties, of the standardization of the Module software, and of special testing functions incorporated in CMP (e.g. numerical Jacobian check) which make the debugging of a Module very reliable.

2.2 The computation master program (CMP)

CMP is splitted into 8 main "activities", each one corresponding to a well defined Phase of a typical study. These Phases can, moreover be grouped according to the following natural classification of the package functions:

(i) Model building (Phase 1-3)

(ii) (Nonlinear) simulation (Phases 4-6)

(iii) Model linearization and analysis of the linearized model (Phases 7-8).

To each Phase there corresponds an executable program (TASK), which is generally interactive and is obtained by linking the corresponding activity of CMP with ML. The different Tasks share the information concerning process topology and physical-geometrical characteristics of the components through a number of permanent files.

An overall description of the logical sequence of the Phases and of their interaction (e.g. information sharing) is given in Fig. 1.

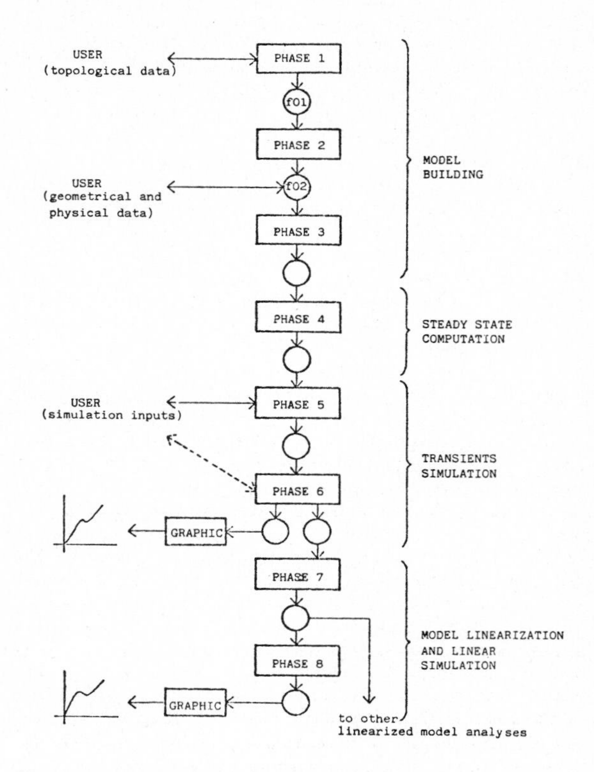

Fig. 1 Logical phases sequence of the CAMP.

Phase 1 is devoted to the definition of the plant topology, in which the user conversationally specifies:

a) Which are the Modules necessary for modelling the various process subsystems (Blocks).

b) The (conventional) names (Labels) of the input and output variables of each Block. Process interconnections are established by Label identities.

During this Phase, the user is aided by CAMP which prints the physical meaning of the variables of the Blocks (see SUB 1 of ML) verifies the formal correctness of the topological data supplied by the user and, finally, stores them in a special permanent file (F01).
Phase 2 is devoted to the preparation (SUB 2 of ML) of a special permanent file (F02) storing:

a) geometrical and physical data relative to the various process subsystems;

b) initial tentative values of the process variables.

Suitable conversational editing programs assist the user in data entering, updating and merging (e.g. to put together data concerning parts of the process at hand). Note that F01 and F02 constitute a complete documentation of the model.
Phase 3 is devoted to the acquisition, elaboration and verification of the geometrical and physical data specified by the user.
Phase 4 is devoted to the computation of process steady states. The equation systems (1) relative to the different Blocks composing the process model are put together and ordered according to the topological connections defined in

Phases 1-3. Then CMP builds up the global system (GS) of algebraic and differential equations (still of the form (1)) which describes the whole mathematical model. The input variables of the global system are all those Block input variables which do not coincide with output variables of any Block. They are the actual physical inputs of the plant. For the computation of a steady state the user must specify which are the process variables assumed to be known and supply the stipulated values of such variables together with tentative values for the remaining state, output and input variables.

A steady state is computed by solving the (nonlinear) algebraic equation system derived from (GS) by setting to nought time derivatives.

In Phase 5 the user specifies the kind of transient to be simulated, by assigning:

- The plant input variables as functions of time, over the whole simulation interval;
- The integration time step to be adopted;
- The tolerance allowed in the solution of the global equations system (as better stated in the next paragraph);
- The set of the process variables to be recorded.

Phase 6 is devoted to the computation, in strict sense, of plant transients. The computation is carried out by solving GS resulting from the assemblage performed in Phase 4.

Since all the user's specifications and options are entered during the preceding Phases, Phase 6 is purely computational and can also be run in batch. However, this CAM package has been designed so as to allow the user to run a process simulation in an interactive way, possibly intervening during the simulation to vary plant inputs in view of the past model evolution.

Since, dealing with large scale processes, the analysis of simulation results requires flexible graphic output, a graphic package has been devoleped for displaying process variables with the following options:

- using different output devices;
- plotting together, for the sake of comparison, records relative to different variables (possibly coming from different runs);
- arbitrarily varying the plot scales;
- plotting additional quantities easily derivable from recorded model variables;
- using the graphic output both for post-mortem analysis and for interactive simulation.

Additional, somehow special, functions allowed by the CAM package are the computation of the linearized system matrices (Phase 7), which can typically be used for control system design or for small variation analysis by direct simulation of the linearized model (Phase 8).

It is worth noticing that the above described package organization, where independent Tasks communicate among them through permanent files as shown in Fig. 1, is of fundamental importance for making the package an effective CAM tool. In fact, the logical sequence of operations as indicated in Fig. 1 can easily be interrupted at any point to repeat some of them according to the user's judgement about the obtained results. This feature is particularly useful for steady state computation where the (possible) redundancy of information about the process is not exploitable within a single computation run.

3. ALGORITMS

The main computation algorithms has been chosen so as to meet the following requirements:

- allowing the user to be involved in numerical integration problems as little as possible, that is using methods essentially robust from the point of view of the numerical stability;
- keeping at the minimum simulation errors due to process nonlinearity;
- automatically solving algebraic loops;
- ability of takling stiffness;
- ability of dealing with complex processes, that is with large scale systems of algebraic and differential equations;
- allowing flexible, cheap and robust computation of the process steady states.

3.1 Steady state computation

The computation of the steady state is performed by a damped Newton-Raphson method. The computation is organized in the following steps:

1) The user specifies the known variables (input, output or states), the only constraint that they square with the equations[1].
2) CMP communicates the initial tentative values of the relevant variables to each Module, which gives back the equations residues and the corresponding part of the Jacobian matrix.
3) CMP performs a Newton iteration and repeats step 2) with the modified values of the plant variables.
4) The iterative procedure involving step 2) and step 3) is stopped when all the equation residues are smaller than a specified threshold value (tolerance).

Since a Newton iteration requires the use of a factorized Jacobian whose computation is very time consuming, the factorized Jacobian computed at a given iteration is retained in the subsequent n iterations (n=5 by default)[2] unless convergence difficulties arise. In this latter case, also a suitable damping of the Newton iteration can be applied.

To avoid imbalance among the tolerances attributed to different equations, each system variable and residue is divided by a suitable normalization value which is specified within each Module (consistently with other Modules which can be interfaced with it).

Moreover, since the inputs of a module are not necessarily coincident with the physical inputs of the corresponding component in any plant configuration, the computation of a steady state can be used also to estimate uncertain process parameter by defining them as possible (constant) Module inputs. They will be threated as unknown inputs during the computation according to the specifications of step 1); of course suitable process outputs has to be assigned as known[1].

3.2 Dynamic simulation

The computation of the process transients is based essentially on the same numerical methods as that of the steady state. It is worked out in two main steps: first, the global system GS of algebraic and differential

[1] More precisely, it is necessary that the equation system be (at least locally) solvable, i.e. the Jacobian matrix at any iteration point be non singular.

[2] Periodic recomputation of Jacobian is adopted during the computation of the steady state because it is expected that the system variables can be subject to consistent variations.

equations is transformed into a system of algebraic equations; then, at each time step, the obtained algebraic system is solved iteratively by a damped Newton Raphson algorithm.
More precisely, let GS be described by:

$$\dot{x} = f(x, u, y) \quad , \quad n \text{ equations,} \qquad (2.a)$$
$$0 = g(x, u, y) \quad , \quad p \text{ equations,} \qquad (2.b)$$

where $x \in R^n$, $y \in R^p$ and $u \in R^m$ are the state, output and input variable vectors of the process model.
Equation (2.b) is maintained unaltered, while equation (2.a) is discretized according to the following trapezoidal formula:

$$x_{k+1} - x_k = .5h\,(f(x_{k+1}, u_{k+1}, y_{k+1}) + f(x_k, u_k, y_k)) \qquad (3)$$

where h is the integration step and the subscript k denotes evaluation at the time kh. Then, noting that x_k, u_k, y_k and u_{k+1} are known values at the (k+1)th integration step, eqs. (3) and (2.b) represent a system of (n+p) algebraic equations for the (n+p) unknown quantities x_{k+1}, y_{k+1}. The solution of the algebraic system (3), (2.b) yields the output of the model at time (k+1)h.
Due to the well known properties of the trapezoidal integration rule [6], the stability of the method essentially depends on the tolerance chosen for the solution of system (3), (2.b) which should be related to the integration step. The experience gained with the use of the package suggested adopting a fixed (quite small) tolerance and possibly acting on the integration step.
A peculiarity of the adopted numerical procedure is that, though the integration method is essentially with fixed step, the integration step can be lengthened when the time derivatives becomes sufficiently small.

3.3 Sparsity

Dealing with large scale systems, that is with systems describing complex processes composed by weekly compled subsystems, important saving of CPU time and memory requirement can be obtained by exploiting the inherent sparsity of system (3), (2.b). This is done in the package by limiting the Jacobian computation to that of the non-zero entries (identified on the basis of the topological information) and adopting an algorithm for sparse matrices [7] for Jacobian factorization and system solution.

4. APPLICATION

4.1 General Aspects

The CAM package described in the preceding sections has been applied to a large-scale real process, that is a 320 MW fossil fired power plant. The plant is equipped with a once-through Universal Pressure boiler [8] and can be splitted into 3 main systems:
(i) boiler and furnace;
(ii) turbine and thermal cycle;
(iii) control and protection system.
Object of the present application is, in particular, system (i) whose scheme is reported in Fig. 2.
The feedwater preheated in the thermal cycle enters the economizer (ECO), where the water is further heated by the flue gas at the end of the gas duct. Then, it is fed to the furnace walls tubes, where radiation heat is released to the fluid by the flame and evaporation takes place. At the outlet of the furnace panels evaporation is completed and

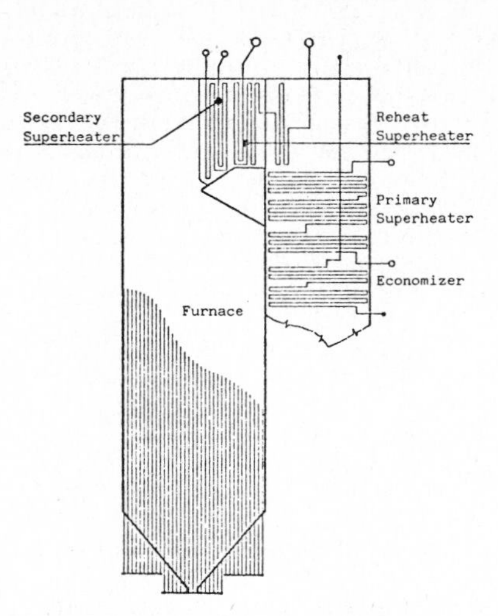

Fig. 2 UP once-through boiler outline.

the generated steam is finally superheated by the hot flue gas in the gas duct enclosure tubes and in the primary and secondary superheater, successively. At the outlet of the secondary superheater, the steam (nominally at 167 bar and 540 °C) is sent to the high pressure (HP) turbine. The HP turbine exhaust steam is superheated again in a single large reheater located in the middle part of the gas duct and then readmitted to the middle low pressure (M-BP) turbine. Two desuperheating sprays are located at the inlet of the final superheater and of the reheater in order to allow steam temperature control.
The hot gas, after going through the gas duct, are partially recirculated, to allow control of the ratio between radiation and convection heat.
The modelling problem has been approached according to the following basic concepts:
- Any system (e.g. the boiler) has to be considered as a connection of simple physical elements (Blocks), each one modelled by a Module, so as to keep the number of different Modules sufficiently small.
- Modules should not be defined so simple as to require too many Blocks in order to limit the complexity of the topology of the overall model.
- Physical properties like the ones depending on the type of fluid or the type of fluid flow are not included in the Module equations, but dealt with in special "general purpose" subroutines.
- The Modules are based on "first principle" non linear lumped parameter equations. A Module can contain either a single lump or a set of similar lumps, whose number and size is parametrically adjustable.

The following Modules have been defined and set up for system, (i):

M1: Fluid Volume, which describes the behaviour of enthalpy and pressure along a fluid flowing inside a tube, allowing different lumping for mass and energy storage.
M2: Tube Wall, which describes the behaviour of temperature within a metal wall in dependence on the heat exchange through the external and internal boundary surfaces.
M3: Gas Volume, which describes the behaviour of temperature within a gas flowing in a duct and exchanging heat with the duct enclosure and banks of tubes located along the gas stream.
M4: Furnace, which describes the temperature distribution along the furnace vertical axis, assuming gas

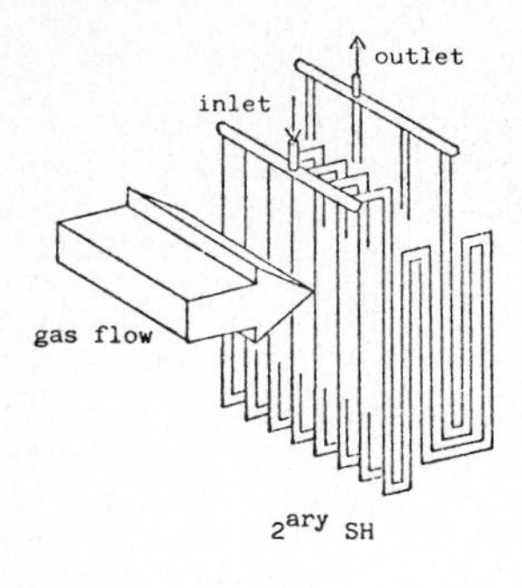

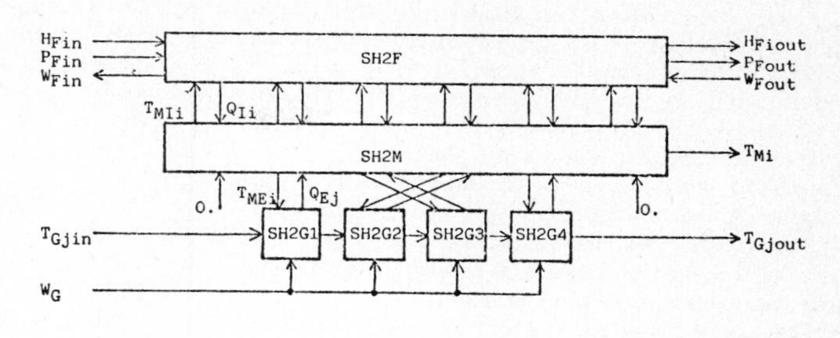

Simbols :

H enthalpy
P pressure
T temperature
W mass flow rate
Q heat exchanged

Subscripts :

F fluid
M metal tubes
I internal surface of tubes
E external surface of tubes
G gas

i ith lump of metal or fluid (i=1,..,6)
j jth gas volume (j=1,..,4)
in inlet
out outlet

Fig. 3 Tubes layout of secondary superheater and correspondent model block scheme.

properties as uniform on each horizontal plane.

M5: Regulating Valve, which computes the mass flow rate as a function of the boundary pressure, inlet enthalpy and valve position.

M6: Attemperators.

M7: Impulse turbine stage.

M8: Reaction turbine stage.

The model of the whole system (i) completed by a part (turbine) of system (ii) has been realized by using about 60 Blocks (with the only 8 Modules described above).

4.2 Illustrative example

For the sake of illustration, the Block scheme (drawn according to CAMP rules) of the secondary superheater is presented in Fig. 3 together with the corresponding plant layout.

The steam side of the secondary superheater is modelled by block SH2F (Module M1) with one lump for mass storage and six lumps for energy storage. Metal of tubes is modelled by block SH2M (Module M2) with six lumps corresponding to the fluid lumps.

Gas side is modelled by a cascade of four blocks SH2G1,...,SH2G4 (Module M3). Connections between Blocks are accounted for by variables shared by different blocks, that is by Q_{Ii}, T_{MIi}, Q_{Ei}, T_{MEi}. The cross connections between gas volumes SH2G2 and SH2G3 and lumps 3 and 4 of SH2M accounts for the counter flow path of steam and gas, resulting from the shape of the superheater bank.

The package ability of solving general algebraic nonlinear systems can be recognized also in this example; in fact, the exchanged thermal power Q_{Ii} and Q_{Ei} and the metal surfaces temperatures T_{MIi}, T_{MEI} are computed by solving the following equations:

- Equation of steam-metal heat transfer (SH2F)

$$g_{QI}\ (Q_{Ii},T_{MIi},H_{Fi},P_{Fi},W_{Fi}) = 0$$

- Equation of the thermal resistances of the two halves of the tube wall (SH2M)

$$g_{TMI}\ (T_{MIi},Q_{Ii},T_{Mi}) = 0$$

$$g_{TME}\ (T_{MEi},Q_{Ej},T_{Mi}) = 0$$

- Equation of gas-metal heat transfer ($SH2G_j$)

$$g_{QF}\ (Q_{Ej},T_{MEi},T_{Gj},W_G) = 0,$$

where the functions g_{QI}, g_{TMI}, g_{TME} and g_{QE} need not to be explicitly solved with respect to any variable, so that general (complex) heat exchange correlation can be easily incorporated. The model of Fig. 3 is completed by a set of geometrical data crudely evaluated on the basis of process design data.

4.3 Steady state computation

The steady state computation has been carried out in the following steps:

(i) The steady state of the process at the nominal (maximum continuous) load was computed by assuming as "strictly" known variables the physical quantities measured with the highest degree of accuracy, e.g. HP and MP turbine inlet steam temperatures and pressures and generated electric power and as tentatively known variables as many measured process quantities, as necessary to make the problem solvable, even in presence of some uncertain parameters (fouling factors, friction factors etc.). The redundancy of process data/measurements has been conveniently exploited by changing/adjusting the set of the tentatively known variables until a satisfactory solution has been obtained.

(ii) Other interesting steady states of the process have been computed by keeping as known the uncertain parameters (intrinsecally valid over the whole load range) computed in step (i) and obviously changing the relevant plant ratings. Validation of the computed steady state at 50% of the maximum load by field measurements was quite satisfactory.

Steady state computation is carried out with particular efficiency by the present package (with respect to more classical simulation packages) as demonstrated by the obtained results. In fact, the model, composed by 270 algebraic equations and 130 differential equations, incorporating a very accurate computation of the heat exchange coefficients (valid over the whole range of operation) and of the steam properties (according to the steam tables described in [9]), took about 100 s of CPU time on a VAX 750 computer for the computation of the 50% steady state starting from the 100% steady state.

4.4 Transient computation

Simulation of plant transients have preliminarly been performed to verify both the modelling criteria and the package performance, by comparing typical step responses with field tests. Though the model, linearized at the nominal steady state, has eigenvalues ranging (in modulus) from less 10^{-3} to more than 10^{3}, the integration of the model equations has been performed with a time step of 4 s.

This was possible because the integration method is stable even with a relatively long integration step provided that the tolerance of the residues is set sufficiently small. Of course, very fast phenomena dominated by the biggest eigenvalues are roughly approximated. They however are not essentially excited in most plant transients.
For the sake of demonstration, simulated responses and recorded field results are compared in Fig. 4, for a step-wise reduction of the fuel flow (about 1.6% of maximum flow rate). Agreement between model and plant is within the accuracy of the industrial measurement devices.
The CPU time spent for simulating the above transient (lasting 720 s of real time) was about 610 s on a VAX 750 computer.

4.5 Comments and Extensions

A useful reference for the results obtained in the application is a preceding work [10] aimed at simulating the full range behaviour of a similar plant. There is shown how comparable simulation accuracy was obtained by using the computer code described in [5], which however is based on the direct integration of partial differential equations and requires a substantially greater computational burden and user's effort.

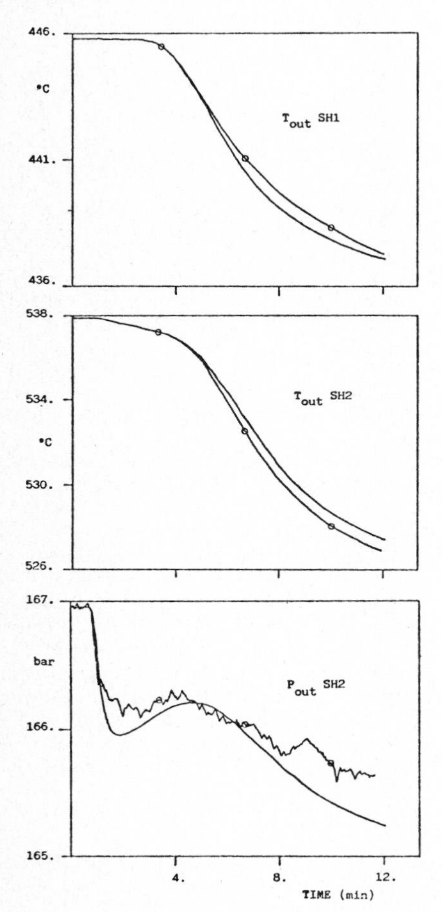

Fig. 4 Comparison between computed and measured (-θ-) transient after a step change in fuel flow rate at maximum load (-1.6% of maximum flow rate at t=45 s).

It is worth noticing that, apart from the work inherent to the separate modelling of additional plant systems (in which a few additional modules appear), the global plant model composed by systems (i)-(iii) can be easily obtained by merging the "data bases" of the individuals models (files F01 and F02) without significant additional work.

5. CONCLUDING REMARKS

The CAM package presented in this paper has been applied to different real case studies, including an experimental solar powered generating plant. The gained experience and the work in progress has demonstrated the undoubted superiority of CAMP with respect to more traditional simulation tools, as for numerical robustness and, primarily, for the flexibility and ease of use. Moreover, since the package has been used for less than three years, it is still under development particularly for what concerns improvements and extensions of the Modules Library.

Aknowledgements. The authors are grateful to A. De Marco and C. Maffezzoni for fruitful discussions concerning the development of the package and the preparation of the paper.

REFERENCE

[1] System/360 Continuos Modelling Program, User's Manual. IBM Corporation Technical Publications White Plains, New York.
[2] A.BRINI, R.FERRARI, S.GOLINELLI, T.MONTAGNA, G.PERNA, J.SZANTO
GOSPEL, an interactive language for continuos system simulation. Summer Comput. Simulation Conf., Washington 1976.
[3] G.K.LAUSTERER, J.FRANKE, E.EITELBERG,
Mathematical modelling of a steam generator. IFAC/IFIP Conference on Digital Computer Application to Process Control, Dusseldorf (FRG) 1980.
[4] O.W.DURRANT
Design, Operation, Control and modelling of pulverized coal fired boilers. Boiler-Turbine Modelling and Control Seminar Sydney 1977.
[5] M.MAXANT, M.PERRIN
Mathematical modelling of two phase flow applied to nuclear power plant. II Multiphase flow and heat transfer Symp., Miami 1979.
[6] H.W. DOMMEL, N.SATO
Fast transient stability solutions. IEEE Winter Power Meeting, New York 1972.
[7] MA28 A set of FORTRAN subroutines for sparse unsymmetric linear equations. United Kingdom Atomic Energy Authority, Harwell AERE R8730 (1979 revision).
[8] "Steam - Its generation and use" 38th Edition BABCOCK & WILCOX, New York.
[9] M.PERRIN
Water and Steam thermodynamic properties, numerical calculation programs (in French). REVUE GENERALE DE THERMIQUE n. 182 February 1977.
[10] P.COLOMBO, R.CORI, A.DE MARCO, G.B.GARBOSSA, G.QUAZZA
Experience with steam power plant modelling at the Italian Electric Power Board. III Power Plant dynamics, Control and Testing Symposium, KNOXVILLE 1979.

Simulation in Engineering Sciences
J. Burger and Y. Jarny (eds.)
Elsevier Science Publishers B.V. (North-Holland)

FLEXIBLE SOFTWARE PACKAGE FOR RAILCAR DESIGN

R.C. White, A.A. Merabet

Department of Electrical Engineering, Arizona State University
Control Systems Laboratory, National Research Council of Canada

Abstract: A design approach for high performance vehicle-guideway-suspension dynamical systems is formulated as a parameter optimization algorithm for the conflicting objectives of stable guidance on tangent and curved track, vibration isolation, and tractive adhesion. As a design tool the simulation must be validated before confidence can be placed in the levels of the quantities predicted by the mathematical procedures used. This validation is performed as a correlation strategy between the theoretically computed results and experimentally obtained or prevalidated analytical values. In this paper a complex vertical/lateral 30 degree-of-freedom lumped parameter full car model simulating the dynamics of a large class of railcar, guideway and suspension subsystems is developed. The mathematical techniques for extraction of the performance criteria are described, and a heuristic design procedure is presented. The correlation strategy is applied to the Metroliner and the Via passenger railcars. The simulation complexity can be reduced by the introduction of two types of simplifications: model reduction achieved by constraining and/or eliminating degrees-of-freedom, efficient algorithms applied in conjunction with effective system parameter values obtained by quasi-linearization or linearization.

1. INTRODUCTION

This paper describes the development of a software package for the simulation and analysis of rail vehicle dynamics. Many rail vehicle simulation and analysis programs exist [1,2], and most are tailored to specific analysis tasks, such as evaluation of vertical ride quality or lateral stability. General purpose simulation and analysis programs [3] offer flexibility, but are more difficult to apply to the specialized problems of rail vehicle dynamics. The aim of the work described here is to construct an analysis package based on one rather comprehensive dynamical model from which models of different types and complexities can be generated by introducing constraints between variables and/or discarding degrees of freedom [4]. The vehicle model software is interfaced to routines for performing functions such as simulating time histories or analyzing stability. Some advantages of this approach to modeling and analysis are:

- nonduplication of software common to different simulation and analysis functions,
- the ability to individually tailor a model to a specific task,
- compatibility of different models of the same vehicle.

This software package is in a development stage. In this paper we describe the comprehensive vehicle-guideway model, the implementation of routines for simulation of time histories and stability analysis, and initial work on model validation.

2. RAIL VEHICLE AND ROADBED MODEL

A brief description of the vehicle model (Figure 2.1) is given here. A detailed description and derivation of the vehicle equations are in [5]. The full model of Figure 2.1 represents a passenger vehicle travelling at constant speed. Constraints between the degrees of freedom can be introduced to simplify the passenger car model or to generate freight car models of varying complexity. The full model comprises the following bodies and degrees of freedom:

- 4 wheelsets: sway, yaw, bounce
- 2 truck frames: sway, yaw, roll, bounce, pitch
- 1 car body: sway, yaw, roll, bounce, pitch, lateral bending, vertical bending, torsional bending.

Car body flexibility is modeled by assuming the body is a uniform beam vibrating at only its lowest bending frequency. The mode shape is taken as that of a uniform beam, although a different mode shape based on experimental data could be specified. The flexible body motions are coupled to other degrees of freedom through the secondary suspension.

The special nature of rail vehicle dynamics arises from the contact geometry and force laws that describe the wheel-rail interaction [6,7]. The contact geometry enters the wheelset equations in terms of constraint functions. For example, the normalized difference in rolling radii of the left and right wheels can be described by:

$$(r_L - r_R)/2a = \lambda(\Delta x)\ [\Delta x/a] \tag{2.1}$$

where r_L and r_R are the rolling radii, 2a is the track gauge, Δx is the lateral displacement of the wheelset relative to the rail, and $\lambda(\Delta x)$ is the conicity. In our model we assume that wheelset displacements are small enough to avoid flange contact with the rail. Then the function above can be adequately represented by $\lambda[\Delta x-\rho_0]/a$, where ρ_0 is a constant. Wheel-rail forces are similarly formulated. These approximations are adequate to describe operation on tangent track and in curves where flange contact is avoided. Situation of flange contact with the rail, such as occurs in severe curving or hunting, requires nonlinear descriptions of the contact geometry and wheel-rail forces.

In the following we remark on some features we believe are unique to this vehicle model.

Elements S_{aL}, S_{bL}, S_{aR}, S_{bR} in Figure 2.1 represent secondary suspension units such as pneumatic springs. These elements are modeled as having non-zero height and the capability of generating roll moments as well as vertical and lateral forces. One such unit is shown in Figure 2.2, where the attachment point on the car body has been displaced from its nominal position by Δx and Δy and rolled by $\Delta\phi$. The forces and moment generated by the spring, found by writing equillibrium equations for the spring, are [8]:

$$F_x=k_x(\Delta x+h_s\Delta\phi/2)+C_x(\Delta\dot{x}+h_s\Delta\dot{\phi}/2)$$

$$F_y=L-k_y\Delta y-c_y\Delta\dot{y} \tag{2.2}$$

$$M_Z=-\tfrac{1}{2}\;(h_sk_x+L)\Delta x+h_s^2k_x\Delta\phi/2$$

$$+\;h_sc_x\Delta\dot{x}\;+\;h_s^2c_x\Delta\dot{\phi}/2$$

Spring models commonly employed in rail vehicle simulations neglect spring height and either ignore the roll moment or assume that the connection point moves on the vehicle. In comparison with the model described here, the former method predicts less body roll and the latter greater body roll.

Another feature of the vehicle model is the representation of lateral and yaw misalignments between the wheelsets and truck frames and lateral and yaw interwheelset misalignments. It has been shown that such misalignments produce lateral offsets of the wheelsets that reduce the effective flange clearance, resulting in increased flange contact with the rail [9].

All inertial, geometric, and suspension parameters can be assigned individually; that is, similar parameters can be assigned different values for, say, each wheelset. This flexibility is useful to evaluate important effects. For example, it has been shown that the wheel profiles of different wheelsets on freight trucks assume different shapes with service wear, and these variations strongly affect vehicle stability [10]. The same study showed that asymmetric loading of the vehicle between trucks influences stability.

The roadbed is straight or curved with variable curvature and superelevation. Superimposed are irregularities in alignment, crosslevel, or vertical profile. The irregularities can be deterministic, such as would represent a crossover, or random. Random irregularities are represented in terms of power spectral densities. The track is modeled as flexible for vertical motions (Figure 2.1) but rigid laterally. The large wheel-rail forces that would require modeling lateral flexibility are inconsistent with our linear model of wheel-rail contact geometry and forces.

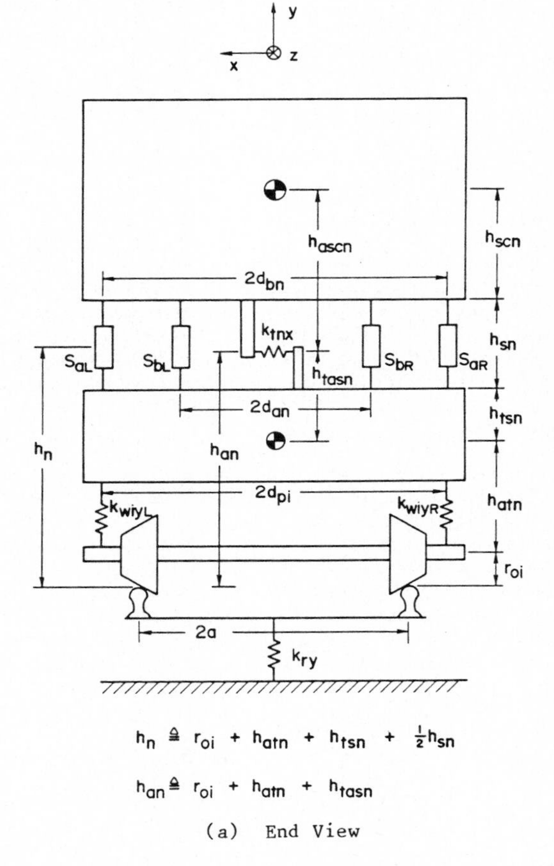

(a) End View

Figure 2.1: Railcar dynamical model

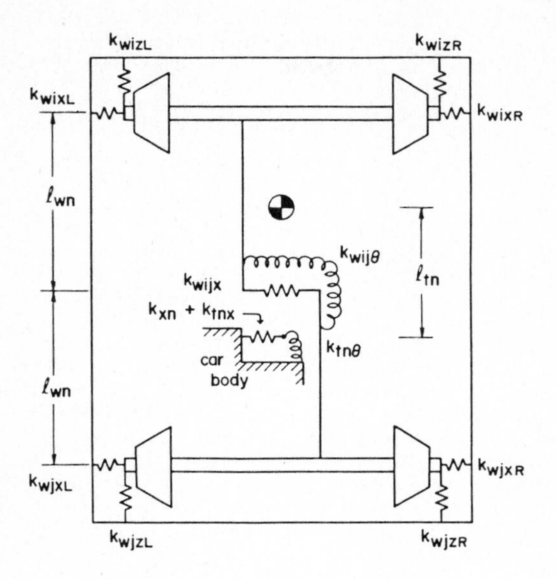

(b) Plan view of truck

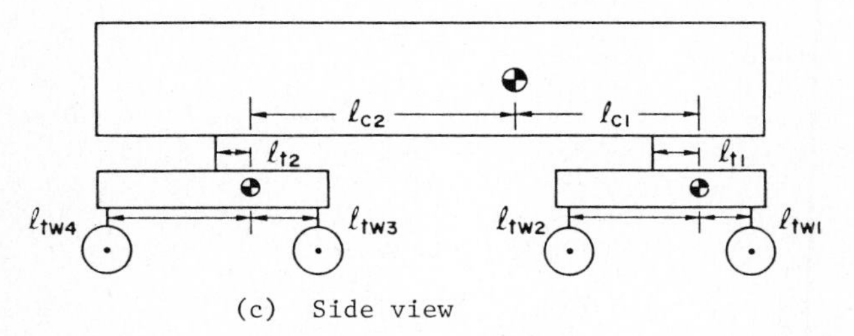

(c) Side view

Figure 2.1 (Cont'd): Railcar dynamical model

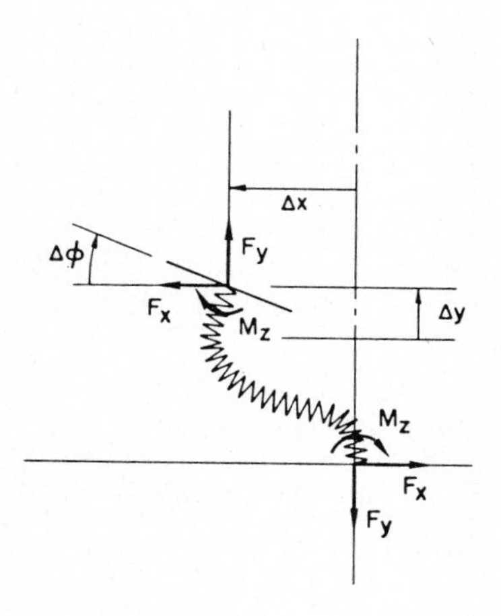

Figure 2.2: Secondary suspension spring model

3. MODEL IMPLEMENTATION

This section presents the simulation of the models dynamics, the time and frequency domain analysis tools applied and a strategy for design of vehicles.

3.1 Equations of Motion

From a lumped parameter vehicle-track-suspension passenger full car structure represented by a set of second order form nonlinear ordinary differential equations, vertical and/or lateral freight car, half car, truck and wheelset sub-models can be extracted by application of kinematic constraints and/or elimination of degrees-of-freedom.

For computational purposes the reduced equations are transformed to a first order form having the displacements and velocities of the wheelset, trucks and carbody as system state, the corresponding track displacements, velocities and accelerations as inputs, and algebraic relations for observation of the vehicle performance, with the coefficients listed in [11].

$$\underline{\dot{x}}(t)\ [A_x]\underline{x}(t) + [B_x]\underline{u}(t)$$

$$\underline{y}(t) = [C_x]\underline{x}(t)+[D_x]\underline{u}(t) \tag{3.1}$$

Two types of nonlineanties are considered: the spring-damper components of the suspension and the wheel-rail geometry of the contact forces. The force-displacement and force-velocity relations of the springs and dampers; and the rolling radii difference, contact angle difference and wheelset roll constraint functions, can be characterized by three constant coefficients describing deadband, linear and stop regions of operations.

3.2 System Inputs

The system may respond to a combination of four types of forcing functions (roadbed geometry, deterministic irregularities, random perturbations and constant biases). Since the vehicle forward speed is assumed constant, the inputs at the trailing wheelsets, trucks and the carbody are obtained by delaying the leading wheelset inputs.

The roadbed geometry is simulated by successive segments representing an entry tangent track, an entry spiral beginning and ending with quadratic transition zones, a steady state curve, an exit spiral with different quadratic transitions and an exit track.

Two different deterministic input forms are generated: a sinusoidal wave, and a pseudo-step functions with a half-cosine transition and the corresponding derivatives for representation of impulses.

The vertical and lateral random irregularities are described by power spectral densities having the form of spectra estimated from track

measurements, and the crosslevel irregularities are obtained from the vertical irregularities by low pass filtering [12].
The suspension misalignments and wheel-rail geometry offsets behave as biases to the vehicle operation and constitute constant input functions to the dynamic model.

3.2 Vehicle Stability

A formal nonlinear stability criteria has not been applied to the system (the Popov test is too conservative). Instead the QR algorithm from the International Mathematical and Statistical Library (IMSL) software package is used to obtain the A matrix eigenvalues for stability analysis and eigenvectors for mode shape representation [13].

The critical velocity where "hunting" begins is obtained iteratively when the operating speed produces eignevalues with positive real parts, and is given by the relation [14]:

$$V_{operating} = 1.2\ V_{critical} \quad (3.2)$$

3.3 Random Response

Only the time domain response is implemented, and uses the Gear (predictor-corrector method suitable for "stiff" systems) algorithm from the IMSL package for integration of equation (3.1).

The track disturbances are obtained in two stages: zero mean gaussian noise sequences are produced from pseudo-random number samples uniformly distributed in the interval (0,1) according to Figure 3.1, "colored" noise with the specified spectra shape is generated by filtering after amplitude matching by the standard deviations of the corresponding processes as shown in Figure 3.2 [15].

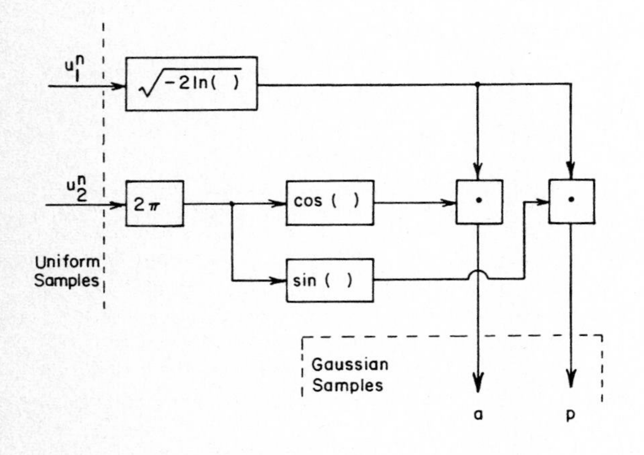

Figure 3.1 Transformation of Uniform to Gaussian Samples

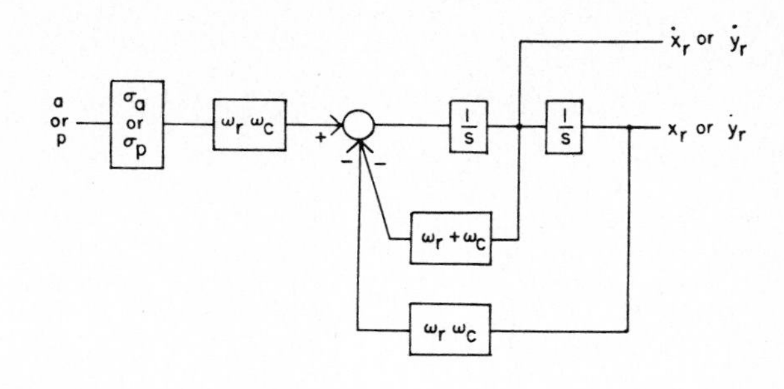

a) Lateral or vertical perturbations

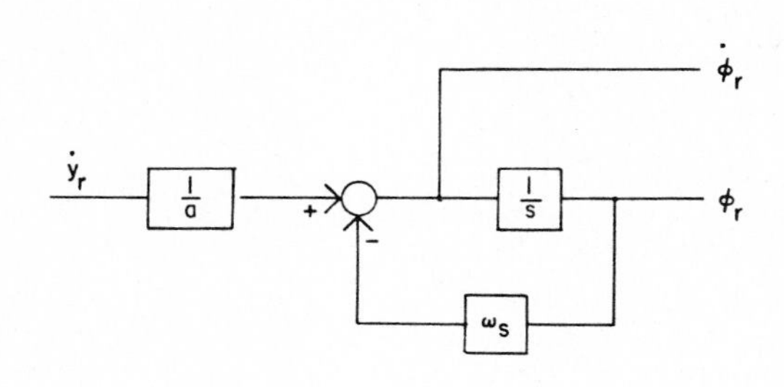

b) Crosslevel perturbation

Figure 3.2: Shaping filters where $\omega_c, \omega_r, \omega_s$, a, σ_a, σ_p characterize the track spectra.

The dynamics of the filters corresponding to equation (3.1) are implicitly integrated using the state transition method.

$$x^n(k+1)=[\Phi^n]\ x^n(k)+[h^n]u^n(k) \quad (3.3)$$

The algorithm used for the calculation of Φ^n and h^n express the required matrix exponential in a finite series expansion.

Optionally, displacements, velocities, peak and rms accelerations for the leading wheelset, leading truck, carbody and passenger; and the wheel-rail relative displacement, the suspension strokes and the lateral/vertical force ratio, are listed and/or plotted.

3.4 Ride Quality

The reduced comfort boundaries for an exposure time of 2.5 hours defined by the International Standard Organization (ISO) are taken as the ride quality performance criteria.

The corresponding rms accelerations over one-third octave bands with center frequencies (f_0) specified by the ISO are obtained from power spectral densities using the relation [16]:

$$a_{rms} = [\int_{f_1}^{f_2} P(f)d_f]^{\frac{1}{2}} \qquad (3.4)$$

where: $f_1 = 2^{-1/6} f_0$, $f_2 = 2^{1/6} f_0$

Evaluation of spectra from time history data is performed using the Fast Fourier Transform (FFT) algorithm and the following discrete approximation:

$$\hat{P}(q) = \Delta f. |Y(q)|^2 \qquad (3.5)$$

where:

$Y(q) = \Delta t.\ FFT\ (y(k))$

$\Delta f = 1/N.\Delta t$

N = Number of samples of y(k)

In order to enhance the accuracy of this spectral estimation method some capability for preprocessing the sampled data and computing certain statistics is incorporated in a software program developed in [17].

3.5 Design Procedure

For rail vehicles, a constrained parameter optimization can be stated as follows:
Criterion function:
Maximum operating speed on a given track.
Optimization variables:
Suspension stiffness and damping values.
Constraints:

- Stable vehicle for all speeds less than the critical speed.
- Acceptable ride quality as defined by the 2.5 hours ISO comfort boundaries.
- Spring and damper strokes limited to avoid hitting suspension stops.
- Wheel-rail displacements limited to avoid flange contact.
- Lateral/vertical force ratio limited to avoid derailment in curved track.

4. MODEL VALIDATION

A confidence building process in three levels of complexity has been applied. First a simulation verification is obtained by qualitatively correlating theoretical predictions and computed trends. The second level of validation entails correlation of a critical value from the analysis with experimental or other prevalidated results. The highest level of system testing involves the correct predicition of the simulated behaviour over a valid frequency band.

4.1 Simulation Verification

This level of validation was performed in three steps, corresponding to the different sources of instabilities of the system equation (3.1). Using known stable vehicle data (Metroliner passenger railcar well below its critical speed) and simple stability tests for reduced models, the A matrix errors were removed. The integration methods were shown to be stable in two ways: the implicit technique was verified by direct calculation of the matrices of equation (3.3), the explicit technique was shown to be converging by submitting the Metroliner model to a set of initial condition and observing that the system returned to its rest position. Finally the verification of the B matrix was performed by exciting the system with step inputs and observing the mode triggered, the shapes of the response and the steady state equilibiriums.

4.2 Critical Speed

Stability analysis was performed with a Metroliner and a Via passenger full car model parameters [11]. The critical speed obtained for the Metroliner (35.6 meter/second) is within 5% of another analytically estimated value [14]. For the Via car (40.3 meter/second) it is higher than the one calculated using an Association of American Railroads (AAR) program.

4.3 Shaker Response

The laboratory shaker at the National Research Council of Canada (NRCC) can be simulated by a lateral slow sine wave input (.0254 meter amplitude) to the vehicle wheelsets. This test was performed for the Via car at its resonnant frequency (1 Hz) and results were comparable (carbody displacement = .03 meter, acceleration = .1 g rms).

4.3 Random Response

This validation stage was attempted in two ways: the computed time histories for a Via car on a class 4 tangent track did not resemble the field recorded data (accelerations presenting hunting type patterns), an effort to duplicate the guideway spectra and passenger comfort boundaries results obtained previously is not completed.

5. CONCLUSIONS

The present model equations describe the lateral-vertical dynamics of a single railcar travelling at constant speed on tangent and curved guideways with vertical flexibility and superimposed irregularities. Wheel-rail interaction is described by linear Kalker creep coefficients and nonlinear geometry constraint functions with offset terms. The vehicle inertial characteristics may be asymetric, the secondary suspension has finite height components, and the nonlinear primary spring-damper arrangements include misalignments. The vertical, lateral and tortional flexibilities of the car body are also incorporated.
From this 30 degrees-of-freedom vehicle, simpler model dynamics may be analysed. The simulation is valid for small wheel-rail contact angles and in the frequency range 0-30 Hz. Different mathematical tools may be applied to the above models for evaluation or optimization of vehicle performance criteria of ride quality, critical speed, curve negociation and guidance. At present not all possible linear, quasi-linear and nonlienar analysis and design techniques have been implemented and applied to all possible submodels. Instead the least restrictive, yet feasible, algorithms have been programmed and tested with different full vehicle models. This "top-down" software engineering approach, the generality of the equations of motion, the model reduction strategies and the simplified analysis algorithms will result in a flexible simulation aided design package.
The validation methodologie is based on a step by step testing of the robustness of the algorithms used for the solution of the system state equations and for the accuracy of the results obtained. Besides the completion of the validation of existing programs, future plans include modelling enhancement, analysis simplification, design formalization and application development.

REFERENCES

[1] Tsai, N.T., Pillkey, W.D., Review and Summary of Computer Programs for Railway Vehicle Dynamics, 10th IMACS World Congress on System Simulation and Scientific Computation (Montreal 1982).

[2] Garg, V.K., Computer Models for Railway Vehicles Operation, Rail International (June 1982).

[3] Billing, J.R., DYNSYS: A General Purpose Computer Program Applicable to Problems of Dynamics, Ontario Ministry of Transportation and Communication (August 1979).

[4] Kortum ,, W., Lehner, M., Richter, R., Multibody Systems Containing Active Elements: Generation of Linearized System Equations, System Analysis and Order Reduction, in Magnus, K., (ed.), Dynamics of Multibody Systems (Springer-Verlag, Berlin, 1978).

[5] White, R.C., Rail Vehicle Equations of Motion, Acorn Associates (December 1981), available in National Research Council of Canada Report 01SX.3155-9-0643 (Appendix).

[6] Hull, R., Cooperrider, N.K., Influence of Nonlinear Wheel-rail Contact Geometry on Stability of Rail Vehicles, Journal of Engineering for Industry.

[7] Kalker, J.J., Survey of Wheel-Rail Rolling Contact Theory, Vehicle System Dynamics 4 (1979) 317-358.

[8] Cooperrider, N.K., White, R.C., Notes on Secondary Suspension Models, Acorn Associates (1982).

[9] Law, E.H., Effects of Suspension Misalignment and Wheel Radius Mismatch on Steady State Lateral Offsets of Rail Vehicle Wheelsets, Clemson University (August 1980).

[10] Tuten, J.M., Law, E.H., Cooperrider, N.K., Lateral Stability of Freight Cars With Axles Having Different Profiles and Asymetric Loading.

[11] Merabet, A.A., Software Package for the Simulation of Guideway-Railcar-Suspension Dynamical Systems, National Research Council of Canada Report 01SX.31155-9-0643 (January 1982).

[12] White, R.C., Limbert, D.A., et al., Guideway-Suspension Tradeoffs in Rail-Vehicle Systems, U.S. Department of Transportation Report DOT-OS-50107 (January 1978).

[13] Hague, I., Law, E.H., Cooperrider, N.K., User's Manual for Lateral Stability Computer Programs for Railway Freight-car Models, U.S. Department of Transportation Report DOT-OS-40018 (April 1980).

[14] Cooperrider, N.K., Law, E.H., Rail Vehicle Dynamics, Lecture Notes Arizona State University and Clemson University.

[15] Merabet, A.A., A Semi-active Suspension Railcar System, M.S.E. Reprot, Dept. of Elect. Eng., Airzona State University (June 1977).

[16] Klinger, D.L., Cooperider, N.K., et al., Guideway-Vehicle Cost Reduction, U.S. Department of Transportation Report DOT-TST-7695 (June 1976).

[17] Fallon, W.J., An Investigation of Railcar Model Validation, M.S.E. Thesis, Arizona State University (May 1977).

Simulation in Engineering Sciences
J. Burger and Y. Jarny (eds.)
Elsevier Science Publishers B.V. (North-Holland)

CATPAC: A SOFTWARE PACKAGE FOR COMPUTER-AIDED CONTROL ENGINEERING

M. Barthelmes, P. Bressler, D. Bünz
K. Gütschow, J. Heeger, H.-J. Lemke

Philips GmbH Forschungslaboratorium Hamburg,
2000 Hamburg 54, F.R.G.

CATPAC is a software system for computer-aided engineering in the area of control instrumentation. Modern methods for process analysis, simulation, and synthesis of control strategies rely on the extensive use of computers. The wide range of algorithms which is required to apply these methods has to be compiled into comprehensive program packages to ensure their convenient application. The CATPAC software system provides offline support control engineering tasks. It enables the control engineer to set up mathematical models of processes, to design control loops on the basis of these models and to simulate the time-behaviour of dynamic systems.

1 INTRODUCTION

If we consider the automation of an industrial process, we can distinguish several phases:

- At the beginning the control engineer needs a mathematical description of the dynamical behaviour of the process.
  Such a model may be calculated from measured values of the input and output signals, if no theoretical model is available.
- To check the validity of the resultant model a simulation can be carried out. E.g. for certain input signals the measured output signals can be compared with simulated output signals.
- The next step is the design of suitable control algorithms based on the resultant model from the process identification and analysis.
- A further simulation can support the assessment of the designed controller.
- The controller models serve as a basis for the hardware and software design of the control system.
- Afterwards the set-up and installation of the instrumentation system has to be realized.
- So finally the initial operation and tests of the complete system may start.

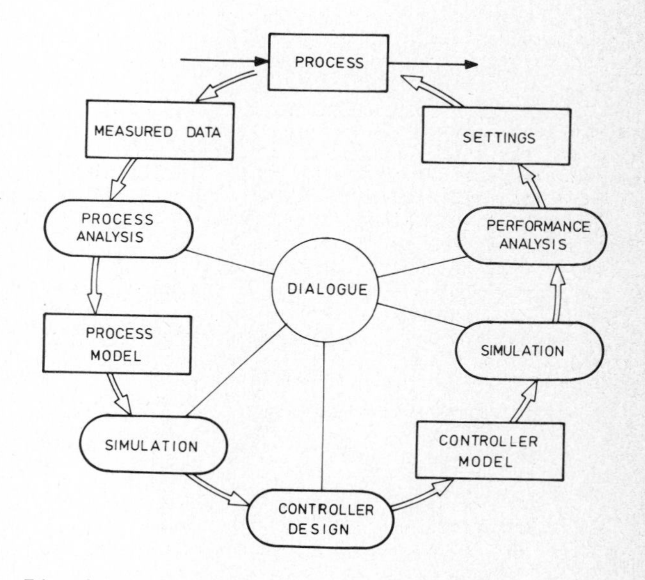

Fig 1: Steps in analysis and design

In practice the sequence of working steps described above may have to be suspended at any stage because of insufficient results.

In this case the sequence will have to be repeated once more, starting at an appropriate previous step.

The CATPAC software facilitates the use of algorithms for varied applications independent of the given process. And it supports the phases of process analysis, controller design and simulation.

2 BASIC CONCEPTION OF THE CATPAC SOFT WARE SYSTEM

CATPAC is an interactive software system for control engineering tasks. The programmed methods for control engineering tasks are uniformly handled as procedures. They need input data and they produce output data, both called models which have a standardized structure.

Due to the interactive mode optional procedures can be activated successively during one CATPAC session.

By providing the possibility to insert additional procedures or to delete existing ones, the performance of the system can be adapted to variable requirements.

The possibility to insert new procedures without effort will be achieved owing to the fact that the procedures are autonomous main programs. Inserting these programs is supported by service routines.

A basic conception of the CATPAC software structure has been developed, see Fig. 2.

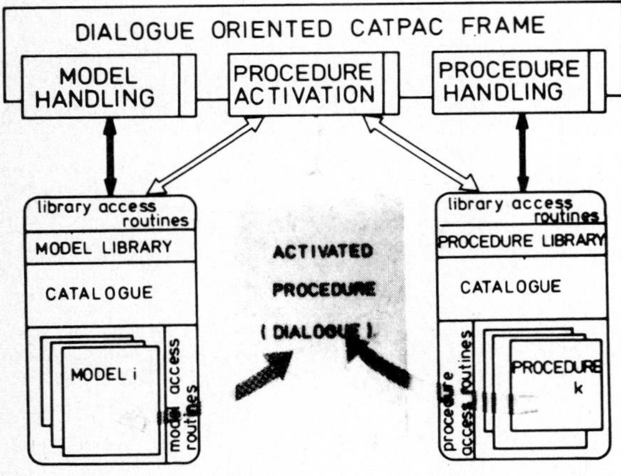

Fig. 2: Basic conception of the CATPAC software system

- The frame program activates procedures and handles procedures and models controlled by interactive user requests. The access to the model catalogue and to the procedure catalogue is realized by a set of access routines.

- All generated models and all available procedures are comprised in a model library and a procedure library. To enable the management of these libraries, there exists a catalogue for each of them containing reference data to any item.

- **Each model is stored on an individual** file. The models have a standardized structure to enable different procedures to access to models.

- The procedures are independent main programs which can be activated by the frame program.

The CATPAC software is available on a process computer.

Further no less important aspects for the conception of the CATPAC software are mentioned in the following.

To facilitate a transfer of the system to other process computers the software package must be portable. Portability is influenced by the chosen programming language and by the interface to the computing system environment.

The system is supposed to be used frequently. For this reason security aspects become very important (e.g. in case of system breakdown, runtime errors).

Readability of program source texts simplifies the design of correct programs, ensures an easy maintenance, and helps the user to become acquainted with the system. Readable programs can be set up by comprehensive documentation as well as by considering their general structure relating to modular design, structured programming, and aspects of high level languages.

3 MODELS

The design phases of a process control system were mentioned in the introduction. Starting with process identification the control engineer needs a mathematical description, which represents the dynamical behaviour of the process. This could be e.g. a set of measured input and output data. Taking these measured values as an input model for a parameter estimation algorithm, the result would be an output model which is a parametric one.

In each step of analysis and design the concerned procedure needs one or more input models and generates one output model.

The dynamic behaviour of a process can be described by measured values, state differential equations, transfer functions etc. The different notations of these models are condensed into three types of models.

- **Parameter sets:** vector differential/difference equations, state differential/difference equations, transfer functions, frequency responses, Fourier coefficients etc.

- **Tables of values:** measured values, frequency characteristics, transfer loci, impulse response functions, step response functions etc.

- **Characteristic values:** gain, dead time, settling time, damping factor, bandwidth average value, variance, standard deviation etc.

To enable a uniform access to these different types of models a standardized structure of model data is a necessary precondition. This is even more evident when one considers that otherwise for each model notation a set of dedicated access routines would be required.

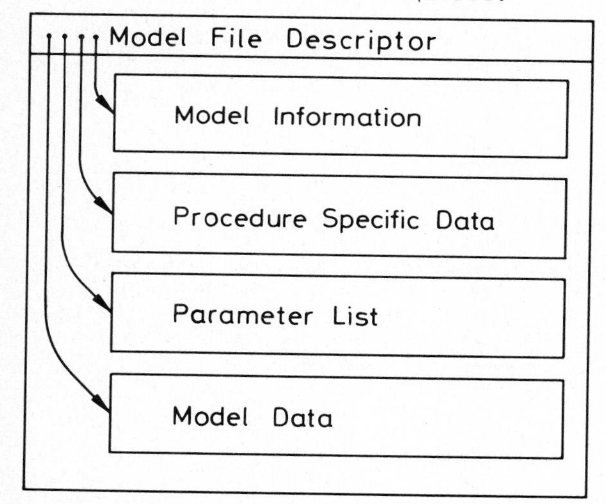

Fig. 3: Structure of the standardized model file

The standardized model file comprises four data areas which are described by the model file description. The data areas are the model information area, the procedure specific data area, the parameter list area and the model data area.

- The model information characterizes the structure, and the meanings of the different model types. It contains a user's comment and a model specification. The model specification block includes data about the corresponding input model and the applied procedure.

- To enable the repetition of a procedure run the procedure specific data are stored. For example when designing a state controller the required weighting matrices are such procedure specific data.

- The parameter list area contains data which cannot be an integral item of the numerical model data (e.g. description of an operating point).

- The model data area is divided into the proper model data and the attached headings.

The main advantages of this standardized model file structure are:

- a unique set of input and output routines can operate on each model file

- all information related to the model such as model data, procedure specific data and user's comment is located in one file.

4 PROCEDURES

The philosophy of the CATPAC package defines procedures as programmed methods for control engineering tasks.

The purpose of procedures which operate on models only, is to process input model(s) and to produce a new set of data, the output model.

The procedures are classified due to control engineering aspects:

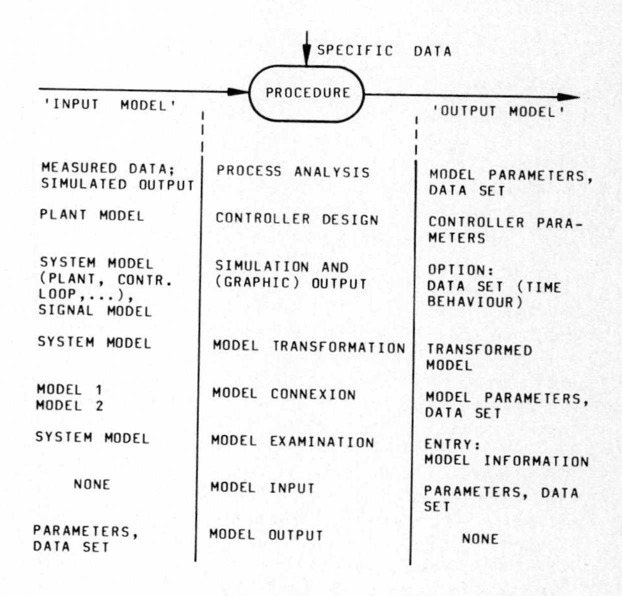

SPECIFIC DATA
PROCEDURE
'INPUT MODEL'
'OUTPUT MODEL'
MEASURED DATA; SIMULATED OUTPUT
PROCESS ANALYSIS
MODEL PARAMETERS, DATA SET
PLANT MODEL
CONTROLLER DESIGN
CONTROLLER PARAMETERS
SYSTEM MODEL (PLANT, CONTR. LOOP,...), SIGNAL MODEL
SIMULATION AND (GRAPHIC) OUTPUT
OPTION: DATA SET (TIME BEHAVIOUR)
SYSTEM MODEL
MODEL TRANSFORMATION
TRANSFORMED MODEL
MODEL 1
MODEL 2
MODEL CONNEXION
MODEL PARAMETERS, DATA SET
SYSTEM MODEL
MODEL EXAMINATION
ENTRY: MODEL INFORMATION
NONE
MODEL INPUT
PARAMETERS, DATA SET
PARAMETERS, DATA SET
MODEL OUTPUT
NONE

Fig. 4: List of procedure classes

The procedure class simulation e.g. can be divided into continuous system simulation and time-discrete system simulation. Among the procedures belonging to the continuous system simulation area there is e.g. a 'Runge-Kutta' procedure.

For each procedure there exists a corresponding information part. This part contains data for the frame program and for the procedure itself. Data for the frame program are: A reference to the program load module, requirements concerning the input and output models and requirements relating to the peripheral units. Information required by the procedure itself is given by default values of procedure specific data which are numerical parameters (e.g. precision parameters) and system design parameters (e.g. weighting matrices).

For creation and implementation of new procedures programming aids are available, e.g.:

- A pool of mathematical standard routines
- Standardized routines for the handling of interfaces between procedures and models, and between procedures and the system respectively.
- Aids for dialogue, graphic, I/O, and for program documentation.

All these service routines located in a module library are made available to the procedure programmer.

6 SIMULATION AND MODEL REPRESENTATION

Besides numerical output of model data, as parameters, or eigenvalues, the graphical representation of the model behaviour is the most important interface to the user. Therefore the procedure class 'simulation' is essential for the assessment of computed results both in modelling and in controller design.

Simulation programs include algorithms for solution of sets of linear differential and/or difference equations. This allows discrete simulation of models of any standardized form as used in control engineering. Model description of other kind, as transfer functions or transient responses, are automatically preprocessed by special transformation programs, available in the procedure pool.

The simulation results, that is a table of values of process variables at discrete time intervalls, can optionally be stored as a new 'model' in the model catalogue. It additionally containes information about the particular simulation run which has generated that data set.
Normally, simulation results are immediately displayed to the user.

A plotter and a graphic display are supplied for graphic output. As required the user can switch over to a device by software commands.

There exists a simplified output version and a more comfortable one. The simplified version provides the representation of one graph only with standard headings and legends of the axes.

The comfortable version offers among other features the possibility to represent more than one graph with individual inscriptions and text for headings.

The information relevant to draw a figure is requested from the user by dialogue and stored for further utilization.

After specification of the desired model an automatic integer scaling is performed if the user does not wish to specify the scaling.

The previously stored information about the desired graphical representation is used to produce the frame with its texts as well as the axes with their tick marks and legends and the graphs with their specified layout.

7 CATALOGUE CONCEPTION

To enable the management of the model library and the procedure library in a fast and systematic way a standardized catalogue conception has been designed which is the same for both libraries.

General Structure

A catalogue is hierarchically structured and consists of three levels. The components on each level are tables which are composed of a header and an arbitrary number of entries. The header consists of text and other data which are important for documentation. Each entry contains user information and on the higher levels a reference to the corresponding table of the subordinate level. On the lowest level this reference is the file name of the respective object. This object is a model or a procedure information.

Access Conception

The possibility to set up a catalogue, to read from it and to modify its contents requires a set of access routines. A hierarchically structured access conception has been developed.

The interface between the frame program and the catalogue is represented by routines of the highest level. They only allow an access to logically combined parts of the catalogues, e.g. user information, text and data for documentation. The information about the realization of such a logically combined part is hidden from the frame program to ensure a modular structure of the software.

Structure of the Model Catalogue

The model catalogue provides a survey of all the models (i.e. all data) within the CATPAC software package which are available for further utilization. In industrial control engineering, where usually many different processes are to be modelled, simulated, and finally equipped with control instrumentation, it is essential to keep information about how the different data sets have been gained.

To this end the structure of the model catalogue reflects the hierarchy of model classification.

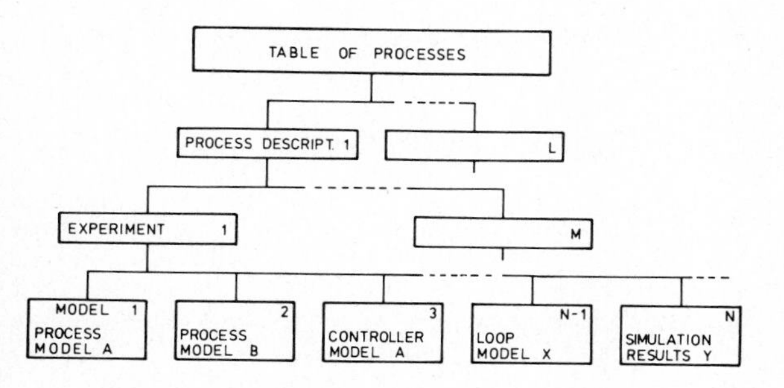

Fig. 5: Model catalogue

On the highest level there exists only one table containing information on all processes which are represented by more detailed data on other catalogue levels.

The next lower level is named the process level. Each table of this level consists of detailed information about the process and a short comment on each corresponding experiment which has been performed to get the necessary data for the process analysis step. The lowest level is called the experiment level. Each of its tables comprises detailed information concerning the performed experiment and short comment on all models which have been derived from the data gained by this experiment.

Since one process model can be used for the design of various controllers and for different simulation runs, it is important to relate all these data to the original model and to provide this information together with the data catalogued.

Structure of the Procedure Catalogue

The procedure catalogue enables all the available procedures within the CATPAC package to be managed. The structure of this catalogue reflects the classification of these procedures.

The tables of the first two levels contain information of the subordinate procedure classes and a short comment on each of them. On the lowest level the tables contain detailed information about these subclasses and for each procedure a reference to the corresponding procedure information which contains among other items the name of the related program.

8 DIALOGUE-ORIENTED PROGRAM FLOW

As shown in Fig. 2 the entire software can be divided into three main parts:

The procedure activation enables the user to apply different procedures in an interactive manner. Therefore it activates the chosen procedure and provides the selected input models.

The procedure handling allows the user to manipulate procedures, i.e. to include new procedures or to write, to alter, or to delete already catalogued ones.

The model handling gives the possibility to manipulate models in an analogous manner as mentioned above.

To ensure an easy and convenient use of these three parts, an interactive program flow has been designed.

On the highest level one of the three operational parts can be selected:

- Procedure activation
- Procedure handling
- Model handling.

Within the dialogue different command

levels are distinguished. Some standardized commands are available for all levels:

EX causes a return to the next higher level or ends the entire CATPAC session

?? causes an output of all possible input commands of the corresponding level.

Procedure Activation

All the different steps of an procedure activation are dialogue-controlled. The most important part of this dialoque is the possible modification of procedure specific data. These data can be read in either from the corresponding procedure information or from an output model which has been generated by the same procedure in a former activation run. After that the user has the possibility to alter the numerical values of particular data.

Another essential step within the procedure activation is the presentation of provisional results by the provided output facilities. On this basis the user should be able to assess the computed results in order to decide, whether the procedure should be started anew with modified parameter values or continued or ended. This is a very important feature of the designed activation scheme, because it enables an iterative handling of algorithms.

Model Handling

As a unified dialogue structure facilitates the use of an interactive software package there exists a sequence very similar to the dialogue for the procedure handling.

An example of a dialogue-sequence for reading a set of (model) data from the model catalogue is shown in Fig. 6. The different levels of the entire sequence can easily be recognized. User input commands are underlined.

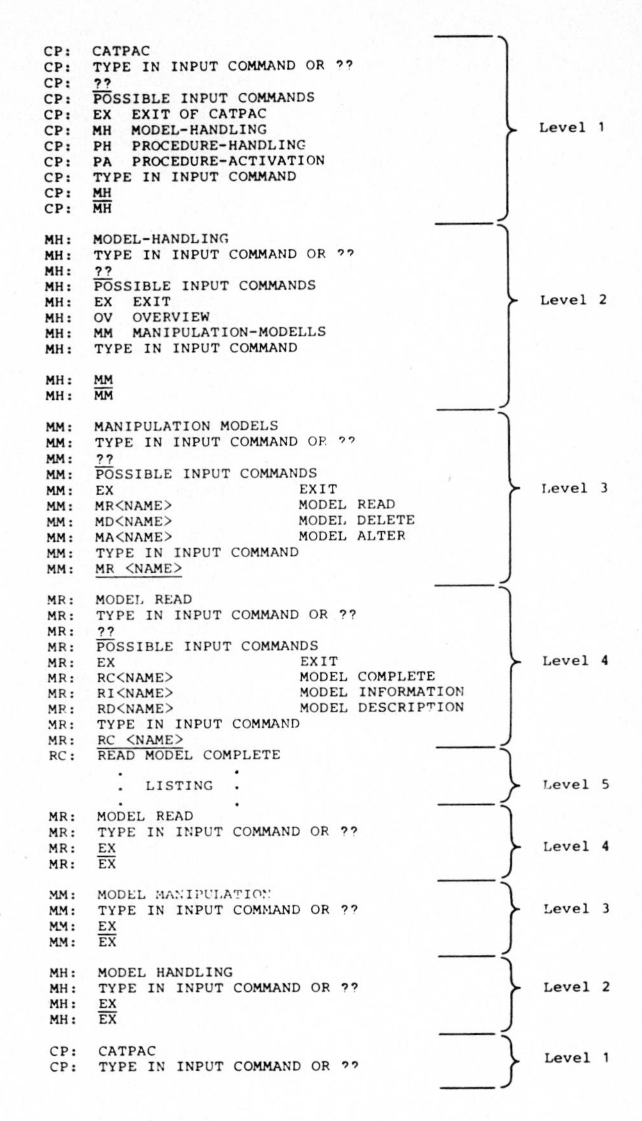

Fig. 6: Example of a dialoque-sequence for model handling

9 CONCLUSION

A brief survey of the CATPAC software has been presented. The idea of this project became apparent during our collaboration in different process automation projects carried out in the past. The gained experience revealed the necessity of advanced computer-aided tools. The CATPAC software is the framework, providing, besides other features, a fast and flexible use of these tools. For these reasons CATPAC is not only an appropriate tool for research work but also for industrial applications.

At present the package is designed for an interactive offline support with no direct link to the concerned process. A future step may be to provide the CATPAC support also in an online manner by data communication with the underlying instrumentation system. In this case the CATPAC software can be seen as a part of the process automation system itself.

II. SIMULATION METHODS – NUMERICAL METHODS

Simulation in Engineering Sciences
J. Burger and Y. Jarny (eds.)
Elsevier Science Publishers B.V. (North-Holland)

SIMULATION OF ON-LINE STATE ESTIMATION FOR DISTRIBUTED DYNAMIC SYSTEMS

A. Maslowski

Technical University of Bialystok
Bialystok, Poland

This paper presents the problem of an approximate state estimation for linear space-distributed systems in the case of incomplete information about the system. Formulation of the problem was based on the model of dynamics and observations of the systems in question as well as on the proposed estimation quality index. Using finite elements concept minimization of the weighted squared index is transformed into large scale dynamic optimization problem in time domain. In the case of low dimensionality the sequential identification algorithm is generated by variational and invariant imbedding methods.

1. INTRODUCTION

During the past decade considerable attention has been made given the development of unified theoretical approach for state approximation of the systems in the case of incomplete information about the system [1] ÷ [6] . This paper presents the investigational technique for state estimation for a class of linear deterministic continuous space-distributed systems in question with the use finite element theory [7] , invariant imbedding [8] and estimation [5]. Incomplete knowledge of the system is specified by unknown vectors which include uncertainty of the mathematical description as well as disturbance acting on the system and output observation. Disturbances can be considered in the spatial domain and on the its boundary, but no statistical assumptions are made concerning its characteristics. The error criterion utilized is the classical least squares formulation. It can be noted that under special hypothesis concerning the disturbances, namely, they be purely random with Gaussian probability density, above classical criterion coincides with the criterion results from extending of maximum-likelihood approach. Using the system dynamics, the output transformation, the error criterion and finite elements, the minimization of the weighted squared index is transformed into large scale dynamic optimization problem in time domain. In the case of low dimensionality the sequential identification algorithm is generated by variational and invariant imbedding methods.

2. STATEMENT OF THE STATE ESTIMATION PROBLEM

We shall consider space-distributed systems described by the following dynamics equation:

$$\frac{\partial}{\partial t}U(t,X) = \mathcal{L}(t,X)U(t,X) + F_D(t,X), \quad X\in\Omega,\ t\in[t_0,T] \quad (1)$$

with boundary condition:

$$\mathcal{B}(t,X)U(t,X) + F_{DB}(t,X) = 0, \quad X\in\partial\Omega,\ t\in[t_0,T] \quad (2)$$

and with initial condition:

$$U(t_0,X) = U_0(X) + F_0(X), \quad X\in\bar{\Omega} \quad (3)$$

and by the following observation equation:

$$Z(t,X) = \mathcal{H}(t,X)U(t,X) + F_0(t,X), \quad X\in\bar{\Omega},\ t\in[t_0,T] \quad (4)$$

In the above equations the following symbols are defined:

- X - three dimensional spatial coordinate vector of the region Ω , $\partial\Omega$ is the boundary and $\bar{\Omega}$ is the closure of Ω ,
- t - time from the observation time domain $[t_0,T]$,
- $U(t,X)$ - p-dimensional state vector,
- $\mathcal{L}(t,X)$ - p×p linear partial differential matrix operator on U, including coefficients which are functions of t and X,
- $F_D(t,X)$ - p-dimensional unknown error vector, including the uncertainty of the mathematical description of the system dynamics as well as unknown disturbance into the spatial domain of the system,
- $\mathcal{B}(t,X)$ - p×q linear differential matrix operator of boundary conditions, which coefficients depends on t and X,
- $F_{DB}(t,X)$ - q-dimensional unknown error vector, including the uncer-

tainty of the mathematical description of the boundary conditions as well as unknown disturbance into boundary domain of the system,

$U_0(X)$ - p-dimensional vector of the a priori estimate of the state at $t=t_0$,

$F_0(X)$ - p-dimensional unknown error vector, including the uncertainty of the mathematical description of the initial condition as well as unknown disturbance into the system at $t=t_0$,

$Z(t,X)$ - r-dimensional vector of observations,

$\mathcal{H}(t,X)$ - p×r matrix operator that relates the state $U(t,X)$ to the observations $Z(t,X)$, which coefficients depends on t and X,

$F_0(t,X)$ - r-dimensional unknown error vector, including the uncertainty of the mathematical description of the output observation as well as unknown disturbance associated with the system observations.

It can be noted that if some coefficients in operator $\mathcal{L}(t,X)$ are unknown, they can be included then into augmented state vector. Because it caused dynamic equation, Eq. (1), to be nonlinear, the method presented can be applied after its linearization. Also in the practical situations it is not always possible to indicate spatial profil of $Z(t,X)$ in the continuous but we shall consider the problem associated with the above vector of observations. Next we can formally report the results obtained for the case of point observations in space and/or time.

The criterion used to the best state identification is choosen to be weighted least squares quality index. If knowledge of the statistics of the random disturbances, which are associated with the system and/or the measurements is available, it can be used to advantage in choosing the weighting functions. As a measure of difference between mathematical model and the real system we define the criterion in the form of the following functional:

$$\beta = \int_{\bar{\Omega}} \|F_0(X)\|^2_{P_0} dX + \int_{t_0}^{T}\int_{\Omega} \|F_D(t,X)\|^2_{Q_D} dXdt + \int_{t_0}^{T}\int_{\partial\Omega} \|F_{DB}(t,X)\|^2_{Q_{DB}} dXdt + \int_{t_0}^{T}\int_{\Omega} \|F_0(t,X)\|^2_{R_0} dXdt \qquad (5)$$

where weighting matrices $P_0(X)$, $Q_D(t,X)$, $Q_{DB}(t,X)$ and $R_0(t,X)$ are positive definite matrices defined respectively for all $X\in\bar{\Omega}$, $(t,X)\in\Omega\times[t_0,T]$, $(t,X)\in\partial\Omega\times[t_0,T]$ and $(t,X)\in\Omega\times[t_0,T]$, and $\|(\cdot)\|^2$ is appropriate squared metric.

The optimal state estimation problem can now be summarized as follows. In order to obtain the least square estimate of the state, we must minimize quality index β, Eq. (5), with respect to $F_0(X)$, $F_D(t,X)$, $F_{DB}(t,X)$, and $F_0(t,X)$, subject to the dynamics equation, Eq. (1), boundary condition equation, Eq. (2), initial condition equation, Eq. (3), and the observation equation, Eq. (4). The method of the solution employed here is based on the reduction of the space-time systems to time-multipoint in the spatial domain by the use of the finite element concept. Then the minimization of the estimation quality index is transformed into large scale optimization problem in time domain only.

3. DECOMPOSITION BY FINITE ELEMENTS

Making use of finite elements theory [7] we will consider the finite element model of the domain $\bar{\Omega}$ which is now the union of E closed and bounded subdomains $\bar{\Omega}_e$. These subdomains $\bar{\Omega}_e$, where $\bar{\Omega}_e$ is the closure of an open subdomain Ω_e, $\bar{\Omega}_e = \Omega_e \cup \partial\Omega_e$, are finite elements and are disjoint:

$$\bar{\Omega} = \bigcup_{e=1}^{E} \bar{\Omega}_e, \qquad \bar{\Omega}_e \cap \bar{\Omega}_f = \emptyset \qquad e \neq f \quad (6)$$

where $\emptyset$ is the null set. The finite elements $\bar{\Omega}_e$ are to be connected together at a number G of nodal points labeled X^Δ, $\Delta = 1,2,\ldots G$. Locally, it is meaningful to label the nodal belonging to element $\bar{\Omega}_e$ by X_e^N, $N=1,2,\ldots N_e$, N_e being the number of nodal points belonging to element $\bar{\Omega}_e$. Assuming the nodal compatibility conditions are satisfied, the connectivity and decomposition of the model are established by respective mappings:

$$X^\Delta = \sum_{N=1}^{N_e} \overset{e}{\Lambda}{}^\Delta_N X_e^N, \ (e \text{ fixed}),\ X_\Delta^N = \sum_{\Delta=1}^{G} \overset{e}{\Omega}{}^N_\Delta X^\Delta \quad (7)$$

where X_e^N is the nodal point in $\bar{\Omega}_e$, and

$$\overset{e}{\Lambda}{}^\Delta_N = \begin{cases} 1 & \text{if node } \Delta \text{ of the connected } \bar{\Omega} \text{ is coincident with node N of element } \bar{\Omega}_e, \\ 0 & \text{if otherwise,} \end{cases}$$

and $\overset{e}{\Omega}{}^N_\Delta$ is simply the transpose of $\overset{e}{\Lambda}{}^\Delta_N$.

Next we consider the construction of the semidiscrete model for the state vector i'th component $U_i(t,X)$, $i=1,2,\ldots p$. If we used so called finite element first order representation of $U_i(t,X)$, then we obtain:

$$U_i(t,X)=\bigcup_{e=1}^{E}U_{ie}(t,X),\quad U_{ie}(t,X)=\begin{cases}0 \text{ if } X\notin\bar{\Omega}_e \text{ for all } t,\\ \sum_{N=1}^{N_{ei}}a_{ie}^{N}(t)\psi_{iN}^{e}(X) \text{ in otherwise,}\end{cases} \quad (8)$$

where $a_{ie}^{N}(t)$ are continuous time-dependent nodal coefficients and $\psi_{iN}^{e}(t)$ are local interpolation functions corresponding to the element $\bar{\Omega}_e$, which are defined so as have the properties:

$$\psi_{iN}^{e}(X)\equiv 0,\quad X\notin\bar{\Omega}_e$$
$$\psi_{iN}^{e}(X^M)=\delta_N^M \quad X^M\in\bar{\Omega}_e \qquad i=1,2,\dots p. \quad (9)$$

where δ_N^M is the Kronecker delta, $M,N=1,2,\dots N_{ei}$.
Globally, we can write:

$$U_i(t,X)=\sum_{\Delta=1}^{G}\alpha_i^{\Delta}(t)\varphi_{i\Delta}(X)=\bigcup_{e=1}^{E}\sum_{N=1}^{N_{ei}}a_{ie}^{N}(t)\psi_{iN}^{e}(X), \quad (10)$$

where the following relations are true:

$$\varphi_{i\Delta}(X)=\bigcup_{e=1}^{E}\sum_{N=1}^{N_{ei}}\overset{e}{\Omega}_{i\Delta}\psi_{iN}(X),\quad a_{ie}^{N}(t)=\sum_{\Delta=1}^{G}\overset{e}{\Omega}{}_{i\Delta}^{N}\alpha_i^{\Delta}(t)$$

The approximated state vector can now be written as:

$$U(t,X)\cong\bigcup_{e=1}^{E}U_e(t,X) \quad (11)$$

and

$$U_e(t,X)=[U_{1e}(t,X),\dots U_{ie}(t,X),\dots U_{pe}(t,X)]^{Tr}=$$
$$=\Big[\sum_{N=1}^{N_{e1}}a_{1e}^{N}(t)\psi_{1N}^{e}(X),\dots\sum_{N=1}^{N_{ei}}a_{ie}^{N}(t)\psi_{iN}^{e}(X),\dots\sum_{N=1}^{N_{ei}}a_{pe}^{N}(t)\psi_{pN}(X)\Big]^{Tr}= \quad (12)$$
$$=\big[A_1^{e^{Tr}}(t)\phi_1^e(X),\dots A_i^{e^{Tr}}(t)\phi_i^e(X),\dots A_p^{e^{Tr}}(t)\phi_p^e(X)\big]^{Tr},$$

where $A_i^e(t)$, $\phi_i^e(X)$, $i=1,2,\dots p$, are vectors of time-dependent nodal coefficients / to be determined / and interpolation functions / assumed to be known /, respectively.
Now, making use of above properties of the finite element approximation we can formulate decomposed relations corresponding to the state estimation problem The quality index - functional β, Eq. (6) - may be replaced by the sum of integrals as follows / taking into account Eq. (1) to Eq. (4) /:

$$\beta=\sum_{e=1}^{E}\beta_e=\sum_{e=1}^{E}\Big\{\int_{\bar{\Omega}_e}\|U_e(t,X)-U_o(X)\|_{P_o}^2dX+\int_{t_o}^{T}\int_{\Omega_e}\|\tfrac{\partial}{\partial t}U_e(t,X)-\mathcal{L}(t,X)U_e(t,X)\|_{Q_D}^2dXdt$$
$$+\int_{t_o}^{T}\int_{\partial\Omega}\|\mathcal{B}(t,X)\|_{Q_{DB}}^2dXdt+\int_{t_o}^{T}\int_{\bar{\Omega}_e}\|Z(t,X)-\mathcal{H}(t,X)U_e(t,X)\|_{R_o}^2dXdt\Big\} \quad (13)$$

Substituting Eq. (12) into Eq. (13) and integrating over spatial domains, we obtain:

$$\beta=\sum_{e=1}^{E}\beta_e=\sum_{e=1}^{E}\Big\{\|A^e(t_o)\|_{\alpha_o^e}^2-2A^{e^{Tr}}(t_o)\beta_o^e+\gamma_o^e+\int_{t_o}^{T}\Big\{\|\tfrac{\partial}{\partial t}A^e(t)\|_{\alpha_D^e}^2+ \quad (14)$$
$$-2\tfrac{\partial}{\partial t}A^{e^{Tr}}(t)\beta_D^e(t)A^e(t)+\|A^e(t)\|_{(\gamma_D^e+\gamma_{DB}^e+\gamma_R^e)}^2-2A^{e^{Tr}}(t)\beta_R^e(t)+\alpha_R^e(t)\Big\}dt\Big\}$$

where

$$A^e(t)=[A_1^{e^{Tr}}(t),A_2^{e^{Tr}}(t),\dots A_p^{e^{Tr}}(t)]^{Tr},$$

and

$$\alpha_o^e=\mathrm{diag}\Big[\int_{\bar{\Omega}_e}\phi_1^e(X)P_{o1}(X)\phi_1^{e^{Tr}}(X)dX,\dots\int_{\bar{\Omega}_e}\phi_p^e(X)P_{op}(X)\phi_p^{e^{Tr}}(X)dX\Big],$$

$$\beta_o^e=\Big[\int_{\bar{\Omega}_e}\phi_1^{e^{Tr}}(X)P_{o1}(X)U_{o1}(X)dX,\dots\int_{\bar{\Omega}_e}\phi_p^{e^{Tr}}(X)P_{op}(X)U_{op}(X)dX\Big]^{Tr},$$

$$\gamma_o^e=\int_{\bar{\Omega}_e}U_o^{Tr}(X)P_o(X)U_o(X)dX,$$

$$\alpha_D^e(t)=\mathrm{diag}\Big[\int_{\Omega_e}\phi_1^e(X)Q_{D1}(t,X)\phi_1^{e^{Tr}}(X)dX,\dots\int_{\Omega_e}\phi^e(X)Q_{Dp}(t,X)\phi_p^{e^{Tr}}(X)dX\Big],$$

$$\beta_R^e(t)=\sum_{i=1}^{r}\Big[\int_{\bar{\Omega}_e}Z_i(t,X)R_{oi}(t,X)\mathcal{H}_{i1}\phi_1^{e^{Tr}}(X)dX,\dots\int_{\bar{\Omega}_e}Z_i(t,X)R_{oi}(t,X)\mathcal{H}_{ip}\phi_p^{e^{Tr}}(X)dX\Big]^{Tr}$$

$$\alpha_R^e(t)=\int_{\bar{\Omega}_e}Z^{Tr}(t,X)R_o(t,X)Z(t,X)dX,$$

$$\beta_D^e(t)=\begin{bmatrix}\int_{\Omega_e}\phi_1^e(X)Q_{D1}(t,X)\mathcal{L}_{11}\phi_1^{e^{Tr}}(X)dX, & \dots\int_{\Omega_e}\phi_p^e(X)Q_{Dp}(t,X)\mathcal{L}_{1p}\phi_p^{e^{Tr}}(X)dX\\ \vdots & \vdots\\ \int_{\Omega_e}\phi_p^e(X)Q_{Dp}(t,X)\mathcal{L}_{p1}\phi_1^{e^{Tr}}(X)dX, & \dots\int_{\Omega_e}\phi_p^e(X)Q_{Dp}(t,X)\mathcal{L}_{pp}\phi_p^{e^{Tr}}(X)dX\end{bmatrix}$$

$$\gamma_D^e(t)=\sum_{i=1}^{p}\begin{bmatrix}\int_{\Omega_e}\mathcal{L}_{i1}\phi_1^e(X)Q_{Di}(t,X)\mathcal{L}_{i1}\phi_1^{e^{Tr}}(X)dX, & \dots\int_{\Omega_e}\mathcal{L}_{i1}\phi_1^e(X)Q_{Di}(t,X)\mathcal{L}_{ip}\phi_p^{e^{Tr}}(X)dX\\ \vdots & \vdots\\ \int_{\Omega_e}\mathcal{L}_{ip}\phi_p^e(X)Q_{Di}(t,X)\mathcal{L}_{i1}\phi_1^{e^{Tr}}(X)dX, & \dots\int_{\Omega_e}\mathcal{L}_{ip}\phi_p^e(X)Q_{Di}(t,X)\mathcal{L}_{ip}\phi_p^{e^{Tr}}(X)dX\end{bmatrix}$$

$$\gamma^e_{DB}(t)=\sum_{i=1}^{q}\begin{bmatrix}\int_{\partial\Omega_e}\mathcal{B}_{i_1}\phi^e_1(X)Q_{DBi}(t,X)\mathcal{B}_{i_1}\phi^{e^{Tr}}_1(X)\,dX, \dots \int_{\partial\Omega_e}\mathcal{B}_{i_1}\phi^e_1(X)Q_{DBi}(t,X)\mathcal{B}_{i_p}\phi^e_p(X)\,dX\\ \vdots \qquad\qquad \vdots\\ \int_{\partial\Omega_e}\mathcal{B}_{i_p}\phi^e_p(X)Q_{DBi}(t,X)\mathcal{B}_{i_1}\phi^{e^{Tr}}_1(X)\,dX, \dots \int_{\partial\Omega_e}\mathcal{B}_{i_p}\phi^e_p(X)Q_{DBi}(t,X)\mathcal{B}_{i_p}\phi^{e^{Tr}}_p(X)\,dX\end{bmatrix}$$

$$\gamma^e_R(t)=\sum_{i=1}^{r}\begin{bmatrix}\int_{\bar\Omega_e}\mathcal{H}_{i_1}\phi^e_1(X)R_{oi}(t,X)\mathcal{H}_{i_1}\phi^{e^{Tr}}_1(X)\,dX, \dots \int_{\bar\Omega_e}\mathcal{H}_{i_1}\phi^e_1(X)R_{oi}(t,X)\mathcal{H}_{i_p}\phi^{e^{Tr}}_p(X)\,dX\\ \vdots \qquad\qquad \vdots\\ \int_{\bar\Omega_e}\mathcal{H}_{i_p}\phi^e_p(X)R_{oi}(t,X)\mathcal{H}_{i_1}\phi^{e^{Tr}}_1(X)\,dX, \dots \int_{\bar\Omega_e}\mathcal{H}_{i_p}\phi^e_p(X)R_{oi}(t,X)\mathcal{H}_{i_p}\phi^{e^{Tr}}_p(X)\,dX\end{bmatrix}$$

where sign Tr denotes the transpose, and P_{oj}, U_{oj}, Q_{Dj}, Q_{DBj}, Q_{DBi_q}, R_{oi_r}, Z_{i_r}, $\mathcal{L}_{i_pj}$, $\mathcal{B}_{i_qj}$, $\mathcal{H}_{i_rj}$, $j=1,2,\dots p$, $i_p=1,2,\dots p$, $i=1,2,\dots q$, $i_r=1,2,\dots r$, are components of P_o, U_o, Q_D, Q_{DB}, R_o, Z, $\mathcal{L}$, $\mathcal{B}$, $\mathcal{H}$, respectively.

Next defining:

$$\zeta^e_R(t)=\gamma^e_D(t)+\gamma^e_{DB}(t)+\gamma^e_R(t)-\beta^{e^{Tr}}_D(t)\,[\mathcal{L}^{e^{-1}}_D(t)]^{Tr}\beta^e_D(t)\,,$$

$$\Gamma^e_o=\mathcal{L}^{e^{-1}}_o\beta^e_o,\quad \Gamma^e_D(t)=\mathcal{L}^{e^{-1}}_o(t)\beta^e_D(t),\quad \Gamma^e_R(t)=\zeta^{e^{-1}}_R(t)\beta^e_R(t),$$

$$\varrho^e=\gamma^e_o-\beta^e_o[\mathcal{L}^{e^{-1}}_o]^{Tr}\beta^e_o+\int_{t_0}^{T}[\mathcal{L}^e_R(t)-\beta^{e^{Tr}}_R(t)[\zeta^{e^{-1}}_R(t)]^{Tr}\beta^e_R(t)]\,dt,$$

we can rewritten Eq. (14) as follows:

$$\beta=\sum_{e=1}^{E}\beta_e=\sum_{e=1}^{E}\Big\{\|A^e(t_o)-\Gamma^e_o\|^2_{\mathcal{L}^e_o}+\int_{t_o}^{T}\|\frac{\partial}{\partial t}A^e(t)-\Gamma^e_D(t)A^e(t)\|^2_{\mathcal{L}^e_D}dt+ \tag{15}$$
$$+\int_{t_o}^{T}\|\Gamma^e_R(t)-A^e(t)\|_{\zeta^e_R}\,dt+\varrho^e\Big\}$$

or:

$$\beta=\sum_{e=1}^{E}\beta_e=\sum_{e=1}^{E}\Big\{\|A^e(t_o)-\Gamma^e_o\|^2_{\mathcal{L}^e_o}+\int_{t_o}^{T}[\,\|f^e_D(t)\|^2_{\mathcal{L}^e_D}+ \tag{16}$$
$$+\|\Gamma^e_R(t)-A^e(t)\|^2_{\zeta^e_R}]\,dt+\varrho^e\Big\}$$

with:

$$f^e_D(t)=\frac{\partial}{\partial t}A^e(t)-\Gamma^e_D(t)A^e(t) \tag{17}$$

Reasuming, Eq. (15) , or Eq. (16) with (17) , constitues decomposed state estimation problem.

4. LARGE SCALE OPTIMIZATION

The estimation problem decomposed in the previous section can now be treated as large scale dynamic optimization problem, that is, minimization of the functional of the form:

$$\beta'=\sum_{e=1}^{E}\beta'_e=\beta-\sum_{e=1}^{E}\varrho^e \tag{18}$$

with respect to $A^e(t)$ without constraints or $f^e_D(t)$ and $A^e(t)$ with equality constraints, Eq. (17) , e=1,2,...E. In both case it is easy to seen that:

$$\min\beta'=\min\sum_{e=1}^{E}\beta'_e \tag{19}$$

because of fundamental property of finite element approximation.

In the case of low dimensionality of the vector $A^e(t)$, e=1,2,...E we can use classical variational approach to find minimum of β' . Employing results of calculus of variations, we define the Hamiltonian function as follows:

$$H=\sum_{e=1}^{E}\|f^e_D(t)\|^2_{\mathcal{L}^e_D}+\|\Gamma^e_R(t)-A^e(t)\|^2_{\zeta^e_R}+\lambda^{e^{Tr}}(t)[\Gamma^e_D(t)A^e(t)+f^e_D(t)] \tag{20}$$

where $\lambda^{e^{Tr}}(t)$ is adjoint vector. Now, the necessary conditions for a minimum of β' to be satisfied are following:

$$\frac{\partial}{\partial t}A^e(t)=\frac{\partial H}{\partial\lambda^e}=\Gamma^e_D(t)A^e(t)+f^e_D(t), \tag{21}$$

$$\frac{\partial}{\partial t}\lambda^e(t)=-\frac{\partial H}{\partial A^e}=2\zeta^e_R(t)[\Gamma^e_R(t)-A^e(t)]-\Gamma^{e^{Tr}}_D(t)\lambda^e(t), \tag{22}$$

$$\frac{\partial H}{\partial f^e_D}=2\mathcal{L}^e_D(t)f^e_D(t)+\lambda^e(t)=0\,, \tag{23}$$

or in the form of canonical equations:

$$\frac{\partial}{\partial t}A^e(t)=\Gamma^e_D(t)A^e(t)-\frac{1}{2}\mathcal{L}^{e^{-1}}_D(t)\lambda^e(t)\,, \tag{24}$$

$$\frac{\partial}{\partial t}\lambda^e(t)=2\zeta^e_R(t)[\Gamma^e_R(t)-A^e(t)]-\Gamma^{e^{Tr}}_P(t)\lambda^e(t), \tag{25}$$

with transversality conditions:

$$\lambda^e(t_o)=2\mathcal{L}^e_o[A^e(t_o)-\Gamma^e_o], \tag{26}$$

$$\lambda^e(T)=0 \tag{27}$$

Eqs. (24) and (25) with conditions, Eq. (26) and Eq. (27) constitue two-point boundary value problem which solution yields seeking nodal coefficients vector $A^e(t)$.

Depending on the instant at which the state estimation is desired, the problem can be associated with finding of

$A^e(t)$ at the current time or past time which is contained in the interval of observations. Because of on-line computing of the problem is also prefered we shall find a sequential solution. With this in view, we will make use of the technique of invariant imbedding and extend some results obtained by Bellman et all [8] .
In the case of current time state identification the differential equations of sequential generating of nodal coefficients vector $A^e(T)$ are obtained in the form:

$$\frac{\partial}{\partial T} P^e(T) = \Gamma_D^e(T) P^e(T) + P^e(T) \Gamma_D^e(T) + 2 P^e(T) \zeta_R^e(T) P^e(T) - \tfrac{1}{2} \alpha_D^{e^{-1}}(T), \quad (28)$$

$$\frac{\partial}{\partial T} A^e(T) = \Gamma_D^e(T) A^e(T) - 2 P^e(T) \zeta_R^e(T) [\Gamma_R^e(T) - A^e(T)], \quad (29)$$

with initial conditions:

$$P^e(t_o) = \tfrac{1}{2} \alpha_o^{e^{-1}} \quad (30)$$

$$A^e(t_o) = \Gamma_o^e \quad (31)$$

where $P(T)$ is a matrix and T is now current time $T \geqslant t_o$.
It can be noted that Eq. (28) is independent upon $A^e(T)$ as well as the terms connected with change in time of the observation vector $Z(t,X)$, that is $\Gamma_R^e(t)$, and can be solved with suitable initial condition without solving for $A^e(T)$.
Thus, solving linear differential equation, Eq. (29) with initial condition Eq. (31) we next obtain the nodal coefficients vector $A^e(T)$.
In the case of past time state identification is required to synthese the state at a fixied instant t_1, where $t_o \leqslant t_1 \leqslant T$, for all $X \in \Omega$, based on knowledge of $Z(t,X)$ in the interval of observations $[t_o, T]$. Determination of the nodal coefficients vector $A^e(t_1, T)$, connected with above problem, can be made in the following manner. At first we are solving Eqs. (28) and (29) with initial conditions Eqs. (30) and (31) till the fixed instant t_1. At this instant the following equations are adjoined to Eqs. (28) and (29) :

$$\frac{\partial}{\partial T} P^e(t_1, T) = P^e(t_1, T) \Gamma_D^{e^{Tr}}(T) + 2 P^e(t_1, T) \zeta_R^e(T) P^e(T), \quad (32)$$

$$\frac{\partial}{\partial T} A^e(t_1, T) = -2 P^e(t_1, T) \zeta_R^e(T) [\Gamma_R^e(T) - A^e(T)], \quad (33)$$

and solved simultaneously with the conditions:

$$P^e(t_1, t_1) = P^e(t_1) \quad (34)$$

$$A^e(t_1, t_1) = A^e(t_1) \quad (35)$$

It is seen that Eq. (32) can be solved with initial condition equation, Eq. (34) without solving for $A^e(T)$ and $A^e(t_1, T)$. Thus, solving Eq. (33) with initial condition Eq. (35) we obtain the nodal coefficients vector $A^e(t_1, T)$.
The generalization of this solution when the state to be identified at a fixied instants set of time $t_1, t_2, \ldots t_n$, where $t_o \leqslant t_1 \leqslant t_2 \leqslant \ldots \leqslant t_n \leqslant \ldots \leqslant T$ is straightforward.

5. DISCUSSION OF ALGORITHM

Now, we can estimate the approximate state vector $U(t,X)$ as follows:

$$U(t,X) \cong \bigcup_{e=1}^{E} \left[A_1^{e^{Tr}}(t) \phi_1^e(X), \ldots A_i^{e^{Tr}}(t) \phi_i^e(X), \ldots A_p^{e^{Tr}}(t) \phi_p^e(X) \right]^{Tr} \quad (36)$$

for determined the nodal coefficients vector $A^e(t)$, for assumed interpolation vector $\phi_i^e(x)$, and for given observation vector $Z(t,X)$. The quality of this approximation depends upon the initial value selected for the state estimate and the stability of the real and estimated systems as well as upon some convergence conditions of finite element method. Detailed development of the relationships between these quantities is still to be accomplished.
From computational point of view it can be noted that substantial restriction for sequential solution is dimensionality of the nodal coefficients vector $A^e(t)$. This is caused by the number of choosen nodal points in each finite element as well dimension of the state vector $U(t,X)$. If sequential generating of nodal coefficients vector cannot be handled by computers at a reasonable speed and cost, we must use to solve large scale optimization problem multilevel optimiazation technique.
Effective realisation of the presented state estimation algorithm is closely connected with the use of advanced software and developed in classical finite element technique, sequential solution of two-point boundary value problems, and multilevel optimization methods.
Particularly useful will application of new computational techniques developed for main finite element method, for example, microprogramming and multiprocessing [10] , [11] .

6. CONCLUSIONS

The problem considered here was concerned with the choice of the method and presentation of algorithm of online state estimation for space-distributed dynamic systems. The / approximate / solution of the problem in question, involving partial differential equations was studed by finite element decomposition of the spatial domain and the state vector. This approach is atractive because sequential estimation,which is suitable in on-line computations, can be obtained by variational approach to optimization and converting an associated boundary value problem to an initial value problem using the method of invariant imbedding.

References

[1] W.H.Ray, D.G.Lainiotis, Eds, Identification, Estimation, and Control of Distributed Parameter Systems, Dekker, New York, 1978.

[2] G.A.Philipson: Identification of Distributed Systems, American Elsevier, Publishing Company, Inc., New York, 1971 .

[3] S.G.Tzafestas, J.M.Nightingle: Maximum Likelihood Approach to the Optimal Filtering of Distributed Parameter Systems, Proc. IEEE, Vol. 116, No. 6, 1969 .

[4] A.Masłowski: On Identification of the Systems with Space-Distributed Parameters, Eng. Transactions of Polish Academy of Sciences, 24, 4, 1975 .

[5] A.Masłowski: State Estimation for Dynamic Systems with Space-Distributed Parameters, IV IFAC Symp. on Identification and System Parameter Estimation, Tbilisi, USSR, Sept., 1976 .

[6] Y.Sawaragi, S.Takeda: Identification Methods in Environmental Pollution Problems, ibidem.

[7] J.T.Oden: A General Theory of Finite Elements, Int. J. num. Meth. Engng, Vol. 1, 1969 .

[8] R.Bellman, R.Kalaba, G.Wing: Invariant Imbedding and the Reduction of Two-Point Boundary Value Problem to Initial Value Problem. Proc. Nat. Acad. of Sc. of U.S.A., Vol. 46, 1960 .

[9] D.A.Wismer, Ed.: Optimization Methods for Large Scale Systems, McGraw-Hill, New York, 1971 .

[10] E.I.Field, S.E.Johnson, H.Stralberg: Software Development Utilizing Parallel Processing, in Structural Mechanics Computer Programs, University Press of Virginia, 1974 .

[11] A.K.Noor, R.E.Fulton: Impact of STAR-100 Computer on Finite Element System, J.Struct. Div. ASCE, 101, 1975 .

Simulation in Engineering Sciences
J. Burger and Y. Jarny (eds.)
Elsevier Science Publishers B.V. (North-Holland)

THE USE OF NUMERICAL SIMULATION TO VERIFY THE EFFICIENCY OF NEW PWM STRATEGIES FOR THE FEEDBACK CONTROL OF A D.C. MOTOR DRIVE

L. Fortuna*, A. Gallo* and M. La Cava**

*Istituto Elettrotecnico, Università di Catania, Italy
**Dipartimento di Sistemi, Università della Calabria, Cosenza, Italy

In this paper the problem of the practical implementation of some particular linearizing PWM feedback control strategies is considered. The study is worked out considering the application of the control law to electrical drives which use as actuator a d.c. motor with armature control. First, new linearizing modulation laws are analyzed, then, the realization of the whole drive is taken into consideration. To this purpose the scheme adopted for the control is that with both speed feedback and current feedback. An automatic computer program has been worked out, which gives the parameters of the linear sections of the regulating device based on linear synthesis considerations. Several simulation tests verify the efficiency of the proposed modulation laws and give useful indications on the choice of the implementation structure for the whole drive.

1. INTRODUCTION

As it is well known pulse-width modulation (PWM) of the manipulating signal is a control mode very widely used in feedback systems for industrial drives. The advantage of such a control mode is connected to the efficiency and reliability of the power amplifier, since it is substantially operated as an on-off device.

Traditionally the analysis of PWM control systems is considered as a very hard task because the model of the feedback system is highly non linear due to the presence of the modulating device. This is particularly true as regards the evaluation of the influence of the project parameters on the dynamic performance [1],[2],[3],[4].

New ideas, introduced in more recent years, consider the possibility of simplifying substantially the synthesis of PWM feedback control systems by making use of some particular types of PW modulators (with nonlinear characteristics) such as to make as linear as possible the behaviour of the whole control system [5],[6],[7]. The interest for this solution has proved to be more and more increasing as a consequence of the extension of microcomputer based hardware for the control law implementation.

In the present paper, first, a rewiew is made of the various modulation laws which linearize exactly the discrete behaviour of the whole feedback control system. These expressions have been proposed in previous papers of the authors [6],[7] and are very easily implementable on a simple microcomputer device. Then, the realization is taken into consideration of the whole drive, which is one using as actuator a d.c. motor with armature control. To this purpose a general scheme is adopted for the control in which both speed feedback and current feedback are used. With reference to such a system, first are chosen both the linearizing modulation laws and the expressions of the linear regulators. Several examples of simulation are worked out, which show both the efficiency of the proposed modulation laws and give useful indications on the choice of the feedback structure.

2. LINEARIZING MODULATION LAWS

Considering the general PWM feedback regulator system given in Fig. 1, P(s) is the controlled process and is completely controllable, completely observable, asymptotically stable and strictly low pass, R(s) is a current regulator of PI type, $R_V(s)$ is a speed regulator of PI type and the block denoted as PWM represents the modulating device.

As regards the control m(t) (the modulators are considered with constant amplitude of the output variable), it has the following form:

$$m(t) = \begin{cases} m_1(kT) \ \forall \ t \ \epsilon (kT, \ kT + \tau_k) \\ m_2(kT) \ \forall \ t \ \epsilon (kT + \tau_k, \ kT+T) \end{cases}$$

where T is the sampling period and τ_k (0,T) is

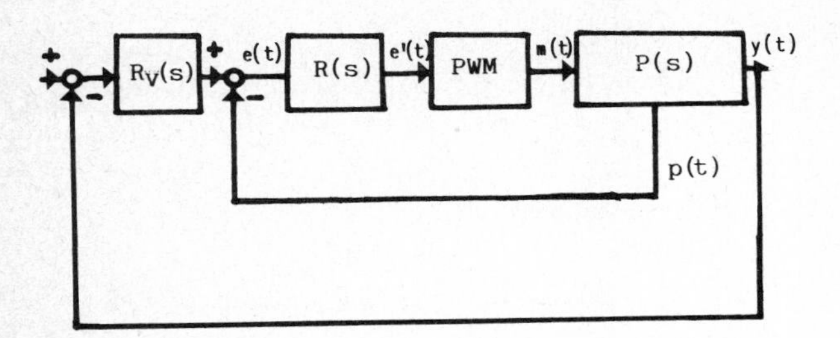

Fig. 1 General regulation system with PWM into internal loop.

the parameter which determines the waveshape of the control variable in correspondence with each sampling interval (kT, kT+T).

The continuous block (R(s) P(s) = G(s)) is described by the following state equations:

$$\begin{cases} \dot{x}(t) = Ax(t) + bm(t) \\ p(t) = c^T x(t) \end{cases} \quad (1)$$

where $x \in \mathbb{R}^n$ is the state vector, $p \in \mathbb{R}^1$ is the output, $m \in \mathbb{R}^1$ is the control input and A,b,c are matrices or vectors of appropriate dimensions.

The linearizing condition of the discretized system (1) is imposed by choosing a particular modulation law [6], [7]:

$$\tau_k = \Psi'' \quad e'(kt)$$

The general linearizing condition is obtained by:

$$\left[m_1(kT) \left(\frac{T^R}{R!} - \frac{(T-\tau_k)^R}{R!} \right) + m_2(kT) \frac{(T-\tau_k)^R}{R!} \right] =$$

$$= K'' \frac{T^R}{R!} \quad e'(kT)$$

where R is the pole-zero difference of G(s) and K" represents the linearized equivalent modulator gain.

The expression of the linearizing modulation laws for uniform sampling modulators are:

a) unipolar lead modulator

$$[1 - (1-\alpha_k)^R] = K'' \frac{|e'_k|}{M}$$

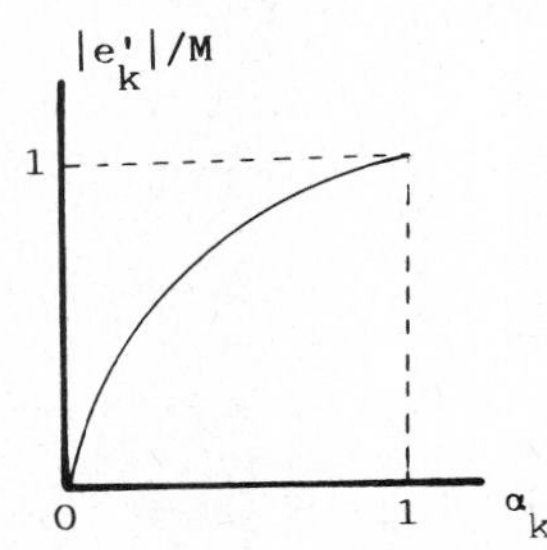

where $\alpha_k = \tau_k/T$ and M is the constant amplitude of the modulated variable.

b) unipolar lag modulator

$$(1 - \alpha_k)^R = K'' \frac{|e'_k|}{M}$$

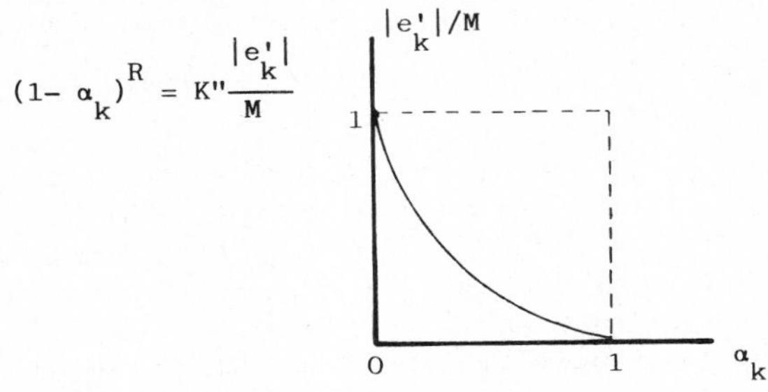

c) bipolar lead modulator

$$[1 - 2(1-\alpha_k)^R] =$$

$$= K'' \frac{|e'_k|}{M}$$

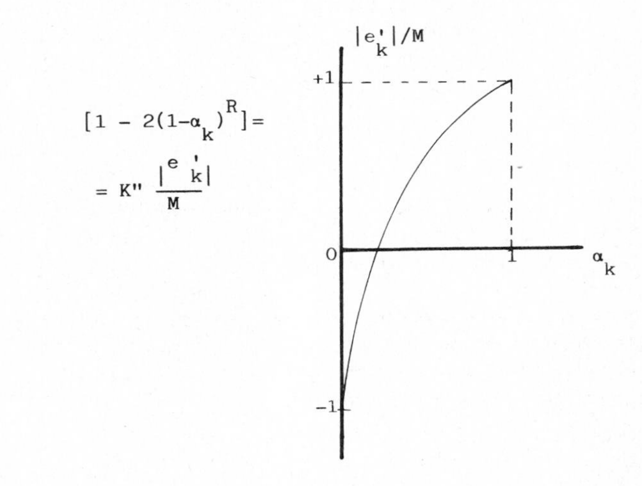

d) bipolar lag modulator

$$[2(1 - \alpha_k)^R - 1] =$$

$$= K'' \frac{|e'_k|}{M}$$

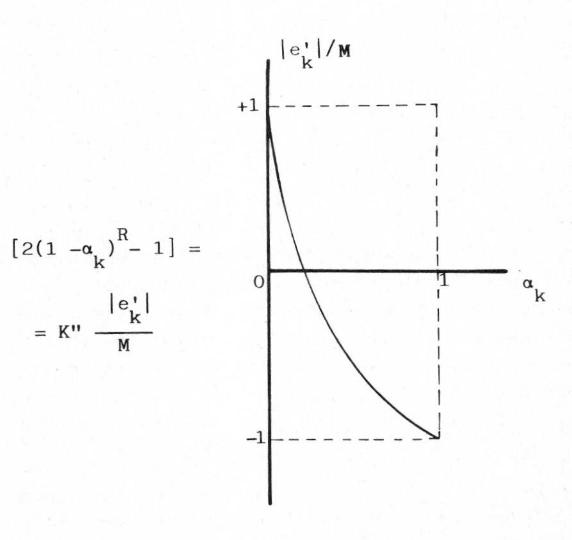

3. CONTROL SYSTEM SYNTHESIS AND SIMULATION

To verify the proposed modulator behaviour, it is considered its application to the speed control of a separately excited d.c. motor. The transfer function between angular speed and supply voltage is characterized by the following transfer function:

$$P(s) = \frac{P}{(s+\beta_1)\,(s+\beta_2)}$$

where: $P = 120.3704$ $(rad/V\ s^3)$
$\beta_1 = 68.747$ $(1/s)$
$\beta_2 = 5.0857$ $(1/s)$.

The motor parameters are:

$R = 2.2\Omega$
$L = 0.03H$
$r = 0.01\Omega$
$l = 3mH$
$C = 2.6Vs$
$M = 380\sqrt{2}\,V$
$J = 0.72\,Nms^2$
$F = 0.36\,Nms$

The d.c. motor speed regulation system consists mainly of two loops (Fig. 1):

- an inner armature current feedback loop with the PWM linearizing scheme;
- an outer speed feedback loop.

Current limitation is obtained by approximately limiting the current loop reference.

The synthesis of the current regulator has been carried out in accordance with the equivalent current loop model shown in Fig. 2. The aim of the R(s) synthesis is to find suitable parameters of the PI regulator and a sampling time T in such a manner to consider instantaneous the sub-system shown in Fig. 2 in comparison with the other regulation devices such as the speed regulator.

The synthesis of the speed regulator $R_v(s)$ is referred to the speed loop model shown in Fig. 3, where K* represents the static gain of the linearized inner armature current loop. The choice of the suitable parameters for the PI regulator used are imposed by an analysis of the dynamic system behaviour. A CAD procedure for an automatic synthesis procedure for the design of K_p and K_i parameters has been used [7] .

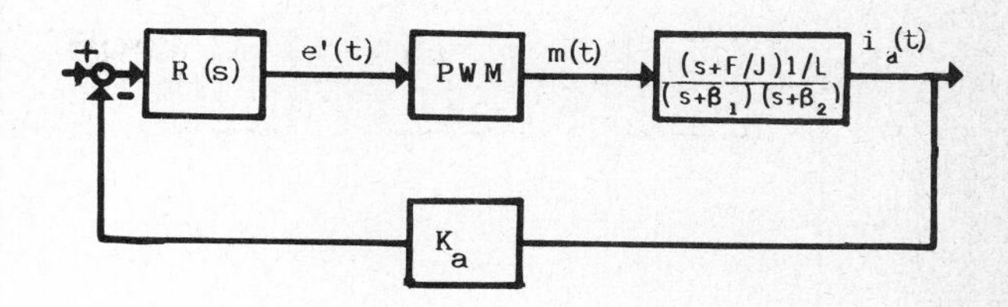

Fig. 2 Model of the current internal regulation loop.

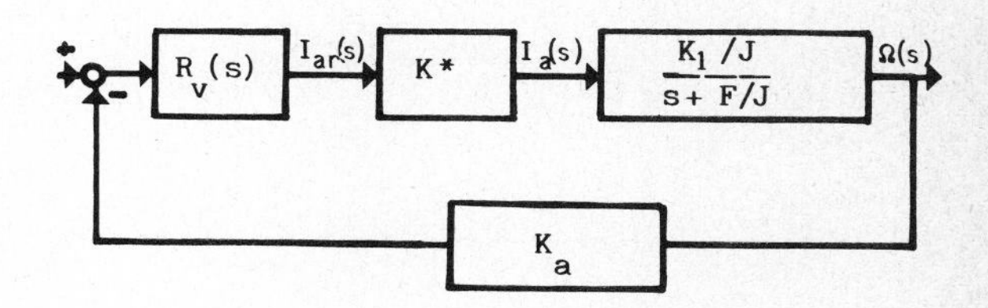

Fig. 3 Model of the speed external regulation loop.

The general scheme for the system simulation is shown in Fig. 4 where the various signals that represent completely the dinamic behaviour of the system are particularized. Considering the corresponding discrete system of Fig. 4, the simulation studies have been carried out according to the discrete system simulation procedures.

Various simulation tests have been made. We have considered the case (B) when the system, without load, has been forced with different speed steps and its successive behaviour when after settling time period it has been loaded with 50 N m torque (that can be considered as a disturbance). Then, we have simulated the case (A) when the system, with a torque load of 50 N m, has been forced with various speed steps, and unloaded after steady-state conditions were reached.

The plots of the angular speed versus time are shown in Fig. 5 and 6 for the considered cases. The suitable properties of the regulators design and the powerful of the linearizing PWM procedure are evident, in particular analyzing the behaviour of the system after the disturbance torque effects. The suitability of the current loop effect is shown in Fig. 7 where

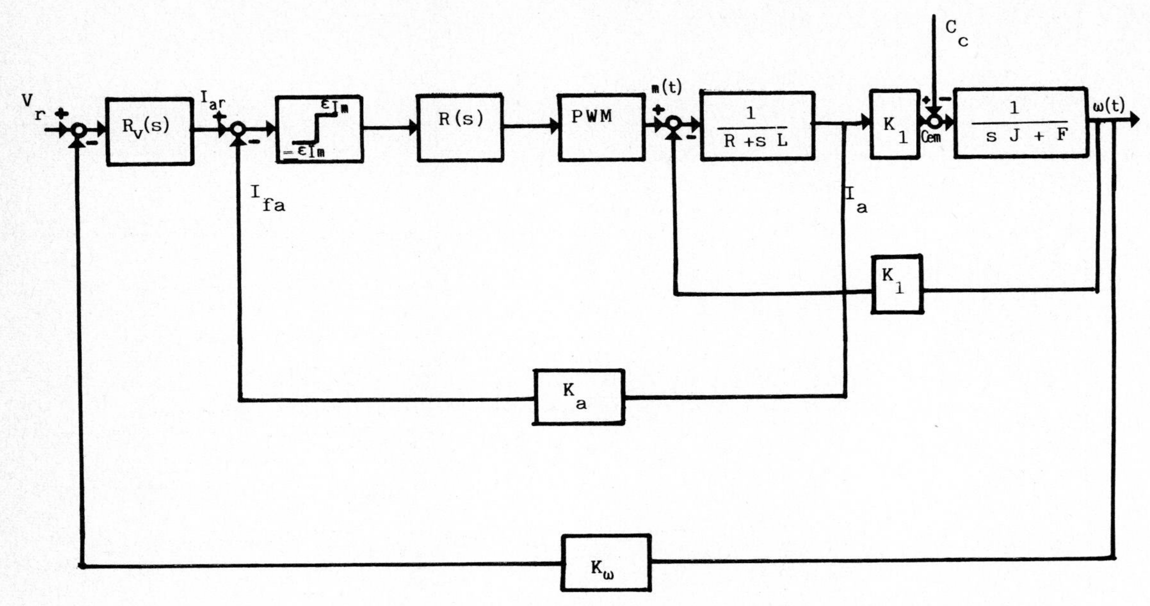

Fig. 4 General scheme for the control system simulation.

the peak current in each sampling period, relative to the B3 speed simulation test, is reported.

4. CONCLUSIONS

A new PWM modulator type is analyzed where the modulation law has been chosen to linearize, within reasonable limits, the whole feedback control system behaviour of a d.c. motor drive. Numerical simulation of the whole controlled system verify the linearizing effect. The efficiency of the modulation law is verified by several tests regarding a wide range of set-point variations. The simulation tests, with different running conditions, have confirmed the good performances of the linearized system where the effect of non-linearity appears negligible in comparison with the traditional PWM systems.

The whole regulation system proposed is suited to be realized using distributed hardware structure such as multimicroprocessor system to implement the modulation law and the discrete speed regulator $R_v(z)$ and the discrete current regulator R(z). This step of realization will be the future development of this work.

ACKNOWLEDGEMENTS

This work has been supported by the Ministero della Pubblica Istruzione (MPI 40%).

NOMENCLATURE

R(s)	current regulator
$R_v(s)$	speed regulator
K_a	current transducer constant
i_{ar}	current reference
i_{fa}	current feedback
εI	current error
Cc	load torque
i_a	motor armature current
Cem	electromagnetic torque
K_1	internal motor current feedback gain
ω(t)	angular speed
m(t)	armature voltage
V_r	speed reference
K_ω	tacho-generator constant
T	sampling time

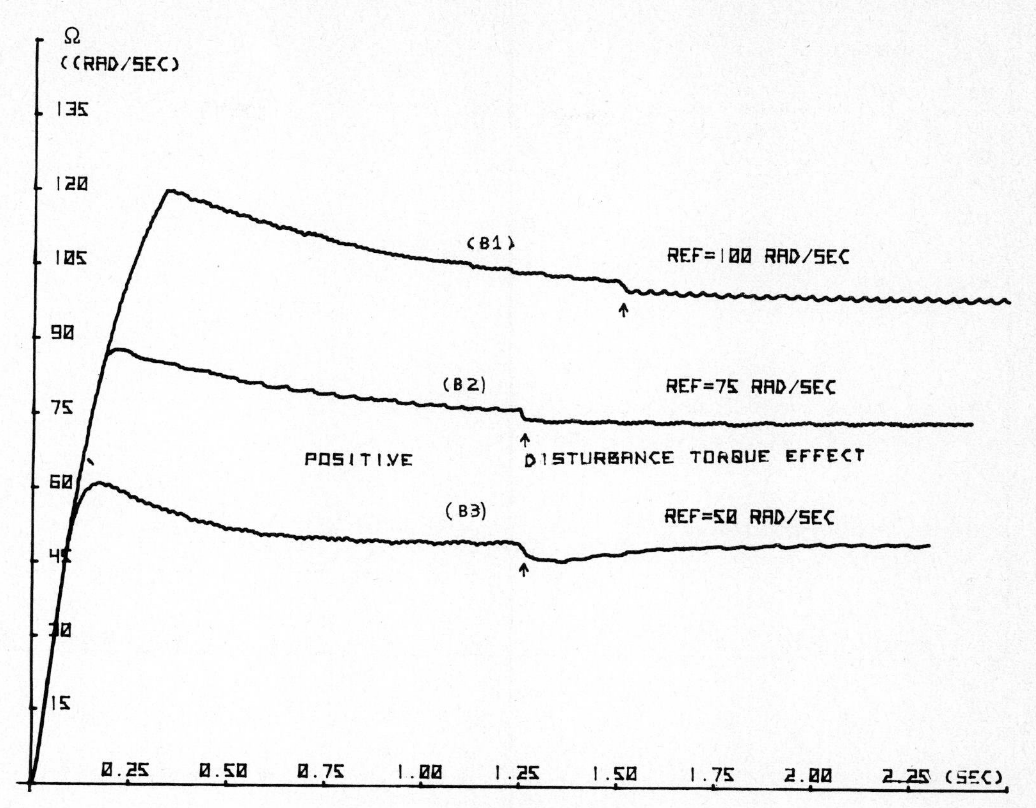

Fig. 5 Simulation test (B).

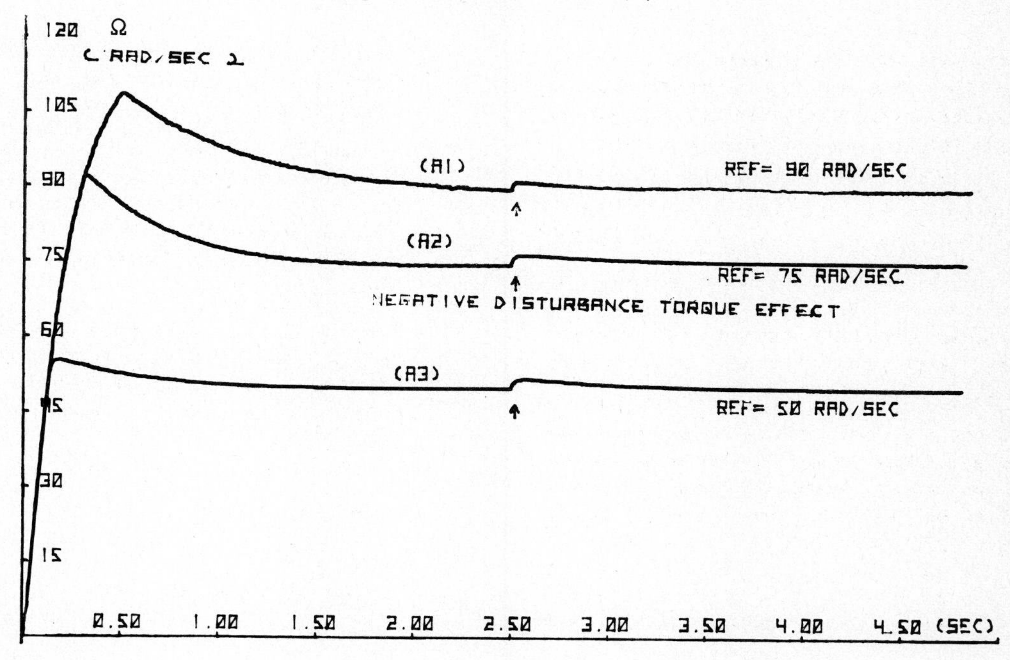

Fig. 6 Simulation test (A).

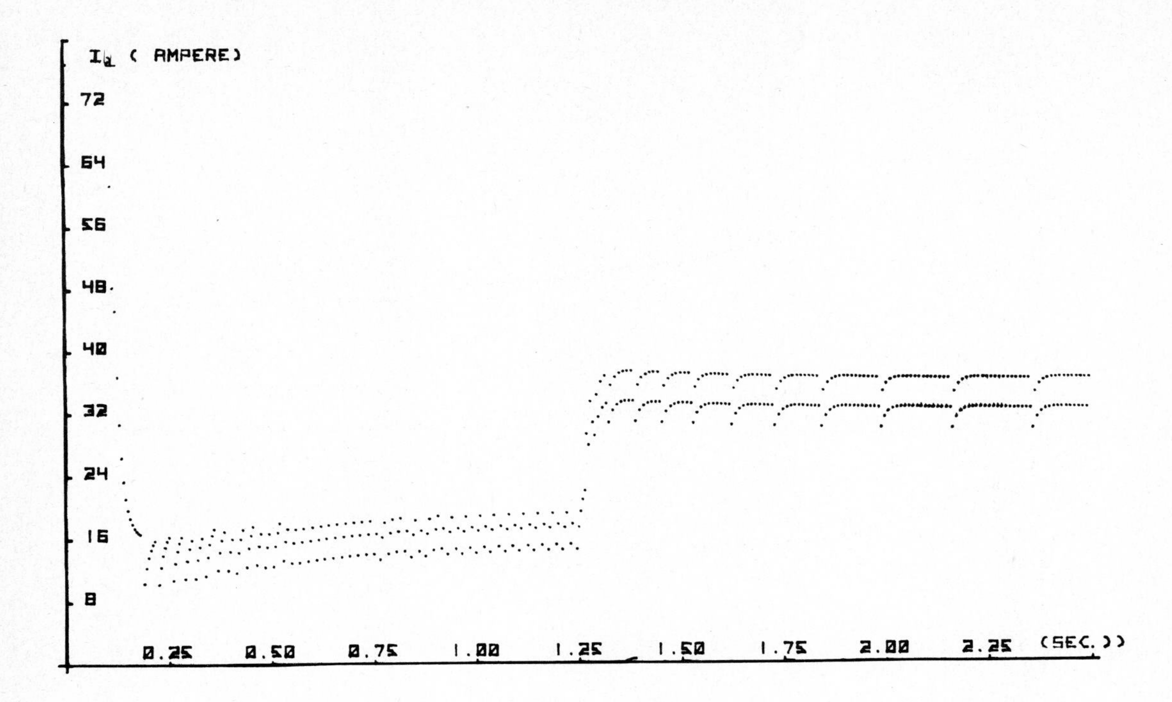

Fig. 7 Peak current trend in each sampling period relative to the B3 speed simulation test.

REFERENCES

[1] Balestrino, A., A. Eisinberg and L. Sciavicco, "Generalized approach to the stability of PWM feedback control system", J. Franklin Inst., 298, 45-58, 1974.

[2] Balestrino , A., A. Eisinberg and L. Sciavicco , "On the stability of pulse modulated control systems", Ric. Automat., 7, 33-43, 1976.

[3] Min, B.J., C. Slivinsky and R. G. Hoft, "Absolute stability analysis of PWM systems", IEEE Trans. Autom. Control, 22, 445-450, 1977.

[4] La Cava, M. and C. Salvino, "Lyapunov stability analysis of PWM feedback control systems with PI regulators", Ric. Autom., 9, 144-160, 1978.

[5] Nougaret, M., "Essential linearization: a synthesis technique for PWM control systems" V IFAC Congress, Paris, 1972.

[6] Fortuna, L., A. Gallo and M. La Cava, "Synthesis of PWM feedback control systems for a d.c. motor drive by a new linearization procedure", XV UPEC (Universities Power Engineering Conference), Leicester, England, paper 5C-2, 1980.

[7] Fortuna, L., A. Gallo and M. La Cava, "A new PWM feedback control system for electrical drives", IFAC VIII Triennal World Congress, Kyoto, Japan, paper 82.2, 1981.

Simulation in Engineering Sciences
J. Burger and Y. Jarny (eds.)
Elsevier Science Publishers B.V. (North-Holland)

DIFFERENTIAL DYNAMIC PROGRAMMING - IMPLEMENTATION OF ALGORITHMS AND APPLICATIONS

J. Lopez-Coronado - L. Le Letty
Département d'Automatique du CERT
2 avenue Edouard Belin -BP 4025
31055 TOULOUSE Cedex - France

This paper presents the principle, the differential equations satisfied by the variation of the cost function and its partial derivatives with respect to the state which are used in differential dynamic programming in order to solve optimization problems in non linear dynamic systems. Starting from the developments of Hamilton Jacobi-Bellman equation the approach leads to several algorithms which are described. Examples of applications and some implementation problems are given.

1 - INTRODUCTION

The work which is presented here concerns the resolution of optimization problems of non linear differential equations based on Dynamic Programming and the associated Hamilton-Jacobi-Bellman equation. Different in its principles of more classical methods of second variation, quasi-linearisation, gradient methods, iteration methods on the final conditions, the approach uses the first and second order developments of the partial differential equations of dynamic programming and is named differential dynamic programming. This approach leads to several iterative algorithms which gives at each iteration, under given general assumptions, an improvement of the optimal cost function with the aid of the differential equations satisfied by first (or second depending on the algorithm) partial derivatives of the cost fonction with respect to the state vector. The approach does not necessitate linearization of the dynamics and leads at each iteration to a local closed loop law of variation about the successive trajectories which satisfy the first order condition on the hamiltonian. Several algorithms are given and have been applied to strongly non linear cases, one of which is an aerodynamic reentry problem which is very sensitive to the input and presents instability zones during the trajectory.

2 - OPTIMIZATION PROBLEMS. FORMULATION AND PRINCIPLES OF DIFFERENTIAL DYNAMIC PROGRAMMING APPROACH

We consider the system of ordinary differential equations :

$$\dot{X} = F(X(t), u(t), t)$$
$$X(t_o) = X_o \qquad (2.1)$$

which determine the evolution of the state vector X as a function of the input vector u. We define

$$t \in [t_o, T] \triangleq \mathcal{T} \subset T$$
$$X : \mathcal{T} \rightarrow x \subset R^n \; ; \; u : \mathcal{T} \rightarrow \mathcal{U} \subset R^m$$
$$S \triangleq \{X(t), u(t), t \} X \in x \; ; \; u \in \mathcal{U}, \; t \in \mathcal{T} \}$$
$$F =: \; S \rightarrow R^m$$

We consider the cost function :

$$J = J(u) \triangleq V^u(X_o, t_o) = \int_{t_o}^{T} L(X(t), u(t), t) dt) + G(X(T), T) \qquad (2.2)$$

where V^u is the cost obtained for the initial conditions (X_o, t_o) and the input u(t), t

$$L : S \rightarrow R \quad , \quad G : R^n \times \mathcal{T} \rightarrow R$$

The optimization problem consists in finding optimal $\hat{u}(t) \in \mathcal{T}$ in the sense of the minimization of (2.2). The final time T is fixed or may be implicitly defined by a final target condition :

$$\psi(X(T), T) = 0, \quad \psi: R^n \times \mathcal{T} \;\; R^p, \; p < n$$

(p independent relations).

For the given optimization problem and the use of differential dynamic programming the following assumptions are made :

1/ The set $\mathcal{U} \subset R^m$ is closed bounded, the class of admissible input functions is defined by $\{u \mid u : \mathcal{T} \rightarrow \mathcal{U}$, u continuous except in a finite number of points.$\}$

2/ The functions F, L, G and their partial derivatives with respect to (X, u) up to the third order exist and are continuous for all (X, u, t) $\in$ S except in a finite number of points t_i where the right and left limits exist.

3/ The partial derivatives F_t, L_t exist and are tontinuous for all (X, u, t) $\in$ S

4/ $||F(X, u, t)|| \leq M (||X||+ 1)$, $M < \infty$, $(X,u,t) \in S$
This condition gives existence and continuty of the cost functionnal $V^u(X(t),t)$ and of its three first partial derivatives with respect to X.

5/ By colling G the dass of piecewise continuous functions of $\mathcal{T} \rightarrow \mathcal{U}$, we compare the effect of 2 inputs u, $\bar{u} \in$ G by defining a distance $d(u,\bar{u})$:

$$d(u, \bar{u}) = \int_{t_o}^{T} || u(t) - \bar{u}(t) || \; dt$$

a distance $d(u,\bar{u}) \leq \varepsilon$ allows strong variations on u.

All these assumptions are not always necessary and are only completely used for some second order algorithms of differential dynamic programming for the development to the second order of Hamilton Jacobi equation and the proof of cost improvement at each iteration for strong variations on u.

The differential dynamic programming (DDP) uses the first or second order developments of Hamilton-Jacobi-Bellman equation of dynamic programming.

For the (2.1), (2.2) problem we recall that the optimal cost function $\hat{V} = V^{\hat{u}}$ satisfies the partial differential equation :

$$\frac{\partial V^{\hat{u}}(X,t)}{t} = \operatorname*{Min}_{u \in \mathcal{U}} H(X,u,V_X^u,t) \qquad (2.3)$$

$$V^{\hat{u}}(X,t)\Big|_{t=T} = G$$

where : $H = L + \langle\lambda, F\rangle$

$\lambda(t) \triangleq V_X = \frac{\partial V(X(t),t)}{\partial X}$ (case of continuous V_X)

and where the adjoint state λ verifies (maximum principle of Pontryagin) the second canonical equation :

$$\dot{\lambda} = -H_X(X,u,\lambda,t) \qquad (2.4)$$

with the transversality conditions.

DPP is a successive approximation method using the development of equation (2.3) in order to obtain a successive improvement for an approximation of the cost variation (called $a(t_o) \triangleq \Delta V$) for input variation $\delta u = u-\bar{u}$ around a nominal input $\bar{u}$.

For two inputs $u(t)$, $\bar{u}(t)$ we known that the cost variation for (2.1) with $X(t_o) = \bar{X}_o$ is given by

$$a(t_o) \triangleq \Delta V \triangleq V^u(\bar{X}_o,t_o) - V^{\bar{u}}(\bar{X}_o,t_o)$$

$$= \int_{t_o}^{T} \left[H(\bar{X},u,V_X^u(\bar{X},t),t) - H(\bar{X},\bar{u},V_X^u(\bar{X},t),t)\right] dt \qquad (2.5)$$

in a neighbourhood $(\bar{X} + \delta X,\ \bar{u} + \delta u)$ of the nominal trajectory $(\bar{u} \rightarrow \bar{X})$.

The principle of the approach is a follows :

- we don't known the optimal control $\hat{u}(t)$, $t \in \mathcal{T}$ but we have at our disposal a nominal solution $(\bar{u},\ \bar{X},\ V\bar{u})$
- we make variations $X = \bar{X} + \delta X$, $u = \bar{u} + \delta u$ where δX and δu become the new variables which are measured with respect to the nominal solution (non necessary weak variation)
- the equation (2.3) becomes :

$$\frac{-\partial V^{\hat{u}}(\bar{X}+ X,t)}{\partial t} = \operatorname*{Min}_{\delta u} \left[L(\bar{X}+\delta X,\bar{u}+\delta u,t)+\langle V_X^{\hat{u}}(\bar{X}+\delta X,t),\ F(\bar{X}+\delta X,\ \bar{u}+\delta u,t\rangle\right] \qquad (2.6)$$

Under the hypothesis of existence of the derivatives of V with respect to (X,t) and under the assumptions 2) and 4) we can develop U up to the second order about the nominal $X=\bar{X}$. Introducing the cost variation :

$\hat{a}(t) = V^{\hat{u}}(\bar{X},t) - V^u(\bar{X},t)$:

$$V^{\hat{u}}(\bar{X}+\delta X,t) = V^{\bar{u}}(X,t)+\hat{a}(t)+\langle V_X^{\hat{u}},\delta X\rangle + \frac{1}{2}\langle\delta X,V_{XX}^{\hat{u}}\delta X\rangle + O(||\delta X||^3)$$

where $O(||\delta X||^3)$ is justified if the condition (5) is verified.

For use in (2.6) one also needs the developments up to the second order of $V^u(X)$

$$V_X^{\hat{u}}(\bar{X}+\delta X,t) = V_X^u(\bar{X},t)+\langle V_{XX}^{\hat{u}},\delta X\rangle+ \frac{1}{2}\langle V_{XXX}^{\hat{u}}\delta X,\delta X\rangle$$

where

$$\frac{-\partial V^{\bar{u}}(\bar{X},t)}{\partial t} - \frac{\partial\hat{a}(\bar{X},t)}{\partial t} - \langle\frac{\partial V_X^{\hat{u}}}{\partial t},\delta X\rangle - \frac{1}{2}\langle\delta X,\ \frac{\partial V_{XX}^{\hat{u}}}{\partial t}\delta X\rangle =$$

$$\operatorname*{Min}_{\delta u} \left[H(\bar{X}+\delta X,\ \bar{u}+\delta u,V_X^{\hat{u}},t)+ \langle V_{XX}^{\hat{u}}\delta X + \frac{1}{2} V_{XXX}^{\hat{u}}\delta X\delta X,\ F(\bar{X} + \delta X,\ \bar{u} + \delta u,t\rangle\right] \qquad (2.7)$$

which contains the first order condition :

$$\frac{-\partial V^u}{\partial t} - \frac{\partial\hat{a}}{\partial t} = \operatorname*{Min}_{\delta u} H(\bar{X}+\delta X,\bar{u}+\delta u,\ V_X^{\hat{u}},t) \qquad (2.8)$$

. In a first step, one strictly minimizes (2.8) for $X+\bar{X}$, then $\tilde{u} = \bar{u}+\delta u$ (by making a development of H in δu up to the second order).

. We then determine a new control $u=\tilde{u}+\delta u$ by using the development of (2.7) for $X=\bar{X} + \delta X$ up of the second order. By neglecting the terms of order greater than 2, we will obtain a linear law $\delta u(t) = K(t)\ \delta X(t)$ in closed loop form for the variations $\delta X(t)$ of the iterative algorithms.

This will give the approxiamtion of $\tilde{a}$, $V_X^{\tilde{u}}$, $V_{XX}^{\tilde{u}}$ and of the differential equations which are satisfied by these quantities after the development of (2.7).

The justification and validity of approximated equations has been demonstrated by MAYNE /4/ with the assumptions which are been introduced at the beginning of the paragraph. A brief explanation can be given as follows starting from the definitions of the quantities $V_X^u(\bar{X},t)$, $V_{XX}^u(\bar{X},t)$ it is not possible to obtain their exact expressions for $u \neq \bar{u}$ because they give rise to higher order terms in the differential equations which determines them. We have indeed (we assume T fixed and no constraint, analog developments can be made in the case T is not fixed) :

$$-\frac{d}{dt} V_X^{\bar{u}}(\bar{X},t) = H_X(\bar{X},\bar{u},\ V_X^{\bar{u}}(\bar{X},t),t)$$

$$V_X^{\bar{u}}(\bar{X},t)\Big|_{t=T} = G_X(\bar{X}(T))$$

$$-\frac{d}{dt} V_X^u(\bar{X},t) = H_X(\bar{X},u,V_X^u(\bar{X},t)+V_{XX}^u(\bar{X},t)\cdot \left[F(\bar{X},u,t)-F(\bar{X},\bar{u},t)\right] \qquad (2.9)$$

$$V_X^u(\bar{X},t)\Big|_{t=T} = G_X(\bar{X}(T))$$

For the second derivative of the cost function we have analog lously :

$$\frac{-d}{dt} V^u_{XX}(\bar{X},t) = H_{XX}(\bar{X},\bar{u},V^u_X(\bar{X},t),t) + F^T_X(\bar{X},\bar{u},t)' V^u_X(\bar{X},t) + V^u_X(\bar{X},t)\ f(\bar{X},\bar{u},t)$$

$$V^u_{XX}(\bar{X},t)\Big|_{t=T} = G_{XX}(\bar{X}(T))$$

$$\frac{-d}{dt} V^u_{XX}(\bar{X},t) = H_{XX}(\bar{X},u,V^u_X(\bar{X},t),t) + F^T_X(\bar{X},\bar{u},t) V^u_X(\bar{X},t) + V^u_X(\bar{X},t)\ F(\bar{X},\bar{u},t) + \sum_{i=1}^{n} V^u_{XXX_i}(\bar{X},t)\ F(\bar{X},\bar{u},t) - F(\mathbf{X},\bar{u},t) \quad (2.10)$$

$$V^u_{XX}(\bar{X},t)\Big|_{t=T} = G_{XX}(\bar{X}(T))$$

The fact of looking for the variations of $V^u_X(\bar{X},t)$ $V^u_{XX}(\bar{X},t)$ along the nominal trajectory $\bar{X}(t)$ introduces higher order terms :

$$V^u_{XX}(\bar{X},t),\ V^u_{XXX_i}(\bar{X},t)$$

which do not appear for $(\bar{u}, \bar{X})$. The integration calculus can be only made by neglecting these higher order terms.

Designing the approximated variables by

$$\tilde{a}(t),\ \tilde{V}^u_X(\bar{X},t) \text{ and } \tilde{V}^u_{XX}(\bar{X},t)$$

one can demonstrate (4) under the assumptions 2 and 4 that :

$$\|a(t) - \tilde{a}(t)\| \leq C_1\ \varepsilon^3 \quad C_1 < \infty$$

$$\|V^u_X(\bar{X},t) - \tilde{V}^u_X(\bar{X},t)\| \leq C_2\ \varepsilon^2, \quad C_2 < \infty \qquad \forall\ t \in \mathcal{T}$$

$$\|V^u_{XX}(\bar{X},t) - \tilde{V}^u_{XX}(\bar{X},t)\| \leq C_3\ \varepsilon, \quad C_3 < \infty$$

3 - DIFFERENTIAL DYNAMIC PROGRAMMING ALGORITHMS

We have seen the second order development of Hamilton-Jacobi-Bellman equation along a nominal trajectory $(\bar{X},\bar{u})$ for a variation δu (we write $\bar{V} = V^{\bar{u}}$, $V = V^u$) =

$$\frac{-\partial\bar{V}(\bar{X},t)}{\partial t} - \frac{\partial\tilde{a}(X,t)}{\partial t} - \frac{\partial V_X(\bar{X},t)^T}{\partial t}\delta X - \frac{1}{2}\delta X^T \frac{\partial V_{XX}}{\partial t}\delta X = \min_{\delta u}\left[H(\bar{X}+\delta X,\bar{u}+\delta u,V_X,t) + (V_{XX}\delta X + V_{XXX}\ \delta X\ \delta X)^T\ F(\bar{X}+\delta X,\delta u+\bar{u},t)\right] \quad (3.1)$$

This development is valid under the assumption of δX sufficiently small, δu may being a strong variation (in the sense of distance d). If that is not the case we can,into the algorithms definition, constraint the application of δu on a small interval, or reduce the step size in the minimization of the term on the right.

3.1 - Second order algorithm of Jacobson

As indicated in (2) the algorithm can be presented in two steps :

- minimize H by the first order condition for $X = \bar{X}$:

$$\frac{-\partial\bar{V}(\bar{X},t)}{\partial t} - \frac{\partial\tilde{a}}{\partial t} = \min_{\delta\tilde{u}} H(\bar{X},\ \bar{u}+\delta\tilde{u},\ V_X(\bar{X},t),t) \quad (3.2)$$

The minimization is obtained by developing the hamiltonien up to the second order with using weah variations ($\tilde{u} = \bar{u} + \delta\tilde{u}$ sufficiently close to $\bar{u}$). We obtain :

$$\delta\tilde{u} = -H^{-1}_{uu}(\bar{X},\bar{u},\ V_X(\bar{X},t),t)H_u(\bar{X},\bar{u},V_X(\bar{X},t),t)$$

we have then the new control :

$$\tilde{u}(t) = \bar{u}(t) + \delta\tilde{u}(t)$$

with the condition

a) H_{uu} positive definite

b) $\tilde{u}$ sufficiently close to $\bar{u}$

. by using the development in δX up to the second order, search for the local linear closed loop law ($\delta u = K(t)\delta X$) which minimizes H by maintaining the first order optimality condition that is :

$$\min_{\delta u}\left[H(\bar{X}+\delta X,\tilde{u}+\delta u,V_X,t) + (V_{XX}\delta X + \frac{1}{2}V_{XXX}\delta X\delta X)^T F(\bar{X}+\delta X,\tilde{u}+\delta u,t)\right]$$

$$= \min_{\delta u}\Big[H + H_u^T\delta u + (H_X+V_{XX}F)^T\delta X + \delta u^T(H_{uX}+F_u^T V_{XX})\delta X + \frac{1}{2}\delta u^T H_{uu}\delta u + \frac{1}{2}\delta X^T(H_{XX}+F_X^T V_{XX} + V_{XX}F_X)\delta X + \frac{1}{2}(V_{XXX}\ \delta X\ \delta X)^T\ F\Big] \quad (3.3)$$

where $\delta u = u - \tilde{u}$ is measured with respect to $\tilde{u}$ of the first order condition and where the function arguments are then taken to be $(\bar{X},\tilde{u})$.

As the first order condition $H_u\big|_{u=\tilde{u}}=0$ has to be verified, the minimization leads to the necessary optimality condition of the brachet :

$$\overset{=0}{H_u} + (H_{uX} + F_u^T\ V_{XX})\delta X + H_{uu}\ \delta u = 0$$

then

$$\delta u = K\delta X = -H^{-1}_{uu}(H_{uX}+F_u^T\ V_{XX})\delta X \quad (3.4)$$

under the assumption $H_{uu} > 0$
with

$$K(t) = -H^{-1}_{uu}(H_{uX}+F_u^T(\bar{X},\tilde{u},t)V_{XX}(\bar{X},t))$$

For sufficiently small δX we then maintain the first order necessary condition about $\bar{X}$:

$$H_u(\bar{X}+\delta X,\ \tilde{u}+\delta u,V_X+V_{XX}\delta X,t) = 0 \quad (3.5)$$

The equations to be solved are then obtained by identifying left and right members of (3.1) and (3.3) with δu given by (3.4) then, with the relations :

$$\frac{d}{dt} V(\bar{X},t) = \frac{\partial}{\partial t}\left(\bar{V}(\bar{X},t)+\tilde{a}(t)\right) + \langle V_X, F(\bar{\bar{X}},\bar{u},t)\rangle$$

$$\frac{d}{dt} V_X(\bar{X},t) = \frac{\partial}{\partial t} V_X + V_{XX}\, F(\bar{X},\bar{u},t) \qquad (3.6)$$

$$\frac{d}{dt} V_{XX}(\bar{X},t) = \frac{\partial}{\partial t} V_{XX}(\bar{X},t) + \frac{1}{2}\left[\frac{\partial}{\partial X} V_{XX}(\bar{X},t) F(\bar{X},\bar{u},t) + F^T(\bar{X},\bar{u},t)\frac{\partial V_{XX}}{\partial X}\right]$$

$$\dot{\bar{V}} = - L(\bar{X},\bar{u},t)$$

we obtain :

$$-\dot{a} = H(\bar{X},\tilde{u},V_X^u(\bar{X},t)t) - H(\bar{X},\bar{u},V_X^u(\bar{X},t)t)$$

$$-\dot{V}_X = H_X + K_u^T H_u + V_{XX}(F - F(\bar{X},\bar{u},t))$$

$$-\dot{V}_{XX} = H_{XX} + F_X^T V_{XX}(\bar{X},\bar{u}) + V_{XX}F_X^T(\bar{X},\bar{u}) + (H_{uX} + F_u^T(\bar{X},t))^T H_{uu}^{-1}(H_{uX} + F_u^T V_{XX}(\bar{X},t)) + \sum_{i=1}^{n} V_{XXX_i}(\bar{X},t)(F(\bar{X},\tilde{u},t) - F(\bar{X},\bar{u},t))$$

Taking into account results given in § 2, we can neglect $V_{XXX}(X,t)$ terms by introducing on $\tilde{a}(t)$ only an error of order ε^3.

The equations of the algorithm are then :

$$\tilde{u} = u + \varepsilon\delta\tilde{u} = \bar{u} - \varepsilon H_{uu}^{-1}(\bar{X},\bar{u},V_X^u(\bar{X},t),t) H_u(\bar{X},\bar{u},V_X^u(\bar{X},t),t)$$

$$-\dot{a}(t) = H - H(\bar{X},\bar{u},V_X^u(\bar{X},t)$$

$$-\dot{V}_X(\bar{X},t) = H_X + K^T(t) H_u + V_{XX}^u(X,t)\,|F(\bar{X},\tilde{u},t) - F(\bar{X},\bar{u},t)|$$

$$-\dot{V}_{XX}(\bar{X},t) = H_{XX} + F_X^T(H_{uX} + F_u^T V_{XX}^u(\bar{X},t)) H_{uu}^{-1}(H_{uX} + F_u^T V_{XX}^u(\bar{X},t))$$

$$K(t) = -H_{uu}^{-1}(H_{uX} + F_u^T V_{XX}^u(\bar{X},t)$$

$$\delta u(t) = K(t)\,\delta X(t) \qquad (3.8)$$

$$u(t) = \tilde{u}(t) + \delta u(t)$$

The corrective term ε is used when necessary to modify vaiations of δX on ther interval $|t_o,T|$ taking into account that these variations are only due to control variations $\delta u + \delta\tilde{u}$ and to the fact that the hamiltonian is assumed locally convex

Algorithms characteristics

- This algorithms is a second order method without linearization of the system dynamics

- It uses the improved control law (3.4) while other second order methods use backwards integrations analogous to (3.6) using the nominal control

- The necessity to have H_{uu} positive definite for all $H(X, u,\lambda,t)$ obtained is a restrictive condition for the algorithm

- If ε is too large at the beginning the differential equations have to be integrated in the backwards sense with a smaller ε.

3.2 - Algorithm of Jacobson Mayne

In this algorithm the weak variation on u is replaced by a strong variation $u^*(t)$ such that :

$$u^*(t) = \arg\min_u H(\bar{X}, u, V_X^u(\bar{X},t),t) \qquad (3.9)$$

This improvement leads to a faster convergence of the algorithm and is also in the fact that convexity of the hamiltonien is only needed in the neighbourhood of u . The new algorithm has the same form as the previous one by changing u by u^* in all arguments. We have to take into account (even in the case where H_{uu} is positive definite) the fact that two cases can invalidate the quadratic development of the criterion in δX:

a) the variation $u = u^* - \bar{u}$ is too large and then the domain of validity $\delta u = u - u^* = K\delta X$ is too much limited. Then if $K(t)$ obtained in function of u^* is very large, δX obtained by the integration of

$$\frac{d}{dt}(\bar{X}+\delta X) = F(\bar{X} + \delta X, u^* + K\delta X, t)$$

$$X(t_o) = X_o$$

becomes too large and invalidates not only the development of the criterion but also that of $F(X,u,t)$ and $L(X,u,t)$. Furthermore $\tilde{a}(t_o)$ is not also a good prediction of cost reduction and becomes very different of the really found value ΔV (the cost becomes greater than at the previous iteration).

b) when the Riccati equation in V_{XX} is divergent This situation may happen even if $H_{uu} > 0$ when existence and unicity conditions are not satisfied, furthermore we have also to take into account the coupling in V_X between the two equations which is a non linear coupling.

In order to solve these two problems Jacobson /2/ uses a step size adjustment method. This method proposes the use of u^* only in an interval $[t_1,T]$ where t_1 is a selected time in function of the ratio $a(t_o)/\Delta V$ of the estimated cost against the actual cost (first case) or is the time for which $a(t)$ takes a value less than a definite limite (in function of the tendancy to divergence of the coupled Riccati Equation).The new control law is then :

$$u(t) = \begin{cases} \bar{u}(t) & t \in [t_o\ t_1] \\ u^*(t) + K(t)\delta X(t) & t \in [t_1, T] \end{cases}$$

This means that we start with a t_1 sufficiently close to T in order that the integration interval and application of the new law be such that the variation of δX stays limited to values where the second order developments are valid.

There are several ways to estimate the value of t_1. The more classical (Jacobson) consists in analyzing the ratio between the cost $\Delta V(t_1)$ and the estimated variation $|a(t_1)|$ given by :

$$\frac{\Delta V(t_1)}{|a(t_1)|} > C \quad , \quad C < 0 \qquad (3.10)$$

where :

$$\Delta V(t_1) = \bar{V}(X_o,t_o) - V(X_o,t_o)$$

$$|a(t_1)| = |\int_{t_1}^{T} [H(\bar{X},u^*,\lambda,t) - H(\bar{X},\bar{u},\lambda,t)]\, dt\, |$$

If relation (3.10) is not satisfied, one has to found another t_1 which is closen to T in order to reduce the integration time and decrease X One can take the formulea :

$$t_{1new} = t_{1old} + \frac{(T - t_{1old})}{2} = t_{ff}$$

There exist other choices for t_1, particularly, in the following method, one fixes for the required accurancy a value of 120% of the nominal cost. When a $(t_1) > \eta$ the backwards integration is stopped and we apply the step adjustment method.

3.3 - First order algorithms

The first order algorithms are easily from second order algorithms ; we have with :

$$H_u(X,u^*,V_X^u(\bar{X},t),t) = 0 \qquad (3.11)$$

$$-\overset{\circ}{a}(t) = H(\bar{X},u^*,V_X^u(\bar{X},t),t) - H(\bar{X},\bar{u},V_X^u(\bar{X},t),t)$$

$$-\overset{\circ}{V}_X(t) = H_X(\bar{X},u^*,V_X^u(\bar{X},t),t) + V_{XX}(\bar{X},u^*,V_X^u(\bar{X},t),t) \cdot \Delta F$$

where

$$\Delta F = F(\bar{X},u^*,t) - F(\bar{X},u,t)$$

This formulation is analogous to that presented in § 2, and we can neglect the term containing the second derivative with respect to the state and we have then :

$$-\overset{\circ}{a}(t) = H(\bar{X},u^*,V_X^u(\bar{X},t),t) - H(\bar{X},\bar{u},V_X^u(X,t),t)$$
$$-\overset{\circ}{V}_X(t) = H_X(\bar{X},u\;,V_X^u(\bar{X},t),t) \qquad (3.12)$$

with the boundary conditions :

$$a(T) = 0$$
$$V_X(T) = G_X(X(T)) \qquad (3.13)$$

The calculus procedure is the same that in the second order method.

Control convergence parameter method (CCP)

This method due to Järmark /5/ introduces a penalization term in the integral part of the criterion. This term which is quadratic in the variation of the control between two successive iterations maintains limited δX in the development of the cost function and makes then valid the development of H-J-H equation without modifying the problem at the optimal solution. With this penalization term we make the hessian of the Hamilton positive definite with respect to the control and avoid convergence problems of the Riccati equation associated with second order DDP, the field of application of DDP first order and second order algorithms is then enlarged.

Let make some remarks and characteristics of this approach :

- the original problem is not modified
- the algorithm is simple and easily implementable on a computer
- large domain of convergence
- fast convergence speed
- if applied to the first order algorithm its sacrifies accurancy for speed and if applied to the second order algorithm its increases accuracy by more calculus.

Let us describe the main characteristics of CCP method in the case of no constraints. The application assumptions of the method are the same as previously given.

The criterion becomes in CCP :

$$V^u(X,t) = \int_t^T [L(X,u,t) = \frac{1}{2}(\delta u)^T C \delta u]\, dt \qquad (3.14)$$

$$\tilde{H}^i = \tilde{H}(X^{i-1},u^i,u^{i-1},V_X^{i-1},C^i,t) = H(X^{i-1},u^i,V^{i-1}, t) + \frac{1}{2}(\delta u^i)^T C^i (\delta u^i)$$

$$\tilde{H} = H(\bar{X},u,\bar{u},\bar{V}_X,C,t) = H(\bar{X},u,\bar{V}_X,t) + \frac{1}{2}(\delta u)^T C(\delta u)$$

$$\bar{H} = H(\bar{X},\bar{u},\bar{V}_X,t)$$

with C diagonal matrix C_{kk}^i , K = 1,...,m and u = u-u.

Jarmark /5/ shows that the penalized problem (3.14) converges to the original problem. The algorithm of Jarmark presents the essential interest in the possibility of giving a new value to C^i in function of the ratio between V^i and $a^i(t_o)$ in successive iteration because this ratio represents a measure of the validity of the first order development of H-J-B equation. The method developed by Jarmark will be used in the simulations. Figure 1 shows an example of the decrease of parameter C with respect to the iterations :

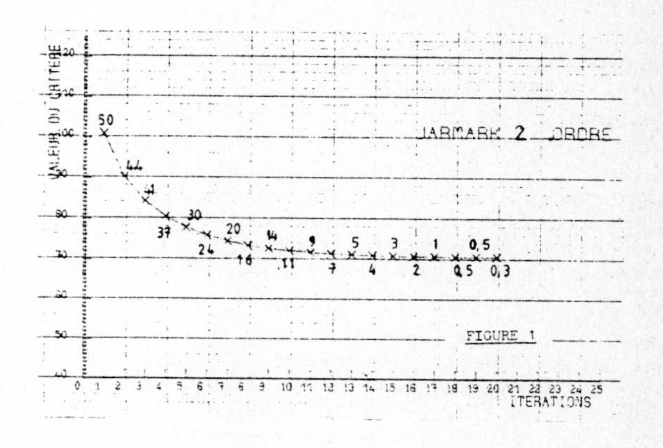

FIGURE 1

4 - APPLICATIONS - SIMULATIONS

Example 1 : Comparative study of algorithms

The example is a non linear system of dimension 3 with only one control variable. The cost function is quadratic and the state equation is highly non linear and we will see that a component of the adjoint state becomes very quickly instable. The problem is without constraint and the final time is fixed. The main interest of the example is due to the fact that around the nominal trajectories used at the beginning of the iterations, the backwards integration on $[T, t]$ of the differential equations in V_x and V_{xx} becomes quickly instable ; this makes inapplicable some other methods as the second variation method for example.

We consider the system :

$$\dot{x}_1 = x_2$$
$$\dot{x}_2 = -x_1 + 1.4\, x_2 - 0.14\, x_2^3 + 4\, u$$
$$\dot{x}_3 = 0.1\, x_1 + 0.3\, (x_1^2 + 0.1\, x_2^2)^{1/2}$$

The cost function to be minimized is

$$J(u) = \int_0^T (x_1^2 + x_3^2 + u^2)\, dt$$

The following figures show the results after application of different algorithms. Figures 2a and 2b show the evolution of the second component (V_{X_2}) of the adjoint state and the optimal trajectory for x_1 which are obtained the second order Jacobson Mayne algorithm. On figure 2 we notice the way of progressing of the iterations in spite of very high values of the adjoint system (V_{X_2} diverges at the beginning of the iterative scheme) and finally converges towards a bounded solution For a precision of 0.001 on the cost the algorithm converges in 21 iterations. We have used the variation formulae for the step :

$$t_1^{i+1} = t_1^i + \frac{t_{FF} - t_1^i}{CH} \qquad \text{with } CH = 2$$

By choosing CH = 5 which corresponds to a new t_1 close to t_{FF} only by 20% we obtain an improvement of convergence speed. On figure 2c we show the adjoint state obtained after 11 iterations for the same precision. Järmark's algorithm has been also used for the same example. The two approach second order and first order methods have been successively applied (results in fig.3 and 4); for example, with an initial control convergence parameter C = 50, convergence has been obtained in 20 iterations for a 0.001 precision on the cost. We note that the number of iterations is a little superior (24) for the first order algorithm, however the number of calculated terms in the pseudo adjoint equation V_X^u and the fact there is no Riccati to integrate makes a gain of a third on the computing time.

We have also implemented a program allowing use of first order or second order algorithms. This has especially allowed to initialize the problem with Jarmark"s first order method and then commute (after 6 iterations) on the Jacobson Mayne second order algorithm, the optimal solution is obtained in 12 iterations. The analysis of the different cases shows that the algorithm with a convergence control parameter although less rapid is is more adapted than the dassical DDP algorithm.

Model 2

The example is a guidance problem in atmospheric reentry in plane movement. The problem consists to minimize the criterion

$$J = \int_{t_0}^T (\dot{Q} + K_7\, a^2)\, dt$$

for the system :

$$\dot{x}_1 = -\, x_2 \sin x_3$$
$$\dot{x}_2 = K_2 \sin x_3 - \frac{D(u)}{K_3}$$
$$\dot{x}_3 = \frac{K_2 \cos x_3}{x_2} - \frac{x_2 \cos x_3}{K_1 + x_1} - \frac{L(u)}{K_3\, x_2}$$

with the constraint and the boundary conditions :

$$|u| \leq 90°$$

$x_1(t_0) = 121.900$ m $\quad (x_1(T) = 76\ 200$ m$)$

$x_2(t_0) = 10\ 970$ m/s $\quad (x_2(T) = 8\ 230$ m/s$)$

$x_3(t_0) = 8.09 \quad (x_3(T) = 0)$

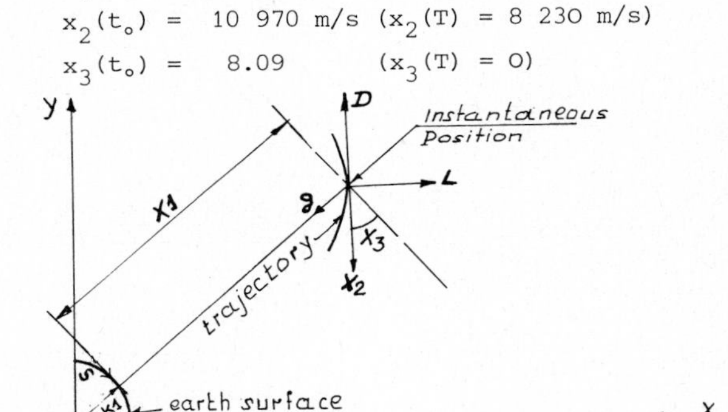

The definition of state variables is :

x_1 vehicle altitude with respect to earth surface

x_2 vehicle velocity

x_3 pitch angle

The control u is the incidence angle

$\dot{Q}$ rate of heat generation due offthe aerodynamic friction

$$\dot{Q} = K_4\, \rho^{1/2}\, x_2^3$$

$$\rho = K_5^2\, e^{K_6 x_1}$$

K_5^2 : air density at sea lavel

K_4 : heating flux constant

a^2 : crew confort parameter

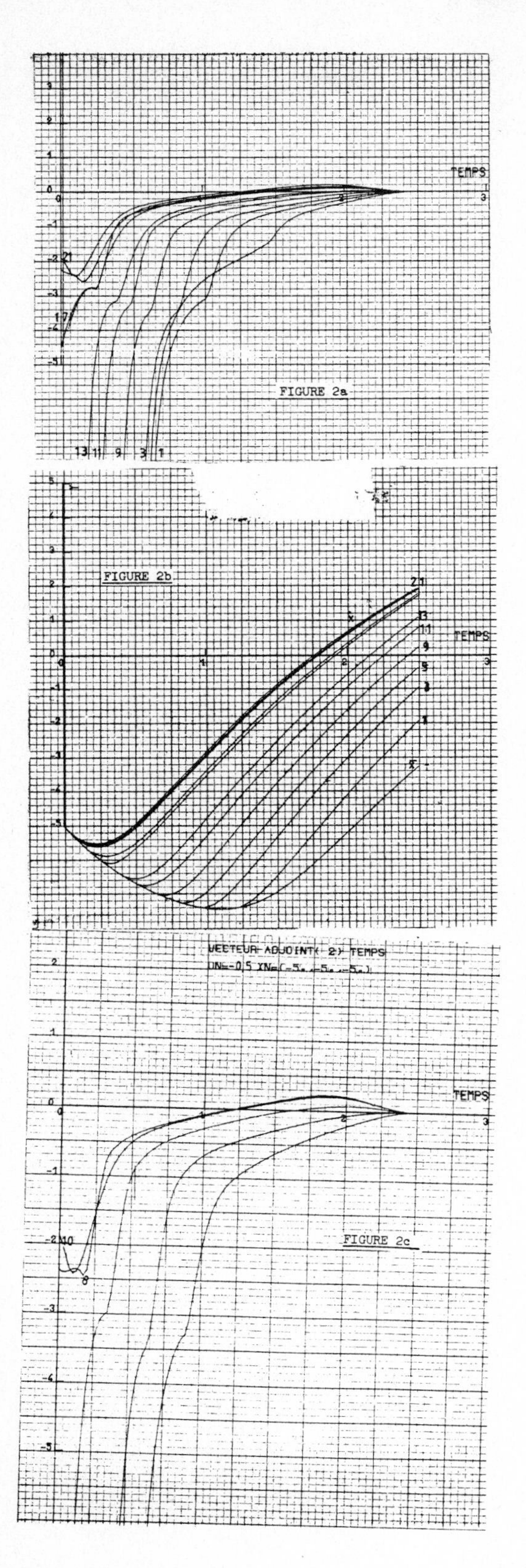

FIGURE 2a

FIGURE 2b

FIGURE 2c

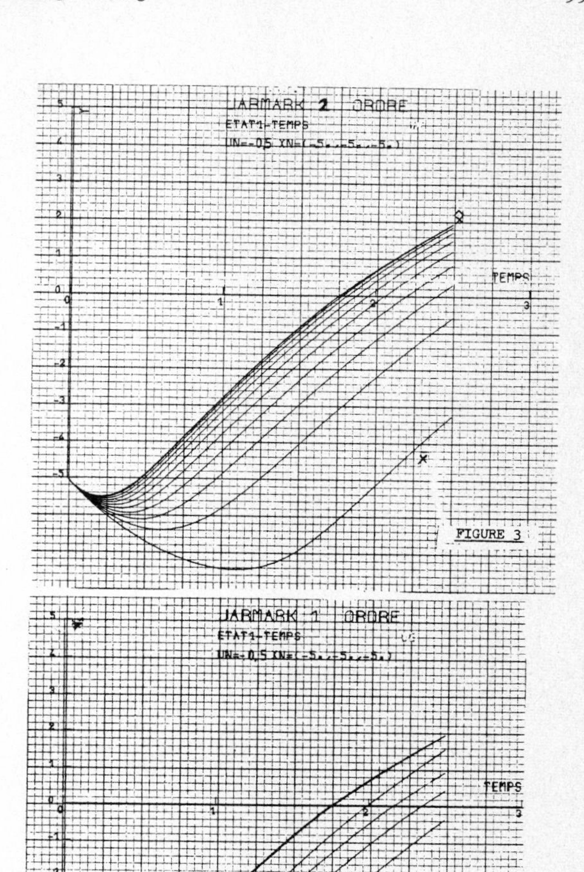

FIGURE 3

FIGURE 4a

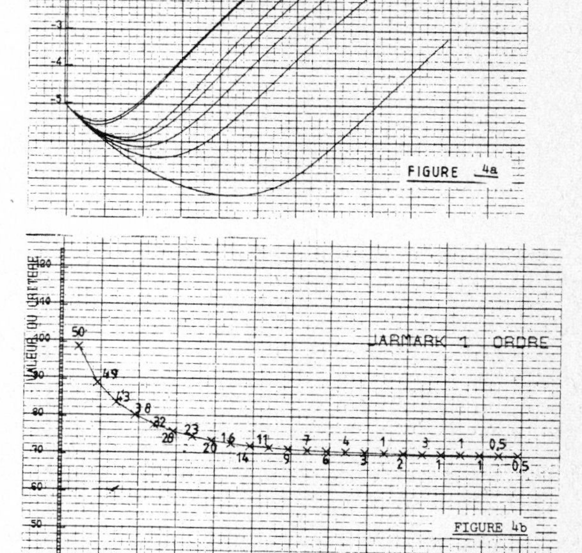

FIGURE 4b

$a^2 = (L^2 + D^2)/K_3^2$

L,D : lift and drag

$L = \frac{1}{2} \rho x_2^2 \; C_L K_{10}, \; D = \frac{1}{2} \rho x_2^2 C_D K_{10}$

C_D and C_L lift and drag coefficients :

$C_L = K_{11} \sin u \cos u,$

$C_D = K_{12} + K_{13} \sin^2 u$

Value of reentry constants

- earth radius : $K_1 = 6370320$ m
- gravity constant $K_2 = 9{,}7536$
- square root of air density at rea level
  $K_5 = 0{,}3768 \; kg^{1/2} \; s/ \; m^2$
- gradient of air density
  $K_6 = -13{,}977 \times 10^{-5} \; m^{-1}$
- parameters of the vehicle
- vehicle mass $K_3 = 150$ kg
  $K_4 = 0{,}6735 \times 10^{-4} \; kg^{1/2} \; s$
- Reference surface
  $K_{10} = 6{,}18 \; m^2$
  $K_{11} = 1.2, \; K_{12} = 0{,}274, \; K_{13} = 1{,}8$
- Weight term between crew confort and rate of heat generation $K_7 = 100$

We consider the criterion

$$J(u) = \int_{t_o}^{T} (\dot{Q} + K_7 \; a^2)\,dt + \sum_{i=1}^{3} b_i (X_i(T) - X_{T_i})^2/2$$

We use Jarmarks"s algorithm =

$$-\dot{a} = H(\bar{X}, u^*, u, \lambda, CJ, t) - H(\bar{X}, \bar{u}, \lambda, t)$$

$$-\dot{V}_X = H_X(\bar{X}, u^*, \bar{u}, \lambda, CJ, t)$$

where

$$H = (\dot{Q} + K_7 \; a^2) + \delta u^T \; CJ \; \delta u + \lambda^T \; F(X,u,t)$$

A good relation of the coefficients b_i and of convergence parameter CJ is obtained after several tests on computer (indeed there is a great sensibility of final precision that can be obtained with respect to these parameters) ; we use :

$$b_1 = 20, \; b_2 = 95 \times 10^5, \; b_3 = 45 \times 10^7$$

We then obtain results which are given on figures 6 The precision is excellent for the desired final state.

REFERENCES

/1/ Breakwell J., Speyer J.L., Bryson A.E. (1963) Optimisation and control of non linear using the second variation. SIAM J.Control, 1.

/2/ Jacobson D.H., Mayne D.Q. (1970) Differential dynamic programming. American Elsevier, N.Y

/3/ Bryson A.E., Ho Y.C. (1969) Applied optimal control. Walthem Mass. Blaisdell.

/4/ Mayne D.Q. (1973) Differential Dynamic programming. A unified approach to the optimization of dynamic systems. Advances in Theory and applications. C.T. Leondes Vol.10

/5/ Mayne D.Q., POLAK E. (1975) First order strong variation algorithms for optimal control. J. Optimization Theory and applicat. 16

/6/ Järmark (1976) Convergence Control in differential dynamic programming applied to air to air combat. AIAA J., vol 14, n°1

/7/ Lopez Coronado J. (1981) Synthèse et applications de la programmation dynamique différentielle. Thèse de Docteur Ingénieur. ENSAE

/8/ Järmark B.S.A. (1978) Some aspects on improving a first order DDP algorithm. SAAB TN 73 Ae., SAAB SCANIA, Sweden

/9/ Järmark B.S.A, Jonsson H.O. (1979) Parameter sensibivity in non linear dynamic systems Proc. Int. Conf. on Inf. Sciences and Systems Patras (Grèce).

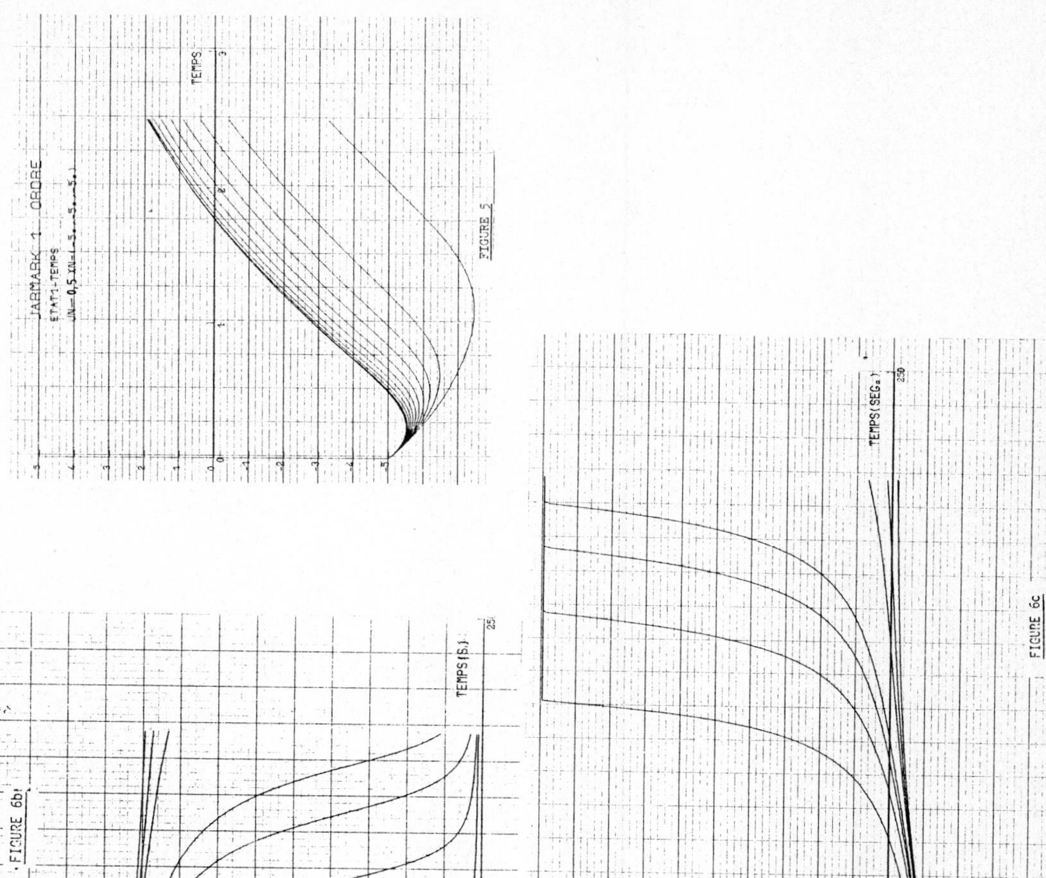
JARMARK 1 ORDRE
ETAT1-TEMPS
TEMPS
FIGURE 5

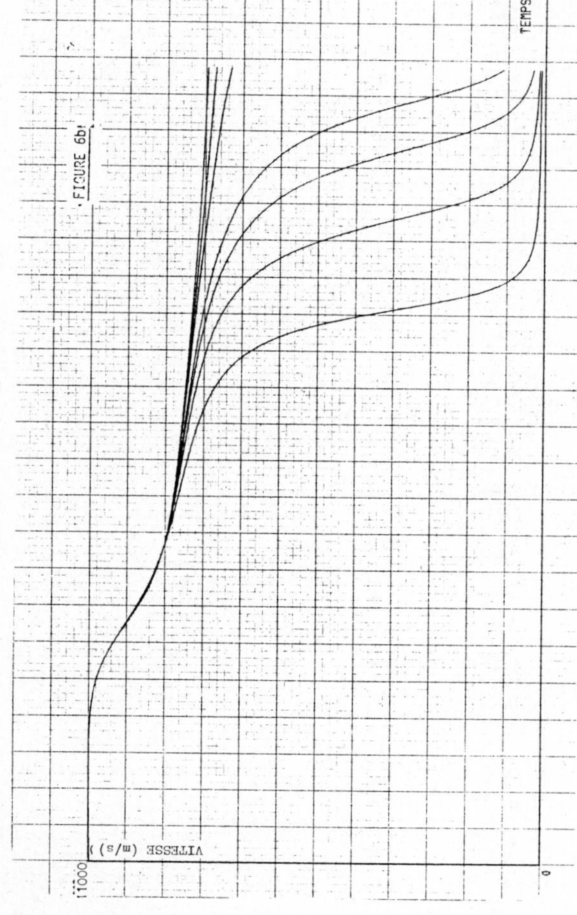
FIGURE 6b
TEMPS(S.)
25
VITESSE (m/s)
1000
0

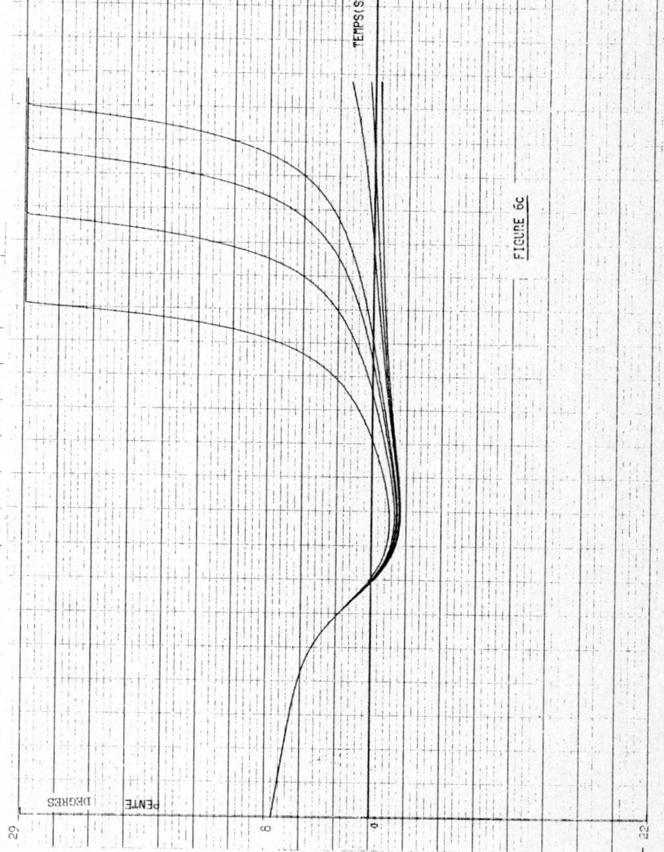
FIGURE 6c
TEMPS(SEG.)
250
PENTE
DEGRES
29
6
0
- 22

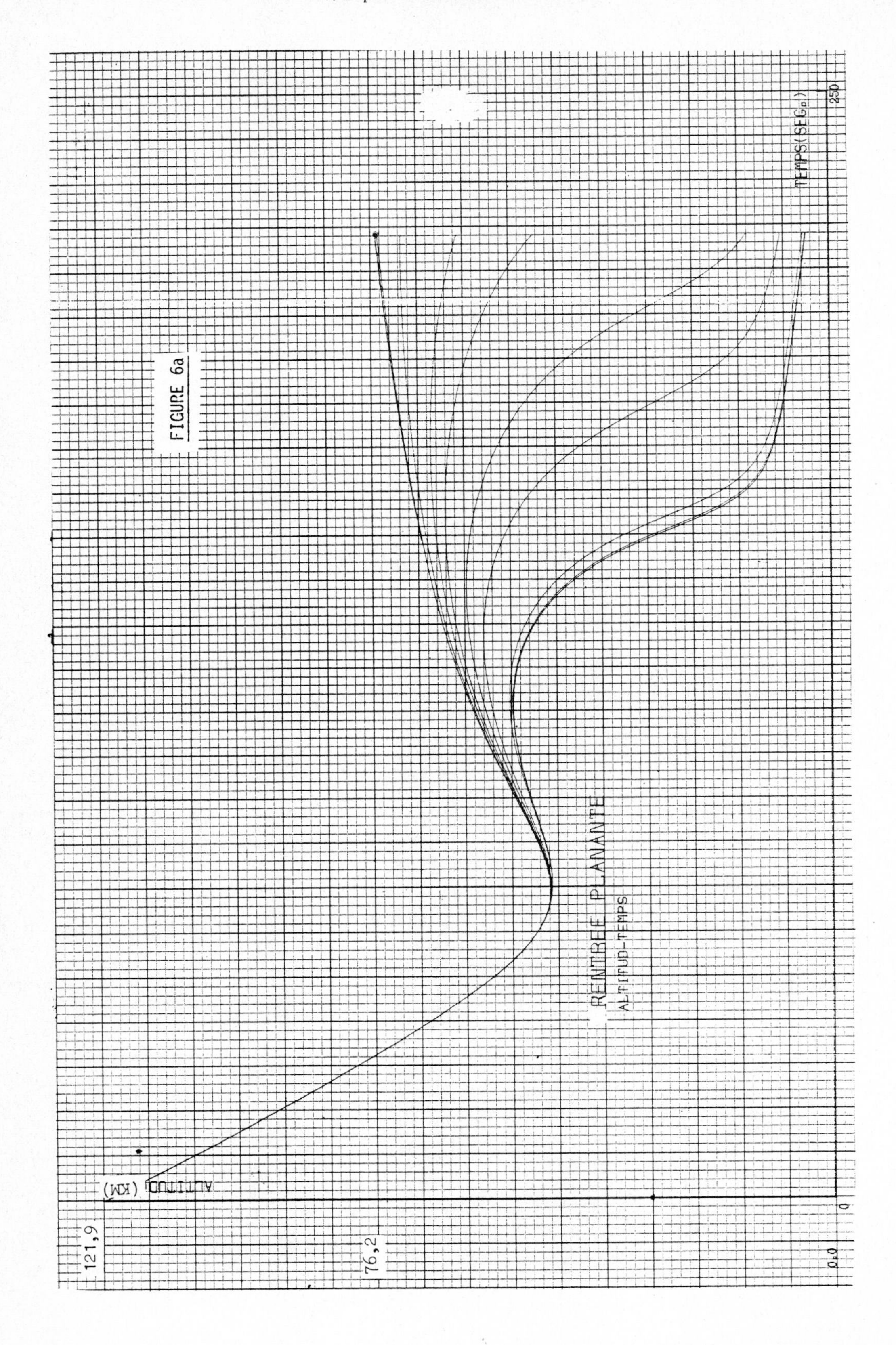
FIGURE 6a
REENTREE PLANANTE
ALTITUD-TEMPS
ALTITUD (KM)
TEMPS(SEG.)
121,9
76,2
0.0
0
250

Simulation in Engineering Sciences
J. Burger and Y. Jarny (eds.)
Elsevier Science Publishers B.V. (North-Holland)

LINEAR APPROXIMATION OF NONLINEAR SYSTEMS BASED ON LEAST SQUARES METHODS

J.G. den Hollander, J.A. Hoogstraten and G.A.J. van de Moesdijk

Delft University of Technology
Department of Aerospace Engineering
Delft, The Netherlands

Linearization is considered in a general sense in approximating nonlinear relations by linear relations while minimizing a sum of weighted squares. The least squares linearization yields a more general concept of transforming nonlinear characteristics into linear relations then the classical linearization. Besides it allows a structurized treatment of multivariate nonlinear systems resulting in generally applicable computer programs.
The basic least squares methodology can be applied to a wide variety of problems related to nonlinear system simulation. Applications treated are: gross linearization of finite dimensional systems, solution of nonlinear equations and nonlinear function optimization.

1. INTRODUCTION

Nonlinearities appearing in control engineering, simulation and optimization considerably hamper a straight forward approach to engineering problems. A tremendous quantity of research has been reported on this subject. However, even when time and specialized knowledge are available to conduct an exhaustive literature search in this rather specialized and in general highly mathematical area, in most cases the final result will still be of limited value to the practizing engineer. For example stability criteria found, often only represent sufficient conditions and as such are unsuitabe for design, yielding results that are both too conservative and are valid only under strictly limiting assumptions.

With the availability of powerful computers, simulation has become applicable to almost any problem. Since in principle a straight forward simulation approach to nonlinear system evaluation has to consider an infinite number of initial conditions, the computer cost for obtaining sufficient information may be prohibitive. In such cases the simultaneous consideration of linearized system equations can significantly reduce the evaluation effort.

The methodology of linear approximation discussed in this paper is based on least squares methods and provides a structurized approach to gross linearization, applicable to a wide class of nonlinear problems. A flexible computer algorithm results, which can be used as a tool in the analysis and evaluation of simulated systems. As such the method does not result in, for instance, exact stability criteria.

Although the method was developed independently the concept shows close relationship to and indeed is a further generalization of the least squares approximation technique [2] such as applied for example in [6]. Besides providing an extension, the method presented encompasses a numerical procedure of considerable generality and flexibility.

The organization of this paper is as follows. First the basic principles of least squares linearization are introduced, followed by a description of the application to finite dimensional system dynamics. Secondly the application to the solution of a set of nonlinear equations is discussed. Finally nonlinear function optimization employing the least squares linearization methodology is described.

2. LEAST SQUARES LINEARIZATION

In the present section the least squares linearization principles are briefly introduced. For a more detailed treatment the reader is referred to a separate report [3].

Consider a set of n nonlinear functional relations formally written as:

$$z(t) = g\{v(t), t\} \tag{1}$$

where $z(t)$ is an n-dimensional column vector of dependent variables, $v(t)$ a m-dimensional column vector of independent variables, t an additional variable such as e.g. time, and $g\{\ \}$ a nonlinear vector valued functional relation.

Let $z_o(t)$ be a solution of (1) for given $v_o(t)$ respectively denoted as the nominal solution and the nominal trajectory. Consequently:

$$z_o(t) = g\{v_o(t), t\},\ t_o \leqslant t \leqslant t_1 \tag{2}$$

To define a least squares linearization for the set of nonlinear relations (1) the following notions are used:

- The perturbations about the nominal solution and the nominal trajectory defined by:

$$\tilde{z}(t) = z(t) - z_o(t) \tag{3}$$

$$\tilde{v}(t) = v(t) - v_o(t) \tag{4}$$

  where $\tilde{z}(t)$ is an n-dimensional column vector of the dependent variable perturbation and $\tilde{v}(t)$ an m-dimensional column vector of the independent variable perturbations.
- The relative frequency density function $f_d(\tilde{v})$ specifying a particular distribution of the independent variable perturbations $\tilde{v}(t)$.
- The m-dimensional independent variable space R_m being an m-dimensional Euclidian space.

The least squares linearization of the set of

nonlinear relations (1) pertaining to a relative frequency density function $f_d(\tilde{v})$ is now defined as:

$$z_L(t) = z_o(t) + \tilde{z}_L(t) \tag{5a}$$

$$\tilde{z}_L(t) = J^*\{v_o(t),\ f_d(\tilde{v}),\ t\}\ \tilde{v}(t) \tag{5b}$$

such that:

$$\int_{R_{\tilde{v}}} (\tilde{z}_L(t)-\tilde{z}(t))^T\ D\ (\tilde{z}_L(t)-\tilde{z}(t))\ f_d(\tilde{v})\ dR_{\tilde{v}}\Big|_{MIN} \tag{5c}$$

Here $\tilde{z}_L(t)$ is an n-dimensional vector of linearized variable perturbations, $J^*\{v_o(t_1),f_d(\tilde{v}),t\}$ is an n * m dimensional matrix function denoted as the generalized Jacobian matrix function, while

$$\int_{R_{\tilde{v}}} (\ )\ dR_{\tilde{v}} = \int_{-\infty}^{+\infty}\int_{-\infty}^{+\infty} \cdots \int_{-\infty}^{+\infty} (\ )\ d\tilde{v}_1\ d\tilde{v}_2\ \cdots\ d\tilde{v}_n$$

and D is an m*m dimensional homogeneity matrix.

The relative frequency density function $f_d(\tilde{v})$ restricts the result given by (5) to a particular region of independent variable perturbations, while furthermore it attributes different weight to the squared errors as a function of the location in the independent variable space. This aspect is illustrated for a simple scalar case in Fig.1. With the relative frequency density function $f_d(\tilde{v})$ the fact can be taken into account that, for instance during operation of a nonliner system, particular combinations of the independent variables occur with different frequency within a certain working region.

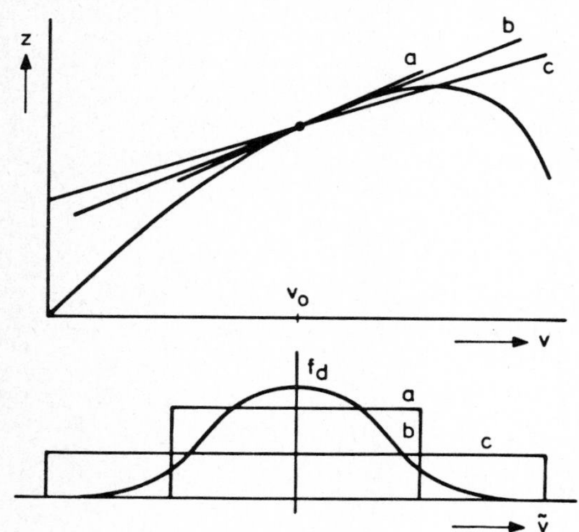

Fig. 1. Three different least squares linearizations pertaining to different relative frequency density functions f_d.

From (5) the following relation for the generalized Jacobian matrix results [3]:

$$J^*\{v_o(t),f_d(\tilde{v}),t\} = \tag{7}$$

$$\int_{R_{\tilde{v}}} f_d(\tilde{v})\tilde{z}(t)\tilde{v}^T(t)dR_{\tilde{v}}\ \left[\int_{R_{\tilde{v}}} f_d(\tilde{v})\tilde{v}(t)\tilde{v}^T(t)dR_{\tilde{v}}\right]^{-1}$$

Note that the homogeneity matrix D has disappeared from (7). As a consequence D only has the formal function in (5c) of matching mutually different dimensions.

The least squares linearization presented above implies $\tilde{z}_L(t) = 0$ for $\tilde{v}(t) = 0$. A generally improved fit to the nonlinear relations, in the sense of the least squares criterium, is obtained by the relaxed least squares linearization as defined in [3].

The fundamental difference between the least squares linearization and the relaxed least squares linearization is depicted for a scalar case in Fig. 2.

The least squares linearization methodology can be interpreted as an elementary form of parameter identification. In this context two conditions are favourable. First the output of the relations is completely deterministic and not hampered by noise sources as usual in the case of system identification. Secondly any desired combination of perturbations of variables can be forced.

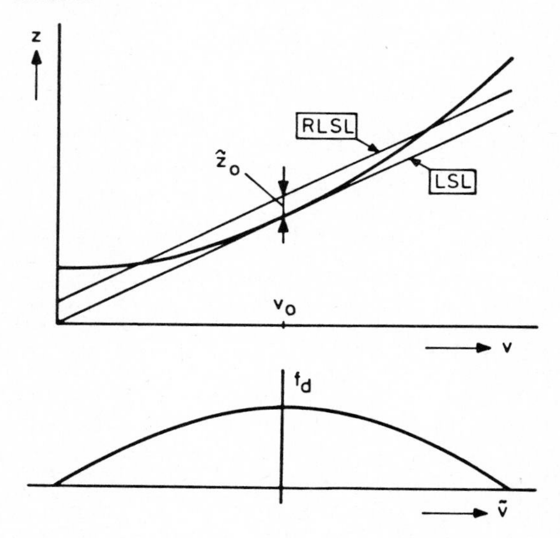

Fig. 2. Least squares linearization LSL compared to relaxed least squares linearization RLSL.

3. COMPUTATIONAL ASPECTS

For the numerical solution of problems described by continuous relations, discretization often results in sufficiently accurate approximation. This approach will here be followed and in the remainder of this paper attention will be focussed on least squares linearization of nonlinear relations which are either t-constant or t-variant relations, discretized with respect to t. Furthermore the linearizations will pertain to perturbation distributions described by discrete (discretized) relative frequency densities.

Consider a sequence of discrete independent variable perturbation vectors:

$$\tilde{v}(i) \; ; \; i = 1(1)\,N \tag{8}$$

In case the sequence of vectors $\tilde{v}(i)$ contains repetitions the sequence can be condensed to:

$$\tilde{v}(j); \; j = 1(1)\,N_1; \; N_1 \leq N; \; r(j) = n_j/N \tag{9}$$

Here $r(j)$ is the relative frequency of occurrence of $\tilde{v}(j)$ in $\tilde{v}(i)$ and n_j the number of repetitions of $\tilde{v}(j)$.

In [3] it is shown that the generalized Jacobian matrix pertaining to a discrete perturbation distribution (9) of the nonlinear relation (1) for nominal trajectory (2) will read:

$$J^*\{v_o(t_o),\, r(j),\, t_o\} = \tag{10}$$

$$= [\sum_{j=1}^{N_1} \{r(j)\tilde{z}(j)\tilde{v}^T(j)\}]\,[\sum_{j=1}^{N_1} r(j)\tilde{v}(j)\tilde{v}^T(j)]^{-1}$$

Introducing the following matrices:

$$\tilde{Z} = col\{\tilde{z}^T(1), \tilde{z}^T(2), \ldots, \tilde{z}^T(j), \ldots \tilde{z}^T(N_1)\} \tag{11}$$

$$\tilde{V} = col\{\tilde{v}^T(1), \tilde{v}^T(2), \ldots, \tilde{v}^T(j), \ldots \tilde{v}^T(N_1)\} \tag{12}$$

$$F = I_{N_1} * col\{r(1), r(2), \ldots, r(j), \ldots r(N_1)\} \tag{13}$$

where I_{N_1} is an $N_1 * N_1$ unit or identity matrix,

Eq. (10) can be reformulated as:

$$J^*\{v_o(t_o),\, r(j),\, t_o\} = \tilde{Z}^T F \tilde{V}\, [\tilde{V}^T F \tilde{V}]^{-1} \tag{14}$$

This basic algorithm can easily be programmed in a general form in which only the evaluation of the dependent variable matrix $\tilde{Z}$ is problem-dependent. Furthermore (14) can be applied to approximate (7) arbitrarily close for t-invariant and with respect to t discretized least squares linearizations.

4. RELATION TO ANALYTICAL JACOBIAN MATRICES

In the present section the relation is discussed to linearizations employing analytical Jacobian matrices, as appearing in the classical linearization based on a truncated Taylor expansion. Furthermore linearizations applied to a simple nonlinear equation will ellucidate different aspects.

Two rudimentary least squares linearizations are identical to currently applied methods in computing numerical approximations of analytical Jacobian matrices.

Consider (14). By selecting $\tilde{V} = \varepsilon I_m$ and $F = I_m/m$, where ε is a scalar, (14) reduces to:

$$J^*\{v_o(t_o),\, \tfrac{1}{m},\, t_o\} = \tilde{Z}^T/\varepsilon \tag{15}$$

which is the well-known finite difference approximation of the Jacobian matrix. It can easily be verified that:

$$\lim_{\varepsilon \to 0} J^*\{v_o(t_o),\, \tfrac{1}{m},\, t_o\} =$$

$$= \left.\frac{\partial g\{v(t),\, t\}}{\partial v^T(t)}\right|_{v_o(t_o),\,t_o} = J\{v_o(t_o), t_o\} \tag{16}$$

where $J\{v_o(t_o),\, t_o\}$ is a Jacobian matrix corresponding to the well-known mathematical definition.

A common extension to the finite difference approximation applies both positive and negative finite differences. This method fits in the least squares linearization methodology by selecting:

$$\tilde{V}^T = [\varepsilon I_m \,\vdots\, -\varepsilon I_m] \; ; \quad F = I_{2m}/2m$$

Then (14) reduces to:

$$J^*\{v_o(t_o),\, \tfrac{1}{2m},\, t_o\} = (\tilde{Z}_1^T - \tilde{Z}_2^T)/2\varepsilon \tag{17}$$

which method will further be denoted as the method of dual finite differences.

Note that the analytic Jacobian matrix (16) can be considered as a limit case of least squares linearizations, pertaining to a particular realization. In this realization only perturbations along the axes of the independent variable perturbation space are considered, while furthermore the set of perturbations proper is restricted to an infinitesimally close neighbourhood of the nominal trajectory.

These aspects will be elucidated considering the well-known Van de Pol equation, reading:

$$\ddot{x} + \mu(x^2 - 1)\,\dot{x} + x = 0 \tag{18}$$

For $\mu = -1$, $\ddot{x} = (x^2 - 1)\,\dot{x} - x$ is linearized around $x = 0$, $\dot{x} = 0$.
The analytic Jacobian matrix as well as the generalized Jacobian matrices based on the single- as well as on the dual finite difference method, both for any value of the finite differences equals: $[-1 \;\; -1]$. Consequently the equivalent linear second order system is stable having a relative damping of 0.5.

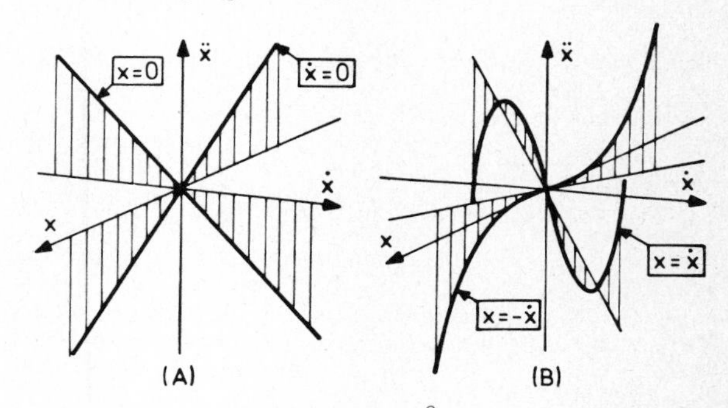

Fig. 3. Function value $\ddot{x} = (x^2 - 1)\,\dot{x} - x$ along axes $x = 0$ and $\dot{x} = 0$ (A) and along axes $x = \dot{x}$ and $x = -\dot{x}$ (B).

The fact that the values of the finite differences are irrelevant in this case might suggest that a system obeying the considered Van de Pol's equation is also stable in the large. However it results from the directions, in which the finite differences are taken, which are in fact arbitrary. Along these directions, i.e. the axes of the independent variable space $(x, \dot{x})$, $\ddot{x}$ is a linear function, as depicted in Fig. 3a. From Fig. 3b it can be seen that for axes rotated about 45°, $\ddot{x}$ turns out to be a nonlinear function, which holds for any other set of directions.

Evaluating nonlinearities by means of least squares linearization such characteristics are readily detected. Consider for instance the simple perturbation distribution:

$$\tilde{V}^T = \begin{bmatrix} \varepsilon_1 & -\varepsilon_1 & -\varepsilon_1 & \varepsilon_1 \\ \varepsilon_2 & \varepsilon_2 & -\varepsilon_2 & -\varepsilon_2 \end{bmatrix}, \quad F = I_4/4 \qquad (19)$$

The resulting generalized Jacobian matrices read:

$$J^* = [-1 \quad -(1-\varepsilon_1^{\,2})] \qquad (20)$$

Although the value of ε_2 does not affect the result, caused by symmetric properties of (19) the nonlinear characteristics are already reflected in the generalized Jacobian J^*. For perturbation ε_1 larger than 1.0 the equivalent linear system becomes unstable. As a consequence stability in the whole can doubted to be a property of the system.

The aspects discussed before illustrate that compared to classical linearization the least squares linearization approach can express essentially different nonlinear characteristics. Furthermore it elucidates that linearization based on finite differences depends on the selected variables and is consequently unsuitable for gross linearization.

5. LINEARIZATION OF FINITE DIMENSIONAL DIFFERENTIAL SYSTEM

Many systems can be classified as finite dimensional differential systems and can be described by a set of simultaneous differential equations:

$$\dot{x}(t) = f\{x(t), u(t), t\} \qquad (21)$$

called the state differential system equations. Here t is the time variable, x(t) is a real n-dimensional time dependent vector denoting the state of the system and u(t) is a real k-dimensional column vector which indicates the input or control variable. The function f{ } is real and vector valued. Often the so-called output variable, i.e. the variable that can be observed or through which the system influences its environment is of interest. In many cases this variable can be expressed as:

$$y(t) = g\{x(t), u(t), t\} \qquad (22)$$

The output variable y(t) is a real ℓ-dimensional column vector.

For least squares linearization it is computationally efficient to group both equations together, yielding:

$$z(t) = g\{v(t), t\} \qquad (23)$$

where: $z(t) = col\{\dot{x}^T(t), y(t)\}$ and $v(t) = col\{x^T(t), u^T(t)\}$, yielding the form of (1). In this form least squares linearization can directly be performed.

The determination of a nominal trajectory $v_o(t)$ might embody the solution of a set of nonlinear relations, which will be discussed in section 6.

Given a nominal trajectory, an independent variable matrix $\tilde{V}$ (12) and a relative frequency matrix F (11) matching the perturbation distribution of interest is specified. Employing (11) the matrix of dependent variables $\tilde{Z}$ is computed using the system equations. The resulting least squares linearized equation then read:

$$\tilde{z}_L = J^* \{v_o(t), r(j), t_o\} \cdot \tilde{v}(t) \qquad (24)$$

Decomposition of this equations result in least squares linearized system equations reading:

$$\dot{\tilde{x}}(t) = A\,\tilde{x}(t) + B\,\tilde{u}(t) \qquad (25a)$$

$$\tilde{y}(t) = C\,\tilde{x}(t) + D\,\tilde{u}(t) \qquad (25b)$$

$$\text{where:} \quad \begin{bmatrix} A & \vdots & B \\ \cdots & \vdots & \cdots \\ C & \vdots & D \end{bmatrix} = J^*\{v_o(t), r(j), t_o\} \qquad (26)$$

For a particular application of the least squares linearization reference is made to a separate contribution to the present conference [4]. Here the least squares linearization is applied on a routine base within a generalized aircraft evaluation and simulation software package [5], in which different comprehensive aircraft subsystem models can be interchanged. In an elementary form the method has already been demonstrated with the Van de Pol equation as discussed in section 4.

6. SOLUTION OF SETS OF NONLINEAR RELATIONS

In the present section application of least squares linearization to the solution of nonliner relations is discussed.

Consider a set of nonlinear relations formally written as:

$$h\{x\} = 0 \qquad (27)$$

where h{ } is a real n-dimensional vector valued functional relation and x a real n-dimensional column vector. Well-known iterative methods to solve such a set of nonlinear relations are the so-called Newton methods. Given particular analytic properties of the equations such an iteration scheme reads:

$$x(k+1) = x(k) - [J\{x(k)\}]^{-1} \cdot h\{x(k)\} \qquad (28)$$

$$\text{where: } J\{x(k)\} = \left.\frac{\partial h\{x\}}{\partial x^T}\right|_{x=x(k)}$$

is a Jacobian matrix. The iteration scheme reflects the following basic principle: Around the current estimate of a solution x(k) the functional relations are considered to be linear and can be described by:

$$h_L\{x\} = h\{x(k)\} + J\{x(k)\}\ \{x - x(k)\} \qquad (29)$$

Equating (29) to zero the improved estimate x(k+1) of the solution (28) results.

By interchanging the analytical Jacobian in (28) with a generalized Jacobian, the solution of nonlinear relations is further extended. For in many cases when analytical Jacobian matrices are no longer defined, still generalized Jacobian matrices exist, e.g. in function discontinuities.

Furthermore the determination of generalized Jacobian matrices only requires function evaluations and is consequently not restricted to analytical functions only, but can be applied without modifications to relations in, for instance, tabular form. Finally an important aspect is that no differentiated functions need to be programmed and consequently especially for complicated functional relations, a considerable reduction in analytical and programming effort results.

For continuous functional relations a finite difference approximation according (17) provides in general an efficient convergence to a solution. A more robust algorithm results applying the method of dual finite differences. This method is used on a routine base within the afore mentioned generalized aircraft evaluation and simulation software package [5] to compute initial conditions matching the aircraft's nonlinear simulation software [4].

7. OPTIMIZATION OF FUNCTIONS

Further extending the application area this section deals by way of introduction with the application of the least squares methodology to function optimization.

Consider a function to be optimized:

$$z = f(x) \qquad (30)$$

where z is a scalar and x a real valued n-dimensional column vector. Within an iteration scheme let x(k) be the k-th iteration to a solution. Around x(k) the following quadratic least squares approximation of (30) can be specified:

$$z^* = z(k) + G^{*^T}(k)\ \tilde{x} + \tfrac{1}{2}\tilde{x}^T\ H^*(k)\ \tilde{x} \qquad (31a)$$

such that:

$$\int_{R_{\tilde{x}}} (z^* - z)^2\ f_d(\tilde{x})\ dR_{\tilde{x}} \Big|_{MIN} \qquad (31b)$$

where $\tilde{x} = x - x(k)$ denotes the perturbations, $G^*(k)$ is an n-dimensional real column vector denoted as the generalized gradient and $H^*(k)$ is an n * n-dimensional real matrix denoted as the generalized Hessian matrix, which can be restricted to a symmetric matrix without loss of generality of (31). Further notation conventions are conform to those used before.

Reformulation of (31a) results in the following relation which is linear in the parameters:

$$z^* = K^{*^T}(k)\ \tilde{v} \qquad (32a)$$

where:

$$\tilde{v} = \text{col}\ \{1,\tilde{x}_1,\ldots,\tilde{x}_n,\tilde{x}_1^{\,2},\ldots,\tilde{x}_n^{\,2}, \tilde{x}_1\tilde{x}_2,\ldots,\tilde{x}_1\tilde{x}_n,\tilde{x}_2\tilde{x}_3,\ldots,\tilde{x}_2\tilde{x}_n,\ldots,\ldots,\tilde{x}_{n-1}\tilde{x}_n\} \qquad (32b)$$

$$K^*(k) = \text{col}\{z(k),g_1^{\,*},\ldots,g_n^{\,*},\tfrac{1}{2}h^*_{1,1},\ldots,\tfrac{1}{2}h^*_{n,n}, h^*_{1,2},\ldots,h^*_{1,n},h^2_{2,3},\ldots,h^*_{2,n},\ldots,\ldots,h^*_{n-1,n}\} \qquad (32c)$$

Since (32) as well as its defining criterium (31) are structurally identical to the least squares linearization (5), a numerically determined least squares approximation for a particular perturbation distribution can be computed with (14). The approximation using the least number of function evaluations results if the number of perturbations used equals the number of unknown coefficients, i.e. $\frac{1}{2}(n+2)(n+1)$. Furthermore (14) then simplifies and the parameter vector K* can be computed as:

$$K^* = V^{-1}\ \tilde{z} \qquad (33)$$

The matrix inversion in (33) reduces to some elementary operations if a properly structurized set of perturbations is selected. Decomposing K* as computed with (33) results in a generalized gradient vector G*(k) and a generalized Hessian matrix H*(k).

The function (30) as well as its local approximation (31) represent hypersurfaces in an n+1 dimensional Euclidian space. In this context the generalized gradient vector represents the slope of the surface in x(k), while the generalized Hessian matrix comprises information concerning the curvature. The orthonormal eigenvectors of this matrix represent the principal axes of the quadric hypersurface (31a), while the corresponding eigenvalues are the second partial derivatives along these principal axes and quantify the curvature.

An algorithm based on these properties is constructed which only employs function evaluation. Such an algorithm yields local quadratic approximations so that during iteration the shape of the function is known along the iteration path, giving insight in the optimization process.

To compare the accuracy and efficiency of the optimization methods with classical minimization methods, experiments were performed on a number of standard test functions. Occasionally the algorithm yielded significantly better results than reported for the classical methods, especially for functions to be minimized which were classified as difficult.

A first example of such a function is the so-called Singular function of Powell [1] reading:

$$f(x) = (x_1 + 10x_2)^2 + 5(x_3 - x_4)^2 + + (x_2 - 2x_3)^4 + 10(x_1 - x_4)^4 \qquad (34)$$

The approach to the minimum x = 0 versus the required number of function evaluations is depicted in Fig. 4. Both accuracy and efficiency compare favourably with results derived from [1].

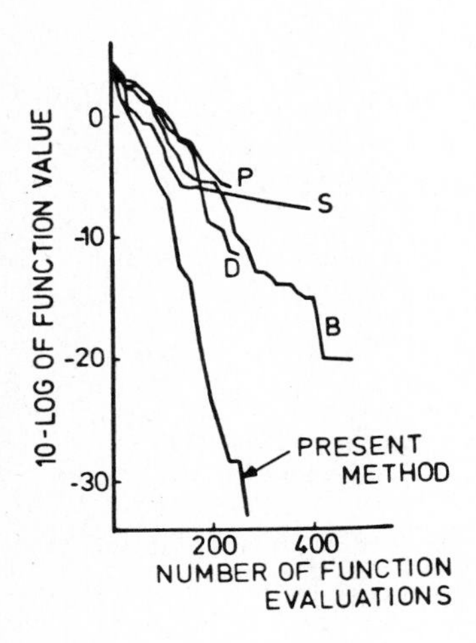

Fig. 4. Minimization performance on singular function of Powell. (P) Powell 1964, (D) Davies ed. 1965, (S) Stewart 1967, (B) Brent 1973 [1].

A second successful example is the minimization of the so-called Watson function [1] reading:

$$f(x) = x_1^2 + (x_2 - x_1^2 - 1)^2 + \sum_{i=2}^{30} \left[\sum_{j=2}^{6} (j-1)\, x_j \left(\frac{i-1}{29}\right)^{j-2} + - \left\{ \sum_{j=1}^{n} x_j \left(\frac{i-1}{29}\right)^{j-1} \right\}^2 - 1 \right]^2 \qquad (35)$$

The best result reported in [1] reads $f(\mu) = 2.2876700 \cdot 10^{-3}$ after 316 function evaluations. The present method passes this minimum after only 198 function evaluations to find the lower minimum $f(\mu) = 2.287346 \cdot 10^{-3}$ after 231 function evaluations.

8. CONCLUSIONS

Least squares linearization of nonlinear relations has been shown to suit various purposes.

Least squares linearizations pertain to well definable perturbation distributions and are optimal as specified for such particular working region.

A numerical method is presented to compute least squares linearization, a method which only requires function evaluations.

The methodology and in particular the resulting numerical method is directly applicable to the linearization of finite dimensional differential systems. A computer program was constructed in which different linearization problems can easily be handled by simply exchanging a subroutine embodying the system dynamics under study. The method does not require an analytical expression for the system dynamics, but can handle all uniquely defined and numerically evaluable relations such as tabular functions or computer simulation programs.

Finite difference linearization methods have shown to be rudimentary least squares linearizations.

Rudimentary least squares linearization applied within modified Newton algorithm results in a robust algorithm suitable to solve nonlinear relations. In the context of nonlinear system simulation this method is applied to initial condition computation. Also here the general concept of the method is a major advantage.

Furthermore rudimentary least squares linearization yields a powerful method when applied to optimization of functions. Favourable optimization characteristics were demonstrated on test functions applied.

For all considered applications the transparency of the method is important. With one basic methodology different problems concerning nonlinear functions can be handled without the need of particular specialized knowledge. However without doubt many problems can be solved more efficiently having more, often highly specialized, knowledge. The present methodology might assist in gaining such specialized knowledge.

REFERENCES

1. Brent, P.B., 1973, Algorithms for minimization without derivatives, Prentice-Hall, Inc., Englewood Cliffs, N.J.
2. Davis, P.J., 1963, Interpolation and Approximation, Waltham, Massachusetts, Blaisdell.
3. Den Hollander, J.G., 1982, Linearization based on least squares approximation with application to finite dimensional system equations, Report LR-364, Delft University of Technology, Delft.
4. Hoogstraten, J.A., Van de Moesdijk, G.A.J., Modular programming structure applied to the simulation of nonlinear aircraft models, in Burger, J., and Jarney, Y., Simulation of Dynamic Systems in Engineering Sciences (North-Holland, Amsterdam, 1983).
5. Hoogstraten, J.A., 1982, A Survey of CASPAR - Control system Analysis and Synthesis Program package for Aerospace Research, LR-336, Delft University of Technology, Delft.
6. Kriechbaum, G.K.L., Noges, E., Suboptimal control of nonlinear dynamical systems via linear approximation, Int. J. Control, Vol. 13, No. 6, 1971, pp. 1183-1195.

Simulation in Engineering Sciences
J. Burger and Y. Jarny (eds.)
Elsevier Science Publishers B.V. (North-Holland)

SIMULATION OF ENGINEERING PROBLEMS USING BOUNDARY ELEMENTS

C.A. Brebbia

Computational Mechanics Centre
Ashurst Lodge, Ashurst, Southampton
SO4 2AA, U.K.

This paper describes in detail the applications of boundary element methods for the solution of engineering problems. The advantages of the new techniques are pointed out, namely i) simple data preparation; ii) more accurate results; iii) definition of system and interpretation of results become easier, which permits a better interface to CAD systems and iv) problems with infinite domains can be solved accurately. Several practical applications help to illustrate the potentialities of the new method and its advantages viz. finite elements.

1. INTRODUCTION

Most engineers are now fully conversant with the finite element method of analysis. This technique which is now nearly twenty years old, has allowed for the solution of problems with complex geometrical shapes, arbitrary supports, different types of loadings, material properties, etc. The possibilities of the technique appeared to be boundless and complex non-linear models and other applications started to be amenable to solution for the first time in engineering history.

Unfortunately the wealth of information required to run a program and the amount of output produced by the computer code itself has the effect of making finite element programs less easy to use and their results more difficult to interpret. An attempt to overcome this problem has been the development of pre- and post processor codes, which require even larger computer facilities and special training.

On the other hand small groups of researchers in engineering started to search for a different theoretical solution to the problem. The basic idea was to develop analysis techniques which were better suited to solve large problems. In general these techniques can be called reduction techniques and can be traced to pre-computer times. Another important development in approximate methods of analysis has been the investigation of mixed principles and the realization that physical problems can be expressed and solved in many different ways in accordance with the equations of the problem that one wishes to approximate. These approximations are of fundamental importance for the computer implementation of the different numerical techniques and can be encompassed under the theory of mixed methods. These methods can be traced to Reissner [1] and more specifically for finite elements to Pian [2]. An excellent exposition of mixed methods in structural mechanics can be found in the book by Washizu [3].

The developments run in parallel with new advances in application of integral equations techniques which were, until recently, considered to be a different type of analytical method, somewhat unrelated to other approximate methods. They became popular in the west mainly through the work of a series of Russian authors, such as Muskhelishvili [4], Mikhlin [5], Kupradze [6] and Smirnov [7] but were not very popular with engineers. A predecessor of some of this type of work was Kellog [8] who applied integral equations for the solution of Laplace's type problems. Integral equation techniques were mainly used in fluid mechanics and general potential problems and known as the source method which is now classified as an 'indirect' method of analysis, i.e. the unknowns are not the the physical variables of the problem. Work on this method continued through the 1960's and 70's in the pioneering work of Jaswon [9] and Symm [10], Massonet [11] Hess [12] and others.

It is difficult to point out precisely who was the first researcher to propose the 'direct' method of analysis. It is found in a different way in Kupradze's book [6]. It seems fair however, from the engineering point of view to consider that the method originated in the work of Cruse [13] in elastostatics and elastodynamics.

Since the early 1960's a small group at Southampton University started working on the applications of integral equations to solve stress analysis problems. Some of this work was reported at the International Conference on Variational Methods in Engineering, held there in 1972 [14]. The conference was convened to discuss different techniques and how they interrelated. A special session of the Conference was dedicated to boundary integral equations although at that time it was not known how to interpret them as a variational or approximate technique. The work at Southampton was continued all through the 1970's through a series of

to the boundary in many cases by an appropriate integration [25]. The fundamental solution is usually the one due to Kelvin for elastostatics, but many other analytical solutions can be used as well. The importance of equation (1) is that the solution of the problem can be reduced by one dimension if the integral in the domain can be treated as shown in reference [25]. Even if the domain needs to be discretized to compute the last integral in (1), the problem unknowns are still those on the boundary only.

The problem described in fig. 1 presents the boundary element method used to describe 1/16 of a rotating machine component. Notice that elements have been defined on the planes of symmetry. Fig. 2 shows instead a thick cylinder divided into curved 9 node elements and presenting three planes of symmetry. Notice that on these planes elements are not required. Well written boundary elements computer programs [26] take symmetry into consideration to save data preparation time and reduce the number of unknowns.

Another important new development in boundary elements has been the use of discontinuous elements. Equation (1) does not require continuity of u or p across interelement boundaries. This implies that the nodes used to define them over the elements do not need to be on the element sides but can be inside. This has important applications for the cases of having a body divided into a series of regions which need then to be combined together. The main advantage is that in this way the cumbersome definition of corners con- ditions is avoided. As a consequence it is al possible to combine easily boundary with fini element matrices. Another advantage of using discontinuous elements is that the surface meshes can be more easily refined, as shown i figure 3, where the mesh can be altered witho regard to inter element continuity.

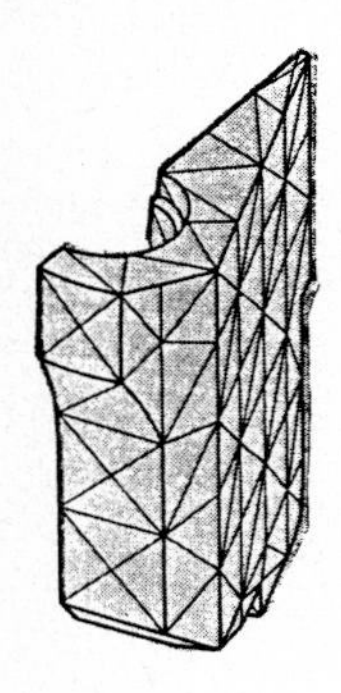

Fig. 1
Three Dimensional
Analysis

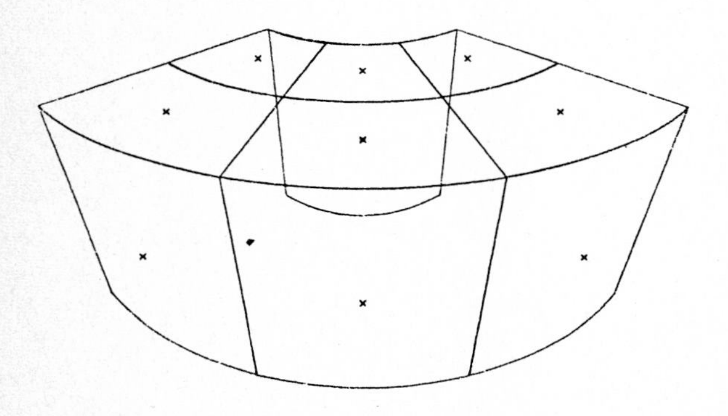

Figure 2 Thick Cylinder with Symmetry Planes

Boundary elements are being extensively used solve potential as well as stress analysis problems. Figure 4 illustrates an electrostatic problem solved for analysing the cathodic protection of a tension leg type platform. Because of symmetry only one quarter of the structure needs to be considered and the figure shows a 700 element mesh. By contrast any domain solution - finite elements or finite differences - used to solve the same problem will easily require thousands of elements. The results obtained in this case were used to design the induced currents protection system for the platform soon to be installed in the North Sea.

One of the most important recent applications of boundary elements has been for the solution of time dependent problems. This has now opened the way for solving more complex problems, which are difficult to attack using domain type techniques.

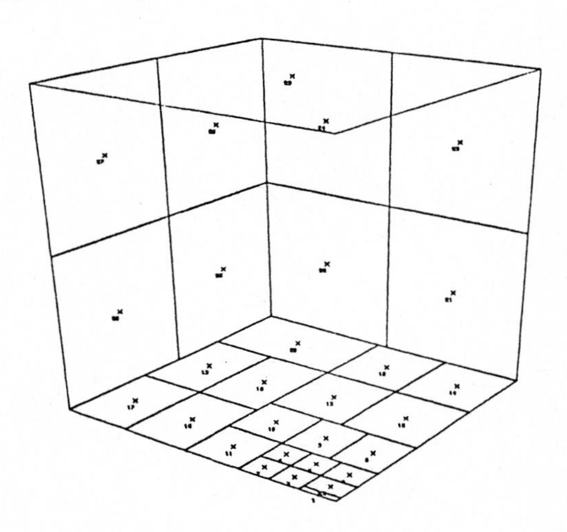

Figure 3 Use of Discontinuous Elements

The diffusion equation can be solved using boundary integral equations in space and a finite difference approximation on time. However, a more elegant and computationally more accurate solution is to use boundary elements in time and space. The approach is described in detail in reference [28] where it is applied to solve complex temperature problems. The use of a time dependent fundamental solution implies that one does not need to perform any finite difference on time and produces an accurate and efficient solution especially when higher order interpolation functions are used.

theses dealing mainly with elastostatics problems. At the same time recent developments in finite elements started to find their way into the formulation of boundary integral equations. and the problem of how to relate the technique to other approximate methods was solved using weighted residuals. This work at Southampton University culminated in 1978, with the first book in which the title "Boundary Elements" was used [16]. The work was then expanded to encompass time dependent and non-linear problems [17] and more recently into a larger book in which the method is described in great detail for a wide range of problems [18].

Four important international seminars were held on the topic of boundary elements in 1978, 1979, 1980, 1981 and 1982. The edited Proceedings of these meetings - the only ones so far in this topic - are now standard references [19][20][21][22]. The Conferences have also produced a series of state of the art books called "Progress in Boundary Element Methods" of which two volumes have been published up to now [23][24]. The series concentrates on the detailed explanation of how to solve certain types of problems using boundary elements. The chapters are written by internationally well known authorities in the field of integral equation techniques.

The boundary element method as understood nowadays, is a reduction technique based on boundary integral equations formulations and interpolation functions of the type used in finite elements. It offers important advantages over "domain" type techniques such as finite elements and finite differences. The main characteristic of the method is that it reduces the dimensionality of the problem by one and hence produces a much smaller system of equations and more important, considerable reductions in the data required to run a problem. In addition, the numerical accuracy of the boundary elements is generally greater than that of finite elements. This is due to using a mixed formulation type of approach for which all the boundary values are obtained with similar degree of accuracy. These advantages are more marked in three dimensional problems. The method is also well suited to problem solving with infinite domains such as those frequently occurring in soil mechanics, hydraulics, stress analysis, etc. for which the classical domain methods are unsuitable. A boundary solution is formulated in terms of influence functions obtained by applying a fundamental solution. If the solution is suitable for an infinite domain no outer boundaries need to be defined.

It is now becoming accepted that the best way of formulating boundary elements for complex engineering problems is by using weighted residual techniques, such as shown in reference [15].

In its simplest form the term boundary elements indicates the method whereby the external surface of a domain is divided into a series of elements over which the functions under consideration can vary in different ways, in much the same manner as in finite elements. This capability marked also an important development as classical integral equations type formulations were generally restricted to constant source assumed to be concentrated at a series of points on the external surface of the boundary.

The main advantages of simulating engineering problems using boundary elements rather than finite elements will become more clear in the examples shown in section 2. They can be summarized as i) simple data preparation; ii) more accurate results are generally obtained; iii) definitions of system and interpretation of results become easier which permits a better interface to CAD (Computer Aided Design) systems and iv) problems with infinite domains can be solved accurately.

2. REPRESENTATIVE APPLICATIONS

The advantages of boundary elements over finite elements are more striking in three-dimensional cases. Figure 1 shows a section of a piece of a centrifugal machine to be analysed. Due to symmetry only a piece needs to be considered. When using finite elements, twenty nodes 'bricks' are used. To define the elements we need the coordinates of all nodes, including internal ones and their connectivity or numbering within each element. By contrast the boundary element solution will require only the surface of the body to be discretized, for instance using 9 nodes elements. The advantages from the point of view of data preparation are obvious.

As the boundary element is a mixed type formulation, results for tractions and displacements are obtained at all nodes, both with a similar degree of accuracy. This is another important advantage as surface tractions in conventional finite elements tend to be poorly given due to the need of finding them from the numerical derivatives of displacements. The boundary element method provides unknowns only on the surface of the domain but based on them one can calculate the stress and displacements at any internal points. The following integral relationship applies for internal boundary points,

$$c_{ij}\, u_j + \int_\Gamma p^*_{ij}\, u_j\, d\Gamma = \int_\Gamma u^*_{ij}\, p_j\, d\Gamma + \int_\Omega u^*_{ij}\, b_j\, d\Omega \qquad (1)$$

where c_{ij} are elements of a unit matrix for a point 'i' inside the domain. For points on the boundary the values of c_{ij} depend on the presence and type of corners. For smooth boundary the c_{ij} - diagonal terms - are equal to $\frac{1}{2}$ and the rest equal to 0. p_j, u_j are the surface tractions and displacements. u^*_{ij} and p^*_{ij} are the surface components in the 'j' direction. b_j are the body forces (or centrifugal forces) components which can be reduced

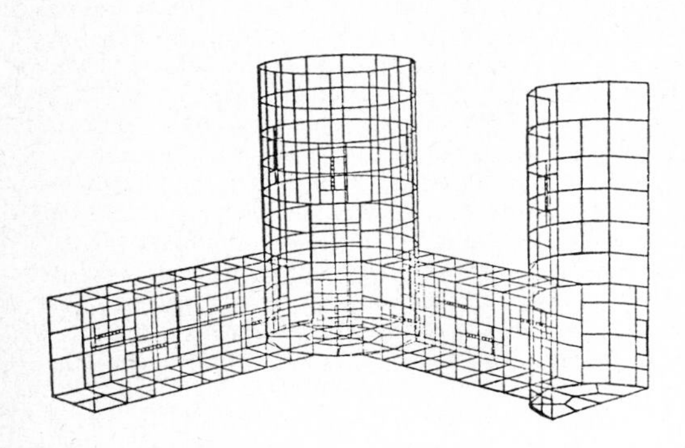

Figure 4 Offshore Platform

The transient example shown in figure 5 illustrates the study of a practical problem with complex time-dependent boundary conditions. The temperature distribution in a turbine disc is sought Although the real structure is axisymmetric, a two-dimensional FEM analysis was provided for comparison purposes [29] and consequently the BEM solution is also two-dimensional. The FEM analysis employs 71 quadratic isoparametric elements and 278 nodes.

The initial temperature of the turbine disc is 295°K and the values of the thermal conductivity density and specific heat are 15 $w/m^{-1}{}^{\circ}K^{-1}$, 8221 kgm^{-3} and 550 $kg^{-1}{}^{\circ}K^{-1}$ respectively. There are 18 different zones along the boundary each with a different set of prescribed nodes for the heat transfer coefficient and the temperature of the surrounding gases and their variation at one of such boundary zones is shown in figure 5. Note that the mathematical representation of the heat transfer coefficient implies the use of mixed boundary conditions of the type $\alpha u + \beta q = \gamma$. No special difficulty is associated with their implementation.

The boundary element discretization employs 90 linear elements in space and time and 106 nodes. (There are 16 double nodes to allow for the discontinuities on the boundary data at the intersection of boundary zones). A stepwise linear variation was prescribed for the boundary temperature. For the boundary flux it was assumed to be linear or quasi-quadratic according to the variation of the heat transfer coefficient (h) and the external temperature within each step. Results (isothermal) at a typical time are plotted in figure 6 for both numerical methods showing excellent agreement.

It is important to point out that the BEM results were obtained using a novel approach of referring the integral equation always to the initial conditions. As the initial conditions are usually everywhere zero this means that one always is solving a boundary only problem, i.e. only boundary integrals need to be computed. This important new idea is described in detail in reference [28]. This technique is most

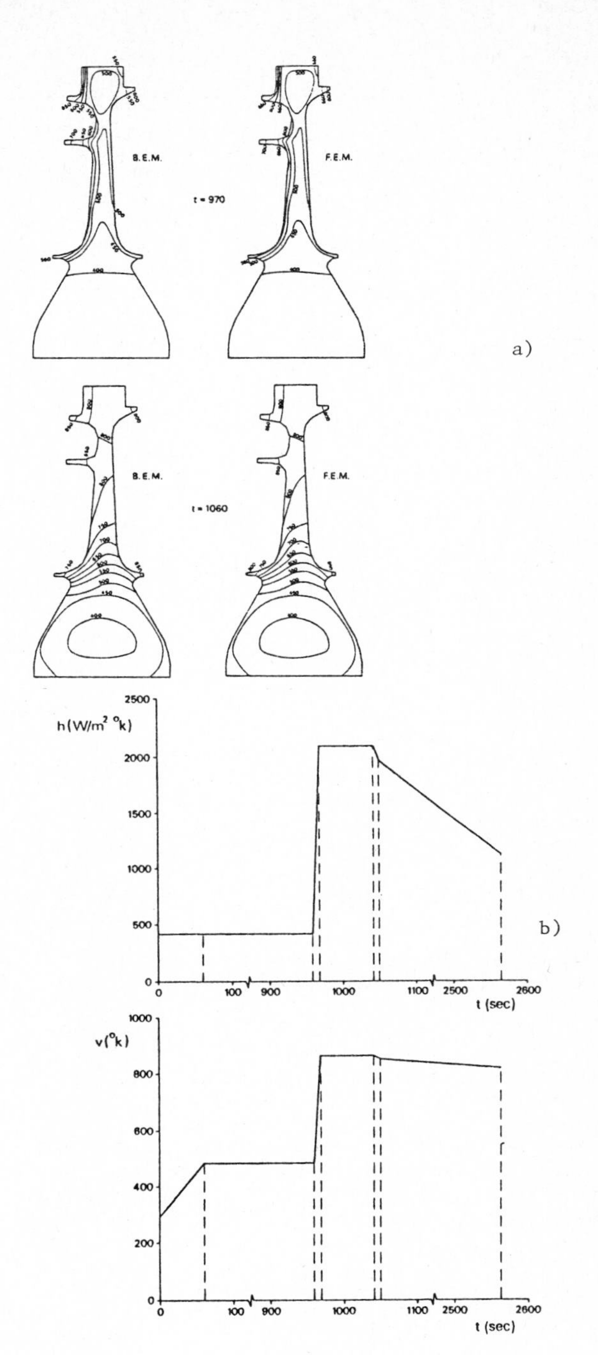

Figure 5 Time Dependent Results for a Turbine Disc a) Isotherms at different times b) Time variation of heat transfer coefficient and temperature of surrounding gas for a typical boundary zone

advantageous for unbounded domains as the engineer does not need to define in any way the extent of the internal zone.

Even more recently the boundary element has

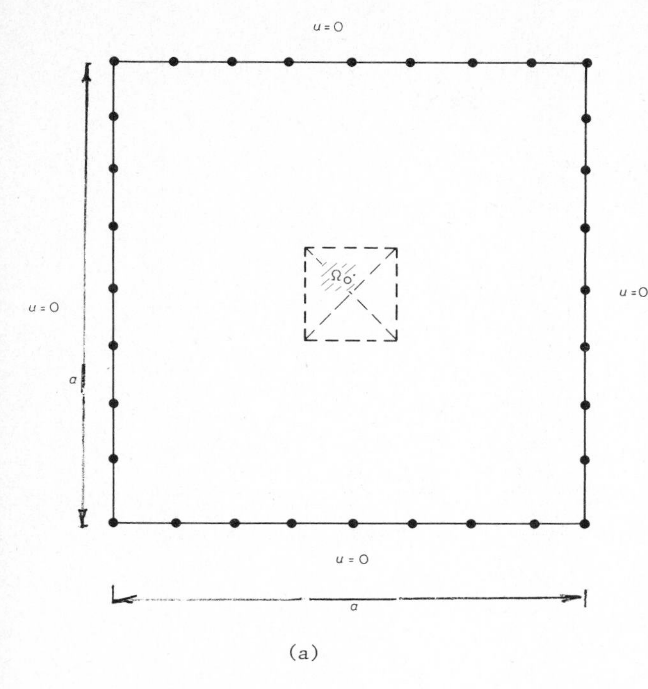

(a)

been applied to solving transient scalar wave propagation problems. These problems especially for two dimensions - are rather complex as they involve the always difficult solution of a hyperbolic equation. References [30][31] deal with this topic in great detail. The problem has been solved in stepwise fashion and using a time and space dependent fundamental solution, with several choices of interpolation functions. The same approach as described previously of referring the integral equation to the initial conditions has been followed, with the result that domain integrals are only required if an initial condition is applied to start the solution, such as is the case over the Ω_o domain shown in the machine described in figure 6, where an initial velocity v_o = c is prescribed.

The boundary was discretized into 32 elements and Ω_o was divided into four cells (see fig. 6). Analytical and BEM results for displacements at certain points and the normal derivative of the displacement were compared.

The values of u and q for $\beta = c\ \Delta t/\ell$, where c is the wave celerity, Δt the time step and ℓ the element length - are plotted in the figure.

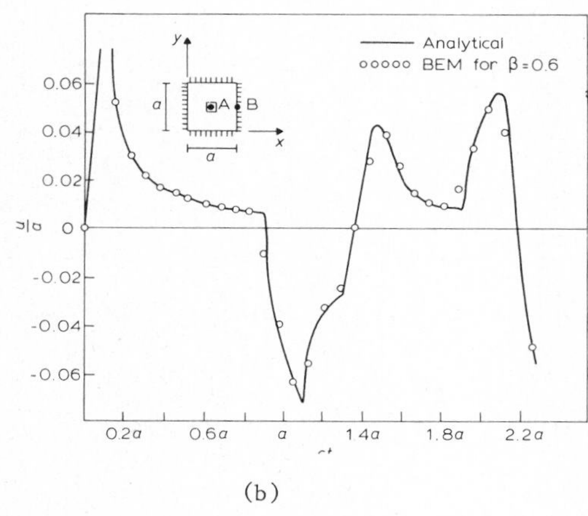

(b)

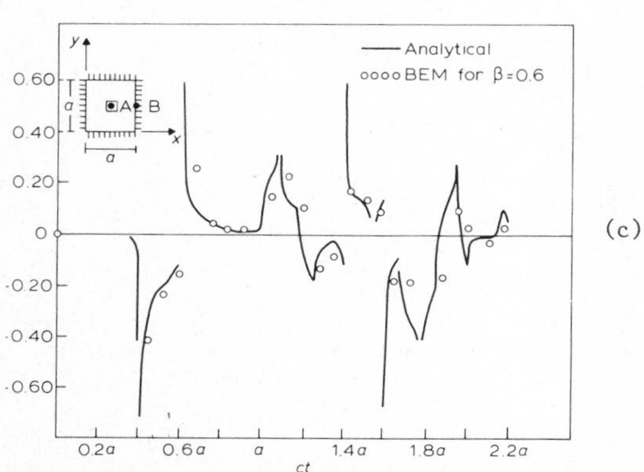

(c)

Figure 6 Results for a Square Membrane a) Machine discretized into 32 elements and 4 cells b) Displacement at point A c) Normal derivative of displacement at point B

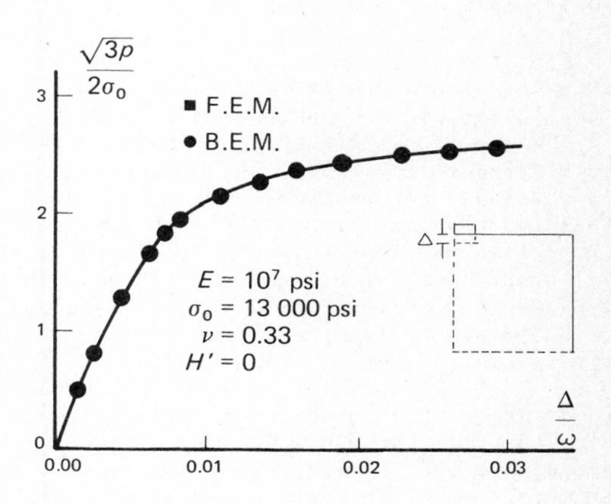

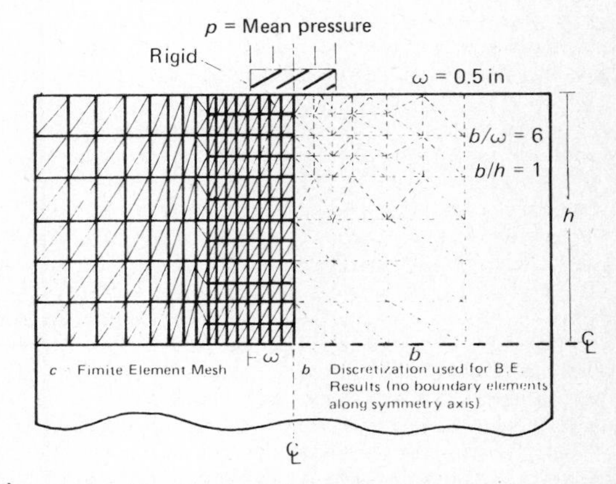

Figure 7 Elastoplastic Analysis of a Square Block a) Geometry of rough punch problems and discretization b) Mean pressure - applied displacement curve for rough punch problem

The agreement for displacements and derivatives is reasonable although better results can be obtained by reducing β (see reference [31] for a discussion of accuracy of results). As a general rule care should be taken in the analysis on the choice of time intervals and boundary discretization in order to avoid contradicting the causality property too far, that is, in each time step waves should not be allowed to travel between nodes far from each other. The simple principle which is required of any numerical solution appears to be frequently violated in practice.

Formulations of boundary elements for plasticity originated early in 1970's but it was not until all the relevant expressions for internal strains and stresses were obtained that the technique could be applied properly. A very comprehensive exposition on how to apply plasticity using boundary elements is given in [32].

In figure 7 the elastoplastic behaviour of a square block compressed by two opposite perfectly rough rigid punches is studied. The problem is analysed under plane strain conditions and the material is considered to be elastic-perfectly plastic obeying von Mises yield criteria.

By using a refined mesh of 274 linear displacement triangular finite elements and 173 nodal points (figure 7) results to this problem are presented by Chen [33][34]. The boundary element analysis was performed with the discretization shown on the RHS of the figure requiring less than one third of the FEM data to run the problem. Notice as well the way in which symmetry conditions have been introduced in the BEM analysis (i.e. no nodes are required on the planes of symmetry).

The indentation process was developed by prescribing the flat punch displacements leading to the average pressure versus displacement curve of figure 7. As can be seen, agreement between the two analyses has been thoroughly obtained, both methods slightly exceeding 4% of the theoretical limit level of $\sqrt{3}p/2J_o=2.5$.

More recently this type of analysis has been extended to solve half-plane problems using the elastic fundamental solution of Melan's. Similarly the Mindlin's solution has been used for three dimensional half space problems. For further details see reference [35].

Comparison has recently been carried out between FE and BE solutions for circular cavities under internal pressure [36]. Lee considered the case of figure 8 for which difficulties arise using finite elements, since it is an inherently infinite problem. As a result the FE outer boundary is taken to be at a distance of ten times the radius of the cavity. The nine noded finite element mesh used is shown in the figure and the material properties were; Young's

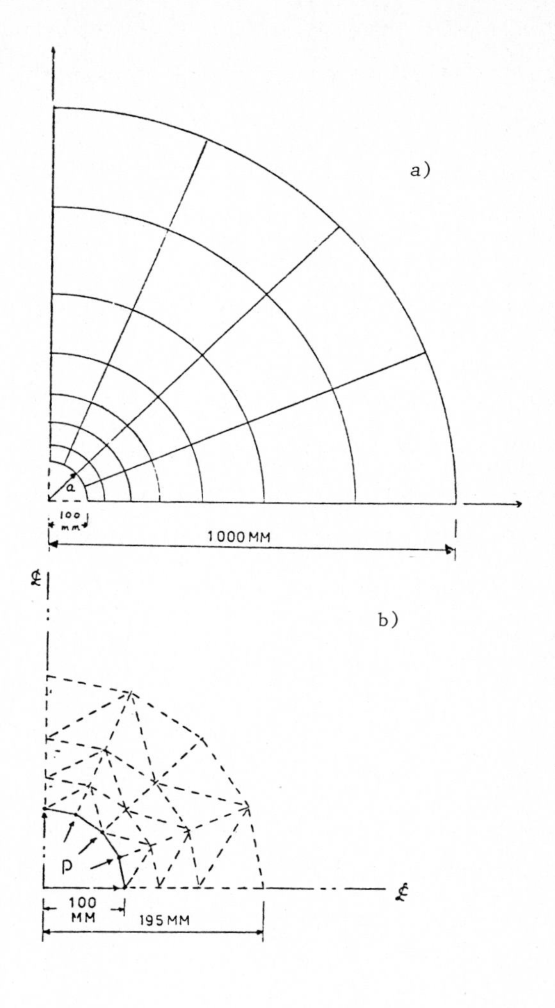

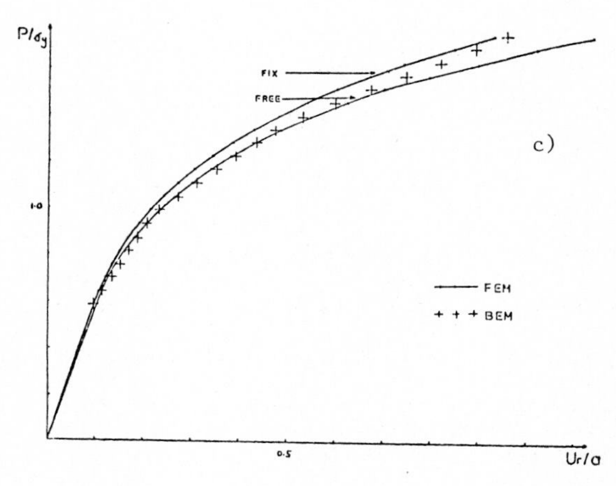

Figure 8 Circular Cavity under Internal Pressure
a) Finite element mesh b) Boundary element mesh c) Pressure - displacement load for elastoplastic behaviour

modulus, E = 7000 kg/m²; Poisson's ratio, ν = 0.2, uniaxial yield stress σ_y = 100 kg/m², strain bending parameter H' = 0.0.

The problem was solved by FEM, first assuming the outer boundary fixed, then free. Theoretically, the radial displacement of the cavity should lie between the displacements found by the two assumptions. Then the problem was solved using BEM for which infinite domains can be properly modelled. The boundary element discretization is shown in the figure and both the FEM and BEM results are presented in the load displacement graph.

From the graph one can notice that for the last increments, boundary element results lie between the two FE cases However, in the early load increments, displacements obtained by BEM are slightly larger than those obtained by the FEM assuming the outer boundary is free. This would be due to the fact that displacements obtained by FEM are always smaller than they should be. When yield zone lies near the cavity, the restraint conditions of the outer boundary have little effect on the radial displacements of the cavity, and the rigidity of the FE dominates the outcome. Further advantages of using BEM can be inferred from table 1. The input problem is especially simplified, the number of unknowns are reduced and substantial amounts of computer time are saved, i.e. the BE technique is very efficient in these cases.

COMPARISON OF BE AND FE SOLUTIONS

Problem	FE or BE	No.of Data Card Input	Total no. of nodes needed	CPU time using ICL2970 (in units)
Circular cavity under	FE (fixed)	201	135	1212.09
internal pressure	FE (free)	194	135	1341.11
	BE	56	20	103.12

Table 1 Comparison of BE and FE Solutions

Other material non-linearities have been studied in reference [37] where viscoplasticity is discussed and in [38] where the problem of non-tension materials is presented. The non-tension problem is especially useful in geomechanics to predict regions of loosening over tunnel roofs and other underground structures. This loosening is usually prevented by covering the tunnel with a thin layer of 'shotcrete' or by prestressing the material surrounding the tunnel. The tunnel shown in figure 9 has been used for a hydroelactric power station and has been studied using finite elements by Valliapan [39]. He analysed as a non-tension problem using 500 constant strain elements and 276 nodes mesh. The figure shows the discretization of this tunnel by the boundary element. Solution for the case of using prestressing forces. A 48 nodes and 39 linear elements boundary mesh was used, the internal cells for initial stress integration are plotted in the same figures. Prestressing using rock bolts has been extensively used in tunnels in place of lining. The prestressing forces are applied at 20 boundary points and 20 internal points which are distributed along a circular path.

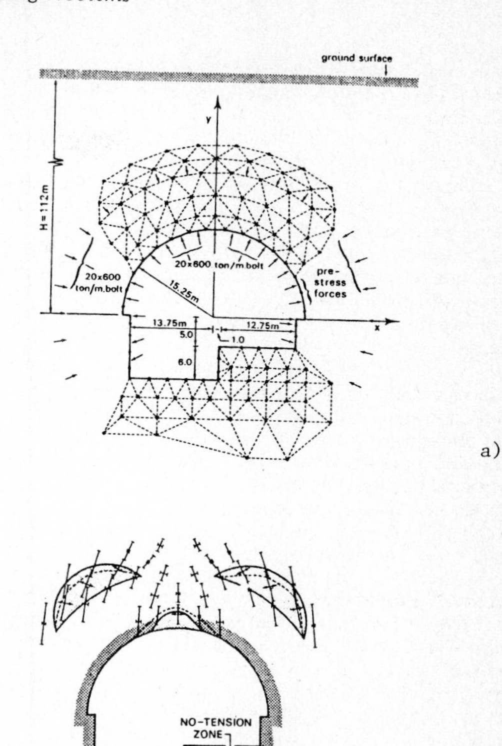

Figure 9 Prestressed Deep Tunnel a) Boundary discretization, internal cells and prestressing forces b) Elastic and non-tension results

The results obtained with the BEM are shown in figure 10 where the initial elastic zone and the final non-tension zone are plotted. As we can see the tension free zone in the roof are very small due to the prestressing effects. These results are in close agreement with the FE solution of reference [39].

3. CONCLUSIONS

This paper presents some important results recently obtained using BEM. Several topics related to boundary elements require further investigation. Boundary element method implies an approximate technique by which the problem dimensions have been reduced by one or in the case of time dependent problems by two. The problem becomes a boundary problem and the surface of the domain can be divided into elements over which some approximate techniques can be applied. What differentiates BE from classical integral equations is its emphasis on elements and compatibility as well as its reliance on general principles such as weighted residuals. This is important as some classical boundary element integral solutions suffer from lack of accuracy and convergence due to poor representation of the geometrical shapes. These problems and the relationship of variational techniques, weighted residuals and boundary integral equations appears not to have been sufficiently investigated.

Although current boundary element practice is heavily dependent on using fundamental solutions based on solving the governing equations under concentrated source and no boundary conditions, it is of the utmost importance to point out that other solutions can be used as well. Analytical functions which have been obtained over a close domain with certain boundary conditions could also be used as fundamental solutions provided that our domain of interest is within the domain of the analytical solutions. This of course can be easily arranged by a change of dimensions. Once these ideas are better understood we shall be able to solve many complex plate, shell and other problems using the wealth of solutions already available as approximate fundamental solutions.

The general field of time dependent and non-linear problems requires further investigation. The method seems well suited to solve problems in time and space. BE is now accepted as a valid and efficient technique for solving parabolic problems and has now started to be applied for the solution of hyperbolic equations. More results however need to be carried out to determine the accuracy and efficiency of the technique for these problems.

In conclusion the BEM method is becoming increasingly popular with practicing engineers as well as other researchers for solving a large variety of practical problems.

REFERENCES

[1] Reissner, E., A Note on Variational Principles in Elasticity, Int. J. Solids and Structures, 1 (1965), 93-95 and 357.

[2] Pian, T.H.H. and Tong, P., Basis of Finite Elements for Solid Continua, Int. J. Numerical Methods Engng., 1, (1969) 3-28.

[3] Washizu, K. Variational Methods in Elasticity and Plasticity, (2nd Edition, Pergamon Press, New York, 1975).

[4] Muskhelishvili, N.I., Some Basic Problems of the Mathematical Theory of Elasticity, (P. Nordhoff Ltd., Groningen, 1953).

[5] Mikhlin, S.G. Integral Equations (Pergamon Press, New York, 1957).

[6] Kupradze, O.D., Potential Methods in the Theory of Elasticity, (Daniel Davey and Co., New York, 1965).

[7] Smirnov, V.J., Integral Equations and Partial Differential Equations, in A Course in Higher Mathematics, IV (Addison-Wesley, 1964).

[8] Kellog, P.D., Foundations of Potential Theory, (Dover, New York, 1953).

[9] Jaswon, M.A., Integral Equation Methods in Potential Theory, I, Proc. R. Soc., A., (1963) 273.

[10] Symm, G.T., Integral Equation Methods in Potential Theory, II, Proc. R. Soc., A., (1963) 275.

[11] Massonnet, C.E., Numerical Use of Integral Procedures, in Stress Analysis, Zienkiewicz, O.C. and Holister, G.S. (Eds.), (Wiley, 1966).

[12] Hess, J.L. and Smith, A.M.O., Calculation of Potential Flow about Arbitrary Bodies, Progress in Aeronautical Sciences, 8, Kuchemann, D. (Ed.) (Pergamon Press, 1967).

[13] Cruse, T.A., Numerical Solution in Three-Dimensional Elastostatics, Int. J. Solids Structures, 5, (1969) 1259-1274.

[14] Brebbia, C.A. and Tottenham, H. (Eds.) Variational Methods in Engineering (2 vols) (Southampton University Press, England, 1973, reprinted in 1975).

[15] Brebbia, C.A., Weighted Residual Classification of Approximate Methods, App. Math. Mod., 2, (1978), September.

[16] Brebbia, C.A., The Boundary Element Method for Engineers (Pentech Press, London, Halstead Press, New York, 1978).

[17] Brebbia, C.A. and Walker, S., Boundary Element Techniques in Engineering, (Butterworths, London, 1979).

[18] Brebbia, C.A., Telles, J. and Wrobel, L., Boundary Elements - Fundamentals and Applications in Engineering, (Springer Verlag, Berlin, 1983).

[19] Brebbia, C.A., (Ed.) Recent Advances in Boundary Element Methods, Proc. 1st Int. Seminar on Boundary Element Methods, Southampton University 1978 (Pentech Press, 1978).

[20] Brebbia, C.A., (Ed.) New Developments in Boundary Element Methods, Proc. 2nd Int. Seminar on Boundary Element Methods, Southampton University, 1980,(CML Publications, Southampton, 1980).

[21] Brebbia, C.A. (Ed.) Boundary Element Methods, Proc. 3rd Int. Seminar on Boundary Element Methods, California, 1981, (Springer Verlag, Berlin 1981).

[22] Brebbia, C.A. (Ed.) Boundary Element Methods in Engineering, Proceedings 4th Int. Seminar on Boundary Element Methods Southampton University, 1982, (Springer Verlag, Berlin 1982).

[23] Brebbia, C.A. (Ed.) Progress in Boundary Element Methods, Vol.1 (Pentech Press, London, Halstead Press, NY, 1982).

[24] Brebbia, C.A. (Ed.) Progress in Boundary Element Element Methods, Vol.2, (Pentech Press London, Springer-Verlag, NY. 1983)

[25] Danson, D. A Boundary Element Formulation of Problems in Linear Isotropic Elasticity with Body Forces, in Boundary Element Methods, Brebbia, C.A. (Ed.), (Springer Verlag, Berlin, 1981).

[26] Danson, D., Brebbia, C. and Adey, R., The BEASY System, Advances in Engineering Software, 4, No.2, (1982), 68-74.

[27] Danson, D., BEASY: A Boundary Element Analysis System, Proceedings of 4th Int. Seminar on Boundary Element Methods, (C. Brebbia, Ed.) Southampton, 1982 (Springer-Verlag, Berlin, 1982).

[28] Wrobel, L. and Brebbia, C., Time Dependent Problems, in Progress in Boundary Element Methods, Vol.1, (Pentech Press, London and Halstead Press, NY 1982).

[29] Rolls Royce, Derby, Private Communication.

[30] Mansur, W. and Brebbia, C., Formulation of the Boundary Element Method for Transient Problems governed by the Scalar Wave Equation, Applied Mathematical Modelling 6, August (1982) 307-311.

[31] Mansur, W. and Brebbia, C., Numerical Implementation of the Boundary Element Method for Two Dimensional Transient Scalar Wave Propagation Problems, Applied Mathematical Modelling, 6, August (1982) 299-306.

[32] Telles, J.C.F. and Brebbia, C.A., Plasticity, in Progress in Boundary Elements Brebbia, C.A. (Ed.) (Pentech Press, London, Halstead Press, NY 1981).

[33] Chen, A.C.T. and Chen, W.F., Constitutive Equations and Punch Indentation of Concrete, Proc. Am. Soc. Civ. Engng., J. Eng. Mech Div., 101, (1975) 889-906.

[34] Chen, W.F., Limit Analysis and Soil Plasticity, (Elsevier Scientific Publishing Co., Amsterdam, 1975).

[35] Telles, J.C.F. and Brebbia, C.A., Boundary Element Solution for Half-Plane Problems, Int. J. Solids Struct., in press, 1981.

[36] Lee, K-H., A Comparison between Finite Element and Boundary Element Methods for the Solution of Elasto-Plastic Problems, M.Sc. Dissertation, Civil Eng. Dept. Southampton University. December 1981.

[37] Telles, J.C.F. and Brebbia, C. Elastic-Viscoplastic Problems using Boundary Elements, Int. J. Mech. Sci., 24, (1982).

[38] Venturini, W. and Brebbia, C.A., The Boundary Element Method for the Solution of No-Tension Materials, in Boundary Element Methods, Brebbia, C.A. (Ed.) (Springer Verlag, Berlin, 1981).

[39] Valliappan, S. Non-linear Stress Analysis of Two Dimensional Problems with Special Reference to Rock and Soil Mechanics, Ph.D. Thesis, University of Wales, March (1968) Swansea.

III. SIMULATION TOOLS – Hardware

Simulation in Engineering Sciences
J. Burger and Y. Jarny (eds.)
Elsevier Science Publishers B.V. (North-Holland)

GRAPHIC MODEL BUILDING SYSTEM -GMBS-

Yoshikazu Yamamoto and Mats Lenngren

Department of Electrical Engineering,
Linkoping University
Linkoping, Sweden

An interactive graphic model building system for discrete event simulation is presented. The GMBS is a completely computer aided software tool which intends to be easy to use even for unexperienced users. The structure and world view of the GMBS are described. We introduce a new concept in which a model is divided into two specifications, that is, structural relations between system components and description of the logical behavior of each component. We also believe that our system can be used for well structured documentations for models.

1. INTRODUCTION

Interactive programming systems have been remarkably improved in recent years[1,2]. However, in case of discrete event simulation, batch processing is still used because of its long execution time. Of course, it is necessary to execute a model program many times in order to obtain statistically significant results. Considering a model building and debugging phase, however, it is certainly convenient if we can use an interactive system. We believe that there are quite large differences between model developing and execution phases in essence. Thus, we divide the whole simulation process into two major phases, that is, the model building phase and the execution phase. In the former phase we utilize an interactive graphic facility, and the latter phase is to be achieved by a special purpose simulation machine.

We have already implemented one experimental parallel simulation machine for efficient execution[3,4,5]. Through the experiments of that project, we learned that the current programming languages for simulation are not sufficient to make model programs in many senses. Thus, we propose in this paper the Graphic Model Building System (GMBS) which is based on a new world view and includes total support tools for accomplishing simulation experiments.

Several papers are already published about interactive model building[1,6] or graphic model representation[7]. However, they are either based on one of the simulation languages or able to treat only a part of the whole simulation process. Our approach is quite different in this aspect. The GMBS is designed to cover the entire simulation process as reported in [8,9,10], and intended to be independent of specific simulation languages as far as possible by making the best use of the graphic capabilities.

The GMBS has several objectives as follows. First, the system should be user friendly and easy to use even for beginners. Secondly, the concept of modeling is as simple as possible, that is, the user should be able to make a model and get results without previous knowledge of specifc simulation languages. The support for structural modeling is the third object, which means the GMBS should support both top-down and bottom-up approaches. Finally, the GMBS assumes a special purpose simulation machine which should take charge of execution of the model program automatically generated by the GMBS.

The remaining paper is divided into six sections. The world view of the GMBS is presented in the next section. The third and fourth sections are provided for descriptions of the software and hardware structure respectively. A small example for structural modeling is presented in the fifth section. The user interface including some features is given in the sixth section. We present the concluding remarks in the last section.

2. WORLD VIEW

There are many simulation languages available and they all have their own world views. The world views applied for discrete event modeling and simulation may be classified into the following three: event oriented, activity oriented and process oriented. When we utilize one of the simulation languages for modeling and programming, it is necessary to transform the inherent concurrency of the objective system into a sequential representation by means of some trick. The form of this transformation is different depending on which language is being used, and it is difficult to do it correctly. It is not too much to say that the modeling process is defined based upon the language you utilize.

From the above consideration we adopt a world view as simple as possible and intend to free the user from the above mentioned mapping problem. We treat an objective system as a pair of physical or logical structural relations between system components and logical descriptions of the behavior of each component. The first part is defined as a graph which consists of any number of circles, each of them representing one or a group of system components, and arrows connecting them. The logical specification of a component behavior may be described by combining predefined procedures or functions and/or user defined procedures.

The GMBS is a system which treats only discrete event simulations so far. The world view of the GMBS is similar to that of GPSS[11], but we do not introduce a large number of blocks to represent actions of a system as in GPSS. Basically, static elements which are used in the GMBS are of only four kinds, that is, Generator, Terminator, General-Server and Submodel. Instead of the transaction of GPSS, we use the term 'packet' to represent dynamic elements of a system. Packets are generated by a Generator and flow through a network of elements and are finally removed from the model at a Terminator. It is, of course, not necessary for packets to correspond to physical components which means that they can be used as signals or to carry information. Each packet may have any number of attributes that are used for many purposes, for example, they may specify some characteristics of a packet and may keep statistics. Integer, real, Boolean, character, string and pointer modes are available for attributes.

Major components of a system can be represented as General-Servers. A General-Server consists of four sequential parts, Packet selection, Queue, Server and Dispatching mechanism. Logical behavior of a General-Server can be specified by giving values to parameters of predefined procedures or defining procedures in the Logic specification part of the GMBS. It is worth noticing that Generator and Terminator are considered as a special General-Server, so that we can also specify logical behavior of them as well as a General-Server. A submodel conceptually corresponds to a subsystem and it can be defined as a set of any number of elements including submodels. It can be also used as a subroutine.

3. SOFTWARE STRUCTURE

The entire software system of the GMBS shown in Figure 1 consists of the following seven submodules: (1) Monitor, (2) Terminal interface, (3) Structural model handler, (4) Logic specification handler, (5) Sequential object program generator, (6) Parallel object program generator and (7) Experimentation support. We briefly discuss each of these submodules in this section.

The GMBS modules are written in Simula except the terminal interface module. The terminal interface module is mainly written in Pascal and the small part of it is written in MACRO.

Though the structure of our system is similar to the model-based simulation (MBS) software system [9,10], our approach is purely based on a software engineering consideration in contrary to the Ören's approach which was based on system theory. The GMBS does not contain a model database, since defined structural models and corresponding logical specifications are saved as ASCII text format files and the management of these files is left to the operating system.

3.1 Monitor

The monitor manages all commands issued by a user and calls an appropriate procedure. All commands in the GMBS are formed as a tree called the command structure (CS) which is generated from a command structure specification (CSS). All nodes of the CS contains either a next level command menu or an indication to a corresponding procedure. Since the leaf nodes of the CS should have no lower command levels, they should be procedure indications. Corresponding to the CS there is a help structure (HS) which also has a tree structure and is generated from a help structure specification (HSS). The HS is used for showing related help messages according to the node in the CS where a user is currently

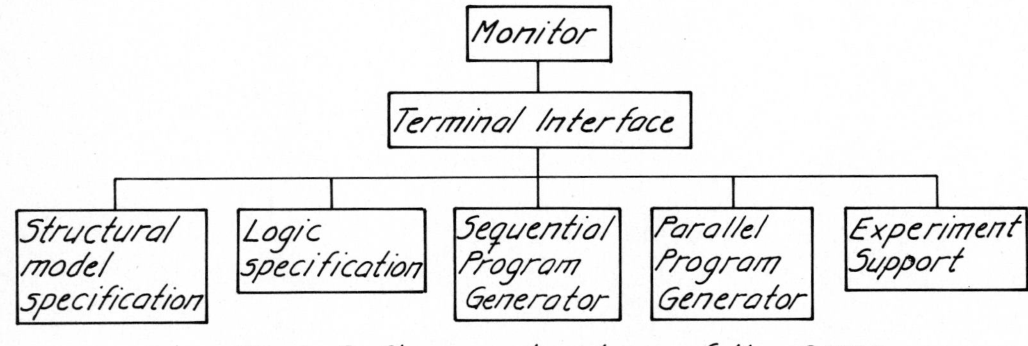

Fig. 1 Software structure of the GMBS

working. An interactive software system such as the GMBS can be well structured by utilizing a hierarchical program structure. Our approach, which is based on the interactive dialogue of a user, can be an excellent software tool because of the fact that the procedures which handle commands in the CS may also have a tree structure from the natural consequence of the CS.

The monitor also has the responsibility for communicating with the operating system, especially for file handling.

3.2 Terminal Interface

We need to use two types of terminals for the GMBS, that is, a conventional character display and a graphic display. To handle these two types of terminals in the same manner from programs, we implement a set of procedures and functions as a terminal interface. This program has the responsibility for sending instructions either to the character display or to the graphic display alternatively.

According to this module, we can use both displays without special treatment from other modules and keep the GMBS transportable to another hardware systems which have different types of displays. In addition to these capabilities, output to a hard copy terminal is also available. The current implementation supports the output to the HP7580 of a single window or a picture (described in the next subsection).

3.3 Structural Model Handler

This module has the responsibility of making up a structural model by utilizing a graphic display and conventional character display terminal. Four different basic elements of the GMBS are represented as circles of different colors.

We call a screen of the graphic display a window. A structural model itself may be a tree structure, and each node of the tree is called a picture. Since a picture may consist of up to 2**15 x 2**15 pixels, it may be regarded as a virtual picture. One picture may have any extent up to the limit, and a user can view any window by issuing a 'V' command. Since the concept of the window is introduced due to the limited size of a display screen, one can scroll the currently viewing window in any size and any direction.

A picture may contain any number of basic elements which are connected to each other by arrows. A General-Server and a Submodel may have any number of input and output lines. A Generator can only have output lines and a Terminator should have no output line.

The tree structure of the model is automatically generated by defining a submodel as an element. Though a submodel may be used as a subroutine, we do not allow a user to define it recursively because it corresponds to a subsystem and an objective system should have some form of concrete shape. Therefore, a tree of pictures defines a single model and there is at least one lowest level picture which has no submodel element. Each submodel can be defined as a single picture which of course may have any number of basic elements including submodels.

By using the above hierarchical tree structure, a user can follow both a top-down and a bottom-up approach to model a system. In case of the bottom-up approach, the user can make a submodel picture first and then save it as a file.

The structural model handler is described in greater detail in section five.

3.4 Logic Specification Handler

Since a Generator and a Terminator can be regarded as a subset of a General-Server as mentioned above, it is enough to explain the logical specification for the General-Server. The General-Server may be divided into the following four sequential parts: (1) Packet selection, (2) Queue, (3) Server and (4) Dispatching mechanism. A user can select a procedure or a function from the predefined sets for each of these parts and define parameter values. The user can leave one or more parameters as variables if the values for these parameters are not constant. These values for variables may be given in the experimentation support part of the GMBS.

3.4.1 Packet selection

If the General-Server has only one input line, the possible specification is a Boolean expression which decides whether the incoming packet is acceptable or not. The Boolean expression may consist of attributes of the packet and predefined state variables. In case of multi-input General-Server, one can define the priority order which is used to select an appropriate packet among the input lines in case of simultaneous arrivals.

3.4.2 Queue

The queue disciplines such as FIFO, LIFO and ranked by some attributes values, are available. Statistics for the queue, for example (current, maximum and average) queue length, (maximum and average) waiting time etc., are automatically updated and can be referred to as state variables.

3.4.3 Server

One can define the capacity of the server, distribution function of the service time, interrupt handling and blocking handling which describes the process when a packet can not enter the next element.

3.4.4 Dispatching mechanism

If the General-Server has more than two output lines, this part specifies the selection procedure. The selection rule includes random, dependent on attribute value and first found available receiver.

In addition to the above predefined specifications, we are preparing a more flexible alternative to define the logical behavior of the element. Each of these four parts or the entire logic of the element can be defined as a dimensional flowchart[12,13,14]. Though it is necessary to refine the dimensional flowchart for generating an executable code, the dimensional flowchart itself is independent of a specific language with the exception of the final refinement level. Thus we can model a system almost completely independent of any languages.

The available attributes of packets and state variables of the General-Server are checked by this module as far as the GMBS knows. The attributes of packets are treated as formal parameters if the General-Server is defined in a submodel picture and the GMBS does not have any knowledge of the actual input packets. Binding between formal and actual parameters is discussed in section 5.

3.5 Sequential Model Program Generator

We noticed from the KDSS project[5] that it is necessary to execute a model program sequentially at least for debugging. This module generates a DEMOS program[15] according to the given structural model, the logic specification and the experimental specification. DEMOS is defined as an external class of Simula and has many convenient tools for debugging such as a trace facility and automatic statistics gathering. Since the DEMOS itself is written in Simula, we can also use SIMDDT[16].

3.6 PARALLEL OBJECT PROGRAM GENERATOR

This module generates an object program for a parallel simulation machine such as KDSS. The module is not implemented yet since we have no way to access to KDSS now.

3.7 EXPERIMENT SUPPORT

This module has two purposes, that is, specification of a single experiment and support for the experimental design including output analysis of the entire simulation experiment. The former part is mainly used for the debugging or verification stage along with the sequential program. It has a similar external view as the experimental frame specification[17]. The latter part, on the other hand, is used for generating necessary execution sequences that can be executed on the parallel machine. The specification includes several alternatives for analyzing and displaying the simulation results.

4. HARDWARE STRUCTURE

We have implemented the GMBS on a hardware structure shown in Figure 2. Almost all software modules except the terminal interface are placed on a host computer (DEC-10 or 20) and the terminal interface is implemented on a front-end computer (LSI 11/23). Both the standard alphanumeric display terminal, the plotter and the color graphics (DEC VSV11) are connected to the front-end computer. A simple protocol is used for communication between the host and the front-end. This means that the Simula software system running on the host can be regarded to be kind of device independent. Since no parallel simulation machine is available now, the interactive execution of a model program has not been realized. The lack of the hardware, however, does not mean that the implementation of a parallel object program generator is impossible. We have already a KDSS simulator written in Simula and can generate data for the simulator completely independent of other modules of the

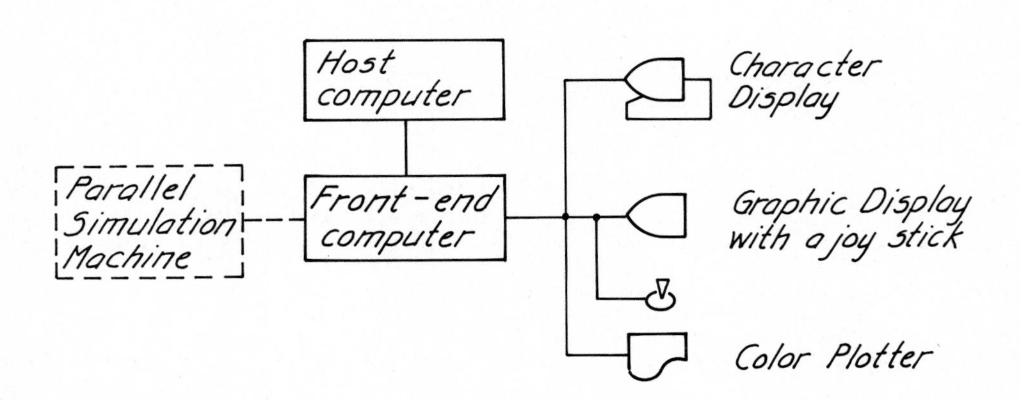

Fig. 2 Hardware structure

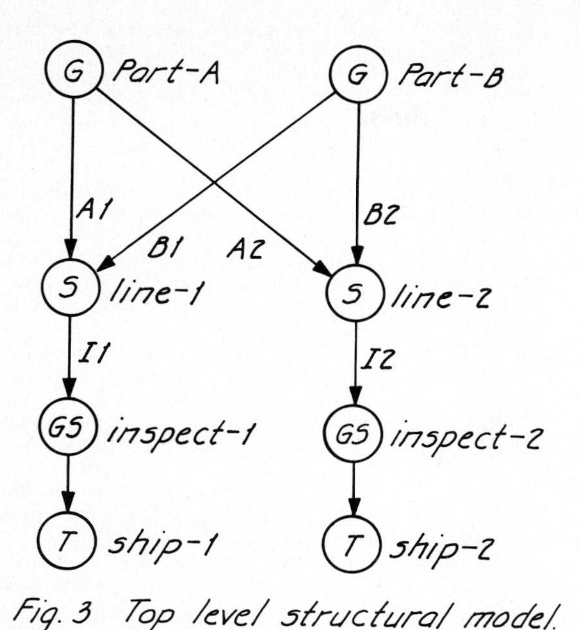

Fig. 3 Top level structural model.

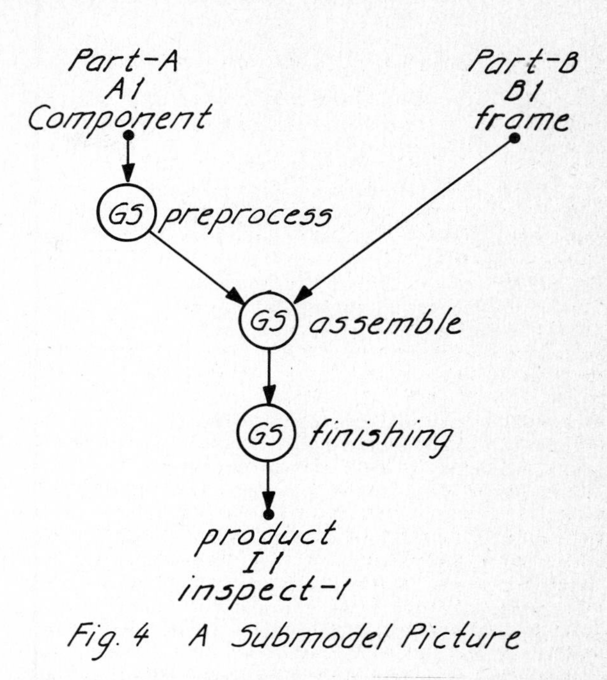

Fig. 4 A Submodel Picture

GMBS. A VAX 11/780 version of the GMBS will also be available.

As mentioned in the former section, a fixed device for the graphic display is out of the question due to the transportable structure of the GMBS software.

5. STRUCTURAL MODEL

A small example is provided here to explain the structural model of the GMBS. Consider a simple job shop which has two assembly lines, two inspection sections and two shipment yards. A product of the job shop consists of two parts, say part-A and B, which are supplied from different sources. The both assembly lines have the same structure and consist of a preprocess of the part-A, an assembly process of two parts and a finishing process.

We can make a structural model of the job shop as shown in Figure 3 by utilizing two submodels, each of them representing an assembly line. Although the actual display and the hard copy can show the types of elements by different colors, we put abbreviated names of types in circles. G, GS, S and T in the circle represent Generator, General-Server, Submodel and Terminator respectively. Since the two submodels have the same structure, we only need to define one structural model as shown in Figure 4. Notice that Figure 4 is defined as a submodel picture called 'line' for the submodel named 'line-1' in Figure 3. Two dots on the top of Figure 4 indicate two input ports to 'line'. Each of the dots has three names, that is, a name of a source element, a name of an input line and a formal name of this input port. The first two names are given automatically when you define a submodel picture, so that you can put a formal name to the input port while keeping the relation between the input line and the input port. One dot at the bottom of Figure 4 indicates the output port. It has also three names but each of them shows a formal name, an output line name and a destination element name from the top respectively.

On the other hand, when utilizing the submodel picture 'line' as a subroutine from 'line-2' in Figure 3, the GMBS shows only the formal name of each input and output port and requests corresponding source and input/output line names.

6. USER INTERFACE

As discussed before, we have two types of I/O devices for interaction between a user and the GMBS. The command menus and the selected command are exchanged via a character display terminal. The pictorial output and an indication of position in a picture are done by a graphic display terminal and a joy stick which is used to move a crosshair cursor on the picture. All commands are designed so that it is possible to identify the first one or two characters. A user can also use a guiding word facility which is invoked by depressing ESC key.

Each node of the CS has the corresponding node in the HS so that the help message is displayed by issuing a command '?' or 'H'. Another important feature of the user interface is an escape facility from a command. It can be activated by '^' command and immediately return to the specified point, usually return to the upper level node in the CS.

Changing of command names or even the command structure is achieved fairly easily by means of the CS and HS structures.

7. CONCLUDING REMARKS

The structural model seems quite useful not only for understanding the structure of a system but also for documentation of a model. By means of the hierarchical structures of both the structural model and logic specification by a dimensional flowchart, one can grasp a model or a system at any level.

The GMBS can also be a useful tool for simulation education since it is independent of any languages so that one does not need to learn troublesome details of a simulation language.

We use the DEC VSV11 color graphics with 4 bits per pixel and 512x512 pixels in resolution, and a conventional character display terminal. It is desirable to use a large color display for mixed presentation of graphics and alphanumeric information, say A4 or A3 size, with high resolution since it is cumbersome to turn our eyes between two displays.

ACKNOWLEDGEMENT

The authors would like to be grateful to Professor Harold W. Lawson Jr. who gave us the chance to do this project and for his useful suggestions. We are also thankful to other members of Telecommunication and Computer Systems group for their contribution on implementing the GMBS.

This project was partially supported by Fukuzawa Foundation.

REFERENCES

[1] Davies, N.R., Interactive Simulation Program Generation, in Zeigler, B.P. et al. (eds.), Methodology in Systems Modelling and Simulation (North-Holland, Amsterdam, 1979) 179 - 200.

[2] Kahan, B.C., Conversational GPSS, Proc. of the 1981 UKSC Conference on Computer Simulation, Harrogate England (May 1981) 108 - 112.

[3] Nakagawa, T. et al., A Multi-Microprocessor Approach to Discrete System Simulation, Proc. of 20th IEEE COMPCON (1980) 350 - 355.

[4] Yamamoto, Y. et al., A Time Advancing Algorithm for Distributed Simulation, Proc. of 12th Annual Pittsburgh Conference on Modeling and Simulation (May 1981).

[5] Yamamoto, Y., On Distributed Simulation, Tech. Report, Institute of Information Science KEIO University (March 1981).

[6] DuBois, D.F., A Hierarchical Modeling System for Computer Networks, Performance Evaluation Review, Vol.11, No.1 (1982) 147 - 155.

[7] Törn, A.A., Simulation Graphs: A General Tool for Modeling Simulation Designs, Simulation Vol.37, No.6 (1981) 187 - 194.

[8] Ören, T.I., Computerization of Model-Based Activities: A Paradigm Shift and New Vistas, in Ören, T.I., et al. (eds.), Simulation & Model-Based Methodologies : An Integrative View (North-Holland, Amsterdam (Forthcomming)).

[9] Zeigler, B.P., System Theoretic Foundations of Modeling and Simulation, same as [8].

[10] Ören, T.I., Concepts for Advanced Computer Assisted Modelling, same as [1], 29 - 55.

[11] Gordon, G., The Application of GPSS V to Discrete System Simulation, (Prentice-Hall, Englewood Cliffs, 1975).

[12] Witty, R.W., Dimensional Flowcharting, SOFTWARE-PRACTICE AND EXPERIENCE, 7 (1977) 553 - 584.

[13] Jönsson, A., Dimensional Flowchart - Some Loose Ends, Internal Report, Dept. of Elect. Eng., Linköping Univ. (Sept. 1981).

[14] Jönsson, A., DIMED Dimensional Flowchart Editor Manual, Internal Report, Dept. of Elect. Eng., Linköping Univ. (Nov. 1980).

[15] Birtwistle, G.M., A System for Discrete Event Modeling on Simula (The Macmillan Press Ltd., London, 1979).

[16] Birtwistle, G.M. et al., DECsystem-10 SIMULA Language Handbook Part II (Swedish Natioanl Defence Institute, Stockholm, 1982).

[17] Rosenblit, J., An Experimental Frame Specification Module, same as [8].

Simulation in Engineering Sciences
J. Burger and Y. Jarny (eds.)
Elsevier Science Publishers B.V. (North-Holland)

THE SOLE SIMULATION PACKAGE IN PASCAL

J.E. Rooda and S. Joosten

Department of Mechanical Engineering
Twente University of Technology
Enschede, the Netherlands

This paper describes the Modular PASCAL version of the SOLE simulation package. We will start to discuss very briefly why a Modular PASCAL version was developed. Then the Modular PASCAL environment will be discussed. Special emphasis will be given to the modular approach and the orthogonal structure of the system. An overview is given on the SOLE system, as it is implemented in Modular PASCAL on PDP/LSI 11 micros of Twente University of Technology.

INTRODUCTION

From 1980 the SOLE simulation package has been developed [Rooda, 1982]. Written in SIMULA and running on DECsystem10 of Twente University of Technology (THT) it offered a friendly environment for mechanical engineers to model their transport and production systems fast and efficiently. However, in the every day practice of transport and production, large computers capable of running SIMULA are not so very common. That is why we considered to adapt SOLE for small computers. With this goal in mind, we have been looking for the right software tools to reprogram the entire package. After studying languages like Algol68, Ada, Forth, and Pascal, the final answer was found at home. At Twente University of Technology, Modular PASCAL [Bron, 1981] has been developed during the past few years.
This paper is built up as follows. After a small philosophical discussion, some attention will be paid to the conception of systems. Then, Modular PASCAL, which is a superset of standard PASCAL will be given the attention it deserves. We will also discuss aspects of programming ease, and how Modular PASCAL helps. Notions as orthogonality and modularity are important in that discussion.
Then we will discuss the basic simulation elements in SOLE: dead and living elements. As SOLE is a process oriented simulation language, we will have a discussion on processes. Particular attention will be paid to the eventlist algorithm. In the end the reader will be confronted with an example, so as to get an impression of SOLE's capabilities.

PHILOSOPHY

In order to understand why decisions have been taken as they are, it is useful to know the system philosophy. We have compressed it in four statements, which must not be seen as dogmas, but as fountains from which thoughts spring.

1. At the beginning of a creation there is philosophy.

This is used by us as an axiom. As a result, we should try to think first, and create later.

2. The thinking frame of the human species is no larger than ten square foot.

It is impossible for human beings to overlook large programs, to comprehend and master them fully. This is why it is so important to be able to subdivide programs into pieces that can be fully understood. This, we think, is essential to a good programming practice.

3. A good algorithm is a trivial algorithm.

No dirty tricks may be used to gain a little time or space. Utmost simplicity is the greatest virtue of an algorithm. It makes programs comprehensible for future programmers, when the original designers have long gone. Furthermore, it enables us to prove the correctness of an algorithm.

4. Man is made to create beauty.

This statement we cannot really explain or defend. Yet we think that programs must be as beautiful as possible, however subjective this may be. Programming is an art, and artist are creators of beauty [Neumann, 1982].

MODULAR PASCAL

Recently Modular PASCAL has become available at Twente University of Technology [Dijkstra 1979; Prins, 1980; Bron, 1981; Joosten, 1982; v. Rossem, 1982]. It has been developed at the department of Computer Science under the supervision of prof. C. Bron. It consists of an operating system, editors, a compiler, a file system and some utilities. At the moment Modular PASCAL runs on PDP/LSI 11 computers. As all software is written in modular PASCAL, portability is very high. A Motorola 68000 version and an M 6809 version have been completed. Modular PASCAL on DEC10 will be ready by the time you read this.

As one might expect, Modular PASCAL programs can be built up in modules. One possible way is to build up a program as one module. It is then similar to an ordinary PASCAL program. In the top-down analysis, one encounters usually several tasks. These tasks can be isolated in modules. Some people think that a module is similar to the SIMULA class concept. This is not true, as the class is very helpful in the bottom-up construction of programs, and modules are used in the top-down analysis. We believe that the two can even be fruitfully combined.

The size of a module should be small enough for a human being to be able to overlook it. In programming this appears to be of great advantage, because a programmer is concerned with only one task at a time, in proportions that are easy to cope with. This results in relatively bug free programming, and a greater programming ease.

Modules are compiled seperately. In changing existing programs, usually only altered modules have to be recompiled.

At runtime, the modular structure appears to be useful too. It is not necessary for all the modules to be in core at the same time. This means that we do no longer suffer from memory shortage, because it is transformed into time usage. This facility makes Modular PASCAL quite suitable for running large programs on small computer systems. As simulation of discrete events in practice is usually quite large, Modular PASCAL is precisely what is necessary to bring our simulation out of the experimental environment on large university computers to the practical environment of transport and production on industrial computers.

Module swapping is implemented quite cheaply in the modular PASCAL environment. From the user, no effort at all is asked. He may only see a somewhat less predictable runtime behaviour, when his program is large. This is improved by an optimal exploitation of locality.

ORTHOGONALITY

Working with Modular PASCAL we have tried to achieve maximum orthogonality. This means that two modules have no interaction with each other. Algorithms must be strictly isolated to guarantee maximum clarity and maximum simplicity. This results in a very maintainable package.

As a result, algorithmic proof for the entire package has come within reach. At the moment work is being done in proving the correctness of the essential algorithms. This makes it possible to guarantee the absence of bugs in the package.

LAYERED STRUCTURE

The environment that is presented to a user is built up of three layers. The bottom layer, the kernel of the operating system, is invisible to the user. Above it, a file system is implemented. This is described in the file system user manual [Bron, 1982; Joosten, 1982], and the user will need it for his input/output other than what is provided by SOLE. The third layer is formed by the SOLE simulation package and offers everything necessary to construct a working model of the transport and production system under study. The user may enhance this by adding layers of his own, thus enlarging the set of tools that he already has.

PROCESSES

SOLE is a "process oriented simulation language" as defined by Nance and Tech [Nance, 1981]. The definition of a process is varying with every publication on simulation languages. Therefore, we have chosen to use a definition widely used in informatics. This definition has the advantage of being more general than most definitions used in simulation literature.

Def: A process is a set of data, describing the state of a process at any time, and actions, describing the time-sequential changes in the state of the process.

This definition is very general and leaves a lot of room for interpretation. Usually processes are not given access to the process data of each other. This is also the case in SOLE. Communication between processes must happen with the aid of semaphores [Dijkstra, 1968], but not by reading or writing each other's process data. This is not restrictive to the possibilities. Because actions are part of a process we can talk about the lifetime or life cycle of a process. There are three important moments in the life of a process:

1. Declaration
   At the moment of declaration the process is defined. Nothing is created within the computer, but only the description of the process is known. This is very much like the declaration of a procedure in languages like ALGOL or PASCAL. Actually, it happens in the same way.

2. Instantiation
   At this moment an instance (occurrence) is made in the computer. The declaration is given a physical realisation in the computer. Of course, more than one instance can be made from one declaration. These are all different processes.
   A process can be active or passive. It is active when it performs actions at current system time. Just after instantiation it is passive, and must be activated explicitly. Processes can be activated and passivated under program control. Activation and passivation can also be done implicitly by SOLE procedures. For instance: hold passivates the current process and activates it after a certain time. We define the lifetime of a process to be the time between instantiation and termination.

3. Termination
   When the actions of a process are performed, the process terminates. The memory space that has represented the process during its lifetime is now returned to the system.

In SOLE the process declaration resembles the ordinary PASCAL procedure declaration. In general it looks like this:

```
PROCESS "proc_name"("proc_parlist");
   VAR "proc_locals";
   BEGIN
     "proc_body"
   END;
```

For example:

```
PROCESS manipulator(input, output: conveyor);
   VAR item: pelem;
   BEGIN
     LOOP item := pick(input);
          handle(item);
          place(item, output);
     END;
   END;
```

The moment of instantiation of a process resembles the call of a PASCAL function. The "call" of a process results in its instantiation. Let us, for example, make three manipulators:

```
r1 := manipulator("Robot 1", conv1, conv3);
r2 := manipulator("Robot 2", conv2, conv4);
r3 := manipulator("Robot 3", conv3, conv2);
```

Here we see that at instantiation the manipulator has to be given a name. That is necessary for tracing purposes. The call of a process reacts just as if an extra formal parameter (the name) has been added by the system at declaration time.
Now the manipulators have to be activated:

```
activate(r1, 0.0);
activate(r2, 3.0);
activate(r3, 1.0);
```

This says that r1 is activated at once, r2 after three time units, and r3 after one time unit.

SYSTEMS

SOLE is built to help modelling and simulation of a system. The way of thinking about systems nowadays, comes from cybernetics. Many definitions exist of the word system. They all have the following in common:
1. A system exists within the universe.
2. A system consists of:
   a. elements, which are characterised by attributes;
   b. relations between the elements within the system;
   c. relations between the elements in the system and the elements outside the system.

Time is a parameter that goes through the system. Changes in the state of the system are related to that parameter.
The state of the system is an inventory of all elements (the values of their attributes) and all relations between elements at a certain moment.

ELEMENTS: DEAD OR ALIVE

The basic simulation element in SOLE is not capable of any actions of its own. Or, more precisely stated, it is not capable of changing the state of the system. We call it a dead element. Dead elements can be created, deleted, put in queues, removed from queues, and so on. They have a cargo-like nature. Examples are: pallets, boxes, crates. A dead element is an information bearer. In the example, at the end of this paper, each letter is a dead element, bearing the information of its destination. When no information was carried, it would have been sufficient to specify how many letters are carried by the postman.
An element can be created alive also. A living element is an extended version of a dead element. Therefore, anyting that we can do with dead elements, can be done with living elements as well. Living elements are capable of changing the state of the system: they can perform action. For this reason, a living element is a process. Examples: manipulators, vessels, cars, aircraft, persons, etc.
Many situations in transport and production require a lot of elements. Most elements however, can be modelled as dead elements. A factory will have a limited number of machines (living elements), but a much larger number of products. Since the representation of dead elements does not require much memory space, they are so cheap that some thousands can be contained in the memory of a microcomputer. Living elements are more expensive, and a few hundred are already quite a lot. Simulation literature has never made a clear distinction between dead and living elements. The basic element is a living element in most simulation languages. This is a pity, because one must stop at, say 300 elements. For this reason, numeric studies in literature seldom consider more than 300 or 400 elements. In SOLE, simulation with some thousands of elements is possible.

AVL TREES

Most simulation languages use an eventlist algorithm based upon some kind of linear list. The eventlist algorithm is regarded as a bottleneck in runtime behaviour. A lot of study has been done on this topic, and many solutions have been thought of. Some have thought of binary trees. This is not such a strange idea, because the number of nodes that have to be passed during the search is at least ${}^{2}\log N$ (N being the total number of nodes) while a linear list needs about one half of the total number of nodes (average number). In practice however, linear lists have appeared to be more successful. This can be understood, when we think that a tree is always broken down on one side: the process with the smallest event-time will be removed from the list. As a result, the tree degenerates to a linear list, and the overhead of the tree remains. A balanced tree is the answer to this problem. We have chosen to implement the AVL tree [Knuth, 1973] which is a balanced tree. Furthermore, each node, corresponding with an instant in future time, in the tree doesn't contain one process, but usually more. All processes that are to be activated at the same time are entered in a queue in the node. This reduces the total number of nodes, and thus the average search time.

RELATIONS

The SOLE Simulation System supports five kinds of relations between elements:

- sync : simple synchronisation
- mutex : mutual exclusion of processes competing for resources
- mesg : synchronisation while passing on a dead element
- coop : synchronisation while passing on a living element
- cond : waiting for some condition to become true

All these relations are based on semaphore synchronisation [Dijkstra, 1968] and are therefore proven correct, and yet efficient structures within the system.
The implementation of the relations is such that they can be used as first in first out structures, unless a given priority decides otherwise.

INPUT GENERATION

When no real-world input is available, input can be generated by random number generators. Several well known distributions are available. Distributions derived from actual experiments in a real-world system can be used as well.

MEASURING INSTRUMENTS (Data Collection)

In order to collect statistical data about the simulated system five types of data collection devices are implemented:

- count : registration of occurrences (i.e. the total number of crates that have come out of a specific machine since a certain moment)
- accu : registration of time dependent data (i.e. the average number of crates in the model)
- tally : registration of time independent data. (i.e. the average number of crates on all pallets)
- pplot : registration of quantities as a function of time (i.e. the level of grain in a silo plotted in time).
- histo : registration with the aid of histograms (i.e. the number of ships that have to wait a particular number of days).

Statistical calculations (mean, standard deviation) are built in. Others can be added with great ease, to suit specific needs.

OTHER FACILITIES

The flow of processes can be traced when desired. A report can be generated on the relations, input generators, and measuring instruments at any moment, giving full statistics on these structures. At initialisation time, the user may decide what relations and input generators will be reported at report time, so as to keep reports small and readable. Measuring instruments will always be reported at report time, because that is what they have been made for in the first place. All elements can be queued in priority queues. These queues are first in first out queues, unless priority decides otherwise. The statistics of these queues can be reported as well.

CONCLUSION

The reasons that SOLE can be regarded as a new simulation language with its own right of existence next to other languages are:

- The most important reason is that a mechanical engineer, having little knowledge of computer science, can be taught quickly how to model his problem. There is no need for him to know implementation details. He will find programming in SOLE very pleasant, as it is very user friendly, and appealing strongly to his background of transport and production systems. He can keep his mind clean for the real work: developing a proper model of reality.
- The distinction between dead and living elements enhances the possibilities of the package significantly.
- The programmer is stimulated to work orderly and efficiently due to the structure of PASCAL and the modularity of his program. This helps him to program quickly, and relatively error free. What is more, it often helps him to understand the problem better.
- SOLE runs in an environment which enables large programs to run on small machines instead of large mainframe computers.

EXAMPLE

The example will be discussed as follows: we will start with a global problem description. Then we will make choices about what is modelled by what structures. Only when that has been done, we can start programming. It illustrates the method of modelling that we think should be used in real world simulations.

PROBLEM

On the westbank of a river a post office is located. On the eastbank, three persons are living in three houses on a street. They are called Gödel, Escher and Bach. They recieve mail from the post office. This mail comes from outside the system. All mail is addressed to Gödel, Escher or Bach. The post office has employed several postmen to bring the mail to its destination. A postman gets his mailbag filled at the post office, brings the mail across a river and delivers the mail at the destination. The only way to cross the river is on a ferry. When his bag is empty, the mailman returns to the post office to get his bag filled. The exact numerical assumptions should be given in this part, but as they speak for themselves, one can find them in the program listing.

SOLUTIONS

Now we will make an inventory of what is needed in our model:

a. The ferry boat is modelled as a process (a living element). It takes at most four waiting passengers on board, where they are left in a queue. The passengers are activated when they are on the other side.
b. The east- and westbank are modelled as coops, as living elements must be passed.
c. The postmen are processes. Each postman has a mailbag.
d. The mailbag is modelled as a queue of dead elements.
e. A piece of mail is a dead element with info on the destination address.
f. The post office is a mail generator, leaving its mail in the mailbag of a postman. Synchronisation is such that there is always exactly one postman being filled. Others can be waiting on a coop. No postman is sent away with less than 10 letters.
g. The residences of Gödel, Escher and Bach are modelled as mail eaters.

In real simulations these specifications will have to be more exact, but for the sake of brevity we have been a bit less precise.

Now we will give the (commented) program listing.

```
PROGRAM mailman(sole);

TYPE destination = (Gödel, Escher, Bach);
     letter      = ELEM dest : destination END;
     mailbag     = queue;
     hook        = ^mailbag;
     bank        = coop;

VAR  post_office : RECORD
                      postbag    : hook;
                      mail_coop  : coop;
                      postman_go : sync;
                    END;
     east, west   : bank;
     street       : ARRAY [destination] OF mesg;

PROCESS mail_gen;
  VAR neg, unif: dist; mail: ^letter;
  BEGIN WITH post_office DO BEGIN
    ini_coop(mail_coop, 'Wait for mail');
    ini_sync(postman_go,'Postman Go');
    activate(take_coop(mail_coop), now);
    ini_dist(neg, 'Letters', negexp, 0.13, 0);
    ini_dist(unif, 'Dests', uniform, 1, 3);
    LOOP
     hold(real_sample(neg));
     mail := new_letter('Letter');
     mail^.dest:=destination(int_sample(unif)));
     into(postbag^, mail^);
     IF (length(postbag^)>=10) AND
        NOT empty_coop(mail_coop)
     THEN BEGIN
            activate(take_coop(mail_coop), now);
            give_sync(postman_go, 1);
          END;
    END;
  END END;

PROCEDURE fill_bag(VAR bag: mailbag);
   BEGIN WITH post_office DO BEGIN
    wait(mail_coop);
    postbag := hook(address(bag));
    take_sync(postman_go, 1);
   END END;

PROCESS postman;
   VAR bag: mailbag;
   BEGIN ini_queue(bag); LOOP
     fill_bag(bag);
     hold(3.13); (* go to the river *)
     wait(west); (* wait, embark on west bank *)
     hold(2.87); (* go to the street *)
     (* while the bag is not empty, deliver *)
     WHILE NOT empty(bag) DO
     BEGIN hold(1.0);
       give_mesg(street[bag.first^.dest],
                 remove(bag));
     END;
     hold(1.91); wait(east); hold(2.09);
   END END;

PROCESS ferry;
   VAR deck: queue;
   PROCEDURE one_way(VAR quai: coop);
      BEGIN hold(2);
        WHILE (length(deck)<4) AND
              NOT empty_coop(quai)
          DO into(deck,take_coop(quai));
        hold(5);
        WHILE NOT empty(deck)
        DO activate(remove(deck), now);
      END;
   BEGIN ini_queue(deck, 'Ferry');
     LOOP one_way(east); one_way(west) END;
   END;
```

```
(* For the ferry it is only necessary to des-
cribe a trip from one side to the other, because
the other way is exactly the same. *)

PROCESS rcvr(VAR box: mesg);
  BEGIN LOOP del_letter(take_mesg(box)) END END;

VAR i: integer; d: destination;
(* Now we will set the system to work in a main
program *)
BEGIN (* main *)
  (* install post office with mail generation *)
  activate(mail_gen('Post Office'), now);
  (* make 15 postmen *)
  FOR i:=1 TO 15 DO
   activate(postman(edit('Postman',i)), i*1.45);
  (* install residences *)
  ini_mesg(street[Gödel ], 'Gödel' );
  ini_mesg(street[Escher], 'Escher');
  ini_mesg(street[Bach  ], 'Bach'  );
  activate(rcvr('Gödel' ,street[Gödel ]), now);
  activate(rcvr('Escher',street[Escher]), now);
  activate(rcvr('Bach'  ,street[Bach  ]), now);
  (* install river banks and ferry *)
  ini_coop(east, 'East Bank');
  ini_coop(west, 'West Bank');
  activate(ferry('Ferry'), now);
  (* Now the main program, which is a process
  too, is held for two hours. Then a report is
  generated, and the program terminates *)
  hold(120);
  report;
END.
```

REFERENCES

[1] Bron, C., Report on the programming language THT PASCAL for the PDP11 series. Revised memorandum nr. 296, dept. of Comp. Sc., Twente Univ. of Techn.(1981).

[2] Bron, C., A concise introduction to the Modular PASCAL Operating System, dept. of Comp. Sc., Twente Univ. of Techn. (May, 1982).

[3] Birtwistle, G.M., Discrete Event Modelling on SIMULA (Macmillan Press Ltd, London, 1979).

[4] Dijkstra, E.W., Cooperation sequential processes, in Genuys (ed.), Programming Languages (Academic Press, London, 1968).

[5] Dijkstra, E.J., Design and implementation of an extended PASCAL dialect for PDP-11 microcomputers, M.Sc. Thesis, dept. of Comp. Sc., Twente Univ. of Techn. (1979).

[6] Jensen, K., and Wirth, N., PASCAL, User Manual and Report (Springer Verlag, Berlin, 1975).

[7] Joosten, S., The Filesystem for the Modular PASCAL Operating System, Bacc. scr., dept. of Comp. Sc., Twente Univ. of Techn. (feb. 1982).

[8] Kaubisch, W.H., Perrott, R.H., Hoare, C.A.R., Quasiparallel Programming, Software-Practice and Experience, 6 (1976) 341-356.

[9] Knuth, D.E., The Art of Computer Programming, volume 3 (Addison-Wesley, Reading (Mass), 1973).

[10] Kriz, J. and Sandmayr, H., Extension of Pascal by Coroutines and its Application to Quasi-parallel Programming and Simulation, Software-Practice and Experience, 10 (1980) 773-789.

[11] Nance, R.E. and Tech, V., The Time and State Relationships in Simulation Modelling, Comm. of the ACM 4 (1981) 173-179.

[12] Neumann, P.G., Psychosocial Implications of Computer Software Development and Use: Zen and the Art of Computing, ACM Software Engineering Notes 7 (April 1982).

[13] Prins, M., PASCAL Operating System for the PDP11, M.Sc. Thesis, dept. of Comp. Sc., Twente Univ. of Techn. (April 1980).

[14] Rooda, J.E., Discrete Event Simulation for the Design and Operation of Logistic Systems, Proceedings of the Society of Logistics Engineers in San Francisco (1981).

[15] Rooda, J.E., Transport- und Produktionssysteme - Modellentwicklung und Simulation, Fördern und Heben 8 (aug 1982), 597-600.

[16] v. Rossem, P.H., A kernel for a distributed network operating system, M.Sc. Thesis, dept. of Comp. Sc., Twente Univ. of Techn. (July 1982).

[17] Stirling, W.D., The Use of a Procedure oriented Language for Process oriented Simulation, Software - Practice and Experience 8 (1978) 137-148.

Simulation in Engineering Sciences
J. Burger and Y. Jarny (eds.)
Elsevier Science Publishers B.V. (North-Holland)

UNISYS - COMPUTER ASSISTED MODELLING AND SIMULATION SYSTEM

Krzysztof A. Grabowiecki

Industrial Institute of Construction Machinery
CAD/CAM Dept.
ul. Kolejowa 57, 01210 Warszawa, Poland

During modelling of dynamic systems, the undetected errors often occures as well as the time which is needed for completing of an overwhole mathematical model is usually pronibitively long. The impact of mentioned features increases as the system to be modelled grow in complexity and size. To minimize that impact, the bond graph formalism had been used in the assisted modelling part of the UNISYS system as a tool for mathematical model generation and its validation. That formalism nad been used also as a tool for linkage of submodels of different formalisms origin (such as finite element, block diagrams etc.). To realize the robustness of simulation segment of UNISYS, the simulation processor was carefully designed to satisfy demands for numerical stability, numerical stiffness and accuracy; the programing segment was designed to reswitch appriopriate integration algorithm. The experimental frames accessible in the UNISYS are described.

INTRODUCTION

Conventional techniques for dynamic analysis of nonlinear systems have shortcomings when applied to modelling of such complex engineering systems as machines of today. The man-computer interface is an inadequate and software designed for modelling and simulation purposes is not robust enough. The features of modern software systems for computer assisted modelling and simulation are well defined up to now [2,7]. Nevertheless, the software which presents those features is still rarity and mostly it is limited to the systems designed for behaviour analysis of the mathematical model. That is why most of systems for system dynamic behaviour analysis is called the simulation systems and not modelling and simulation systems. Still there is few software packages which give the user an opportunity to process the computer assisted modelling (the model generation and the analysis of model acceptability). The problem is particulary easy to notice, when nonlinear deterministic model is the subject of research. Other premises also direct toward the design of a suitable software. Growing complexity of machines together with the variety of possible formalisms for modelling of different machine components (construction, power transmission components, control systems, environmental influence etc.) turn efforts to generate the software facilities for subdividing of physical model into submodels, while they are modelled by different formalisms, and linking them into an overwhole entirety in a man-independent way. That feature would allow to organize model hierarchically and it would speed the construction of new models from the existing ones. That idea, as well as software portability, was pointed out a few years ago in different software research centres [1,3,7].

Below there are presented efforts towards creating of software which would present some features of computer assisted modelling and simulation. The special attention was paid to the computer assisted modelling realization. As a result, the UNISYS system (Unified Simulation System) was worked out.

GENERAL OUTLINES OF THE UNISYS STRUCTURE

To locate the place of software for modelling assistance in the system for computer aided modelling and simulation, it seems to be usefull to look how totally the presented software - the UNISYS - is build up. From the point of view of software engineering the UNISYS software system was build in the modular way, where autonomic programs operate on the common data base. Thus the data base technique is used for data storage and retrival. The data base consists of two files: - file of models and file of

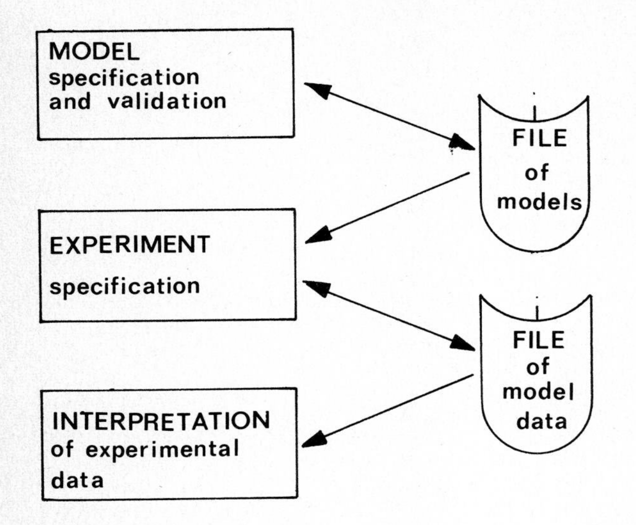

Figure 1.

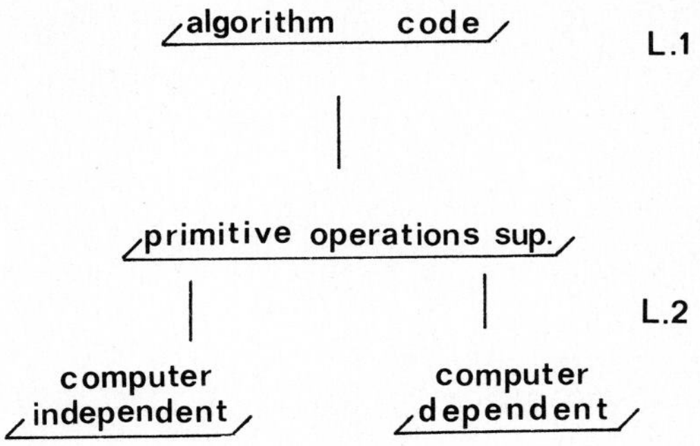

/algorithm code/ L.1

/primitive operations sup./ L.2

computer independent/

computer dependent/

Figure 2.

model data (model parameters, experiment data etc.) - Figure 1. Program modules which allow to process operations defined in the Figure 1 may work independently or in the sequential way, as they have parallel input both from the data base (where are stored the results of processing of previously used modules) and directly from the user defined input data stream, when one wants to work in the local mode.

It seems, that the most important feature of that approach is the saving of computer time and it´s necessary capacity on the level of mathematical model generation and validation. Moreover, becouse of data base technique application the model formulation processing is independent of solution processing (simulation), which is usually an expensive one and demanding the computer power (particulary, when an integration of nonlinear differential equations is needed).

When the general purpose software system is under design, particulary in the software hause which do not cooperate with the computer company, the crucial is to satisfy its portability. To achive it, the UNISYS software was written in the two-level way; the programming modules were written in FORTRAN IV standards, however all specific operations on bits, bytes, characters, strings as well as access to files realization which vary from one computer system to another, there were addressed to the second level of software where those primitive operations are coded - Figure 2.

The second level of software is subdivided into two segments; one of them should be exchanged when the computer is changed. Above described software structure is guaranty for the easy implementation of software on different types of computers; only a small subset of procedures has to be exchanged in the way similar to the exchange of chips in the electronic devices.

UNISYS FACILITIES FOR THE COMPUTER ASSISTED MODELLING

It seems that very few general purpose simulation systems present support for modelling. That unique oportunity is presented by the UNISYS. Most of simulation systems have the input data stream organisation which demands to documentate model with aid of ODE oriented formulations. That formulation is tedious one and sensitive for undetectable errors introduction, when the model grows in the size and complexity. That could be omitted when variables identification could be processed on the level of model documentation rather then the solution approximation. The significant feature of graph-theoretical models is the inherent, simple and systematic maner in which method formulates the system equations directly from the description of the physical system. These automatic formulation procedures render the models extremly computer worthy.

Facilities for subdividing the model into submodels and for organizing them hierarchically to form an overwhole model are necessary to speed the construction of new models from existing ones. It also makes model creating more straigthforward as it is much more easy to work on the simple submodel then on the overwhole system. It is obvious, that graph theory provides with possibility of invoking of piecewise approach.

As it was stated previously, the equal importance for man-computer interface is to satisfy the condition of polydescriptiveness on the stage of model formulation. It seems to be crucial to give an opportunity to the user for job in the most familiar to him modelling formalism. It is even more important if one would try to analyse the behaviour of machinery which consists of different components such as construction which commonly is modelled with the finite element method, control systems modelled with transfer functions and at last the machine interacts with the environment what may be represented for example by set of ordinary differential equations.

The most easy way would be to model those model components in particular formalisms and to link them together into one entirety. What is more, as a rule there exist perfect models documentated in particular formalisms which could be used for an overwhole model analysis, but... there is no tool to connect them with the rest of system under investigation. As the result, the model has to be created once more in the "governing formalism" for given analysis.

Once more the graph provides a unifying approach for developing numerical meodels for lumped parameter systems as well as for continuum problems [8]. Thus there may be found an appropriate representation of given modelling formalism by the graph means. The computer´s role (or rather the software role) here will be to translate the description of model whatever its format into a standardized representation suitable for undergoing the manipulations of the model file as well as for model structure validity and acceptability analysis. In the UNISYS that standardized representation is the graph.

As during the UNISYS design procedure it was assumed, that the system should accept nonlinear systems for analysis, the bond graph [6] was choosen for an internal representation of different formalisms acceptable by that software. Its main features are:

- possibility of identification and isolation of source of nonlinearities and placing them into the constitutive properties (these originating from nonlinear primitive element characteristics e.g. from a nonlinear i=f(E) dependency for an electrical resistance R etc. - it defines terminal equation; additionaly there may be defined so called structural nonlinearity which results when two variables are interdependent through circuit and terminal equations and at least one of terminal equations is nonlinear through constitutive nonlinearity);
- possibility of an input-output dependencies (system causality) checking which may give informations whether physical system is properly modelled from the point of view of "througn" and "accross" variables causal interdependencies.

Those features and some other described elsewhere [6], together with the regular graph properties were the reasons why bond graph formalism had been choosen as a target language for computer assisted modelling in the UNISYS. Actually the user may model his submodels in finite element technique, block diagrams, explicit set of ordinary differential (ODE) or differential-algebraic equations (DAE) and straigth on in the target language which is bond graph. When any of those formulation is the input into the system, first the consistency with respect to the choosen modelling formalism is checked. In the case of success the physical data (coefficients of terminal equations) are loaded into the model data file. Then, an appropriate for modelling formalism graph structure is generated and it´s properties are analysed. The resulting bond graph structure is stored in the model file. In the case when the model was composed of several submodels prepared with different formalisms, an overwhole model structure is build up. The concatenation procedure is based on the input-output or boundary conditions definitions. The overwhole model composed in this way (which is "de facto" the bond graph model) is analysed to check its structure, correctness of power flow through the system and causality assignments. Finally, the mathematical model given by the set of ordinary nonlinear differential or differential-algebraic equations is achived. It is expressed in the form of FORTRAN subprogram generated in a way which satisfies demands for standard interface for ODE/DAE solvers [5]. An advantage of this approach is that it makes the language available to user as

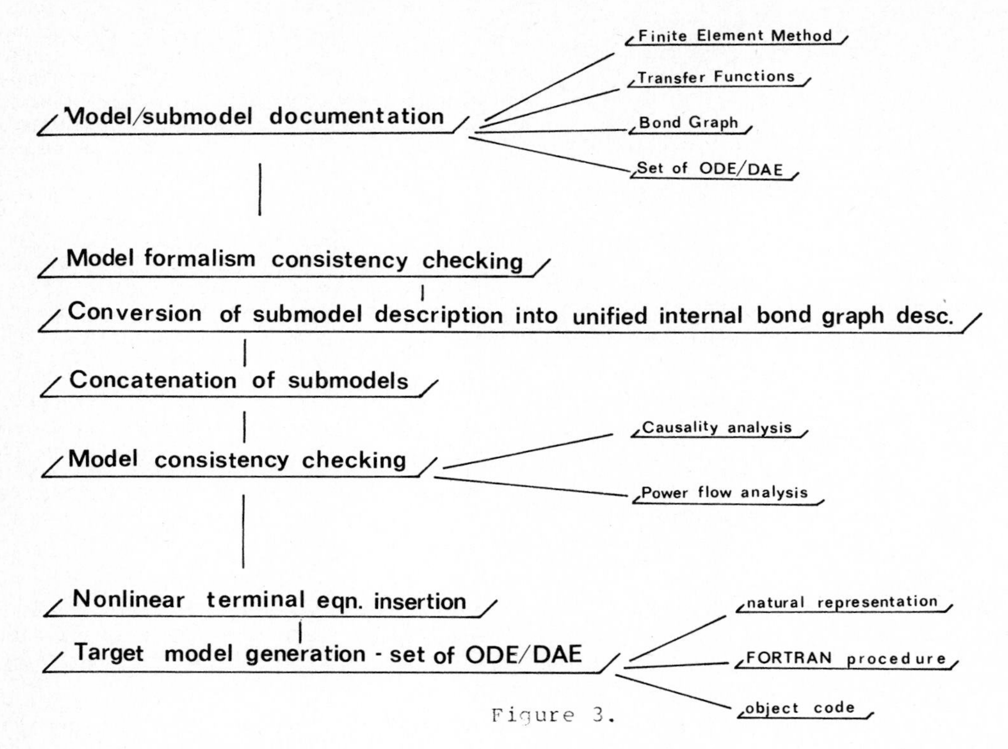

Figure 3.

soon as possible and that it permits (relatively inexpensive) modifications of the language specifications based on user´s experience. Besides the appriopriate low level target language is under design; in that case the ODE/DAE solver standard interface is prepared in the form of model structure data.

UNISYS ASSISTANCE IN THE BEHAVIOUR GENERATION

Despite of the fact, that during the UNISYS design the main efforts were pointed to achive the support for modelling process, it is necessary to mention about realization of its simulation module. The main goal during the design of that part of system was to satisfy some aspects of software robustness [2]. Particulary the attention was paid to the effectiveness of numerical integration in scope for solving of nonlinear numerically stiff initial value problem.

The available simulation software offers a comprehensive selection of integration algorithms. It does not however tell the user which would be the most appropriate procedure for solving of his particular nonlinear problem. The conditions of solution may also vary in the time domain and in that case it is particulary difficult for user to interfere. For this reasons, the selection of integration algorithms in the UNISYS system is optionally automated. The criteria for automation are: relative accuracy, numerical stability and corresponding to it numerical stiffness. In the integration algorithms library there were included extremly carefull and sophisticated Hindmarsh implementation of Gear and Adams algorithms (multistep metnods) and Runge Kutta Fenlberg method of fourth-fifth order implemented by Shampine [9]. If necessary there is also in the disposition algorithms for solving of DAE (based on the Gear´s method).

As an additional facilities the UNISYS gives an opportunity to linearize achived model and to process the regular stability analysis (time and frequency domain behaviour and eigenvalues problem solution combined with transfer function generation) which is commonly used in the control system analysis.

EXPERIMENTAL FRAME ACCESSIBLE IN THE UNISYS

The combination of possibility of computer assisted modelling and data base technique as well as an efficient software for integration of achived state equations formulation make possible at least three level analysis of problem being under consideration - Figure 4.

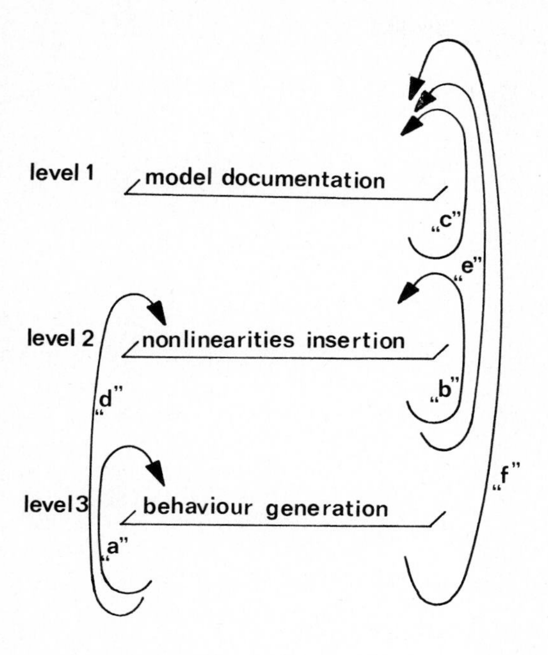

Figure 4.

Let assume that the model is generated and its behaviour generation is used for parameter sensitivity analysis - the loop "a". As the mathematical model is stored in the model file, one needs only to change physical parameters and to continue observation of output variables; the repetition of model generation process (loop "f") is not needed in that case. If there is decision, that parameter changing is not sufficient to achive proper results, one may decide just to change constitutive nonlinearities (i.e. from square to the cubic function on the resistance). Then there is need to come back to the level 2 - (loop "d"), to retrive from the model file an appropriate model structure and to fill it with the new constitutive nonlinearity. Then the parameter sensitivity analysis may be followed to achive the success. If there is still no luck one has to come back to the model structure documentation and to change it (loop "f"). It is obvious, that such procedure is straigthforward. Loops "b", "c" and "e" are connected with the analysis of model consistency. Loop "c" allows to define the premises for proper structure of model on the base of causality and power flow directions which exist in that model. Loops "b" and "e" mark the process of checking and reassignment of causality arising in the model and assumed when constitutive nonlinearity is established (i.e. whether assumed independent variable in the constitutive equation conforms to that achived during system causality analysis).

Described analysis of "b", "c" and "e" loops allows designer to deduce whether input-output (boundary condition) is realizable in the model which is assumed to copy the real system. That organisation of the tool for dynamic analysis also promises possibility to include into experimental frame the optimization algorithm (which seems to be not very difficult in the case of parameter sensitivity analysis, but it may be exciting on the model structure analysis level).

THE UNISYS APPLICABILITY

So far the UNISYS has been used with success in the number of design processes in the machine industry as a part of computer aided design software for system dynamics analysis. Presently the routine use of that system is in the analysis of dynamic of telescopic cranes (power transmission, control system and construction interactions), and electrohydraulic systems and devices. It was also used to solve problems of industrial robot dynamics. The system may be used in other cases, where mixed formalisms (lumped-distributed parameter models) are usefull; for instance, recently, efforts aimed at using it for biological systems (an artifical kidney-patient system dynamics model) have been made.

ACKNOWLEDGEMENTS

This project was made possible due to a grant from Marie Sklodowska-Curie Fund, a joint undertaking of the US and Polish Goverments administrated by the US National Science Foundation and the Polish Academy of Science.

REFERENCES

[1] Bossak, M., Grabowiecki, K.A., Rosenberg, R.C., Zgorzelski, M., Analysis of physical dynamics system using a polydescriptive approach: bond graphs and finite elements, Proc. of II Int. Conference on Engineering Software (Pentch Press 1981).

[2] Cellier, F.E., Combined continuous/discrete system simulation by the use of digital computer: technique and tools (Ph.D

dissertation, Swiss Federal Institute of Technology, Zurich 1979).

[3] Grabowiecki, K.A., Stepniewski, Wl., Zgorzelski, M., POLSYAS - A polydescriptive, polyalgorithmic simulation system for nonlinear continuous problems, Proc. of I Int. Conference on Engineering Software (Pentch Press 1979).

[4] Grabowiecki, K.A., Margolis, D., Finite element method bond graph representation - A way to incorporate finite element subsystem into overall dynamic system model (to be published).

[5] Hindmarsh, A.C., LSODE and LSODI, Two New Initial Value Ordinary Differential Equations Solvers, ACM SIGNUM Newsletter, Vol. 15, No. 4 (1980) 10-11.

[6] Karnopp, D.C., Rosenberg, R.C., System Dynamics: A Unified Approach, (Wiley 1975).

[7] Ören, T.I., Zeigler, B.P., Concepts for advanced simulation methodologies, Simulation, Vol. 32, No. 3 (1979), 69-82.

[8] Savage, G.J., Kesavan, H.K., The Graph - Theoretical Field Model - Modelling and Formulation, J. of Franklin Institute, Vol. 307, No. 2 (1979).

[9] Shampine, L.F., Watss, H.A., Practical Solution of Ordinary Differential Equations by Runge Kutta Methods, SAND 76 - 0585, Sandia Lab., Albuquerque, New Mexico (1976).

[10] van Dixhoorn, J.J., Simulation of bond graphs on minicomputers, Trans. ASME, J. Dyn. Systems, Measurement and Control, Vol. 99, No. 1 (1977).

Simulation in Engineering Sciences
J. Burger and Y. Jarny (eds.)
Elsevier Science Publishers B.V. (North-Holland)

DYNAMIC SYSTEM SIMULATION IN DESIGNING COMPUTER PERIPHERALS

Martin H. Dost

International Business Machines Corporation
San Jose, California, USA

Numerous IBM products, especially computer peripherals and memory devices, contain complex automatic control systems or other dynamic systems which have been designed with the aid of digital computer simulation. An important design tool has been the continuous system simulation language, DSL. Over many years it has evolved into a very powerful, yet user-friendly language. Its application to a paper motion control system of the IBM 3800 Printer is shown in detail to illustrate the use of DSL for time domain simulation and frequency domain analysis.

INTRODUCTION

The advent of more and more application-oriented, high-level programming tools, coupled with powerful graphic display of computer results, has encouraged many engineers and scientists to use computers for help in solving problems which were deemed impractical for simulation only a few years ago. Unlike the aerospace or process industries, where life threatening and very high cost situations have for years forced designers to use simulation techniques to study the performance of systems under consideration, the business machines industry started to make heavy use of simulation only relatively recently.

Several reasons account for this change. Among them are the availability of suitable hardware and software for easy modeling and coding of problems, rapid execution, and immediate graphic output. The decreasing cost of computing, the training in computer sciences of recent graduates, and the scarcity of good experimental technicians are other reasons. But the increasing demand for ever higher performance, greater reliability and miniaturization make development of new machines in reasonable time frames and with the limited manpower impossible without use of really practical computational tools.

With higher performance requirements, more attention must be given to dynamic aspects of systems. Increasingly, feedback control techniques have to be employed to meet specifications and, of course, these closed-loop systems have to be consistently stable, requiring much analysis in time and frequency domains. Whether for motion control of magnetic heads over disks, of checks to be microfilmed, of electron beams in vacuum columns, of robots in manufacturing processes, etc., analysis of chemical reactions or physical processes, the simulation language, DSL, is likely to have played an important role in the design of IBM's peripheral equipment.

The first version of the language, DSL/90 (1), written in FORTRAN for the IBM 7090 was widely distributed free of charge in the mid-60's. Also, it was the basis of the program product CSMP/360 (2) which, in turn, had a strong influence on guidelines for such "Continuous System Simulation Languages" defined by an SCi* committee in 1967 (3). Since then, DSL has undergone many evolutionary changes. First, DSL/360 was written for the System/360 computers in 1967. It was drastically restructured, mainly to avoid numerical difficulties due to discontinuities, introducing DERIVATIVE and SAMPLE segments of code in addition to the DYNAMIC segment in which structure statements were traditionally coded.

In the early 70's a graphic post-processor was written (4), which gave DSL users much more flexibility in graph generation. Along with new features, facilitating frequency domain analysis and automatic double precision computation, it became part of DSL/dp in 1977, used internally at IBM until now in batch mode. Subsequently, an interactive version, DSL/VM, was written to run DSL in the foreground under VM/CMS (5). When combined with graphic output via the IBM 3277 Graphic Attachment (6) in 1980, it proved to be a winning combination of hardware and software for control engineers and others concerned primarily with dynamic system analysis or design.

All these program developments took place in response to needs by users of DSL, out of a small service group within the IBM San Jose development laboratory, which is the cradle of the disk file and many other products. No company-wide mission was ever given to develop DSL as a program product, yet within IBM it gradually spread to almost all operating units.

---

*Simulation Councils, Inc., the former name of the SCS (Society for Computer Simulation).

It is now even being used by biologists in Brazil to combat destructive fruit flies.

SIMULATION STUDIES OF A HIGH PERFORMANCE PRINTER

A particularly complex machine that was designed with the aid of simulation by DSL is the IBM 3800 Printing Subsystem (7) which employs an electrophotographic process similar to that used in IBM copiers. Figure 1 shows schematically the process components, many of which were studied by computer simulations: A large rotating drum, holding an organic photoconductor (OPC) web on its outside and supply reels of OPC material on its inside, had to have well-regulated speed despite its unbalance. The electrical charge and discharge phenomena were investigated through modeling and simulation of an OPC surface element on its way through the process (8). Exposure to laser light via a rotating mirror and to xenon flash light through forms overlay was studied (sensitometric optimization and mirror dynamics). Developer, cleaner, and several corona charge units were modeled to investigate physical and geometric aspects of these components. The air supply system, though static in nature, was also designed with the aid of DSL, as were temperature control systems and power supply circuits. But, most of all, paper motion control systems (speed, position and lateral steering) at the transfer station and through the fuser rolls were extensively simulated. This enabled the designers to do much of the experimentation on the computer, rather than in the laboratory, where other processes depended on consistent paper movement.

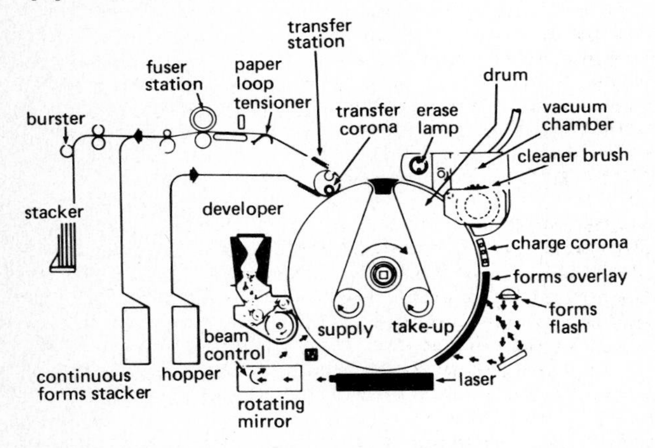

Figure 1. Process components of the IBM 3800 Printer, using transfer-electrophotography.

PAPER MOTION CONTROL SYSTEM DESIGN

The 3800 Printer was required to move many different types of fanfolded paper at printing speeds of 0.82m/sec, while maintaining position accuracy within 0.2mm. Start/stop had to be accomplished within 6.5mm to accommodate various sheet lengths without printing across perforations or to stop the paper while the OPC gap of the drum passes by. At the transfer station, where toned images are transferred from OPC to paper (see Figures 1 and 2), sprocket wheels are propelling the paper, while at the fuser the paper passes through two rolls to fix the images by heat and pressure. At both stations d-c motors drive the loads, and shaft angle encoders track the paper. However, due to large inertia differences a significant difference in response exists, which necessitated a slack loop for the paper between stations.

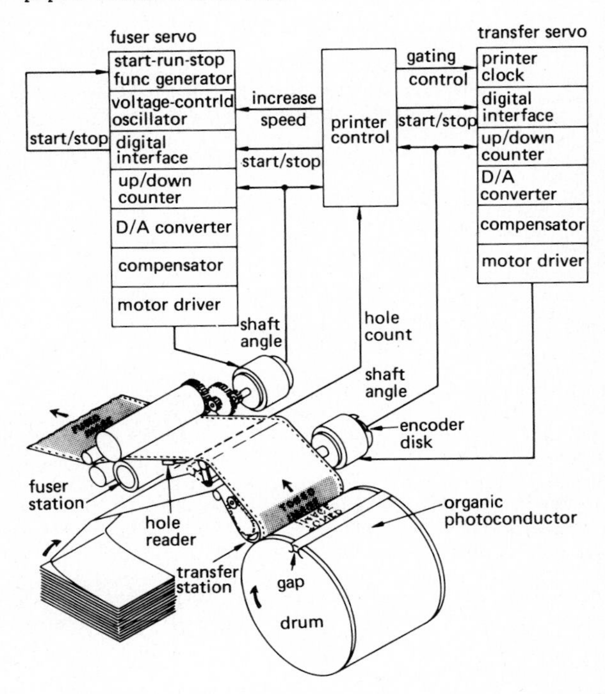

Figure 2. Paper motion control system schematic for the IBM 3800 Printer.

As stated in more detail in Reference 9, the design of these servomechanisms underwent several major changes during the development of the printer. An early control scheme suffered from a parameter sensitivity problem that was only recognized after extensive modeling and simulation in time and frequency domains. One of the transfer servos, though built seemingly just like all others, was vibrating badly; inordinate frequency response diagrams, experimentally obtained, showed the instability. The problem had no simple solution.

As it turned out, the source of the instability was not found until a minor loop was accurately modeled in the frequency domain. It contained a resetting integrator whose function was to provide fill for a shaft-encoder generated position signal (to make a continuous ramp out of a discrete staircase).

While the simulation language DSL is intended primarily for transient analysis, it is also easily usable for frequency response analysis. Since FORTRAN is a subset of DSL, complex arithmetic is readily carried out. Therefore, it becomes a simple matter to evaluate the effect of complex systems even if they contain elements involving time delay, such as the resetting integrator whose transfer function is

$$R(s) = \frac{1}{s}\left(1 - \frac{1 - e^{-sT}}{sT}\right).$$

Table 1 is shown as example of a simple program to study a resetting integrator with DSL in time and frequency domain (Figure 3).

Table 1. Sample DSL code for analysis of a resetting integrator in frequency and time domain.

```
TITLE FREQUENCY AND TIME DOMAIN ILLUSTRATION
TITLE SAMPLE AND HOLD - RESETTING INTEGRATOR
PARAM FS=1, FINTIM=10
COMPLEX S, SH, RI
INTGER I
INITIAL
      T=1./FS
      DELS=T
DERIVATIVE
      V=TIME-2.*RAMP(5.)
      X=INTGRL(0.,V)
DYNAMIC
      FILL=X-STAIR
SAMPLE
      STAIR=X
TERMINAL
      DO 9 I=1,100
      LOGF=I*0.02-1.
      F=10.**LOGF
      W=2.*PI*F*FS
      S=CMPLX(0.,W)
      SH=(1.-CEXP(-S*T))/(S*T)
      RI=(1.-SH)/S
       MAGN=20.*GAIN(RI)
       PHAS=RADEG*PHASE(0.,RI)
    9 CALL SAVE(FR)
SAVE  0.02, X, STAIR, FILL
GRAPH (DE=TEK618,SC=5,NB=5,TI=0.8) ...
      TIME(NB=5) FILL, STAIR(PO=5) X(AX=OMIT)
LABEL SAMPLE AND HOLD, FILL- TRANSIENT RESPONSE
SAVE  (FR) W, F, MAGN, PHAS
*LIST (/FR) W, F, MAGN, PHAS
GRAPH (IN/FR,DE=TEK618,OV,PO=0,5.5,NB=4) F(AX=LOG) ...
      MAGN(SC=10,UN=DB) PHAS(SC=25,UN=DEG,PO=5)
LABEL (IN) RESETTING INTEGRATOR - FREQUENCY RESPONSE
END
STOP
```

The control scheme that was finally adopted for the transfer servo is shown schematically in Figure 4. Rather than using feedback from the armature circuit to produce fill, the system uses additional "feed-forward" signals (FF) which were designed to overcome inertia and back-emf effects known a priori. This system had continuous as well as discrete elements, several time-varying nonlinearities, a fourth-order compensator and multiple inputs, one of which was based on table look-up. Even if all nonlinearities had been ignored, it could not have been studied analytically as a sampled data system, because the sampling rate was a function

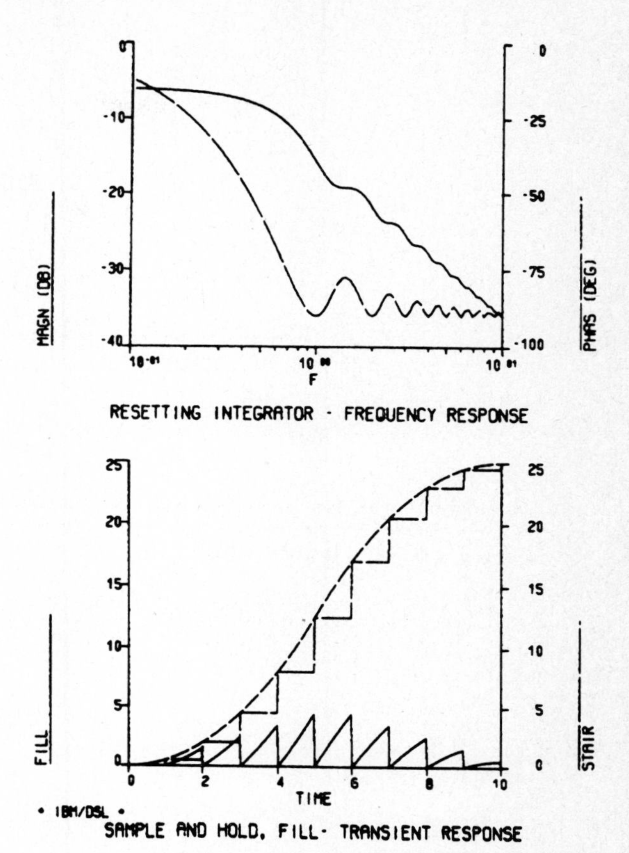

Figure 3. Graphic output from program of Table 1.

of shaft speed. But simulation by DSL was straightforward and enabled the responsible engineers to design the system with a minimum of experimentation.

Table 2 is the program used to code the model depicted in Figure 4, specifically to study the effect of deadspace in the digital encoder on system performance. A representative plot of armature current I, position error XE, paper velocity V and desired velocity VR are shown in Figure 5, as produced for the second of three runs on a Tektronix 618 with 4631 hard copy unit.

By inspection of Table 2, the reader can see the traditional division of code into data, structure and control statements (including graph commands). The structure statements are, however, grouped into DERIVATIVE, DYNAMIC and SAMPLE segments, not merely one DYNAMIC region.

The SAMPLE segment serves to code only those events occurring at sample times, as encountered in sampled data systems. The interval between samples, DELS, may, in general, be made variable. This feature was used to simulate the discrete

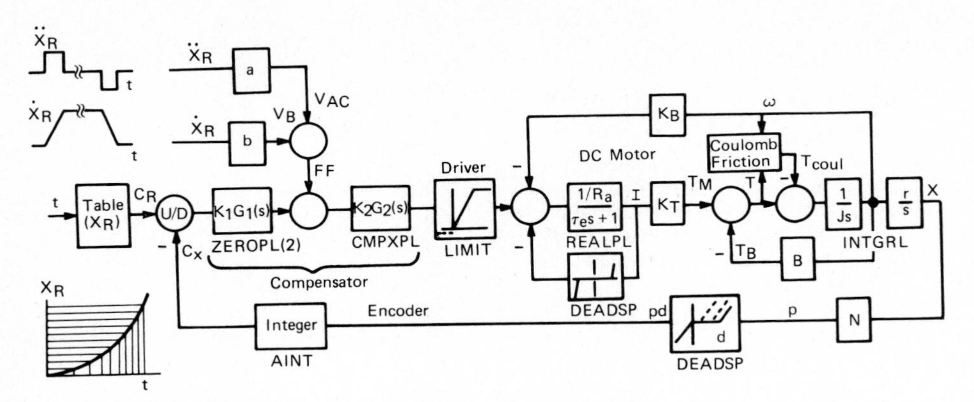

Figure 4. Block diagram of the 3800 transfer servo.

Table 2. DSL program for simulation of 3800 transfer servo (depicted schematically in Figure 4).

```
TITLE TRANSFER SERVO - DEADSPACE STUDY
INTGER M
STORAGE DM(28)
TABLE DM(1-28)=8,7,6,5,3*4,3,4,4*3,2,3,3,2,3,3*2,3,6*2
CONST K1=1.5, KL=5.25, KD=2, N=144, VP=32.3, R=0.478
CONST Z1=104, P1=17.7, Z2=329, P2=8000, WN=4000, Z=0.5
CONST RA=0.47, L=5E-5, KT=0.765, KV=0.0864, J=0.75E-3
PARAM B=7.14E-3, XA=0.1875, IC=5, STROKE=1, D=0
INITIAL
        TCOAST=(STROKE-2.*XA)/VP
        TD=1./(2.*N*VP)
        TA=2.*XA/VP
        T1=TA+TCOAST
        T2=T1+TA
        CR=0.
        M=0
        A=VP/TA
        AB=KV/(KD*R)
        AC=A*J*RA/(R*KT*KD)
        TC=IC*KT
        V2=IC*RA/KD
        V1=V2/KL
        PRE=V1/K1
        W2=WN*WN
        DELS=21.*TD
DERIVATIVE
        VR=A*(TIME-RAMP(TA)-RAMP(T1)+RAMP(T2))
        XR=INTGRL(0.,VR)
        VIN=VDAC+PRE
        VLG=ZEROPL(V1,Z1,P1,VIN*K1*P1/Z1)
        VLD=ZEROPL(V2,Z2,P2,VLG*KL*P2/Z2)
        VC=VLD+VAC+VR*AB
        VN=CMPXPL(V2,0.,Z,WN,W2*VC)
        VD=LIMIT(LOW,LIM,VN*KD)
        DV=VD-50.*ID-KV*W
        I=REALPL(0.,L/RA,DV/RA)
        LIM=20.-I/6.
        ID=DEADSP(-25.,25.,I)
        T=KT*I-B*W
        DT=T-TCOUL
        W=INTGRL(0.,DT/J)
        X=INTGRL(0.,R*W)
DYNAMIC
        VAC=AC*(1.-STEP(TA)-STEP(T1)+STEP(T2))
*       DELPLT=SWITCH(TIME.LT.0.01, 5E-5, 0.)
        LOW=SWITCH(TIME.LT.T1, 0., -3.)
        TCOUL=SWITCH(ABS(W).LT.0.1, LIMIT(-TC,TC,T),...
                                    SIGN(TC,W))
        V=W*R
        PD=DEADSP(0.,D,N*X)
        XE=XR-PD/N
        CX=AINT(PD)
        VDAC=CR-CX
SAMPLE
        IF(TIME.EQ.0.) RETURN
        IF(TIME.LT.TA) GO TO 10
        IF(TIME.LT.T1) GO TO 20
        GO TO 30
10      M=M+1
        DELS=DM(M)*TD
        GO TO 50
20      DELS=2.*TD
        GO TO 50
30      M=M-1
        IF(M.LT.1) DELS=1.
        IF(M.LT.1) GO TO 50
        DELS=DM(M)*TD
50      CR=CR+1
        VDAC=CR-CX
TERMINAL
        IF(D.LT.1.) CALL RERUN
        D=D+0.5
CONTROL FINTIM=0.05, DELMAX=1E-5
PRINT 0.0005, VR, V, VDAC, I, XE, X, VLG, VLD, VN
*SAVE  (DETAIL) VDAC, CX, CR, PD
*GRAPH (/DETAIL,DE=TEK618,SEP,LO=-1,SC=2) ...
*       TIME(TI=0.7) CX, PD, VDAC, CR
SAVE  0.0001, VR, V, XE, I
GRAPH (G1,DE=TEK618,SEP,NB=4) TIME(UN=SEC) I(UN=A)...
 XE(UN=IN) V(LO=0,SC=10,UN=IPS,PO=0,4.5,LI=1) ...
 VR(SC=10,AX=OMIT,PO=-1.5,4.5,LI=2)
LABEL (G1) POSITION ERROR, CURRENT AND VELOCITIES
END
STOP
```

events surrounding the encoder and up/down counter which served as the error detector in this system.

The DYNAMIC segment contains statements which express discontinuous phenomena, slowly changing variables and variables generated only for plotting or tabulation. They are executed once at the completion of every time step DELT, the size of which varies between DELMIN and DELMAX. DELT is controlled by a fifth-order Runge-Kutta integration method and a precise system clock that accomodates all communication requests and predictable events of discrete nature. Several other integration methods, both fixed and variable step size, are optionally available as well.

The remaining structure statements, typically integral statements, linear transfer functions and nonlinearities without discontinuities, as well as algebraic and trigonometric expressions, are coded in the DERIVATIVE segment. They are automatically sorted for optimal sequence. These statements get evaluated at all time steps, including fractions of DELT, as needed by the chosen integration algorithm.

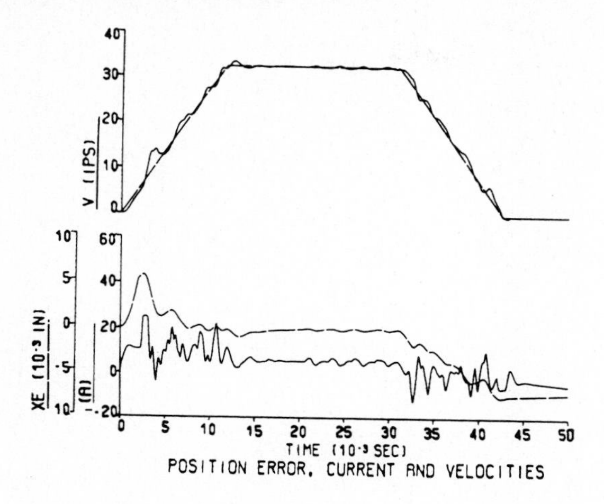

Figure 5. Graphic output from program of Table 2.

One-time calculations, to be carried out at the beginning or at the end of a run, are stated in the INITIAL and TERMINAL segments, respectively. In the example described here, a family of runs is desired and a test is made at the end of each run (FINTIM) to determine whether more runs are desired.

By a few lines of high-level code, a graphic post-processor is invoked to produce plots of the computer results, such as Figures 3 and 5, on the device of your choice. Under VM a printer-plot may be displayed directly at the video display terminal. A high-resolution graph may be seen on a separate screen (for example, the Tektronix 618, attached directly to an IBM 3277 terminal) and printed out on a 4631 hard copier. If improvements in appearance are desired, changes in graph commands are easily made without need for rerunning the job. Finally, when high-quality graphs are needed, a colored pen plot may be obtained by a device such as the IBM XY/750.

CONCLUSIONS

Technologies in computer hardware and software for highly efficient simulation of most dynamics problems encountered today are well proven. At IBM they are being used on a daily basis to aid in development and research work toward products of all kinds. The simulation language DSL is one such tool that is particularly useful for analysis and design of automatic control systems, as illustrated by a paper motion servomechanism for the 3800 Printer.

Tools like DSL allow engineers and scientists to use mathematical modeling and simulation techniques to prove their ideas and to optimize their designs without much need for programming expertise and with a minimum of time and need for experimentation in the real world.

ACKNOWLEDGEMENTS

The programming tool described here and the sample problem are the product of many dedicated individuals. Two men stand out, both from the San Jose development laboratory of IBM's General Products Division: Mr. Wai Mun Syn, who developed most of the DSL language, and Mr. T. Jay Cameron, who was responsible for the 3800 transfer servo design.

REFERENCES

[1] Syn, W. M. and Linebarger, R. N., "DSL/90 - A Digital Simulation Program for Continuous System Modeling," 1966 Spring Joint Computer Conference, April 26-28, 1966, pp. 165-187.

[2] "System/360 Continuous System Modeling Program (360A - CX-16X), User's Manual (M20-0367)," IBM Corporation, White Plains, NY, 1967. Superseded by CSMP III in 1972 (5734-XS9), General Information Manual (GH19-7000) and Graphic Feature (SH19-7001-2), IBM World Trade Corporation, 821 United Nations Plaza, New York, NY 10017.

[3] "The SCi Continuous System Simulation Language (CSSL)," Simulation, Vol. 9, No. 6, December 1967, pp. 281-303.

[4] Dost, M. H., Syn, W. M. and Turner, N. N., "High-Level Control Language for Scientific and Engineering Graph Generation," 1975 Winter Simulation Conference December 18-19, 1975 in Sacramento, CA, pp. 595-602.

[5] IBM Virtual Machine Facility/370: Introduction, Order No. GC20-1800. Terminal User's Guide, Order No. GC20-1810.

[6] IBM 3277 Graphic Attachment RPQ (7H0284) General Information Manual: GA33-3039.

[7] Findley, G. I., Leabo, D. P. and Slutman, A. C. "Control of the IBM 3800 Printing Subsystem," IBM Journal of Research and Development, January 1978, Vol. 22, No. 1, pp. 2-12.

[8] Vahtra, U. and Wolter, R. F., "Electrophotographic Process in a High-Speed Printer," IBM J. Res. Develop. 22, pp. 34-39, January 1978.

[9] Cameron, T. J. and Dost, M. H., "Paper Servo Design for a High-Speed Printer Using Simulation," IBM J. Res. Develop. 22, pp. 19-25, January 1978.

IV. SIMULATION AND CONTROL METHODS AND TECHNIQUES

Simulation in Engineering Sciences
J. Burger and Y. Jarny (eds.)
Elsevier Science Publishers B.V. (North-Holland)

APPLYING A SIMULATION TOOL TO THE DESIGN OF WARPED SURFACES : NOTION OF INTERPOLATING PROCESS

R. HAJ NASSAR**, D. MEIZEL*, P. BIELEC*

* Laboratoire d'Informatique Industrielle et d'Automatique
INSTITUT INDUSTRIEL DU NORD B.P. 48
59651 VILLENEUVE D'ASCQ CEDEX - FRANCE

** Laboratoire de Systématique
UNIVERSITE DES SCIENCES ET TECHNIQUES
U.E.R. d'I.E.E.A
59655 VILLENEUVE D'ASCQ CEDEX - FRANCE

This paper develops a new approach in the problem of interpolating a set of points by a general spline function. The interpolation curve is regarded as the trajectory generated by a dynamical system fed by the interpolation data. Such a system is called the "interpolating process" and its determination is shown to be more simple than others algorithms previously developped for spline functions. Surfaces are investigated as tensor-products of previously defined functions.

1. INTRODUCTION

Computer-assisted design & manufacturing of mechanical objects require algorithms to define geometric shapes in two or three dimensions.

In addition of simple primitive shapes such as straight lines, plans, circles, ..., the conception of car bodies emphasizes the need to design warped curves and surfaces issued from aesthetical and/or aerodynamical considerations. With this aim in mind, interpolation algorithms give a powerful tool to transform a "sketch" defined by a net of points with some tangency conditions into the viewing of the continuous shape created from those discrete data.

In the midst of CAD problems, the convenience of a CAD operator requires the shortest lap of time as possible between the submission of a sketch and the final display of the corresponding surface on the CRT.

This point emphasizes rapidity as a key-property for such used algorithms.

The proposed work defines an original approach in the problem of modelization of warped surfaces defined in the purpose of their conception and tooling.

By this way, a surface is defined by a family of warped curves representing the trajectories followed by a graphical screen-pointer or a tool.

Those trajectories are interpreted as the output generated by a linear stationary dynamical system, implemented either by an analog-circuit or by a simulation program. This system is fed by a control input whose explicit expression is given from the imposed interpolation specifications.

The system and its control input is called "interpolating process".

The specifications imposed to interpolations curves are the same as those that define the interpolation SPLINES functions. Thus, the curves generated • are continuous until 2q-2 order (for any $q > 0$), • are analyticaly defined between two points, • minimize a shape-functional (for example in order to get the smoothest possible variations between the interpolated points).

By the use of the optimal control theory of linear processus, we propose a simple solution to compute the "interpolating process".

Indeed, it is shown that this "interpolating process" is represented by a continuous linear process whose initial conditions are computed from the initial data of the problem (net of points, conditions of tangency, ...).

After having defined the properties that specify the generative and directive interpolation curves, we propose an original formulation of the problem, leading to the complete definition of the "interpolating process" generating those curves.

The interpolation by surfaces is investigated following this technique.

2. DEFINITION OF INTERPOLATION CURVES DEFINED BY A FUNCTION $z = f(x)$

Let a net of points :

$$\{P_i = (x_i, z_i) \in \mathbb{R}^2 \ ; \ i = 1, \ldots, n\}$$

We propose to determine a curve that passes successively through the points $P_1, P_2, \ldots, P_n$.

By limiting the study to the nets ordered by the following relation :

$$(1) \qquad x_1 < x_2 < \ldots < x_i < x_{i+1} < \ldots < x_n$$

we are led to search for an interpolation of equation $z = f(x)$.

So as to simplify the study of the method, we

develop the determination of order 2 SPLINE and SPLINE UNDER-TENSION interpolations [5] [3]. The work can easily be extended for any order SPLINE functions. Those curves are defined by the following properties :

P1) $z = f(x)$ and its first and second derivatives are continuous on $[x_1 , x_n]$.

P2) For $i = 1, \ldots, n$; $f(x_i) = z_i$. In addition, one or the other tangent in the endpoints may be given independently.

P3) $f(x)$ minimizes the following $\phi(g)$ functional of form :

$$\phi(g) = \int_{x_1}^{x_n} (g''(x))^2 + \sigma^2 (g'(x))^2 . dx \qquad (\sigma > 0)$$

Remark : $\sigma = 0$ corresponds to the particular case of polynomial splines functions of order two for which the proposed method has been developped in [1] and [4]. In this case, we search for the "smoothest" possible interpolation. The term introduced with $\sigma \neq 0$ realizes a compromise between the properties of minimal length and smoothness of the interpolation. Reduction of the length of the interpolating curve eliminates extraneous critical points [3].

We propose to determine the analytical expression of the function $f(x)$ corresponding to properties (P1, P2, P3), by regarding it as the trajectory generated by a dynamical system whose representative point evoluates with respect to the variable $x \in [x_1 , x_n]$.

3. NOTION OF "INTERPOLATING PROCESS"

So as to explicit all the terms involved in the definition of the searched function (P1, P2, P3), we propose to represent it under the following block-diagram (Fig. 1).

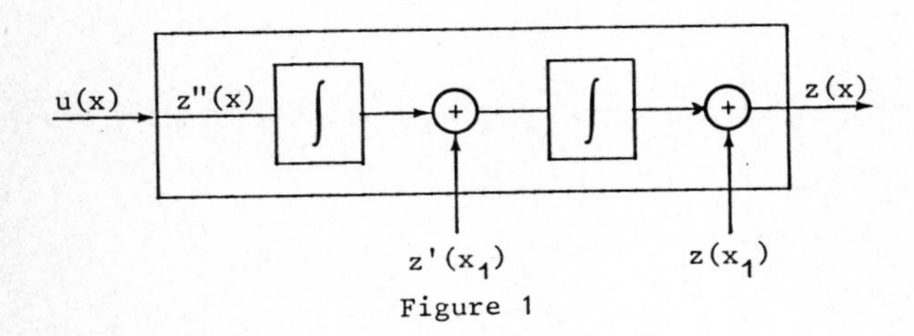

Figure 1

The integrators represented in this scheme are defined with respect to the variable $x \in [x_1 , x_n]$.

The system can be represented by the following state-space equation (3.1) :

$$(3.1) \quad \begin{cases} \dfrac{d}{dx} (\bar{z}(x)) = A\, \bar{z}(x) + b\, u(x) \\ z(x) = c^T . \bar{z}(x) \end{cases}$$

$$\bar{z}^T(x) = [z(x), z'(x)]$$

$$b^T = [0 , 1] \; ; \quad A = \begin{bmatrix} 0 & 1 \\ 0 & 0 \end{bmatrix}$$

$$c^T = [1 , 0]$$

The computation of the interpolation is thus reduced to the determination of the continuous input $u(x)$ (P1) of the previous linear system (3.1) that produces the trajectory passing by the given points (P2) and minimizes the functional $J(u(.) , \bar{z}(.))$ (P3).

$$J(u(.),\bar{z}(.)) = \int_{x_1}^{x_n} (u^2(x) + \bar{z}^T(x) . \begin{bmatrix} 0 & 0 \\ 0 & \sigma^2 \end{bmatrix} \bar{z}(x)) dx \qquad (3.2)$$

We propose to call the previous dynamical system and its control $u(x)$: "Interpolating Process".

This last formulation allows to resolve the problem by the use of the quadratic optimal control theory for which we propose to recall the principal results.

4. OPTIMAL CONTROL OF LINEAR SYSTEMS : A BRIEF REVIEW

Consider a system described by the following state equation (4.1) :

$$(4.1) \quad \begin{cases} \dfrac{d}{dx} (\bar{z}(x)) = A.\bar{z}(x) + b\, u(x) \\ z(x) = c^T . \bar{z}(x) \end{cases}$$

with $u(x), z(x) \in \mathbb{R}$

$\bar{z}(x) \in \mathbb{R}^q$

$A \in \mathbb{R}^{q \times q}$; $b, c \in \mathbb{R}^q$

A specific trajectory $\{\bar{z}(x) \; ; \; x \in [x_1 , x_n]\}$ is caracterized by :

. some given components of $\bar{z}(x)$ in some fixed points $x_i \in [x_1 , x_n]$,

. the fact that the function $z(x)$ ($x \in [x_1 , x_n]$) minimizes a quadratic functional of both the state and the control-law (4.2) :

$$(4.2) \quad \begin{cases} J(u(.),\bar{z}(.)) = + \dfrac{1}{2} \displaystyle\int_{x_1}^{x_n} (\bar{z}^T(x).P.\bar{z}(x) + u^2(x)) dx \\ P \in \mathbb{R}^{q \times q} \; ; \; P \geq 0 \end{cases}$$

This trajectory is denoted optimal trajectory (in the sens of the criterium (4.2)).

It is shown [6], that in every point x of an interval $[x_1 , x_n]$ which does not contain a point x_i for which there exists a constraint on the trajectory, the control $u(x)$ that produces the optimal trajectory is the one that maximizes the HAMILTONIAN $H(u(x), z(x), \Psi(x))$ (4.3) at this point.

$$H(u(x),\bar{z}(x),\Psi(x)) = -\frac{1}{2}(u^2(x) + \bar{z}^T(x)\,P.\bar{z}(x)) + \Psi^T(x).(A\bar{z}(x) + b\,u(x)) \quad (4.3)$$

Without any constraint on $u(x)$, the previous result gives then :

$$u(x) = b^T.\Psi(x) \quad (4.4)$$

The evolution of the adjoint vector $\Psi(x) \in R^q$ at this point satisfies the following system of differential equations (4.5) :

$$\frac{d}{dx}(\Psi(x)) = -A^T.\Psi(x) + P.\bar{z}(x) \quad (4.5)$$

Transversality conditions :

The variation δJ of the functional (4.2) which is due to an admissible variation $\delta\bar{z}(x_1)$ of the initial value $\bar{z}(x_1)$ in the endpoint x_1, is given, for its first order, by the following relation :

$$\delta J = \Psi^T(x_1).\delta\bar{z}(x_1) \quad (4.6)$$

As J is minimal with respect to the control induced by $\Psi(x_1)$, it comes for example, that if one component of $\bar{z}(x_1)$ is not fixed, the corresponding component of $\Psi(x_1)$ is then equal to zero (4.6).

The searched control $u(x)$ appears then as the output generated by a dynamical system (4.4), (4.5) from initial conditions obtained from transversality conditions (4.6).

5. EXPLICIT DETERMINATION OF THE INTERPOLATING PROCESS

We propose, at first, to explicit the structure of the interpolating process, by use of the formulation of the problem mentioned previously, treated as an optimal-control problem (§ 3 & 4).

Further, we propose to clear the computation of the parameters involved in the structure from the problem data (i.e. net of interpolating points).

5.1 Interpolating process structure

The results recalled in the previously section as well as the state-space formulation of the interpolation problem initialy set (§ 2) show that in each interval $]x_i, x_{i+1}[$, which does not contain datas, the searched curve is given by the following scheme (Fig. 2) :

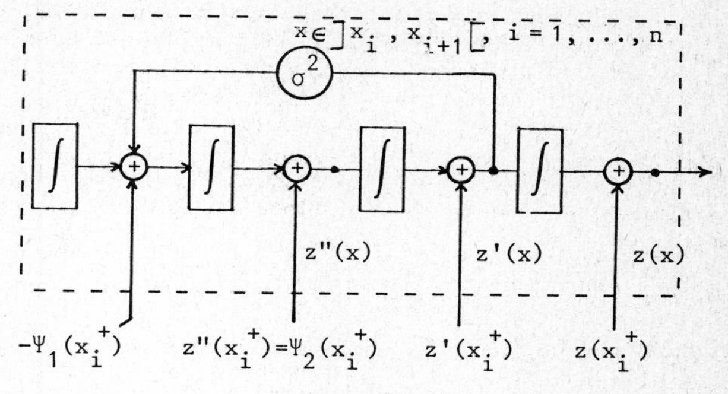

Figure 2

The state equation of this dynamical system is obviously given by :

$$\frac{d}{dx}(v(x)) = \tilde{A}.v(x) \quad (5.1.1)$$

with $v^T(x) = [z(x), z'(x), z''(x) ; -\Psi_1(x)]$

$$\tilde{A} = \begin{bmatrix} 0 & 1 & 0 & 0 \\ 0 & 0 & 1 & 0 \\ 0 & \sigma^2 & 0 & 1 \\ 0 & 0 & 0 & 0 \end{bmatrix}$$

The equation which defined numericaly $z(x)$ on $]x_i, x_{i+1}[$ is thus obtained by integrating the previously state equation (5.1.1) from the value of $v(x_i^+)$.

$$(5.1.2) \quad \begin{cases} \forall h \in]0, x_{i+1} - x_i[\\ v(x_i + h) = M(h).v(x_i^+) \\ M(h) = \exp(\tilde{A}.h) \end{cases}$$

$M(h)$ is respectively given by the following expressions :

$$(5.1.3) \quad \sigma = 0 \rightarrow M(h) = \begin{bmatrix} 1 & h & h^2/2! & h^3/3! \\ 0 & 1 & h & h^2/2! \\ 0 & 0 & 1 & h \\ 0 & 0 & 0 & 1 \end{bmatrix}$$

$$(5.1.4) \quad \sigma \neq 0 \rightarrow M(h) = \begin{bmatrix} 1 & \frac{\mathrm{Sh}\,\sigma h}{\sigma} & \frac{\mathrm{Ch}\,\sigma h - 1}{\sigma^2} & (\frac{\mathrm{Sh}\,\sigma h}{\sigma} - h)\frac{1}{\sigma^2} \\ 0 & \mathrm{Ch}\,\sigma h & \frac{\mathrm{Sh}\,\sigma h}{\sigma} & \frac{\mathrm{Ch}\,\sigma h - 1}{\sigma^2} \\ 0 & \sigma\,\mathrm{Sh}\,\sigma h & \mathrm{Ch}\,\sigma h & \frac{\mathrm{Sh}\,\sigma h}{\sigma} \\ 0 & 0 & 0 & 1 \end{bmatrix}$$

The function $z(x)$ generated by the dynamical system represented (Fig. 2) is then completely defined from (n-1) initial conditions $v(x_i^+) \in \mathbb{R}^4$ ($i = 1, \ldots, n-1$).

We now complete the interpolating process by associating to the proposed dynamical system a computation-block that gives the initial conditions $v(x_i^+)$ $(i = 1, \ldots, n-1)$ from the initial data (the set of points to interpolate).

The interpolating process structure can then be represented in the following way (Fig. 3).

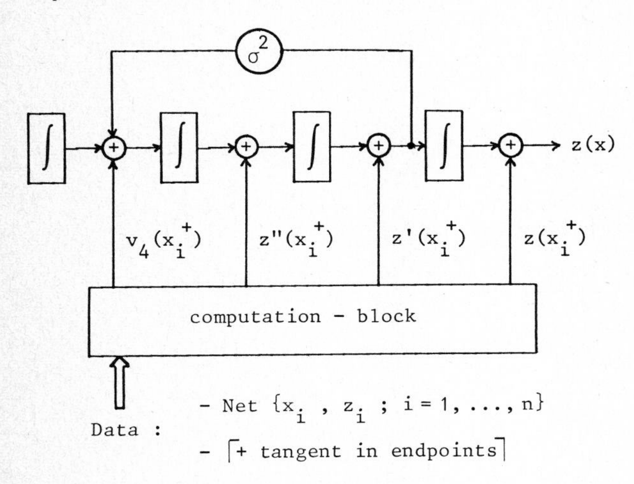

Figure 3

We propose to explicit the computation-block.

5.2 Computation of interpolation parameters

We first show that $v(x_{i+1}^+)$ can be obtained from $v(x_i^+)$ by an affine relation. The computation of $v(x_1^+)$ will come to conclude this method.

5.2.1 Relation between $v(x_{i+1}^+)$ and $v(x_i^+)$

The advance function (5.1.2) allows to compute $v(x_{i+1}^-)$ from $v(x_i^+)$ by :

$$v(x_{i+1}^-) = M(x_{i+1} - x_i).v(x_i^+) \tag{5.2.1.1}$$

Given that $z(x)$ is continuous up to the order 2 on $[x_1 , x_n]$, the first three components of $v(x_{i+1}^-)$ equate the corresponding components of $v(x_{i+1}^+)$. The fourth component is given from the data $z(x_{i+2}) = z_{i+2}$, by the first line of the advance equation (5.2.1.1) on interval $[x_{i+1} , x_{i+2}[$

$$\begin{aligned} z_{i+2} &= [1,0,0,0].M(x_{i+2} - x_{i+1}).v(x_{i+1}^+) \\ &= [1 ; m_{12}(i+2) ; m_{13}(i+2) ; m_{14}(i+2)] .v(x_{i+1}^+) \end{aligned} \tag{5.2.1.2}$$

The relation between $v(x_{i+1}^+)$ and $v(x_i^+)$ can then be represented under the following matrix form (5.2.1.3) :

$$\begin{bmatrix} 1 & 0 & 0 & 0 \\ 0 & 1 & 0 & 0 \\ 0 & 0 & 1 & 0 \\ 1 & m_{12}(i+2) & m_{13}(i+2) & m_{14}(i+2) \end{bmatrix} .v(x_{i+1}^+) =$$

(5.2.1.3)

$$\begin{bmatrix} 1 & 0 & 0 & 0 \\ 0 & 1 & 0 & 0 \\ 0 & 0 & 1 & 0 \\ 0 & 0 & 0 & 0 \end{bmatrix} M(x_{i+1}-x_i).v(x_i^+) + \begin{bmatrix} 0 \\ 0 \\ 0 \\ z_{i+2} \end{bmatrix}$$

This expression is shortened into the following form :

$$N(i+2).v(x_{i+1}^+) = Q.M(x_{i+1}-x_i).v(x_i^+) + P(i+2)$$

The matrix $N(i+2)$ is invertible and its inverse is easily computable. This gives the following relation between $v(x_{i+1}^+)$ and $v(x_i^+)$:

$$v(x_{i+1}^+) = A(i+2).v(x_i^+) + b(i+2) \tag{5.2.1.4}$$

with
$$A(i+2) = N^{-1}(i+2).Q.M(x_{i+1}-x_i)$$
$$b(i+2) = N^{-1}(i+2).P(i+2)$$

We must now initialize this previous induction relation by the computation of $v(x_1^+)$.

5.2.2 Computation of $v(x_1^+)$

The first component of $v(x_1^+)$ is, by the definition of $v(x)$ (5.1.1) equal to z_1.

The second component is equal to $z'(x_1)n$ which can be one of the problem data.

If it is not, the transversality condition (4.6) at the point x_1 establishes the third component $(z''(x_1))$ of $v(x_1^+)$ equal to zero.

Two components are then directly known. There fails two relations between the components of $v(x_1^+)$.

The first relation is obtained from the advance equation (5.2.1.2) not used in the previous step (5.2.2.1).

$$z_2 = [1 , 0 , 0 , 0] .M(x_2-x_1).v(x_1^+) \tag{5.2.2.1}$$

The missing relation comes from the knowledge of $z'(x_n)$ or the nullity of $z''(x_n)$. This relation is the second (resp the third) line of the equation which joins $v(x_n^-)$ to $v(x_1^+)$ obtained from the previous relations :

(5.2.2.2)
$$v(x_n^-) = M(x_n - x_{n-1}).(\prod_{k=3}^{n} A(k)).v(x_1^+) + M(x_n - x_{n-1}).\left(\sum_{k=3}^{n}(\prod_{e=k+1}^{n} A(e)).b(k)\right)$$

5.2.3 Constitution of the computation-block

From the subprograms that define the matrix $M(x_{i+1} - x_i)$ (5.1.3) (5.1.4) and A(i) (5.2.1.4), as well as the vector b(i) (5.2.1.4) directly from the data, the computation of the interpolation parameters is summed up to the computation of the affine relation (5.2.2.2), then to the resolution of a twice order linear system.

To emphasize the simplicity of those computations we may compare them to the solution proposed in [2] in the case of $\sigma = 0$, which requires the solution of a tridiagonal system of linear equations whose order is equal to the number n of points in the net. Otherwise Pilcher [3] resolves the case $\sigma \neq 0$ by resolving a system of equations of order 4n-4.

The simplicity of the computation-block we propose induces thus a high treatment speed, which answers to the specifications formulated in the introduction.

An example of proposed algorithm is represented below (Fig. 4) for different values of σ.

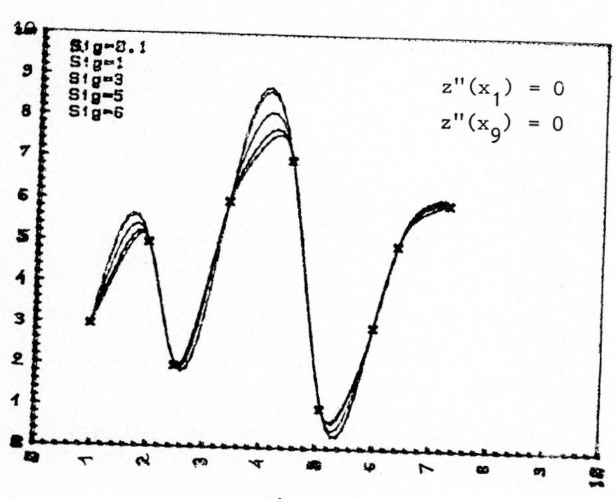

Figure 4

6. SURFACE GENERATION

So as to simplify, we propose to limit the statement to the case of surfaces defined by a function z(x , y) over a rectangular net.

$$x \in [x_1 , x_m] \; ; \; y \in [y_1 , y_n]$$

The problem data are constituted of the knowledge of the function value in the points of the net, with eventually some tangent planes fixed in the endpoints.

(6.1)
$$\forall i \in 1, \ldots, m \; ; \; \forall j \in 1, \ldots, n \qquad z(x_i, y_j) = z_{ij}$$

The surface that interpolates those data must be continuous up to the order 2.

We propose to define its form from those of the generatives :

(6.2) $\quad z(x , y_0 = Cst \in [y_1 , y_n])$

or directives curves :

(6.3) $\quad z(x_0 = Cst \in [x_1 , x_m] , y)$

This approach can be motivated by the fact that the evolution of two "smooth" curves generally called generative and directive [7], along the x and y axes generate a surface having the same property of "smoothness".

Each curve is specified by the problem of the previous chapter (§ 2, Properties P1, P2, P3) and can be generated by an "interpolating process".

The way to define a surface is then the following :

1) Interpolation of m curves defined by the m unidimensionnal nets :

$$\mathcal{R}_i = \{y_j , z_{ij} \; ; \; j=1,\ldots,n \; ; \; i=1,\ldots,m\}$$

2) From the knowledge of the m directives $z(x_i , y)$, we can compute any generative

$$z(x , y_0 = Cte \in [y_1 , y_n])$$

The problem is then resolved according to the following scheme :

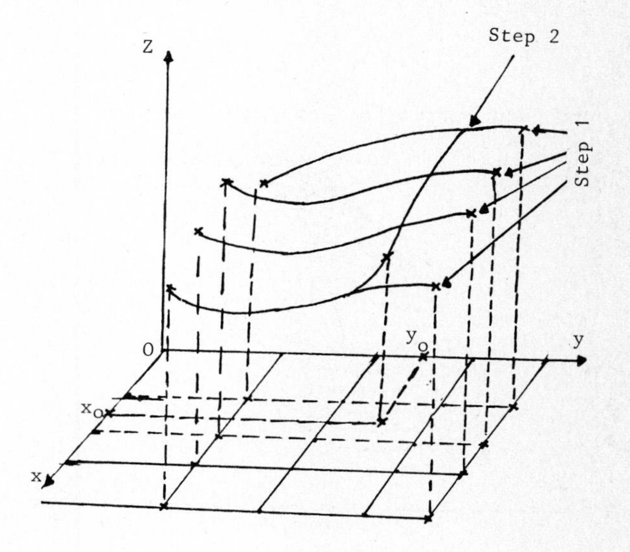

Figure 5

We propose now to show that the volume of computation needed by step 2 can become independant of the number of generatives we want to draw.

6.1 Analysis of step 2

After step 1, we know the value of $z(x_i, y)$ for each $\tilde{y} \in [y_1, y_n]$.

The interpolation of the net $\{x_i ; y_o ; z(x_i, y_o)$ $i = 1, \dots, m\}$ is executed by an interpolating process whose computation block determines the components of the following vector (6.1.1).

$$(6.1.1)\quad v^T(x_1,\tilde{y}) = \left[z(x_1,\tilde{y}) ; \frac{\partial z}{\partial x}(x_1,\tilde{y}) ; \frac{\partial^2 z}{\partial x^2}(x_1,\tilde{y}) ; \frac{\partial^3 z}{\partial x^3}(x_1,\tilde{y}) \right]$$

Two components of $v(x_1,\tilde{y})$ are explicit at once : for example when $\partial z(x_1,\tilde{y}) / \partial x$ is not fixed, $z(x_1,\tilde{y})$ is determined from step 1 and $\partial z^2 / \partial x^2$ $(x_1,\tilde{y})$ is zero. The computation of the two other components are obtained by the solution of a twice order linear system the second member of which is affine in values of $z(x_i,\tilde{y})$ computed at step 1.

It comes the following relations :

$$\begin{cases} v_1(x_1,\tilde{y}) = z(x_1,\tilde{y}) \\ v_3(x_1,\tilde{y}) = 0 \\ \begin{pmatrix} C_{11} & C_{12} \\ C_{21} & C_{22} \end{pmatrix} \begin{pmatrix} v_2(x_1,\tilde{y}) \\ v_4(x_1,\tilde{y}) \end{pmatrix} = \begin{pmatrix} z(x_2,\tilde{y}) - z(x_1,\tilde{y}) \\ \sum_{i=1}^{m} \gamma_i . z(x_i,\tilde{y}) \end{pmatrix} \end{cases} \quad (6.1.2)$$

The computation of C_{11}, C_{12}, C_{21}, C_{22}, γ_i $(i = 1, \dots, m)$ is executed only once (Fig. 5).

6.2 Example

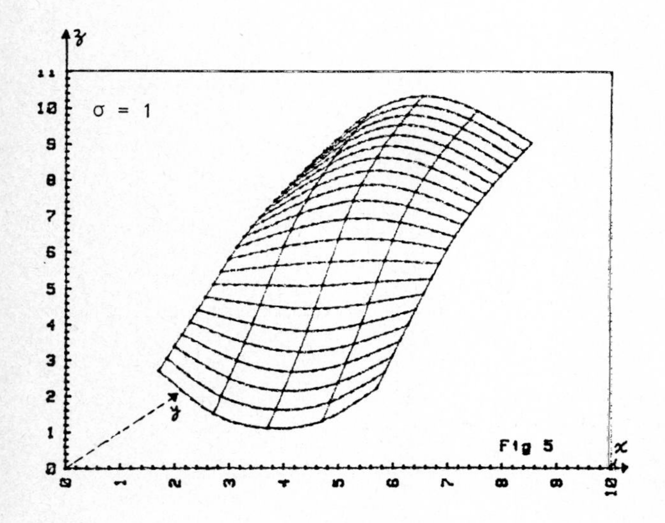

Figure 5

7. CONCLUSION

The definition of the "interpolating process" gives adequate solution to the problems of modelization of warped curves and surfaces.

The exposed method on spline under tension of twice order can easily be generalized to an order q and to other types of splines functions (trigonometric, ...).

BIBLIOGRAPHY

[1] R. HAJNASSAR, D. MEIZEL, P. BIELEC
"Une procédure rapide de calcul et de tracé des fonctions spline d'ordre 2"
Actes du Congrès IASTED ICD 82, Tunis, septembre 1982.

[2] G. MARCHOUK
"Méthodes de calcul numérique"
Editions Mir, Moscou, 1980.

[3] D. PILCHER
"Smooth approximation of parametric curves and surfaces"
Ph. D. Thesis, Mathematics Department, University of Utah, 1973.

[4] D. MEIZEL, R. HAJNASSAR, P. BIELEC
"Une procédure rapide d'interpolation de données ponctuelles"
Actes du Congrès IASTED ICD 82, Tunis, Septembre 1982.

[5] P.J. LAURENT
"Approximation et optimisation"
Collection Enseignement des Sciences, 13, Hermann, Paris, 1972.

[6] L. PONTRIAGUINE, V. BOLTIANSKI, GAMKRELIDZE, MITCHENKO
"Théorie mathématique des processus optimaux"
Editions Mir, Moscou, 1974.

[7] P. BEZIER
"Procédé de définition de courbes et surfaces non mathématiques ; Système UNISURF"
Automatisme, Vol. 13, Mai 1968.

Simulation in Engineering Sciences
J. Burger and Y. Jarny (eds.)
Elsevier Science Publishers B.V. (North-Holland)

SIMULATION - AID TO PROCESS INTERACTION

D. De Buyser, L. De Wael, G.C. Vansteenkiste

University of Ghent
Ghent, Belgium

Simulation support to process control is discussed in the paper with particular emphasis to fermentation processes. The computational tools used fulfill three purposes : education, modelling and process monitoring.
A description and justification is given of the complete set-up. A highly instrumented pilot-fermentor is coupled via a data-acquisition system to a mini-computer under control of an intelligent terminal of the CDC-PLATO network. The different functions of the units are described as there are instructional task, user-process interaction, information display, data collection and processing.

1. INTRODUCTION

A prerequisite to process interaction is to have some knowledge about the process itself. Sources of this knowledge are a priori information, process data and models. The support to the control of a fermentation process from an education, process monitoring and modelling standpoint are discussed in this publication.

2. THE PROCESS UNDER STUDY

Figure 1 gives an overview of the present computer coupled fermentation installation.

It is used for the study of aerobic fermentations in general. The production of the oligopeptide antibiotic gramicidin S, produced by certain Bacillus brevis strains, is the subject of intensive research.

From the point of view of process control it seems that the formation of gramicidin S is mainly influenced by the oxygen level and pH. Therefore the instrumentation is available for controlling these parameters by the computer.

Because one of the main bottlenecks in computer-controlled fermentation is the availability of adequate sensors, the system is also used for

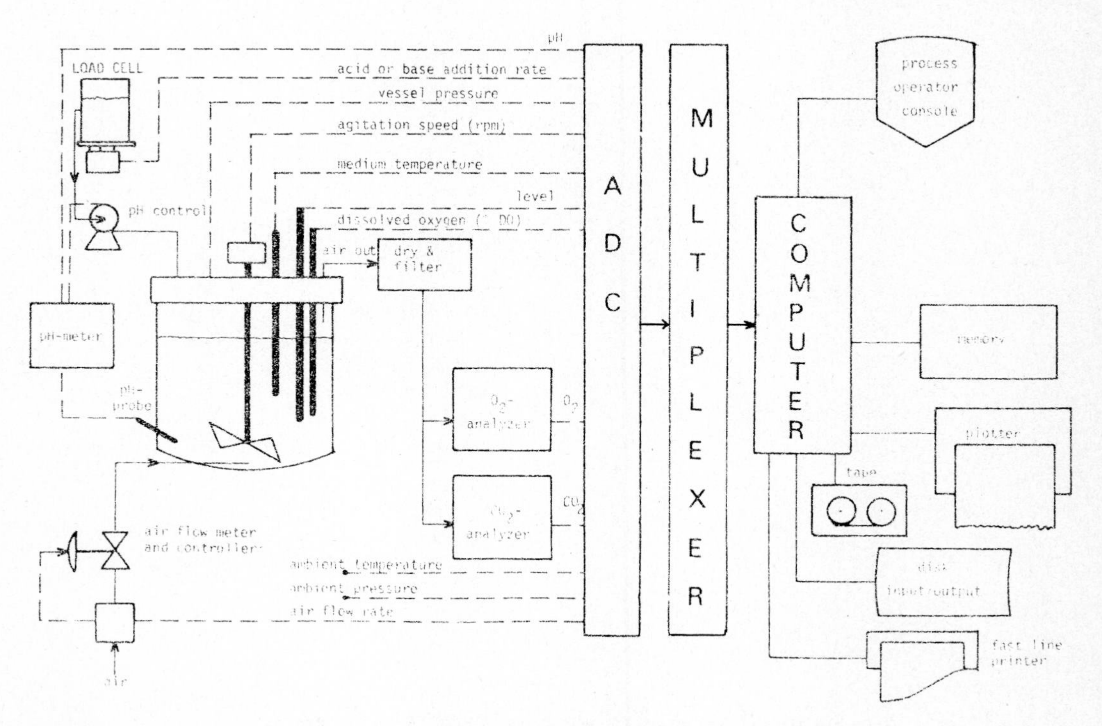

Figure 1 : Fermentor + measured variables

the development of 'net-effect' sensors. These provide immediate information on the physiological and biological state of the process, in contrast with the measurable physical (fermentorpressure, fermentortemperature, agitationspeed, ...) and chemical variables (% O_2, % CO_2 in exhaust gas, dissolved oxygen).

3. THE COMPUTER CONFIGURATION USED

The complete set-up is given in figure 2. To start with an overview of the different components will be given. Then some of them will be discussed in more detail.

Physical variables from the fermentor are converted in numbers by the data-acquisition system. By a transducer the variable (pressure, temperature, concentration,...) is transformed into a voltage. This voltage is converted in a 12 bit integer by the ADC.

Some physical variables cannot be measured online. They are measured on samples taken from the fermentatorculture at regular intervals. The procedures contain different steps, take some time and are duplicated. Examples are dry cell weight, antibiotic (bio-assay) and protein determination, gaschromatographic analysis (substrate : fructose, glycerol, inositol, ...; products : volatile and non volatile fatty acids, carbonic and tricarbonic acids, ...) and amino-acid analysis.

These raw data are put in the computer by the operator and stored in the database and processed (e.g. the antibiotic concentration is calculated by the use of standards measured with the samples).

Data acquisition is supervised by a PDP 11/34 minicomputer. The raw data are filtered, processed and stored in a database. The measured physicochemical variables don't give accurate information about the physiological state of the fermentation process itself. They are used as inputs to programs ("net-effect sensors") that calculate on-line process state variables such as biomass, substrate consumption and product formation. These computed variables are also stored in the database.

Another task of the minicomputer is process control. To reduce the load on the computer setpoint control is used. Agitation speed, pH and inlet air flow are controlled by computerset analog controllers. The purpose of the control is to optimize yield, but the optimal conditions depend heavily on the micro-organism used and the products formed.
Several tools help the experimenter to find this optimal trajectory.

Information provided by the process specialist (a microbiologist) is delivered by the CDC-PLATO computer based education system. This system is accessed via the PLATO-terminal linked to a mainframe by a special datatransmissionline. The PLATO system functions as a knowledge exchange vehicle between the microbiologist and the control engineer. A priori information about the process is given in an informal way by means of text, diagrams, graphs and pathways. These data are stored in a lessondatabase on system discs.

This a priori information together with data from experiments are used to build models to describe process knowledge in a formal way.

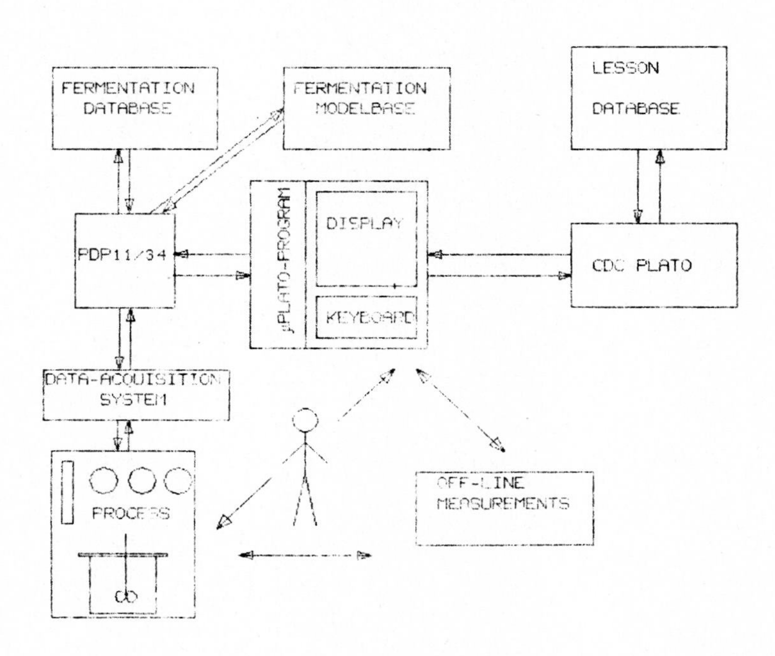

Figure 2 : Complete set-up of fermentation system

These models are stored in a modelbase.

An interface between the process computer and the mainframe is provided by microprogramming the terminal. Real data can be used in an instructional lesson as example or demonstration.

4. THE COMPUTER BASED EDUCATION SYSTEM (CBE)

4.1. Introduction

As an alternative to the traditional education some computer producers have introduced CBE . Computer based education can be separated in two main parts : computer aided instruction (CAI) and computer managed instruction (CMI).

In CAI the instructor programs a computer to transfer knowledge to the student. Because the student follows a lesson in front of a terminal this type of education can be individualized. The original and new acquired knowledge of the student can be probed by tests and depending on the results he can be routed through different parts of the lesson. In this way a student acquainted with the material only gets a brief summary. Somebody not familiar with the new knowledge receives a step for step development of the new topics.

CMI is more general than CAI. In CMI the computer indicates to the student the different learning materials he has to use. These can be on several media : CAI, video, slides, textbooks. After the student has worked through this material, he has to complete tests. Depending on the results he can pass to a new topic, or in case of bad performance he has to restudy the old material or gets assigned new learning materials for the old topic.

4.2. The PLATO system-courseware

The programs written in the TUTOR language that deliver instruction to the user compose the courseware (lessons).

There are four distinct types of lessons. Drill and practice lessons learn the student new concepts by putting a sequence of simple questions and giving appropriate feedbacks. Questions that are answered incorrectly are repeated at the end of the session. This type of instruction is particularly suited for language apprehension.

Tutorial lessons introduce new knowledge to the student. Such a lesson consists of a presentation of new facts, followed by a test to check if the student has grasped the new material. If this is the case the lesson continues. If not the student is given a review of the material. A lesson written in this way can ressemble very much to a book. The PLATO-system however offers the possibility to present material in a stepwise manner, to accentuate and repeat important topics. It also requires a more active participation of the user because he has regularly to answer questions, without these answers he cannot proceed.

New concepts can also be introduced by use of simulation. In this way the course of real processes can be speeded up or delayed. Only the dominant aspects of the process can be shown and highlighted. Examples are experiments from physics, chemistry. It is also possible to give the student the possibility to control the conditions of the experiment.

On the system already exists a vast amount of lessons covering a wide range of disciplines. To find his way in this wood of lessons the user disposes of a catalog of lessons in which all published lessons are summarized.

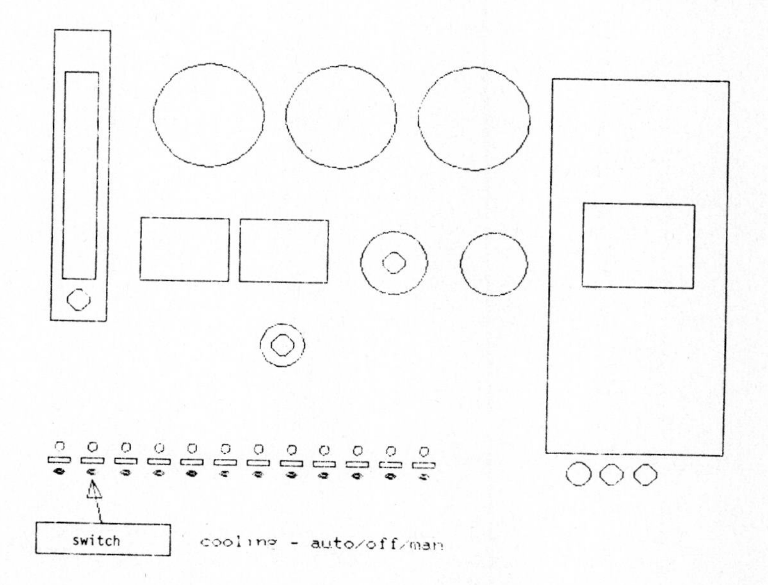

Figure 3 : Frontview of fermentorpanel

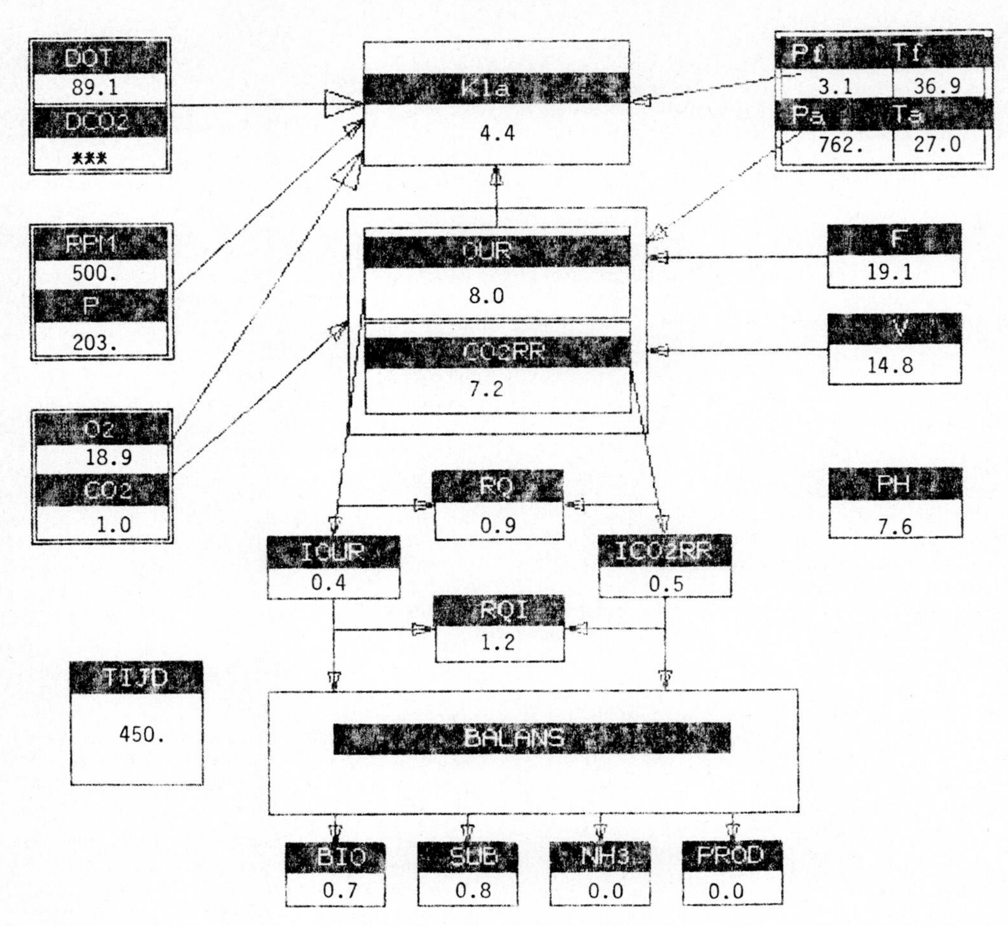

Figure 4 : Snapshot of process

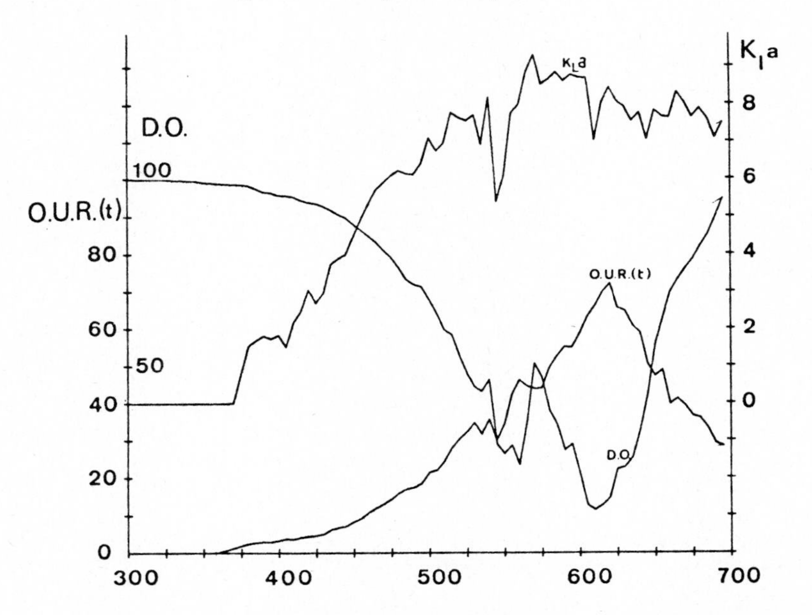

Figure 5 : Time course

4.3. The PLATO-system as an educational tool in fermentation

The fermentor with its associated instruments and facilities forms a fairly complex device. A lesson has been written that has two main objectives. The student should understand the reason and the use of the different valves and switches on the fermentor and know the instructions for the instruments and appliances. Secondly he should be able to conduct a total fermentation process consisting of the following steps : sterilisation, instrument calibration, inocculation, fermentation, separation of the product and termination of the process.

In this way the PLATO-system functions as a guide for experimentation. The lesson consists of three distinctive parts. The first is tutorial. It transfers the knowledge from the process specialist (microbiologist) to the student. Besides a brief introduction to fermentation processes and some generalities it consists of an online manual of the different devices. A typical display from this part is given in figure 3. The frontview of the fermentorpanel is shown. The function of the different switches is explained in a stepwise manner.

The second part of the lesson makes use of the simulation capabilities of the PLATO-system. The PLATO-terminal is programmed to provide an interface between the system and the process. To get acquainted with the course of a fermentation process the user can observe the trajectories of the different variables. Data can be displayed in a numerical or a graphical form. Measurements from different fermentations can be compared with each other and the influence of the distinct environmental conditions can be assessed. The impact of the interventions of the operator on the process can be evaluated.

The third part of the lesson consists of a computer guided execution of the distinct steps of the fermentation process. In this stage of the lesson the PLATO-terminal is programmed to provide an interface between the PLATO-system and the real-time process.

The student is questioned about the different steps he has to perform on the installation to initiate the different parts of the process. Then he is asked to execute those actions on the real process. It is not possible to check the settings of switches and potentiometers and the opening of valves directly on the process, but their effects can be derived from the measured variables. If these variables are not between acceptable limits, the student is given appropriate feedback to check all over again. If unacceptable conditions occur an alarm is set and intervention of an experienced operator is required.

5. PROCESS MONITORING

On the process console the user can have a display of the instantaneous values of the distinct variables in a diagram form (figure 4). This display gives a snapshot of the process and illustrates the relationship between the variables. Not only the physical variables are displayed but also the important physiological variables that give the operator a better view of the actual state of the process.

It is also possible to get a graphical output of the time course of several variables between two specified moments. Figure 5 illustrates the time course of dissolved oxygen (D.O.), oxygen uptake rate (O.U.R.) and volumetric transfertcoefficients (k_1a).
The operator has also the possibility to input the raw data from the off-line processing. The different manual actions carried on the process are also to be documented.

When variables go outside acceptable ranges an alarm is set to alert the operator.

6. MODELLING AND SIMULATION

Another important tool in the process interaction is the use of simulation.
The user can select a parametrized model from the model database and specify the parameters and the experimental frame. This experimental frame specifies the conditions under which the simulation has to be run : initial conditions, termination conditions, trajectories of the inputs to the model, and output variables to be calculated and stored. As an illustration a simple model consisting of the Monod equation for growth and the Brown equation for product formation is given in figure 6. The initial conditions and parameters are specified in the beginning.

```
MUMAX = 0.98

M = 0.1

KS = 34

KP = 0.133

YS = 6.5

XO = 0.01

SO = 60.

TO = 2.

C    DIFFERENTIAL EQUATIONS.
C    -----------------------

X' = X * MUMAX * S / (KS + S)

S' =  - YS * X' - M * X

P' = KP * X'(T-TO)
```

Figure 6 : Simple model for product formation

The parameters of the process and the experimental frame can be changed interactively.

Modelling can be an aid in experimental design. A new model can result in new questions about the system. This can lead to new experimental frames and new experiments.
The ultimate goal is to find a model that is accurate enough to use as a process model for optimal control.

7. CONCLUSION

Because the grayness of biological system there are until now no general accepted ways for process control. Questions that remain to be answered are : which are the important process variables, how can they be measured accurately, which are the optimal conditions for maximum yield.
Because the answer to those questions also depends very heavily on the process under study a lot of experimental work remains to be done.
To help the scientist in his experimental design powerful tools are needed. Adequate knowledge transfer, process monitoring and simulation can be of great help.

Simulation in Engineering Sciences
J. Burger and Y. Jarny (eds.)
Elsevier Science Publishers B.V. (North-Holland)

DIGITAL SIMULATION OF TWO LEVEL SUBOPTIMAL CONTROL SYSTEMS

Andrzej Świerniak

Department of Automatic Control,
Silesian Technical University,
Gliwice, Poland

A two level linear control system is considered with the direct control lower level realizing a stabilization task with a prescribed degree of stability and the higher level setting an optimal static set point for the lower one. It is assumed that a discrete-time controller in the lower level is designed on the basis of a reduced order state model and has the sample interval much shorter than that of the higher level interventions. The simulation results are very promissing.

1. INTRODUCTION

Two-level systems with the lower direct control level realizing a stabilization task and the higher static optimization one |1| are examples of suboptimal solutions of complex optimal control problems. In contrast to |2| we assume that digital controllers are used on both levels. A sample interval for the lower level is "small" enough that from the point of view of the higher level the lower layer may be treated as dynamic plant. However the higher layer decision is made on the grounds of a static model. The losses resulting from the use of that model are to be estimated. On the other hand the lower level treats the reference signal as being constant during a control interval and its changes are considered as the new initial state conditions. The sample interval of the higher level is long enough in comparison to that of the lower level as well as its "time constants" so that an infinite control interval is considered for design purposes. Moreover a controller is designed using a reduced order discrete-time model and assures a prescribed degree of stability. All considerations presented in the paper are approximate because of the assumptions mentioned above, but they seem to be useful from the system designer's point of view. In other words we connect vertical division of tasks |1| with time-scale and regulator-servo separation. The simulation is provided to check the validity of the assumptions and analytic estimations.

2. DESIGN OF DIRECT CONTROL LEVEL

Consider the system described in the state-space form by the continuous time model (for the perturbations from the steady state):

$$\dot{\underset{\sim}{x}}(t) = \underset{\sim}{\Phi}\underset{\sim}{x}(t) + \underset{\sim}{\Theta}u(t)$$

that is to be controlled digitally using synchronous input actuation (in the presence of the zero ordered hold) and output sampling of sample interval T_i. The sampled input and state vectors $u_k = u(kT_i)$, $x_k = x(kT_i)$ are related by a discrete-time model:

$$x_{k+1} = Fx_k + Gu_k;\ k = 0,1,\ldots,N_i \qquad (1)$$

x_o - given

where $\dim u_k = r$, $\dim x_k = n$. We assume

that the pair (F,G) is controllable.
In order to simplify the simulation the state equation (1) is assumed to be given in Jordan canonical form. It may be achieved by the use of the state transformation: $x = V\tilde{x}$, where V is the nonsigular matrix built of the eigenvectors of the matrix Θ. Furthermore we have:

$$F = e^{\Phi T_i}, \quad G = \Phi^{-1}(I - e^{\Phi T_i})\Theta$$

where Φ is built of the eigenvalues of Φ and the following conditions hold:

$$V\tilde{\Phi} = \Theta V, \qquad \Theta = V\tilde{\Theta}$$

The objective of the designer is to stabilize with a given degree of stability the m-dimensional combination of the state components in their transient states caused by changes of a set-point given by the higher level. Assuming that a sampling interval of the higher level T_i is much longer than that one of the lower level T_i i.e.

$$T_u = N_i T_i, \quad N_i \gg 1$$

the performance index to be minimized may be proposed in the form:

$$J = \frac{1}{2} \sum_{i=1}^{\infty} \alpha^{2i} [x_i^T H^T Q H x_i + u_{i-1} R u_{i-1}] \qquad (2)$$

being a discrete analogue of the modified quadratic criterion |3|. We assume that $Q = \Gamma^T\Gamma$ and R are positive difinite matrices of proper dimensions, H is of dimension $m \times n$; $\alpha \geq 1$ plays an opposite role to a discount factor |4| and is equal to $e^{\xi T_i}$ where ξ is a desired degree of stability.

Assuming that the controller has m inputs and r outputs we will design a control law on the basis of a reduced order model.

In order to do that we apply a method of dominant eigenvalues |5|.

Divide the matrix F:

$$F = \begin{bmatrix} F^m & 0 \\ 0 & F^s \end{bmatrix} \qquad (4)$$

in such a way that F^s contains Jordan blocks for the eigenvalues lying inside the unit circle far from it, after all lying within the disque of a radius $\frac{1}{\alpha}$. F^m consists of other m eigenvalues, so it is necessary at last for n-m eigenvalues to have their gains less than $\frac{1}{\alpha}$.

A reduced order model is described by:

$$x_{i+1}^m = F^m x_i^m + G^m u, \quad i = 0, 1, \ldots \qquad (5)$$

$$x_o^m = M x_o$$

where

$$M = [I^m \; O^m] \qquad (6)$$

(I^m unit matrix of dim $m \times m$, O^m node matrix of dim $m \times (n-m)$)

$$F^m M = MF \qquad (7)$$

$$G^m = MG \qquad (8)$$

Then the performance index we consider is:

$$J^m = \frac{1}{2} \sum_{i=1}^{\infty} \alpha^{2i} [x_i^{mT}(C^m+P)^T Q (C^m+P) x_i^m + u_{i-1}^T R u_{i-1}] \qquad (9)$$

where

$$C^m = HM^T \qquad (10)$$

and P ensures the equality of the steady states (if one eigenvalue is equal to unity then a state variable which is connected with it is not taken into account)

$x_\infty = (C^m + P)x_\infty^m$ when the same controls are applied.

Hence:

$$H(I^n - F)^{-1}G = (C^m + P)(I^m - F^m)^{-1}G^m \qquad (11)$$

and P may be obtained from the equation:

$$[h_{m+1} \; \dots \; h_n](I^{n-m} - F^s)^{-1} \begin{bmatrix} g_{m+1} \\ \vdots \\ g_n \end{bmatrix} = $$
$$= P(I^m - F^m)^{-1} \, G^m$$

(h_i - ith column of H, g_i - ith row of G).

Since (7) may be written as:

$$F^m M = MF \tag{12}$$

then M is an aggregation matrix |6|.

We assume that the pair $(F^m, (C^m + P)\Gamma)$ is observable. The optimal control for the model has the following form |7|.

$$\begin{aligned} u_i^m &= -(R + G^{mT}K^m G^m)^{-1} G^{mT} K^m F^m x_i^m = \\ &= -R^{-1} G^{mT} W^m F^m x_i^m = -R^{-1} G^{mT} K^m x_{i+1}^m = \\ &= -R^{-1} G^T M^T K^m x_{i+1} \end{aligned} \tag{13}$$

where K^m is a symmetric, positive definite solution of the equation:

$$K^m = \alpha^2 F^m W^m F^m + (C^m + P)^T Q (C^m + P) \tag{14}$$

and W^m is given by:

$$K^m = W^m (I^m + G^m R^{-1} G^{mT} K^m) \tag{15}$$

The solution exists because of the model controllability assured by the system equations controllability. If the control (13) is applied to the plant the value of the performance index (2) is given by:

$$J^r = \frac{1}{2} x_o^T L x_o \tag{16}$$

where L is the symmetric positive definite solution of the equation:

$$\begin{aligned} L &= H^T Q H + \alpha^2 F^T M^T W^m M G R^{-1} \, G^T M^T W^m M F + \\ &+ \alpha^2 F^T (I^n - G R^{-1} G^T M^T W^m M)^T \cdot \\ &\cdot L (I^n - G R^{-1} G^T M^T W^m M) F \end{aligned} \tag{17}$$

The most important result is given by the following theorem |8|.

<u>Theorem 1:</u> The obtained closed-loop system has all eigenvalues lying inside the circle of the radius $\frac{1}{\alpha}$.

3. EVALUATION OF THE HIGHER LEVEL

From the point of view of the higher level the lower one may be treated as a dynamic plant sampled with a sample interval T_u. The sampled input (a decision of the upper layer) $w(k) = w(kT_u)$ and output $y(k) = y(kT_u)$ are related by a model:

$$z(k+1) = Az(k) + Bw(k), \qquad k = 0, 1, \dots N; \; z(0) = z \tag{18}$$

$$y(k) = Cz(k) \tag{19}$$

where z is a state vector of dim n (We omit disturbances in the model but one may consider the measurable disturbances in the similar way as in |2|).

The decision variable w(k) is found on the base of the static model, then the state may be chosen in any convenient form. We use a canonical Jordan form. The matrix A consists of Jordan blocks connected with the eigenvalues of a matrix:

$\underset{\sim}{A} = \{(I - GR^{-1}G^T W^m M)F\}^{N_i}$ which lie inside the circle of the radius $(\frac{1}{\alpha})^{N_i}$

z is connected with x in the following way

$$z = \underset{\sim}{V} x - Dw \tag{20}$$

where $\underset{\sim}{V}$ is a matrix built of eigenvectors of the matrix $\underset{\sim}{A}$ and D is connected with B by:

$$AD - D = B \tag{21}$$

(In that sense x may be interpreted as specially defined error of the overall system under assumption that w is constant).

The static model used has a form:

$$y = C(I - A)^{-1} Bw \tag{22}$$

(for simplicity in this section we use

I instead of I^n), and the objective is to have y to be equal to the given signal s in discrete time instants kT_u. To make the problem as simple as possible consider y as a single variable and measure the performance by the average square index:

$$\phi = \frac{1}{N+1} \sum_{i=0}^{N} |y(i) - s(i)|^2 \qquad (23)$$

where N is assumed to be a large number. The decision of the higher level is given by:

$$w(k) = \frac{s(k)}{a} \qquad (24)$$

$$a = C(I - A)^{-1} B$$

Assuming that s is changing slowly we denote by b the upper bound of the velocity of s i.e $\underset{t}{Max}|\dot{s}(t)| \leq b$

Thus
$$\underset{k}{Max}|s(k+1) - s(k)| \leq bT_u \qquad (25)$$

(The reference signal s may be for example a result of a global optimization of a system divided into subsystems connected only by common resources. One of the subsystems is under consideration). Hence we have:

$$y(k) = CA^k z(0) + C \sum_{i=1}^{k} A^{k-1}B \frac{s(i-1)}{a} =$$
$$= CA^k z(0) + C(I-A)^{-1}B \frac{s(k)}{a} - C(I-A)^{-1}A^k B \frac{s(0)}{a}$$
$$-C \sum_{i=1}^{k} A^{k-i}(I-A)^{-1}B \frac{s(i) - s(i-1)}{a}$$

The error may be estimated by:

$$|y(k)-s(k)| \leq |CA^k[z(0) - (I-A)^{-1}B \frac{s(0)}{a}]| +$$
$$+ |C(I-A)^{-1} (I-A^k)A(I-A)^{-1} B|\frac{bT_u}{a}$$
$$\leq |CA^k[z(0)-(I-A)^{-1}B \frac{s(0)}{a}]| +$$
$$+ |C(I-A)^{-1} A(I-A)^{-1}B|\frac{bT_u}{a} \qquad (26)$$

Assuming for simplicity only distinct eigenvalues p_i (in other cases the perturbation technique may be used) and omitting members dependent on the initial conditions (because of the assumptions of N being large and "good" system stability), we have:

$$\phi \leq \frac{N}{N+1} \sum_{i=1}^{n} \left(\frac{p_i}{1-p_i} \frac{|a_i|}{a} \cdot b\right)^2 T_u^2$$

where $|a_i| = |\frac{c_i b_i}{1-p_i}|$; c_i, b_i are entries of the vectors C and B; but $|p_i| \leq (\frac{1}{\alpha})^{N_i}$ so the index may be estimated by:

$$\phi \leq D \cdot n \left(\frac{bT_u}{(\alpha)^{N_i} - 1}\right) \qquad (27)$$

where

$$\frac{1}{n} \leq D = \frac{\sum_{i=1}^{n} |a_i|^2}{a^2} \leq \frac{n(\max_i a_i)^2}{a} \qquad (28)$$

Note that use of the dynamical model of the lower layer to find the input variable w would require the knowledge of s in the whole control interval NT_u.

4. SIMULATION PROCEDURE

During the simulation following procedures have been used:

1) EIGEN - the procedure calculates eigenvalues and eigenvectors of the given matrix
2) DIGIT - the procedure calculates the parameters of the discrete model of the lower level
3) ORDER - the procedure calculates the parameters of the lower order model
4) RICCA - the procedure calculates the gain of the feedback control law
5) INDEX - the procedure calculates the value of the performance index
6) MODEL - the procedure simulates the behavior of the overall systems and calculates the mean square error

7) GENER - the procedure generates functions bt, $1-e^{-bt}$, sin bt and stores their values at kT_u.

The block diagram for the simulation may be illustrated in the following form:

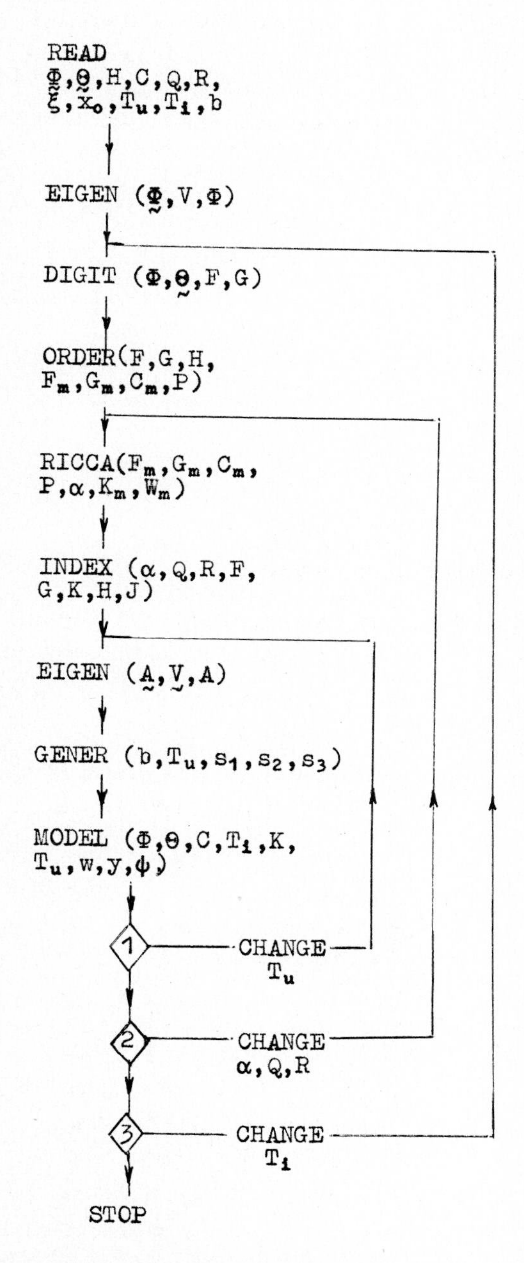

Because of lack of space we are not able to present the simulation results. However for the Nicolson's 9 x 9 model of boiler |5| the simulation results are in agreement with assumptions and analytic estimations for $N_u \geq 10$, and $bT_u < 1$ even for 3 x 3 reduced order model.

5. CONCLUSIONS

The simulation procedure has been proposed in the paper for the two-level discrete-time control system with different sampling interval short on the lower level and "long" on the higher one. The controller of the lower level in spite of using the reduced order model of the system assures the prescribed degree of stability. The higher level changes the set point on the basis of a static model of the lower one. The average square error is found to depend on the ratio of maximum increments of the reference signal and the degree of stability (from the point of view of the higher level) of the controlled system. The simulation results are in agreement with that statement. The analysis has however an approximate characteristic based on the hypothesis of the time-scale separation and the regulator-servo separation. These hypothesis have been verified in a great number of simulation examples. A part of the assumptions may be weakened without difficulty, for example the finite control interval in performance index (2) in part 2 may be considered without difficulty, as well as the multivariable problem in part 3. The assumptions in the paper has been made to neglect the main ideas.

REFERENCES:

|1| Singh, M.G. and Titli, A., Systems

Decomposition, Optimisation and Control (Pergamon Press, Oxford, 1978)

|2| Świerniak A., 'Performance deterioration in two-layer control suboptimal system', Proceedings of X IMACS World Congress, Montreal, 1982

|3| Anderson, B.D.O. and Moore,J.B., Linear Optimal Control, (Englewood Cliffs N.J. Prentice Hall, 1981)

|4| Bertsekas, D.P., Dynamic Programming and Stochastic Control, (Academic Press, New York, 1976

|5| Davison, E.J., 'A method for simplifying linear dynamic systems, IEEE Trans. on Automatic Control, AC-11, 93-101, (1966)

|6| Aoki, M., 'Control of large scale dynamic systems by aggregation', IEE Trans. on Automatic Control, AC-13, 246-253, (1968)

|7| Bryson, A.E. and Ho, Y.C., Applied Optimal Control, (Blaisdell Publishing Comp., Massachusetts, 1969)

|8| Świerniak A., A suboptimal discrete-time two-level control system, O.U.E.L. Report No 1141/82

Simulation in Engineering Sciences
J. Burger and Y. Jarny (eds.)
Elsevier Science Publishers B.V. (North-Holland)

MEAN SQUARE STABILITY AND FEEDBACK CONTROL OF A DISCRETE MACRO ECONOMIC DYNAMIC SYSTEM WITH SMALL STOCHASTIC PERTURBATIONS IN PARAMETERS

M.E.Elaraby* and A.A. EL-Kader[x]
*Prof.and[x] Doctoral student Mechanical
Design and Production Dept.,Cairo University.Egypt

The present paper is an exploratory work to formulate aprescriptive simulation model of a developing country national economy using stochastic optimal control theory. The prescriptive simulation model incorporates a set of linear difference equations with small stochastic perturbations in parameters. These are expressed as Gaussian white noise random matrix sequences. In a prescriptive approach optimal stochastic economic controls are generated such that a quadratic social welfare functional is minimized. To explore the average behavior of the system under-consideration the difference equation that govern the mean square matrix of the solution sequence is derived. The stability of this equation leads directly to the mean square stability of the macro-economic model.

1. INTRODUCTION

High rates of economic growth and price stability are the most accepted goals of a developing country market economy. The increased public acceptance of a managed economy have created considerable interest on the problem of simulating economic control policies to achieve specified economic goals. Prescriptive and predictive econometric models have attained a considerably sophisticated level in terms of complexity. |1| and |2| . By the aid of these simulation models the governmental authorities are able to test hypothetical economic policies which can be used finally to produce the desired economic objectives.

The performance of a macroeconomic dynamic system is in fact a function of the economic parameters that characterize the system. Generally, estimation of macroeconomic model parameter values making use of the past economic data, resulted in significant variations in parameter behavior for different time intervals. In reality economic parameter values vary randomly about a deterministic central value. This is mainly due to :

(a) Inherent inaccuracies in the past economic data.
(b) Effects of environmental disturbances.

Accordingly one of the major goals of this work is to simulate the economic policy of a developing country taking into account the uncertainities associated with the economic parameters. Next the effect of these parameter uncertainities is invistigated on the average behavior of the macroeconomic model (specially the mean square matrix) both from quantitative and qualitative stand points.

DESCRIPTION OF THE MACROECONOMIC SIMULATION MODEL

The model consists of eleven linear first order random coefficient difference equations with inhomogenious random parts. These are describing the relationships between all aggregated flow and stock variables characterizing the performance of the national economy. The total demand for the gross domestic product is expressed as the sum of its components and is given by :

$$G.D.P(i) = C_g(i)+C_p(i)+I(i)+E(i)-M(i)+\Delta S(i) \quad (1)$$

where:

- G.D.P. is the gross domestic product
- C_g is the governmental consumption
- C_p is the private consumption
- I is the total investiments and is given by :

$$I(i) = A_r(i)+N_d(i) + R_c(i) + S_e(i) \quad (2)$$

where:

- $A_r(i)$ is the agricultural investiments
- $N_d(i)$ is the industrial investiments
- $R_c(i)$ is the residental construction investiments
- $S_e(i)$ is the services investiments
- $\Delta S(i)$ is the change in inventories
- $E(i)$ is the total exports
- $M(i)$ is the total imports
- i is a quarterly sampling times index.

The governmental consumption simulation model is based upon a linear disposable income function that relates G.D.P and disposable income. This linear function is believed to be satisfactory for the tax structures in most of the developing countries. Accordingly we have

$$C_g(i+1) = A_g\, G.D.P(i) + W_1(i) \quad (3)$$

where:

$A_g = \delta C_g(i+1)/\delta\, G.D.P(i)$

$W_1(i)$ is an additive stochastic environmental noise term.

On the other hand the private consumption is simulated using also a linear disposable income function and an additional term to allow for the influence of population growth, then C_p is given by :

$$C_p(i+1) = A_p\, G.D.P(i) + B_p L(i) + W_2(i) \quad (4)$$

where:

$A_p = \delta C_p(i+1)/\delta\, G.D.P(i)$ and $B_p = \delta C_p(i+1)/\delta L(i)$
$L(i)$ is the labour force
$W_2(i)$ is an additive stochastic environmental noise term.

The total investimates as given by eq.(2) is modelled by :

$$I(i+1) = A_i G.D.P(i) + B_i L(i) + C_i K(i) + D_i r(i) + W_3(i) \quad (5)$$

where:

A_i, B_i, C_i and D_i are defined as in the previous equations
$K(i)$ is the agricultural, industrial and residental construction stock
$r(i)$ is the interest rate which is chosen as a control variable
$W_3(i)$ is an additive stochastic environmental noise term

The second term of this model allows for the effect of the labour force on investiment opportunities. The third term simulates a flexible accelerator mechanism, while the fourth term allows for the effect of the monetary system resulted from supply of and demand for money.

In a developing country it is believed that there is no definite relationship to correlate the total exports with the other economic state variables hence it is expressed by :

$$E(i+1) = H_e . E(i) + E_g(i) + W_4(i) \quad (6)$$

where:

$E_g(i)$ is a governmental control to be determined such that a quadratic social welfare function is minimized.
$W_4(i)$ is an environmental noise term.

The total imports is on the other hand given as:

$$M(i+1) = A_m . G.D.P(i) + M_g(i) + W_5(i) \quad (7)$$

in this simple model we have expressed the total imports as a function of a lagged G.D.P term and another governmental policy (control) variable. $M_g(i)$.

Change in inventories was simulated by the following identity which, like the model of the total investiments, incorporates a flexible accelerator mechanism.

$$\Delta S(i+1) = A_s . G.D.P(i) - D_s r(i) - G_s S(i) + W_6(i) \quad (8)$$

such that A_s, D_s and G_s are defined as in the previous equations and $S(i)$ is the inventory, which is given simply by :

$$S(i+1) = S(i) + G_{\Delta s} \Delta S(i+1) \quad (9)$$

The stock variable, K is given such that the stock level at the end of quarter , i+1 equal to the stock level at the end of quarter, i, in addition to an expression to simulate either consumption or addition of stock due to expenditures in quarter (i+1) and allowances for depreciation during the same quarter, hence we have :

$$K(i+1) = (1-C_k)\, K(i) + I(i+1) \quad (10)$$

where:

C_k is the annual stock replacement rate.

The labour force in the present simulation model was used to allow for the influence of population growth on the national economy. Accordingly it was assumed that the labour force grows exponentially with time as follows:

$$L(i+1) = (1-B_\ell+B_t)L(i) + F_g(i) + W_g(i) \quad (11)$$

in this equation $B_t - B_\ell$ stands for the labour force growth rate and is linearly related to the overall domestic population growth rate. $F_g(i)$ is a governmental policy to control the labour market such that the quadratic social welfare function is minimized.

Fig.(1) Shows a signal flowgraph of the above described macroeconomic simulation model.

IDENTIFICATION OF MACROECONOMIC PARAMETERS

In general the performance of the above described macroeconomic model will depend on the values of system parameters. To compensate for the inherent inaccuracies associated with the economic data and to simulate the parameters uncertainities, we express the economic parameters as follows :

$$P = P_o + \varepsilon(P_{1W} + P_{1C}) \quad (12)$$

where:

P is a model economic parameter , P_o is a deterministic leading term of the parameter P_{tW} is a stationary Gaussian white noise sequence with zero mean and covariance $q_{1W}\delta_{1W}$, P_{1C} is a stationary Gaussian colored noise with zero mean and covariance $q_{1C}\, \alpha^{|i-j|}$ and ε is a positive small constant.

Now substituting eq(1) in to eqs.(3) through (11) and making use of eq.(12) for all parameters, we can express the above econometric formulation in a random coefficients state space format as follows.

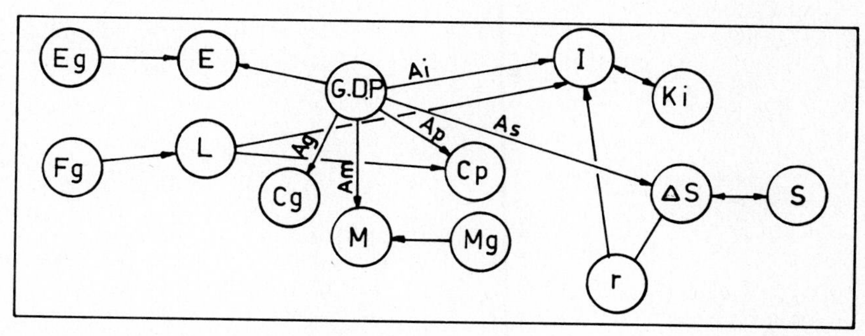

Fig.1.
Simulation model state space construction

$$X(i+1)=F_o X(i)+G\ U(i)+\varepsilon(R_{1w}+R_{1c})X(i)+W(i) \qquad (13)$$

$$i=0,1,\dots,N_s \quad \text{and } X(0) \text{ is given}$$

where:

$$X^T(i)=(C_g(i),C_p(i),I(i),E(i),M_o(i),\Delta S(i),S(i),K(i),L(i)) \qquad (14)$$

$$F_o = \begin{bmatrix} A_{go} & A_{go} & A_{go} & A_{go} & A_{go} & A_{go} & o & o & o \\ A_{po} & A_{po} & A_{po} & A_{po} & A_{po} & A_{po} & o & o & B_{po} \\ A_{io} & A_{io} & A_{io} & A_{io} & A_{io} & A_{io} & o & C_{io} & B_{io} \\ o & o & o & H_{eo} & o & o & o & o & o \\ -A_{mo} & -A_{mo} & -A_{mo} & -A_{mo} & -A_{mo} & -A_{mo} & o & o & o \\ A_{so} & A_{so} & A_{so} & A_{so} & A_{so} & A_{so} & -G_{so} & o & o \\ G_{\Delta so}A_{so} & G_{\Delta so}A_{so} & G_{\Delta s}A_{so} & G_{\Delta so}A_{so} & G_{\Delta so}A_{so} & G_{\Delta so}A_{so} & 1-G_{\Delta so}G_{so} & o & o \\ A_{io} & A_{io} & A_{io} & A_{io} & A_{io} & A_{io} & o & 1-C_{Ko}+C_{io} & B_{io} \\ o & o & o & o & o & o & o & o & 1-B_\ell+B_t \end{bmatrix} \qquad (15)$$

$$R_{1w} = \begin{bmatrix} A_{qlw} & A_{qlw} & A_{qlw} & A_{qlw} & A_{qlw} & A_{qlw} & o & o & o \\ A_{plw} & A_{plw} & A_{plw} & A_{plw} & A_{plw} & A_{plw} & o & o & B_{plw} \\ A_{ilw} & A_{ilw} & A_{ilw} & A_{ilw} & A_{ilw} & A_{ilw} & o & C_{ilw} & B_{ilw} \\ o & o & o & H_{eiw} & o & o & o & o & o \\ -A_{mlw} & -A_{mlw} & -A_{mlw} & -A_{mlw} & -A_{mlw} & -A_{mlw} & o & o & o \\ A_{slw} & A_{slw} & A_{slw} & A_{slw} & A_{slw} & A_{slw} & -G_{slw} & o & o \\ G_{\Delta so}A_{slw} & G_{\Delta so}A_{slw} & G_{\Delta so}A_{slw} & G_{\Delta so}A_{slw} & G_{\Delta so}A_{slw} & G_{\Delta so}A_{slw} & 1-G_{\Delta so}G_{slw} & o & o \\ A_{ilw} & A_{ilw} & A_{ilw} & A_{ilw} & A_{ilw} & A_{ilw} & o & 1-C_{klw}+C_{ilw} & B_{ilw} \\ o & o & o & o & o & o & o & o & 1+B_{lw} \end{bmatrix} \qquad (17)$$

$$G = \begin{bmatrix} o & o & o & o \\ o & o & o & o \\ D_i & o & o & o \\ o & 1 & o & o \\ o & o & -1 & o \\ -D_s & o & o & o \\ -G_{\Delta s}D_s & o & o & o \\ D_i & o & o & o \\ o & o & o & 1 \end{bmatrix}$$

such that $E(R_{1w})=o$ and $E(R_{1w}.R^T_{1w})=Q_{1w}.\delta$ .Matrix R_{1C} is defined similarly.

$\bar{W}(i)$ is 9x1 dimensional Gaussian white noise vector sequence such that $E(w(i) = o$ and $E(W(i).W^T(J)) = Q\ \delta(i-J)$.

The elements of matrix F_o were estimated based on national economic data and using a least square curve fitting algorithm, and a regression analysis to identify the most significant parameter simulation model.

Due to the insufficient statistical data concerning either parameters uncertainities or environmental noise we consider the optimal stochastic control problem for the case of $R_{1C} = 0$.

OPTIMAL STOCHASTIC FEEDBACK CONTROL

The random coefficient state space macroeconomic model described in the above sections is used here for synthesizing a stochastic optimal feedback controller for the national economic policy. In this case the macroeconomic state vector is regarded as perturbations from a potential economic state. The social welfare functional to be minimized contains two terms, the first of which is the expectation of a quadratic penalty function on the terminal deviations and the second term is the expectation of a summation of quadratic forms in economic states and controls, thus we have :

$$J=\min_U E\Big|1/2\ X^T(N_s)S(N_s)\ X(N_s) + 1/2\Big(\sum_{i=o}^{N_s-1}(X^T(i)AX(i) + U^T(i)BU(i)+X^T(i)NU(i) + U^T(i)N^TX(i))\Big)\Big| \qquad (18)$$

where:

N_s is the total number of quarters contained in the minimization process.
$S(N_s)$and A are 9x9 positive-Semi definite matrices, B is 4x4 positive definite matrix and N is 9x4 matrix.

The problem now is to determine the optimal feedback economic control sequence U(i) that minimizes the functional (18) such that the macro economic constraints eq.(13) are satisfied. In the present work the discrete dynamic programming method is used. The fundamental lemma of stochastic optimal control implies that we can use optimal economic control policy for i=o up to N_s-2, then for the last stage we find $U(N_s-1)$ such that :

$$J_{(N_s-1)} = \min_{U(N_s-1)} E\Big|1/2X^T(N_s)S(N_s)X(N_s) + 1/2(X^T(N_s-1)AX(N_s-1)+U^T(N_s-1)BU(N_s-1) +X^T(N_s-1)NU(N_s-1)+U^T(N_s-1)N^TX(N_s-1))\Big| \qquad (19)$$

Substituting for $X(N_s)$ from eq.(13), rearanging, regrouping terms and performing the expectation operation we get :

$$J_{(N_s-1)} = \min_{U(N_s-1)} \Big|1/2\{X^T(N_s-1)(F_o^TS(N_s)F_o + \varepsilon^2 S_{NW} + A)X(N_s-1)+ X^T(N_s-1)(F_o^TS(N_s)G+N)U(N_s-1)+U^T(N_s-1)(G^TS(N_s)G+B)U(N_s-1)+U^T(N_s-1)(G^TS(N_s)F_o+N^T)X(N_s-1)+Tr(S_NQ)\}\Big| \qquad (20)$$

where $S_{NW}=E|R^T_{1w}S_NR_{1w}|$.Performing the minimization operation w.r.t $U(N_s-1)$ we get :

$$U(N_s-1) = -|G^TS(N_s)G+B|^{-1}|G^TS(N_s)(F_o)+N^T|X(N_s-1) = -C(N_s-1)X(N_s-1) \qquad (21)$$

Substituting eq.(21) in eq.(20), we get after after manuplation :

$$J^*_{(N_s-1)} = 1/2|X^T(N_s-1)S(N_s-1)X(N_s-1)+Tr(S_NQ)| \qquad (22)$$

where we have :

$$S_{N-1}=F_o^TS(N_s)F_o-C^T(N_s-1)(G^TS(N_s)G+B)^{-1}C(N_s-1)+\varepsilon^2 S_{Nw} \qquad (23)$$

The last term in eq.(23) allows for the effect of the stochastic perturbations in macroeconomic system parameters. Next we assume that we have used optimal control policy for i=oup to $i=N_s-3$, then we proceed determine $U(N_s-2)$ such that;

$$J_{(N_s-2)}=E\ 1/2(J^*_{(N_s-1)}+X^T(N_s-2)AX(N_s-2)+X^T(N_s-2).NU(N_s-2)+U^T(N_s-2)BU(N_s-2)+U^T(N_s-2)N^TX.(N_s-2)| \qquad (24)$$

making use of eq.(22)in eq.(24), substituting for $X(N_s-1)$ from eq.(13), performing the expectation and the minimization w.r.t $U(N_s-2)$ we get :

$$U(N_s-2)=-(G^T S(N_s-1)G+B)^{-1}(G^T S(N_s-1)(F_o) + N^T)X(N_s-2)=-C(N_s-2)X(N_s-2) \quad (25)$$

which is identical to eq.(21) except for the indices. The same procedure applies until the zeroth stage, the terms $\varepsilon^2 S_{iw}$, $i = 0,1,\ldots,N_s$ indicate the quantitative effect of the stochastic parameter perturbations on the optimal economic control policy.

AVERAGE BEHAVIOR OF THE MACROECONOMIC MODEL

Obviously the average behavior of an optimally controlled national economic policy is a powerful tool which can be used to qualify the generated control sequences for different expressions of the national welfare functional from a quantitative point of view. On the other hand the average behavior is used to investigate the qualitative influence of the parameters stochastic perturbations by means of stochastic noments stability. The mean equation is easily derived, by taking the expectation of eq (13), hence we get :

$$X(i+1) = |F_o-G(C(i)|X(i), \; X(o) \text{ is given} \quad (26)$$

where we have made use of, $U(i) = -C(i)X(i)$
Next to get the equation governing the mean square matrix sequence of the solution sequence, we multiply eq(13) by its transpose and take its expectation we get :

$$X_{m.s}(i+1)=(F_o-GC(i))X_{m.s}(i)(F_o-GC(i))^T+Q+\varepsilon^2 X_r(i), \; X_{m.s}(o) \text{ is given} \quad (27)$$

where we have $X_r(i)=E|R_{1w}X(i)X^T(i)X^T(i)R_{1w}^T|$

The condition for asymptotic stability in the mean sense is derived from eq.(26), it is satisfied if and only if all the eigen values of the matrix $(F_o-GC(i))$, have moduli strictly less than 1. To explore the influence of parameters stochastic perturbations on mean square stability of the economic policy we rewrite eq.(27) in the form :

$$X_{m.s.t}(i+1) = F_{og}(i) \;\; X_{m.s.t}(i)+Q_v \quad (28)$$

where, $X_{m.s.t}(I) = X_{m.s}(J,K)$ I= 1 to 45, J=1 to 9 and K = J to 9 for $i = o,1,\ldots,N_s$.

In the present work we have assumed no cross correlation between parameters stochastic perturbations, accordingly only the diagonal elements of matrix F_{og} will contain terms depending on the covartance of these perturbations as well as "ε". Asym ptotic stability in the mean square sense is achieved if and only if all the eigen values of matrix F_{og} have modeli strictly less than 1.

NUMERICAL RESULTS

Due to insufficient data concerning the statistical parameters of environmental white noise vector, W(i), and the random coefficient matrix, R_{1w}, these parameters were estimated according to a parametric study. Economic Data during 1973 to 1979 were used for this study. Statistical parameters of the deviations between actual and the unperturbed simulation model i.e. (for ε = o) trajectories were taken as the statistical parameters of W(i). On the other hand economic data during the same period from different sources were used to determine the statistical properties of parameters stochastic perturbations based upon the deviations between the economic parameters of each data source. Figs.(2) and (3) show some of the results of this study Fig.(2) illustrates some of the elements of the optimal feedback gain matrix . C(i) during the period 1973-1979, Fig.(3) illustrates the average behavior of the controlled national economic policy.

REFERENCES

|1| Sakwa, Y. and Ueda, Y. , The optimal investiment for the control of environmental pollution and economic growth, Int. Jrnl. of systems science 9, 2 (1978).

|2| Erikson, L. and Norton, E., Application of sensitivity constraned optimal control to national economic policy formulation , in Leondes, C. T.,Advances in control systems (Academic press , London, 1973).

V. BIOLOGICAL SYSTEMS

Simulation in Engineering Sciences
J. Burger and Y. Jarny (eds.)
Elsevier Science Publishers B.V. (North-Holland)

RESOLUTION, IDENTIFICATION D'UN MODELE STOCHASTIQUE DE TRANSIT CEREBRAL

Moïse Sibony

Faculté des Sciences, Laboratoire de Mathématiques
Parc de Grandmont
37200 Tours, France

Des modèles stochastiques circulatoires ont été introduits et étudiés par K.L. Zierler[7], B. Hoop, R.G. Ojemann, G.L. Brownell[3], T. Planiol, R. Itti, R. Floyrac[4] et de nombreux autres auteurs sur le plan expérimental. Les auteurs "calculent" les courbes de transit pour différentes valeurs des paramètres choisis à l'avance, pour que l'allure de la courbe obtenue soit assez proche de celle des courbes observées chez les sujets normaux. Notre méthode envisage la détermination numérique de tous les paramètres, distribués chacun dans le seuil de ses contraintes physiques, ainsi que la détermination des courbes de transit. Nous généralisons ensuite le modèle au cas de dimension 2 ou plus. Suivent des expériences numériques.

Soit $x \in [a,b]$. On cherche à minimiser la fonction coût optimal

$$(1)\quad J(D,v) = \int_a^b |f(x) - \int_0^T C(x,t)|^2 dx$$

lorsque les paramètres D et v se trouvent dans l'ensemble des contraintes admissibles

$$(2)\quad \mathcal{U}_{ad} = \{(D,v) \mid |D| \le \alpha, v \in [\beta,\gamma] \text{ avec } \alpha>0, \beta>0\}$$

pour a,b,α,β,γ et f donnés, $C(x,t)$ étant solution du problème aux limites :

$$(3)\begin{cases} \dfrac{\partial C}{\partial t} + D\dfrac{\partial C}{\partial x} - v\dfrac{\partial^2 C}{\partial x^2} = 0 \text{ ; équation de diffusion} \\ C(a,t) = C(b,t) = 0 \text{ ; conditions aux limites} \\ C(x,0) = C_0(x),\ C_0 \text{ donné ; condition initiale.} \end{cases}$$

Nous envisageons l'identification optimale des paramètres D,v et $C(x,t)$ successivement dans les cas suivants :

i) où D,v sont des constantes (modèle initial)

ii) où D,v sont des fonctions de $x \in [a,b]$

iii) Nous généralisons au cas de la dimension 2 d'espace : D,v sont constants et $C(x,y,t)$ régit par une équation d'évolution en dimension 2, d'espace. Les cas ii) et iii) offrent implicitement un champ d'applications plus vaste. Suivent des expériences numériques sur ordinateur.

1-Identification optimale des coefficients v,D constants. Détermination de la fonction C(x,t) et du coût optimal J(D,v)

A partir de la relation (1), nous calculons le gradiant $(J'_D, J'_v) = J'_\mu$ pour tout couple $\mu=(D,v)$ et nous cherchons la solution optimale $\tilde{\lambda}=(\tilde{D},\tilde{v})$ réalisant les contraintes (2),(3) et caractérisée par l'inégalité

$$(4)\quad (J'(\lambda), \mu-\lambda) \ge 0,\ \forall\ \mu = (D,v) \in \mathcal{U}_{ad}$$

(,) désignant le produit scalaire dans $\mathbb{R}^2$ (Cf. [1],[6]).

La méthode de la Mire (Cf.[6]) permet de minimiser directement la fonctionnelle (1), sous les contraintes (2),(3). Une deuxième méthode itérative permet de résoudre le problème (4) équivalent sous les contraintes (2),(3). Nous nous proposons de comparer les résultats fournis successivement par la méthode itérative et la méthode de la Mire.

Notations. Posons $Q =]a,b[\times]0,T[$, $T>0$, $V = H_0^1(]a,b[)$ muni de la norme $\|C\| = \|\frac{\partial C}{\partial x}\|_{L_2(a,b)}$. On munit $L_2(0,T;V)$ de la norme $\|C\|_{L_2(0,T;V)} = (\int_0^T \|C\|^2 dt)^{1/2}$.

On note par

$$W = \{\varphi \mid \varphi \in L_2(0,T;V),\ \frac{\partial \varphi}{\partial t} \in L_2(0,T;V'),\ V' = H^{-1}(a,b)\},$$

muni de la norme

$$[\varphi]_W = (\|\varphi\|^2_{L_2(0,T;V)} + \|\frac{\partial \varphi}{\partial t}\|^2_{L_2(0,T;V')})^{1/2}$$

W est un espace de Hilbert.

Considérons l'espace $\{\mathcal{G}\}$ des opérateurs linéaires continus de V dans V' définis par

$$(5)\quad \mathcal{G}\varphi = D\frac{\partial \varphi}{\partial x} - v\frac{\partial^2 \varphi}{\partial x^2}, \text{ avec } (D,v) \in \mathcal{U}_{ad}.$$

Nous démontrons alors les résultats suivants :

Proposition 1

Soit $\mathcal{A}\in\{\mathcal{A}\}$; il existe $\beta,\gamma>0$ tels que

(6) $\beta\|\varphi\|^2 \leq (\mathcal{A}\varphi,\varphi) \leq \gamma\|\varphi\|^2, \ \forall\ \varphi\in V$

(7) $(\mathcal{A}\varphi,\psi)\leq M\|\varphi\|.\|\psi\|, \ \forall\varphi,\psi\in V\ ;\ M=\beta(b-a)+\gamma$

ce qui assure l'existence et l'unicité de $C\in W\cap C(0,T;V)$, solution du problème (3).

Considérons l'application $\Phi:\mathcal{A}\rightsquigarrow C=\Phi(\mathcal{A})\in W$, C étant la solution de (3). Φ est dérivable au sens de Gateau et sa dérivée $\Phi'(\mathcal{A})$ est définie en tout point $\mathcal{A}\in\{\mathcal{A}\}$ par l'application $\Phi'(\mathcal{A})\ :\ \delta\mathcal{A}\in\{\mathcal{A}\}\to\delta C\in W$, δC étant solution du système

$$(8)\begin{cases}\frac{\partial}{\partial t}\delta C+\mathcal{A}\delta C=-\delta\mathcal{A}.\Phi(\mathcal{A})=-\delta\mathcal{A}.C\\ \delta C(0,x)=0\end{cases}$$

Proposition 2

Soient $\lambda=(D,v)\in\mathcal{U}_{ad}$ et $y:W\to L_2(a,b)$ l'opérateur d'observation (linéaire continu) défini par

$$(9)\quad y(C)=\int_0^T C(x,t)dt$$

Pour $f\in L_2(a,b)$ donné, on définit sur $\{\mathcal{A}\}$ une fonctionnelle des moindres carrés

$$(10)\ J(\mathcal{A})=\|y(\Phi(\mathcal{A}))-f(x)\|^2_{L_2(a,b)} = \int_a^b \left|\int_0^T C(x,t)dt-f(x)\right|^2 dx$$

1°) Alors J est dérivable sur $\{\mathcal{A}\}$ et sa dérivée $J'(\mathcal{A})$ en $\mathcal{A}\in\{\mathcal{A}\}$ est définie par $\delta\mathcal{A}\in\{\mathcal{A}\}\to J'(\mathcal{A})\delta\mathcal{A}=\int_0^T(\delta\mathcal{A}C(t),p(t))_{V',V}dt$

où $C\in W$ est solution du système direct

$$(11)\begin{cases}C\in W,\ \frac{\partial C}{\partial t}+D\frac{\partial C}{\partial x}-v\frac{\partial^2 C}{\partial x^2}=\frac{\partial C}{\partial t}+\mathcal{A}C=0\ ;\\ C(0,x)=C_o(x)\in L_2(a,b)\end{cases}$$

et p est solution du système adjoint

$$(12)\begin{cases}p\in W,-\frac{\partial p}{\partial t}-D\frac{\partial p}{\partial t}-v\frac{\partial^2 p}{\partial x^2}=-2\left(\int_0^T C(x,\tau)d\tau-f(x)\right);\\ p(T,x)=0\ ;\ p(a,t)=p(b,t)=0\end{cases}$$

2°) On notera alors $J'(\lambda)=J'(\mathcal{A}(\lambda))$; $\delta\mathcal{A}=\delta D\frac{\partial}{\partial x}-\delta v\frac{\partial^2}{\partial x^2}$; $Q=]a,b[\times]0,T[$. On en déduit

$$(13)J'(\lambda)d\lambda=J'_D\delta D+J'_v\delta v=\left(\int_Q p\frac{\partial C}{\partial x}dxdt\right)\delta D - \left(\int_Q p\frac{\partial^2 C}{\partial x^2}dxdt\right)\delta v$$

$J'(\lambda)=(J'_D,J'_v)\in\mathbb{R}^2$; $\|J'(\lambda)\|^2_{\mathbb{R}^2}=|J'_D|^2+|J'_v|^2$.

Démonstration

1°) Nous avons successivement

$$J'(\mathcal{A})\delta\mathcal{A}=2\left(\int_0^T C(x,t)dt-f(x),\int_0^T\delta C(x,t)dt\right)_{L_2,L_2(a,b)}$$

$$(14)J'(\mathcal{A})\delta\mathcal{A}=2\int_0^T\left(\int_0^T C(x,t)dt-f(x),\delta C(x,t)\right)dt_{L_2,L_2}$$

où δC est solution de

$$(15)\ \frac{\partial}{\partial t}\delta C+\mathcal{A}\delta C=-\delta\mathcal{A}\Phi(\mathcal{A})=-\delta\mathcal{A}.C$$

En introduisant l'état adjoint par (12), (14) devient

$$J'(\mathcal{A})\delta\mathcal{A}\ -\int_0^T\left(-\frac{\partial p}{\partial t}-D\frac{\partial p}{\partial x}-v\frac{\partial^2 p}{\partial x^2},\delta C\right)_{V,V'}dt$$

en intégrant par parties

$$J'(\mathcal{A})\delta\mathcal{A} = -\int_0^T\left(p,\frac{\partial}{\partial t}\delta C+\mathcal{A}\delta C\right)_{V,V'}dt$$

compte tenu de (8)

$$J'(\mathcal{A})\delta\mathcal{A} = \int_0^T(\delta\mathcal{A}\ C,p)_{V,V'}\ dt$$

2°) Posons $\lambda=(D,v)$, $\delta\lambda=(\delta D,\delta v)$, $\delta\mathcal{A}=\delta D\frac{\partial}{\partial x}-\delta v\frac{\partial^2}{\partial x^2}$. On a alors

$$J'(\lambda)d\lambda=J'(\mathcal{A}(\lambda))d\lambda=\left[\int_Q p\frac{\partial C}{\partial x}\right]d\delta-\left[\int_Q p\frac{\partial^2 C}{\partial x^2}\right]\delta v =J'_D\delta D+J'_v\delta v.$$

Proposition 3

1°) J atteint un minimum en $\bar\lambda=(\bar D,\bar D)$ caractérisé par l'inégalité

(16) $<J'_D(\bar\lambda),\mu-\bar\lambda>\geq 0,\ \forall\ \mu\in\mathcal{U}_{ad}$

$<,>$ désignant le produit scalaire dans $\mathbb{R}^2$.

2°) De plus la méthode itérative

$$(17)\ \lambda_{n+1} = P_{\mathcal{U}_{ad}}(\lambda_n-\rho_n J'(\lambda_n))$$

converge pour $\rho_n>0$ convenablement choisi en fonction des itérations ; $P_{\mathcal{U}_{ad}}$ étant l'opérateur de projection sur $\mathcal{U}_{ad}$. Plus précisément on peut extraire de (λ_n) une sous-suite (λ_{n_k}) qui converge vers $\bar\lambda\in\mathcal{U}_{ad}$.

3°) Le problème aux limites (3) est discrétisé suivant un schéma implicite :

$$(18)\frac{1}{k}(C_h^{s+1}-C_h^s)+\mathcal{A}_h C_h^{s+1}=0, C_h^o=C_{oh};s=0,\ldots,S$$

où $\mathcal{A}_h$ désigne l'opérateur de discrétisation en $h\to0$ à 5 points. Le système (18) étant résolu numériquement par la méthode itérative

$$(19)\ u_h^{m+1}=u_h^m-\rho(B_h u_h^m-g_h),\ u_h^o=v \quad \text{avec}$$

$g_h=C_h^S$; $B_h u_h^m=u_h^m+k\ \mathcal{A}_h u_h^m$.

Pour $\rho>0$ convenablement choisi, on a $\lim_{m\to\infty} u_h^m = C_h^{s+1}$ et $\lim_{(h,\varepsilon)\to 0} q_{hk} C_h^s = C$ dans $L^2(0,T;V)$; q_{hk} étant un opérateur linéaire continu convenable, C étant la solution du problème aux limites (3). De plus on a $\lim_{(h,k)\to 0} J_{hk}(\lambda_{hk}) = J(\tilde{\lambda})$, J_{hk} étant la discrétisation de J.

Démonstration

1°) Pour $\mathfrak{a}\in\{\mathfrak{a}\}$, $f\in L_2(a,b)$, la solution C de (11) vérifie la majoration $\|C\|_{L_2} \le \|C_o\|_{L_2}$ pour tout $t\in]0,T[$.

Il en résulte que

$$J(\mathfrak{a}) = \|\int_0^T C(x,\tau)d\tau - f(x)\|_{L_2}$$

reste borné indépendamment de $\mathfrak{a}\in\{\mathfrak{a}\}$ = espace fermé. J est continu de $\{\mathfrak{a}\}$ dans $\mathbb{R}+$, donc atteint un minimum en au moins $\tilde{\mathfrak{a}}\in\{\mathfrak{a}\}$. A $\tilde{\mathfrak{a}}$ nous associons $\tilde{\lambda}=(\tilde{D},\tilde{v})\in\mathcal{U}_{ad}$ solution de l'inégalité (16).

2°) D'après (13) nous avons

$$J'(\lambda)=(J_D', J_v')\in\mathbb{R}^2; \|J'(\lambda)\|^2_{\mathbb{R}^2} = |J_D'|^2+|J_v'|^2$$

alors la méthode itérative

$$(20)\begin{cases}\lambda_o\in\mathcal{U}_{ad}\\ \lambda_{n+1} = P_{\mathcal{U}_{ad}}(\lambda_n - \rho_n\omega_n)\end{cases}$$

avec $\omega_n = J'(\lambda_n)/\|J'(\lambda_n)\|_{\mathbb{R}^2}, \|\omega_n\|_{\mathbb{R}^2}=1$, est convergente pour $\rho_n>0$ convenablement choisi. En effet, nous avons successivement

$$\Delta J_\rho = J(\lambda_n) - J(\lambda_n - \rho\omega_n)$$

$$\Delta J_{\rho_n} = \rho_n \langle J'(\lambda_n - \bar{\rho}\omega_n), \omega_n\rangle_{\mathbb{R}^2\times\mathbb{R}^2} = \rho_n J'_{\bar{\rho}}$$

avec $0<\bar{\rho}<\rho_n$.

De l'uniforme continuité de J' on tire $|J'_{\bar{\rho}} - J'_o| \le \bar{\rho}.C \le C_2\, J'_o$ pour ρ_n tel que

$$\frac{C_1}{C}\|J'(\lambda_n)\|_{\mathbb{R}^2} \le \rho_n \le \frac{C_2}{C}\|J'(\lambda_n)\|_{\mathbb{R}^2}.$$

On en déduit

$$(21)\begin{cases}\Delta J_{\rho_n} > 0\\ \lim_{n\to\infty}\Delta J_{\rho_n} = 0 \Rightarrow \lim_{n\to\infty}\|J'(\lambda_n)\|_{\mathbb{R}^2} = 0\end{cases}$$

La suite $(\lambda_n)\in$ à un fermé borné de $\mathbb{R}^2$ et l'on peut extraire $(\lambda_{n_k}) \to \bar{\lambda}\in\mathcal{U}_{ad}$.

J étant positive bornée sur $\mathcal{U}_{ad}$ et J' étant continue on a $J'(\bar{\lambda})=0, J_D'(\bar{D},\bar{v}) = J_v'(\bar{D},\bar{v})=0$ avec $(\bar{D},\bar{v})\in\mathcal{U}_{ad}$.

3°) Nous démontrons successivement les assertions suivantes :

(22) $(\mathfrak{a}_h v_h, v_h)_h \ge \beta\|v_h\|_h^2$, $\forall\, v_h\in V_h$ = espace de dimension finie de discrétisation.

(23) $(\mathfrak{a}_h v_h, w_h)_h \le M\|v_h\|_h\|w_n\|_h ; \forall\, v_h, w_h\in V_h$

Nous posons alors d'après (19)

(24) $u_h^{m+1} = T_\rho u_h^m = u_h^m - \rho(B_h u_h^m - g_h)$

et nous montrons que

(25) $\|T_\rho w\|^2 \le (1-\rho)^2\|w\|^2$ pour $\rho = \dfrac{\beta}{\beta + \frac{4k}{h^2}M^2}$

ce qui assure la convergence des itérations (24) ou (19) et l'on a $\lim_{m\to\infty} u_h^m = C_h^{s+1}$.

4°) On note

u_m^* = limite de la suite $(u_h^m)_m$ définie par le schéma (19) et $\varepsilon_h^m = u_h^m - u^*$.

C_h^s = approximation de u^* obtenue en arrêtant l'algorithme (19) à la m_s-ième itération et $\varepsilon_s = C_h^s - u^*$.

Si $(w)_i$, $i=0,\ldots,S$ est la solution du système (18), w_s vérifie

(26) $w_s - w_{s-1} + k\,\mathfrak{a}_h\, w_s = 0$

et $u^*\in V_h$ vérifie

(27) $u^* - C_h^{s-1} + k\mathfrak{a}_h u^* = 0$

Par différence on a

(28) $(u^* - w_s) + k\mathfrak{a}_h(u^* - w_s) = (C_k^{s-1} - w_{s-1})$

L'erreur $e_h^s = C_h^s - w_s$ vérifie alors au temps $t=sk$

(29) $e_h^s - \varepsilon_s + k\,\mathfrak{a}_h(e_h^s - \varepsilon_s) = e_h^{s-1}$

ce qui donne

(30) $(e_h^s - e_h^{s-1}, e_h^s)_h + k(\mathfrak{a}_h e_h^s, e_h^s)_h = (B_h\varepsilon_s, e_h^s)_h$

Posons $\gamma_o = k/h^2$; $\delta = \beta/(b-a)$. En utilisant les inégalités (22),(23), il vient

$$\frac{1}{2}\{|e_h^s|_h^2 - |e_h^{s-1}|_h^2 + |e_h^s - e_h^{s-1}|_h^2\} + k\delta|e_h^*|_h^2 \le (1+4\gamma_o M)|\varepsilon_s|_h|e_h^s|_h$$

ce qui donne

$$(31)\{|e_h^s|_h^2 - |e_h^{s-1}|_h^2 + |e_h^s - e_h^{s-1}|_h^2\} + k\{2\delta - (1+4\gamma_o M)\xi\}|e_h^s|_h^2 \le \frac{1}{\xi k}(1+4\gamma_o M)|\varepsilon_s|_h^2, \xi>0$$

On prend $\xi = \delta/(1+4\gamma_o M)$ et on somme (31) pour $s=1,\ldots,n$ avec $n=1,\ldots,S$; il vient

(32)

$$|e_h^n|_h^2 + \delta.k\sum_{s=1}^n |e_h^s|_h^2 \le \frac{1}{\delta.k}(1+4\gamma_o M)^2\sum_{s=1}^n|\varepsilon_s|_h^2.$$

On note $\eta=1-\rho$ où ρ est donné par (25) ce qui permet la majoration

$$(33)\quad |\varepsilon_s|_h^2 \leq \frac{\eta^{2m_s}}{(1-\eta^{m_s})^2}(2|e_h^s|^2+2|w_s|_h^2)$$

Or $w_s \in V_h$ vérifie

$$(34)\begin{cases}(w_s-w_{s-1},w_s)_h+k(\mathcal{C}_h w_s,w_s)_h=0 \\ \qquad\qquad\text{pour } s=1,\dots,S \\ w_o=C_{oh}\end{cases}$$

En sommant (34) de $s=1,\dots,n$ avec $n=1,\dots,S$, il vient en utilisant (22),(23)

$$(35)\begin{cases}k\sum_{s=1}^{n}|w_s|_h^2\leq\frac{|C_{oh}|^2}{2\delta} \quad \text{pour } n=1,\dots,S \\ |w_n|_h \leq |C_{oh}|_h\end{cases}$$

Posons $m=\underset{s=1,\dots,S}{\text{Min}}\ m_s$, $L=(1+4\gamma_o M)^2/\delta^2$.

En reportant (33),(35) dans (32) on obtient

$$|e_h^n|_h^2\leq\frac{L}{k^2}\frac{\eta^{2m}}{(1-\eta^m)^2}\exp\{4\delta L\frac{\eta^{2m}}{(1-\eta^m)^2}\frac{T}{k^2}\}$$

Pour $k/h^2=\gamma_o=c^{te}$ et si m vérifie

$$(36)\quad \lim_{h,k\to0}\frac{\eta^m}{k}=0,$$

ce qui est réalisé pour $m\geq c^{te}|\text{Log}k|$, on aura :

$$(37)\quad \lim_{h,k\to0}\|q_{hk}e_h\|_{L^\infty(0,T;L_2(a,b))}=0$$

où $q_hk=q_k\cdot q_h$ avec $q_k\varphi=\sum_{s=0}^{S-1}\chi_s(t)\varphi^s$, $\chi_s(t)$ étant la fonction caractéristique de $[sk,(s+1)k[$ et $\varphi=(\varphi^s)_{s=0,\dots,S-1}$. A partir des relations (35),(37) on démontre la convergence $\lim_{h,k} q_{hk}C_h^s=C$, C étant l'unique solution du problème (3).

(5°) Pour $\lambda=(D,v)\in\mathcal{U}_{ad}$ nous avons

$$J(\lambda)=\|\int_0^T C(x,t)dt-f(x)\|_{L_2(a,b)}$$

C étant la solution du système (3). Pour h,k fixé nous notons

$$J_{hk}(\lambda)=\|q_h\int_0^T q_kC_{hk}dt-f(x)\|_{L_2(a,b)}$$

C_{hk} étant la solution du système (18). Soit $\varepsilon_{hk}>0$ une suite destinée à tendre vers 0 lorsque $h,k\to0$. On note λ_{hk} une solution du problème de minimisation modulo ε_{hk} :

$$(38)\quad J_{hk}(\lambda_{hk})\leq J_{hk}(\lambda)+\varepsilon_{hk},\ \forall\,\lambda\in\mathcal{U}_{ad}$$

On montre alors que si $\chi=(D,v)\in\mathcal{U}_{ad}$ est un élément minimisant $J(\lambda)$ sur $\mathcal{U}_{ad}$ alors

$$(39)\quad \lim_{h,k\to0}J_{hk}(\lambda_{hk})=J(\tilde\lambda)$$

2 - Applications

1°) Cas de coefficients D,C constants

Pour les données suivantes :

$[a,b]=[0,1]$; $T=1$; $C_o(x)=\sin\pi x$; $f(x)=(1-e^{-T})\sin\pi x$; $\mathcal{U}_{ad}=[0,2]\times[10^{-6},2]$

la fonctionnelle J atteint son minimum $J(\bar\lambda)=0$ pour $C(x,)=e^{-t}\sin\pi x$; $D=0$; $v=1/\pi^2\simeq 0.101$.

Si l'on se donne des pas de discrétisations en espace et temps, $h=1/10$, $k=1/10$, on obtient les résultats du tableau n°1 donnant l'évolution des valeurs de $J_{hk}(D,v)$ de D et v en fonction des itérations. On peut constater que $J_{hk}(\tilde D,\tilde v)=0.5910^{-9}$; la vraie valeur est $J=0$
$\tilde D=0.110\times10^{-3}$; la vraie valeur est $D=0$
$\tilde v=0.109$; la vraie valeur est $v=0.11$

L'emploi d'une méthode directe d'optimisation (Cf.[6]) aboutit à des résultats plus précis pour un nombre restreint de points $h=k=1/10$ comme le tableau n°2 :

$J(\tilde D,\tilde v)=0.12\times10^{-17}$; $\tilde D=0.95\times10^{-9}$; $\tilde v=0.109$

Mais la méthode devient nettement plus onéreuse en temps machine pour un nombre de variables plus grand.

Tableau n° 1 — Tableau n° 2

Nbre d'itérations	$J_{hk}(\tilde D,\tilde v)$ (Tableau n° 1)	$\tilde D$ (Tableau n° 1)	$\tilde v$ (Tableau n° 1)	$J_{hk}(\tilde D,\tilde v)$ (Tableau n° 2)	$\tilde D$ (Tableau n° 2)	$\tilde v$ (Tableau n° 2)
1	0.12×10^{-1}	0.00	0.201	0.13×10^{-1}	0.10×10^{-2}	0.51×10^{-1}
5	0.22×10^{-4}	0.871×10^{-3}	0.110	0.13×10^{-4}	0.72×10^{-2}	0.107
10	0.25×10^{-7}	0.785×10^{-3}	0.109	0.71×10^{-6}	$0.1\ \times10^{-2}$	0.109
30	0.68×10^{-8}	0.359×10^{-3}	0.109	0.11×10^{-11}	0.36×10^{-5}	0.109
40	0.15×10^{-8}	0.173×10^{-3}	0.109	0.43×10^{-14}	0.26×10^{-6}	0.109
50	0.59×10^{-9}	0.110×10^{-3}	0.109	0.12×10^{-17}	0.95×10^{-9}	0.109

2°) Cas de coefficients variables

On suppose l'existence d'une constante $\alpha>0$ vérifiant pour $D\in C^1(a,b), v\in C^2(a,b)$ l'inégalité

$$(40)\quad \operatorname*{Inf}_{x\in[a,b]} v(x) - \frac{(b-a)^2}{2} \operatorname*{Sup}_{x\in[a,b]} (D'(x)-v''(x)) \geq \alpha.$$

On envisage alors la détermination du couple $(D(x),v(x))$ dans l'ensemble des contraintes admissibles :

$$(41)\quad \mathcal{U}_{ad}=\{(D,C)\in C^1\times C^2 \ ;\ 0<\alpha<v(x)\ ;\ D'(x)\leq v''(x),\ \forall\ x\in[a,b]\}$$

minimisant la fonctionnelle $J(D,v)$ où C est solution du problème aux limites

$$(42)\begin{cases} \dfrac{\partial C}{\partial t} + D(x)\,\dfrac{\partial C}{\partial x} - v(x)\,\dfrac{\partial^2 C}{\partial x^2} = 0 \\ C(0,x)=C_o(x), \text{ pour } C_o, f \text{ donnés dans } L_2(a,b). \end{cases}$$

L'ensemble des assertions de la proposition 3 restent valables et l'on peut résoudre le problème (41),(42), après discrétisations sur les variables d'espace et de temps et approximations par une méthode itérative du type (19), l'équation (42) étant approchée par la méthode implicite (18).

Exemple 1 de résolution numérique : $D(x), v(x),\ x\in[a,b]$

On se donne un problème modèle dont on connait la solution exacte, que l'on retrouve par les schémas d'approximations énoncés plus haut. Plus précisément, on pose

$C_o(x)=x^2(x-1); f(x)=C_o(x)(1-e^{-T}); x,t\in[0,1]$

$D(x)=(3x+1)/9; v(x)=x/9$

$C(x,t)=x^2(x-1)e^{-T}$.

Alors le minimum atteint par la fonctionnelle $J=0$, comme l'unicité de la solution n'est pas assurée, en fixant l'un des coefficients, D ou v à sa valeur exacte, on retrouve l'autre et on obtient $J=10^{-8}$ après 20 itérations seulement.

Exemple 2 : Modèle en dimension 2 d'espace

Cette fois les coefficients D et v sont fonctions de deux variables $(x_1,x_2)\in\Omega =]0,1[\times]0,1[$. Plus précisément on pose :

$$(43)\quad C_o(x_1,x_2) = \sin\pi x_1 . \sin\pi x_2\ ;$$
$$f(x_1,x_2)=C_o(x_1,x_2)(1-e^{-T})$$
$$\mathcal{U}_{ad}=[-2,+2]\times[10^{-6},2]$$

où C est solution du système : $C\in W$

$$(44)\begin{cases} \dfrac{\partial C}{\partial t}+D(x_1,x_2)\displaystyle\sum_{i=1}^{2}\frac{\partial C}{\partial x_i}-v(x_1,x_2)\sum_{i=1}^{2}\frac{\partial^2 C}{\partial x_i^2}=0 \quad \text{p.p. sur }]0,T[^1 \\ C(x_1,x_2,0)=C_o(x_1,x_2) \end{cases}$$

La fonctionnelle J atteint son maximum $J(\overline{D},\overline{v})=0$ sur $\mathcal{U}_{ad}$ pour $\overline{D}=0$, $\overline{v}=1/2\pi^2 \simeq 0.507\times10^{-1}$.

On opère encore comme précédemment : au bout de $n=12$ itérations

$\overline{D}=0.45\times10^{-2}$, $\overline{v}=0.529\times10^{-1}$; $J=0.18\times10^{-5}$

alors que par la méthode directe de la Mire (Cf.[6]) :

$\overline{D}=0$, $\overline{v} = 0.529\times10^{-1}$; $J = 0.15\times10^{-9}$

Conclusion : Les méthodes et algorithmes précédents ont abouti à la résolution numérique complète, avec une très grande précision, de l'ensemble du problème (1), (2),(3), dans les cas de coefficients constants, variables, en dimension 1 et 2, en évitant la constitution onéreuse de données expérimentales moins précises et aléatoires. Nous avons présenté les résultats numériques pour des problèmes modèles de type analytique, mais nos méthodes fonctionnent normalement lorsque les données proviennent de mesures expérimentales.

3 - Méthode du Hamiltonien-Algorithme

La définition du Hamiltonien associé aux problèmes (1),(2),(3) permet la formulation de l'état adjoint et le calcul analytique du gradient de $J=(\frac{\partial J}{\partial D},\frac{\partial J}{\partial v})$ pour aboutir à un algorithme général d'approximation numérique des coefficients (D,v) du système et de la fonction coût J. Nous explicitons brièvement les étapes de cette méthode générale d'optimisation, dans le cadre de notre problème d'identification de coefficients (D,v) par la méthode du gradient projeté. Plus précisément le Hamiltonien du problème (1),(2), (3) est défini par

$$(45)\quad \mathcal{H}(C,p,D,v)=\int_0^1 \left|f(x)-\int_0^T C(x,t)dt\right|^2 dx \ \dots -2\int_Q p\left(\frac{\partial C}{\partial t}+D\frac{\partial C}{\partial x}-v\frac{\partial^2 C}{\partial x^2}\right)dxdt$$

où $Q=]0,1[\times]0,T[$, p=multiplicateur de Lagrange de l'équation d'état. Si le multiplicateur de Lagrange (p) associé au problème initial existe, alors $\mathcal{H}(C,p,D,v)$ est minimum sur $\{C,D,v\}$ au point $(\overline{C},\overline{D},\overline{v})$ sur l'ensemble

$$(46)\quad X=\{C \mid C(x,0)=C_o(x), C(0,t)=C(1,t)=0\} \times \mathcal{U}_{ad}.$$

Plus précisément on aura

$$(47)\quad \mathcal{H}(\overline{C},p,\overline{D},\overline{v}) \leq \mathcal{H}(C,p,D,v),\ \forall(C,D,v)\in X$$

Proposition 4

La dérivée de Gateaux de $\mathcal{H}$ est définie par

$$(48)\quad \mathcal{H}'_C=\lim_{\gamma\to0} \mathcal{H}(C+\gamma z,p,D,v)-\mathcal{H}(C,p,D,v)$$
$$=-2\int_a^b\{f(x)-\int_0^T C(x,t)dt)\int_0^T z dt\}dx$$

$$= -2\int_Q p\left(\frac{\partial z}{\partial t}+D\frac{\partial z}{\partial x}-v\frac{\partial^2 z}{\partial x^2}\right)dxdt$$

pour $\begin{cases} z(x,0)=0 \\ z(0,t)=z(1,t)=0. \end{cases}$

La condition $\mathcal{H}'_C=0$ donne le système adjoint

$$(49)\begin{cases} -\frac{\partial p}{\partial t}-D\frac{\partial p}{\partial x}-v\frac{\partial^2 p}{\partial x^2}=\int_0^T C(x,t)dt-f(x) \\ p(0,t)=p(1,t)=0 \\ p(x,t)=0 \end{cases}$$

Démonstration

La condition (48) est immédiate. De $\mathcal{H}'_C=0$ on déduit :

$$(50)\quad -\int_Q p\left(\frac{\partial z}{\partial t}+D\frac{\partial z}{\partial x}-v\frac{\partial^2 z}{\partial x^2}\right)dxdt + \int_Q z(x,t)\left\{\int_0^T C(x,t)dt-f(x)\right\}dxdt = 0$$

L'égalité (50) devient alors

$$(51)\quad \int_Q z(x,t)\left\{\int_0^T C(x,t)dt-f(x)\right\}dxdt + \int_Q \left(z\frac{\partial p}{\partial t}+Dz\frac{\partial p}{\partial x}+vz\frac{\partial^2 p}{\partial x^2}\right)dxdt - \int_0^1 p(x,t)z(x,t)dx+v\int_0^T\left\{p(1,t)\frac{\partial z}{\partial x}(1,t)-p(0,t)\frac{\partial z}{\partial x}(0,t)\right\}dt=0$$

pour tout z tel que $z(x,0)=0$, $\forall x$; $z(0,t)=z(1,t)=0$, $\forall t$. On pose $z=\tilde{z}(x)\varphi(t)$ avec $\varphi\in C^\infty(0,T)$; $\varphi(0)=\varphi(T)=0$ et on interprète (51) pour en déduire (49).

Nous démontrons alors la

Proposition 5

$$\lim_{\gamma\to 0}\frac{\mathcal{H}(z,p,D+\gamma\delta D,v+\gamma\delta v)-\mathcal{H}(z,p,D,v)}{\gamma} = -2\int_Q p\left(\delta D\frac{\partial z}{\partial x}-\delta v\frac{\partial^2 z}{\partial x^2}\right)dxdt.$$ D'où

$$(52)\quad \operatorname{grad} J(D,v) = (J'_D, J'_v) = \left(-\int_Q p\frac{\partial z}{\partial x}dx,dt, \int_Q p\frac{\partial^2 z}{\partial x^2}dxdt\right)$$

D'où l'algorithme d'approximation des coefficients D,v : On se donne $D^0,v^0\in\mathcal{U}_{ad}$. Si D^n,v^n sont déterminés, on calcule D^{n+1},v^{n+1} comme suit :

1°) z^{n+1} est déterminé par

$$(53)\begin{cases} \frac{\partial}{\partial t}z^{n+1}+D^n\frac{\partial}{\partial x}z^{n+1}-v^n\frac{\partial^2}{\partial x^2}z^{n+1}=0 \\ z^{n+1}(0,t)=z^{n+1}(1,t)=0 \\ z^{n+1}(x,0)=z_0(x) \end{cases}$$

Le problème (53) admet une solution unique $z^{n+1}\in L_2(0,T;V)$.

2°) p^{n+1} est solution de :

$$(54)\begin{cases} -\frac{\partial}{\partial x}p^{n+1}-D^n\frac{\partial}{\partial x}p^{n+1}-v^n\frac{\partial^2}{\partial x^2}p^{n+1} = \int_0^T z^{n+1}(x,t)dt-f(x) \\ p^{n+1}(1,t)=p^{n+1}(0,t)=p^{n+1}(x,t)=0,\forall x \end{cases}$$

On pose $\mathcal{B}p=D\frac{\partial p}{\partial x}+v\frac{\partial^2 p}{\partial x^2}$ et l'on vérifie que l'opérateur $\mathcal{B}$ satisfait les conditions (6) et (7), ce qui assure l'existence et l'unicité de p^{n+1} solution de (54).

3°) Nous obtenons alors (D^{n+1},v^{n+1}) par le schéma itératif :

$$(55)\quad (D^{n+1},v^{n+1})=P_{\mathcal{U}_{ad}}\left(D^n+\rho_1\int_Q p^{n+1}\frac{\partial}{\partial x}z^{n+1}dxdt, v^n-\rho_2\int_Q p^{n+1}\frac{\partial^2}{\partial x^2}z^{n+1}dxdt\right)$$

ou encore

$$(D^{n+1},v^{n+1})=P_{\mathcal{U}_{ad}}(D^n-\rho_1 J'_{D^n}, v^n-\rho_2 J'_{v^n})$$

avec $J'_{D^n}=-\int_Q p^{n+1}\frac{\partial}{\partial x}z^{n+1}dxdt$

$$J'_{v^n}=\int_Q p^{n+1}\frac{\partial^2}{\partial x^2}z^{n+1}dxdt$$

Pour ρ_1 et $\rho_2>0$ convenablement choisis (comme dans (25)) on a enfin

$$\lim_n(D^n,v^n) = (D,v).$$

Connaissant (D,v) et $C=\lim_n z^n$ on peut alors donner une approximation du coût J à l'aide de la formule (1).

BIBLIOGRAPHIE :

[1] J.Cea, Optimisation, théorie et algorithme, Dunod, 1971.

[2] C.Chavent, Analyse fonctionnelle et identification de coefficients répartis dans les équations aux dérivées partielles, Thèse, Univ. Paris, 1971.

[3] B.Hoop, R.G.Ojemann, G.L.Browell, A stochastic model of regional cerebral circulation, J.Nucl.Med. 12, 8, 540(1971).

[4] T.Planiol, R.Itti, R.Floyrac, Essai d'interprétation théorique des tracés de γ-angioencéphalographie, Ann. Phys. Biol. et Med., 5, 173-185, 1971.

[5] T.Planiol, R.Itti, R.Floyrac, L'intérêt des modèles théoriques pour l'interprétation des courbes de transit cérébral, Dynamic studies with radioisotopes in medecine 1971, Vol.2, Vienna, 1974

[6] M.Sibony, Sur l'approximation d'équations et inéquations aux dérivées partielles non linéaires de type monotone J.Math.Anal.Appl., 34, 3, 502-564, 1973.

[7] K.L.Zierler, Equations for measuring blood flow by external monitoring of radioisotopes, Circ. Res., 16, 4, 305 (1965).

Simulation in Engineering Sciences
J. Burger and Y. Jarny (eds.)
Elsevier Science Publishers B.V. (North-Holland)

NEURODYNAMICAL SIMULATION OF PERISTALTIC MOVEMENT

Yoji Umetani, Norio Inou

Tokyo Institute of Technology
Tokyo, Japan

The small intestine is an active transport system which carries intraluminal contents by peristalsis. The authors propose a neurodynamical simulation model of the system, and clarify the principle of movements to transport semifluid contents as well as solid ones. They also represent that this simulation method will usefully be applicable to the analysis of dynamical distributed parameter system for a tubular shaped active object.

1. INTRODUCTION

Peristalsis, as generally be seen in the digestive organs of animals, has two notable features. One is that the intestines can transport any intraluminal contents by peristalsis whether they are liquid or solid. The other is that the intestines almost autonomously move without innervation of extrinsic nerves. Although this peristaltic movement has been studied by many physiologists, its neural controlling mechanisms have not yet been made clear.

The authors are making biomechanical study on this subject for many years, and have showed that a neural network model preliminarily proposed could simulate the peristaltic movement for transporting only solid contents.[1] But this model was not always successful in transporting non-solid contents.

In this paper, a new model for transporting both semifluid and solid contents is proposed.

2. CHARACTERISTICS OF PERISTALSIS

The authors in the first place, started to grasp the peristaltic movement under intact state. Thus, in order to observe it in vivo, experiments on canine small intestinal movements with an abdominal-window-technique[2] was carried out. As the result, two different types of peristaltic movements were discovered as shown in Fig. 1(a) and (b). The first type of peristalsis generally occurs when the intestine is transporting liquid contents such as soup. On the contrary, the second type solid contents such as meat. The movement of the first type as Fig 1(a), is observed to have a series of wider rings of constriction and faster propagation velocity of the rings, than respective those of the second type.

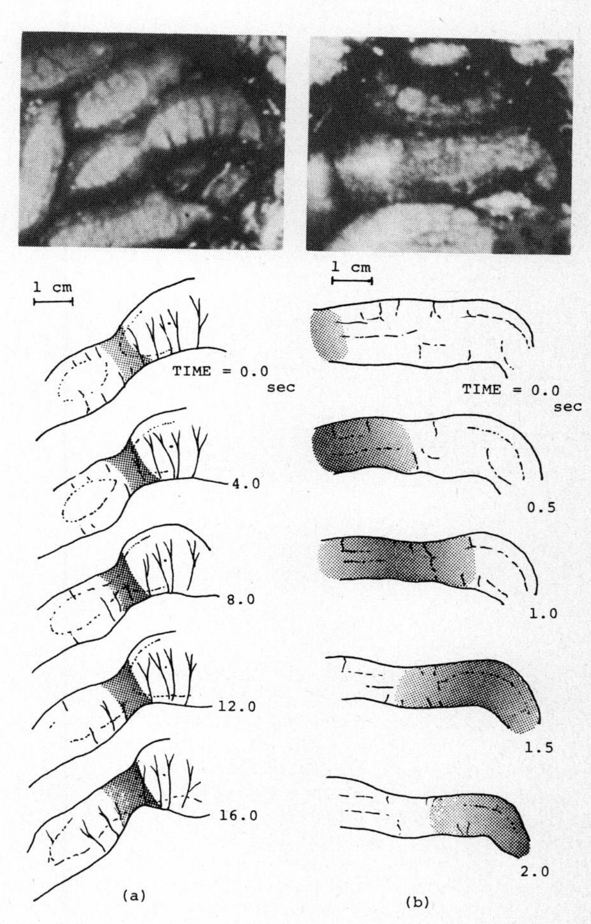

Fig.1 Typical figures of peristaltic movement. (traced from photographs of canine small intestine) Dotted portion denotes contracting.
(a) transporting a lump of meat at rate of ≈4 cm/sec.
(b) transporting soup of meat at rate of ≈0.1 cm/sec.

These results suggest that the movement of peristalsis is affected by the property of intraluminal contents. Examining many references[3,4,5,6] related to the deformation in peristaltic movements, the authors lead to presumption that the configuration of intestines varies due to the physical consistency of the contents.

3. MODELING OF NEURAL CONTROLLING MECHANISM

The small intestine is mainly composed of longitudinal and circular muscle layers, Auerbach's and Meissner's nerve plexus and mucosa as schematically shown in Fig. 2.

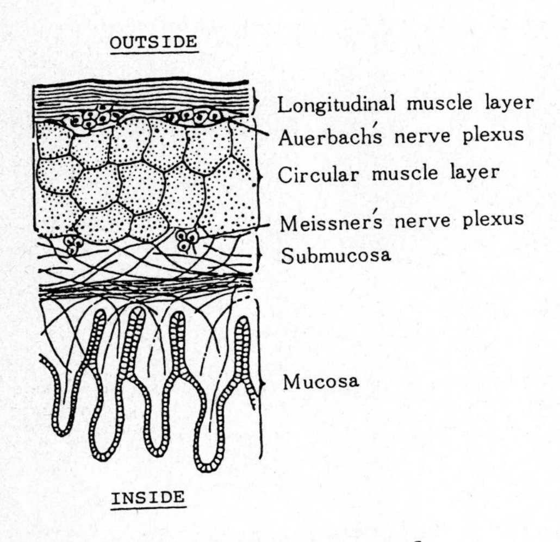

Fig.2 Schematic diagram of longitudinal wall section of small intestine.

The circular muscle is responsible for varying the diameter of the small intestine. Auerbach's nerve plexus is scattered in the thin space between the two muscle layers, and each of them is a fine conglomerate of neurons packed up closely. This plexus is supposed to play an important role in controlling muscular movement. There should exist, probably in the mucosa, the mechanical receptor[7] which is responsive to intraluminal pressure, not to the stretch. This property of the receptor has been confirmed by computer simulation.[1] Moreover, it is inferred that the other type of receptor responding to stretch of intestinal muscle tissue exsists in the muscle.[8]

Thus, we set forth in the modeling that the most fundamental structural elements of the small intestine are circular muscle, Auerbach's plexus, and two types of mechanical receptors. In spite of many physiologists'efforts[9,10], they are not as yet successful to find out the connecting relationship of these elements enough to build up the simulation model of peristaltic movement. On the contrary, some engineers have proposed its neurodynamical model[11], and succeeded in simulating it on a digital computer restricting that the intraluminal contents are perfect fluid.

The authors propose a new model, as shown in Fig.3, which enables any consistent

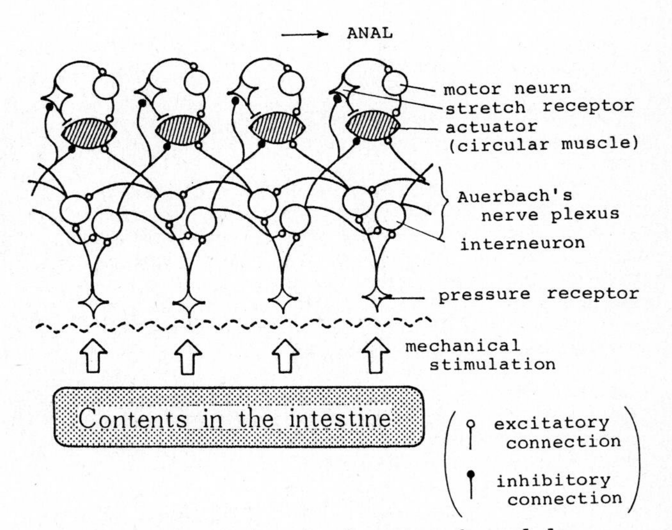

Fig.3 Proposed neural network model for peristaltic movement.

intraluminal contents to transport. This model is linearly arrayed with segments which are discretized along the longitudinal axis. The model is composed of mechanical receptors, a series of neural network corresponding to the Auerbach's plexus, and actuators that function as circumferencial contractile muscles. As the overall performance of this model, we can have an insight that it behaves irreversibly because of its non-symmetrical connection between each segment.

The model and its elements of Fig. 3 work as follows. Stimulation of the contents causes two kinds of receptors activate; the first is to activate the pressure receptor directly by the contents, and the second the stretch receptor by stretch of the intestinal wall. The actuator contracts or relaxes when excitatory or inhibitory input respectively is applied. A minor loop composed of an actuator, a stretch receptor and a motor neuron is provided as to work similary to the myotatic stretch-reflex arc. Namely, the stretch receptor works like a modulator where the inhibitory input applied from the neuron decreases the sensitivity of the loop gain.

4. DIFINITION OF ELEMENTS IN THE MODEL

To simulate peristaltic movement of the new model on a digital computer, the characteristics of elements constructing the model and physical property of intraluminal contents in deformation are defined as follows.

°Pressure receptor
The pressure receptor is defined to work as a pulse-frequency-modulator, that is, the signal S_1,

$$S_1 = k_1 \cdot P \qquad (1)$$

is transduced to a unit pulse train which frequency is proportional to S_1, where P = intraluminal pressure, k_1 = const.

°Stretch receptor
The stretch receptor is designed to work as a pulse-frequency-modulator, that is, the input S_2 is transduced to a unit pulse train of which frequency is proportional to S_2, where

$$S_2 = \begin{cases} k_2 \cdot (R-R_0), & \text{if } R \geq R_0 \\ 0 \quad , & \text{if } R < R_0 \end{cases} \qquad (2)$$

R is radius of the intestine.
R_0 is undeformed radius.
k_2 is variable.

° Neuron
The neurons, both the interneuron constructing a neural network and the motor neuron, are defined to have such characteristics as Reiss' model as in Fig. 4. Thus we describe the output $Y(x_n)$ of neurons in discrete form as below:

$$Y(x_n) = \begin{cases} 1 & \text{if } x_n \geq \Theta_n \\ 0 & \text{if } x_n < \Theta_n \end{cases} \qquad (3)$$

where x_n denotes the state variable of a neuron, Θ_n the threshold level. In order to accord the overall behavior of the model with the real one, value of the threshold Θ_n and weighting of the two inputs coming out of an adjacent interneuron and a pressure receptor can be adequately determined.

°Actuator
The actuator is an element which works to squeeze and propel the intraluminal contents by contraction. The contraction is mostly determined by strength of excitatory or inhibitory inputs. With reference to some physiological results[12,13] the authors define there exists such an apparent state x_a in the actuator as shown in Fig. 5. The movement of actuator can

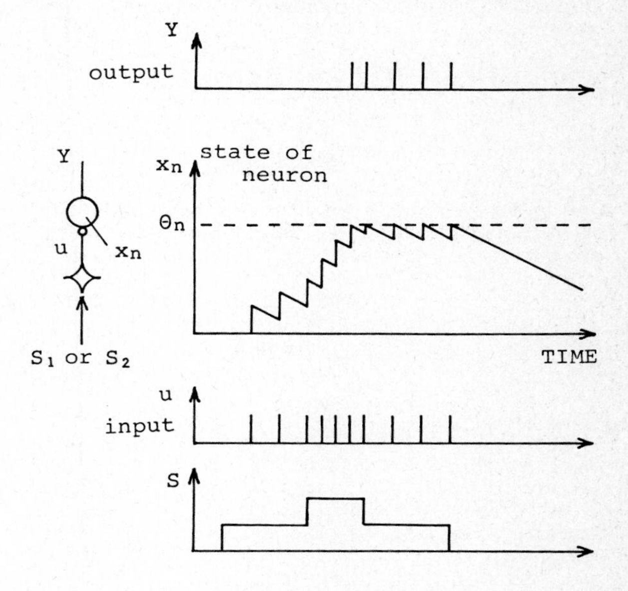

Fig.4 Behavior of neuron.

Fig.5 Behavior of actuator.

be classified into three modes, namely contracting mode, resting one and relaxing one with respect to the value of state variable x_a. In regard to the force F exerted by the actuator, we assume the three forces corresponding to the modes,

$$F = \begin{cases} F_c & : \text{contracting mode} \\ F_r & : \text{resting mode} \\ F_\ell & : \text{relaxing mode} \end{cases} \quad (4)$$

where F_c varies between F_r and F_{max} in accordance with the consistency of the contents, F_r is kept constant $0.2 \cdot F_{max}$ if $R \geq R_0$, otherwise $F_r = 0$ if $R < R_0$, and F_ℓ is definitely 0. In view of dynamical behavior of the actuator, the contracting velocity v should be defined. We approximately assume it as

$$v = k_3 \cdot (F_{max} - F_L) \quad (5)$$

where F_L is a reacting force mainly exerted by the contents in the contracting mode, and corresponds to yield stress of contents τ_E which will be difined later. In case of the relaxing mode, the actuator is passively distended by the contents thrust away from the oral side. In case of the resting mode, the actuator always, has a tendency to recovering to undeformed radius R_0.

°Contents
Intestinal contents are generally rheological, so called "chyme". As such a material is too difficult to formulate or calculate the actual deformation, the behavior of contents is assumed to be of idealized plasticity, that is, it flows like a fluid if the applied force exceeds a critical value τ_E, otherwise it behaves like a solid.

The fluid-dynamical condition of continuity is provided as a matter of course. The flowing mass of the contents squeezed out from a segment should be distributed successively along the longitudinal axis as much as proportional to the facility of distension of each segment.

5. SIMULATION

5-1 Preparatory consideration

The object of simulation experiment is to verify the following two items using the neurodynamical model shown in Fig. 3.

(1) Similarity of characteristic configurations of intestinal movement caused by the model to those observed in the animal experiment.

(2) Capability of propulsion of intraluminal contents with various consistency.

The simulation model is composed of segments which are structured in closed feedback systems where the neural network and the contents are interactive to each other. We can regard the system inputs as both intraluminal pressure P and intestinal distention $\Delta R = R - R_0$, and the system output as the diameter of intestine $2 \cdot R$, in standing upon the neural network side. The diameter of intestine for every segment is calculated by state variables of all elements included within the segment iteratively at every discretized moment. The time interval in discretion is determined as short as contracting time for the radius of intestine to move 0.25% in case that the actuator moves at maximum rate. Referring to the time scale of simulation, the unit time interval is to be estimated as 0.001 second corresponding to the real time scale. The mechanical model of intestine supposed in the experiment is a two-dimensional channel which total length is 50 and $R_0 = 4$, where these values are normalized with that of the width of a segment. As mentioned hitherto, it is linearly arrayed with 100 pieces of segment. This model is supposed to be equivalent to a canine intestine of about 15 centimeter length with the radius of about 0.8 centimeter.

Main parameters of the elements for the simulation are as follows.

°Pressure receptor

$k_1 = 0.5$ in eq(1)

°Stretch receptor
k_2 varies in the domain [0,0.1] in such a way as it is decreased from the normal state ($k_2 = 0.1$) by inhibitory signal as 30% less as per one pulse of the signal.

°Neuron
Threshold level $\Theta_n = 10$.
Weighting of the two inputs is determined so that the input from interneuron is 3.5 times as strong as that from pressure receptor.

°Actuator
Threshold levels indicated along the axis of state variable x_a in Fig.5 are

$\Theta_{ac} = 10$, $\Theta_{ar} = -10$.

In eq(5),

$F_{max} = 1.0$, $k_3 = 0.01$,

where $F_{max} = 1.0$ is approximately equivalent to the maximum contracting force of canine small intestine, 50 gram-force.

5-2 Result of simulation experiment

In carrying out simulation experiments, we settle on some initial conditions; the intraluminal contents are square-like shaped with a side length of 10, and they are laid at a medium place inside the intestine. Three kinds of consistency of the contents are provided. Therefore, every simulation experiment starts from a circumstance that a square-shaped object has suddenly been stuffed into the intestine. The model was simulated on a digital computer (HITAC M-200). The configurational change of the intestinal wall as well as the contents and transport of the contents are traced to print out on a plotter in a cycle which corresponds to 5 seconds in real peristalsis.

Fig. 6(a),(b) and (c) represent a set of results, where the normalized consistency $C_s(=\tau_E/F_{max})$ = 10, 0.6, 0.4 respectively.

Since, in case of Cs=10, the contents are not deformed by the contracting forces of actuators, they are transported undeformably to the anal direction as shown in Fig.6(a). In this case, a norrow constriction occurs at close oral side of the contents, and width of the constriction does not extend while the contents are transported.

In case of C_s=0.6 and 0.4, as in Fig.6 (b) and (c), since the contents are fluid and deformable, they are propelled to the anal direction being deformed by constriction which width is being extended as much as the contents are squeezed out. In regard to the propulsive rate, we observe that the lower the consistency is, the faster the contents are transported.

6. CONCLUDING REMARKS

In regard to result of our simulation experiments, we conclude that the simulation model in Fig. 3 represents a fundamental system for realizing peristaltic movements. That is:

(1) Characteristic configurations of peristalsis shown in Fig.6 are similar to those of actual peristaltic movement as shown in Fig.1, in case of different consistency of the intraluminal contents.

(2) It has been made clear that the contents are transported to the anal direction even if they are deformable like chyme. As described in Introduction, the authors had verified that a neural network model proposed previously could not transport any deformable contents. Therefore, we can regard it reasonable that a minor feedback loop composed of stretch receptor and motor neuron functioning like myotatic stretch-reflex arc is additionally supplied to the former model.

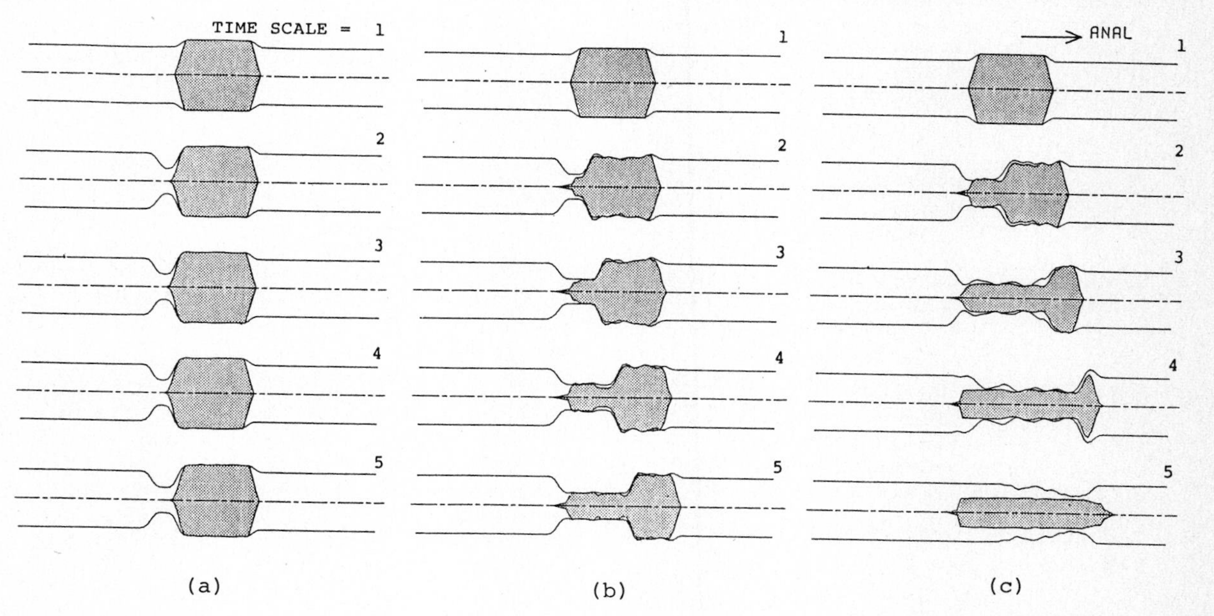

Fig.6 Consecutive process of simulated configuration of intestinal peristalsis at various consistency. (a) Cs=10.0, (b) C_s=0.6, (c) C_s=0.4 . One unit of time scale is set up corresponding approximately 5 seconds.

For further discussion on the proposed model, we can indicate some imperfect matters. One of them is that the Auerbach's plexus is too much simplified, and another one is the function of longitudinal muscle is neglected.

It should be emphasized that this simulation method will usefully be applicable to the analysis of dynamical distributed parameter system for a tubular shaped active object.

ACKNOWLEDGEMENT

The authors wish to thank Dr. T. Watanabe for his cooperative works on zoological experiments, and thanks are also due to Mr. N. Matsuhira for the computational work. This research was partly supported by the Grant in Aid for Scientific Research (1981), the Ministry of Education, Japan.

REFERENCES

[1] Umetani, Yoji and Inou, Norio, Biomechanical Study of Peristalsis --- Neural Control Mechanism of Transport in Gastrointestines, The Trans. of the Society of Instrument and Control Engineers(SICE), Vol.17-1, (1981), 133-138,(in Japanese)

[2] Hukuhara, T.,Die normale Dünndarmbewegung(Mit Hilfe der Bauchfenstermethode und Kinematographie), Pflüger Archiv Physiologie, Vol.226,(1931), 518-542, (in German)

[3] Yanagiya, I. and Ohkubo, Y., Effect of the Viscosity of Intestinal Contents on the Progressing Rate of Peristalsis, J. Physiol. Soc. Japan, Vol.20, (1958), 469-475,(in Japanese)

[4] Nakayama, S., Movements of the Small Intestine in Transport of Intraluminal Contents, Jap. J. Physiol., Vol.12,(1962), 522-533

[5] Costa, M. and Furness, J. B., The Peristaltic Reflex : An Analysis of the Nerve Pathways and Their Pharmacology, Naunyn-Shmiedeberg's Arch, Pharmacol.,Vol.294,(1976), 47-60

[6] Pescatori,M.,Grassetti,F.,Ronzoni,G., Mancinelli, R., Bertuzzi, A. and Salinari, S. , Peristalsis in Distal Colon of the Rabbit : An Analysis of Mechanical Events, Am. J. Physiol., Vol.236, (1979), E464-E472

[7] Hukuhara, T. Yamagami, M. and Nakayama, S., On the Intestinal Intrinsic Reflex, Jap. J. Physiol., Vol.8, (1958), 9-20

[8] Iggo, A., Gastro-intestinal Tension Receptors with Unmyelinated Afferent Fiberes in the Vagus of the Cat, Quart.J.Exp.Physiol., Vol.42,(1957), 130-143

[9] Yokoyama, S. and Ozaki,T., Functions of Auerbach's Plexus, Jap. J. Smooth Muscle Res., Vol.14, (1978), 173-183

[10] Hirst, G.D.S. and Holman, Mollie E., Functions of Enteric Nerve Cells in Relation to Peristalsis, Jap. J. Smooth Muscle Res., Vol.14,(1978), 189-200

[11] Bertuzzi, A., Mancinelli, R., Pescatori, M. and Salinari, S., An Analysis of the Peristaltic Reflex, Biological Cybernetics, Vol.35,(1979), 205-212

[12] Bortoff, Alex, Slow Potential Variations of Small Intestine, Am. J. Physiol., Vol.201, (1961), 203-208

[13] Bortoff, Alex, Intestinal Motility, The New England Journal of Medicine, Vol.280, (1969), 1335-1337

Simulation in Engineering Sciences
J. Burger and Y. Jarny (eds.)
Elsevier Science Publishers B.V. (North-Holland)

SIMULATION SUR MICRO-ORDINATEUR DE SIGNAUX ELECTRO-PHYSIOLOGIQUES

D. GITTON et C. DONCARLI

Laboratoire d'Automatique de l'E.N.S.M.
1, rue de la Noë
44072 NANTES CEDEX - FRANCE

L'analyse par ordinateur des signaux d'électromyographie (E.M.G.) recueillis grace à une électrode aiguille concentrique fait l'objet de travaux dans de nombreux laboratoires de physiologie et d'automatique. La simulation de tels signaux est un outil indispensable au développement, à la mise au point et à la comparaison des algorithmes de traitement du signal. La modélisation utilisée part de la description physiologique du signal, et les paramètres de la simulation seront fixés par le médecin en fonction de son objectif de traitement. Les paramètres sont utilisés pour l'élaboration d'un fichier chronogramme, dont dépend enfin la génération du signal lui-même. Le travail présenté a été implanté sur un micro-ordinateur de série, et les signaux sont transférables sur cassette analogique ou vers d'autres calculateurs par une ligne assynchrone.

I - INTRODUCTION

L'électromyographie (E.M.G.) est une technique qui consiste à enregistrer grace à une électrode aiguille (ou de surface), les phénomènes électriques générés dans les muscles [1-6]. On s'intéresse à l'analyse des signaux recueillis lors de contractions faibles, et on présente les données physiologiques de tels tracés : le signal est constitué par les potentiels de décharge des fibres musculaires excitées par le même motoneurone. Ces fibres constituent une unité motrice (U.M.) dont l'activité électrique est appelée Potentiel d'Unité Motrice (P.U.M.). Le recrutement et la fréquence de battement des U.M. dépend de la force développée par le muscle. L'électrode aiguille concentrique permet d'enregistrer l'activité du petit nombre d'U.M. proches du capteur. Après un examen minutieux, on peut isoler les différents P.U.M. et ainsi calculer divers paramètres (tels que l'amplitude, la fréquence de battement, etc...) utiles au diagnostic de certaines maladies neuromusculaires.

L'idée d'un traitement automatique de ces enregistrements n'est pas récente, mais le problème est complexe et de plus en plus de recherches sont faites sur le dépouillement automatique des enregistrements E.M.G. [7-14]. Les logiciels de traitement sont sophistiqués et délicats à mettre au point. La simulation de tels signaux est donc un outil indispensable à la mise au point et aux tests des algorithmes.

II - MODELISATION DU SIGNAL E.M.G.

II - 1 - Généralités

La modélisation utilisée est issue de la description physiologique du signal. L'enregistrement est généré à partir d'un nombre restreint de formes élémentaires. Chaque forme élémentaire est associée à une loi qui détermine les instants d'apparition et l'amplitude lors de chaque apparition. Le signal est la somme des différentes manifestations des formes élémentaires auxquelles on ajoute des perturbations correspondant d'une part au bruit d'instrumentation (bruit blanc) et d'autre part au bruit musculaire (potentiels d'unitées motrices éloignées du capteur).

Les formes élémentaires sont des formes extraites d'enregistrements E.M.G. réels. On établit un catalogue des formes élémentaires normalisées en amplitude. Dans les différentes apparitions d'une forme choisie, l'amplitude pourra varier, mais la durée est fixe. La constitution de ce catalogue est préliminaire à toute simulation, il doit contenir une variété suffisamment représentative des diverses formes des potentiels normaux et pathologiques.

Les lois déterminant les apparitions et l'amplitude des formes, ainsi que l'élaboration du bruit musculaire, sont un compromis entre nos connaissances du signal E.M.G., les conditions cliniques d'examen, et la puissance des algorithmes utilisés à ce jour. Nous avons gardé toutes les difficultés que peut rencontrer un algorithme de traitement du signal dans le cas d'enregistrements réels, l'avantage de la simulation étant de pouvoir moduler ces difficultés, et de valider ensuite les résultats obtenus.

II - 2 - Définition d'un P.U.M.

Un P.U.M. est composé d'une forme principale et éventuellement de une ou deux formes secondaires liées. Chaque forme est choisie dans le catalogue des formes élémentaires. L'amplitude de chaque forme, principale ou liée est une variable aléatoire dont la distribution est une loi normale, définie par la moyenne et l'écart-type.

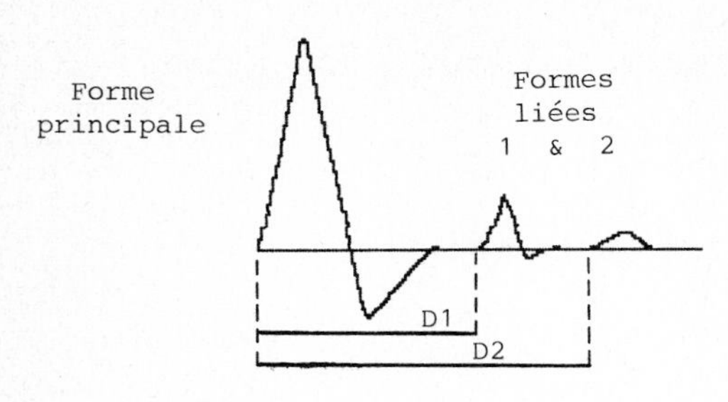

Figure 1 : Différentes formes composant un P.U.M.

II - 3 - Loi d'apparition de la forme principale d'un train de P.U.M.

Pendant la durée de l'enregistrement, les faibles variations de l'effort musculaire introduisent des variations dans le recrutement des U.M. et dans la fréquence de battement des U.M. recrutées. Ainsi un P.U.M. battra dans des fenêtres d'activité, caractérisées par leur date de début (t0 et t2) et de fin (t1 et t3). Nous avons modélisé la variation des Intervalles Inter-Potentiels (I.I.P.) par une loi triangulaire, décroissante pui croissante, et symétrique par rapport au centre de la fenêtre. Aux bords de chaque fenêtre, la fréquence de battement est la fréquence minimale (I.I.P. maximum) à laquelle peut être sollicité l'U.M., l'I.I.P. max est le même pour toutes les fenêtres d'activité du P.U.M. Enfin, cette loi peut être entachée d'une incertitude dont il faut définir un coefficient (voir figure 2).

II - 4 - Loi d'apparition des formes liées

Les formes secondaires liées d'un P.U.M. apparaissent après la forme principale avec un temps de latence D (voir figure 1), caractérisé par une moyenne et un écart-type. Ces formes liées pouvant apparaître par intermittence dans certaines pathologies, on définit une probabilité d'apparition pour chaque forme liée (voir figure 3).

II - 5 - Perturbations du signal

Nous avons considéré deux types de perturbations :
- le bruit d'instrumentation qui est un bruit blanc gaussien, centré dont on définira l'écart-type.

- le bruit musculaire qui correspond aux potentiels des U.M. éloignées de l'aiguille, ces potentiels sont d'amplitudes trop faibles pour être détectés utilement. Nous avons choisi de représenter ce bruit par une forme du catalogue, l'intervalle entre deux apparitions de cette forme, dépendant de la moyenne locale des intervalles entre les formes principales de P.U.M. consécutives. L'amplitude est la valeur absolue d'une loi gaussienne, centrée et caractérisée par un écart-type qui est très faible vis à vis des formes utiles.

III - PARAMETRES DE LA SIMULATION

L'outil développé doit permettre de choisir entre diverses configurations réalistes de tracé E.M.G., mais aussi de choisir la difficulté proposée à l'algorithme de traitement envisagé par ailleurs. On propose donc à l'utilisateur de fixer un lot de paramètres de simulation, en connection directe avec les données physiologiques.

Paramètres d'ordre général :
- la période d'échantillonnage
- la longueur du signal
- la caractéristique du convertisseur simulé (codage sur 8, 10 ou 12 bits)
- les caractéristiques du bruit blanc d'instrumentation
- les caractéristiques des potentiels éloignés, soit :
  * *le numéro de la forme*
  * *l'écart-type sur l'amplitude*
  * *un coefficient indiquant la densité des potentiels éloignés par rapport aux potentiels utiles*
  * *un coefficient d'incertitude temporelle*

On peut créer jusqu'à 16 P.U.M., et pour chaque P.U.M., il faut définir :
- le numéro de la forme principale
- l'amplitude moyenne et l'écart-type

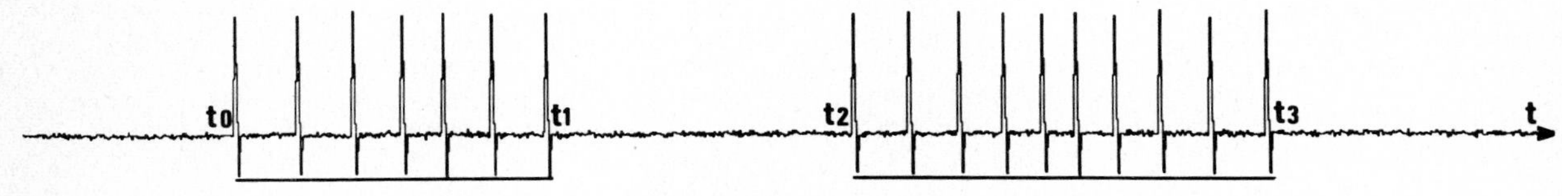

Figure 2 : Fenêtres d'activité d'un P.U.M.

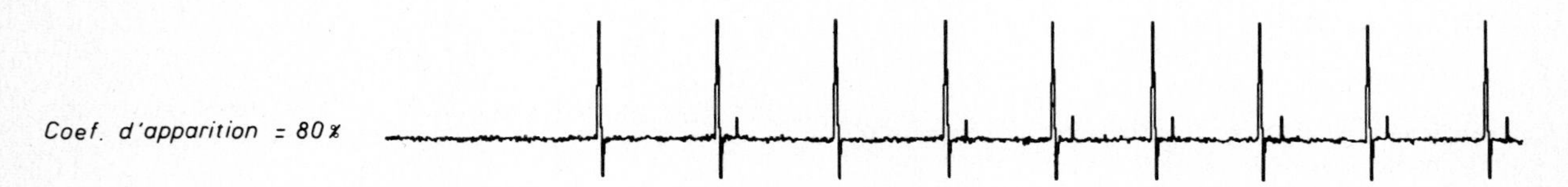

Figure 3 : Forme liée d'un P.U.M.

- l'I.I.P. maximum et le coefficient d'incertitude par rapport à la loi triangulaire des I.I.P.
- les caractéristiques de chaque fenêtre soit :
  * les dates de début et de fin
  * l'I.I.P. minimum
- les caractéristiques des potentiels liés s'il y a lieu, soit :
  * le numéro de la forme
  * l'amplitude moyenne et l'écart-type
  * le temps de latence moyen et l'écart-type
  * la probabilité d'apparition

L'utilisation de ces paramètres conduit donc à l'élaboration du fichier chronogramme.

IV - CREATION DU CHRONOGRAMME

A partir des paramètres de définition, on construit la liste des apparitions des formes élémentaires constituant le signal. Le fichier chronogramme est la description exhaustive du signal, et il donne à l'utilisateur du signal généré toutes les informations nécessaires aux tests.

Dans la suite nous noterons RAND un tirage d'une variable aléatoire uniformément répartie entre 0 et 1, et Ln un tirage d'une variable aléatoire dont la loi de distribution est normale, réduite et centrée.

IV - 1 - Calcul des amplitudes

Pour chaque apparition d'une forme utile, l'amplitude A est calculée par : $A = Am + s.Ln$; Am et s étant la moyenne et l'écart-type.

Pour le bruit musculaire, l'amplitude est : $A = s.|Ln|$

IV - 2 - Calcul des dates d'apparitions des formes principales

Pour chaque P.U.M. et pour chaque fenêtre, td et tf étant les dates de début et de fin de la fenêtre, l'I.I.P. est décrémenté jusqu'au milieu de la fenêtre, puis incrémenté jusqu'à la fin. Notons Imax et Imin le maximum et le minimum, et δI l'incrément ($1 \geqslant \delta I > 0$), soit N le nombre d'apparitions de la forme principale dans la fenêtre, nous avons approximativement :

$$N = \frac{Imax + Imin}{tf - td}$$

$$\delta I = \frac{Imax - Imin}{N - 1}$$

Le calcul des dates est séquentiel :

$t1 = td$ et $I1 = Imax$
$t2 = t1 + I1$ et $I2 = I1 - \delta I$
etc...

On peut introduire une perturbation temporelle dont le coefficient d'incertitude est st, et l'expression récursive devient :

$t1 = td$ et $I1 = Imax$

$$\begin{cases} ti+1 = ti + Ii.\ (1 + st.\ Ln) \\ Ii+1 = Ii - \delta I \text{ si } ti+1 < \frac{tf + td}{2} \\ \text{ou} \\ Ii+1 = Ii + \delta I \text{ si } ti+1 \geqslant \frac{tf + td}{2} \end{cases}$$

Fin si $ti+1 \geqslant tf$

IV - 3 - Dates d'apparition des formes liées

A chaque apparition de la forme principale, un tirage de la fonction RAND indique si la forme liée apparaît ou non. Soit P sa probabilité d'apparition :

si $RAND > P$: pas d'apparition
si $RAND \leqslant P$: apparition de la forme à la date tl : $tl = t + D + sl.Ln$; où t est la date d'apparition de la forme principale, D le temps de latence moyen, et sl l'écart-type.

IV - 4 - Elaboration du bruit musculaire

Le bruit musculaire est calculé à partir du chronogramme des formes utiles. Notons Tk la date d'apparition de la kième forme utile du signal. On définit la fonction en escalier E qui représente une "moyenne locale" de l'intervalle entre deux formes utiles consécutives dans l'enregistrement.

$\forall\ t \in]\ Tk,\ Tk+1\ ]$

$$E(t) = E(Tk+1) = \alpha.E(Tk) + (1 - \alpha).(Tk+1 - Tk)$$
$$\alpha \simeq 0,4$$

Les dates tj d'apparition de la forme représentant le bruit musculaire sont calculée séquentiellement :

$tj+1 = tj + \frac{E(tj)}{C}.\ (1 + s.Ln)$; le coefficient C est la densité des potentiels éloignés par rapport aux potentiels utiles et s un coefficient d'incertitude.

IV - 5 - Fichier chronogramme final

Le fichier chronogramme est une suite de quadruplets : (ti, Ai, $\mathcal{F}i$, $\mathcal{C}i$), ti étant la date de l'apparition, Ai le facteur d'amplitude, $\mathcal{F}i$ le numéro de la forme dans le catalogue et $\mathcal{C}i$ un code indiquant la fonction de cette forme (forme principale ou liée d'un P.U.M., ou potentiel éloigné). Le fichier est classé dans l'ordre croissant des dates.

V - CREATION DU SIGNAL

Le signal est généré à partir du chronogramme. Chaque forme $\mathcal{F}i$ correspond à un tableau de longueur li dans le catalogue des formes : fi(1), fi(2)...,fi(li). A l'instant t, l'amplitude S(t) du signal est donc :

$$S(t) = sb.Ln + \sum Ai\ .\ fi(t - ti)$$
$$\forall\ i : t \in [ti,\ ti + li]$$

où sb est l'écart-type du bruit blanc d'instrumentation.

VI - SYNOPTIQUE DU LOGICIEL ET MATERIEL UTILISE

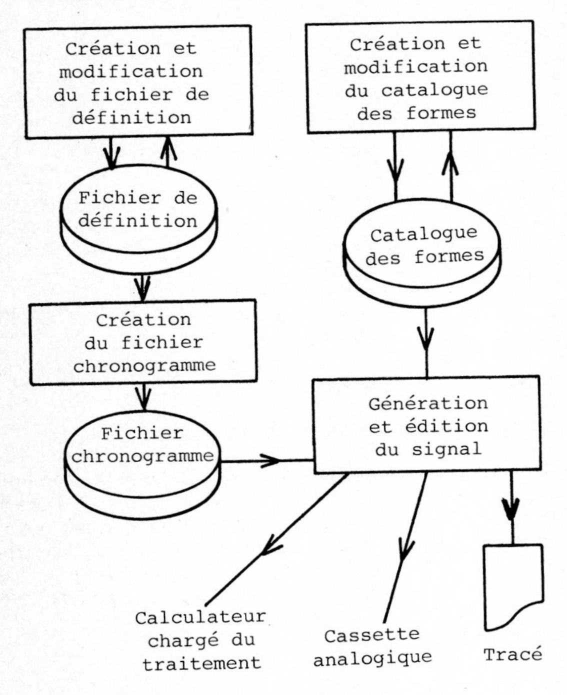

Figure 4 : Synoptique du logiciel

Le matériel de base est un micro-ordinateur APPLE II 48K plus, équipe de deux lecteurs de disquettes 5" et d'une imprimante disposant du mode graphique. L'entrée de courbes a été facilitée par l'utilisation d'une tablette graphique. Le signal généré est stocké sur disquette et peut être enregistré sur cassette analogique par l'intermédiaire d'une carte de conversion numérique-analogique, ou transféré vers un autre calculateur par un ligne assynchrone.

VII - EXEMPLES

On présente d'abord un tracé E.M.G. réel, obtenu pour une contraction faible (figure 5). Les deux exemples suivants sont des tracés simulés correspondant à une contraction faible (3 P.U.M.), seul le bruit d'instrumentation diffère entre les deux tracés (figures 6 et 7). On présente enfin un tracé simulé correspondant à une contraction moyenne, il se compose de 15 P.U.M. Les multiples superpositions des potentiels rendent impossible toute reconnaissance de forme, ce tracé est dit interférentiel (figure 8).

VIII - CONCLUSION

La simulation présentée dans ce papier, permet d'obtenir des signaux dont les caractéristiques coïncident au mieux avec les signaux réels normaux et pathologiques. Il s'agit donc d'un outil permettant de tester et de mettre au point les algorithmes (détection, filtrage, reconnaissance de forme, etc...) destinés à fournir une aide au diagnostic.

Figure 5 : Enregistrement E.M.G. réel

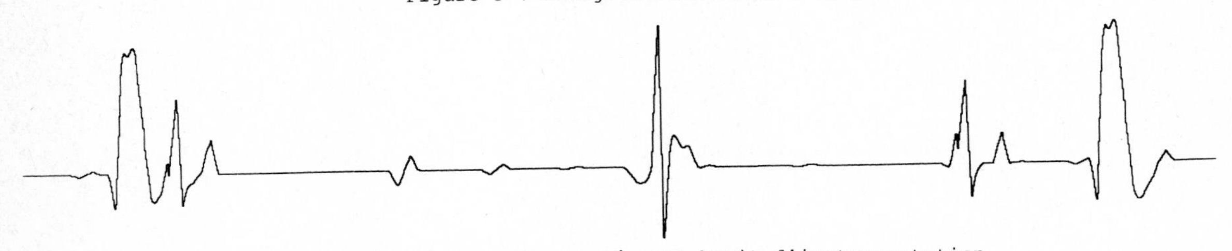

Figure 6 : Enregistrement simulé sans bruit d'instrumentation

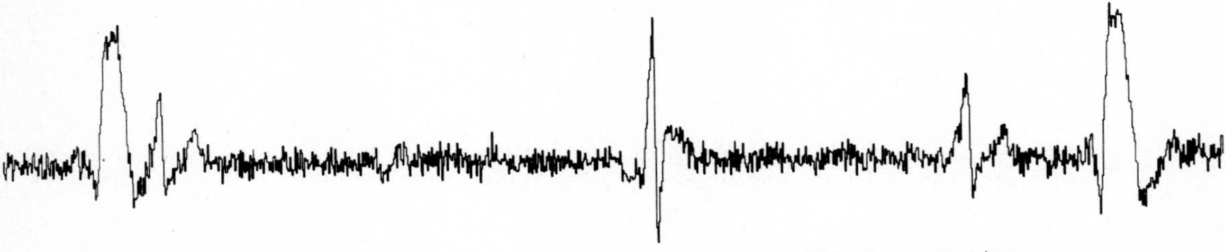

Figure 7 : Enregistrement simulé avec bruit d'instrumentation

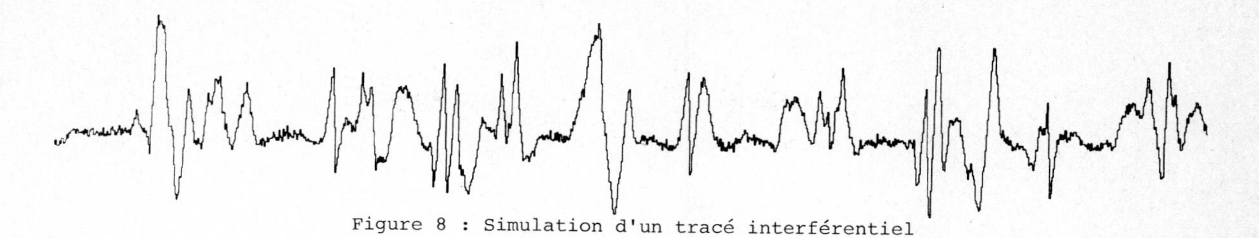

Figure 8 : Simulation d'un tracé interférentiel

[1] F. BUCHTAL - C. GULD - P. ROSENFALCK "Action potential parameters in normal human and their dependance on physical variables". Acta Physiol. Scand., Vol. 32-33, pp 200-218, 1954-1955.

[2] F. BUCHTAL - P. PINELLI - P. ROSENFALCK "Action potential parameters in normal human muscle and their physiological determinants" Acta Physiol. Scand., Vol. 32-33, pp 219-229, 1954-1955.

[3] C.J. DE LUCA "Physiology and mathematics of myoelectric signals". IEEE Trans. Biomed. Eng., Vol. BME26, pp 311-325, 1979.

[4] H. KRANZ - G. BAUMGARTNER "Human alpha motoneurone discharge, a statistical analysis". Brain Res., Vol. 67, pp 324-329, 1974.

[5] B. MATON - S. BOUISSET "Motor unit recruitment during movment in normal man". Excerpta Medica, Vol. 1, pp 312-318, 1972.

[6] S. ANDREASSEN "Interval pattern of single motor units" Ph. D. dissertation, Tech. Univ. of Denmark, 1977.

[7] J.D. DILL - P.C. LOCKEMANN - K.I. NAKA "An attempt to analyse multi unit recordings". EEG Clin. Neurophysiol., Vol. 28, pp 79-82, 1972.

[8] R. KADEFORS "Myoelectric signal processing as an estimation problem". In New Developments in Electromyography and Clinical Neurophysiology. J.E. DESMEDT, Ed. BASEL, KARGER, 1975, Vol. 1, pp 519-532.

[9] H.A. BOMZE - B.A. EISENSTEIN "Decomposition method of analysing nerve potentials" Proc., 21 s^t ACEMB conf. HOUSTON TX, 1968.

[10] J.F. FELDMAN - F.A. ROBERGE "Computer detection, classification and analysis of neuronal spike sequences". Inform., Vol. 9, 1971.

[11] D.H. FRIEDMAN "Detection of signals by template matching" Baltimore M.D. Johns Hopkins, 1968.

[12] F.M. GLASER - W.B. MARKS "The on-line separation of inter-leaved neuronal pulse sequences". In Proc. Rochester Conf. Data Acquisition Biology, Medicine, pp 137-156, 1966.

[13] E.M. SCHMIDT "Unit activity from peripheral nerve bundles utilizing correlation techniques". Med. Biol. Eng., Vol. 9, pp 665-674, 1971.

[14] R.S. LE FEVER - C.J. DE LUCA "A procedure for decomposing the myoelectric signal into its constituent action potentials". Part I : technique, theory and implementation IEEE Trans. Biomed. Engin., Vol. BME 29 n°3, pp 149-157, 1982.

VI. SIMULATION AND CONTROL METHODS AND TECHNIQUES

Simulation in Engineering Sciences
J. Burger and Y. Jarny (eds.)
Elsevier Science Publishers B.V. (North-Holland)

CONTROLLABILITY AND OBSERVABILITY OF LINEAR TIME DELAY SYSTEMS

Vladimir M.Marchenko

Byelorussian S.M.Kirov Technological Institute
Minsk, USSR

The first part of the paper deals with a problem of pointwise controllability and observability which generalizes the well-known problem of relative controllability and conditional observability of delay-differential systems. Then effective criteria of controllability, "controllability-observability" duality principle and formula for the calculation of the minimum number of inputs for which the system is controllable are given. In the second part we consider a problem of modal controllability and a stabilization problem for linear systems with delays. In conclusion we apply the obtained results to the investigation of a problem of control of the flight of a fluing apparatus

0. INTRODUCTION

A problem of controllability of delay-differential system was formulated [1] first by Krasovskii in 1963 as a problem of total quieting (controllability to zero function). In a parallel way Kirillova and Churakova [2] stated and solved the problem of relative controllability of such systems. Various aspects of these problems were investigated in [3]-[6] and today a wide literature exists on the subject of controllability and observability of delay-differential systems (for the entire collection see also the survey papers [7]-[10]). In 1967, Vogt and Cullen [11] formulated and solved the problem of calculation of the minimum number of inputs of systems without delay. Some generalization of this problem to systems with delays are in [12], [13]. Problems of pointwise controllability and observability of time lag systems were considered [13] - [15]. The problem of modal controllability (theory and applicatios) was detailly considered for systems without delay in the book [16]. Generalization of this problem to the linear stationary systems with many delays are given in [17]. The paper [17] contains also an application of the modal control problem to a problem of control of an automatic reostat regulator of voltage.

1. POINTWISE CONTROLLABILITY AND OBSERVABILITY

1.1 Controllability

Let us suppose that a control object is described with the following delay-differential system

$$\dot{x}(t) = \sum_{j=o}^{l} (A_j x(t-h_j) + B_j u(t-h_j)),$$
$$t > 0, \tag{1.1}$$

with the initial conditions

$x(t) = g(t)$, $u(t) = 0$, $t \in [-h,0]$,
$x(+0) = g_o$,

where $x' = (x_1,\ldots,x_n)$, $u' = (u_1,\ldots,$

u_r), $0 = h_o < \ldots < h_l$ are constant delays; the coefficients A_j, B_j, $j = 0,\ldots,l$, are the constant matrices of the corresponding dimensions; g is a piecewise continuous vector function, $g_o \in R^n$; the stroke (') denotes transposition.

The system (1.1) is called: 1) pointwisely controllable in the points $s_o,\ldots,s_N$, $0=s_o<\ldots<s_N$, if there exists a time moment t_1, $t_1 > s_N$, such that for any initial data g, g_o and for any n-vectors c_i, $i=0,\ldots,N$, there exists a piecewise continuous control $u(t)$, $t \in [0,t_1]$, such that the corresponding solution $x(t)$, $t \in [0,t_1]$, of the system has the property $x(t_1-s_j) = c_j$, $j=0,\ldots,N$; 2) z-pointwisely controllable for $z \geq 0$ if the system is pointwisely controllable in any points $s_o,\ldots,s_N$ such that $0=s_o<\ldots<s_N \leq z$; 3) pointwisely controllable if the system is z-pointwisely controllable for all z, $z \geq 0$.

Let $X_k(t)$, $k \geq 1$, $t \geq 0$, be the solution of the determining equation [5] :

$$X_{k+1}(t) = \sum_{j=o}^{l} (A_j X_k(t-h_j) + B_j U_k(t-h_j))$$

where $X_k(t) = 0$ if $k \leq 0$ or $t < 0$; $U_k(t) = 0$ if $k^2 + t^2 \neq 0$, $U_o(0) = I_n$, I_n is the identity $n \times n$ matrix. Introduce the set

$$S = \Big\{ s_o,\ s_1,\ s_2,\ldots \ \Big|\ 0 = s_o < s_1 \ldots,\ s_k = \sum_{k_1=1}^{l_1} h_{j_{k_1}} - \sum_{k_2=1}^{l_2} h_{j_{k_2}},\ j_{k_1} \in \{0,1,\ldots,l\},\ j_{k_2} \in \{0,\ldots,l\},\ l_1 = 1,2,\ldots;\ l_2 = 1,2,\ldots \Big\}.$$

Then the following statements are valid

Theorem 1.1
The system (1.1) is pointwisely controllable in the points $s_o, \ldots, s_N$ if and only if the condition $\operatorname{rank} X(s_o,\ldots,s_N) = n(N+1)$ holds where by the symbol $X(s_o,\ldots,s_N)$ we denote the matrix which formed by the columns of the matrix

$$\begin{bmatrix} X_k(t-s_o) \\ X_k(t-s_1) \\ \vdots \\ X_k(t-s_N) \end{bmatrix}$$

when t run from 0 to $s_N + (n-1)h$ and $k = 1,\ldots,n$.

Theorem 1.2
The system (1.1) is z-pointwisely controllable if and only if the equality $\operatorname{rank} X(s_o,\ldots,s_j) = n(j+1)$ holds for $s_j \in S$ such that $s_j \leq z$ but $s_{j+1} > z$.

Theorem 1.3
The system (1.1) is pointwisely controllable if and only if there exists a real number m^*, $m^* \geq 0$, such that the system

$$\dot{x}(t) = \sum_{j=o}^{l} (m^*)^{h_j} A_j x(t) + \sum_{j=o}^{l} (m^*)^{h_j} B_j u(t)$$

is controllable in Kalman's sense.

The proof of the theorems 1.1 - 1.2 can be given using the technique of the determining equation by analogy with the proof of the criteria of relative controllability [5]. The proof of the theorem 1.3 can be found in [13] (see also [15]).

Remarks

Since $X_k(t) \neq 0$ not more than for finite points t from $[0,s_N+(n-1)h]$ then the matrix $X(s_o,\ldots,s_N)$ has finite dimensions and its number of rows is equal to $n(N+1)$.

We can state that if the system (1.1) is z-pointwisely controllable then it is also q-pointwisely controllable for $q \leq z$. The inverse statement is not true in general but if the system (1.1) is z^*-pointwisely controllable for $z^* = \max_{j\in\{0,\dots,l\}} \min\{h_j,\ h_j \text{ rank } B_j\} + (n-1)(n-2)h/2$ then it is also pointwisely controllable

1.2 Observability

Consider the system

$$\dot{x}(t) = \sum_{j=0}^{l} A_j' x(t-h_j), \quad t > 0, \tag{1.2}$$

with the output

$$y(t) = \sum_{j=0}^{l} B_j' x(t-h_j), \quad t > 0, \tag{1.3}$$

where the matrices A_j, B_j and the numbers h_j, $j=0,\dots,l$, are the same that in (1.1).

Let $s_0,\dots,s_N$ be real numbers such that $0 = s_0 < \dots < s_N$. In each interval $[s_j-h, s_j]$ we take the initial conditions for the system (1.2) as follows $x(t) = g_j(t)$, $t\in[s_j-h, s_j]$, $x(s_j+0) = x_j$. The corresponding solution of the system (1.2) we denote as $x_j(t)$, $t > s_j$.

Let $Y_j(t) =$

$$\sum_{i=0}^{l} B_i' x_j(t-h_i), \quad t \geq 0,$$

$$x_j(t) = \sum_{i=0}^{l} \int_{s_j-h}^{s_j} K(t-s-s_i) A_i g_j(s)\,ds, \quad t < s_j,\ j = 0,\dots,N,$$

where $K(t)$ is defined by equation

$$\frac{dK(t)}{dt} = \sum_{j=0}^{l} A_j' K(t-h_j)$$

with initial conditions $K(+0) = I_n$, $K(t) \equiv 0$ for $t \leq 0$.

Consider the following functional F_{t_1}:

$$F_{t_1}(Y_j, g_j) = \int_0^{t_1} v'(t) Y_j(t)\,dt + \sum_{i=0}^{l} \int_{s_j-h_i}^{s_j} q_i'(t) g_j(t)\,dt,$$

where the components of the vector functions v, q_i, $i = 0,\dots,l$, are piecewise continuous functions.

The system (1.2),(1.3) is said to be: 1) observable in the points $s_0,\dots,s_N$ if there exists a number $t_1 > 0$ such that for any n-vectors p_j, $j = 0,\dots,N$, there exist vector function $q_i(t)$, $t\in[-h, s_N]$; $v(t)$, $t\in[0,t_1]$; such that $F_{t_1}(Y_j, g_j) = p_j' x_j$ for all $x_j \in R^n$ and piecewise continuous vector functions g_j, $j = 0,\dots,N$; 2) z-pointwisely observable if it is z-pointwisely observable in every points $s_0,\dots,s_N$ such that $0 = s_0 < \dots < s_N \leq z$; 3) pointwisely observable if the system is z-pointwisely observable for all z, $z \geq 0$. The process observation is shown schematically in Fig.1

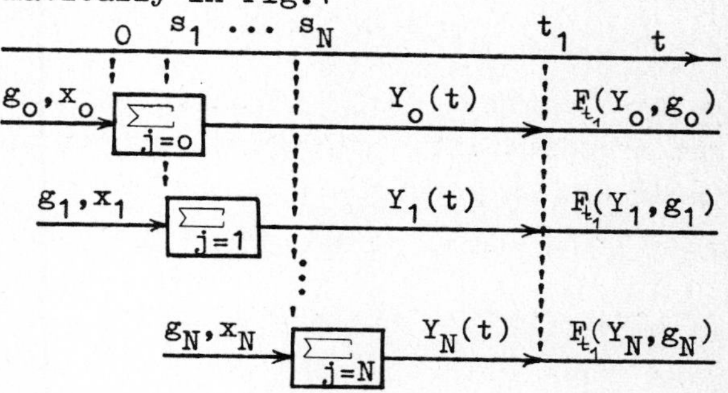

Figure 1 : Pointwise observation

If we connect this scheme with a physical system for which the vector functions Y_j, $j = 0,\dots,N$, are outputs of concrete physical models then the sense of the pointwise observation is contained in the construction of single-type observers $F_{t_1}(\cdot,\cdot)$ regenerating the initial data x_j, $j = 0,\dots,N$, by means of measurements of the output Y_j and initial data g_j, $j=0,\dots,N$. It should be noted that the final moment t_1 is the same for all models.

We can state the following [15]

Theorem 1.4 (duality principle)
The system (1.1) is controllable in the points $s_0,\dots,s_N$ (z-pointwisely controllable, pointwisely controllable) if and only if the system (1.2),(1.3) is observable in $s_0,\dots,s_N$ (z-pointwisely observable, pointwisely observable).

1.3 Calculation of minimun number of inputs

Consider the system (1.1) with $B_j = 0$, $j = 1,\dots,l$. The problem is to find the least number r_o of columns of the matrix B_o for which the system is pointwisely controllable.
From theorem 1.3 and [11] we have

Theorem 1.5
The minimum number r_o of inputs of the pointwisely controllable system is defined by

$$r_o = \min_{m \geqslant 0} \rho\left(\sum_{j=o}^{l} m^{h_j} A_j\right)$$

where the symbol $\rho(D)$ denotes the number of nontrivial (different from unit) invariant polynomials of p-matrix $pI_n - D$.

2. MODAL CONTROL AND STABILIZATION

Consider now the system of neutral type

$$\dot{x}(t) = \sum_{j=o}^{l}(D_j\dot{x}(t-h_j) + A_j x(t-h_j) + b_j u(t-h_j)), \quad t > 0, \qquad (2.1)$$

where D_j, A_j are $n \times n$ constant matrices, $b_j \in R^n$, $j = 0,\dots,l$; $D_o = 0$. For these system the state concept includes also an information on derivative $\dot{x}(t)$, $t > 0$. Therefore consider a regulator of the form

$$u(t) = \sum_{i=o}^{N} \sum_{j=o}^{M} q'_{ij}\, x^{(i)}(t-w_j) \qquad (2.2)$$

The system (2.1) is said to be modally controllable by the regulator (2.2) if for any real numbers $0=s_o<\dots<s_k$; r_{ij}, $i=0,\dots,n$; $j=0,\dots,k$; $r_{no} = 0$, there exist numbers $w_1,\dots,w_M$ and n-vectors q_{ij}, $i=0,\dots,N$; $j=0,\dots,M$, such that

$$\det\left[pI_n - \sum_{j=o}^{l}(D_j p e^{-ph_j} + A_j e^{-ph_j} + b_j e^{-ph_j}\left(\sum_{i=o}^{N}\sum_{j=o}^{M} q'_{ij} p^i e^{-pw_j}\right))\right] \equiv$$

$$p^n + \sum_{i=o}^{n}\sum_{j=o}^{k} r_{ij}\, p^i e^{-ps_j}.$$

Theorem 2.1 [13]
If the identity

$$\det\left[b(m) \,\vdots\, A(m,p)b(m) \,\vdots \dots \vdots\, (A(m,p))^{n-1} b(m)\right] \equiv \text{constant} \neq 0 \qquad (2.3)$$

holds for any complex p and real m, $m \geqslant 0$, then the system (2.1) is modally controllable by the feedback (2.2).
Here we denote

$$b(m) = \sum_{j=o}^{l} m^{h_j} b_j,$$

$$A(m,p) = p\sum_{j=o}^{l} m^{h_j} D_j + \sum_{j=o}^{l} m^{h_j} A_j.$$

Corollary 2.1 [13]
The system (2.1),(2.2) with $D_i = 0$, $q_{vj} = 0$, $i=0,\dots,l$; $v=1,\dots,N$, is modally controllable if and only if the identity (2.3) holds for $p = 0$ and for all real m such that $m \geqslant 0$.

The problem of stabilization is a partial case of the problem of modal control, i.e. every modally controllable system can be also stabilized. Below we consider some problem of stabilization for the simplest system with delay

$$\dot{x}(t) = A\, x(t) + A_1\, x(t-h) + b\, u(t) \qquad (2.4)$$

with the output feedback

$$\sum_{j=o}^{N} z_j\, u(t-jh) = \sum_{j=o}^{N} q'_j\, x(t-jh) \qquad (2.5)$$

where A, A_1 are $n \times n$ constant matrices, $b \in R^n$, $q_j \in R^n$, $z_j \in R$, $j=0,\dots,N$, $z_o \neq 0$, $h > 0$.

The system (2.4),(2.5) is said to be stabilized if there exist parameters z_j, q_j, $j=0,\dots,N$; $z_o \neq 0$ such that the system (2.4) closed by the regulator (2.5) is asymptotically stable, i.e. the roots of the equation

$$\det\left[\sum_{j=o}^{N} z_j\, e^{-pjh}(pI_n - A - A_1 e^{-ph}) -\right.$$

$$- b \sum_{j=o}^{N} q_j' e^{-pjh} \Big] = 0$$

has negative real parts.

We have

Theorem 2.2
If the all roots of the equation

$$\det \left[b \vdots (A + mA_1)b \vdots \ldots \vdots (A + mA_1)^{n-1}b\right] = 0$$

belong to the ring $0 < |m| < 1$ then the system (2.4),(2.5) is stabilized.

3. APPLICATION

Consider a plane longitudinal motion of a fluing apparatus for which the vector of velocity of the center of gravity belong to the vertical symmetry plane.

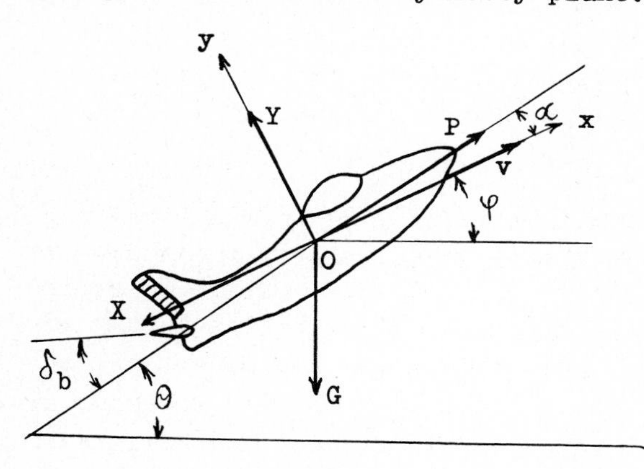

Figure 2 : Longitudinal motion of fluing apparatus

Under some assumptions this motion is described [18] by the equations

$$\dot{v} + n_{11}v + n_{12}\alpha + n_{13}\Theta = n_p \delta_p + f_1,$$
$$-n_{21}v + \dot{\alpha} + n_{22}\alpha - \dot{\Theta} - n_{23}\Theta = f_2,$$
$$n_{31}v + n_o\dot{\alpha} + n_{32}\alpha + \ddot{\Theta} + n_{33}\dot{\Theta} = -n_b\delta_b + f_3, \quad (3.1)$$

where (see also Fig. 2):

v - velocity of the fligt
Y - body force
X - resistance force
G - gravity force
P - propulsive force
Θ - pitch angle
φ - angle of inclination of trajectory
α - angle of attack
f_1, f_2, f_3 - components of disturbing forces
δ_p, δ_b - angles of deviation of rudders
n_{ij}, n_p, n_b - coefficients characterizing effectiveness of control organs and characteristics of fluing apparatus

Assume that the components f_1, f_2, f_3 are such that $f_i(t) = a_i\, v(t-h)$ where a_i, $h > 0$, are constants, $i=1,2,3$. Setting $x' = (v, \alpha, \Theta, \dot{\Theta})$ we have

$$\dot{x}(t) = A^*x(t) + A_1^*\, x(t-h) + Bu(t) \quad (3.2)$$

where $A^* =$

$$\begin{bmatrix} -n_{11} & -n_{12} & -n_{13} & 0 \\ n_{21} & -n_{22} & n_{23} & 1 \\ 0 & 0 & 0 & 1 \\ -(n_{31}+n_o n_{21}) & n_o n_{22}-n_{32} & -(n_{33}+n_o n_{23}) & -n_o \end{bmatrix}$$

$$A_1^* = \begin{bmatrix} a_1 & 0 & 0 & 0 \\ a_2 & 0 & 0 & 0 \\ 0 & 0 & 0 & 0 \\ a_3 & 0 & 0 & 0 \end{bmatrix}, \quad B = \begin{bmatrix} n_p & 0 \\ 0 & 0 \\ 0 & 0 \\ 0 & n_b \end{bmatrix}$$

$$u' = [\delta_p, \delta_b].$$

To ensure the stable given flight (under action of disturbances) the pointwise controllability of the system (3.2) is required. Assume that $n_{21} + m\, a_2 \neq 0$, $m \geq 0$. Then we have rank $\left[B \vdots (A^* + mA_1^*)B \vdots (A^* + mA_1^*)^2 B \vdots (A^* + mA_1^*)^3 B\right] = 4$ for some $m \geq 0$. By the theorem 1.3 the system (3.2) is pointwisely controllable. In this situation we want to know minimum number of inputs necessary for the pointwise controllability of the system.

Applying the rown and column elementary operations to the p-matrix $pI_n - A^* - mA_1^*$ we obtain

$$r_o = \min_{m \geqslant 0} \rho(A^* + m A_1^*) = 1$$

if $n_{12} + m a_2 \not\equiv 0$, $m \geqslant 0$, and, in particular, it is possible to take the vector $(n_p,0,0,0)'$ or $(0,0,0,n_b)'$ as the required matrix of inputs.

Conclusion

For the control of the plane horisontal flight of a fluing apparatus given by the system (3.1) it is sufficiently to take one control parameter:

δ_b - angle of deviation of height rudders or

δ_p - situation of control organ of motor.

REFERENCES

[1] Krasovskii, N.N., Optimal processes in lag systems, in Proc. of the Second IFAC Congress, 2 (Nauka, Moscow, 1965).

[2] Kirillova, F.M. and Churakova, S.V., Dokl. Acad. Nauk SSSR, 174 (1967) 1260-1263.

[3] Kurzhanskii, A.B., Prikladnaya Matematika i Mekhanika, 6 (1966) 1121-1124.

[4] Krasovskii, N.N., Theory control by movement (Nauka, Moscow, 1968).

[5] Gabasov, R. and Kirillova, F.M., Qualitative theory of optimal processes (Nauka, Moscow, 1971).

[6] Marchenko, V.M., Dokl. Acad. Nauk SSSR, 236 (1977) 1083-1086.

[7] Gabasov, R. and Kirillova, F.M., Automatika i Telemekhanika, 9 (1972) 31-62.

[8] Gabasov, R. and Kirillova, F.M., Mathematical theory of optimal control, in Itogi Nauki i Tekhniki. Matematicheskii Analis, 16 (VINITI, Moscow, 1979).

[9] Manitius, A. and Triggiani, R., SIAM J. Control and Optimization, 16 (1978) 549-552.

[10] Gabasov, R. and others, Problems of control and observation for infinite-dimensional systems, Prepr. of Institute of Mathematics of AN BSSR (to appear, 1982).

[11] Vogt, W.G. and Cullen, G.A., IEEE Trans. Automat. Contr.,12 (1967) No3.

[12] Marchenko, V.M., Izvestiya Vuzov. Matematika, 4 (1978) 42-52.

[13] Marchenko, V.M., Controllability problems for time lag systems, Institute of Mathematics of PAN (Prepr. No 234, 1981).

[14] Minyuk, S.A., Vestnik Belorusskogo universiteta, ser.1, 1 (1972) 8-11.

[15] Marchenko, V.M., To the theory of controllability and observability of linear systems with retarded arguments, in Gabasov and Kirillova (eds), Problems of Optimal Control (Nauka i Tekhnika, Minsk, 1981).

[16] Porter, B. and Grossley, R., Modal control. Theory and applications, (Taylor and Francis, London, 1972).

[17] Marchenko, V. and Asmykovich, I., On the problem of modal control in linear systems with delay, in Tzafestas (ed),Simulation of Distributed Parameter and Large-Scale Systems (North - Holland, Amsterdam, 1980).

[18] Bodner,V.A., Theory of automatic control of the flight (Nauka,M.,1964).

Simulation in Engineering Sciences
J. Burger and Y. Jarny (eds.)
Elsevier Science Publishers B.V. (North-Holland)

FRACTIONAL ORDER POSITION CONTROL AND COMPARISON WITH THE ADAPTIVE CONTROL

A. OUSTALOUP

E.N.S.E.R.B. - University of Bordeaux I
351, cours de la Libération - 33405 TALENCE CEDEX - FRANCE

This note is based on the fact that fractional order systems avoid the usual compromise between stability and static precision. In the first part, we define fractional order and show how it may be achieved.

Given that fractional order feedback systems with orders between 1 and 2 are very precise and have damping related essentially to the order, the second part of this note applies this remarkable property to a position control problem. The dependence between inertia and damping is removed. This dependence is an important limitation to the dynamic control of manipulators used on industrial robots.

The third part describes the performances achieved in an analog simulation of a fractional order position control process. Comparison is made with a traditional control and with an adaptive control with a reference model.

1. INTRODUCTION

While the problem of fractional order was seriously tackled from the stand-point of automation in 1975, work on fractional orders is still largely unknown to some researchers in this field, particularly those who are unconditional supporters of integer orders. This work has shown that linear feedback systems of fractional order i) lead to a logarithmic phase (for orders between 0 and 1) ; ii) are not subject to the compromise between stability and precision (orders between 1 and 2) ; iii) obey the conditions for linear voltage-frequency conversion (orders over 2).

This note is based on the fact that fractional orders escape the usual compromise between stability and precision. Fractional order feedback systems with orders between 1 and 2 have high static precision and damping determined essentially by a nearly linear dependence on the order (7). Their field of application is wide (7), including robots where they naturally eliminate the dependence between inertia and damping, a most remarkable property for a feedback system. This dependence is a major limitation to the dynamic control of manipulators used on industrial robots, particularly those with several degrees of freedom.

The purpose of this paper is to propose a flexible dynamic control system which minimises this problem while giving identical damping performances in very different conditions of use.

2. CONCEPT OF FRACTIONAL ORDER

Transfer functions of the form :

$$F_n(p) = F_{o_n}(1 + \tau_n\, p)^{-n}, \qquad 2.1$$

with n an integer ($n \geq 1$), are obtained from linear differential equations of order n, to which correspond polynomial equations of degree n in p where all the coefficients have the same sign :

$$1 + \sum_{i=1}^{n} \frac{1}{i!}(n-i+1)(n-i+2)\ldots(n)\tau_n^i\, p^i = 0 \qquad 2.2$$

Less common transfer functions of the form :

$$F_{n'}(p) = F_{o_{n'}}(1 + \tau_{n'}\, p)^{-n'}, \qquad 2.3$$

where n' is a fraction ($0 < n' < 1$), lead to linear differential equations of infinite order, to which correspond polynomial equations of infinite degree in p where the terms constitue alternating series :

$$1 + \sum_{i=1}^{\infty} \frac{(-1)^i}{i!}(i-1-n')(i-2-n')\ldots(-n')\tau_{n'}^i\, p^i = 0. \qquad 2.4$$

The first type of transmittance can be obtained physically with active and passive components, but the second type are idealised transfer functions, physically unobtainable. However, one can in practice produce approximate forms in certain frequency domains. In fact, a fractional order may be achieved only by distributing the parameters (6-7).

Although n' is a fraction, it plays the same role as n in determining the asymptotic frequency behaviour. This leads us to arbitrarily fix the order of a system as the coefficient n or n' characteristic of this behaviour, particularly since this definition takes account of more physical considerations than does that previously given, where the order of a system is defined as that of its differential equation.

3. PRINCIPLES OF FRACTIONAL ORDERS

3.1 Fractional order systems with local parameters

The passive components-resistances, condensers and induction coils can be combined in eight types of elementary cell-four containing resistances and condensers and four with resistances and coils.

The first four give low-pass networks with fractional order (7). These networks are composed of many elementary cells with ratios of α^{-1} and of η^{-1} respectively between the resistances and the capacitances of successive cells, α and η are greater than unity.

The second type give high-pass networks with fractional order (7). These are composed of many elementary cells with ratios of α^{-1} and η^{-1} respectively between the resistances and the inductances of successive cells, α and η are greater than unity.

The theoretical impedance of an infinite chain of cells is given by :

$$Z_{n'}(p) = Z_{o_{n'}}(1 + \tau_{n'} p)^{\pm n'} \quad 3.1$$

The exposant is positive or negative for high-pass or low-pass filters respectively (7).

3.2 Fractional order systems with distributed parameters

The value of distributing the parameters is that one thus obtains a better approximation of a fractional order. Any segment of a distributed parameter device corresponds to an infinite number of elementary cells, so that 3.1 is no longer an approximate relation in a given frequency domain. We have used this property to study cylindrical and plane symmetrical systems with distributed parameters which may be characterised by fractional order impedances (6-7). This work was illustrated by the making of thick layer devices of cylindrical or plane symmetry with orders 0.2 (cylindrical) or 0.5 and 0.8 (plane) (7). A model of a cylindrical thermal corrector of order 0.55 has been made in our Laboratory. It is designed to achieve the optimal order (1.45) for the temperature feedback system for which it is intented.

4. WHY USE FRACTIONAL ORDER CONTROL ?

In this paragraph, we justify the use of fractional order control, particularly adaptive controls.

Adaptive control with a reference model is characterised by compensating variations of the gain of the direct line of the feedback. The invariance of damping with changes in inertia leads to a greater sensitivity of the eigenfrequency. By considering an electro-mechanical process (servo-motor) we can overcome this problem if the device has a current limiting loop, no matter whether there is a speed controller. When the error signal of the adaption loop controls the gain of the feedback chain relative to the current limiting loop, both the damping and the undamped eigenfrequency are unsensitive to changes of inertia.

Simplicity is the rule for fractional order controls, both in the system itself and in its application. This kind of control amounts in fact to introducing a simple high-pass corrector of fractional order in the direct feedback circuit. This correction does away with the need of a reference model, an adaption loop comparator, a decision unit, a current limiting loop and a voltage controlled amplifier (V.C.A.). Moreover, use of a fractional order gives a systematic value of the damping : to such and such fractional order corresponds such and such a damping factor and such and such overshots. Furthermore, this solution does not accentuate the variations of the eigenfrequency when the damping is unsensitive.

5. FRACTIONAL ORDER POSITION CONTROL

5.1 Principle and operating conditions

A fractional order position control may be obtained by integrating a fractional order (between 0 and 1) correction in the direct line of a second order fundamental position control. The transmittance of this high-pass corrector is

$$F_{n'}(p) = (1 + p/\omega_{n'})^{n'} \text{ avec } 0 < n' < 1 , \quad 5.1$$

where the transitional frequency $\omega_{n'}$ is chosen to verify

$$\omega_{n'} = \omega_m , \quad 5.2$$

ω_m is the transitional frequency of the servo-motor corresponding to optimal inertia (cf relation 5.5).

The transmittance in open circuit is of order $n = 2 - n'$:

$$\beta_n(p) = (1 + p/\omega_m)^{n'-1} \omega_1 p^{-1} \quad 5.3$$

where ω_1 is the frequency related to integration in open circuit use. ω_1 and ω_m verify :

$$\omega_1 = \frac{\mu}{N} \frac{A A_V \Phi_0}{A B \Phi_0^2 + f(R_g + R)} \quad 5.4$$

$$\text{and } \omega_m = \frac{A B \Phi_0^2 + f(R_g + R)}{J(R_g + R)} , \quad 5.5$$

where the parameters are :

- μ : transfer factor of the feedback loop
- A_V : voltage gain of the power amplifier of the direct loop
- A and B : constants for a given motor, A being determined by the torque and B by the back-e.m.f.
- Φ_0 : magnetic flux under one pole at constant excitation
- R_g, R : output resistance of the power amplifier and resistance of the windings
- N : step-down ratio of the speed reducer
- J and f : moment of inertia and coefficient of viscous friction of the rotor.

The operating conditions must ensure an asymptotic frequency behaviour over a wide range around the frequency at unit again in open circuit ω_{u_n} (6). This condition is satisfied when :

$$\omega_{n'} \text{ and } \omega_m << \omega_{u_n} \qquad 5.6$$

5.2 Relative change, with the inertia, of the frequency at unit gain in open circuit

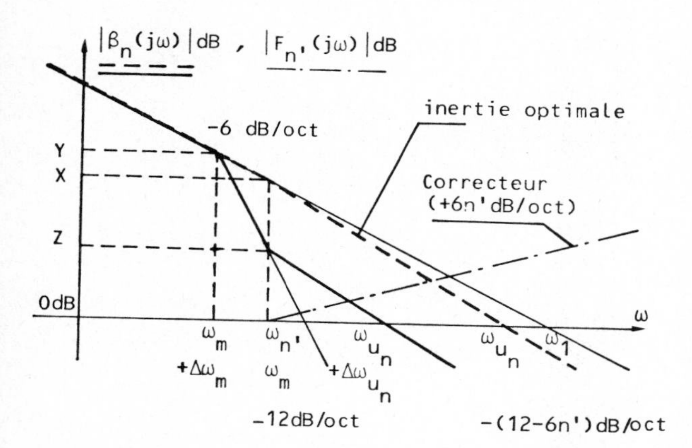

Figure 1 - Deformation of the asymptotic gain in open circuit, due to a change in the inertia

The relative change of ω_m with inertia is deduced from 5.5 :

$$\Delta\omega_m/\omega_m = \left(1 + \frac{\Delta J}{J}\right)^{-1} - 1 . \qquad 5.7$$

In figure 1, we have :

$$Y = X(1 + \Delta\omega_m/\omega_m)^{-1} \qquad 5.8$$

$$Z = Y(1 + \Delta\omega_m/\omega_m)^2 \qquad 5.9$$

$$Z = X(1 + \Delta\omega_m/\omega_m) \qquad 5.10$$

$$\text{and } \Delta\omega_{u_n}/\omega_{u_n} = (Z/X)^{1/n} - 1 \qquad 5.11$$

whence, using 5.10 and 5.7 :

$$\Delta\omega_{u_n}/\omega_{u_n} = \left(1 + \frac{\Delta J}{J}\right)^{-1/n} - 1 \qquad 5.12$$

5.3 Analytical form of the response to elementary tests and the corresponding performance

The response to a step function or to a delta or a ramp excitation have been derived elsewhere. Taking account of the presence of the integration in open circuit.

$$s(n,t) = k_n \, E \, u(t) \left[1 - e^{-\zeta(n)\omega_x t} \cos g(n)\, \omega_x t\right] \qquad 5.13$$

$$s(n,t) = k_n \, A \, u(t) \, \omega_x \, e^{-\zeta(n)\omega_x t} \cos\left(g(n)\omega_x t - \text{arc cos } \zeta(n)\right) \qquad 5.14$$

and $s(n,t) = k_n \, \alpha \, u(t) \left[t - \zeta(n)\omega_x^{-1} + \omega_x^{-1} e^{-\zeta(n)\omega_x t} \sin\left(g(n)\omega_x t - \text{arc sin } \zeta(n)\right)\right]$, 5.15

where ω_x is given by :

$$\omega_x = \omega_{u_n}(1 + \Delta\omega_{u_n}/\omega_{u_n}) \qquad 5.16$$

Using 5.12, we have :

$$\omega_x = \omega_{u_n}(1 + \Delta J/J)^{-1/n} \qquad 5.17$$

The last equation can be used to determine the reduced expressions for the performances:

$$f'_p = f_p/f_{u_n} = (-n^2+4n-3)^{1/2}(1+\Delta J/J)^{-1/n} \; ;$$

$$\theta'_r = \theta_r \cdot f_{u_n} = 0{,}25(-n^2+4n-3)^{-1/2}(1+\Delta J/J)^{1/n} \; ;$$

$$\theta'_R = \theta_R \cdot f_{u_n} = \frac{3}{2\pi}(2-n)^{-1}(1+\Delta J/J)^{1/n} \; ;$$

$$\varepsilon'_{sp} = \varepsilon_{sp} \cdot f_{u_n}/k_n\alpha = \frac{2-n}{2\pi}(1+\Delta J/J)^{1/n} \; ;$$

$$\theta'_{sp} = \theta_{sp} \cdot f_{u_n} = \frac{2-n}{2\pi}(1+\Delta J/J)^{1/n} . \qquad 5.18$$

The performances are plotted in figures 2 to 5 as functions of order at different values of the relative change of inertia. It will be noted that at any given order, the variations for positive changes of the inertia are smaller in magnitude than those for negative changes.

6. ANALOG SIMULATION WITH OPERATIONAL AMPLIFIERS OF TRADITIONAL, ADAPTIVE, FRACTIONAL ORDER AND ADAPTIVE FRACTIONAL ORDER CONTROLS

The results presented here concern mainly the optimal order n = 1.45, although some simulations at other fractional orders were done. This value of n minimises the criteria accounting for rigidity, drag error, the transient state energy and Tupicyn's criterion (7).

The analog simulator designed in this Laboratory (7) simulates a current limited position command with speed correction. A fractional order corrector in the action line, as well as a fractional order adaption loop with a reference model are incorporated (Fig. 7).

The error signal controls the amplification of the feedback line relative to the current limiting loop. The control is in fact applied via a system comprising a proportional integrator corrector, an analog commutator with transfer factor ± 1, depending on the logical state of an upstream decision making unit whose role is to give the error signal of the adaptive loop the sign of its first half wave. This causes double rectification with commanded polarity.

The simulator uses 23 operational amplifiers. The logical unit is composed of a forming circuit, a monostable, a memory and a comparator. The gain controlled amplifier is composed of an operational amplifier and a field effect transistor used as a variable resistance between the drain and the source.

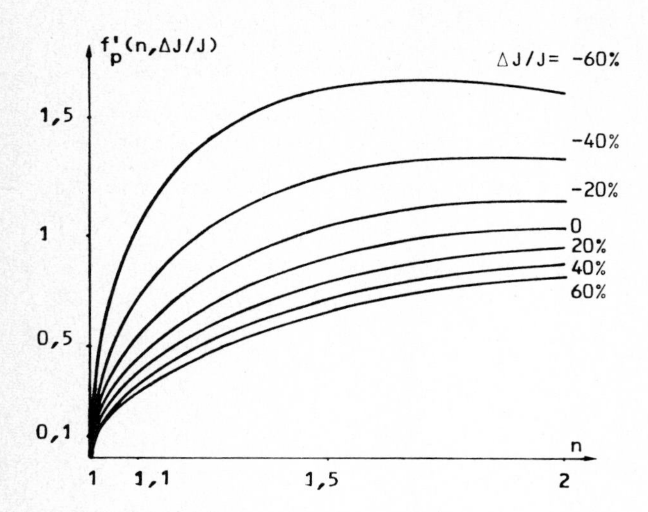

Figure 2 : Variation of the reduced eigenfrequency with the order, for different relative changes of inertia.

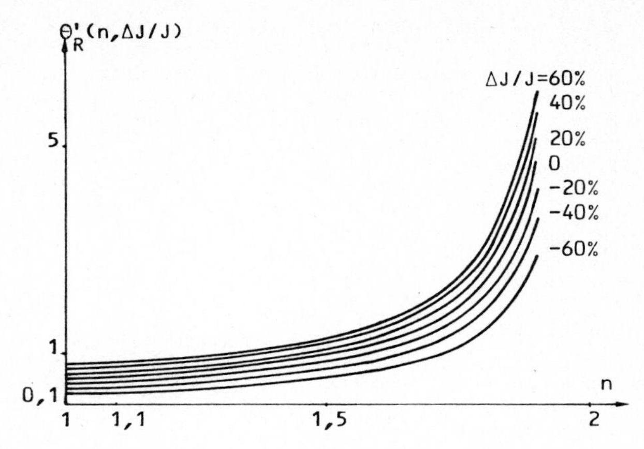

Figure 4 : Reduced response time vs order at different values of the relative change of inertia

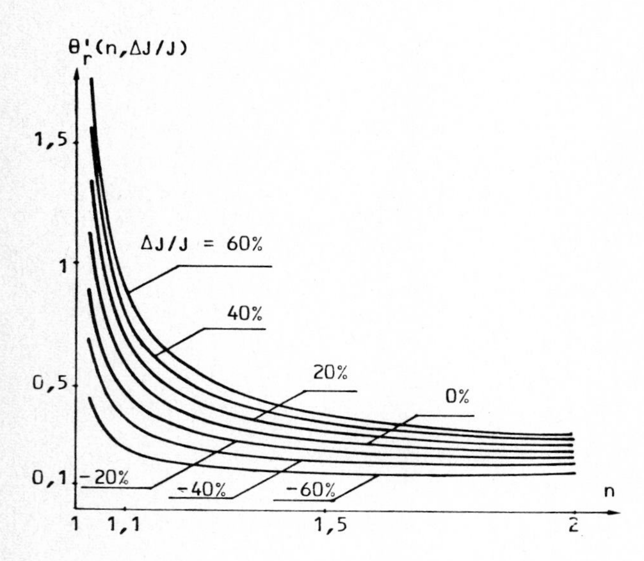

Figure 3 : Reduced time characteristic of the rigidity as a function of order, for different relative changes of the inertia.

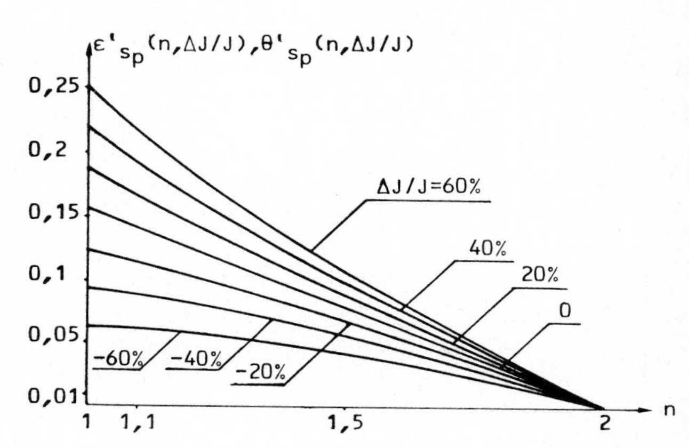

Figure 5 : Reduced drag error and final delay vs order at different values of the relative change in inertia.

Switches can be set to produce any of the following four cases, namely a position control system with current limitation and speed control :

a) without adaption loop
b) with adaption loop (adaptive command)
c) with fractional order correction and no adaption loop (fractional order command)
d) with fractional order control and adaptive loop (adaptive, fractional order control).

The frequencies f_1 related to integration in open circuit are fixed at 7 Hz for all operating regimes. The transitional frequencies in open circuit f_m are 5 Hz without fractional order correction ($f_1/f_m = \sqrt{2}$) and 0.078 Hz with correction. The transitional frequencies f_n' of the fractional order correctors must be set equal to or preferably smaller than the transitional frequency f_m (7). An arbitrary value of 0.039 Hz was used ($f_n'/f_m = 0.5$).

Changes in inertia were simulated by a variable resistance R in an integrator. R varied between 100Ω and 100 KΩ, a factor of 1000 between the extreme values of the simulated inertia. Figure 6 shows the variation of the first reduced overshot of the response to a slep function vs. The normalised inertia, for the four cases described above. A relative change of 100 between the extreme values of the inertia leads to changes of 70% without adaptive loop and 28% with adaptive loop.

The same relative variation of the inertia produces variations of only 16% with a fractional order corrector and only 9% when both adaptive correction and a fractional order correction are employed. A ratio of 1000 leads to equally promising changes of only 22% and 13% for the last two cases.

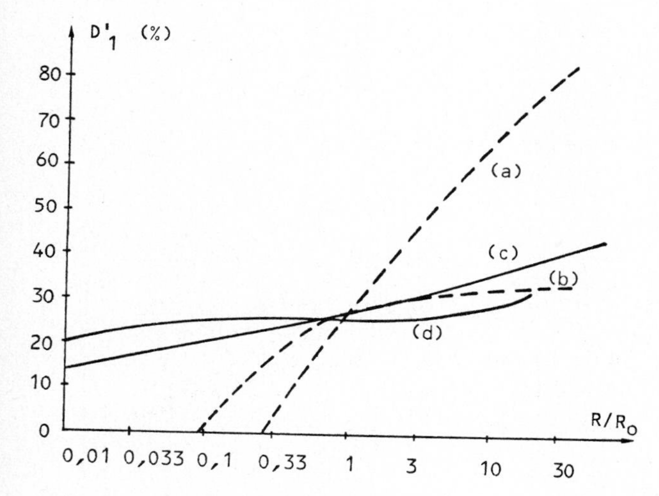

Figure 6 : Reduced first overshot of the response to a slep function vs normalised inertia. R_o is the value of the resistance corresponding to identity between the model signals and those :

a) without adaption loop
b) with adaption loop
c) with fractional order command
d) with adaptive fractional order command.

7. CONCLUSION AND PERSPECTIVES

In this paper, we have reviewed the idea of fractional order and how it may be achieved in systems with localised or distributed parameters. We present the performance of a fractional order position control, not forgetting to justify its use compared to that of an adaptive control.

The performance of this type of control, up to the present, have been compared only with those of traditional uncorrected controls, in order to treat impartially fractional and integer orders, and thus bring out the natural damping properties of fractional orders.

While numerical simulation has not been completed, the results of analog simulation, particularly those on adaptive fractional order controls, promise interesting future research. We plan, within the frame of our work, to study an adaptive fractional order control for a manipulator with three degrees of liberty using the structure of an inclinable polar table. The inertia of this system varies over a wide range.

BIBLIOGRAPHY

Some earlier work by the author :

(1) OUSTALOUP, A.
L'Onde Electrique, 1979, vol. 59, n°2, pp. 41-47

(2) OUSTALOUP, A.
European Conference on Circuit Theory and Design, Warsow, Poland, September 2-5, 1980

(3) OUSTALOUP, A.
I.E.E.E. Transactions on Circuits and Systems, 1981, vol. cas 28, n° 10, pp 1007-1009

(4) OUSTALOUP, A. and BEN HAFSIA
L'Onde Electrique, 1981, vol. 61, n° 3, pp 31-37

(5) OUSTALOUP, A.
I.E.E.E. International Symposium on Circuits and Systems, Chicago, Illinois, April 27-29 1981

(6) OUSTALOUP, A.
Thèse de Doctorat d'Etat ès Sciences, Université de Bordeaux I, 1981.

(7) OUSTALOUP, A.
Systèmes asservis linéaires d'ordre fraction naire, Ed. MASSON, 1982

(8) COIFFET, Ph.
Les Robots, Hermes publishing, 1981

(9) LANDAU, Y.
Adaptive control, Control and System Theory Series, vol. 8

(10) NAJIM, K.
Commande adaptative des processus industriels Ed. MASSON, 1982

(11) OUSTALOUP, A.
Brevet d'invention n° 78 357 28 INPI, 1978

(12) OUSTALOUP, A.
L'Onde Electrique, 1979, vol. 59, n° 6-7, pp 61-68

(13) OUSTALOUP, A.
L'Onde Electrique, 1979, vol. 59, n° 8-9, pp 102-107

(14) OUSTALOUP, A.
L'Onde Electrique, 1980, vol. 60, n° 4, pp 40-44

(15) OUSTALOUP, A.
Conférence "Application de la Modélisation et de la simulation"
LYON, FRANCE, 7-11 septembre 1981

(16) OUSTALOUP, A., PISTRÉ, J-D. and MORA, A.
IASTED Conference, Davos, Switzerland, march 2-5, 1982

(17) OUSTALOUP, A., MORA, A. and PISTRÉ, J-D.
Conference "Modelling and Simulation", Vallée de Chevreuse, Paris-Sud, France, 1-3 juillet 1982.

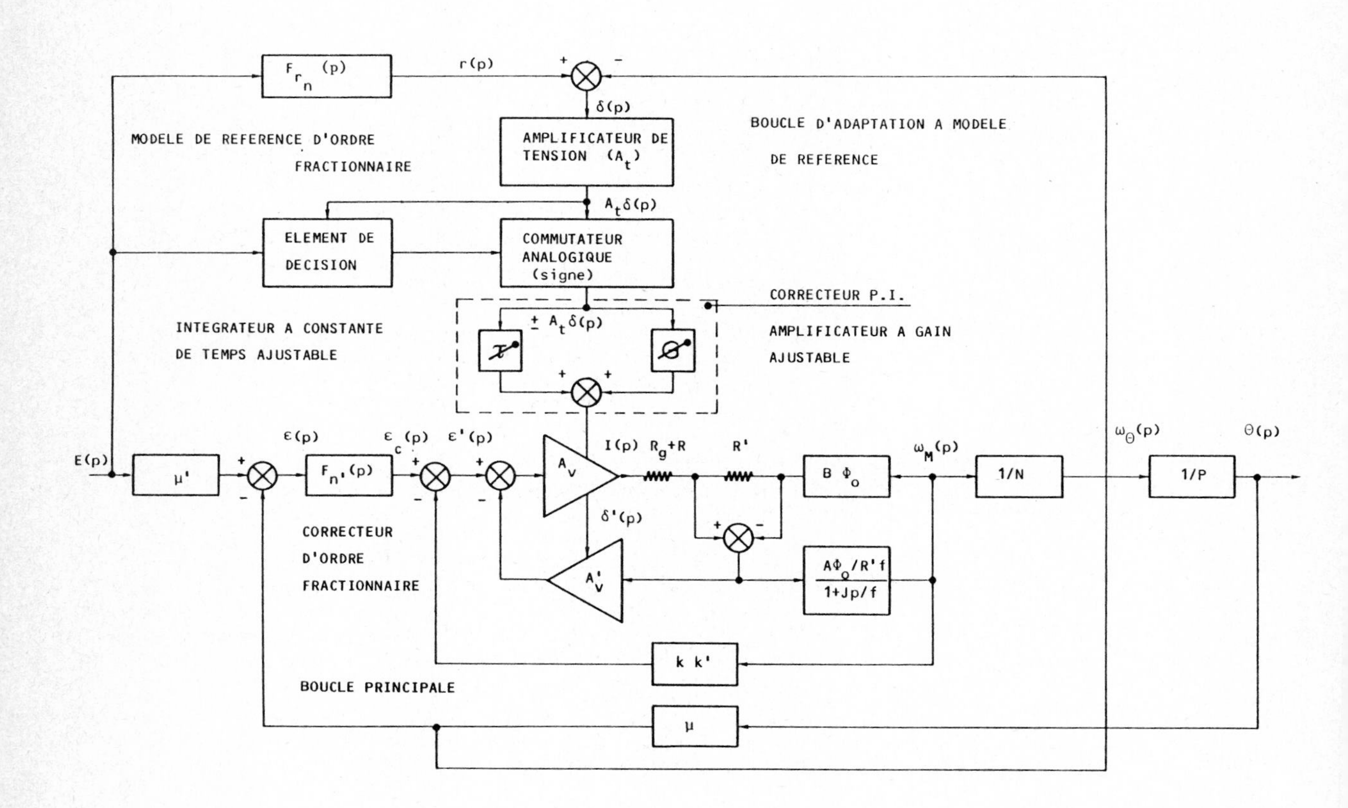

Figure 7 : Block diagram of the simulator

μ' : transfer factor of the display unit
kk' : transfer factor of the feedback line of the speed control loop. k is the transfer factor of the tachometric generator
A'v : voltage amplification at controlled gain VCA
R' : resistance put in the winding circuit to detect the image of the winding current (current limitation)
$F_{n'}(p)$: transmittance of the fractional order (n') corrector, $0 < n' < 1$
$F_{r_n}(p)$: transmittance of the fractional order (n) reference model, $n = 2 - n'$.

APPENDIX : DETERMINATION OF THE OPTIMAL ORDER $n = 1,45$

An empirical model established by identification with the experiment (6), gives a step response of the form :

$$s(n,t) = K_n\ E\ u(t)\ \left[1-e^{-\zeta(n)\omega_{un}t} \cos g(n)\ \omega_{un}\ t\right], \quad A.1$$

in which E is the height of the step and where :

$$\zeta(n) \simeq 2 - n \quad \text{for} \quad 1 \leqslant n < 2$$

$$\text{and } g(n) = \omega_p\ (n,\omega_{un})/\omega_{un} \simeq \left(1 - \zeta^2_{(n)}\right)^{1/2} \quad A.2$$

This model leads to simple expressions for the caracteristics of the responses to the step function $(e(t) = E\ u(t))$, the delta function $(e(t) = A\ \delta(t))$ and the slope function $(e(t) = \alpha t\ u(t))$. In particular, the reduced eigenfrequency is

$$f'_p(n) = f_p/f_{un} = g(n) = (-n^2 + 4n - 3)^{1/2}\ ; \quad A.3$$

the first time θ_r that the response reaches its steady state value and the response time θ_R verify the relations :

$$\theta_r \cdot f_{un} = \theta'_r(n) = 0,25\ (-n^2+4n-3)^{-1/2} \quad A.4$$

$$\text{and } \theta_R \cdot f_{un} = \theta'_R(n) \simeq 3/2\ \pi\zeta = (3/2\pi)(2-n)^{-1}\ ; \quad A.5$$

the delay error ε_{sp} and the corresponding final delay θ_{sp} are given by :

$$\varepsilon_{sp} \cdot f_{un}/K_n\ \alpha = \varepsilon'_{sp}(n) = (2-n)/2\pi \quad A.6$$

$$\text{and } \theta_{sp} \cdot f_{un} = \theta'_{sp}(n) = (2-n)/2\pi . \quad A.7$$

The energy error in steady state operation, defined by :

$$\varepsilon_{wp} = \int_{-\infty}^{+\infty} \left(s^2(n,\infty)\ u(t) - s^2(n,t)\right)dt\ , \quad A.8$$

obeys the relation :

$$\varepsilon_{wp} \cdot f_{un}/s^2(n,\infty) = \frac{1}{8\pi}\ \frac{7n^2 - 28n + 27}{2 - n}\ ; \quad A.9$$

the final delay θ_{wp} is given by :

$$\theta_{wp} \cdot f_{un} = \frac{1}{8\pi}\ \frac{7n^2 - 28n + 27}{2 - n}\ . \quad A.10$$

In steady state operation ($t \geqslant \theta_R$), the true energy dissipation W(t) is a function of the desired dissipation $W_d(t) = s^2(n,\infty)t$:

$$W(t) = W_d(t) - \varepsilon_{wp}\ , \quad A.11$$

whence, using A.9 and A.10 :

$$W(t) = s^2(n,\infty)(t - \theta_{wp}). \quad A.12$$

The energy involved in the transient follows immediately :

$$W(\theta_R) = s^2(n,\infty)(\theta_R - \theta_{wp}).$$

Replacing θ_R and θ_{wp} by their values given by A.5 and A.10, we have :

$$W(\theta_R) \cdot f_{un}/s^2(n,\infty) = W'(n) = \frac{1}{8\pi}\ \frac{7n^2-28n+15}{n-2} \quad A.13$$

Two criteria were defined. The first takes into account the first passage time (cf. above), the delay error and the energy involved in transient state operation. Its reduced form is :

$$C'_1(n) = \theta'_r(n) \cdot \varepsilon'_{sp}(n) \cdot W'(n)\ ,$$

$$\text{i.e. : } C'_1(n) = \frac{(-3+4n-n^2)^{-1/2}(-15+28n-7n^2)}{64\pi^2}\ ; \quad A.14$$

the minimum occurs at $n_1 = 1.662\ (\zeta = 1/\sqrt{7})$.

The second criteria includes Tupicyn's criterion. Its reduced expression is of the form :

$$C'_2(n,k) = C'_1(n) \cdot T'(n,k)\ ,$$

$$\text{where } T'(n,k) = \frac{f_{un}}{s^2(n,\infty)} \int_{-\infty}^{+\infty} \left\{ \left(s(n,\infty)\ u(t) - s(n,t)\right)^2 + (k^2/\omega^2_{un}) \left[\partial\left(s(n,\infty)u(t)-s(n,t)\right)/\partial t\right]^2 \right\} dt\ ,$$

$$\text{i.e. : } C'_2(n,k) = \frac{1 + k^2}{512\ \pi^2}\ (-3+4n-n^2)^{-1/2}\left[\frac{13}{2-n} - 44 + 78n + 7n^2(6+n)\right]\ ;$$

for $k = 0$, the minimum occurs at $n_2 = 1.276$ ($\zeta = 0,723$).

Both criteria are roughly constant in the interval $1,4 < n < 1,5$. Given that we want the best order, it seems natural to choose the mid-point, $n = 1.45$. The resonance factor is then equal to 1.315 and the phase margin to 49.5°, which is additional justification.

Simulation in Engineering Sciences
J. Burger and Y. Jarny (eds.)
Elsevier Science Publishers B.V. (North-Holland)

NETSY: NETWORK SYSTEM SIMULATOR (FROM NETSY TO NESSY - A CASE-STUDY)

D. Ambrózy

Central Research Institute for Physics
H1525 Budapest 114, P.O.B.49, Hungary

In order to investigate the monitoring problems of computer networks, in the KFKI Research Institute for Measurement and Computing Techniques a simulator has been construed. The simulator is modular and interactive; the parameters of the simulation as well as of the traffic to be simulated is modifiable while running the program. The report deals with the system and the more important events of its history. On basis of experience the further development of the simulator is under way.

Introduction

Early in 1977 our Institute became interested in the problem of measuring computer networks. Since at these times no networks of the needed dimensions were available, we had to choose an appropriate way of modeling. However, the established models known at that time turned out to be of general nature and they failed to give account of the particular problems involved in the design and installation of measuring systems. Therefore, we decided to build a simulator.

By the fall of 1977 the following objectives were agreed upon:

a. design of a simulator, the performance of which corresponds to the statistics [1,2,3] of the computer networks

b. the simulator should be modular with respect to the configuration of the simulated network as well as with respect to the applied strategies (routing, forwarding, etc., technics)

c. the parameters of the simulation as well as those of the traffic should be adjustable

d. the system should qualify as part of the measuring device (as a traffic-generator, as a source of outside disturbances, etc.,) and as part of the system to be measured (in cases of the tuning of prospective gateways, hosts, or monitors)

e. the system should work on a minicomputer with a possibility of its later real time realisation of a µP parallel-processor system.

These objectives hold good up till now.

The short history of NETSY

NETSY's name is an acronym for network simulation system. Its development started in November 1977. In 1978 it was in working order as a basic realisation.

After a series of pilot-runnings and evaluations, in the fall of 1978 it was furnished with additional, interchangeable modules and in 1979 it obtained a provisory monitor.

Meanwhile, quite unexpectedly, NETSY became employable as a research-tool in the investigation concerning a game-theoretical approach to computer network measurement technics [4].

Early in 1980 several hundred simulations were performed on the standard-NETSY for a better understanding of the working of a network with monitor. In 1981 other experiments were conducted, too. One series of simulations concerned the problem of a single-server system with selfadjusted service-time.

Based on experience, some ideas of the system's modification emerged. One of them concernes the language, for which originally FORTRAN was chosen, because minicomputers with FORTRAN compilers were accessible during the time of the development, and FORTRAN proved suitable.

Our future plans include the realisation of the parallel µP version of the modified NETSY, which implies the switching to an appropriate language.

The structure of NETSY

The structure of the modular simulator is depicted in Fig. 1.

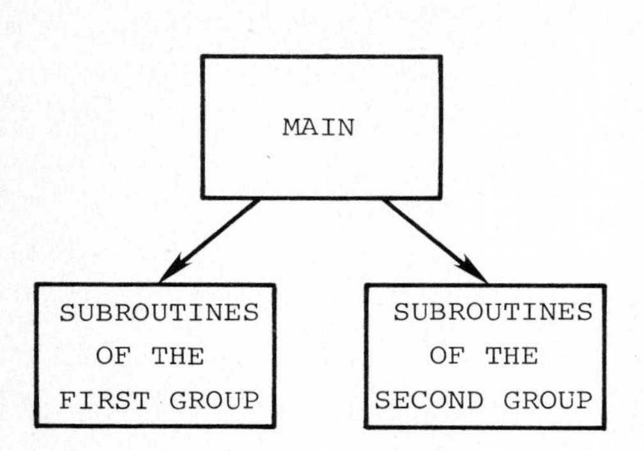

Fig.1.

The framework of the whole system is the main-segment of the FORTRAN program. The chief duties cf the main segment are

- to ensure the communication between the system and the experimenter
- to generate the objects to be transported by the simulated system (these objects could be anything that may travel on a traffic-network; in our special case these are messages)
- to time the internal events which occur in the simulated system; to activate the subroutines which are responsible for the execution of these events
- to deliver the reports, one part of which contains the periodic report of the traffic, the other contains the event-driven snapshot of the actual status of the simulated system and the optional tracing of the individual messages
- to terminate the simulation at some predetermined time or occurence.

The main segment is thus the driver and the executive of the whole model; it may conduct any kind of traffic-simulation, depending on the performing subroutines.

The hitherto realized subroutines may be divided in two groups.

To the first group belong the segments that perform the system's communications with the experimenter and in the future with the network.

The second group contains the programs that execute the traffic of the objects generated by the main segment and those performing some special events (e.g. break-down of some unit of the simulated system, etc.).

Since the subroutines may be changed one-by-one, with NETSY we have practically as many simulators as there are workable combinations of the subroutines of the second group.

The simulated system

At present NETSY simulates computer networks.

The simulated network is an arrangement of stations: hosts, nodes, (possibly a monitor, too,) in which messages originate, stay for some time and reach their destination. The propagation of the messages from station to station forms the traffic of the network.

The stations are represented by lists and tables. The propageting messages may form queues in the lists of the stations, according to the technical data of these stations, enumerated in the tables. (Buffer-size, number of living connections etc.). These technical data are intaken at the first phase of the running of the program in the followin form:

```
1.  buffersize   number of connections
2.      "                 "
...
N       "                 "
```

The integer numbers 1 to N are the identifiers of the stations.

The tables contain some other values, too, which are updated after every event. These are: length of queue (if any), the time of the next event concerning the station, the positions of its pointers (handling the lists of the queues) and the data of the housekeeping of the station (sum of transfers, sum of failures).

The interconnection between the stations (i.e. the topology of the arrangement) is contained in another table (tables) input after the technical data. The description-format of the topology depends on the adopted routing strategy. At present three of them do work: a fixed one (where a single and only a single sequence of stations leads from any source to any destinations), a split-routing (where before every skip from one station to the next one the more favourable is chosen out of two possibilities) and a fully random one (where the next station is chosen by the Monte Carlo method out of the existing connections).

The intake of these data is the first action of NETSY.

The process of simulation

After obtaining the initial description, the network is in its initial state - i.e. it is totally empty.

The first event which happens after the beginning of the simulation proper is the generation of the first message by the main segment.

Messages are represented by vectors, with their serial number as identifier for them. Their fixed properties are stochastically ordered to them as components, but once decided on, they prevail until the arrival of the message at its final destination, or until it reaches its age-limit and is discarded. The properties are:

identifier of source of origination
identifier of final destination
value of size.

The messages have variable properties too, which obtain their values in consequence of the events of the simulation

their actual age (elapsed time since their generation)
identifier of their actual residence.

Message are generated by stochastic, uniformly distributed time intervals. The expected value of the intervals is defined by the experimenter in the following way. Hundred tacts make a simulation-period. (The period is a purely technical time-unit for the benefit of the user, who may set the time-span of the whole simulation as an arbitrary number of periods.) Among other, hitherto unmentioned uses of this measure, one is the facility of setting simulation- and traffic-parameters with its help, as in the case of defining the intensity of the message flow, as follows.

After the input of the technical and topology-description, the experimenter answers the question:

Massage/period?

The answer may be a real number fixing the expected value of messages per period, (which is the parameter of the resulting Poisson-process of the entering of messages into the system.)

The new message is enlisted as first in the waiting queue of the first station of its route.

The duration of the residence of a message in a station depends on the forwarding strategy in use. At present, two different strategies are working (both of them may be used with any of the routing strategies).

In the simpler one, the message's stay in any station depends on the queue waiting before it (on FCFS basis) and on its size. Since the service time of the message depends also on their respective sizes, the waiting time for any of them in station n is

$$T_n \quad \sum_{i=1}^{m} t_i - t_{el}$$

where m-1 denotes the number of messages waiting for service prior of message$_m$ under consideration, t_i is their individual service-time, T_n the sum of these times including the service-time of message$_m$ while t_{el} means the time elapsed between the beginning of servicing t_1 and the arrival of message$_m$, which is to be subtracted from the sum.

The skip of a message is represented by enlisting its identifier (i.e. serial numer) into the appropriate queue (list) and by the updating of the concerned tables: that of the stations (length of the queue, pointer-handling, counting of the traffic, decision on its next event-time) and of the message-vector (marking the actual residence). After every period the ages of the messages are incremented accordingly.

When a message reaches its final destination, it is counted, its delay is recorded (which is identical with its age), eventually other statistics are performed. After this the system destroys the message, and its serial number becomes free.

When the congestion of the station-queues impedes the propagation of a message, it is possible that its age reaches an upper limit, after which NETSY annihilates it - doing first the necessary book-keepings, - and announces this event to the experimenter.

All the happenings after the generation of a message discussed hitherto were of a regular nature. The properties governing them were all fixed from the beginning of the process, either stochastically but following well-defined distributions with parameters set by the experimenter, or given directly by him (e.g. the technical description of the system). All the events and the caused changes in the states of the model could have been calculated - which in fact the computer did - but for the starting of new messages.

Now we to discuss the three remaining topics: the irregular events, the problems inherent in the message-size and the question of the monitor.

The irregular, or random events have impact either on the topology of the system, or on the traveling of the individual messages.

Temporary break-downs can be caused of communication-lines. These may happen, according to the experimenters request, either stochastically, with exponential distribution (the parameter of which is given by the experimenter just after defining the parameter of the message-flow) or between any two chosen neighbouring stations. When its line is broken, NETSY decrements the number of living lines in the tables of the concerned stations. When a station loses its last line, it is closed down.

As chance event the faulty delivery of individual messages is introduced. After such failure the sending station has to repeat the attempt of forwarding the message.

The failure-rate is determined by the experimenter, as the main number of succesful deliveries for one failure. The sparsing of the repetitions among the transfers is uniformly distributed.

<u>The message-size</u>

To every message an integer number is ordered by the main segment, which is characteristic of its size. It may be regarded either as its length in bytes, or as the time needed for servicing it, being these two quantities roughly proportional. (Hence the notation t in (1).)

According to present theories these integers should have exponential distribution, and were developed accordingly until 1978 fall. However, a deeper investigation made clear that the assumption is not justifiable; since although packets with about zero netto-length may exist, messages needing about zero buffer and zero service-time are unimaginable. Therefore NETSY attaches to the exponentially formed message-lengths a constant tag, the value of which is defined by the experimenter at the initial dialogue. The value may range from Ø to 2ØØ (tacts as service-time).

Taking f(t) as the distribution function of the original sizes, the addition of the constant k to the sizes yields the new distribution function

$$f(t^*) = f(t-k) \times 1(t-k) \qquad (1)$$

1(t-k) being the unit step-function /giving Ø when (t-k)<Ø.

The procedure shifts the whole function to right on the t axis, with the deterioration of the distribution as result, but since μ (the parameter of the original exponential distribution) is not affected by the manipulation, the assumption of the constant tag is justifiable, (as well as consistent with the hitherto established models when $t << \frac{1}{u}$.)

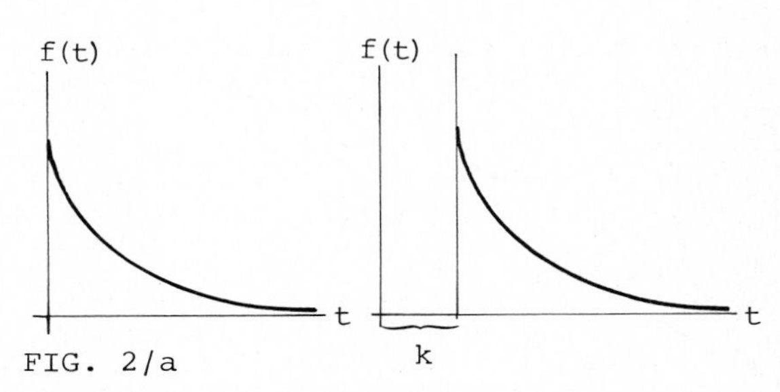

FIG. 2/a

density function of f(t) — density function of f(t*)

It may be shown, the character of the Erlang distribution resulting from the servicings with time-durations proportional to the sizes of the messages is not affected by the right-shift of the distribution of the sizes. The parameter γ was originally $\lambda \cdot \frac{1}{\mu}$. Now it becomes

$$\gamma = \lambda(\frac{1}{\mu} + k)$$

The constant tag accounts for the time- and buffer-requirements of the protocollar processes. (Fig. 2/a).

<u>NETSY with monitor</u>

In the summer of 1980 NETSY was supplied with a monitor.

Its first realisation was built as a simplified model of the CIGALE failure-protection system [5]. The monitor is sending probe packets as regular intervals, these are echoed by the destination-stations. The frequency and length of the probes is given by the experimenter in the initial dialogue. The experimenter chooses the monitor-station from among the others.

The remaining are standard nodes. The monitor-station becomes dedicated to the task of sending and receiving the probes - meanwhile it also forwards the standard messages of other senders when their path happens to lead through it. However, the monitor-station is no gateway to a source-or destination-host.

The periodically transmitted probes distort the original Poisson-process of the message-flow into the system in a way resembling to that observed in connection with the constant tag of the messages. Let again f(m) be the distribution of the incidences of messages, m being the number of incidences in a given time-interval. (Fig. 2/b).

The new distribution becomes, after the addition of k periodic messages (constant over time)

$$f(m^*) = f(m-k) \times 1(m-k) \qquad (2)$$

which is a discreate version of (1).

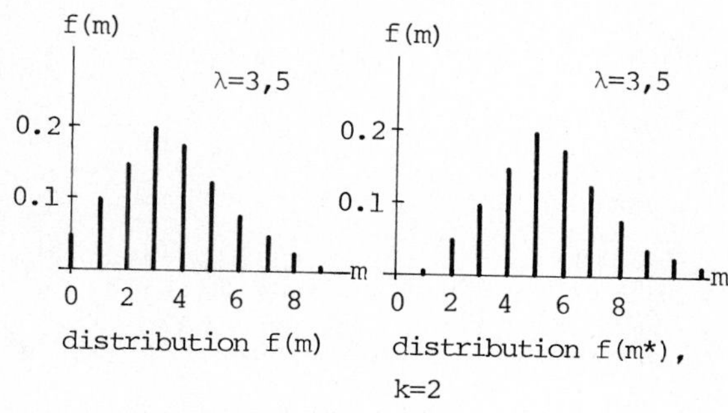

Fig. 2/b

This applies, however, only to the process of the starting of messages.

The pilot-run with monitor showed that the arrival-process in the stationary case a Poisson-one, (as it had been expected). (The measurement was suggested by T.L. Török.) The missing of consequences of the distortion of the starting process is due to the nature of the subsequent delays which the probes, too, undergo.

The distributions of the delays show a wide variety according to the topology of the simulated network, the traffic conditions etc. This, too, is well known from literature [6].

The probes may be sent - when the experimenter gives his order thus - stochastically or event-driven, too.

Simulation results

As pilot-runs, we simulated the configurations recommended as etalons by Keinrock, Price, and others (Fig.3).

The Price-network was simulated as a ten-node arrangement, the others were run with 5 to 15 nodes. Special attention was paid to the effects of the assignement of the monitor-function to different nodes of a given arrangement.

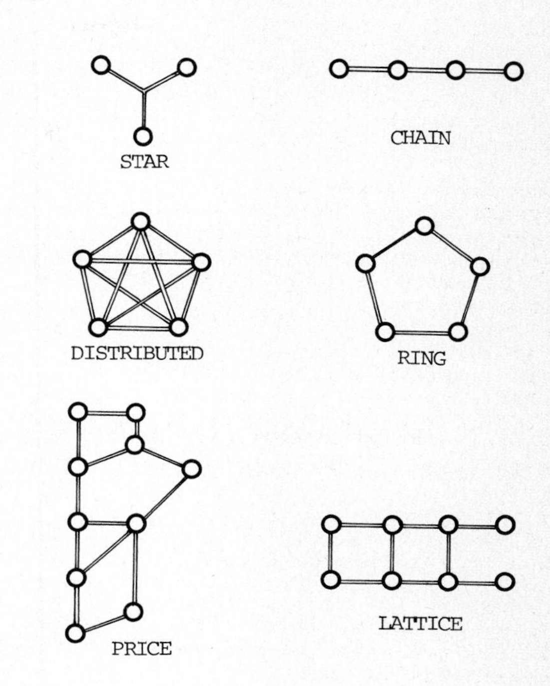

FIG. 3 THE ETALION-CONFIGURATIONS

When both the source and the destinations are drawn with equal chances, the number k of possible pairs is the sum of the arithmetical series from 1 to N (N being the number of nodes). The number n of the possible j lengths of routes between any two of them, measured by the skips of the traveling messages, is N-1.

The probability of any route-length j. (j = 1...,n) is

$$p_j = \frac{N-1}{k}, \quad \text{being } k = \sum_{i=1}^{N-1} i \qquad (3)$$

and the expected value of the route-length

$$j = \sum_{i=1}^{n} p_i j_i = \frac{1}{k} \sum_{i=1}^{N-1} i(M-i) \qquad (4)$$

However, when a node becomes a monitor, its messages are echoed back from every destination, and therefore the condition of equal chances of drawing is not met; the probes, traveling their routes twice, double the relative frequency of their route-lengths. Moreover, the monitor sends its probes periodically, while standard messages start stochastically which renders the calculation difficult. Even more complicated situation evolves when the distribution of the message-delay is investigated, since the service-time also of the probe is proportional to its size, which is kept constant, whereas the standard messages are devoloped according to (1).

The simulation showed cleary the difference between the two arrangements shown in Fig.4. This difference is especially sharp regarding the respective traffics of the nodes in the different arrangements.

MONITOR MONITOR

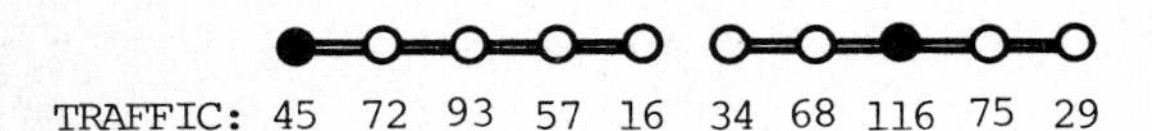

TRAFFIC: 45 72 93 57 16 34 68 116 75 29

Fig.4. ALLOCATION OF MONITOR

Other interesting experiment is the play-back of the hypothesis, that fixed routing may give the same mean values of delays between node-pairs [7]. Here the experiments were conducted using the more elaborate forwarding strategy. Thus far the hypothesis seems to hold only in the cases when the network is loaded moderately (under 0.6 Erlag/node on the average) but above this, quite unexpectedly, the fully random routing performed superior to the fixed one. The experiment is not yet closed; correct evaluations will follow.

Evaluation of experience

Up till now a few houndred sessions, i.e. some thousan runs were performed with NETSY giving the following results:

a. NETSY fulfills the first requirement stated on page 1: its performance corresponds to the hitherto observed behaviour of computer networks

b. the second requirement is fulfilled, too: the experience with interchangeable strategies is quite statisfactory as was shown

c. the interactive realisation supports the easy tuning of the simulator as well as of the simulated system

d. The real-time operation necessary for the application of NETSY needs further study

e. which - together with the above mentioned facts - not merely allows, but rather calls for the realisation on a dedicated parallel-processing system.

Final remarks

Computers are considered as the most complex systems of the technics of our days - computer-networks are of an even higher level complexity. Their modeling is quite impossible without a coordinated endeavour in every possible area of the art: from the abstract mathematical descriptions to the concreteness of measurement. In the integration of these efforts NETSY - expecially in its future version - can be a usable tool. Its continuous development is a necessary condition - and that is made possible by its modular structure.

Acknowledgement

The author is indebted to Ms L. Emmi Kovåcs (MTA-KFKI) for her valuable help, to A. Gåspår (MTA-SZTAKI) and L.T. Török (MTA-KFKI) for the englightening discussions and especially to Ms Katalin Tarnay (MTA-KFKI) for her suggestions broadening the scope of this report and of the underlying research.

Literature

[1] Kleinrock, L. & Naylor, W.E., On measured behaviour of the ARPA network AFIPS Conf. Proc. 1974, Vol.43, p.: 767-780.

[2] J.R., Jackson, Job-shop-like queueing systems Managements Sci. Vol. 10, pp.: 131-142 (1963).

[3] Tobagi, F.A. & al.: Modeling and measurement techniques in packet communication networks, Proc. IEEE, Vol. 66, No 11. pp.: 1423-1447.

[4] Tarnay, K., Ein Matrix-Spieltheoretisches Modell für Rechnernetzwerkmessungen, DREZDA, 1979.

[5] Pouzin, L., CIGALE, the packet-switching machine of the CYCLADES computer network, IFIP Congr. Stockholm, 1974 Aug. pp.: 155-159.

[6] Kleinrock, L., Queuing system, II. Wiley, 1976 New York.

[7] Schoemaker, L., Simulation of Computer Networks, North Holland, 1979.

Simulation in Engineering Sciences
J. Burger and Y. Jarny (eds.)
Elsevier Science Publishers B.V. (North-Holland)

NUMERICAL SIMULATION OF THE CONTROL OF THE SPACE-TIME THERMAL PROFILE IN AN EPITAXIAL REACTOR

J. Souza Leao*, J.P. Gouyon, J.P. Babary

Laboratoire d'Automatique et d'Analyse des Systèmes,
7, avenue du Colonel Roche - 31400 Toulouse, France

Liquid phase epitaxy is one method of manufacturing electronic components, requiring the utilization of a cylindrical reactor in which the temperature must be controlled during the various heating and cooling phases. The latter corresponds to a critical phase in the manufacturing process. In a control context, it may be posed as a problem of tracking a space-time thermal profile.
The modelling of the thermal behaviour of the reactor leads to a parabolic non linear distributed parameter system. This system may be linearized about a nominal profile. The problem of controlling the heating coils of the five heating zones is resolved by using the Porter-Bradshaw method for finite-dimensional systems.
The purpose of this paper is to present a comparative study of two numerical methods for simulating the controlled system :
- the first method (FORTRAN) is adopted for the problem of controlling the non linear distributed parameter system (partial differential equations),
- the second method (CSMP III) is adopted for the problem of controlling the linearized and approximated system.
The comparative study is carried out with respect to both the characteristics and performance of the simulation methods and the obtained results.

1. INTRODUCTION

Among the different methods of manufacturing electronic components, such as diffusion, epitaxy, ionic implantation, the liquid-phase epitaxy is a process which on one hand presents many advantages from the viewpoint of component quality, but, on the other hand, requires particular care and attention in relation to the optimal control of the thermal behaviour of the reactor.

This liquid-phase epitaxy process can be decomposed into five phases :

1 - preheating up to 750°C necessary for the purification of the atmosphere in the reactor.
2 - heating up to 800°C (approximately) in minimum time, without overshooting this temperature (tolerance of 1°).
3 - regulation of a thermal profile along the reactor.
4 - slow cooling in one or several prescribed intervals, with controlled space-time thermal gradients.
5 - natural cooling (stoppage of the reactor).

2. PHYSICAL AND MATHEMATICAL DESCRIPTION OF THE REACTOR.

The reactor consists of two concentric quartz cylinders heated by five coils distributed along the total length of the reactor (Figure 1). The main charge is located in the central zone and is composed of a partitioned graphite crucible, with a sliding lid for multilayer deposits. The crucible contains a gallium solution saturated with a polycrystalline Ga As solution.

The dynamic thermal behaviour of the reactor is modelled via a balance of the heat exchanges between charge, gas (circulating at low velocity), quartz tubes, and heating elements, under a certain number of simplifying hypotheses [1]. This model, a system of three non linear partial differential equations, enables the evolution of the thermal profile in the different constituent, to be determined, given the evolution of the heating power profile (model 1) :

$$\left\{\begin{aligned}
\frac{\partial Y_1}{\partial t} &= A_1 \frac{\partial^2 Y_1}{\partial x^2} + A_2(Y_2^4 - Y_1^4) + A_3 (Y_2 - Y_1) \\
\frac{\partial Y_2}{\partial t} &= A_4 \frac{\partial^2 Y_2}{\partial x^2} + A_5(Y_2^4 - Y_1^4) + A_6(Y_3^4 - Y_2^4) + A_7(Y_2 - Y_1) \\
\frac{\partial Y_3}{\partial t} &= A_8 \frac{\partial^2 Y_3}{\partial x^2} + A_9(Y_3^4 - Y_2^4) + A_{10}(Y_3 - T_0) + A_{11}\, P(x,t)
\end{aligned}\right. \tag{1}$$

* UFRJ-COPPE, Rio de Janeiro, Brazil

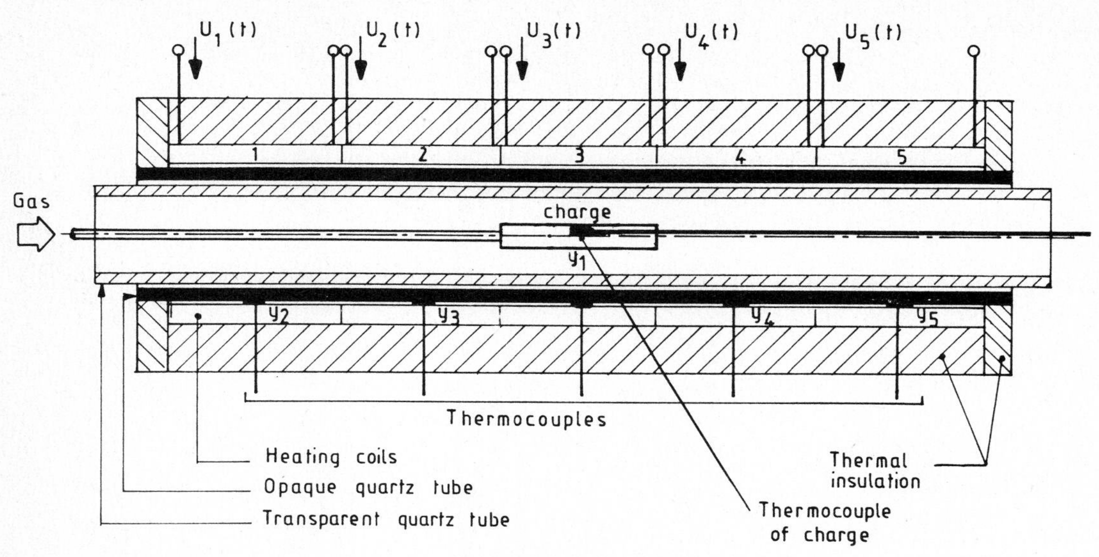

Fig. 1. Scheme of the reactor

where $Y_i(x,t)$, i=1,2,3 represent the thermal profile of the charge, quartz tubes and heating elements, respectively. The coefficients A_i are constant. The boundary conditions, at x = 0 and x = 1, are assumed constant. T_0 is the ambient temperature and P(x,t) is the heating power profile.

Since the range of temperature is relatively small during phases 2 - 3 - 4, the model 1 is linearized about the steady state obtained for phase 3, by setting :

$$\begin{cases} Y_i(x,t) = Y_{ic}(x) + y_i(x,t) \\ P(x,t) = P_c(x) + p(x,t) \end{cases} \tag{2}$$

Simulation has shown that the heat transfer dynamics across the quartz tubes are significantly faster than in the other constituents, hence the linearized model may be reduced to a system of two partial differential equations of the form (model 2) :

$$\begin{cases} \frac{\partial y_1}{\partial t} = [f_1(x) + A_1 \frac{\partial^2}{\partial x^2}] y_1 + f_2(x) \cdot y_3 \\ \frac{\partial y_3}{\partial t} = f_3(x) . y_1 . [f_4(x), + A_8 \frac{\partial^2}{\partial x^2}] y_3 + A_{11} \; p(x,t) \end{cases} \tag{3}$$

with the boundary conditions :

$$y_1(0,t) = y_1(1,t) = 0$$

$$y_3(0,t) = y_3(1,t) = 0$$

An approximation to model 2 is obtained by the Galerkin method [2] by setting :

$$y_k(x,t) = \sum_{i=1}^{N} a_{ki}(t) \cdot \varphi_i(x) \quad k=1,3 \tag{4}$$

The N basis functions are orthonormal on [0,1]:

$$\varphi_i(x) = \sqrt{2} \sin i\pi x \qquad i=1,\ldots,N$$

The $a_{ki}(t)$ are state variables; the control is of the form :

$$p(x,t) = \sum_{j=1}^{5} h_j(x) \; u_j(t)$$

with $h_j(x)=1$ on the jth zone and $h_j(x)=0$ outside this zone.

By calculation, a linear finite dimensional (2N) system is obtained, which may be written in matrix form as (model 3):

$$\begin{bmatrix} \dot{a}_1 \\ \dot{a}_3 \end{bmatrix} = \begin{bmatrix} A_{11} & A_{12} \\ A_{31} & A_{32} \end{bmatrix} \begin{bmatrix} a_1 \\ a_3 \end{bmatrix} + \begin{bmatrix} \mathbb{O} \\ B_3 \end{bmatrix} u \tag{5}$$

where $a_1 = a_1(t)$ and $a_3 = a_3(t)$ are N-dimensional vectors

u = u(t) is a 5-dimensional vector.

The initial conditions are given by :

$$a_{ki}(0) = \int_0^1 y_k(x,0) \, \varphi_i(x) \, dx \quad k=1,3; \; i=1,\ldots,N \tag{6}$$

3. STATEMENT OF THE PROBLEM

For the purpose of this paper, we have selected

as an example the problem of controlled cooling of the charge (Phase 4). The objective is to determine a control u(t) whichs enables the main charge temperature (situated in the central zone : $0.4 \leq x \leq 0.6$) to decrease linearly as a function of time; thus, for model 1 :

$$Y_{1d}(x,t) = Y_{1c}(x) + \alpha t \text{ for } \begin{bmatrix} 0,4 \leq x \leq 0,6 \\ 0 \leq t \leq T \end{bmatrix} \quad (7)$$

α corresponds to the desired temporal thermal gradient of cooling ($\alpha < 0$).

For model 2 :

$$y_{1d}(x,t) = \alpha t; \quad 0.4 \leq x \leq 0.6 \text{ and } 0 \leq t \leq T \quad (8)$$

For model 3, the desired state $a_{1d}(t)$ may be calculated on the basis of the approximation :

$$y_{1d_N}(x,t) = \sum_{i=1}^{N} a_{1d_i}(t) \cdot \varphi_i(x) \stackrel{\Delta}{=} \sum_{i=1}^{N} c_i \alpha t . \varphi_i(x) \quad (9)$$

The coefficients c_i are determined by minimizing the mean square error between y_{1d} and y_{1d_N} on the interval $0.4 \leq x \leq 0.6$.

Thus, the problem is to determine the control u(t) such that, on a given time interval T, the state vector $a_1(t)$ follows the reference vector v(t) with components :

$$v_i(t) = a_{1d_i}(t) = c_i \alpha t \qquad i=1,2,\ldots,N \quad (10)$$

Writing system (5) more concisely :

$$\dot{a}(t) = A\, a(t) + B\, u(t) \quad (11)$$

with the ouput vector defined at 6 measurement points (1 thermocouple at the centre of the charge and 5 thermocouples equi-spaced along the heating coils) :

$$y_s(t) = [y_1(0.5,t) \mid y_3(0.1,t) \mid y_3(0.3,t) \mid y_3(0.5,t) \mid y_3(0.7,t) \mid y_3(0.9,t)]^T$$

The method developed by Porter and Bradshaw [3] consists of constructing, on the basis of (11), an augmented system by adjoining a double integration of the difference between the actual state $a_1(t)$ and the desired state $a_{1d}(t)$:

$$\begin{cases} \dot{a}(t) = A\, a(t) + B\, u(t) \\ \dot{z}_1(t) = v(t) - a_1(t) \\ \dot{z}_2(t) = z_1(t) \end{cases} \quad (12)$$

with the initial conditions a (0) as in (6) and

$$z_1(0) = z_2(0) = 0$$

A control of the following form is now defined :

$$u(t) = K_1\, a(t) + K_2\, z_1(t) + K_3\, z_2(t) \quad (13)$$

where the matrices K_1, K_2, K_3 are such that a quadratic cost of the form :

$$J = \int_0^{\infty} (a^{*T} Q\, a^* + u^T R u)\, dt \quad (14)$$

is minimized, where $a^*(t)$ represents the augmented state vector $[a \mid z_1 \mid z_2]$.

The optimal solution is then :

$$u_{opt}(t) = - R^{-1} B^{*T} \mathbb{P}\, a^*(t) \quad (15)$$

avec : $B^* = \begin{bmatrix} B \\ \mathbb{O} \end{bmatrix}$ (matrix of dimension 4N x 5)

and $\mathbb{P}$, a matrix of dimension 4N x 4N, is the solution of the algebraic Riccati equation, associated with the augmented system (12) and the quadratic cost (14).

$\mathbb{P}$ was calculated by using the Kleinmann method [4].

The choice of the numerical values of the elements of the weighting matrices Q and R, while arbitrary, was made via a sequence of numerical simulations of the controlled system, in order to obtain a satisfactory system dynamics.

4. SIMULATION METHODS

Two methods were adopted to simulate the dynamic behaviour of the system :

- a specific method : CSMP III (Continuous System Modelling Program)
- a general FORTRAN-based method.

-1- C.S.M.P. III simulation [5]

The control problem for the linearized system (model 3) was solved by numerical simulation, using C.S.M.P. III language. This language exhibits a considerable advantage for the user over more classical programming techniques based on the use of languages such as FORTRAN... It allows a more simple description of the mathematical model; it enables simulations to be carried out under different conditions and it offers greater flexibility at the input - output level.

A TSO control procedure, called #CSMP, has been developed at L.A.A.S., which enables CSMP III to be used interactively, via a Tektronix 4014 graphics terminal. It allows several simulations to be performed sequentially by modifying the model, or preserving the model in a load-module executable form and modifying the execution parameters (execution time, integration method, etc...), or, again, by displaying successively different aspects of the same simulation, stored simultaneously during execution.

-2- FORTRAN simulation program

This program is used for simulating the system of nonlinear partial differential equations (model 1) under closed-loop control. Since the control law (15) is given as a function of the state $a^*(t)$ of the system and not of the output $y_s(t)$, an observer was constructed on the basis of temperature measurements taken at the 6 thermocouples.

Mathematically, the system simulated is system (12) in which the differential equation for a(t) is replaced by the equations of model 1, together with the observation equation (determined on the basis of the linear system) and of the form :

$$\overset{\circ}{\hat{a}}(t) = A\,\hat{a}(t) + B\,u(t) + N\,(y_s(t)-C\,\hat{a}(t)) \qquad (16)$$

All these equations were discretized in space (Δx = 0,02) and/or in time (Δt = 0,001 hr.) according to the following schemes :

$$\frac{\partial y}{\partial t} \Leftrightarrow \frac{y_j^{n+1}-y_j^{n}}{\Delta t} \quad \text{and} \quad \frac{\partial^2 y}{\partial x^2} \Leftrightarrow \frac{\theta(\delta^2 y)_j^{n+1}+(1-\theta)\,(\delta^2 y)_j^{n}}{(\Delta x)^2}$$

with θ = 0.8, this corresponds to a stable discretization method.
The graphical results were obtained using a BENSON digital plotter.

5. SIMULATION RESULTS

All numerical simulations were carried out on the IBM 3033 computer of the Centre National Universitaire Sud de Calcul (C.N.U.S.C.) at Montpellier from alphanumeric and graphics terminals at L.A.A.S.

-1- The C.S.M.P. based simulation of the controlled linear system has been found to be convenient for the determination of the weighting matrices Q and R. Figures 2 and 3, respectively, represent the evolution of the spatial profile and the evolution of the temperature difference with respect to the nominal profile at the centre of the charge for a cooling defined by a temporal gradient of α = - 1 degree/minute, on a time interval of T = 36 minutes.

These figures clearly show the maintenance of a virtually zero spatial gradient in the central zone (figure 2) and the accurate "tracking" of the desired cooling (figure 3).

This simulation was carried out in 8.47 seconds and required 512 K bytes of memory.

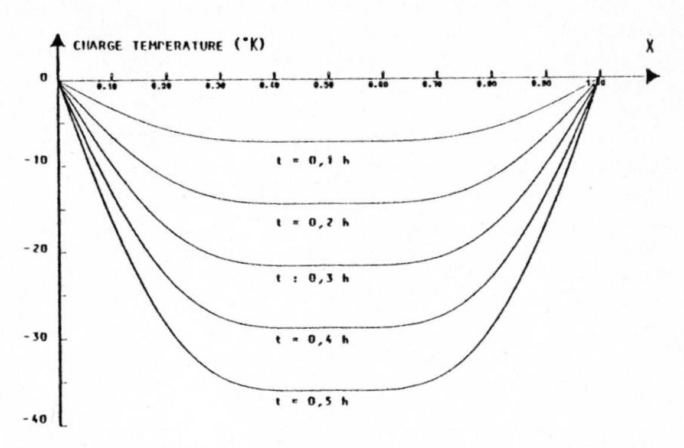

Fig. 2. Evolution of the thermal profile in the charge (linear system)

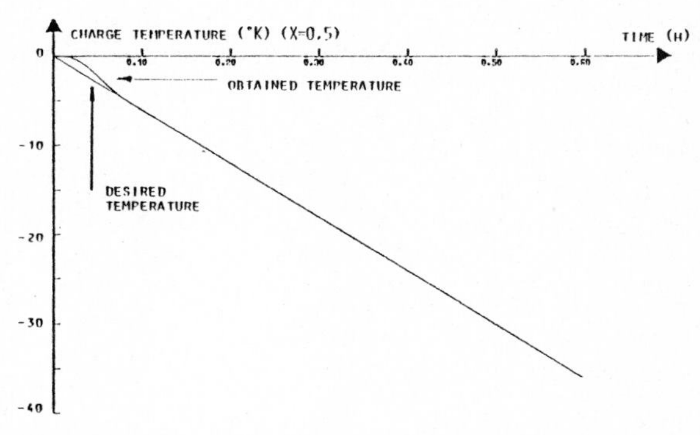

Fig. 3. Evolution of the temperature in the middle of the charge (linear system)

-2- Figure 4 and 5 refer to the simulation (FORTRAN program) of the controlled nonlinear system (model 1).

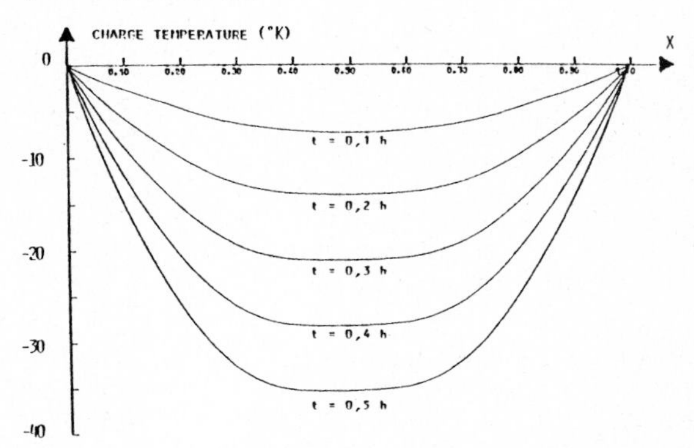

Fig. 4. Evolution of the thermal profile in the charge (non-linear system)

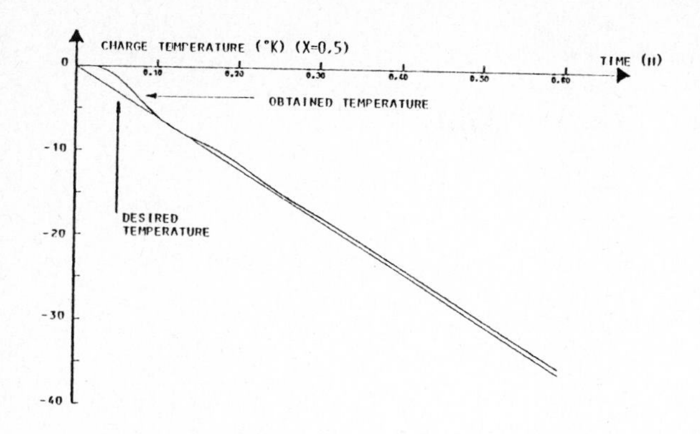

Fig. 5. Evolution of the temperature in the middle of the charge (non-linear system)

This simulation was carried out with the same loop gains as for the C.S.M.P. simulation and for the same desired spatio-temporal thermal profile. It is remarked (figure 4) that the spatial gradient is again maintained at zero for the first 20 degrees of cooling, then a slight degradation is evident.

With regard to the temporal profile (figure 5), a tracking error of the order of 2% may be noted which is due to the fact that there is a "modelling error" between the nonlinear model 1 and the linear approximated model 3 and that the control and observer were calculated on the basis of the linear approximated model 3.
This simulation was carried out in 3.54 seconds and required approximately 224 K bytes of memory.

6. CONCLUSION

From the comparative simulation study presented in this paper, the following conclusions may be drawn :

- the C.S.M.P. simulation method is simpler from the point of view of its programming and utilization; but the execution time and memory required to carry out a simulation would seem in general to be somewhat greater (simulation of the same controlled linear system was also carried out, using the FORTRAN program developped for the nonlinear system).

- because of the load-module form, CSMP III allows more frequent re-runs than in FORTRAN, with associated economies in compilation time.

- the most significant time saving is that associated with the storing of all variables during a simulation; the model behaviour can be analysed by displaying successively several curves, which is inexpensive in computing time, since it is not necessary to perform several runs.

7. REFERENCES

[1] J.P. Babary, A. Benhammou, B. Dahhou, Optimization problems associated with the thermal behaviour of an epitaxial reactor, 3rd IFAC Conf. on Systems Approach for Development, Rabat (Maroc) 24 - 27 Nov. 1980.
[2] B.A. Finlayson, L.E. Scriven, The method of weighted residuals, a review. Appl. Mech. Rev. Vol. 19, n° 9, Sept. 1966.
[3] B. Porter, A. Bradshaw, Design of linear multivariable continuous-time tracking systems, Int. Journal of Systems Science, vol. 5, n° 12, 1974.
[4] D.L. Kleinman, On an iterative technique for Riccati equations computations. IEEE Trans. AC vol. AC 13, Février 1968.
[5] J.P. Gouyon, Utilisation du logiciel C.S.M.P. III à partir d'un terminal graphique interactif, Note Technique LAAS-SIS n° 82.0.42, Août 1982.
[6] J.E. Doucet, Programme de simulation d'un four à diffusion. Note Interne LAAS n° 75.I.34 1975.

Simulation in Engineering Sciences
J. Burger and Y. Jarny (eds.)
Elsevier Science Publishers B.V. (North-Holland)

SIMULATION OF SYSTEMS WITH MOBILE POWER SOURCES

E.P. Chubarov and V.A. Kubyshkin

Institute of Control Sciences
Moscow, USSR

Approximate models of systems with mobile power sources are discussed. Such models enable reducing nonlinear systems to linear. Examples of solving programmed control and feedback control problems with the aid of such models are given. Computer simulation of such systems and results in modeling feedback control systems are also considered.

1. INTRODUCTION

Mobile power sources such as electron, ion, and laser beams, plasma fluxes, and electric arcs now find their way into iron and steel, mechanical, and many other industrial systems where the control inputs moves over a spatially distributed plant.

In a general for a system with a mobile power source 1 - 3 is described in the following way 3

$$A_Q[Q(x,t),x,t] + u(t)\Psi[x-s(t),t,P(t)]=0,$$
$$x \in D \subset R^n, \quad t \geqslant t_0 \qquad (1)$$

where x is a coordinate in an n-dimensional Euclidian space; $x=(x_1,\dots,x_n)\in R^n$ unually $n = 1,2,3$; t is time; $Q(x,t)$ is the state of the plant which usually takes up a region D; A_Q is a differential, integral, or any other system operator; $u(t)$ is the power of the mobile source; $\Psi(x,t,p(t))$ is a normalized function which dictates the distribution of the source power density relative to its center of gravity, or the source shape; $s(t)$ represent the motion of the source center; and $p(t)$ is the vector of parameters which decine the source shape and rotation.

In real life very significant is the case where the operator describes the temperature distribution in the object of heating (under some boundary and initial conditions). We will concentrate on the case.

A classification of control problems for systems with mobile sources of impact has been given elsewhere 1 - 5 .

Such systems are nonlinear with respect to the control parameters $s(t)$ and $P(t)$, which complicates calculations and determination of control function. Consequently, it is necessary to develop approximate models of such systems so as to simplify their investigation. Furthermore, methods for investigation of nonlinear systems. Computer simulation is difficult too. The paper will discuss ways to overcome these difficulties and report the results in modeling programmed control and feedback control systems.

2. ASYMPTOTIC MODELS OF SYSTEMS WITH MOBILE SOURCES INPUTS

Asymptotic models of plants subjected to mobile sources of input signals have long been used in the engineering practice. The heat conductivity operator is simplified in most cases in that with fast movement of the source i one direction along a straightline the heat conductance along the motion direction is neglected 6 .

This simplification is, however, inapplicable, if the source can change the direction or scan the object of heating fast. In this case time-averaged models are effective tools [7,8] whereby the nonlinearity of the control parameter in the right-hand side of the plant equation is eliminated.

Let us take up replacement of an accurate equation by an averaged equation of a plant which is described by a one-dimensional heat conductivity equation

$$\frac{\partial Q}{\partial t} - a\frac{\partial^2 Q}{\partial x^2} + bQ = u(t)\Psi[x - s(t), p], \qquad (2)$$

$$x \in D \subset R^1, \; t > 0,$$

$$\alpha Q - \beta \frac{\partial Q}{\partial x}\Big|_{x=x_0} = 0, \; \alpha Q + \beta \frac{\partial Q}{\partial x}\Big|_{x=x_1} = 0, \qquad (3)$$

$$Q(x,0) = Q_0(x), \qquad (4)$$

where a, b, α, β and p are consist factors, $D = (x_0, x_1)$ is an interval of the numerical axis R^1, and $Q_0(x)$ is the initial plant state. For specificity the source shape is assumed to be time-invariant. The following assertion enables using an averaged equation. Let $Q(x,t)$, $x \in D$, $t \geq 0$ be a solution to equations (2) - (4) and $\bar{Q}(x,t)$, $x \in D$, $t \geq 0$, a solution of the same equations but with a signal of the form

$$\bar{F}(x,t) = \sum_{n=1}^{\infty} \bar{F}_n(x)\chi_n(t) \qquad (5)$$

$$\bar{F}_n(x) = \frac{1}{T_n}\int_{I_n} u(\tau)\Psi(x - s(\tau), p)\,d\tau, \qquad (6)$$

where $s(t)$ is a piecewise-continuous function; the semi-axis $t \geq 0$ is divided into semi-intervals (cycles) $I_n = [t_{n-1}, t_n)$ by points $0 = t_0 < t_1 < \ldots$, $T_n = t_n - t_{n-1}$ are cycle durations; and $\chi_n(t)$ is a characteristic function of I_n. Then for any $\eta > 0$ one can found $\varepsilon > 0$ such that with $\max_n T_n < \varepsilon$

$$\| Q(x,t) - \bar{Q}(x,t) \|_{L_2(D)} < \eta$$

over an interval $0 < t < \infty$.

This assertion is proved with the aid of the Bogolyubov theorem on averaging [9] as in [8]. Consequently, use of averaged equations enables studying systems with an averaged distributed signal (8) which is linear in the plant object, rather than with a mobile signal. This is found to be very useful in control, computer simulation, etc. Let us consider several examples of solving control problems with a mobile source where averaged models are effective.

Example 1. Obtaining and maintaining a specified temperature field. In a plant (2) - (4) it is required to find control functions $u(t)$ and $s(t)$ so as to obtain and maintain a temperature field which would be close to the specified one, $Q^*(x)$, $x \in D$.

The averaged temperature field $\bar{Q}(x,t)$ is time invariant if it satisfies the stationary equation

$$-a\frac{\partial^2 \bar{Q}}{\partial x^2} + b\bar{Q} = \bar{F}(x), \; x \in D, \qquad (7)$$

$$\alpha \bar{Q} - \beta\frac{\partial \bar{Q}}{\partial x}\Big|_{x=x_0} = 0, \; \alpha\bar{Q} + \beta\frac{\partial Q}{\partial x}\Big|_{x=x_1} = 0. \qquad (8)$$

In this case $\bar{F}(x) = \bar{F}_n(x)$, $n = 1, 2, \ldots$. Assume that $Q^*(x)$ satisfies the boundary conditions (8). (If $Q^*(x)$ does not, let us change it somewhat in the vicinity of its ends so that it does). Then the control $\bar{F}^*(x)$ which insures for (7) and (8) a state which is close to the desired one, $Q^*(x)$ is obtained by substituting $Q^*(x)$ into equation (7). Then control functions $u(t)$ and $s(t)$ should be chosen. To determine them, use the integral equation

$$\bar{F}^*(x) = \frac{1}{T_n}\int_{I_n} u(\tau)\Psi(x - s(\tau), p)\,d\tau.$$

If, for instance $u(t) = U = \text{const}$ and $s(t)$ is in each cycle I_n a monotone (increasing or decreasing) function such that $s(t_{n-1}) = x_0$, $s(t_n) = x_1$ then by replacing $s(t)$, $t \in I_n$ we have

$$\bar{F}^*(x) = U\int_{x_0}^{x_1} \Psi(x-\sigma, p)\, w_n(\sigma)\, d\sigma,$$

$$w_n(x) = \frac{[s_n^{-1}(x)]'}{T_n} = \frac{1}{T_n V_n(x)}, \quad (10)$$

where $s_n^{-1}(x) = t_n(x)$ is a function inverse of $s(t),\ t \in I_n$, $V_n(x)$ is the rate of motion of the source center in each point of the plant. Assuming that $\Psi(x) = \delta(x)$ (this assumption is sound in many important actual cases) we obtain the power and the motion function from the formula

$$U = \int_{x_0}^{x_1} \bar{F}^*(x)dx, \quad t_n(x) = \frac{T_n}{U}\int_0^x \bar{F}^*(\xi)d\xi. \quad (11)$$

Figure 1 shows an example of computed control functions $F^*(x)$ and $s(t)$ which insure a state close to $Q^*(x)$ in steady state operation.

Example 2. Development of a feedback control function for offsetting unobservable disturbances.

Include into the right-hand side of equation (2) an additional term $g(x,t)$ which stands for exogenous disturbance. It is required to find control functions $u(t)$ and $s(t)$ as functions of the plant state $Q(x,t)$ which would insure maintenance of a state close to $Q^*(x)$ in the presence of a disturbance $g(x,t)$. A detailed description of the solution is given in 8 where it was obtained thanks to using an averaged equation of a plant with distributed control which was to minimize the rate of decrease of the Lyapunov functional

$$J(t) = \frac{1}{2}\int_D [\bar{Q}(x,t) - Q^*(x)]^2 dx.$$

The control function of the source motion (or velocity) which is obtained for this condition has the form

$$\tilde{w}_n(x) = -w_{max}\, \mathrm{sign}\Big[\tilde{Q}(x,t_{n-1}) - \int_D \tilde{Q}(\sigma,t_{n-1})\, w^*(\sigma)\, d\sigma\Big],$$
$$n = 1, 2, \dots,$$

where $w^*(x)$ is programmed control which insures the state $Q^*(x)$ which is obtained as in Example 1; $\tilde{Q} = \bar{Q} - Q^*$, $\tilde{w}_n(x)$ is a variation of the programmed control $w^*(x)$, w_{max} is the maximal admissible value of the programmed control variation. The control $w_n(x) = w^*(x) + \tilde{w}(x)$ is then replaced by the velocity $V_n(x)$ or the motion function $s(t), t \in I_n$ in compliance with formula (10).

3. MODELS OF A TOTALITY OF MODES

Let us considere the specifics of modal representation of systems with movable sources of exogenous signals with the plant (2) - (4) as an illustration. The state of the plant is known to be representable as a series of a special form

$$Q(x,t) = \sum_{K=1}^{\infty} q_K(t)\, \varphi_K(x),$$

where $\varphi_K(x)$ are spatial modes of the plant (eigen functions of the plant operator) and $q_K(t)$ are time modes with $q_K(t)$ described by the equations

$$\dot{q}_K + \lambda_K q_K = u(t)\int_D \Psi(x - s(t), p)\, \varphi_K(x)\, dx, \quad (12)$$

$$q_K(0) = C_K, \quad K = 1, 2, \dots,$$

where λ_K are eigen values of the plant operator and C_K are coefficients of $Q_0(x)$ series expansion for $\varphi_K(x)$.

The system (12) is nonlinear for the control $s(t)$, which is especially obvious if $\Psi(x) = \delta(x)$. Assume that the plant (2) - (4) should be moved from the initial state $Q_0(x) = 0$ to the desired state $Q^*(x)$ within time t^*. Assuming for specificity that $\Psi(x) = \delta(x)$ we arrive at the following problem of moments, nonlinear for $s(t)$,

$$d_K = \int_0^{t^*} e^{-\lambda_K(t^*-t)} u(t)\varphi_K[s(t)]dt, \quad K=1,2,\ldots, \tag{13}$$

where d_K, $K = 1,2,\ldots$, are coefficients for expansion of $Q^*(x)$ into a series of eigen functions $\varphi_K(x)$.

Important results in the theory of the nonlinear problem of moments (finite dimensional and infinite dimensional) have been obtained in 10,11 . If the system is described by a heat conductivity equation, approximate solutions can be obtained by considering a truncated, finite dimensional problem of moments. A convex set theoretic approach analogous with that used in the theory of the linear problem of moments has proved fruitful. It reduces the determination of a solution, as in the linear problem, to a conditional extremum problem for a convex function which depends on a finite number of variables.

4. COMPUTER SIMULATION OF SYSTEMS WITH MOBILE ENERGY SOURCES

Numerous dufficulties arise in computer simulation of systems with movable sources, especially where the sources moves along (scans) a certain path on the surface. For illustration let us take up the plant (2) - (4) where the shape function looks like a Gaussian distribution, $\Psi(x,p) = (p/\sqrt{\pi})\, e^{-px^2}$

and the spurce moves in cycles from the point x_0 to the point x_1 and back. The cause of the difficulties is that on each time stratum (with solution of equations (2) - (4) by the finite difference method) a function of the form $\Psi(x,p)$ should be approximated fairly accurately as a network function. Therefore with fairly large coefficients p (sharp function of the form) the network step along a space coordinate should be made small. Besides, steps in time τ should be chosen such that the translation of the source $h = s(t+\tau) - s(t)$ in a time step stay below a certain admissible value h_{max}.

Numerical experiments have revealed that the best network for calculations of systems with mobile energy sources has a variable time step and a constant space coordinate step. The time steps are chosen so that the shape function center on each time stratum coincide with the network node along the spatial coordinate.

Let us present some results of investigating a close-loop control system with a movable source of signals that were obtained with the aid of a digital system.

The system to be controlled is described by the equations

$$\frac{\partial Q}{\partial t} - a\frac{\partial^2 Q}{\partial x^2} + u(t)\Psi(x - s(t), p) + g(x,t), \tag{14}$$
$$x \in D = (0, L), \quad t > 0,$$

$$Q(0,t) = Q(L,t) = \mu_0, \tag{15}$$

$$Q(x,0) = 0, \tag{16}$$

where $g(x,t)$ is a disturbance; $\Psi(x,p)$ has form of a Gaussian distribution. The source is assumed to move in cycles from point 0 to point L and back. The relation between the motion function $s_n(t)$, $t \in I_n$ in each cycle and the velocity $V_n(x)$ in each point (or distributed control $w_n(x)$) in the cycle is specified by formulae (10) and (11).

It is required to stabilize the specified temperature distribution $Q^*(x)$ in the system (14) - (16) by feedback control and in the face of unmeasurable exogeous distrubances $g(x,t)$ which are offset by varying the source power and velocity given in the form

$$V_n(x) = V^*(x) + \Delta V_n(x), \tag{17}$$

$$\tau_u \frac{du(t)}{dt} + u(t) = u^*(t) + \Delta u(t), \tag{18}$$

where the coefficient τ_u allows for the delay in the power actuator; $u^*(t)$ and $V^*(x)$ are functions which maintain the specified $Q^*(x)$ with zero disturbance (programmed control); and $\Delta u(t)$ and $\Delta V_n(x)$ are source power and velocity variations introduced by the feedback. The signals $\Delta V_n(x)$ and $\Delta u(t)$ are defined as functions of system deflection from the specified distribution.

The following constraints are imposed on the source velocity and power

$$0 < V_{min} \leqslant V^*(x) + \Delta V_n(x) \leqslant V_{max}$$

$$0 < u_{min} \leqslant u^*(t) + \Delta u(t) \leqslant u_{max}$$

The plant state is estimated from the cycle-avaraged deflection in each

$$\Delta Q(x,n) = \frac{1}{T_n}\int_{I_n} Q(x,t)\,dt - Q^*(x)$$

or from the mean deflection at the end of the source motion cycle.

The specified distribution $Q^*(x)$ was represented as a parabolic distribution

$$Q^*(x) = -A(x-0{,}5L)^2 + B, \quad x\in[0,L]$$

which satisfies the conditions

$$Q^*(0) = Q^*(1) = \mu_0 .$$

To obtain this distribution, the programmed control should be specified in the form

$$u^* = 2A, \quad V^*(x) = \frac{L}{T_n}, \quad x\in[0,L].$$

The source shope was specified as a Gaussian function. The disturbance was specified as a waveform $g(x) = G_0 \sin(2\pi/L)x$ and a half-waveform $g(x) = G_0 \sin(\pi/L)x$ of the sine line; G_0 is a constant.

The following control algorithms were modeled.

Algorithm 1.

$$\Delta V_n(x) = K_1 \Delta Q(x,n-1),$$

$$\Delta u(t) = -K_2 \int_D \Delta Q(x,n-1)\,dx,$$

where K_1 and K_2 are factors.

Algorithm 2.

$$\frac{1}{T_n V_n(x)} = \frac{1}{T_n V^*(x)} - K_1 \Delta Q(x,n-1)$$

$$w_n(x) = w^*(x) - K_1 \Delta Q(x,n-1)$$

and the power variation is as in Algorithm 1.

The advantage of Algorithm 2 over Algorithm 1 is that the power which is transferred from the source to the plant at each point due to velocity variations is directly proportional with the deflection in this point, $\Delta Q(x,n)$.

Algorithm 3, proportional integral

$$\Delta V_n(x) = K_1 \Delta Q(x,n-1) + K_3 \sum_{j=1}^{n-1} \Delta Q(x,j)$$

$$\Delta u(t) = -K_2 \int_D \Delta Q(x,n-1)\,dx - K_4 \sum_{j=1}^{n-1} \int_D \Delta Q(x,j)\,dx$$

Figure 2 shows the computed dependence $\Delta Q(G_0)$ of the error in steady state condition on the disturbance G_0 for the proportional algorithm. The dependence $\Delta Q^0(G_0)$ shows the static error value in an open-loop system. The curve of $\Delta Q(G_0)$ breaks when the source velocity reaches its constraints. In algorithms 1 and 2 the static error cannot be reduced to zero because, as computations show, with the gain exceeding a certain value the system becomes unstable. Algorithm 3 has no such disadvantage. The $\Delta Q_{PI}(G_0)$ curve in Fig. 2 shows that the static error in the system is zero until the source velocity reaches the constraints.

REFERENCES

1 Butkovskiy A.G., Pustyl'nikov L.M. Theory of Mobile Control of Systems with Distributed Parameters. Nauka. Moscow, 1980 (in Russian).

2 Chubarov E.P. Measurement and Control with Mobile Local Inputs. Energiya. Moscow, 1977 (in Russian).

3 Chubarov E.P. Mobile Control Systems. 3-rd Symposium "Control of Distributed Parameter Systems". Preprints. Tolouse (France), 1982.

4 Butkovskiy A.G. Some New Results in Distributed Parameter System Control. 3-rd Sympositum "Control of Distributed Parameter Systems".

5 Distributed control by mobile inputs. Collection of papers ed A.G. Butkovskiy. Nauka. Moscow, 1979. (in Russian)

6 Butkovskiy A.G. Theory of Optimal Control for Distributed Systems. Nauka. Moscow, 1965 (in Russian)

7 Kubyshkin V.A. Mathematical Modeling of Control Problems with a Mobile Multi-Cycle Inputs. In: Distributed Control by mobile inputs. Nauka. Moscow, 1979 (in Russian).

8 Breger A.M., Butkovskiy A.G., Kubyshkin V.A., Utkin V.I. Use of Sliding regime to mobile multicycler inputs Avtomatika i telemekhanika, No.3, 1980.

9 Bogolubov N.N., Mitropolskiy Yu.A. Asymptotic Methods in Nonlinear Oscillations. Fizmatgiz. Moscow, 1963 (in Russian)

10 Danilov V.Ya., Fedorchenko I.S., Tsitritskiy O.Ye. On Solvability of the Nonlinear Problem of Moments. In: Distributed Control by Mobile Inputs. Nauka. Moscow, 1979 (in Russian)

11 Butkovskiy A.G., Kubyshkin V.A., Pustyl'nikov L.M., Sharfarets B.P. Investigation time optimale mobile control. Avtomatika i telemekhanika, No.9, 1980 (in Russian).

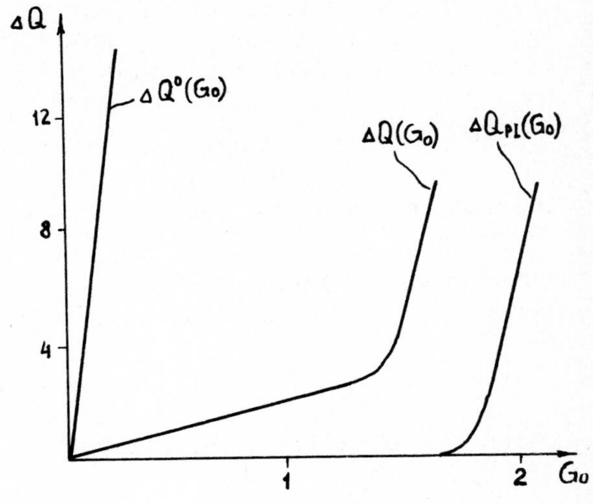

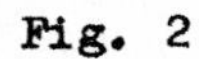
Fig. 2

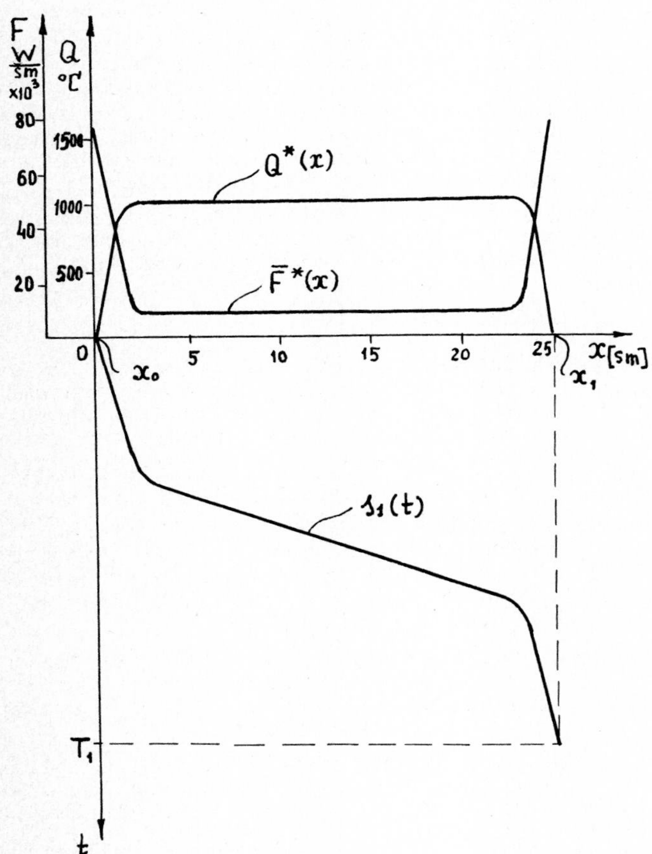

Fig. 1

VII. ENERGY SYSTEMS

Simulation in Engineering Sciences
J. Burger and Y. Jarny (eds.)
Elsevier Science Publishers B.V. (North-Holland)

MODELLING AND CONTROL OF AN UNDERGROUND GASIFICATION WELL

R. Gorez and C. Duqué

Université Catholique de Louvain
Unité d'Automatique et Analyse des Systèmes
B-1348 Louvain-la-Neuve Belgique

A model for the study of the thermal state in an exhaust well of an underground gasification plant is set up and simulated on a digital computer. Simulation allows to predict the transient responses to disturbances, to define a control strategy to be followed by the operators and to gain experience in the control of the plant. Simulation problems and solutions are discussed, and results are presented and analyzed.

1. INTRODUCTION

At present time, underground gasification experiments are carried out in Belgium. Gas is produced by the combustion of coal layers located about 1000 m beneath the ground surface.

The gas is collected by a 865 m depth well, consisting of two concentric pipes. The pipes temperatures must be limited for safety; this is achieved by injecting water in the annular space between the two pipes. This water should be transformed into overheated vapor before getting to the bottom of the well; otherwise, the gas generator inside the coal layer would be flooded. The overheated vapor will be carried away by the gas stream, the mixture outflowing upwards through the inner tubing.

The water flow rate must be such as to obtain a volume of overheated vapor of appropriate heigth (20 to 50 m) in the bottom of the well. Then, it is necessary to monitor the thermal state in the bottom of the well by means of thermocouples and to keep it within acceptable limits by varying the water flow (in normal operating conditions, the gas flow rate should be kept time-invariant in order not to disturb the operation of the gas generator; but, the gas temperature is not known and can be time-varying, then, it must be considered as an external disturbance).

The next section of this paper deals with a description of the plant and the set up of a mathematical model allowing to predict the state of the fluids and the temperature distributions along the well. This model is used in digital simulation programs, allowing to analyse the steady-state operation of the system or its transient response to variations of the gas temperature or of the mass flow rate of gas or water. The programs and the problems related to the simulation of the system are presented in the third section of the paper.

In the following sections, simulation results

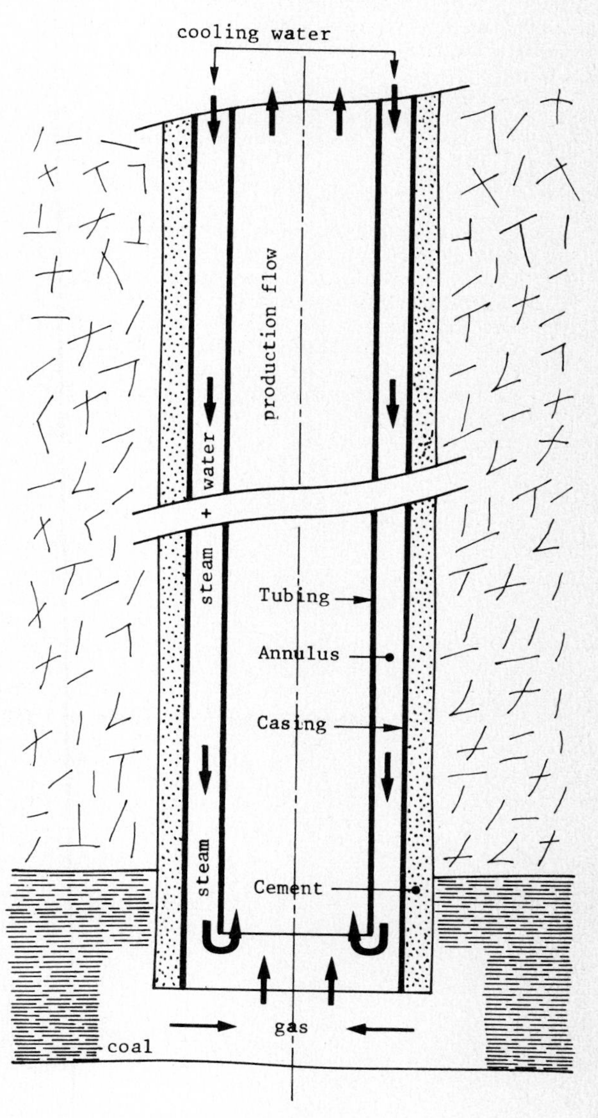

Fig. 1

are given and discussed. This allows to define a control strategy which should be followed by the operators; the resulting closed-loop system is simulated, and conclusions are derived in the last section.

2. DESCRIPTION AND MODELLING OF THE PLANT

Fig. 1 shows a sketch of the well. The depth of the well is 865 m, the tubing and casing are steel tubes, 5 and 7 inches diameter respectively.

Heat and mass transfers should be described by a distributed parameters mathematical model, taking into account the axial and radial distributions of the state variables : temperatures of the pipes, temperatures, pressures and composition of the fluids. This model has been converted into a lumped parameters model as follows :

1) in the axial direction, the well is divided into N equal length cells, marked by a subscript i ranging from $i = 0$ (bottom of the well) to $i = N$ (top of the well); average values of the parameters and variables are defined and mass and energy balance equations are written for each cell in turn;
2) radial variations of the state variables inside the fluid flows and of the temperatures through the pipewalls are neglected; then, the thermodynamical state in the ith cell can be represented by 1) the pressure in the upwards flow through the inner tubing : p_{mi}, 2) the vapor composition of the mixture of vapor and liquid flowing down the annular region tubing and casing : x_i, or if the vapor is overheated, its pressure : p_{wi}, 3) five temperatures : T_{mi}, T_{ti}, T_{wi}, T_{ci}, T_{oi} (subscripts m, t, w, c, o, refer to the mixture of gas and vapor, the tubing, the mixture of vapor and liquid water, the casing and the surrounding ground, respectively);
3) thermal capacitances of fluids are neglected, being very small compared to those of the pipewalls, and only radial heat transfers and axial mass transfers are considered, axial conductive and convective heat transfers being negligible.

All these assumptions result in the analog electrical model of Fig. 2, where each cell is represented by two capacitances, connected by conductances to each other and to current generators providing the enthalpy fluxes removed by fluids from and to the adjacent cells, Φ_w and Φ_m respectively.

Tubing

$$C_{t,i}\,\dot{T}_{t,i} = \Phi_{mt,i} - \Phi_{wt,i}$$

Casing

$$C_{c,i}\,\dot{T}_{c,i} = \Phi_{wc,i} - \Phi_{o,i}$$

Gas-vapor mixture

$$0 = \Phi_{m,i} - \Phi_{m,i+1} + \Phi_{mt,i} + F_m g\,\Delta L$$

Water (liquid-vapor)

$$0 = \Phi_{w,i+1} - \Phi_{w,i} + \Phi_{wt,i} - F_w g\,\Delta L$$

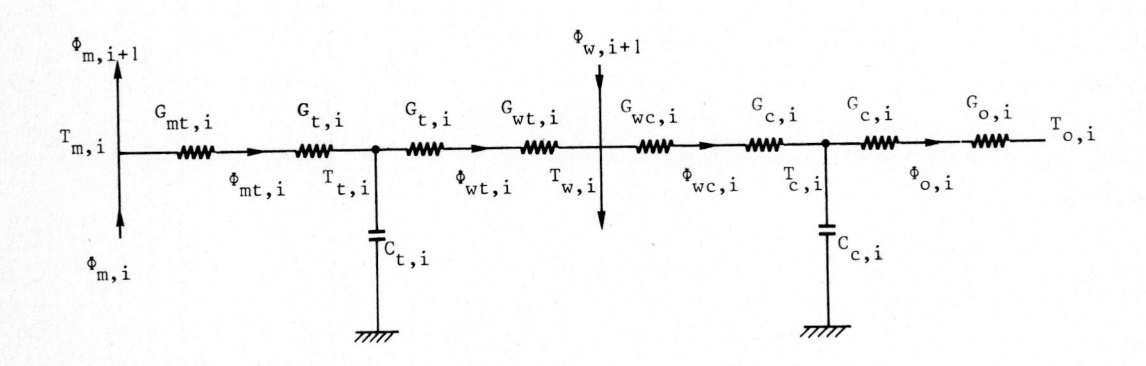

Fig. 2 - Analog model of cells

where T, F, Φ are used for temperature, mass flow rate and enthalpy flux respectively, C_t and C_c are the thermal capacitances of the tubing and the casing, and g is the gravitational constant (because of the well depth, potential energy related to gravity cannot be neglected in energy balances). Enthalpy balances yield a system of $2N$ ordinary differential equations and $2N$ algebraic equations; two other algebraic equations are given by mass and enthalpy balances in the mixing of gas and steam at the bottom of the well :

$$F_m = F_g + F_w ,$$

$$\Phi_{m,1} = \Phi_{g,o} + \Phi_{w,1}$$

Enthalpy fluxes can be calculated by the following relationships :

$$\Phi_{mt,i} = (T_{m,i} - T_{t,i})/(G^{-1}_{mt,i} + G^{-1}_{t,i})$$

$$\Phi_{wt,i} = (T_{t,i} - T_{w,i})/(G^{-1}_{wt,i} + G^{-1}_{t,i})$$

$$\Phi_{wc,i} = (T_{w,i} - T_{c,i})/(G^{-1}_{wc,i} + G^{-1}_{c,i})$$

$$\Phi_{o,i} = (T_{c,i} - T_{o,i})/(G^{-1}_{o,i} + G^{-1}_{c,i})$$

$$\Phi_{m,i} = F_m h_{m,i} , \quad \Phi_{w,i} = F_w h_{w,i}$$

where h is used for the enthalpy per unit mass of fluid, G_t^{-1} and G_c^{-1} are the half thermal resistances through the walls of the tubing and the casing, G_{mt}, G_{wt}, G_{wc} are thermal conductances related to the transfer coefficients between fluids and walls. Those and enthalpy are nonlinear functions of several variables : temperatures, pressure, mass flow rate, vapor composition, and also thermal flux density for the transfer coefficients between vapor and walls [1]. Moreover, pressures must be calculated taking into account friction losses and gravitational potential energy :

upwards flow (gas-vapor mixture)

$$p_{m,i+1} = p_{m,i} - (g + 2\lambda V^2_{m,i}/D_t)\rho_{m,i} \Delta L$$

downwards flow (liquid-vapor mixture)

$$p_{w,i+1} = p_{w,i} - (g - 2\lambda V^2_{w,i}/(D_c - D_t))\rho_{w,i} \Delta L$$

where D is used for pipes diameters, ρ for mass per unit volume, V for velocity, ΔL is the cells length, and λ is a friction coefficient depending on diameters and Reynolds number [1,2].

In the upwards flow, friction and gravity pressure drops are added up. In the downwards flow, friction losses are nearly compensated by gravity effects, resulting in a quasi-constant pressure; this is due to the fact that a fraction of the injected water is vaporized when coming in the well. Then, on the longest part of the well, water is flowing in the form of thin liquid films gliding along the pipewalls, the rest of the annular space between tubing and casing being filled by a gigantic vapor bubble [3]. Temperature $T_{w,i}$ will be the liquid-vapor equilibrium temperature in this part of the well, the overheated vapor temperature in the lowest part.

3. SIMULATION PROBLEMS AND PROGRAMS

The mathematical model described in the previous section is strongly nonlinear : enthalpy and transfer coefficients are complicated function of several state variables. Especially, the transfer coefficient between vapor and walls may be four times smaller or ten times larger than in the case of liquid only, according as vapor is overheated or saturated; in case of mixture of vapor and liquid, it will depend on the vapor composition. Thereafter, the position of the transition point between the part of the well where vapor is overheated and the upper part must be determined accurately : the cell inside which the transition occurs must be found, and the position of the point inside this cell must be calculated. Computations must be done iteratively, the number of iterations required for given accuracy being smaller if cells are shorter. The first simulations have shown that 1 m length was adequate, requiring only a single iteration ; accordingly, the dimension of the set of equations is pretty large (N = 865).

Then, two computer programs have been written. The first one, called STATIC, is used for calculating the distributions of temperatures and other state variables in steady-state operation of the plant, and also for determining the water mass flow rate which is required for equilibrium. The equations of this model are those of the previous section, where time derivatives have been nullified and the temperatures of the water, the casing and the surrounding ground are equal. The following boundary values may be choosen :

- temperature, pressure and mass flow rate of the gas inflow at the bottom of the well (the composition of the gas and its physical and thermodynamical properties are known)
- temperature of the water inflow at the top of the well
- overheating rate (difference between the temperature of the overheated vapor and the vapor-liquid equilibrium temperature at the bottom of the well).

As it is a nonlinear boundary values problem, it must be solved iteratively. The steady-state solution is reached after a few iterations only, with errors smaller than 1 %. Fig. 3 gives the temperature distributions along the well for mass flow rates of 2500, 5000 and 10 000 m^3/h and an overheating rate of 300°. A systematic use of this program yields curves relating the transition point position to the overheating rate and the gas flow rate (for given pressure and temperature of the gas). Such curves are useful for selecting operating conditions. Besides, simulations have shown that relative variations of some coefficients (viscosity, friction, ...) were very small, in the same way as pressure variations in the downwards flow (less than 10^5 *Pa* compared to an inlet pressure of $25\ 10^5$ *Pa*); they have shown also that all the state variables variations may be neglected at levels upper than 200 m above the well bottom, if the operating conditions are such as the distance from the transition point to the well bottom is less than 100 m.

The second program, called DYNAM, is used for obtaining the transient responses to variations of the gas temperature or of the mass flow rate of gas or water. Initial conditions are those of steady-state equilibrium and must be computed first by the program STATIC. Moreover, as the thermal properties of the ground are not known and the conductivity of cement is very low, it has been assumed that the cement coating achieves a perfect thermal insulation between casing and surrounding ground ($\Phi_{o,i} = 0$). The structure of DYNAM is as follows [4] :

INITIATING SECTION
- Read initial values of state variables from a data file created by STATIC
- Select disturbance signal

→ State variables values

DERIVATIVE SECTION
- Test for end ($t = t_{max}$)
- Display results on a CRT terminal or store them in an output file
- Compute right-hand sides of differential and auxiliary equations

end ← time derivatives of state variables

INTEGRATION SECTION
Predict and correct state variables values at $t + \Delta t$

The integration section calls a fixed step second order Runge-Kutta integration routine. Several values of the integration interval have been tried; it appears that a time interval of 10 s was the best trade-off taking into account accuracy and computer time.

Simulation of the complete model, including 865 cells and computations of pressures and coefficients at each time step, on an IBM 370 computer, resulted in a CPU time of 10 min for a transient length of 1000 s. Also, it has confirmed conclusions of the steady-state analysis; then, the dynamic model has been simplified, resulting in a saving of CPU time by a factor of 5 :
- state variables are calculated along the first 200 m from the bottom; beyond, they are held at fixed values
- pressures and friction coefficients are computed at each 10 time steps.

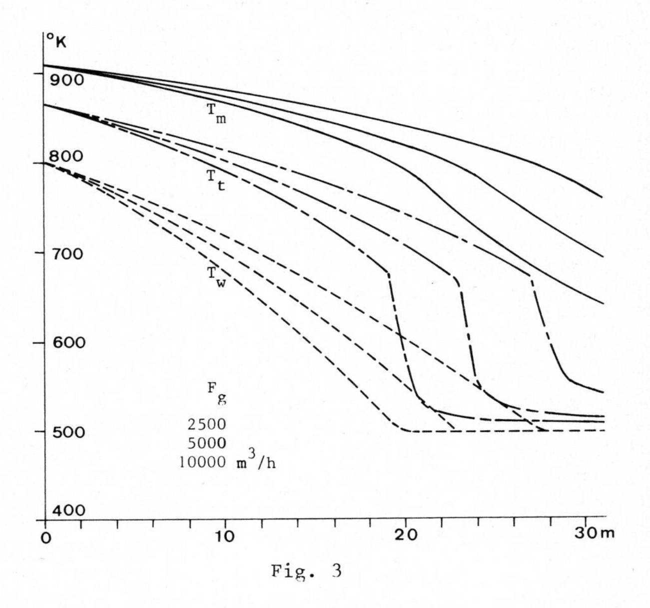

Fig. 3

Then, simulation of transients of more than 1 h length requires a computer time of 10 to 15 min in batch processing or 20 to 30 min in on-line time-sharing (On-line processing with sampling and display of relevant information on a CRT terminal and entry of data such as the water flow rate at given time intervals allows the simulation of the control of the thermal state in the well by a human operator).

4. ANALYSIS OF SIMULATION RESULTS

A lot of experiments consisting in step disturbances of the gas temperature or the flow rate of gas or water have been simulated in order to obtain the step responses of the well thermal state to the input variables.

Fig. 4 and 5 give the displacement of the transition point and the variations of temperatures of vapor at given locations in the lower part of the well, for various steps of the gas temperature and flow rates respectively. They show that transients can be decomposed into two parts. The first part of the transient responses is related to the fitting of the temperature distributions of the walls, with time-constants of approximately 1 min.; the duration of this part is less than 5 min. Afterwards, the time-variations of the temperatures and the displacement of the transition point are roughly linear, at least as long as the length of the overheated zone exceeds a few meters; below, transients are accelerating. In the same way, it appears that variations are faster when they are related to a downwards rather than an upwards displacement of the transition point. Also, it can be seen that equal relative variations of the water and gas flow rates yield the same results, in opposite directions, the effects of a 20 % flow rate variation being approximately the same as those of a gas temperature variation of 100°.

The previous results agree with these of the analysis of an oversimplified model, derived from a global enthalpy balance :

$$C\dot{T} = F_g c_g (T_g - T_w) - F_w r ,$$

where r is the latent heat of water, c_g is the specific heat of the gas and C is the effective thermal capacitance of the walls, which is approximately equal to $(c_t + c_c)L_s$, where c_t and c_c are the specific heat per unit length of tubing and casing and L_s is the distance from the transition point to the bottom of the well.

5. CONTROL OF THE THERMAL STATE

From the simulation results, it appears that the thermal process is not self-regulatory; then, control of the thermal state at the bottom of the well is necessary. Otherwise, a 100° decrease of the gas inlet temperature or a 20 % excess of the water flow rate versus the gas flow rate, could result in flooding of the gas generator in less than 5 min. Disturbances in the opposite sense would lead to an upwards displacement of the transition point, resulting in excessive thermal expansions of the pipes. Then, the cement coat could be damaged, and part of the casing would be hanged up freely in the well, with risks of clash. As for the tubing, its downwards expansion is limited by the bottom of the well; excess overheating would induce thermoelastic stresses in the fixing devices; then, there would be a risk of damage either at the ground surface, or in the bottom of the well.

In the first stages of the experiments, automatic control will not be achieved; control will be done by human operators, allowed to manipulate the flow rates. In normal operating conditions, the only maniputed variable will be the water flow rate, the gas flow rate should be kept at a fixed value in order not to disturb the process of combustion and gasification in the coal layers. As a consequence of the film type flow in the major part of the well, a variation of the water flow rate resulting from opening or closing the control valve comes

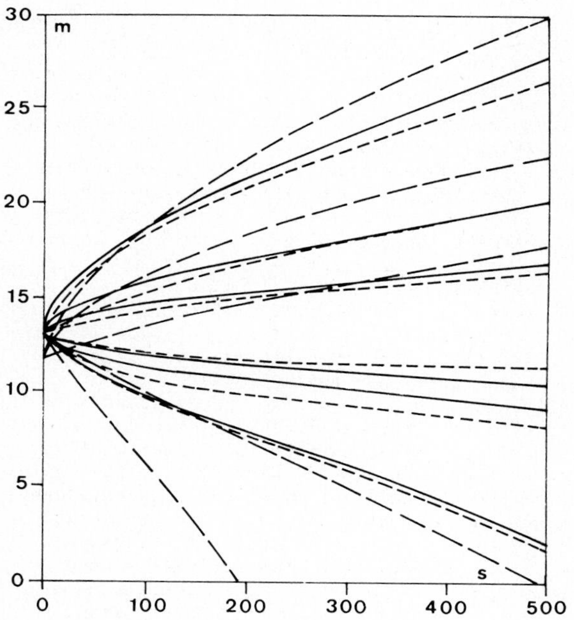

Fig. 4

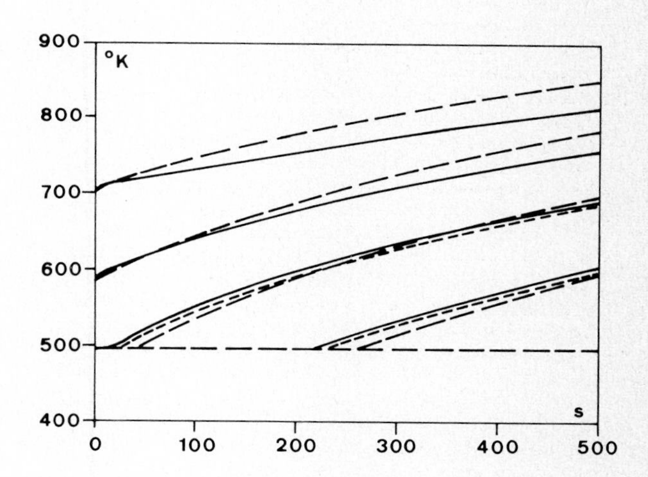

Gas flow rate : 5 000 m^3/h

Full lines : $\Delta F_w/F_w = \pm 5, \pm 10, \pm 20$ %

Dotted lines : $\Delta F_g/F_g = \pm 5, \pm 10, \pm 20$ %

Broken lines : $\Delta T_g = \pm 100, \pm 200, + 400°$

Fig. 5

down the well at the velocity of the film. Preliminary theoretical and laboratory experiments [5] have shown that in normal operating conditions, this velocity should be between 1.2 and 1.6 m/s; it results in a delay of approximately 10 min., in the control of the process.

The thermal state of the well will be monitored by several thermocouples, located on the outside surface of the tubing, providing measurements of the vapor temperature. Thermocouples signals will be processed by a minicomputer in the control room, and temperatures values will be displayed on a CRT terminal. In view of not overloading the operator's task, and also, because of the delay mentioned above, it is assumed that operators could sample data and manipulate the control variable at 5 min. intervals. Several control strategies have been tested by simulation, using the program DYNAM, in various operating conditions, for different thermocouples locations.

Finally, the control strategy which is proposed, is based on the following relationship :

$$F_w(k)=F_w(k-1)+K_o[T_m(k) - T_d]+K'_o[T_m(k) - T_m(k-1)]$$

where T_d is the desired value of the temperature at a point located 15 m above the bottom of the well, and T_m is the measurement provided by a thermocouple at this point; $T_m(k)$, $T_m(k-1)$, $F_w(k)$, $F_w(k-1)$ denote the values of the measured temperature and the manipulated water flow rate at times t_k and t_{k-1}, where $t_k - t_{k-1}$ = 5 min. Simulations have shown that T_d should have the same value in all operating conditions, selected in view of achieving a steady-state position of the transition point between 25 and 30 m from the bottom of the well. (This position moves upwards for higher gas flow rate; this improves the safety margin at the same time as the gain of versus disturbances has been increased). Fig. 6 shows the displacement of the transition point following a disturbance of the gas temperature, when using this control strategy. Other control strategies had led to better results, but they are very sensitive to the exact value of the delay; then, they cannot be recommended before experiments have been done and have provided a better knowledge of the plant parameters.

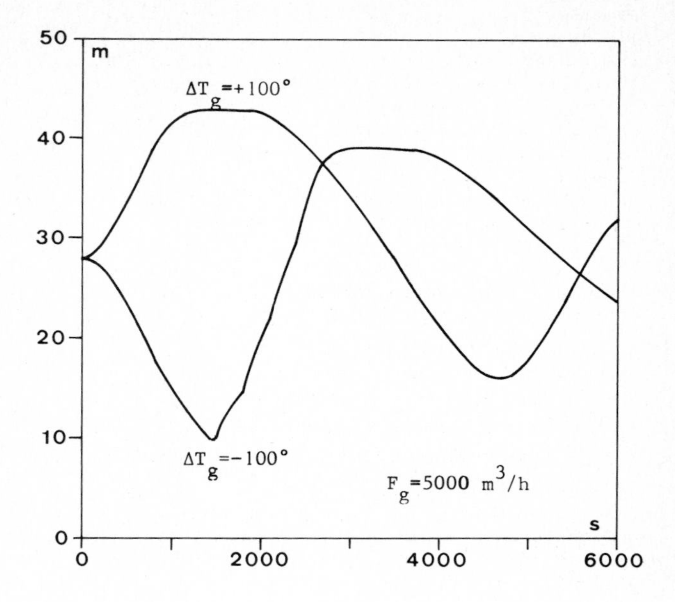

Fig. 6

6. CONCLUSIONS

Simulation has yielded the transient responses of the considered process to various disturbances. Those have shown that the process is not self-regulatory and that monitoring and control are necessary. From the results of simulation, it has been possible to define some operating conditions of the process and to select and try a control strategy; then new simulation experiments have allowed to gain experience in the control of the process. This has resulted in recommendations which should be followed by the plant operators. Moreover, simulation has allowed to determine the most appropriate locations of thermocouples for the monitoring of the process. In this way, simulation has proved to be an indispensable tool in the design of the operating conditions of the process and in the training of the operators.

REFERENCES

[1] Houberechts A., La thermodynamique technique (Vander, Leuven, 1976)

[2] Van Simaeys C. & Giot M., Etude du refroidissement du gaz de gazéification souterraine par injection d'eau dans l'espace annulaire situé entre le casing et le tubing, Int. Report, Dept of Thermodynamics, Univ. of Louvain (1979)

[3] Renaux J.P., Patigny J., Giot M., Recherche sur le refroidissement des sondages de gazéification, Int. Report, Dept of Thermodynamics, Univ. of Louvain (1981)

[4] Franks R.G.E., Modeling and Simulation in Chemical Engineering (Wiley, New York, 1972)

[5] Renaux J.P., Recherche sur le refroidissement des sondages de gazéification : étude des écoulements en film, Int. Report, Dept of Thermodynamics, Univ. of Louvain (1982).

Simulation in Engineering Sciences
J. Burger and Y. Jarny (eds.)
Elsevier Science Publishers B.V. (North-Holland)

MODELLING ANALYSIS OF SOLAR ENERGY SYSTEMS

L. Fortuna*, A. Gallo*, M. La Cava** and A. Barbarino*

*Istituto Elettrotecnico, Università di Catania, Italy
**Dipartimento di Sistemi, Università della Calabria, Cosenza, Italy

The purpose of this paper is the working out of several suitable approximate models of a experimental solar energy system which are successively compared on the resultant approximation. Distributed and lumped parameter mathematical models from the third to the fifth order have been developed. Initially a distributed non-linear 5th order model, derived from basic physical balance equations, is analysed. This model-building is assumed as the standard, from which both reduced and approximate models are obtained. Various kinds of lumped models are, successively, considered on the basis of theoretical and experimental hypoteses. The various models are tested by simulation comparing their outputs with experimental data.

1. INTRODUCTION

To work out optimal control solutions for the thermal behaviour of a solar energy circuit, process typically characterized by successively varying conditions, an accurate procedure is necessary to build up appropriate models suited to develop the required control law.

Distributed parameter models represent obviously the most accurate tool for the solution of control problems in solar energy systems [1] [2]. These models allow one to evaluate overall system performance and interactions between the state vectors of the various sections of the plant with sufficient accuracy. Various difficulties arise, however, usually both for system simulation and control implementation: computer runs should be in general very costly and the task of the controller design too complicated to be easily implemented. The modern control theory approach utilizes largely simplified lumped parameter models which are able to hold the fundamental dynamic relationships and to allow simulation with low computer time and control implementation with substantially reduced cost.

The physical process consists of several distinct distributed parameter elements, mathematically represented by non linear partial differential equations with space and time as independent variables. Lumped parameter approximations may be used to formulate finite-dimensional state-space models based on specific assumptions of physical character. This approach has been shown adequate in modelling various solar plants by experimental verifications [3]. More accurate models of the same kind may be obtained by applying suited mathematical procedures to the definition and working out of the relative parameters.

In this paper, after we have assumed as standard a distributed non-linear parameter model-building of an experimental-pilot solar circuit, derived from basic physical equations, reduced and approximate models are obtained: first a lumped model is derived simply from physical considerations; then, a lumped time-invariant model obtained by time discretization is estimated using normal operating data; finally, a lumped model is derived analytically from the distributed one by means of the well known method of Galerkin. The various approximate models are compared one with another by computing the output behaviour of each of them in response to the same inputs.

2. SYSTEM DESCRIPTION

The solar plant here considered is a simple experimental-pilot circuit (Figure 1). It consists of a solar flat plate single glass collector (cupper tube and aluminium plate type), a boiler, a heat exchanger, a generalized load and two circuit control devices. This circuit was designed both for instantaneous collector efficiency (NBS standard) and for mean daily efficiency (calorimetric method) measuring; the experimental conditions were referred to summer in Sicily. For our studies we have considered only the performance with energy accumulation in the day. The smallest time constant has been estimated to about 300 s for the tube and plate collector and 10 hours for the boiler. A set of experiments, described in detail in [4], have been designed according to the NBS specifications; various measure tests have been performed and the main variables of the typic multivariable solar process have been automatically recorded from a data acquisition system; for each variable a number of 360 samples has been considered at interval of 90 s. The data are converted into physical units and processed by means of a package of FORTRAN programs implemented on off-line computer. In Figs 2, 3 the experimental evolutions of the solar radiation, the collector fluid temperature, the boiler fluid temperature and the ambient temperature in the primary circuit, related to a summer day, are shown.

3. DEVELOPMENT OF THE MODELS

Various distributed and lumped mathematical models, based on a priori physical knowledge of the process, from the third to the fifth order have been developed to study the solar heating system. The complex structure of the process leads to the multivariable model. Therefore special attention was devoted to the problem of structure determination and, particularly, to the decoupling of the multivariable system into suitable scalar subsystems. The order depends essentially on the scheme of the process; if this is the complete scheme comprising the primary and secondary circuits of Fig. 1, the order is five; if this is only the primary circuit (which is the proper solar energy system) the order is three; if this is only the secondary circuit (which is the user load system) the order is two.

A modelbuilding simplified in the structure, but accurate in the parameters, is obtained in accordance with the following procedure. First, the most appropriate definition of the model structure is derived by using basic energy balance techniques; then a lumped parameter approximation was used to formulate a finite dimensional-state space model. In a third step suited identification algorithms allow one to determine from experimental data values of the parameters of the reduced model which are closer to physical phenomena in spite of adopted simplifications. Finally a lumped model is derived analytically reducing the partial differential equations to a finite number of ordinary differential equations with time as the independent variable.

3.1 Distributed Parameter Model

In the complete scheme of the system we have five state variables relating to the five energy storage processes (i.e. the glass, collector, primary and secondary fluid and boiler) and two input variables (i.e. the solar radiation and the exernal ambient temperature). To represent with high accuracy the spatial and time evolution of the complete system, so as to have a standard model to which to refer all the other simplified models, we choose a fifth order state model with a distributed time-varying non linear structure. To build this model we consider five energy balance equations which are derived for a generic zone in the collector and boiler. The following assumptions were made in deriving these equations: negligible thermal losses in the carter and at the borders;

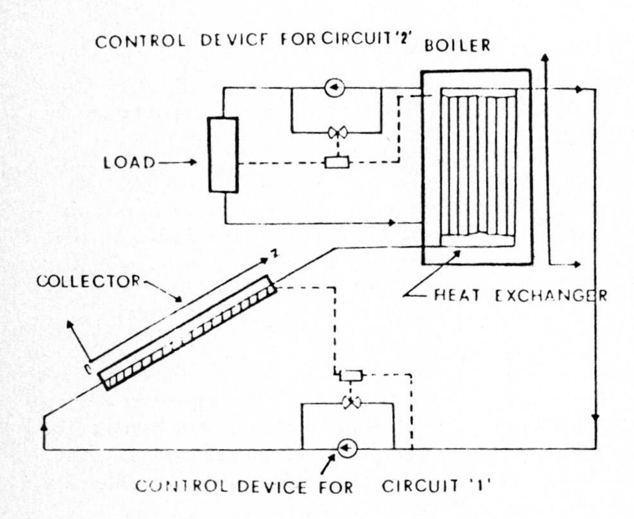

Fig. 1 Experimental solar circuit.

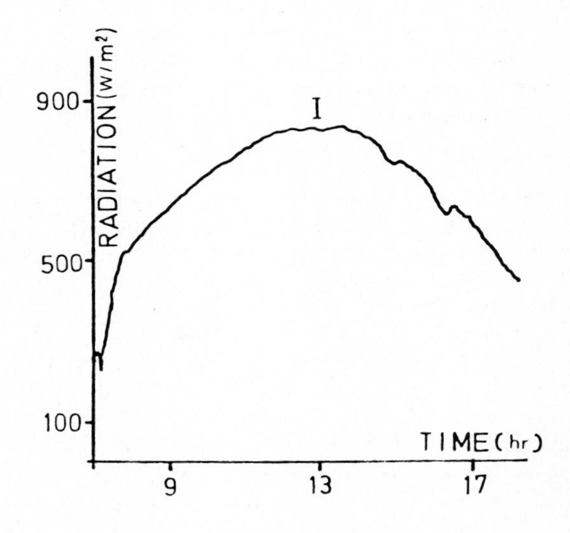

Fig. 2 Experimental solar radiation.

negligible thermal capacity of the air between the glass and the absorber plate; insulated ducts; negligible transmittance of the glass at high wave-lenghts (>3 μm); constant thermal capacities with the temperature. These typical modelling equations are described in Appendix. In a second step the equations are arranged in the following standard state-space form:

$$\frac{\partial y}{\partial t} = A\ y + B\ \frac{\partial y}{\partial z} + C\ u \qquad (1)$$

$$z \in [0,1],\ t \geq 0$$

where z denotes the abscissa in the collector and boiler. The initial and boundary conditions are: $y(z,0)= y_0(z)$ and $y(0,t) = y_b(t)$.
The input vector u and the state vector y are:

$$u'=[I\ \ T_a],\quad y'=[T_1\ T_2\ T_3\ T_4\ T_5]$$

where the apex ' denotes the transpose vector. A, B, C are matrices with defined structure.

3.2 Simplified Model

The simplified model adopted is given by three energy balance equations respectively for the collector fluid, the absorber plate and the glass cover. They are derived by assuming valid the single-cell approximation and suitably linearizing the terms due to thermal losses. Therefore the lumped parameter approximation is used to formulate a finite-dimensional state space model in the following form:

$$\dot{y} = D\ y + E\ u \qquad (2)$$

where $u'= [I\ \ T_a]$, $y'=[T_1\ \ T_2\ \ T_3]$.

The elements of the matrices D and E, with defined structure (Appendix), are related in a very strict way to the coefficients of the standard model regarding the proper solar energy subsystem.

3.3 Identified Model

To represent in a suited way an expression relating model parameters and measured data, it is convenient to work out a linear structure in dicrete form according to a priori knowledge. Therefore the process is assumed to be given as a mathematical model whose structure is known because it is derived from the state space model (2). Every equation is discretized by using the classical finite difference approximation. In effecting the required algebraic manipulation, one obtains a discrete time-invariant model, which has the following form:

$$y_{j,k} = a_j\ y_{j,k-1} + \sum_{i=1}^{m_j} b_{ji}\ u_{ji,k-1} \qquad (3)$$

$$j=1,2,3$$

where k denotes the discrete time, j represents the equation index, $y_{j,k}$ is the process output T_j temperature according to the corresponding jth equation at the sampling instant kth, m_j is the input number according to the jth equation; a_j and b_{ji} represent the estimating parameters, that are strictly related with the physical coefficients.

Many practical methods of system identification were investigated , utilizing the solar experimental data. In particular the least squares (LS) and generalized least square (GLS) have been tested to identify the parameter vector:

$$\theta_j' = [a_j\ ,\ b_{j1}\ ,\ b_{j2}\ \ldots\ b_{jm_j}]$$

Identification experiments have been performed by using a FORTRAN package of identification and simulation algorithms developed for on-line or off-line procedures. Using classical off-line GLS procedure [4] the daily means values of the parameters are obtained. With LS and GLS on-line recursive methods [5] the intrinsic present parameters are identified.

3.3 Approximate Model by Galerkin Method

The well-known Galerkin's method is applied to the model (1) with the aim to obtain a suitable approximate lumped model. Considering only the primary circuit, the standard state model is reduced to three differential distributed equations (Appendix), and can be expressed in the following form:

$$P\ \frac{\partial y}{\partial t} = R\ y + Q\ \frac{\partial y}{\partial z} + S\ u\ ,\ z\epsilon[0,1],\ t \geq 0 \qquad (4)$$

where now the state vector y, of n=3 elements, is $y_i' = [T_1\ \ T_2\ \ T_3]$ and the input vector, of m=2 elemnets, is $u' = [I\ \ T_a]$. The P, R, and Q are (3x3) square matrices, and S is a (3x2) matrix. The initial and boundary conditions are $y(z,0) = y_0(z)$, $y(0,t) = y_b(t)$.
We assume the following approximate solutions:

$$\tilde{y}_i = \sum_{j=1}^{n_1} x_{ij}(t)\ \omega_j(z) + T_{io},\quad i=1,2,3 \qquad (5)$$

where $\omega_j(z)= z^j$ is the generical basis function and n_1 is the number of terms of the solution.

Each approximate solution $\tilde{y}_i$ is replaced in the original system (4) and in a such a manner the residual vector r_E of n=3 elements is given. Applying the weighted residuals procedure by Galerkin method, we have:

$$\int_0^1 r_{E_i}\ z^k\ dz = 0,\quad i=1,2,3,\quad k=1,2,\ldots n_1$$

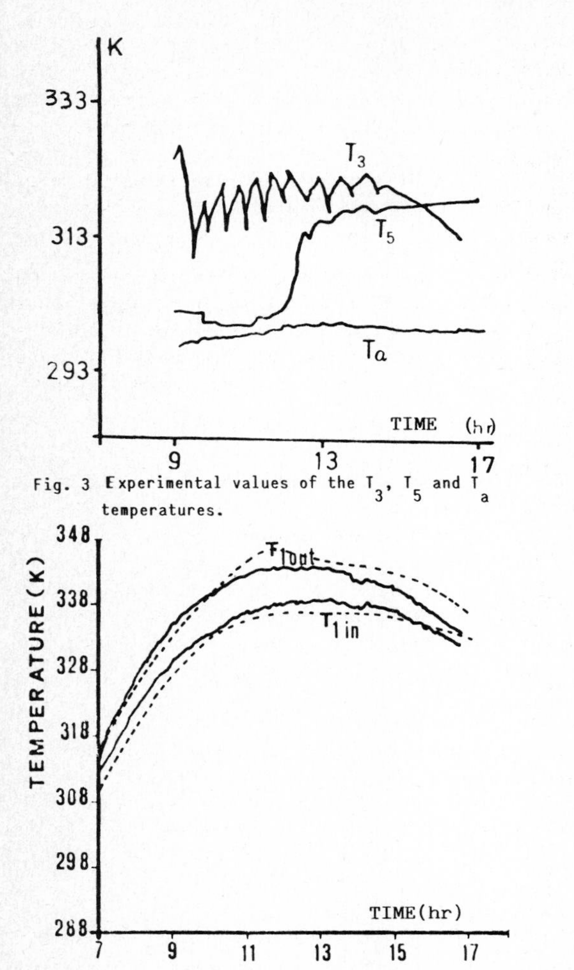

Fig. 3 Experimental values of the T_3, T_5 and T_a temperatures.

Fig. 4 Comparison between the T_1 temperature of the standard (continuous) and approximate (dashed) models (in=input,out=output zones of the collector).

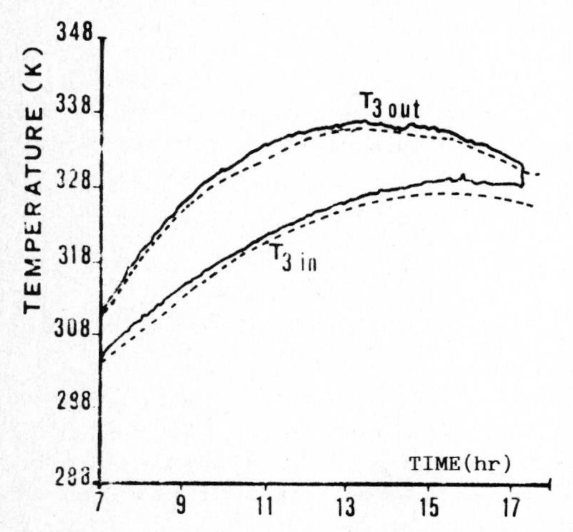

Fig. 5 Comparison between the T_3 temperature of the standard (continuous) and approximate (dashed) models.

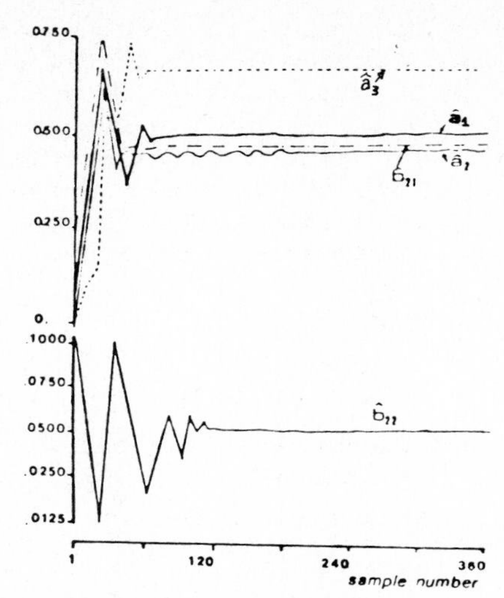

Fig. 6 Parameter estimation.

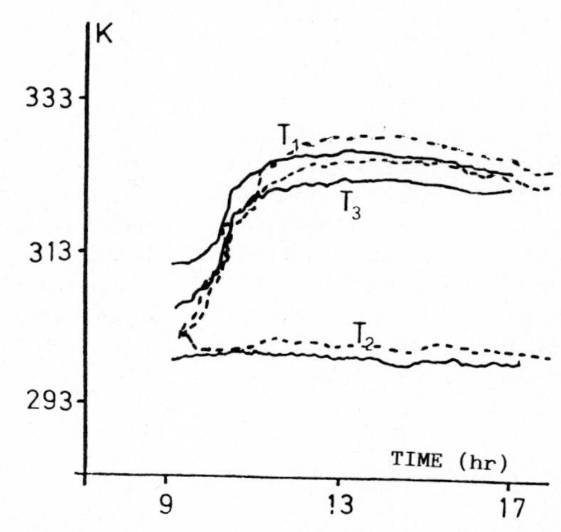

Fig. 7 Simulated outputs T_1,T_2,T_3 relative to the identified (continuous) and simplified (dashed) models.

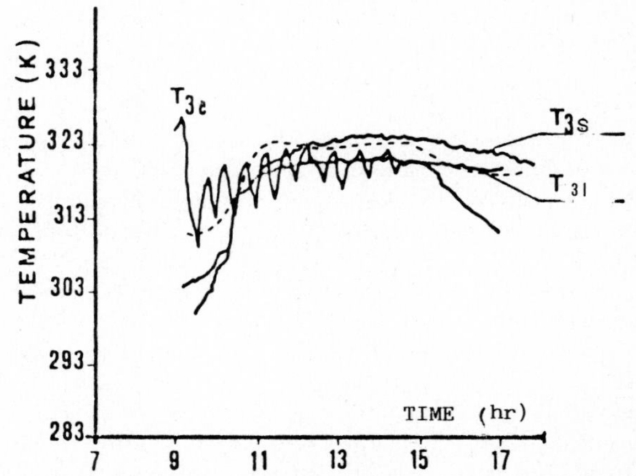

Fig. 8 Comparison among the temperatures of the simplified (T_{3s}), identified (T_{3i}) and approximate (dashed) models with experimental values (T_{3e}).

A system of $n_2 = 3xn_1$ differential equations, with n_2 unknown functions $x_{ij}(t)$ is derived. This system in compact form can be written as follows:

$$L\dot{x} = [M + J]x + Ku + HT_o \quad \text{where}$$

$$x' = [x_{11}x_{21}x_{31}x_{12}x_{22}x_{32} \cdot \cdot x_{ij} \cdot \cdot x_{1n_1}x_{2n_2}x_{3n_3}]$$

and (n_2xn_2) L, M, J and (n_2x2) K , (n_2x3) H are matrices with defined structure.

Inverting the L matrix,

$$F = L^{-1}[M + J] \quad \text{and} \quad G = L^{-1}K ,$$

the system (4) assumes the lumped approximate form:

$$\dot{x} = Fx + Gu + L^{-1}HT_o \tag{6}$$

called "Galerkin's state equation".

4. SIMULATION AND CONCLUSIONS

The proposed models can be used to describe the system behaviour and to evaluate its performance when it is subjected either to simulated inputs, or to experimental inputs derived from the historical data. With reference to the experimental plant presented in Fig. 1, we have simulated both the standard and the simplified models, fixing the parameter values on the basis of the design parameters values in this plant and of physical considerations about the energetic balances.

The simulation of standard and simplified models has been carried out by means of the finite difference method. The selection of the simulation step has been very important in the simulation because it determined the precision of the results. In general the solution of the discretized system obtained by means of a step by step procedure (finite difference method) has been integrated with other numerical method (as the predictor-corrector method) to achieve a good precision of the results.

In the distributed simulation the collector and the boiler have been subdivided into three zones which we consider the best spatial subdivision to have acceptable computer time and suitable accuracy. The theoretical values of the coefficients of the standard model are derived in the central zone at midday.

The simulation of approximate model by Galerkin method has been developed by the solution of the equation (6) and successively, after a discretization procedure, by the output transformation (5). In this manner it is possible to calculate the physical variables (in particular temperatures) of each zone of the collector instant for instant. A number of basis functions $n_1 = 6$ has been chosen in such a manner to obtain a great accuracy. The approximate solution (5) has been assumed in this form to verify the boundary conditions.

In Figs 4 and 5 the simulation results for the standard model are shown; the collector and fluid temperatures, relative to the input and output zones of the collector, are compared with the trend of the simulated temperatures (dashed) of the approximate model by Galerkin.

The trend versus time of some parameters by GLS estimated is shown in Fig. 6. It is evident the variability of some parameters which depend mainly from the dynamic of the solar process; in particular the behaviour of the physical coefficients is manifest: in fact the parameter b_{22} depends on the heat removal factor F_r and the total heat loss coefficient of the collector U_L depending also on the absorber radiativity, then the parameters b_{21} and a_2 are closely depending on the thermal transmittance k_1 between plate and primary fluid.

In Fig. 7 the simulation results for the simplified and identified models are shown. These results do not differ much among them and also agree acceptably with the experimental data. In Fig. 8 the comparison among the fluid temperatures relative to the simplified, identified and approximate by Galerkin (for a mean zone) models with experimental values is shown.

The results show the validity and the reliability of the applied methods for the modelling of the solar circuit. Therefore we note that the simplification has allowed us to reduce the non-linear standard model (distributed fifth order) to lumped (third order) model and this by identification to the linear time-invariant discrete model having mean parameter values. The Galerkin equation is most convenient for the solar circuit optimization. In fact we obtain the simulation at each desidered zone with reference to the standard model where the values are related to fixed zones, and with reference to the simplified and identified models where the simulation variables are related to means values in the whole collector. Moreover the approximate model is appropriate to save computing time in controlling procedure by modern optimization theories. The on-line identification algorithms are mainly suitable to study the behaviour of some physical parameters which are classically derived from empirical or simplified formulas. These parameters, estimated at each instant, can be used, in second step, in the simple Galerkin's state equation which gives the space-time evolution.

The application of optimal control techniques to verify the accuracy of the chosen approximate model in complex solar plants will be the future development of this work.

5. REFERENCES

[1] Dorato, P., "A rewiev of the application of modern control theory to solar energy system", Proc. IEEE Conf. on Decision and Control, USA, 1979.

[2] Rorres, C, Orbach, A., and Fischl, R. "Optimal and suboptimal control policies for a solar collector system", IEEE Trans., AC-25, 1085-1091, 1980.

[3] Ray, A., "Non-linear dynamic model of a solar steam generator", Solar Energy, 26, 297-306, 1981.

[4] Arcidiacono, P., Cammarata, G., Faro,A., and Gallo, A., "Identification of solar energy systems", Proc. 5th IFAC Symp. on Ident. and System Par. Est.,F.R.Germany,1979

[5] Fortuna, L, Gallo, A, Cammarata, La Cava, "Microcomputer system allow a real-time parameter estimation for the control of a solar plant", 3rd IFAC Conf. on System Approach for Development, Morocco, 1980.

APPENDIX

Distributed parameter equations:

$$b\,dz\,F_1 I - h_1 b\,dz\,(T_1 - T_2) - \sigma_1 b\,dz\,(T_1^4 - T_2^4) - k_1 (T_1 - T_3)\,p_3\,dz = \rho_1 c_1 s_1 \frac{\partial T_1}{\partial t} \quad (1)$$

$$b\,dz\,F_2 I + h_1 b\,dz\,(T_1 - T_2) + \sigma_1 b\,dz\,(T_1^4 - T_2^4) - h_2 b\,dz\,(T_2 - T_a) - \sigma_2 b\,dz\,(T_2^4 - T_a^4) = \rho_2 c_2 s_2 \frac{\partial T_2}{\partial t} \quad (2)$$

$$k_1 p_3\,dz\,(T_1 - T_3) = \gamma_1 c_1 \dot{M}\,dT_3 + \rho_3 c_3 n_3 \left(\frac{d_3}{2}\right)^2 dz \frac{\partial T_3}{\partial t} \quad (3)$$

$$\gamma_1 c_1 \dot{M}\,dT_4 = k_4 p_4\,dz\,(T_4 - T_5) + c_3 \rho_3 p_4\,dz \frac{\partial T_4}{\partial t} \quad (4)$$

$$k_4 p_4\,dz\,(T_4 - T_5) - k_5 \pi D\,dz\,(T_5 - T_a) - \gamma_2 c_3 \dot{M}'\,dT_5 = \rho_5 c_5 \pi \frac{D^2}{4} dz \frac{\partial T_5}{\partial t} \quad (5)$$

Structure of the matrices D and E of the system (2):

$$D = \begin{bmatrix} d_{11} & d_{12} & d_{13} \\ d_{21} & d_{22} & 0 \\ d_{31} & 0 & d_{33} \end{bmatrix}, \quad E = \begin{bmatrix} e_{11} & 0 & e_{13} \\ e_{21} & e_{22} & 0 \\ 0 & 0 & e_{33} \end{bmatrix}$$

Simplified distributed parameter equations:

$$b\,dz\,F_1 I - U_{pg}\,b\,dz\,(T_1 - T_2) - k_1 (T_1 - T_3)\,p_3\,dz = \rho_1 c_1 s_1 \frac{\partial T_1}{\partial t} \quad (1)$$

$$b\,dz\,F_1 I + U_{pg}\,b\,dz\,(T_1 - T_2) - U_{ga} (T_2 - T_a)\,b\,dz = \rho_2 c_2 s_2 \frac{\partial T_2}{\partial t} \quad (2)$$

$$k_1 p_3\,dz\,(T_1 - T_3) = \gamma_1 c_1 \dot{M}\,dT_3 + \rho_3 c_3 n_3 \pi \frac{d_3^2}{4} dz \frac{\partial T_3}{\partial t} \quad (3)$$

NOMENCLATURE

T_a	ambient temperature
T_1	collector-plate temperature
T_2	glass-cover temperature
T_3	fluid temperature in the collector
T_4	fluid temperature in the heat exchanger
T_5	stored fluid temperature
F_1	radiation fraction absorbed in the plate
F_2	radiation fraction absorbed in the cover
h_1	convective coefficient between plate and cover
h_2	convective coefficient from cover to ambient air
σ_1, σ_2	emissivity plate-cover and cover-air
k_1	transmittance plate-fluid
k_4	transmittance heat-exchanger-thermal storage
k_5	transmittance storage-ambient air
p_3	heat transmission perimeter from plate to fluid
p_4	heat transmission perimeter from heat exchanger to thermal storage
I	solar radiation
dz	thickness of the collector strap
U_{pg}	overall thermal factor (plate-glass)
U_{ga}	overall thermal factor (glass-ambient)
b, D	collector width, storage diameter
s_1, s_2	thickness of the plate and of the cover
$\rho_1\ \rho_2$	specific mass of the plate and cover
$\rho_3\ \rho_5$	primary,secondary fluid specific mass
γ_1	pump function of the primary circuit
γ_2	pump function of the secondary circuit
d_3, n_3	diameter and number of the fluid tubes
$\dot{M}, \dot{M}'$	primary, secondary fluid mass flow-rate
c_1, c_2	collector plate, glass specific heat
c_3, c_5	primary,secondary fluid specific heat

Simulation in Engineering Sciences
J. Burger and Y. Jarny (eds.)
Elsevier Science Publishers B.V. (North-Holland)

Simulation of the fuel elements, for their optimization

Ph.Dr.Eng.Dănilă Nicolae, Eng.Stan Nicolae, Ionescu Dan Cezar

The article attempts a simulation of the fuel element of the nuclear reactor in order to determine the limit operating values, in view of selecting the maximum admissible power which can be generated, taking into account the uncertainties influencing the nominal parameters.

1. INTRODUCTION

In the thermal design of the fuel elements of the nuclear reactors it is necessary that their generated power and geometrical sizes should be selected in such a way that :
- there should exist a very low probability of temperature overshoot and critical stresses ;
- the coefficients of safety selected from the technological point of view should not be very penalizing.

During the steady state of the core region the temperature distribution can be calculated for nominal or overpower conditions.
The distribution and the maximum values of the fuel temperatures are often different from the expected ones because of the various uncertainties (hydraulic, thermal, manufacturing tolerances, accuracy of the calculation models, the operating conditions, a.s.o.).
In order to determine the limit operating values of the fuel elements, in view of selecting the maximum admissible power which can be generated, it is necessary to calculate the temperature field in the fuel under "hot spot" conditions characterized by the consideration of the uncertainties influencing all nominal parameters.
The article attempts a simulation of the fuel elements in view of evaluating the temperatures and the risks of overshooting the critical temperatures, taking into account the uncertainties influencing the nominal parameters, as a function of the power level selected for the operation of the nuclear reactor.

2. INFLUENCE OF UNCERTAINTIES ON THE TEMPERATURE FIELD

The uncertainties influencing the determination of the temperature field in the fuel can be divided into /1/:
- manufacturing uncertainties (tolerances);
- uncertainties of modelling (temperature and neutron calculation) and of the knowledge of the physical parameters (thermal conductivity,coefficient of heat exchange, a.s.o.) ;
- operation uncertainties.

Every uncertainty is characterized by its own law of distribution and standard deviation from its nominal value.
These uncertainties can appear at different levels in the calculation of the temperatures field :
1. oversall uncertainties common to all fuel elements within the core region considered;
1.1. region uncertainties (density, enrichment, a.s.o.);
1.2. uncertainties of the channel (the path among the fuel elements, a.s.o.);
1.3. uncertainties of the unit (calibration of the inlet opening of the cooling agent);
1.4. uncertainties of the core region (coefficients of the heat transfer).
2. local uncertainties which are random functions along the axis of the technological channel and which are called the "specific standard deviation"

Table 1 presents the main uncertainties which influence the determina - tion of the temperature field in the fuel element.
The temperature in the centre of the fuel element is given by the relation:

$$t_c = t_f + \Delta t_{ft} + \Delta t_t + \Delta t_g + \Delta t_c = t_f + \sum_{i=1}^{4} \Delta t_i = t_f + \Delta t_{max}.$$

or

$$t_c = t_f + q_l \left[\frac{1}{\pi d_e \alpha} + \frac{d_e - d_i}{\pi \lambda_t (d_e + d_i)} + \frac{1}{\pi \alpha_g d_i} + \frac{1}{4 \pi \lambda_c} \right] \quad (2)$$

Taking into consideration the uncer - tainties influencing the determination of t_c, we can consider it a random variable. If we term x_1 the parameters on which t_c depends, then it follows :

$$t_c = f(x_1, x_2, \ldots x_i, \ldots x_n) \quad (3)$$

As t_c is a linear function of parameters x_1, the average value of the random variable t_c will be :

$$\mu\{t_c\} = f\left[\mu\{x_1\}, \ldots, \mu\{x_n\}\right] = t_{c\ nom.} \quad (4)$$

and the standard deviation will be calculated by means of the relation :

$$V\{t_c\} = \sigma_{t_c}^2 = \left(\frac{\partial f}{\partial x_1}\right)_\mu^2 \sigma_{x_1}^2 + \ldots + \left(\frac{\partial f}{\partial x_n}\right)^2 \sigma_{x_n}^2 \quad (5)$$

We apply (5) in Table 2 which presents the calculation formulas for the standard deviation of the main uncertainties characterizing the temperature field in the fuel and the shell.
In order to define the specific standard deviations, the function g (r) will be considered space variable. If we want to make a statistical examination of the average volumetric value of g (r) on the volume $_i$ then :

$$\mu\{g_i\} = \frac{1}{\Sigma_i} \int_{\Sigma_i} g(\vec{r})\, dV \quad (6)$$

the average and the variance of /2/ are defined by :

$$\mu\{g_i\} = \mu\{g\} \quad V\{g_i\} = V\{g\} / \Sigma_i \quad (7)$$

and in a analogue way the values on a surface or a length.
The combined effect of the direct uncertainties which have a unitary probability of occurence determines the probable temperature in the centre of the fuel element :

$$t_{c\ prob} = \mu\{t_f\} + \sum_{i=1}^{4} a_i \mu\{\Delta t_{i\ max}\} \quad (8)$$

By combining the statistical uncertainties we can obtain the maximum temperature as a function of the selected confidence :

$$t_{c\ max.} = t_{c\ prob.} + \lambda \sqrt{V\{t_c\}} \quad (9)$$

For the design calculations it is necessary to define a factor of temperature safety $F_{ST} = F_{ST}$ (degree of temperature safety) which has to take into account the uncertainties influencing the temperature design of the core region, so that the temperature value :

$$t = \mu\{t_f\} + F_{ST} \cdot \mu\{\Delta t_{max.}\} \quad (1o)$$

should not be exceeded in any spot of the considered component (shell or fuel). Starting from this definition, if is the safety degree imposed in the temperature design, $t_{crit.}$ is the value of the critical temperature which must not be exceeded, then the maximum temperature difference $t_{max.}$ can be calculation by means of the relation:

$$\mu\{\Delta t_{max.}\} = \frac{t_{crit.} - \mu\{t_f\}}{F_{ST}(\lambda)} \quad (11)$$

Having $\mu\{\Delta t_{max.}\}$ determined from the relation (2) we can also determine the linear power value q_e for which the safety conditions of the core region are satisfied.

3. SIMULATION OF THE FUEL ELEMENT IN VIEW OF DETERMINING THE TEMPERATURE FIELD

The numerical simulation of the fuel element was performed by means of the interactive system CD-4oo shown in Figure 1.
The operating system utilized was RSX-11M which is a system primarily used for the development and running of programmes of scientific importance. By means of this interactive system, the values used in the optimization of the maximum linear power could be easily varied.

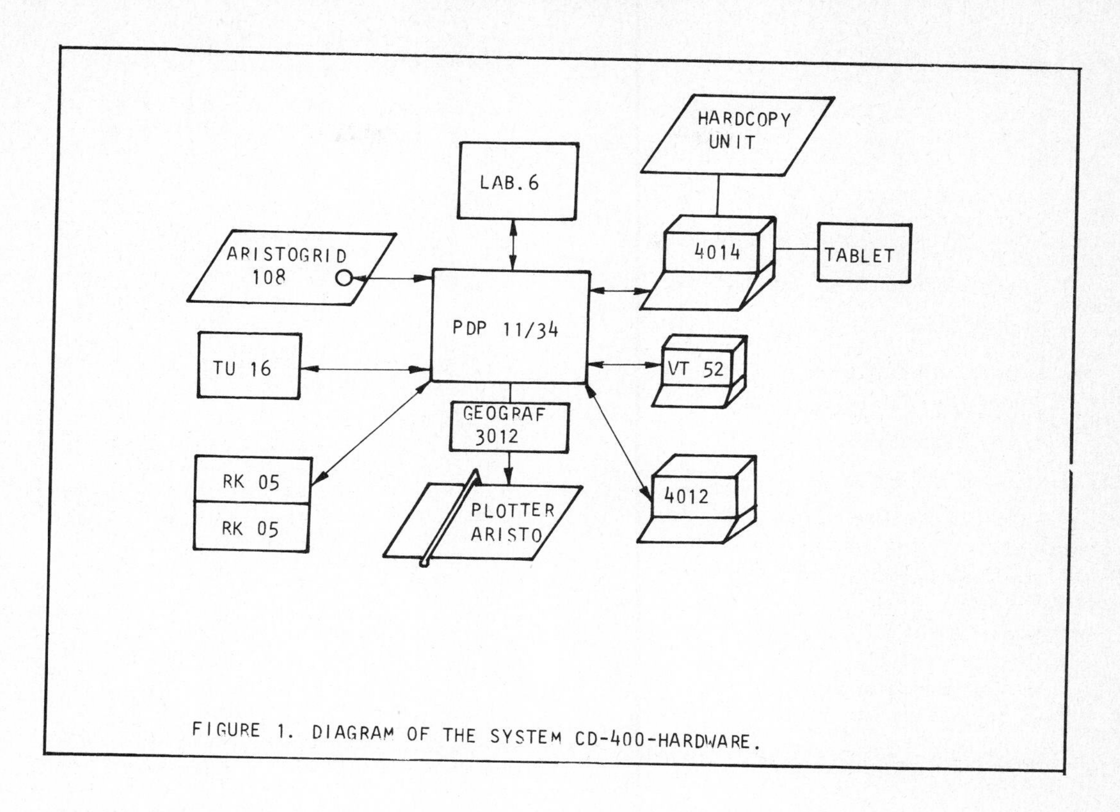

FIGURE 1. DIAGRAM OF THE SYSTEM CD-400-HARDWARE.

4. RESULTS OBTAINED

The obove-presented model was used in the determination of the limit values of the operating power of the fuel elements.
Figure 2 presents the variation of the temperature fields in a fuel element during a power variation as well as the probable temperature and the maximum temperature /3/.
Figure 3 shows the variation of maximum temperatures in the fuel as a function of the probability of occurence for the nominal power and for 1.o5 of the nominal power.
Figure 4 indicates the variation of the maximum and probable temperatures in the centre of the fuel element as a function of the linear power for different values of the confidence λ , enabling the determination of the power value for which, with a certain probability, the critical temperature is not overshooted.
Figures 5 and 6 show the influences of three of the main uncertainties on the factor of temperature safety and the maximum temperature in the fuel.

5. CONCLUSIONS

The present paper makes a description of a simulation method of the fuel elements in view of a statistical determination of temperatures in the centre of the fuel used for :

- defining the limit operating conditions of the fuel elements;
- straining efforts towards the determination of the optimum operating power by studying the influence of every parameter (uncertainty) on temperature, therefore on geometry and power.

NOTATION

t_c - temperature in the centre of the fuel ;
t_f - temperature of the fluid ;
Δt_{ft} - temperature drop between fluid and can;
Δt_t - temperature drop in the can ;
Δt_g - temperature drop in the gap ;
Δt_c - temperature drop in the fuel;
q_l - linear power ;
α - coefficient of convection ;
d_e - outer diameter ;
d_i - inner diameter ;
λ_t - thermal conductivity of the can
λ_c - thermal conductivity of the fuel ;
α_g - gap conductivity ;
$\mu\{x_i\}$ - average value of the variable x x_i ;
σ^s - specific standard deviation.

REFERENCES

1. RUSSO,S. - "Traitement statistique de la thérrmique du combustible des reacteurs rapides"
2. DANILA,N., STAN,N., IONESCU, D.C. - "Modèles stochastiques pour déterminer le champ de témperature dans les élements combustibles des reacteur nucleaires" Proceeding International AMSE Conference, Paris, 1982
3. STAN,N. - "Contribuţii la studiul proceselor termice tranzitorii în reactoare nucleare". Ph.D.Thesys, Institutul Politehnic Bucureşti, 1982

Table 1

Statistical uncertainties	Hot spot in the fuel	Hot spot in the can	Hot channel
Density	x	x	x
Enrichment	x	x	x
Inner diameter of the can	x	x	x
Outer diameter of the can	x	x	x
Thickness of the can	x	x	
Asymmetry of enrichment	x		
Asymmetry of density	x		
Eccentricity pellet-can	x	x	x
Lattice path	x	x	x
Core length	x	x	x
Calibration of diaphragms	x	x	x
Neutron flow	x	x	x
Measurement of power	x	x	x
Inlet temperature	x	x	x
Critical temperature of the can		x	
Critical temperature of the fuel	x		
Shell conductivity	x	x	
Fuel conductivity	x		
Coefficient of convection	x	x	
Coefficient of heat exchange in the gap	x		
Specific heat of natrium	x	x	x

Table 2

Parameter	Relative standard density	σ_j^s, Δt C°	
		Fuel	Shell
Density	σ_γ^s	$\sigma_\gamma^s \cdot \Delta t_{fc}$	$\sigma_\gamma^s \Delta t_{ft} \frac{1}{\sqrt{l_s}}$
Asymmetry of density	$\sigma_{\gamma a}^s$	$\sigma_{\gamma a}^s \cdot \Delta t_{fc}$	-
Enrichment	$\sigma_{\hat{\imath}}^s$	$\sigma_{\hat{\imath}}^s \cdot \Delta t_{fc}$	$\sigma_{\hat{\imath}}^s \cdot \Delta t_{ft} \frac{1}{\sqrt{l_s}}$
Asymmetry of enrichment	$\sigma_{\hat{\imath}a}^s$	$\sigma_{\hat{\imath}a} \cdot \Delta t_{fc}$	-
Axial flow	$\sigma_{\phi a}^s$	$\sigma_{\phi a}^s \cdot \Delta t_{fc}$	$\sigma_{\phi a}^s \cdot \Delta t_{ft}$
Eccentricity fuel can	σ_e^s	-	$\sigma_e^s \cdot \Delta t_{ft}$
Coefficient of convection	σ_α	$\frac{q_l}{\pi d_e \alpha^2} \sigma_\alpha \cdot \Delta t_{fc}$	$\frac{q_l}{\pi d_e \alpha^2} \sigma_\alpha \cdot \Delta t_{ft}$
Thickness of the can	σ_t	$\frac{2q_l}{\pi \lambda_t (d_e+d_i)} \sigma_t \cdot \Delta t_{fc}$	$\frac{2q_l \cdot \sigma_t}{\pi \lambda_t (d_e+d_i)} \cdot \Delta t_{ft}$
Fuel conductivity	$\sigma_{\lambda c}$	$\frac{q_l}{4\pi \lambda_c^2} \sigma_{\lambda c} \cdot \Delta t_{fc}$	-
Gap conductivity	σ_g	$\frac{q_l}{\pi \alpha_g^2 d_i} \sigma_g \cdot \Delta t_{fc}$	-

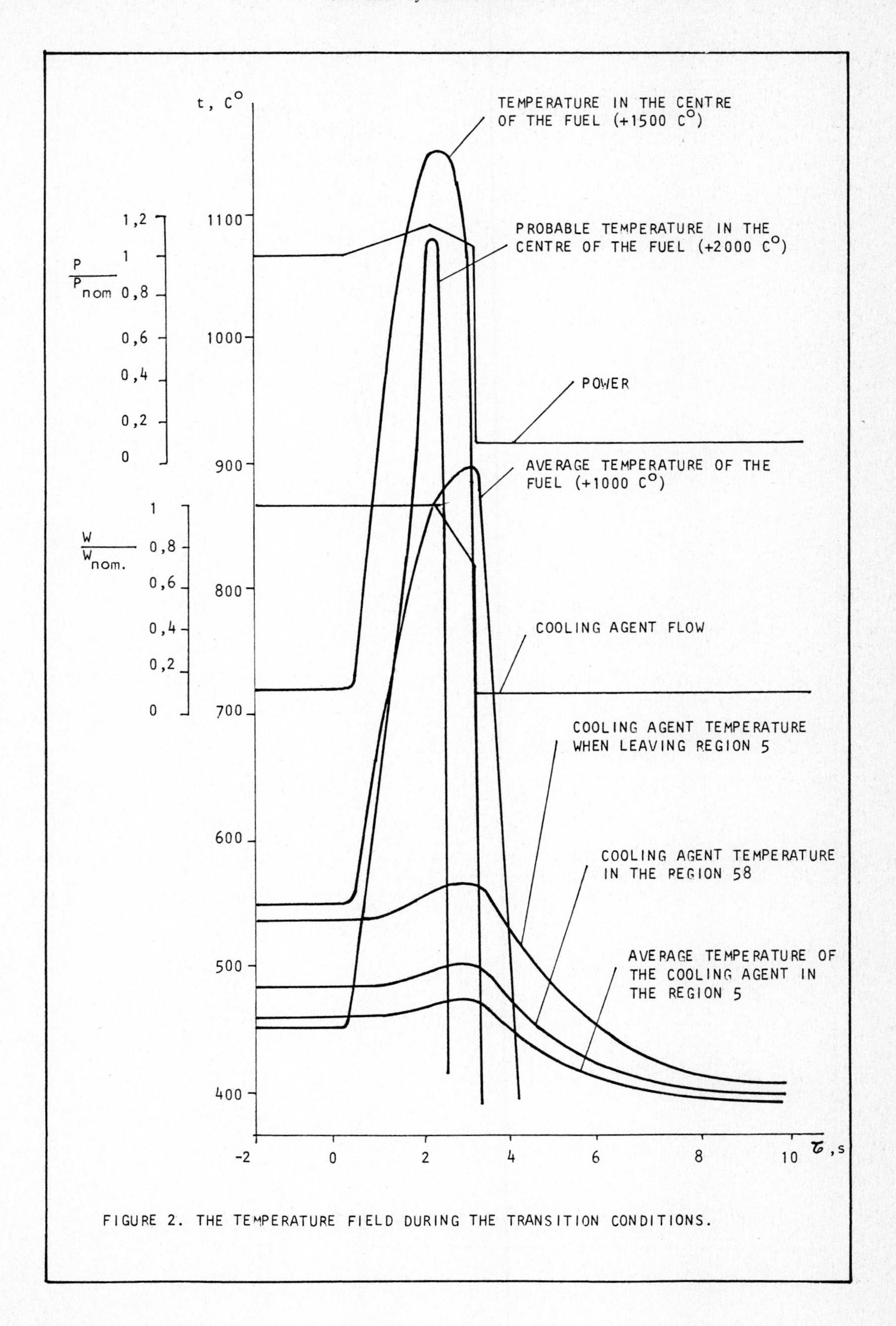

FIGURE 2. THE TEMPERATURE FIELD DURING THE TRANSITION CONDITIONS.

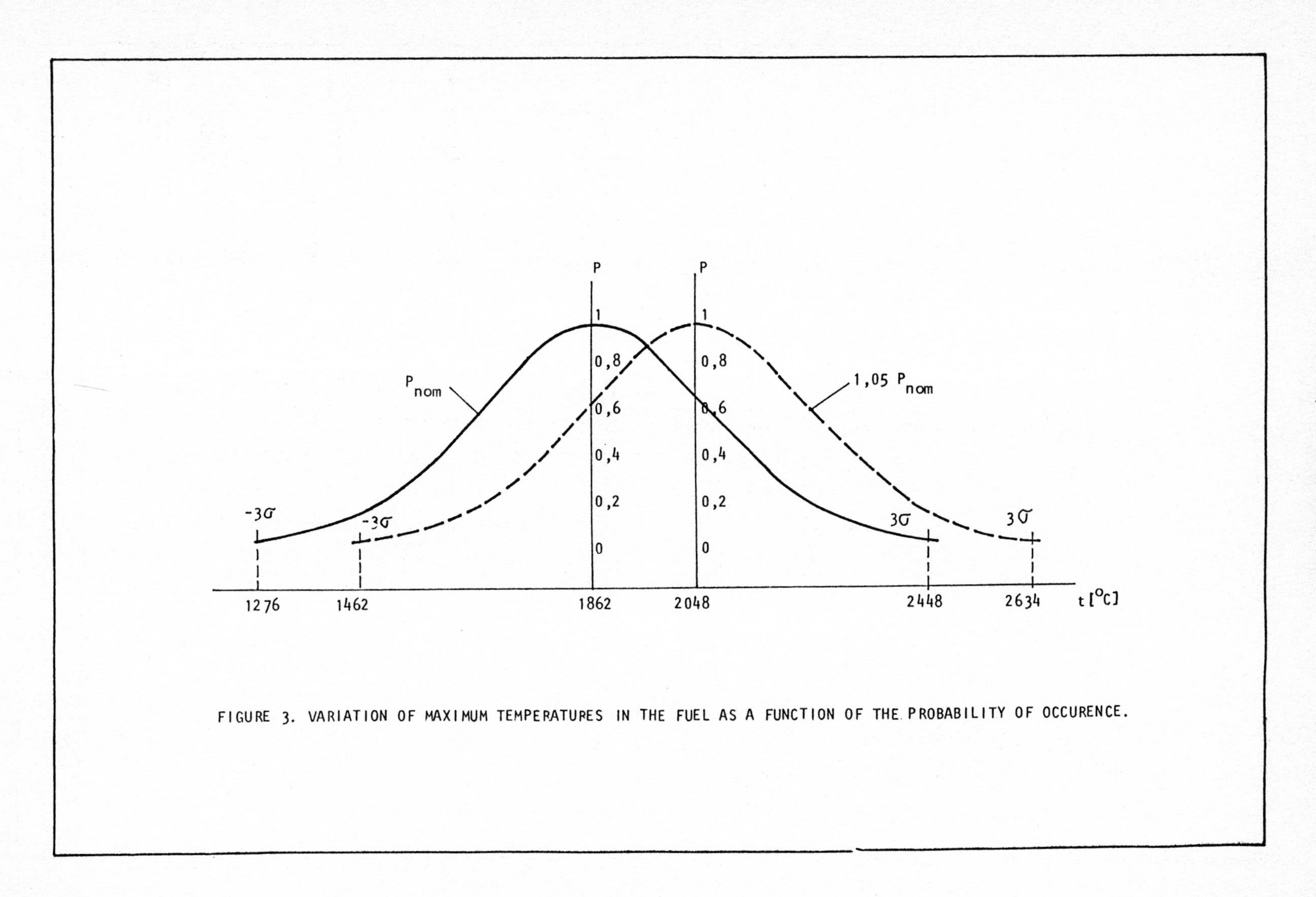

FIGURE 3. VARIATION OF MAXIMUM TEMPERATURES IN THE FUEL AS A FUNCTION OF THE PROBABILITY OF OCCURENCE.

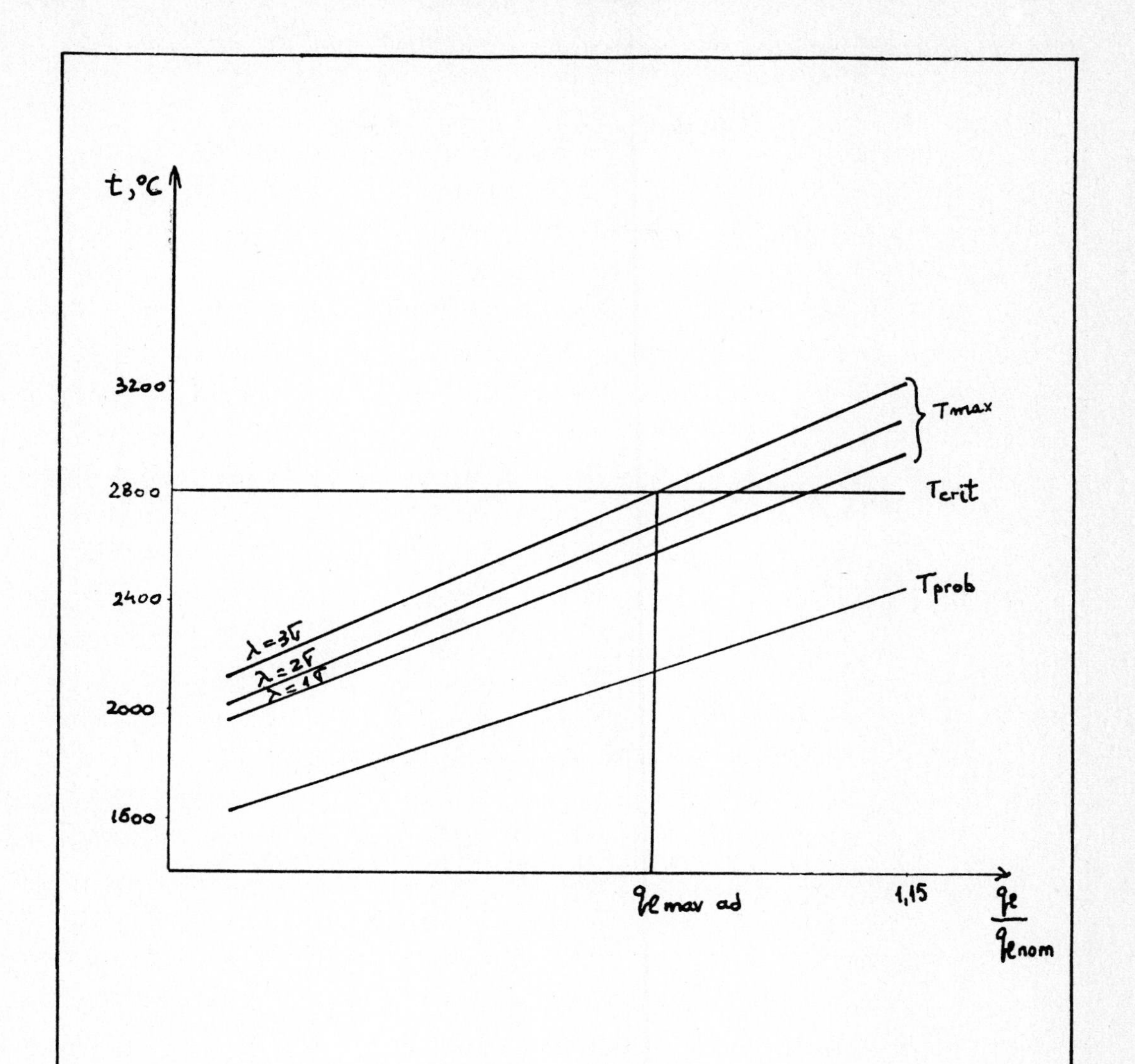

Figure nr. 4 Variations of temperatures vs linear power

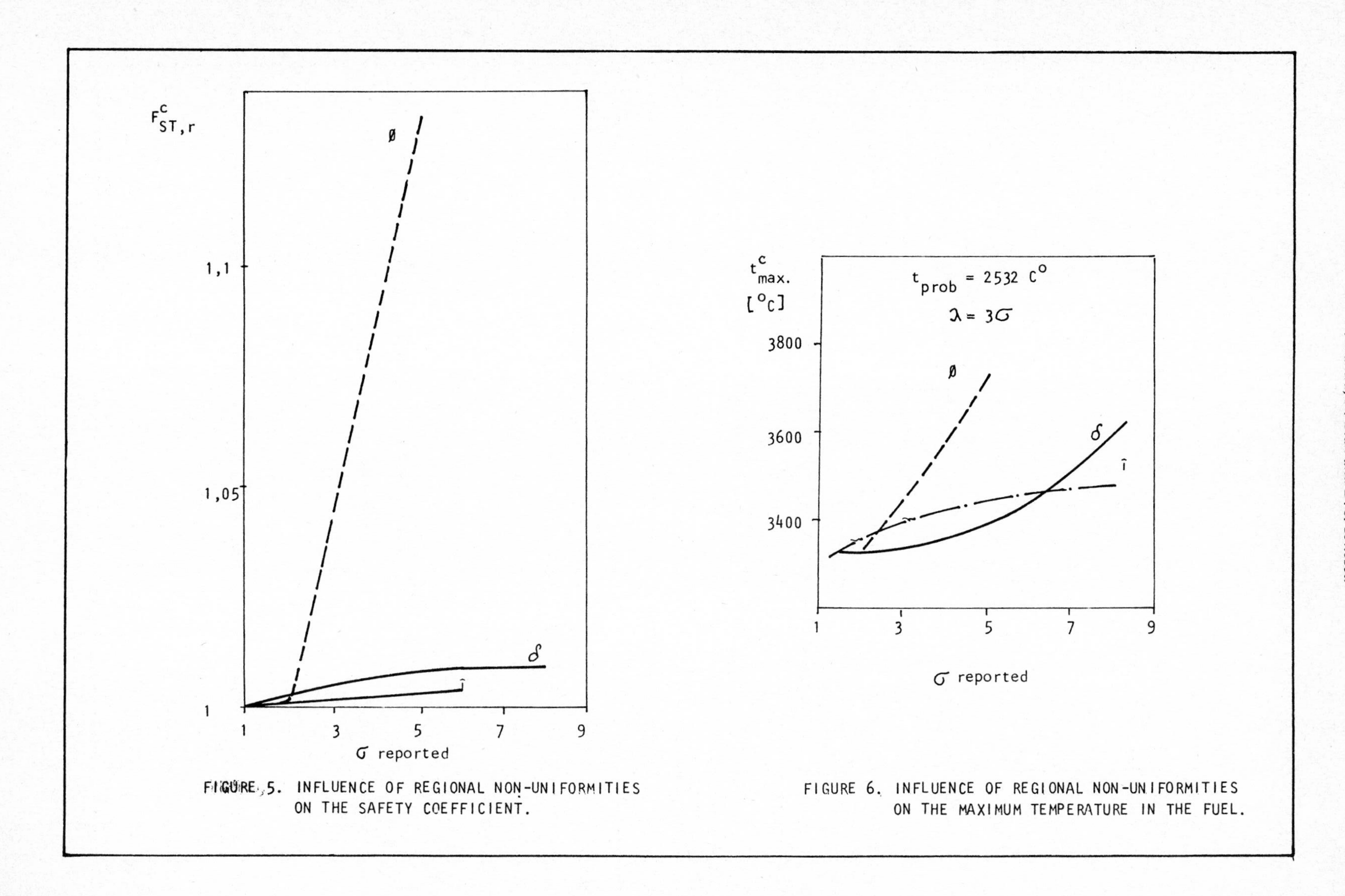

FIGURE 5. INFLUENCE OF REGIONAL NON-UNIFORMITIES ON THE SAFETY COEFFICIENT.

FIGURE 6. INFLUENCE OF REGIONAL NON-UNIFORMITIES ON THE MAXIMUM TEMPERATURE IN THE FUEL.

Simulation in Engineering Sciences
J. Burger and Y. Jarny (eds.)
Elsevier Science Publishers B.V. (North-Holland)

OPTIMIZATION OF OPERATING POLICIES IN HYDRO ELECTRIC POWER SYSTEMS

H. Habibollahzadeh & J. Bubenko
Energy Systems Laboratory,
Royal Institute of Technology,
S-100 44 Stockholm, Sweden

D. Sjelvgren
Swedish State Power Board
S-162 87 Vällingby,
Sweden

N. Andersson & G. Larsson
Krångede AB
S-103 93 Stockholm,
Sweden

This paper presents the development of a solution method that reforms the problem into a network formulation and employs the Network Flow Algorithm. The network flow algorithm is modified to incorporate hydro reservoir head variation and piecewise linear nature of cost function. Dantzig-Wolfe Decomposition Method is employed for the additional constraints with non-network structure.

1. INTRODUCTION

This paper contains the result of the research work done at the Energy Systems Laboratory of the Royal Institute of Technology in Stockholm on the development and computer programming of a new effective method for optimal weekly hydro scheduling. The optimization model has a nonlinear function which is well approximated as a piecewise linear function, linear constraints with network structure, and linear constraints that do not exhibit this nature. A combination of network programming and Dantzig-Wolfe decomposition method has been used in this study. The computation time of the computer program developed for this method is very good in comparison with other methods.

2. NETWORK FORMULATION

The dynamics of hydro reservoir for the system considered in this study are described by the following difference equation,

$$x_i(k+1) - x_i(k) + u_i(k) - u_{i-1}(k) = b_i(k)$$

$$\underline{x_i} \le x_i(k) \le \overline{x_i} \qquad (1)$$

$$\underline{u_i} \le u_i(k) \le \overline{u_i}$$

where

$x_i(k)$ = the content of the i:th reservoir at the beginning of period k

$u_i(k)$ = the discharge from the i:th reservoir during period k

$b_i(k)$ = the natural inflow to the i:th reservoir during period k

(i-1) represents the upstream plant

The following assumptions are implicit in the formulation:

(i) the traveling time of water flowing from one reservoir to the next is less than the one hour time step of the model,

(ii) no reservoir has more than one reservoir directly down- or upstream from it, and

(iii) no spillage is allowed at the plants.

The topological structure of the system is shown in Figure 1. The final destination of the water is a sea which is also included, because it will be used in the solution method later.

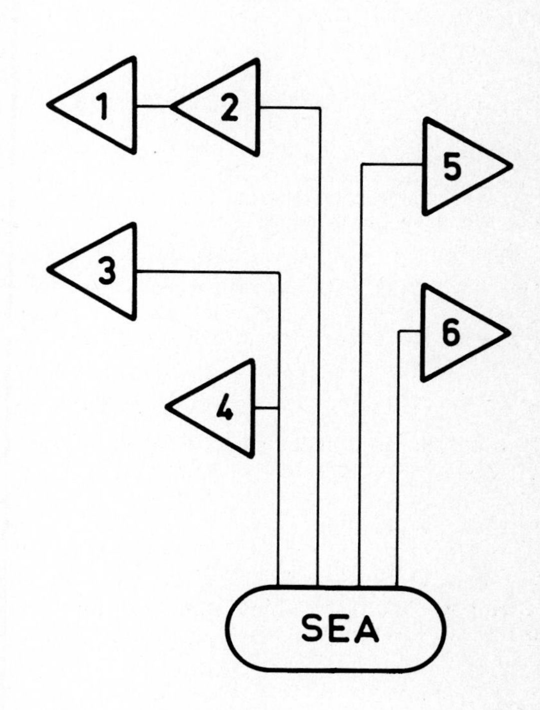

Figure 1 : A six reservoir system

The power system output for a hydro station, station i, at constant head during hour k is,

$$P_i(k) = \sum_{j=1}^{K} n_{ij} \, u_{ij}(k) \tag{2}$$

where this equation represents the piecewise approximation of a hydro production curve with K line segments, Figure 2.

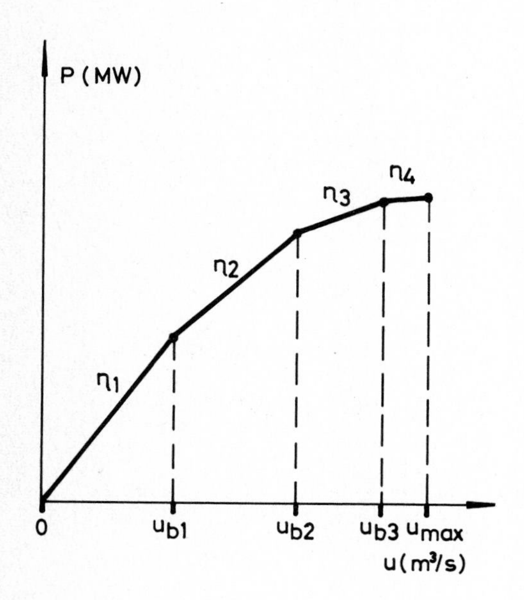

Figure 2 : Hydro production characteristics

n_{ij}'s are the slops of different segments ($n_{i1} > n_{i2} > \ldots > n_{ik}$) and $u_{ij}(k)$'s are elements of $u_i(k)$, each corresponding to a line segment in the hydro production curve.

$$u_i(k) = \sum_{j=1}^{K} u_{ij}(k) \tag{3}$$

The energy production for a hydro station at hour k is then

$$e_i(k) = T \, P_i(k) \tag{4}$$

The total energy production at hour k, e(k), is the sum of energy production at each plant

$$e(k) = \sum_{i=1}^{M} e_i(k) \tag{5}$$

where M is the total number of hydro plants in the system. The hydro plants benefit function can now be written as

$$f = \sum_{i=1}^{N} e(k) \, c(k) \tag{6}$$

where N is the number of time intervals in the optimization period and c(k) is the marginal benefit at hour k.

Letting [x] be the vector of all state variables, [b] a vector of the natural inflows to the reservoirs, [D] the vector of unit benefits from each state variable, the system equations (1) through (6) can be written as

$$\max \quad [D]^t[x] \tag{7}$$

$$\text{subject to} \quad [A]\,[x] = [b] \tag{8}$$

$$[\underline{x}] \leq [x] \leq [\bar{x}] \tag{9}$$

The matrix [A] is the node-arc incidence matrix of a network. It must be noticed that equations (7)-(9) are for a constant head and have a network programming structure.

For simplicity of explanation, we would consider the first two cascade plants in Figure 1. The expanded topological structure of this plant over the period of optimization, one week (168 hours), is drawn in Figure 3. This graph has one node for each reservoir at a specific hour and one node for the final destination of the water, the sea. The arcs either carry the water in the reservoirs from one hour to the next or carry the water flows from one reservoir to the next. The [A] matrix for these two reservoir systems is the node-arc incidence matrix of its expanded topological structure graph, Figure 3. This matrix is an mxn matrix, where m and n are the number of nodes and arcs respectively, and its elements are as follows,

$$A_{ij} = \begin{cases} +1 & \text{if arc j is incident on node i and directed away} \\ -1 & \text{if arc j is incident on node i and directed towards} \\ 0 & \text{if arc j and node i are not incident} \end{cases}$$

Each column of matrix [A] has exactly two non-zero entries, one being a + 1 and the other -1. So matrix [A] does not have a full rank and its rank is one less than the number of its rows. This is the special feature of a coefficient matrix in a linear programming problem.

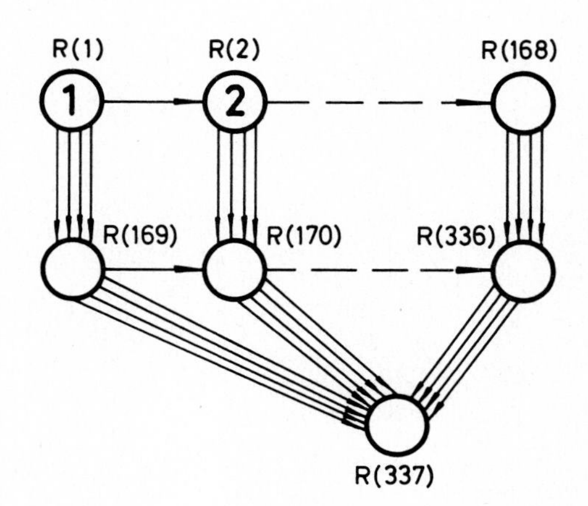

Figure 3 : Expanded topological structure of two cascade plants over a period of 168 h

3. SIMPLEX METHOD ON A GRAPH

In this section we review the simplex method on a graph for solving network programs such as equations (7)-(9). This method is of a great importance as it completely eliminates the need for carrying and updating the basis inverse. In fact, the method can be performed directly on a network diagram such as Figure 3. The method is very efficient and about one hundred times faster than general algorithms.

The constraint matrix [A] in a network program does not have a full rank. The network program can be reformulated by adding a linearly independent column (i.e. variable) having an upper-bound of zero, so that the constraint matrix has a full rank. The network program can be written as follows,

$$\min \quad [c]^t[x] \tag{10}$$

$$\text{subject to} \quad [A][x] + a[e^r] = [b] \tag{11}$$

$$0 \leq [x] \leq [u] \tag{12}$$

$$0 \leq a \leq 0 \tag{13}$$

Column e^r is a new variable (new arc) which is added at node r (in our case the node is corresponding to sea) and is incident only on this node. Node r is called the root node and the graph corresponding to equations (10)-(13) is called rooted graph.

The matrix $[A:e^r]$ can be partitioned as follows,

$$[A:e^r] = [B:N]$$

where matrix [B] is a nonsingular matrix and a basis of matrix $[A:e^r]$. In network programs, matrix [B] is triangular or can be made triangular by proper row or column interchanges. The arcs of the rooted graph corresponding to matrix [B] are the basic arcs and generated a spanning tree for the graph. The arcs corresponding to matrix [N]are the non-basic arcs. The steps of the simplex method on a graph are described in the following algorithm.

ALG. 1 Simplex method on a graph

Step 1. Initialization

Let $[x^B:x^N]$ be a basic feasible solution with the basis tree T_B. The quantities of dual variables or node potentials are given by $[\pi] = [c^B][B^{-1}]$. This is the solution to a linear system of linear equations $[\pi][B] = [c^B]$. Since B is triangular, $[\pi]$ may be obtained by solving the last component first and backward substitution to iteratively obtains all components.

Step 2. Pricing

In this step, we search for non-basic arcs with negative reduced costs to enter the basis. Assume that the non-basic arc j has the following characteristics:

$\pi_{S(j)}$ = potential of starting node of arc j

$\pi_{E(j)}$ = potential of end node of arc j

c_j = unit cost along arc j

Then arc j is a candidate to enter the basis if,

1 - $\pi_{S(j)} - \pi_{E(j)} - c_j > 0$ and arc j is at the lower-bound

2 - $\pi_{S(j)} - \pi_{E(j)} - c_j < 0$ and arc j is at the upper-bound

The set of all such non-basic arcs is called the candidate set. If this set is empty, then we terminate with $[x^B:x^N]$ an optimum; otherwise the non-basic arc, k, with the largest reduced cost $(\pi_{S(k)} - \pi_{E(k)} - c_k)$ is the best candidate to enter the basis tree T_B.

Step 3. Ratio test

If the best candidate arc is added to the existing basis tree, a cycle is generated, say C. The flow of the candidate arc is increased by Δ if it is at the lower-bound and decreased by Δ if it is at upper-bound. Consequently, the flow of the basic arc in cycle C will increase by Δ if they are in the same direction as Δ and decrease by Δ otherwise. Let S_1 be the set of basic arcs in cycle C that are decreasing, then find,

$$\Delta_1 = \min_{i \in S_1} \{x_i, \infty\}$$

Let S_2 be the set of basic arcs in cycle C that are increasing

$$\Delta_2 = \min_{i \in S_2} \{u_i - x_i, \infty\}$$

Then the maximum amount of increase in Δ, before an arc in cycle C assumes one of its bounds, can be found from Δ_1, Δ_2 and upper-bound of candidate arc (u_k).

$$\Delta\max = \min \{\Delta_1, \Delta_2, u_k\}$$

The arc assuming one of its limits at Δmax is called the blocking variable.

Step 4. Update flow

The flow of the arcs in cycle C is adjusted by Δmax. If the blocking variable is the non-basic candidate we proceed with step 2.

Step 5. Update tree and duals

When the blocking variable is a basic arc, it is replaced by the entering non-basic arc and we return to step 2.

4. MODIFIED PRIMAL SIMPLEX METHOD ON A GRAPH

In the primal simplex method on a graph, the basis tree changes from one iteration to the next. Thus, in the hydro system model equations (7)-(9)

this will result in head variation and consequently there will be a new cost coefficient vector in each iteration. The problem would actually be as follows,

$$\begin{array}{ll} \min & [c(h)]^t[x] \\ \text{subject to} & [A][x] + a[e^r] = [b] \\ & 0 \leq [x] \leq [u] \\ & 0 \leq a \leq 0 \end{array}$$

Since the constraints are the same in all iterations, the spanning tree from one iteration will be feasible for the next iteration. Only the dual variables must be modified in accordance with the new cost coefficient vector. So the modified primal simplex method on a graph for the above case can be stated as follows:

ALG. 2 *Modified primal simplex method on a graph*

Step 1. Initialization

Let $[x^B:x^N]$ be a basic feasible solution with basis tree T_B.

Step 2. Cost coefficient vector and dual variables determination

Evaluate the cost coefficient vector c with respect to the new heads and correct the dual variables $[\pi]$ accordingly.

The remaining steps of this algorithm are the same as steps 2-5 in ALG. 1.

5. APPLICATION OF THE METHOD

A computer program has been developed for application of this method to hydro system short-term scheduling. For the initial basic feasible solution, an all initial start such as Figure 4 is used. Each arc of this tree corresponds to a node and connects it to the root node. The flow of each arc is equal to the absolute value of the requirement at the corresponding node, and it is directed away from this node if the requirement is negative and directed towards the node if the requirement is positive. The cost and upper-bounds of these arcs are set to infinity.

In pricing step, since the size of candidate set can be very large at early iterations, a maximum size is appointed to this set. Once a candidate set is generated, its elements are entered into the basis tree one by one with respect to their reduced cost. A new candidate set is generated, when there are no more elements with negative reduced cost in the existing candidate set. Labeling basis tree techniques are employed for ratio test and updating, which are the most efficient data structure for the physical context of the problem.

The program has been tested on Krångede hydro system. The computational time and memory requirements are excellent.

An interesting feature of the optimal tree in this case is that a major part of non-basic arcs represents the discharges at different hours. This means that the discharges are at lower-bound, upper-bound or one of the best operating points.

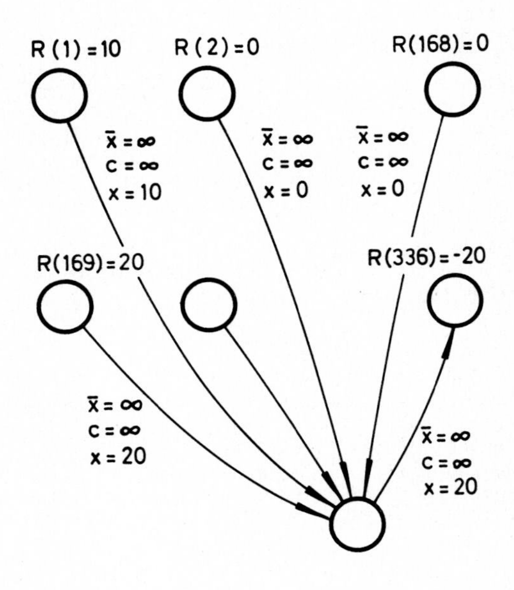

Figure 4 : An all artificial initial solution

6. SYSTEM CONSTRAINTS WITH NON-NETWORK STRUCTURE

There are usually other constraints in hydro system such as reserve and participation factor requirements that do not exhibit network structure. Each of these constraints is for a single interval. Solving the optimization problem with these constraints ignored, it is observed that these constraints are fulfilled for most of the time intervals. To comply with these constraints at the few hours that they are not met, Dantzig-Wolfe decomposition method can be used.

7. CONCLUSION

The network programming method used in this study is a very efficient method. The time requirement is very low in comparison with conventional methods. For example, the time required for the solution of the two cascade systems over 168 hours was 10 seconds on Vax-11/780 computer. The computation time for the Dantzig-Wolfe decomposition method is not comparable with network programming and it depends on the number of hours that the external constraints are not met. Since this number is usually very low in comparison with the total number of time intervals, the overall computation time is also very good in comparison with other methods.

8. REFERENCES

[1] Bakes, M.D., Solution of special linear programming problem with additional constraints, Operation research quarterly. Vol. 17, No. 4.

[2] Bubenko, J.A. and Waern, B.M., Short-range hydro scheduling, IEEE Conference 1972, paper C 72456-2.

[3] Chen, S. and Saigel, R., A primal algorithm for solving a capacitated network flow problem with additional linear constraints, Network. Vol. 7 (1977).

[4] Cunninghan, W.H., A network simplex method, Mathematical programming, 11-2 (1976).

[5] Dantzig, G.B. and Wolfe, P., The decomposition algorithm for linear programs, Econometrica. Vol. 29, No. 4 (1961).

[6] El-Hawary, M.E., Optimal economic operation of electric power systems (Academic Press, New York, 1979).

[7] Glover, F., Klingman, D. and Stutz, J., Augmented threaded index method for network optimization, Infor. Vol. 12, No. 3 (1974).

[8] Habibollahzadeh, H., Optimal short-term hydro scheduling, Energy Systems Laboratory, Royal Inst. of Techn., Stockholm (July 1982).

[9] Kennigton, J.L. and Halgan, R.V., Algorithms for network programming (John Wiley & Sons, New York, 1981).

[10] Lasdon, L.S., Optimization theory for large systems (Macmillan, London, 1970).

[11] Sjelvgren, D. and Dillon, T.S., Modeling and organization of production planning system for hydro-thermal regulation in the Swedish State Power Board, (PSCC VII, Lausanne, 1981).

[12] Sjelvgren, D. and Dillon, T.S., Seasonal planning of a hydro-thermal system based on the network flow concept, (PSCC VII, Lausanne, 1981).

[13] Turgeon, A., Optimal short-term hydro scheduling from the principal of progressive optimality, Hydro-Quebec, Canada (1981).

[14] Turgeon, A. Optimal short-term scheduling of hydro plants in series - A review, Hydro-Quebec, Canade (1981).

Janis A. Bubenko Sr (M'46 - SM'56 - LM'80) was born in Valmiera, Latvia, on October 1, 1911. He has been associated professor at Chalmers University of Technology, Gothenburg, Sweden, 1950-54 and the head of the computer division at the Swedish State Power Board, 1955-65. From 1965 he is professor at the Royal Institute of Technology, Stockholm, Sweden, heading the Energy Systems Laboratory.

Hooshang Habibollahzadeh was born in Ardebil, Iran, on September 15, 1951. He received his B.Sc. and M.Sc. degrees in electrical engineering from the University of Washington, Seattle, USA, in 1973 and 1974 respectively. He has been lecturer at the Iran University of Science and Technology from 1975 to 1980. He has also six years of experience in power system engineering field and design. Since 1980 he has been a doctoral student at the Royal Institute of Technology in Stockholm, Sweden.

Denis Sjelvgren was born in Karleby, Finland, on September 4, 1947. He received his M.Sc. and Ph.D. degrees at Chalmers University of Technology in 1972 and the Royal Institute of Technology in 1976 respectively. He has been working at the Swedish State Power Board since 1977. He is also a senior member of the Energy Systems Laboratory of the Royal Institute of Technology, Stockholm.

Nils L. Andersson was born in Ljungby, Sweden, on September 14, 1944. He graduated in electrical engineering from Chalmers University of Technology in 1969. In 1970 he joined the Krångede Power Company, where he became the head of the Department of operation in 1979.

O. Gunnar Larsson was born in Vänersborg, Sweden, on April 21, 1948. He graduated in electrical engineering from Chalmers University of Technology in 1972. In 1977 he joined the Krångede Power Company, where he works for the Department of operation.

VIII. MECHANICAL SYSTEMS

Simulation in Engineering Sciences
J. Burger and Y. Jarny (eds.)
Elsevier Science Publishers B.V. (North-Holland)

SIMULATION AND CONTROL OF A NONLINEAR ELECTROMAGNETIC SUSPENSION SYSTEM

P.K. Sinha and A.J. Hulme

Department of Engineering, University of Warwick, Coventry, CV4 7AL, England.

Abstract. This paper presents a brief simulation analysis of an electromagnetic suspension system with linear and nonlinear magnet force-distance characteristics. This analysis is then used to develop a real-time adaptive control algorithm for single-degree of freedom suspension system with nonlinear force characteristics. The algorithm is realised through a simulated microcomputer on a PRIME-550 digital computer. Effectiveness of this method of controlling the nonlinear system is demonstrated through computer simulation and some practical aspects of implementing this control scheme briefly discussed.

1. INTRODUCTION

The possibility of using controlled direct-current electromagnets to support large weights was demonstrated by Kemper in 1932. However, it was only in the mid-1960's, with the advent of high power solid-state devices, the full potential of electromagnetic suspension was realised. Because of the exciting possibility of overcoming mechanical friction, the potential of magnetic suspension in transportation is being explored in several countries [1,2]. This paper presents a brief account of controlling a single-point (single-degree of freedom) suspension system with particular reference to a novel method of reference-following adaptive control. Responses presented here were obtained by simulating a nonlinear d.c. suspension system and the reference model by using simulation language ACSL†; the adaptive control algorithm was realized through a floating point arithmetic subroutine. All of these were programmed on a PRIME-550 digital computer. Implementation of the adaptive algorithm on a Rockwell-6502 microcomputer will form a part of the experimental work now being undertaken by the authors. A computer-aided-design framework is being developed in this paper.

The paper first briefly examines the dynamics of a linearized suspension system (Section 2). The deterioration in transient performance due to square-law nonlinearity and the limits of linear feedback control are highlighted through simulation results in Section 3. In Section 4, a novel model-reference adaptive control method is described. The control algorithm is implemented by a floating point algorithm for adaptive state feedback control of the nonlinear suspension system. The improvement in force disturbance rejection that can be obtained through adaptive gains is illustrated through simulation results in Section 5. The final section of the paper discusses some of the key issues of implementing this adaptive algorithm on a microcomputer.

†Advanced Continuous Simulation Language.

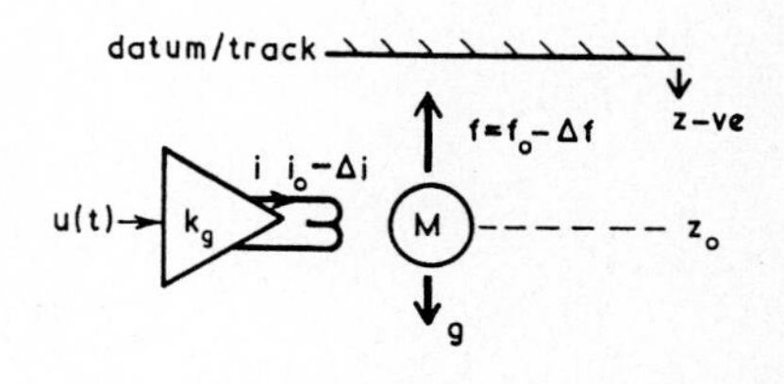

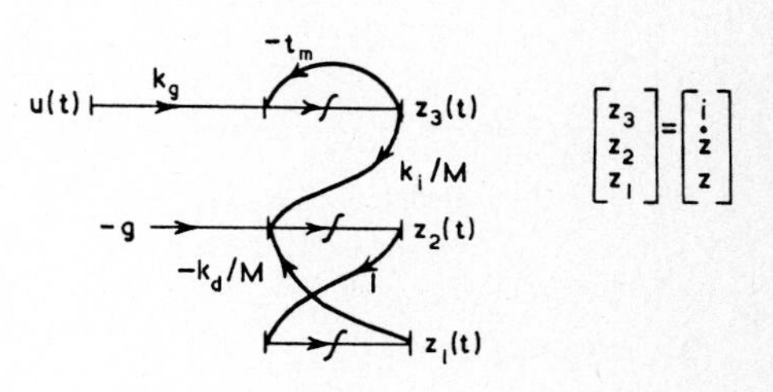

Figure 1. Single-point electromagnetic suspension system.

2. LINEAR SYSTEM: DYNAMICS AND CONTROL [3]

The open-loop dynamics of an electromagnetic suspension system can be easily derived by using the force-distance and force-current characteristics and Newton's second law. A small perturbation model is obtained on the assumption that at any nominal point (z_o, i_o) the electromagnet generates an attraction force $f_o = mg$ to provide a position of unstable equilibrium. Thus, if k_d and k_i are the slopes at (z_o, i_o), dynamics of the open-loop system may be represented by Figure 1, k_g and t_m being the gain and time-constant of power amplifier-magnet coil combination. It is apparent that stability cannot be achieved by increasing either k_i or k_d.

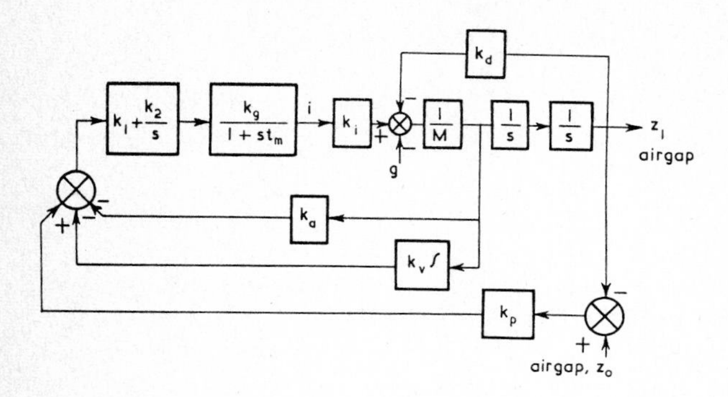

Figure 2. Closed-loop linearized system.

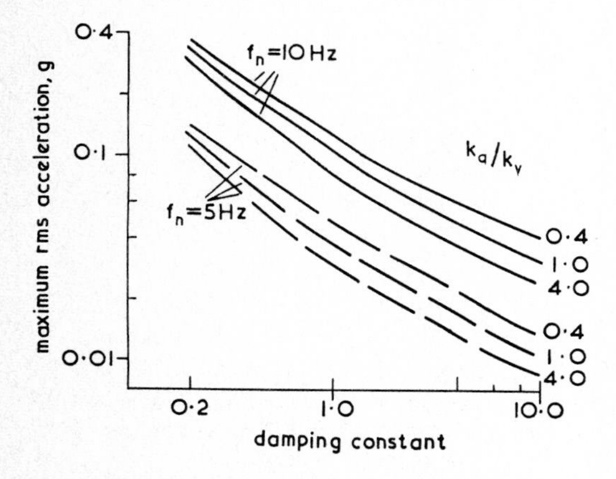

Figure 3. Effect of k_a and k_v on vertical acceleration.

Various methods of deriving feedback control laws to satisfy stability and several performance requirements are now available. The general configuration* is shown in Figure 2; position feedback gain k_p controls suspension stiffness, velocity feedback gain k_v controls the ride-quality through damping, and force (or acceleration) feedback gain k_a may be used to improve stability as well as ride quality (vertical acceleration levels). Influence of these parameters on system performance are shown in Figure 3 for a 20 cm U magnet (data in Section 10).

3. NONLINEAR SYSTEM: LINEAR FEEDBACK CONTROL

The assumptions of small perturbations and linearity in the preceding section are not strictly valid in general operational mode where the airgap variations may be as much as 100%.

*Time constants associated with various transducers are discounted in this analysis.

Apart from saturation effects, the non-linearities which have dominant influence on the stability of an electromagnetic suspension system are magnet force-distance characteristics, effect of eddy-current on moving magnets and hysteresis in magnet core. Of these non-nonlinearities, the force-distance nonlinearity is known to have most significant effect on suspension stability. Constraints imposed on linear feedback gains by force nonlinearity (Figure 4) are highlighted in computer

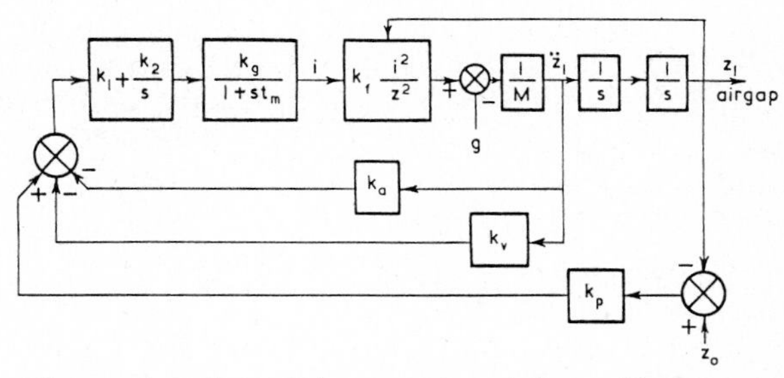

Figure 4. Suspension system with nonlinear force-distance characteristics and linear feedback law.

simulation responses in Figure 5. The general observation is that the introduction of non-linearities reduces the effective damping and dynamic stiffness; this is consistent with experimental results [4]. The effect of non-linearity on step responses are shown in Figure 5; the sharp contrast highlights the limitation of linear theory in the design of magnetic suspension systems.

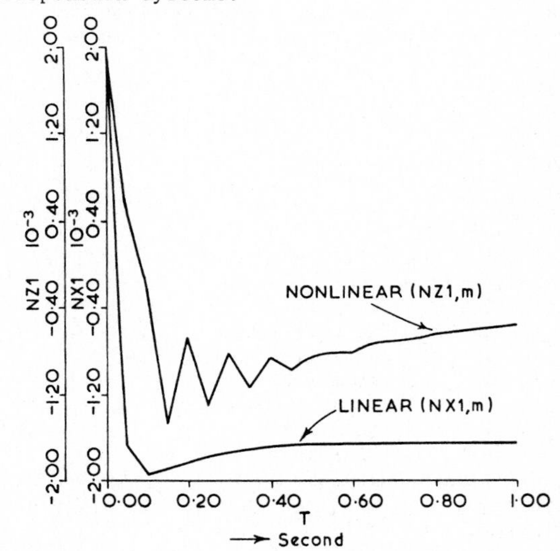

Figure 5. Step responses of the linear (Figure 2) and nonlinear (Figure 4) systems; k_a = 0.2, k_v = 270.0, k_p = 2.0.

The feedback gains in the linear and nonlinear models, needed to achieve similar degrees of transient stability is very much dependent on the intrinsic parameters of a particular system. Some of these are magnet/amplifier time-constant, values of k_i, k_d, k_f and the weight of the suspended mass. Formulation of a generalised method of modifying the linear feedback gains to match the performance of the experimental (nonlinear) system with that of a linear model is therefore not practicable. One plausible method of obtaining this similarity is to derive the feedback gains in the nonlinear system from a model-following adaptive control algorithm. Such a scheme has a significant practical advantage in that the adaptive algorithm can usually be expressed in terms of the parameters of a reasonably accurate* reference model. A mechanism of deriving adaptive feedback gains based on the method of stable maximum descent is described in the following section.

4. MODEL-REFERENCE ADAPTIVE CONTROL [5,6]

An outline derivation of the adaptive control method used to control the nonlinear suspension system is given below. The requirements in respect of desired system behaviour are embodied in a reference model. The same inputs are applied to both the system and the model, the difference between their respective outputs being used to derive the adaptive algorithm. The system (S) and the reference model (S_r) are described (with usual notations, $x \in X^n$, $x_r \in x_r^{\,n}$) by

$$S: \dot{x}(t) = Ax(t) + bu(t) + h_o \tag{1}$$

$$S_r: \dot{x}_r(t) = Ax_r(t) + b_r u(t) + h_r \tag{2}$$

From the above representations, an error state variable equation

$$\dot{e}(t) = A_r e(t) + Gx + h_o \tag{3}$$

may be derived, where $e = x_r - x$, $G = A_r - A$ and $h = h_r - h_o$.

The problem of adaptive control is regarded here as the minimization of a performance function e_M given by

$$e_M = \tfrac{1}{2} \, || E_1 e + E_2 \dot{e} ||^2 \tag{4}$$

* Many 'finer' details may be included in the (non necessarily linear) reference model. Relationship between the nature of the reference model and performance of the nonlinear system is currently under investigation.

where E_1 and E_2 are square matrices of order n. The rule of adjusting an element α of A, by using the method of stable maximum decent may then be derived as

$$\frac{d\alpha}{dt} = + \frac{\partial e_M}{\partial \alpha} \tag{5}$$

where the dynamic adjustment of α relies upon the sensitivity of e_M to α. Continuing equations 3, 4 and 5

$$\frac{d\alpha}{dt} = \left[E_1 e + E_2 \dot{e}\right]^t \left[E_1 \frac{\partial e}{\partial \alpha} + E_2 \frac{\partial \dot{e}}{\partial \alpha}\right]$$

$$= \left[E_1 e + E_2 \dot{e}\right]^t \left[E_1 \frac{\partial e}{\partial \alpha} + E_2 \{A_r \frac{\partial e}{\partial \alpha} + \frac{\partial A_r}{\partial \alpha} e + G \frac{\partial x}{\partial \alpha} + \frac{\partial G}{\partial \alpha} x + \frac{\partial h}{\partial \alpha}\}\right]$$

$$\simeq \left[E_1 e + E_2 \dot{e}\right]^t \left[E_1 \frac{\partial \alpha}{\partial \alpha} + E_2 \{A_r \frac{\partial e}{\partial \alpha} + \frac{\partial G}{\partial \alpha} x\}\right]$$

$$= \left[E_1 e + E_2 \dot{e}\right]^t \left[E_2 \frac{\partial G}{\partial \alpha} x\right] \tag{6}$$

where

$$E_1 = -E_2 A_r \tag{7}$$

Equation 6 represents the adaptive controller for any parameter α in A of the system in equation 1.

Similarly the controller for each disturbance correction can be derived as

$$\frac{dh_{ri}}{dt} = \left[E_1 e + E_2 \dot{e}\right]^t \left[E_2 \frac{\partial h}{\partial h_{oi}}\right] \tag{8}$$

Analytical investigation of stability with this method of adaptive control is not considered, but demonstrated through computer simulation results presented in the following section.

5. ADAPTIVE STATE FEEDBACK CONTROL

The method of adaptive scheme described above is used here to control the nonlinear system in Figure 4. The closed-loop linear system in Figure 2 is taken as the reference model. The synthesis objective here is to generate the feedback gains so that the nonlinear system follows the reference model. A schematic of the overall system is shown in Figure 6, where the effective feedback gains in the nonlinear system are adk_a, adk_v and adk_p. These, generated by using equation 6, are given by

$$adk_a = (e_1 + \frac{1}{\Delta}\alpha_1)(\frac{1}{\Delta}\frac{k_g a_g k_d}{m} Z_{ir}$$

$$+ (e_3 + \frac{1}{\Delta}\alpha_2)\ (\frac{1}{\Delta}\frac{k_i}{m} Z_{ir})$$

$$adk_v = (e_1 + \frac{1}{\Delta}\alpha_1)\ (\frac{1}{\Delta}\frac{k_g v_g k_d}{m} Z_{2r})$$

$$+ (e_3 + \frac{1}{\Delta}\alpha_2)\ (\frac{1}{\Delta}\frac{k_i}{m} Z_{2r})$$

$$adk_p = (e_1 + \frac{1}{\Delta}\alpha_1)\ (\frac{1}{\Delta}\frac{k_g p_g k_d}{m} Z_{3r})$$

$$+ (e_3 + \frac{1}{\Delta}\alpha_2)\ (\frac{1}{\Delta}\frac{k_i}{m} Z_{3r}) \qquad (9)$$

where

$$\Delta = (t_m + k_a a_g k_g)\frac{k_d}{m} + \frac{k_i}{m} k_g k_p p_g$$

$$\alpha_1 = \frac{k_d}{m}\dot{e}_1 - k_g k_p p_g\ \dot{e}_2 + k_g k_v v_g \frac{k_d}{m}\dot{e}_3$$

$$\alpha_2 = \frac{k_i}{m}\dot{e}_1 + (t_m + k_a a_g k_g)\dot{e}_2 + k_g k_v v_g \frac{k_i}{m} e_3$$

$$\left.\begin{array}{l} e_i = Z_i - Z_{ir} \\ \dot{e}_i = \dot{Z}_i - \ddot{Z}_{ir} \end{array}\right\} \quad i \in 1,\ 2,\ 3.$$

a_g, v_g and p_g being the gains of the corresponding sensors.

To examine the effectiveness of this adaptive control algorithm, the closed-loop system in Figure 6 was simulated using ACSL. Some of

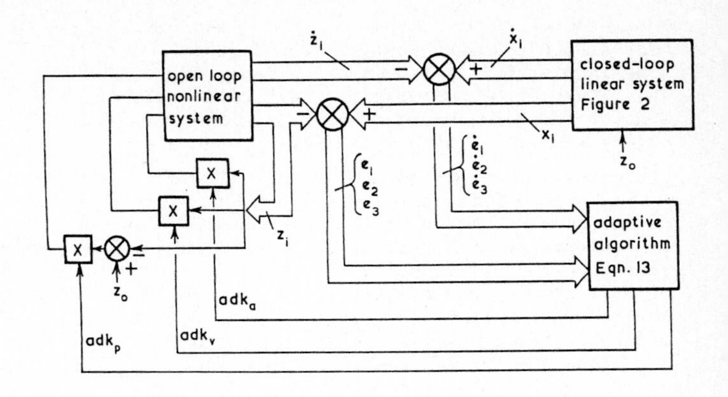

Figure 6. Model-reference adaptive control schematic

the responses obtained from this simulation study are shown in Figure 7. Although several detailed features are yet to be investigated (for example, the initial values of k_a, k_v, k_p in the adaptive algorithm has been observed to have significant influence on the transient

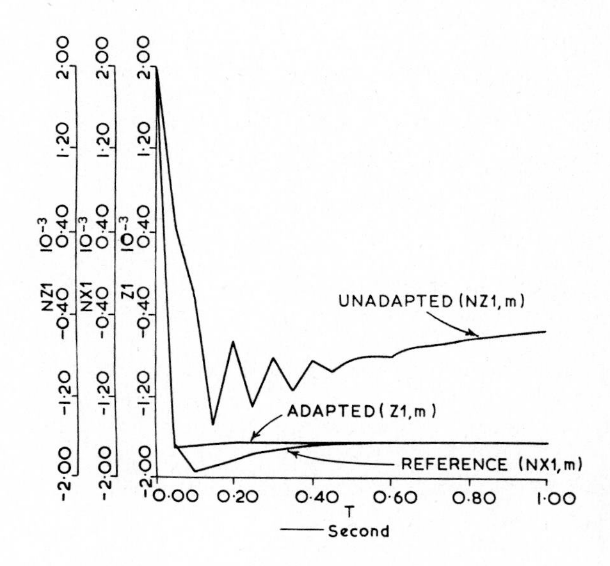

Figure 7. Step responses of the nonlinear system (unadapted from reference responses as in Figure 5).

response of the nonlinear system), the response in Figure 8 and other results obtained so far demonstrate the feasibility of the proposed control strategy.

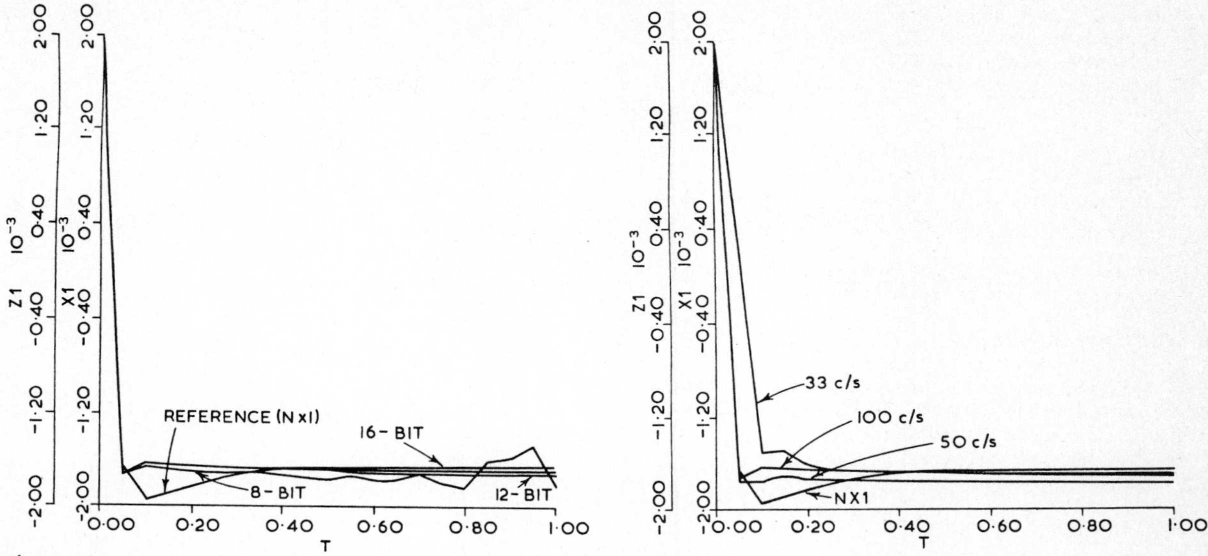

Figure 8. Effect of input sampling frequency (a) and adaptive algorithm word length (b) on transient response.

The following section examined some of the practical aspects of implementing the algorithm in equation 9 on a microcomputer.

6. MICROPROCESSOR SIMULATION

The adaptive algorithm in equation 9 was realized through a simulated microprocessor program on PRIME-550 using floating point arithmetic. This 'theoretical' study of a microprocessor-based adaptive control scheme is considered to be useful in identifying the effects of certain key parameters associated with the use of 'micros' in real-time control. Although this work is yet to be concluded, results obtained so far suggest that despite the fast transient response of the suspension system (Figure 8), adaptive technique is amenable to microprocessor control, though care is needed in studying the sampling frequency and the word length of the ADC and DAC; figures 8a and b highlight the effects of these parameters on system step responses. The effect of microprocessor computing time on system stability is currently being studied.

7. CONCLUDING COMMENTS

This paper has concentrated on presenting a novel method of controlling nonlinear electromagnetic suspension systems. The simulation results demonstrate the effectiveness of this scheme. This single-degree of freedom system is now being studied on an experimental system with a view to developing a near accurate digital model. Subject to satisfactory agreement between the theoretically predicted results and experimental responses, adaptive control of nonlinear magnetic suspension system will have significant influence in the control of multimagnet suspension systems [2]. Some recent results on this simulation project will be presented at the Conference.

8. ACKNOWLEDGEMENTS

This work forms part of a magnetic suspension research at the University of Warwick with financial and computing support from the UK Science Research Council.

9. REFERENCES

[1] Sinha, P.K., and Jayawant, B.V. 'Electromagnetic wheels', Electronics & Power, October, pp. 723-727 (1979).

[2] Sinha, P.K., Yu K.C. and Hulme, A.J. 'Computer-aided design of magnetically suspended vehicles', Submitted to Automatica.

[3] Jayawant, B.V., and Sinha, P.K. 'Dynamics and ride quality of a magnetically suspended low-speed vehicle', Automatica, Vol. 13, pp. 605-610 (1977).

[4] Jayawant, B.V., Sinha, P.K., Wheeler, A.R., and Willsher, J. 'Control and dynamics of magnetically suspended vehicles', Preprints 7th IFAC World Congress, Helsinki, Vol.3, pp. 1325-1332 (1978).

[5] Donaldson, D.D. 'The theory and stability of a model referenced parameter tracking technique for automatic control systems', Ph.D. thesis, University of California, (1961).

[6] Green, J.W., and Walker, P.H. 'Stable maximum descent; an improved method of reference model adaptive control', Measurement and Control, Vol.7, pp. 425-432 (1974).

[7] Kalman, R.E., and Bertram, J.E. 'Control System Analysis and design via the "Second Method" of Liapunov', ASME.J. Basic Engineering, pp. 371-393 (1960).

[8] Ogata, K. State Space Analysis of Control Systems, Prentice-Hall (1967).

10. SYSTEM PARAMETERS

20 cm U magnet design to operate at 3 mm airgap and has copper enamelled coils on each limb which are connected in series.

Weight = 8.75 kg

Pole length = 20 cm

Pole width = 0.95 cm

Number of turns = 280

Coil cross sectional area = 959 cm

Coil resistance at 20°C = 0.47 Ω

Time-constant = 85 ms

Transducer and amplifier

Accelerometer: Scheavitz EM inductive type; range: ± 1 g, output 2.3 v per 1 g. Velocity signal is derived by integrating the accelerometer output.

Position sensor: Hall-probe type; range: 0-10 mm: bandwidth: 1 kHz.

Power Amplifier: Switching (4 kHz) transistor amplifier; bandwidth: in excess of 250 Hz.

Simulation in Engineering Sciences
J. Burger and Y. Jarny (eds.)
Elsevier Science Publishers B.V. (North-Holland)

HYBRID SIMULATOR ASSISTED DESIGN OF A FEED-BACK CONTROL TO RESONANCE FOR A VIBRATING SOIL COMPACTOR

Guy Garcin[x], Michel Guesdon[x], François Degraeve[x]

[x] Vibration and Simulation Laboratory
Centre Technique des Industries Mécaniques
Senlis, France

[xx] Direction of Research
Société ALBARET
Rantigny, France

Recent tests have shown the benefits of compacting soil at resonant frequency. The principle admitted for such a performance is to control the phase between the compacting force and the movement of the vibrating part, by the phase measured at resonant frequency.
One way to develop such a system is to use the possibilities of hybrid calculus. The soil compactor is simulated with an analog computer whereas the feed-back system is represented by the digital computer.
Tests on the compactor have established the validity of the simulation model, and permitted to define the coefficients. The settings of the PID controller were determined by simulation.
The next step consisted in locating the control program of the compactor into a microprocessor, and to check its performance on an analog computer, before mounting into its working place.

1. SOIL COMPACTION BY VIBRATIONS

The Albaret Company currently fabricates soil compactors whose vibrations are generated by rotating unbalanced masses actuated at variable speed by a fixed displacement hydraulic motor coupled to a variable displacement pump, the latter fixed at the end of the crankshaft of the Diesel engine. Control of the displacement of the pump and therefore of the speed of the unbalanced masses, is by an hand-operated servo-positioner.

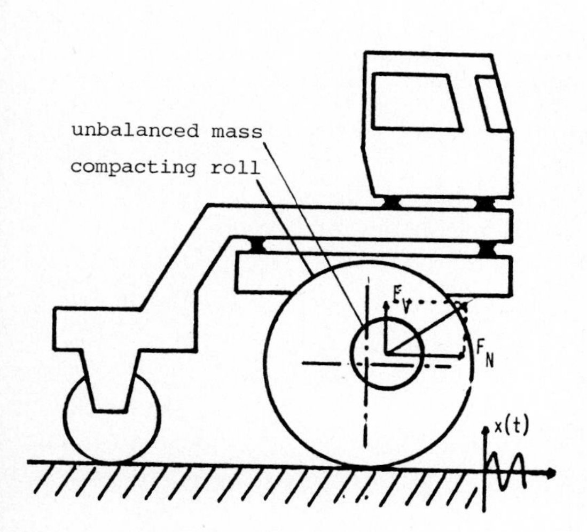

Figure 1 : Sketch of the compactor

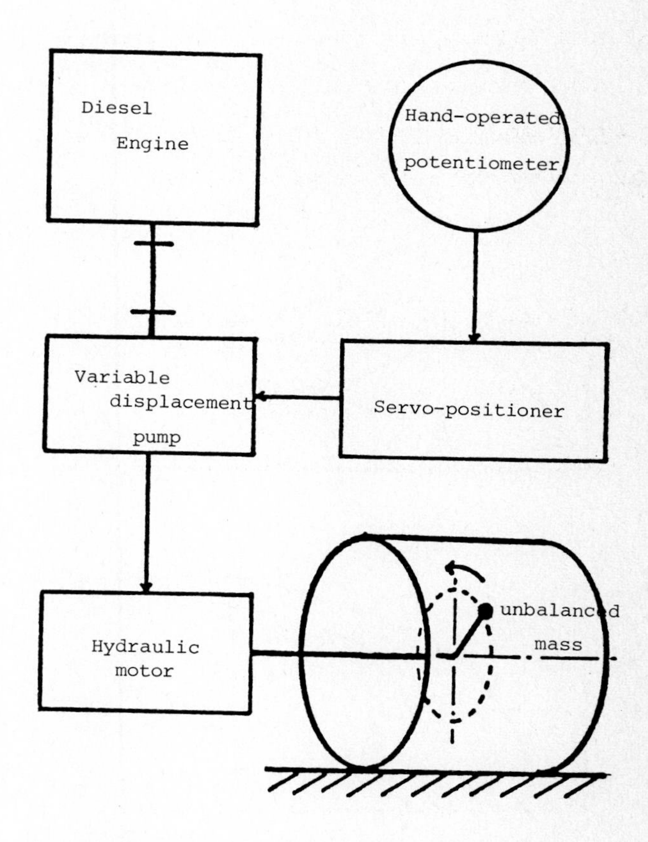

Figure 2 : Block-diagram of the rotation control of the unbalanced mass.

Quite a lot of research and experiment carried out on outdoor sites as well as in laboratories have evidenced the effect of a good selection of the working parameters of soil compactors, frequency and amplitude of vibrations being major factors.

The frequency is compactor dependent, and may be adjusted at will by variating the displacement of the hydraulic pump, whereas the vibration amplitude depends upon the compacting machine (weight of the unbalanced masses, ratio of the suspended mass to the vibrating mass) and either upon the vibrating frequency of the soil which varies with its mechanical properties during compaction.

Tests on different soils have clearly shown that there is an interest in compacting with high amplitude vibration, hence at resonant frequency. Compacting is then more regular and, above all, yield may be increased up to 40 % in comparison with current practice.

In order to get compacting related to the resonant frequency of the system mass-spring represented by the compacting roll and the elastic soil, the Albaret Company has developed a system whose principle has been patented, based upon a continuous readjustement of the phase shift of the vibrating force of the unbalanced masses in relation to the movement of the compacting roll. The phase shift previously mentioned must continuously be maintained at the value measured at resonant amplitude on a proving ground. This phase shift would have a value of 90° for a perfect linear system.

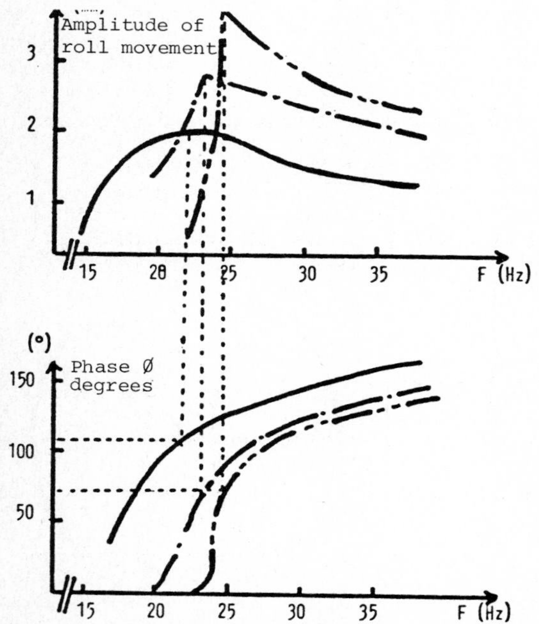

Figure 3 : Reaction of the compacting roll during tests at variable frequency on three different soils

2. SCOPE OF RESEARCH

The problem is to automatically maintain at a value Øc set by the operator, the phase shift Ø between the excitation force and the movement of the compacting roll. The solution is to develop a closed loop control system which continuously cancels the difference between Ø and Øc.

Besides the function of automatic regulation, the control system of the compactor has to carry out supplementary tasks such as :

- Automatic search of the value of the reference phase during an appropriate test,
- Detection of abnormal operation and automatic shifting to an emergency procedure,
- Starting and stopping the unbalanced masses,
- Allow manual operation.

Under abnormal operation, we understand particularly galoping which is a movement of bouncing of the compacting roll on the ground whith loss of contact. This phenomenom is rapidly harmful to the equipment, and the urgency procedure requires a quick modification of the rotation frequency of the unbalanced masses, and a progressive return to the main loop.

Others cases of abnormal operation involve lack of information from one of the transducers and the urgency procedure implies manual operation only.

Due to the number and complexity of tasks to be performed, the choice goes to the development of a microprocessor based numerical control system allowing the control of numerous variables and allowing further developments without major changes in the electronic circuits.

The control algorithms are previously developed on a hybrid computer. The process of continuous mode, that is the dynamic behaviour of the system servo positioner-pump-motor-unbalanced masses-compacting roll-soil is simulated on the analog part of the computer, while the control system is simulated on its digital part. Doing so makes design easier and faster.

The behaviour of the Diesel engine has not been simulated, being considered as an infinite power source.

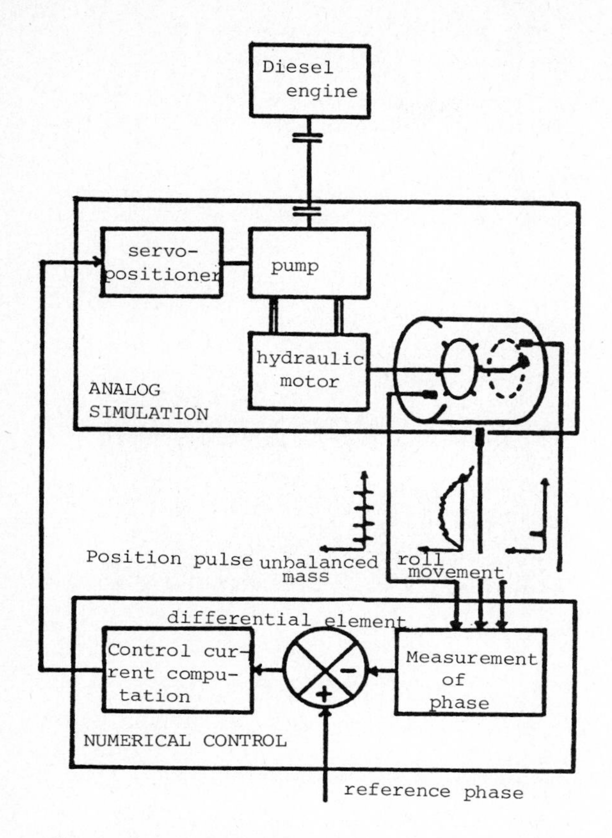

Figure 4 : System simulation

3. MODELLING OF THE COMPACTOR

From the point of view of its functions, the compactor may be represented by the assembly of three blocks as shown by fig. 5

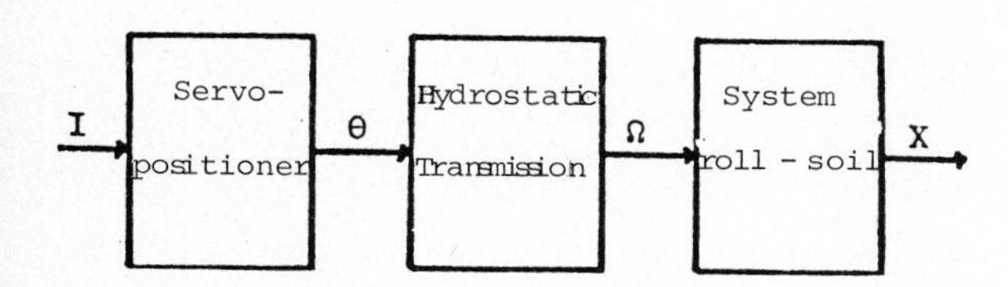

Figure 5 : Blocks diagram of the compactor
I : control current

θ : Tilting angle of the cam-plate of the hydraulic pump

Ω : rotational speed of the unbalanced masses

X : movement of the compacting roll.

The servo-positioner is characterized by a second order transfer function as follows :

$$\frac{\Theta(p)}{I(p)} = \frac{K}{\left(\frac{p}{\omega_{01}}\right)^2 + 2\frac{\xi_{01}}{\omega_{01}}p + 1}$$

Where : p is the Laplace variable
K is the static gain
ω_{01} is the angular frequency,
ξ_{01} is the reduced damping ratio

The hydrostatic transmission (hydraulic pump and motor) may also be modelized by a second order transfer function :

$$\frac{\Omega(p)}{\Theta(p)} = \frac{KTH}{\left(\frac{p}{\omega_{02}}\right)^2 + 2\frac{\xi_{02}}{\omega_{02}}p + 1}$$

Where : ω_{02} is the engine angular frequency
ξ_{02} is the reduced damping ratio
KTH is the static gain as explicited by following relation

$$KTH = \frac{N\ pump}{\theta\ max} \times \frac{V\ pump\ max}{V\ motor}$$

Where : N pump is the rotational speed of the pump
θ max is the maximal angle of tilt of the cam-plate of the hydraulic pump
V pump max is the maximal rotational speed of the pump.
V motor is the rotational speed of the motor.

The shaft bearing the unbalanced masses is directly connected to the hydraulic motor. It generates a centrifugal force acting on the mass-spring system represented by the compacting roll and the soil. The mechanical model considered linear is represented schematically in fig. 6

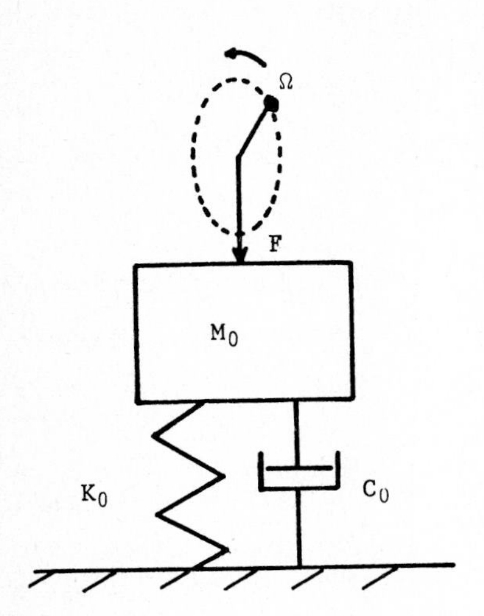

Figure 6 : Equivalent mechanical system "unbalanced masses-compacting roll-soil"

The vertical component of the centrifugal force may be written :

$$F = m\, e\, \Omega^2 \sin(\Omega t)$$

where :

e : the offset of the unbalanced masses

m : the total weight of the unbalanced masses

The transfert function from the exciting force to the compacting roll displacement reads :

$$\frac{X(p)}{F(p)} = \frac{\frac{1}{K_0}}{\left(\frac{p}{\omega_{03}}\right)^2 + 2\frac{\xi_{03}}{\omega_{03}}p + 1}$$

Where :

Ko : the equivalent spring stiffness of the soil

ω_{03}: $\frac{Ko}{Mo}$

where : Mo : equivalent mass of the compacting roll and the soil

$$\xi_{03} = \frac{C_0}{2\sqrt{K_0 M_0}}$$

Co : the equivalent viscous damping of the soil

4. IDENTIFICATION OF SOME PARAMETERS OF THE MODEL

Some parameters involved in modelling the compactor are not well known. This is the case, for instance, for the eigenfrequency and the reduced damping of the transmission. It has been necessary therefore, to identify them by means of a full scale test.

Figure 7 : Measurements on the compactor

- Flow rate in the hydrostatic transmission
- Rotational speed of the Diesel engine
- Speed of displacement and acceleration of the compacting roll

have been measured for different shapes of the control signal of the servo positioner.

The measurements have shown the stability of the speed of the Diesel engine, hence the validity of the assumption of infinite power availability.

However, they do not allow a separate determination of the characteristics of the servo-positioner and the hydraulic transmission, the cam-plate of the pump being inaccessible. They have to be determined as a whole.

The corresponding parameters have been determined by a standard graphic method from the measurements of flowrate in the hydrostatic transmission during one step of the control signal.

Comparing the results with the data given by the manufacturer of the servo-positioner has shown that the dynamics of the hydraulic drive was mainly restricted by the capacity of the servo-positioner. The hydrostatic transmission has therefore a mere effect of static amplification in the pass-range of the servo-positioner.

Control current of the
servo-positioner

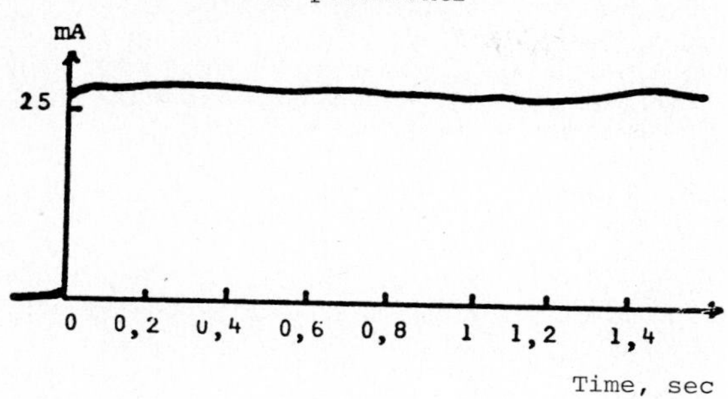

Discharge rate

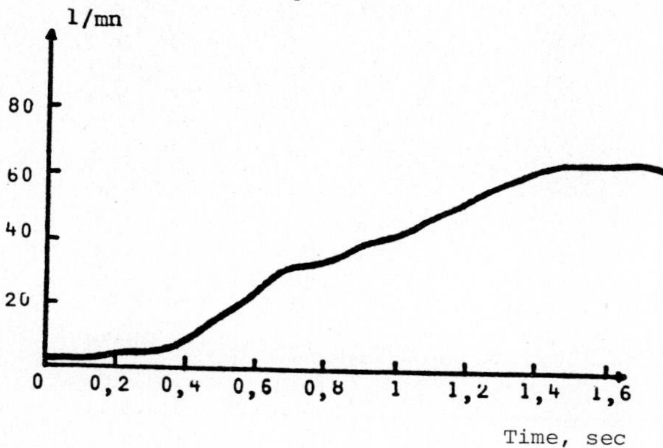

Figure 8 : Flowrate of the pump during one step of the control signal

Regarding the system compacting roll-soil, there is no absolute requirement to have it déterminèd. In fact, the values of the coefficients are depending upon the type of soil, and only current values for the eigenfrequency and the damping ratio have been retained. For the simulation of differents soils the parameters are just to be altered.

The simulated system is written as follows :

$$\frac{\Theta\ (p)}{I\ (p)} = \frac{0{,}36}{1 + 0{,}18\ p + 10^{-2}\ p^2}\ \text{degré/mA}$$

$$\frac{\Omega\ (p)}{\Theta\ (p)} = \frac{13}{1 + 7.10^{-3}\ p + 1{,}5.10^{-5}\ p^2}\ \text{rad/s/degré}$$

ω_{03} : 126 rad/s

ξ_{03} : 0,1

$(K_0)^{-1}$: 2,6 10^{-8} m/N

5. PRINCIPLE OF PHASE MEASUREMENT

When passing in front of a fixed tranduced, a marker on the unbalanced masses triggers the sampling of the rotation signal of the compacting roll versus time at a sampling frequency determined by an encoder wheel with a number n teeth, connected itself to the unbalanced masses (fig. 4).

An harmonic analysis of the sampled data for a lapse of time of two revolutions of the unbalanced masses allows the determination of :

- The phase shift of the exciting force in regard to the roll movement at fundamental frequency,
- The amplitude at harmonic 0,5 in view of detecting galoping.

For the encoding wheel, a number of 6 teeth has been selected in order to pay attention to the more elevated frequencies which might be included in the signal, and prevent, therefore, aliasing in the spectrum.

6. DESIGN OF THE CONTROLLER

The error signal $\varepsilon = \emptyset_c - \emptyset$ is entered into a PID controller. The current controlling the servo-positioner which is déterminèd by the error signal ε is written.

$$I\ (t) = Kp\ \left(\varepsilon(t) + \frac{1}{Ti} \int_0^t \varepsilon(t)\ dt + T_D \frac{d\varepsilon(t)}{dt} \right)$$

Where Kp, Ti and T_D are the parameters of the controller to be defined.

As the feed-back has to be carried out by a microprocessor, this control has been transcripted into the numerical mode by discretizing the equation. When applying the trapezoid equation, we are getting the control algorithm in integral form as below :

$$I_k = I_{k-1} - A\varepsilon_k + B\varepsilon_{k-1} - C\varepsilon_{k-2}$$

$$A = \left(1 + \frac{\Delta(k)}{2\ Ti} + \frac{T_D}{\Delta(k)}\right) Kp$$

$$B = \left(1 - \frac{\Delta(k)}{2\ Ti} + \frac{T_D}{\Delta(k)} + \frac{T_D}{\Delta(k-1)}\right) Kp$$

$$C = \frac{T_D}{\Delta(k)}\ Kp$$

$\Delta(k)$ is the time elapsed between sampling (k) and (k-1).

Considering the inertia of the system and the number of teeth of the encoding wheel, we write

$\Delta(k) = \Delta(k-1)$ and in this case

$$B = \left(1 - \frac{\Delta(k)}{2\,Ti} + \frac{2\,T_D}{\Delta(k)}\right) Kp$$

The computation steps will be as follows (see fig. 9) :

- At each pulse generated by the encoding wheel the vibration signal is sampled digitized, and the sine and cosine terms of the Fourier series are summed. Besides, the control current is computed and visualized.

- After 12 time-generator pulses i.e two complete revolutions of the unbalanced masses, the program checks for galoping, and afterwards computes the revolution period. The true phase Ø is then visualized.

The computation speed is such that the program is run between two pulses of the time-generator, even at the highest rotational speed of 35 Hz of the unbalanced masses.

7. TESTING THE NUMERICAL CONTROL OF THE SOIL COMPACTOR ON AN HYBRID SIMULATOR

Analog simulation of the model of the compactor has been used first for the determination of the PID controller coefficients.

Figures 10 and 11 show the simulated behaviour of the compactor when starting the rotation of the unbalanced masses, the control being either wrong (fig. 10) or right (fig. 11). The segmental aspect of the curve of phase Ø (t) is due to the discretization of measurement with time.

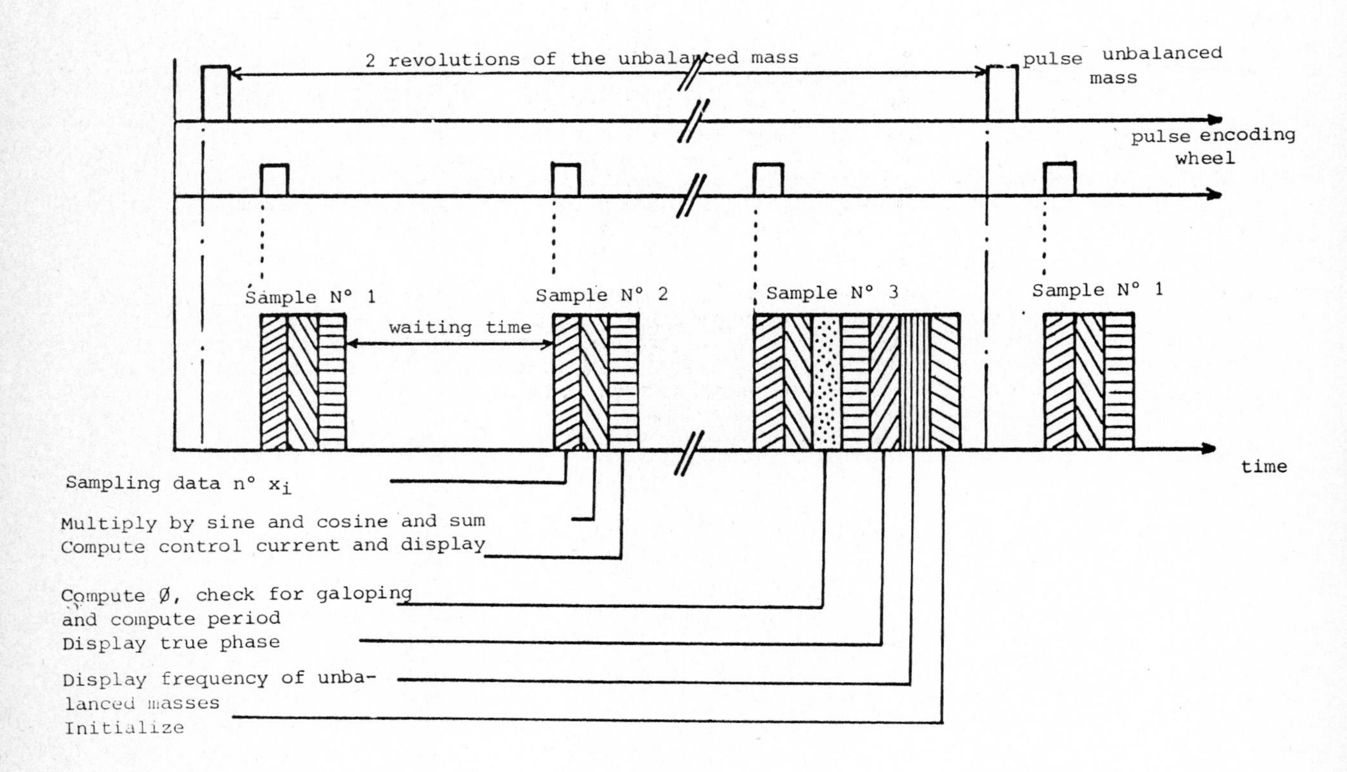

Figure 9 : Sequencing control computation

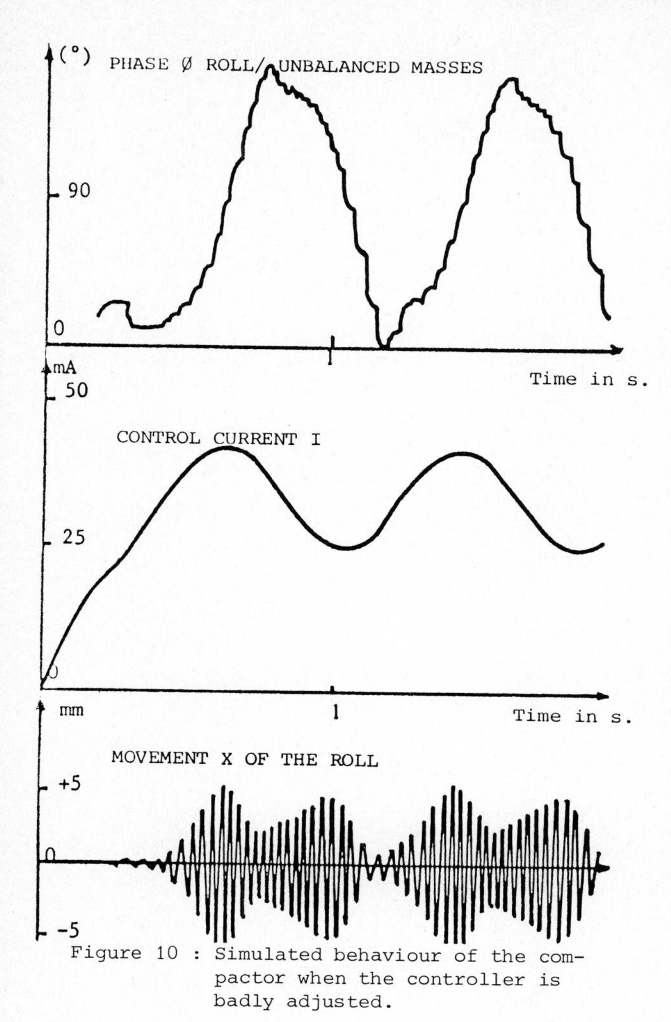

Figure 10 : Simulated behaviour of the compactor when the controller is badly adjusted.

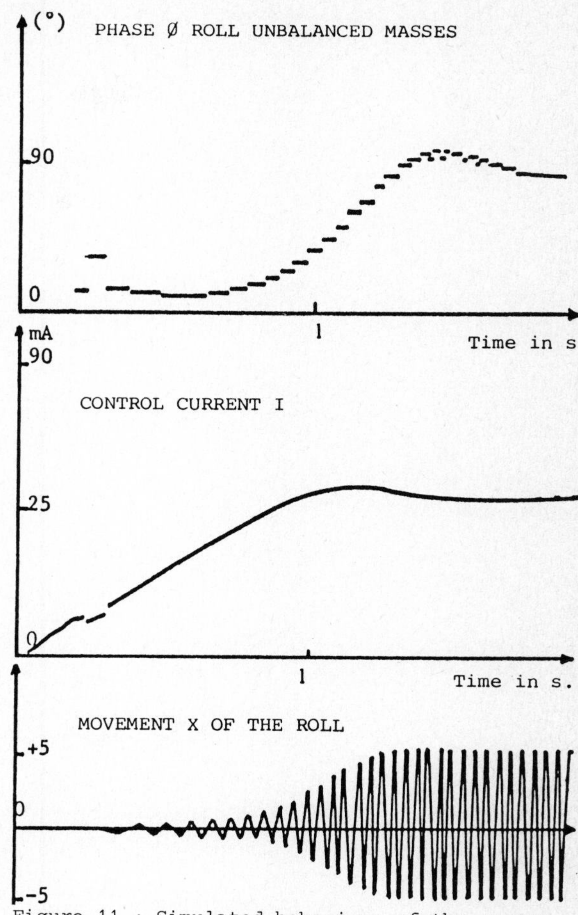

Figure 11 : Simulated behaviour of the compactor when the controller is well adjusted.

In the same way, the simulation model allowed testing of following cases :

- Sudden variation of soil stiffness,
- Urgency procedure at occuring of galoping,
- Automatic testing for the determination of Ø c.

8. CONSTRUCTION AND ADJUSTEMENT OF A PROTOTYPE

After locating the algorithms of the control system in the simulator, a detailled specification has been set up.

This specification was handed over to a specialized laboratory to have a prototype located into a microprocessor.

Before experimenting on site, the prototype controller has been connected to the simulator to make sure that its behaviour was exactly that of the model. The flexibility of the simulation methods permitted to eliminate within two days the unavoidable programming errors of the microprocessor. This is very short in comparison with the duration of testing on the compactor.

Considering the very good results in simulation it has been possible to carry out some simplifications in manufacturing the industrial controller, i.e using a time generator at four times the rotational frequency of the unbalanced masses in place of the encording wheel and reducing sampling to 4 times per revolution.

After solving the programming problems, the controller has been put into place on the compactor and adjusted.

The final tests showed good results, and the system has been adopted in practice.

Simulation in Engineering Sciences
J. Burger and Y. Jarny (eds.)
Elsevier Science Publishers B.V. (North-Holland)

SIMULATION OF FLEXIBLE STRUCTURES WITH MULTIPLE MODES FOR A WHITE NOISE INPUT

Victor Florin Poterasu, Mario Di Paola and Giuseppe Muscolino

Polytechnic Institute of Jassy,Department of Mechanics,Romania
Istituto di Scienza delle Costruzioni,Facoltà di Ingegneria,Università di Palermo
Italia

An identification method for a simulated flexible structure with many degrees of freedom in frequency domain is given.To this aim,we use the PSD(power spectral density) and FFT(fast Fourier transform),the input of the system being a white noise(Dirac's impulse).We studied two cases in certain hypotheses.Finally,a numerical example is given concerning a plane frame with four levels excitated by a Dirac's impulse to the foundation.A very good agreement exists between the calculated and simulated response.

1. INTRODUCTION

For the simulated flexible structures in vibrational state are given many identification techniques in the last years.The structure with many degrees of freedom is simulated in their motion by a system of linear differential ordinary equations with unknown coefficients.These coefficients(or structural parameters)are determined by two principal techniques:optimum parameters determined in the temporal domain [3], [6], [8], [9], [10] and in the frequency domain [1], [2], [3], [11].

The aim of this paper is the determination of the system coefficients in the frequency domain for a simulated structure,in motion,with many degrees of freedom.Some parameters are determined by imposing the minimum of the difference between the PSD calculated and measured and others by the FFT in the same manner.This parameter identification method is very easy because the problem is linear,therefore it is well adapted to the engineering technique.The input function is the Dirac's impulse.

2.THEORETICAL CONSIDERATIONS

The response of a linear system considering the modal contributions can be written in matricial form as

$$\{X\} = [\Phi]\{Y\} \tag{1}$$

where $\{X\}$ is the displacement knots vector, $\{Y\}$ is the displacement vector in the modal coordinates, $[\Phi]$ is the modal matrix.

The relation for the p-th knot is

$$X_p(t) = \sum_{i=1}^{m} \Phi_{pi} Y_i(t) \tag{2}$$

where m are the number of the modes, $Y_i(t)$ is the solution for an elementary oscillator which simulates the elastical structure with multiple modes of vibrations

$$\ddot{Y}_i + a_i \dot{Y}_i + b_i Y_i = -c_i \ddot{Z}_g(t) \tag{3}$$

-the i-th equation.

Here, a_i, b_i, c_i are the damping,stiffenes and partecipation parameters, $\ddot{Z}_g(t)$ is the support acceleration (ξ_i= the critical damping coefficient = $a_i/2b_i$).

The partecipation coefficient is given by

$$\{c\} = [\Phi]^T [M] \{1\} \tag{4}$$

where $[\Phi]$ is the modal matrix orthonormal with respect to the masses matrix $[M]$: $[\Phi]^T[M][\Phi]=[I]$; $\{1\}$ being the vector of 0 and 1 representing the activated knots by the support acceleration $\ddot{Z}_g(t)$

Multiplying the differential equation by and introduising the notation

$$X_{pi}(t)=\Phi_{pi} Y_i(t) ; c_{pi} = \Phi_{pi} c_i \tag{5}$$

relations (2) and (3) are

$$X_p(t) = \sum_{i=1}^{m} X_{pi}(t) \tag{6}$$

$$\ddot{X}_{pi} + a_i \dot{X}_{pi} + b_i X_{pi} = -c_{pi} \ddot{Z}_g(t) \tag{7}$$

The response in the (6) and (7) form simplifies very much the identification problems because for the calculation of the Φ_{pi} terms it is not necessary to obtain an approximate response.
For the frequency domain problem, all the records of the response have a finite duration. Therefore it is necessary to study a finite Fourier transform of relation (7) in $0 \div T$ interval.

$$X_{pT}^{(i)}(f) = -c_{pi} H_i(f) Z_T(f) - H_i(f)[\dot{X}_{pi}(T) - \dot{X}_{pi}(0)] - H_i(f)(a_i + j2\pi f) \quad (8)$$

where $Z_T(f)$ is the input Fourier transformation $\dot{X}(T), \dot{X}(0), X(T), X(0)$ the velocity and the displacement for the final and initial period, $H_i(f)$ the transfer function in the i-th mode

$$H_i(f) = \frac{(b_i - 4\pi^2 f^2) - j2\pi f a_i}{(b_i - 4\pi^2 f^2)^2 + 4\pi^2 f^2 a_i^2} \quad (9)$$

From the practical point of view, the easiest way is to use the relative or absolute accelerations of the response. Relation (8) is written for the p-th mode in relative acceleration form as

$$A_{pT}(f) = \sum_{i=1}^{m} 4\pi^2 f^2 c_{pi} H_i(f) Z_T(f) + \sum_{i=1}^{m} (b_i + jfa_i) H_i(f) v_{pi} + \sum_{i=1}^{m} jf b_i H_{pi}(f) d_{pi} \quad (10)$$

where the v_{pi} and d_{pi} parameters are the difference between the knot velocities respectively the knot displacements in $0 \div T$ interval:

$$v_{pi} = \dot{X}_{pi}(T) - \dot{X}_{pi}(0); \quad d_{pi} = X_{pi}(T) - X_{pi}(0) \quad (11)$$

3. IDENTIFICATION ALGORITHM

The problem consists in the parameters identification $a_i, b_i, c_{pi}, v_{pi}, d_{pi}$ so that the difference between the approximate response (6) - (10) and the recorded or simulate response is minimum. Concerning the proposed problem, we make the following remarks.
If in the first case, the measurements are considered in infinite time, the coefficients for the terms v_{pi} and d_{pi} lose their meaning and the parameters a_i, b_i, c_{pi} are determined by the simultaneous conditions:

a) the resonance amplitude of the approximate response (the peak) is the same with the amplitude recorded or simulated response in the PSD form;

b) this amplitude corresponds to the same damping frequency f_{di};

c) the energy comprised in the significant band of the approximate response peak (from the resonance) is the same with that for the recorded or simulated response.

From the above conditions, we obtain a system with $3 \times i$ unknowns (i-the number of modes considered). From the mathematical standpoint, the conditions for a_i, b_i, c_{pi} can be written in the form

$$G_{pi}(f_i) = S_{pi}(f) \quad \text{for } f = f_{di}$$

$$\frac{\partial G_{pi}(f)}{\partial f} = 0 \quad \text{for } f = f_{di} \quad (12)$$

$$\min \left\{ J = \int_{f_{di}-\varepsilon_i}^{f_{di}+\varepsilon_i} \left[G_{pi}(f) - S_{pi}(f)\right]^2 df \right.$$

where

$$G_{pi}(f) = 16\pi^4 f^4 c_{pi}^2 |H_i(f)|^2 \; G_{zz}$$

is the output PSD and the input G_{zz}, ε_i is the i-th half- length band appropriately defined in the i- th interval of eigenfrequency of the structure and $S_p(f)$ represents the PSD expressed in recorded acceleration.
If the input is a white noise (in thetemporal domain of Dirac's $\delta(t)$), relations (12) give the values

$$b_i = 4\pi^2 f_{id}^2 \sqrt{1 - c_{pi}^2 / S_{pi}(f_{id})}$$

$$a_i = \sqrt{\frac{4\pi^2 f_{id}^2 c_{pi}^2}{S_{pi}(f_{id})} - \frac{(b_i - 4\pi^2 f_{id}^2)^2}{4\pi^2 f_{id}^2}} \quad (13)$$

$$c_{pi}^2 = \int_{f_{di}-\varepsilon_i}^{f_{di}+\varepsilon_i} \left[S_{pi}(f) / G_i(f)\right] df.$$

The above system consisting of the nonlinear equations is solved by an iterative method. In the system (13), $S_{p\,max}^{(i)} = S_{pi}(f_{di})$ is the PSD maximum for the i-th damping frequency.

Relations (12) and (13) give a quasi-exact coincidence for PSD in the interval near to the damping frequency of the structure. We remark that the c_{pi} sign is defined by the response curve form and the integrals (12) and (13) are simbolical.

Giving up the above hypotheses, we determine the $2 \times i$ parameters v_{pi}, d_{pi} from the minimum of J

$$J = \sum_{\ell_{min}}^{\ell_{max}} [A_{pT}(\ell\Delta f) - A_{eff}(\ell\Delta f)][\bar{A}_{pT}(\ell\Delta f) - \bar{A}_{eff}(\ell\Delta f)] \quad (14)$$

where A_{eff} is the recorded or simulated response Δf is a sample, $l_{min}\Delta f \div l_{max}\Delta f$ is the recording interval (the under-symbol shows the conjugate complex). With $\{\delta_p\}^T = \{\{v_p\}^T\{d_p\}^T\}$, the necessary condition for the minimization is

$$\left\{\frac{\partial \mathcal{F}}{\partial\{\delta_p\}}\right\} = 2\sum_{\ell_{min}}^{\ell_{max}}\left\{Re\left(\frac{\partial A_{pT}}{\partial\{\delta\}}\right)Re(A_{pT}) + Jm\left(\frac{\partial A_{pT}}{\partial\{\delta\}}\right)Jm(A_{pT})\right. \quad (15)$$
$$\left. - Re\left(\frac{\partial A_{pT}}{\partial\{\delta\}}\right)Re(A_{eff}) - Jm\left(\frac{\partial A_{pT}}{\partial\{\delta\}}\right)Jm(A_{eff})\right\} = \{0\}.$$

In the appendix we show that the relation (15) is a linear system of algebraic equations for the direct determination of $\{\delta\}$.

The algorithm has the following steps: first are determined the parameters a_i, b_i, c_{pi}, the PSD for the p-th mode record; second, the resolution of a linear algebraic equation system for the determination of parameters v_{pi}, d_{pi} minimize the difference between the recorded or simulated response and the approximate response.

4. NUMERICAL APPLICATION

The identification method is showed for a plane frame excited by a Dirac's $\delta(t)$ impulse to the support. The system characteristics are given in fig. 1. The calculated response is determined by the theoretical results.

We identify the system parameters for the response in terms of relative acceleration of the 4-th level.

Figure 2 shows the PSD for the effective and approximate response. The parameters a_i, b_i, c_{pi} are calculated by condition (13).

Figures 3 and 4 give, respectively, the real and immaginary parts of the Fourier transformation of the response, comparing those effective (or simulated) with those approximated, imposing satisfaction of either the conditions on a_i, b_i, c_{pi} or those on v_{pi}, d_{pi}.

The coefficients are given in table 1.

Table 1.

modes	a_i	b_i	c_{pi}^2	ξ_{di}	ω_{di}	c_{pi}	v_{pi}	d_{pi}
1	1.085	309.09	1.558	3.09%	17.581	-1.248	0.0431	0.0719
2	1.269	2770.82	0.142	1.20%	52.638	0.377	0.0678	-7.15 10
3	1.080	7376.62	0.023	0.63%	85.887	-0.011	8.30 10	1.538 10
4	1.745	12308.33	2.90 10	0.78%	110.94	0.054	0.0205	-5.592 10

$c_1 = c_3 = c_4 = 5$ daN s/cm

$c_2 = 8$ daN s/cm; $m = 5.097$ daN s /cm

$K_p = (30 \times 30)$ cm, $K_t = (30 \times 50)$ cm; $h = 300$ cm

$E = 310000$ daN/cm ; $L = 400$ cm.

$\omega_1 = 17.583$ rad/s, $\xi_1 = 3.08\%$

$\omega_2 = 52.691$ rad/s, $\xi_2 = 1.14\%$

$\omega_3 = 85.853$ rad/s, $\xi_3 = 0.58\%$

$\omega_4 = 111.159$ rad/s, $\xi_4 = 0.54\%$

Some parameters are calculated by Dirac's impulse and the others taking into account the actual recording of the support.

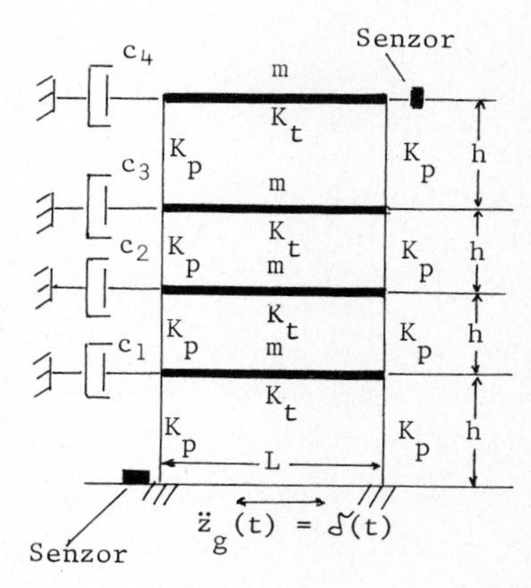

Fig. 1

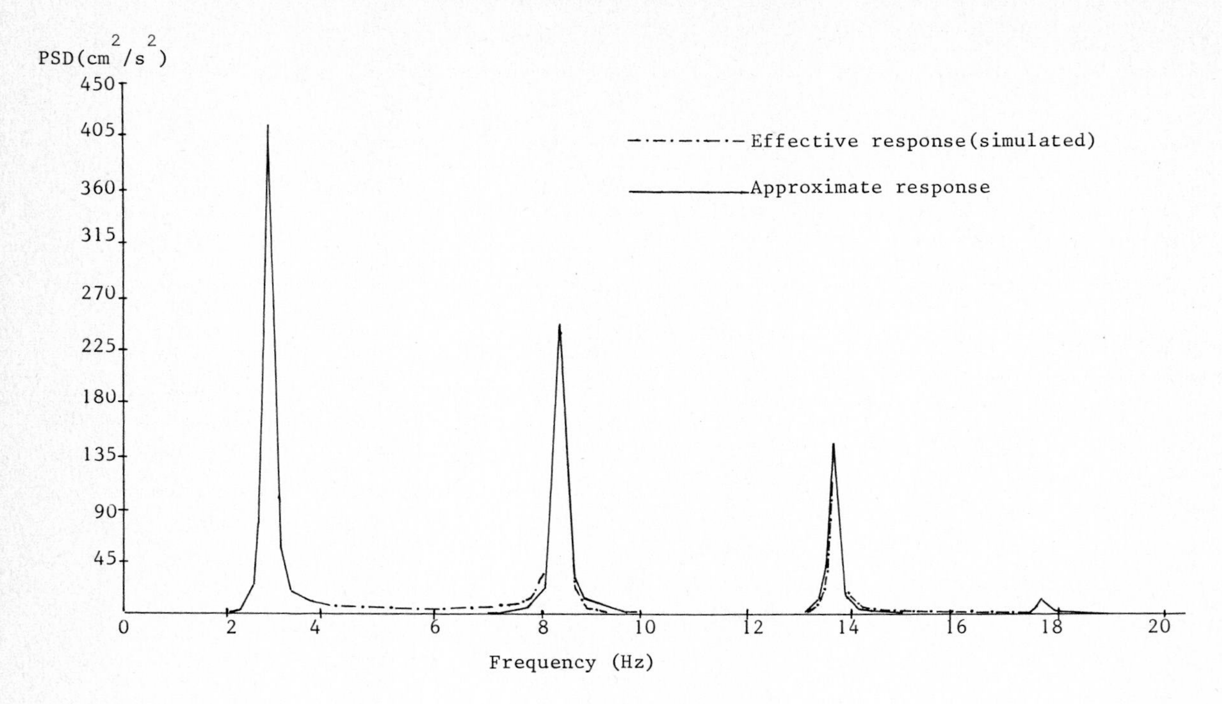
PSD(cm2/s2)
450
405
360
315
270
225
180
135
90
45
0
2
4
6
8
10
12
14
16
18
20
Effective response(simulated)
Approximate response
Frequency (Hz)

Fig. 2

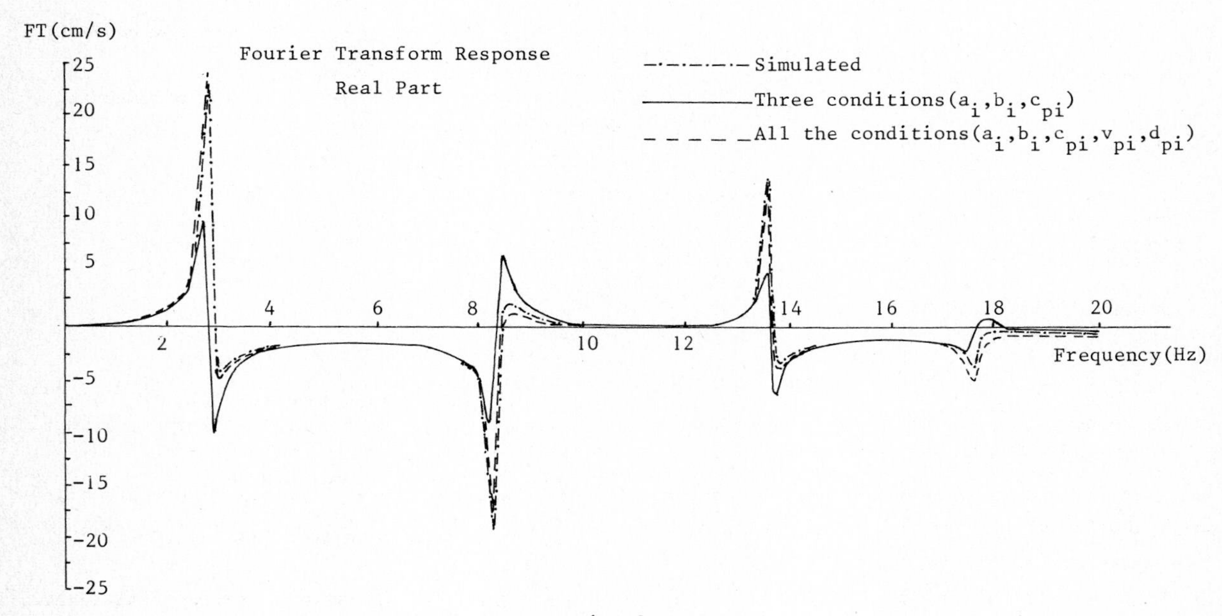
FT(cm/s)
Fourier Transform Response
Real Part
Simulated
Three conditions(a_i,b_i,c_{pi})
All the conditions(a_i,b_i,c_{pi},v_{pi},d_{pi})
25
20
15
10
5
-5
-10
-15
-20
-25
2
4
6
8
10
12
14
16
18
20
Frequency(Hz)

Fig. 3

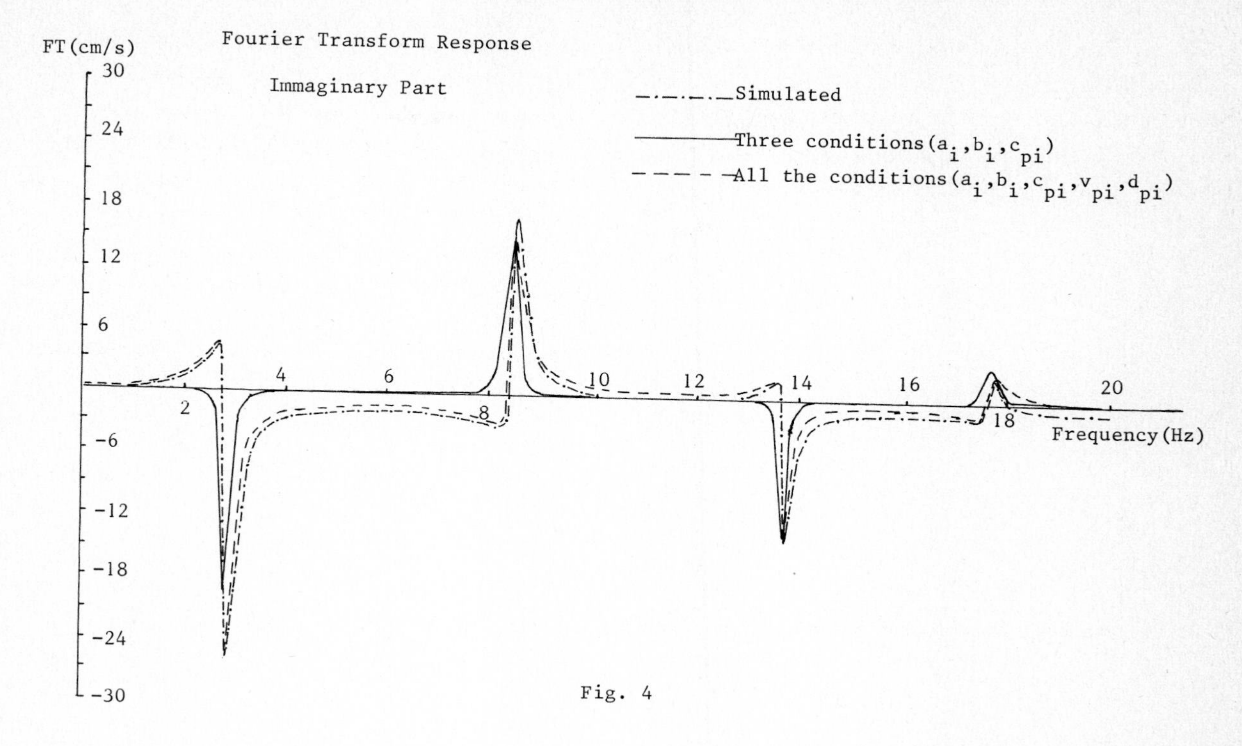

Fig. 4

5. APPENDIX

Equation (15) is written in matrix form as

$$\left[\sum_{\ell_{min}}^{\ell_{max}}\left([\alpha(\ell\Delta f)]+[\delta(\ell\Delta f)]\right)\right]\{\delta\}=\left\{\sum_{\ell_{min}}^{\ell_{max}}\left(\beta(\ell\Delta f)\{\tilde{\alpha}(\ell\Delta f)\}+\right.\right.$$
$$\left.\left.+\,\varepsilon(\ell\Delta f)\{\tilde{\delta}(\ell\Delta f)\}\right)\right\}$$

where the matrices $[\alpha(\ell)]$ and $[\delta(\ell)]$ are of the order $2m \times 2m$ and the vectors $\{\tilde{\alpha}(\ell)\}$, $\{\tilde{\delta}(\ell)\}$ of the order $2m \times 1$, $\{\delta\}$ being the vector of unknowns.

$$[\alpha(\ell\Delta f)]=\{\tilde{\alpha}(\ell\Delta f)\}\{\tilde{\alpha}(\ell\Delta f)\}^T \quad \text{and}$$
$$[\delta(\ell\Delta f)]=\{\tilde{\delta}(\ell\Delta f)\}\{\tilde{\delta}(\ell\Delta f)\}^T$$

where
$$\tilde{\alpha}_i = b_i Re[H_i(\ell\Delta f)] - 2\pi f a_i Jm[H_i(\ell\Delta f)]$$
$$\tilde{\alpha}_{i+m} = -2\pi f b_i Jm[H_i(\ell\Delta f)]$$
$$\tilde{\delta}_i = 2\pi f a_i Re[H_i(\ell\Delta f)] + b_i Jm[H_i(\ell\Delta f)]$$
$$\tilde{\delta}_{i+m} = 2\pi f b_i Re[H_i(\ell\Delta f)]$$

The sufficient condition of the minimum is ensured by the positively defined matrix

$$\sum_{\ell_{min}}^{\ell_{max}}\left([\alpha(\ell\Delta f)]+[\delta(\ell\Delta f)]\right).$$

REFERENCES

[1] McVerry G.H.,Structural identification in the frequency domain from earthquake records, Earthq.Eng.Struct.Dyn.,8(1980)161

[2] DiPaola M.,Muscolino G.,Un metodo numerico per l'analisi dei sistemi elastici dissipativi, 5° Congresso AIMETA Palermo,(1980)

[3] Caravani P.,Thompson W.T.,Identification of damping coefficients in multidimensional linear systems,Jrnl.Appl.Mech.,(1973)

[4] Thomson W.T.,Calkins T.,Caravani P.,A numerical study of damping,Earthq.Eng.Struct.Dyn., 3(1974)97

[5] Hart G.C.,Yao T.P.,System identification in structural dynamics,Proc.ASCE,Jrnl.Eng.Mech.Div. 6(1972)1089

[6] Beck J.L.,Jennings P.C.,Structural identification using linear models and earthquake records,Earthq.Eng.Struct.Dyn.,8(1980)145

[7] Eykhoff P.,System identification(John Wiley, New York,1974).

[8] Poterasu V.F.,Vieru D.,Uniqueness results related to damping matrix identification in building structures,Bull.Inst.Polyt.Jassy 2,V (1981)7.

[9] Udwadia F.E.,Marmarelis P.Z.,The identification of building structural systems.I.The linear case,Bull.Seism.Soc.Am.,66(1976),121.

[10] Jurukowski D.,Mathematical model formulation of a two-storey steel frame structure using parametric system identification and shaking table experiments,7-th WCEE,6(1980)569.

[11] Loh C.H.,Mau S.T.,A preliminary study on a multi-variate system identification technique for building seismic records,7-th WCEE,6(1980) 41.

IX. VEHICLES – TRANSPORTATION (1)

Simulation in Engineering Sciences
J. Burger and Y. Jarny (eds.)
Elsevier Science Publishers B.V. (North-Holland)

THE MODELLING AND CONTROL OF RC HELICOPTER

Katsuhisa Furuta, Yasuhiro Ohyama, Osamu Yamano

Department of Control Engineering
Tokyo Institute of Technology
Oh-Okayama, Meguroku, Tokyo, Japan

A mathematical model of an RC helicopter in hovering is derived assuming that the type of the main rotor hub is elastic see-saw-flapping type and that the direction of the total main rotor thrust approximately coincides with that of normal line of tip path plane. The dynamics of two kinds of sensor, rate gyro and inclinometer are investigated. Finally those reliability and validity are confirmed by using simple gimbal equipment.

1. INTRODUCTION

A helicopter is widely used in practice for goods transportation or information media due to its convenient hovering capability. However its dynamical instability causes difficulty in controlling its flight without the "Stability Augmented System". Meanwhile, the Radio Control (RC) helicopter has not yet been used in practice because of the following two reasons, in spite of its various advantages due to small size and wireless operation.

1. The handling is very difficult, since the response of an RC helicopter motion is very fast for its small size.
2. The payload capacity is restricted by the small power device.

In order to handle an RC helicopter desirably, some automatic fligh control technique has to be investigated. In this paper, the mathematical model of the RC helicopter in hovering and its control are investigated.

Reference 1 and 2 state about the mathematical model of the actual helicopter, however they are much complicated and the reliability of the model is difficult to investigate in RC helicopter, because of the difficulty in handling and sensing.

In this paper, the modelling is described. The rotor dynamics is derived from the dynamics of one blade, since a full articulated hinged rotor is usually used in an acrual helicopter. An RC helicopter usually uses a see-saw-flapping-hub with two blades, so the rotor dynamics can be simply derived by approximating the rotor as one rigid stick. Moreover, the total thrust due to the main rotor is calculated from the blade element lift, and it is also assumed that the total thrust direction coincides with that of tip path plane's normal line. In the next, the reliability of thus derived model is investigated by using the gimbal equipment that has three free axes, rolling, pitching and yawing, but has no freedom of horizontal and virtical motions. At the same time, two kinds of sensors, rate gyro and inclinometer are investigated.

2. MODELLING

2.1 Assumptions

The following assumptions are mainly used to derive the linear mathematical model in hovering.

a) Main rotor is composed of two blades without dragging motion against the rotor shaft rotation, and each blade has homogeneity, enough rigidity characteristics and has no twist and no taper.

b) The vehicle mass center is located under the rotor shaft, and the longitudinal axes, the lateral axes and the rotor shaft correspond to the main inertial axes respectively. The air drag force of fuselage is negligible.

c) In hovering, the rotor angular velocity is constant, and the changes of fuselage velocity pitch angle and flapping angle of main rotor and their derivatives are very small.

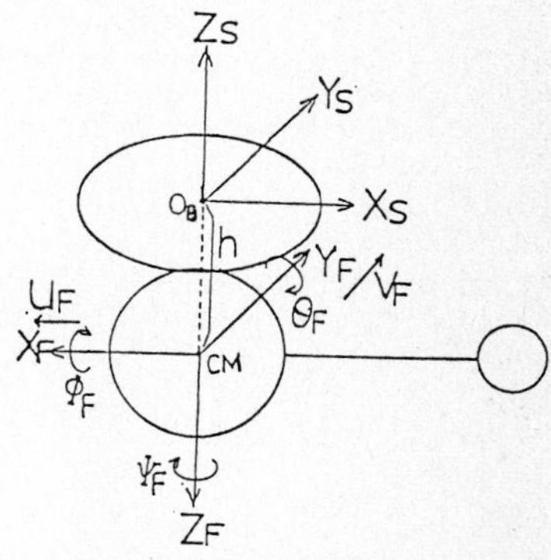

Fig.1 Standard Axes

The fuselage and main rotor standard axes depicted in Fig.1 are introduced and the auxiliary main rotor axes are shown in Fig.2.

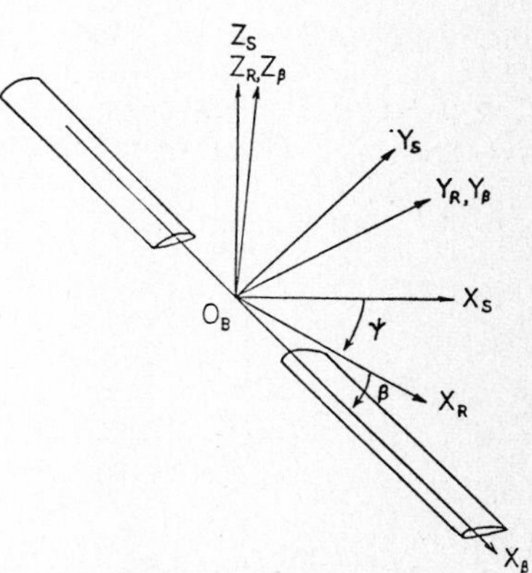

Fig.2 Main Blade Axes

2.2 Dynamics of main rotor

The rotor dynamic equation is usually derived from one blade motion, as the rotor blade is attached to the hub by flapping hinge in an actual helicopter. On the other hand, in an RC helicopter the blades are attached directly to the hub, therefore the rotor can be approximately modeled as one rigid stick with spring like Fig.3. The corning angle is used only for getting blade lift and drag force. By using the law of conservation of angular momentum related to the hub center,

$$\frac{d}{dt}\underset{\beta}{\mathbb{L}}_B + \underset{\beta}{\Omega}^{\beta-I} \times \underset{\beta}{\mathbb{L}}_B = \underset{\beta}{\mathbb{M}}_B + \underset{\beta}{\mathbb{M}}^*_B + \mathbb{K}_B \quad \text{---(1)}$$

$\mathbb{L}_B$: angular momentum of hub with blades
$\mathbb{M}_B$: moment due to one blade lift
$\mathbb{M}^*_B$: moment due to another blade lift
$\mathbb{K}_B$: moment due to hub spring

is derived. The flapping motion is represented by Y component of (1).

$$[\underset{\beta}{\mathbb{L}}_B]_y = I_B(-\dot{\phi}_F \sin\theta + \dot{\theta}_F \cos\psi + \dot{\beta}) \quad \text{---(2)}$$

$$[\underset{\beta}{\Omega}_B^{\beta-I} \times \underset{\beta}{\mathbb{L}}_B]_y = I_B\Omega(-\phi_F \cos\psi - \theta_F \sin\psi + \beta\dot{\psi}_F + \beta\Omega) \quad \text{----(3)}$$

The lift of the blade element is calculated from the velocity resolutions depicted in Fig.4,

$$L_B = \frac{1}{2}\rho ac(u_T^2\theta - u_T u_P) \quad \text{----(4)}$$

and the induced velocity in hovering is assumed constant over the rotor and the average induced velocity determined from the momentum theory is

$$v = \sqrt{\frac{L}{2\rho\pi R^2}} \quad \text{----(5)}$$

The moment by the blade lift is derived by integrating the element lift over the blade as follows.

$$[\underset{\beta}{\mathbb{M}}_B]_y = -\int_0^{BR} r\frac{1}{2}\rho ac(u_T^2\theta - u_T u_P)dr \quad \text{----(6)}$$

$$[\underset{\beta}{\mathbb{M}}^*_B]_y = [\underset{\beta}{\mathbb{M}}_B]_y : \psi = \psi + \pi \quad \text{----(7)}$$

$$[\underset{\beta}{\mathbb{K}}_B]_y = -2K_B \cdot \beta \quad \text{----(8)}$$

The pitch angle of the blade may be approximated by

$$\theta = \theta_0 + \theta_S \sin\psi + \theta_C \cos\psi \quad \text{----(9)}$$

where θ_0: collective pitch angle

and the flapping angle is approximated by its constant and first harmonic content,

$$\beta = \beta_0 + \beta_S \sin\psi + \beta_C \cos\psi \quad \text{----(10)}$$

where β_0: corning angle

Substituting (2)-(10) into (1) and setting the coefficients of sin ψ, cos ψ to zero, the following main rotor dynamic equations are obtained.

$$\begin{aligned}&-\ddot{\phi}_F - 2\dot{\theta}_F\Omega + \ddot{\beta}_s - 2\dot{\beta}_c\Omega \\ &= \frac{-2C_{LB}}{I_B}\big[(\tfrac{2}{3}B^3R^3\Omega\theta_0 - \tfrac{1}{2}B^2R^2 v)(u_F - h\dot{\theta}_F) + \\ &+\tfrac{1}{3}B^3R^3\Omega\beta(v_F + h\dot{\phi}_F) + (\tfrac{1}{2}B^4R^4\Omega\theta_0\beta_0 - \tfrac{1}{3}B^3R^3 v\beta_0) \\ &\cdot\dot{\theta}_F - \tfrac{1}{4}B^4R^4\Omega\dot{\phi}_F + \tfrac{1}{4}B^4R^4\Omega(\dot{\beta}_s - \beta_c\Omega) + \tfrac{1}{4}B^4R^4\Omega^2\theta_s\big] \\ &-\frac{2K_B}{I_B}\beta_s \end{aligned} \quad \text{----(11)}$$

$$\begin{aligned}&\ddot{\theta}_F - 2\dot{\phi}_F\Omega + \ddot{\beta}_c + 2\dot{\beta}_s\Omega \\ &= \frac{-2C_{LB}}{I_B}\big[(\tfrac{2}{3}B^3R^3\Omega\theta_0 - \tfrac{1}{2}B^2R^2 v)(v_F + h\dot{\phi}_F) + \\ &+\tfrac{1}{3}B^3R^3\Omega\beta(u_F - h\dot{\theta}_F) + (\tfrac{1}{2}B^4R^4\Omega\theta_0\beta_0 - \tfrac{1}{3}B^3R^3 v\beta_0) \\ &\cdot\dot{\phi}_F - \tfrac{1}{4}B^4R^4\Omega\theta + \tfrac{1}{4}B^4R^4\Omega(\dot{\beta}_c + \beta_s\Omega) + \tfrac{1}{4}B^4R^4\Omega^2\dot{\theta}_c\big] \\ &-\frac{2K_B}{I_B}\beta_c \end{aligned} \quad \text{----(12)}$$

2.3 Dynamics of fuselage

The force on fuselage is obtained by averaging the blade lift and drag over one rotation. By using the law of conservation of angular momentum related to the vehicle mass center,

$$\frac{d}{dt}I_F \underset{F}{\Omega}^{\beta-I} = \underset{F}{\mathbb{M}}_B + \underset{F}{\mathbb{M}}_{KB} + \underset{F}{\mathbb{M}}_T \quad \text{----(13)}$$

I_F : fuselage inertial moment
$\mathbb{M}_{KB}$: moment due to hub spring
$\mathbb{M}_T$: moment due to tailrotor thrust

is derived and by the law of conservation of momentum related to the vehicle mass center,

$$\frac{d}{dt}m_F \underset{F}{\mathbb{V}}^{F-I} = \underset{F}{\mathbb{F}}_B + \underset{F}{\mathbb{F}}_t - m_F\mathbb{G} \quad \text{----(14)}$$

$\mathbb{V}^{F-I}$: vehicle velocity
$\mathbb{F}_B$: force due to mainrotor thrust
$\mathbb{F}_t$: force due to tailrotor thrust
$\mathbb{G}$: acceleration of gravity

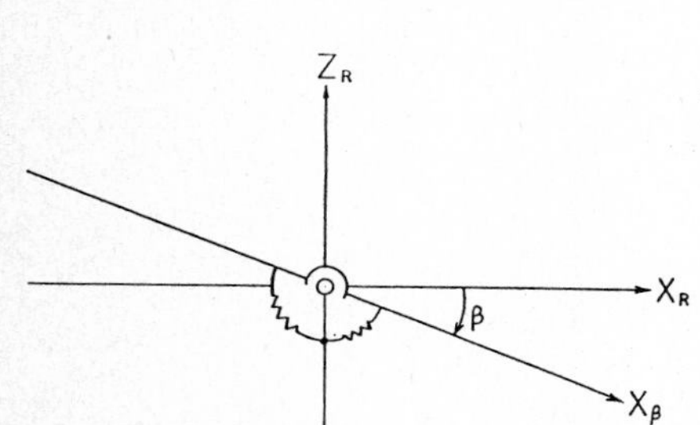

Fig.3 Main Rotor Model

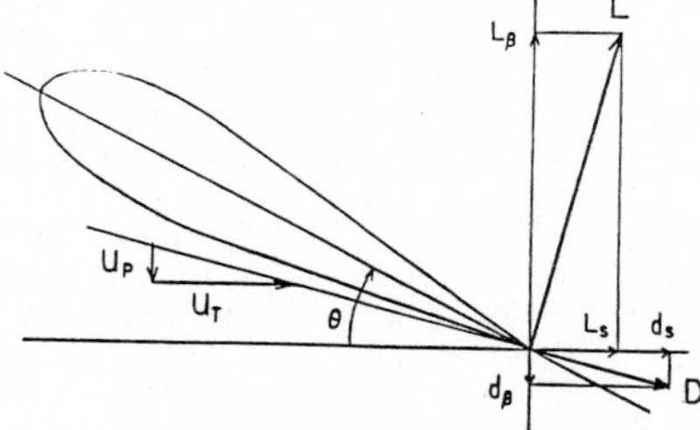

Fig.4 Schematic Blade Element Velocity Resolutions

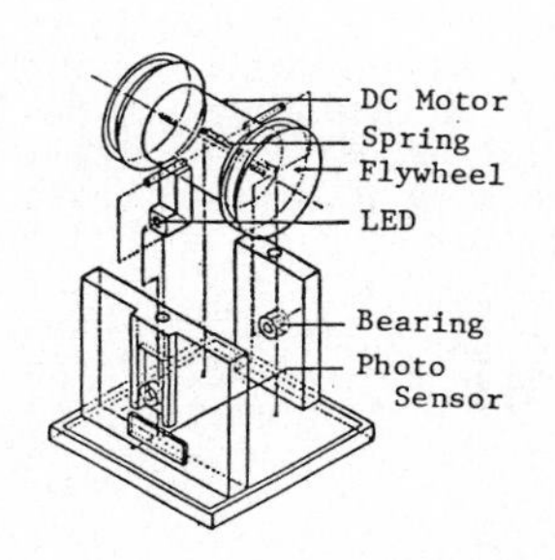

Fig.5 Structure of Rate Gyro

is derived.

The rotor force on fuselage is calculated complicatedly from X and Y components of blade lift.(1) However, the direction of the average rotor thrust is considered approximately to be that of normal line of tip path plane, and the main rotor flapping anlges β_s, β_c, decide the angle between that normal line and rotor shaft. Then the equations of fuselage motion are simply obtained by using average rotor thrust F_{zo} and flapping angle β_s, β_c.
The average mainrotor thrust F_{zo} and the average tailrotor thrust F_{to} are derived as follows.

$$F_{zo}=\frac{2}{2\pi}\frac{1}{2}\rho ac\int_0^{2\pi}\int_0^{BR}(u_T^2\theta_0-u_Tu_p)drd\psi$$

$$=2C_{LB}\{\tfrac{1}{3}B^3R^3\theta_0-\tfrac{1}{2}B^2R^2\Omega v\} \qquad ----(15)$$

$$F_{to}=\frac{2}{2\pi}\frac{1}{2}\rho a_tc_t\int_0^{2\pi}\int_0^{Rt}(u_{tT}^2\theta_t-u_{tT}u_{tP})drd\psi$$

$$=2C_{Lt}\{\tfrac{1}{3}R^3n^2\Omega^2\theta_t-\tfrac{1}{2}R_t^2n\Omega v_t-\tfrac{1}{2}R_t^2n\Omega h_t\dot{\psi}_F\} \qquad ----(16)$$

And the dynamic equations of rolling and pitching motion

$$I_{\phi F}\ddot{\phi}_F=-F_{zo}\beta_sh-2K_B\beta_s \qquad ----(17)$$

$$I_{\theta F}\ddot{\theta}_F=F_{zo}\beta_ch+2K_B\beta_c \qquad ----(18)$$

are is derived and those of lateral and longitudinal motion,

$$m_F\dot{u}_F=-F_{zo}\beta_c-m_Fg\theta_F \qquad ----(19)$$

$$m_F\dot{v}_F=-F_{zo}\beta_s+m_Fg\phi_F+F_{to} \qquad ----(20)$$

are derived.

Meanwhile, the yawing motion is relatively decoupled from the rolling and pitching motion, and is caused mainly by the main rotor drag force and tail rotor thrust. The moment due to the profile drag force is derived from Fig.4 ,

$$[\mathbb{M}_B]_z=\frac{2}{2\pi}\int_0^{2\pi}\int_0^{BR}r(ds+Ls)drd\psi$$

$$=2C_{LB}(\tfrac{Cd}{a}\tfrac{1}{4}B^4R^4\Omega-\tfrac{1}{2}B^2R^2v+(\tfrac{Cd}{a}\tfrac{1}{2}B^4R^4\Omega-\tfrac{1}{3}B^3R^3v\,\theta_0)$$

$$\dot{\psi}+\tfrac{1}{3}B^3R^3\Omega v\,\theta_0+(-\tfrac{2}{3}B^3R^3v-\tfrac{1}{4}B^4R^4\Omega)\,\dot{\beta}_0+B^2R^2v\,w_F) \qquad ----(21)$$

then the yawing dynamic equation

$$I_{\psi F}\ddot{\psi}_F=[-2C_{LB}(\tfrac{Cd}{a}\tfrac{1}{2}B^4R^4\Omega-\tfrac{1}{3}B^3R^3v\,\theta_0)-C_{Lt}R_t^2n\Omega h_t^2]\dot{\psi}_F+(\tfrac{2}{3}C_{Lt}R_t^3n^2\Omega)\,\theta_t \qquad ----(22)$$

is derived.

The vertical motion is also decoupled from other motions. The vertical thrust F_z is derived as follows ,

$$F_z=\frac{2}{2\pi}\frac{1}{2}\rho ac\int_0^{2\pi}\int_0^{BR}(u_T^2\theta-u_Tu_p)drd\psi$$

$$=2C_{LB}\{\tfrac{1}{3}B^3R^3\Omega\,\theta_0+(\tfrac{2}{3}B^3R^3\Omega\,\theta_0-\tfrac{1}{3}B^2R^2v)\dot{\psi}_F+\tfrac{1}{3}B^3R^3\Omega\,\dot{\beta}_0+\tfrac{1}{2}B^2R^2w_F\} \qquad ----(23)$$

and the dynamic equation of the vertical motion

$$m_F\dot{w}_F=-F_z \qquad ----(24)$$

is derived.

3. Sensor

The doppler radar, airspeed indicator, accelerometer, directional gyro and rate gyro are used for an actual helicopter, however these sensors can't be used for RC helicopter because of their heavy and large structure. In this paper, rate gyro and inclinometer are investigated to be used on an RC helicopter.

Fig.5 depicts the structure of a model rate gyro. The DC motor drives two flywheels at high speed, and the inclined angle of the gimbal, which is balanced by the gyro-precession and springs, is detected by using LED and Photo-Sensor. The dynamics of a rate gyro is generally given by second order differential equation.(4) However, the main blade rotation causes strong fuselage vibration, whose frequency component must be filtered out from the rate gyro output, the overall transfer function including the low pass filter is given by

$$\frac{0.45}{1+0.09s} \qquad \frac{[\;V\;]}{[rad/s\;]} \qquad ----(25)$$

This rate gyro is used to be angular acceleration meter by neglecting its time constant, 0.09 sec. .

A small inclinometer depicted in Fig.6 is constructed of a weight bar and damping oil. This transfer function is given by

$$\frac{[\;5.23,0.53]}{1+0.09s} \qquad \frac{[\;V\;]}{[rad,m/s^2]} \qquad ----(26)$$

and the inclined angle and the accelelation to this inclinometer are detected at the same time.

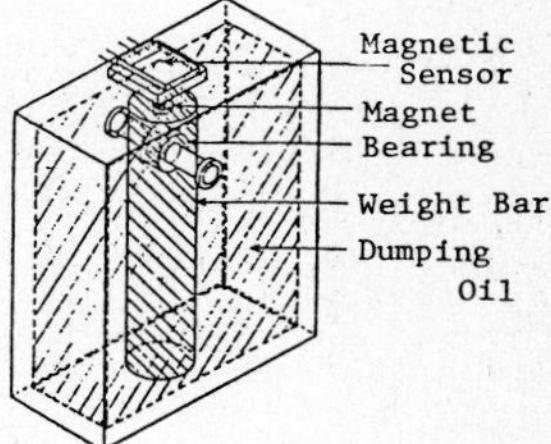

Fig.6 → Structure of Inclinometer

4. GIMBAL EQUIPMENT

It is desired to transform the eight differential equations, (11), (12), (17), (18), (19), (20), (23), (24), to the tweleve first order differntial equations of the state

$$\dot{x}=Ax+Bu \qquad ----(27)$$

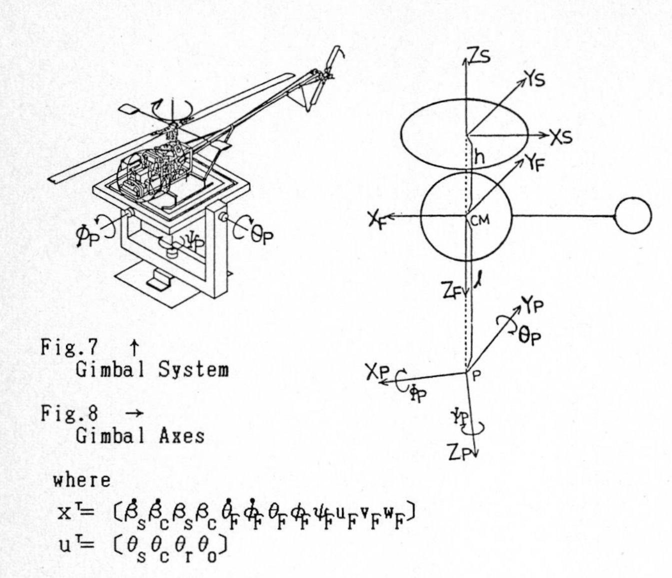

Fig.7 ↑
Gimbal System

Fig.8 →
Gimbal Axes

where

$$x^T = [\dot{\beta}_s \dot{\beta}_c \beta_s \beta_c \dot{\theta}_F \dot{\phi}_F \theta_F \phi_F \psi_F u_F v_F w_F]$$
$$u^T = [\theta_s \theta_c \theta_r \theta_o]$$

To investigate the reliability of the mathematical model and two kinds of sensors, the stabilizing control system of the gimbal equipment as Fig.7 is designed and implemented by the minicomputer. Fuselage rolling, pitching and yawing motion are measured by three potentiometers and fuselage holizontal and vertical motions are restricted. In this helicopter system, the main rotor is driven by two electric motors and its angular velocity is controlled to be constant by single loop servo control system.

The gimbal axes are depicted in Fig.8 and fuselage angles and velocities are substituted for gimbal angles as follows,

$$\theta_F = \theta_p \qquad ----(28)$$
$$\phi_F = \phi_p \qquad ----(29)$$
$$u_F = -l\theta_p \qquad ----(30)$$
$$v_F = l\phi_p \qquad ----(31)$$

The first order mathematical model is derived as follows.

$$\frac{d}{dt}\begin{bmatrix}\beta_s\\ \beta_c\\ \dot{\theta}_p\\ \dot{\phi}_p\\ \theta_p\\ \phi_p\end{bmatrix} = \begin{bmatrix} A_{11} & A_{12} & A_{13} & A_{14} & 0 & 0\\ -A_{12} & A_{11} & A_{14} & A_{13} & 0 & 0\\ 0 & A_{32} & 0 & 0 & A_{35} & 0\\ A_{41} & 0 & 0 & 0 & 0 & A_{46}\\ 0 & 0 & 1 & 0 & 0 & 0\\ 0 & 0 & 0 & 1 & 0 & 0 \end{bmatrix}\begin{bmatrix}\beta_s\\ \beta_c\\ \dot{\theta}_p\\ \dot{\phi}_p\\ \theta_p\\ \phi_p\end{bmatrix} + \begin{bmatrix} B_{11} & B_{12}\\ -B_{12} & B_{11}\\ 0 & 0\\ 0 & 0\\ 0 & 0\\ 0 & 0\end{bmatrix}\begin{bmatrix}\theta_s\\ \theta_c\end{bmatrix} \qquad ----(32)$$

where A_* , B_* are given in Appendix B.

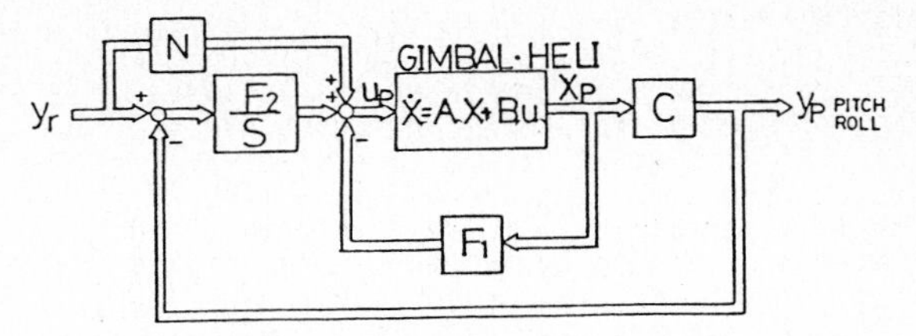

Fig.9 Configuration of servo control system

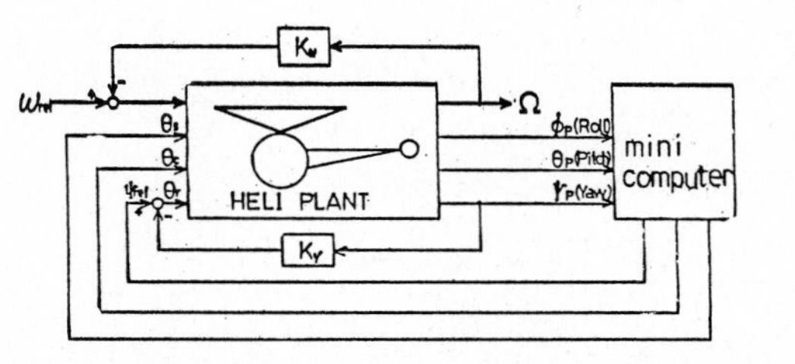

Fig.10 Configuration of total control system

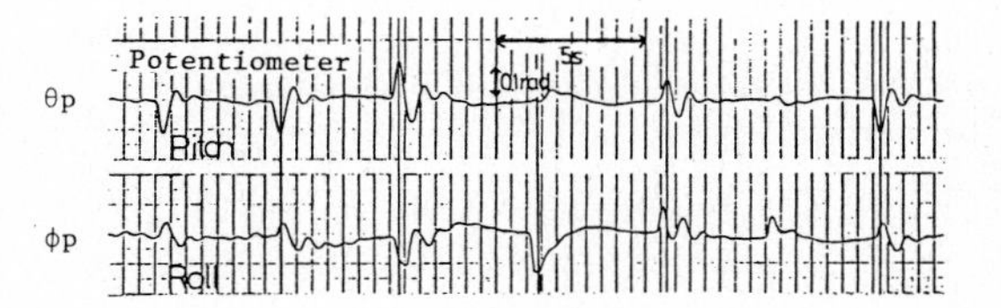

Fig.11 Control result by potentiometer

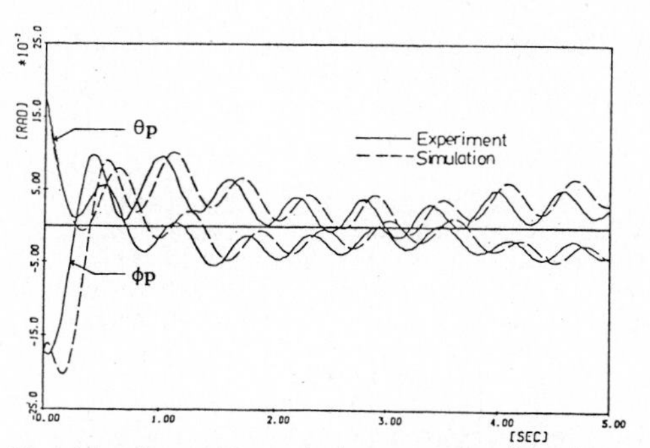

Fig.12 Comparison between experiment and simulation

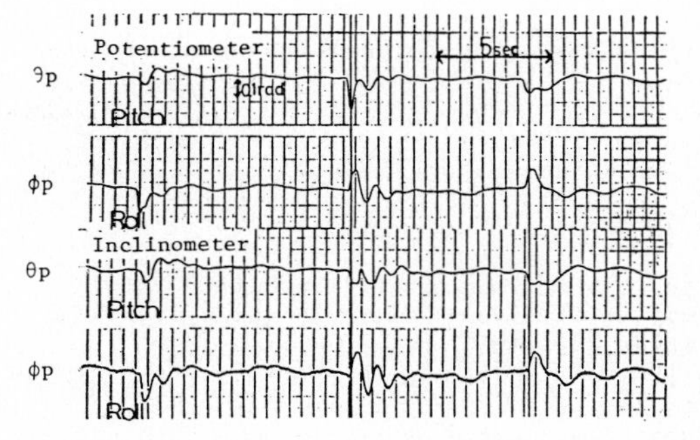

Fig.16 Control result by Inclinometer and Rate Gyro

However, the vertical motion and yawing motion are neglected because they are considered to be decoupled from other motions. The second derivatives of angle in flapping motion are also neglected because they are so small.

Table 1 shows its parameters, and this gimbal system is found to be more unstable compared with hovering helicopter because unstable poles with positive real part are more unstable than those of hovering helicopter, (Table 2). In Table 2, the fast mode(5) is flapping mode of the main rotor.

5. CONTROL

5.1 Control by potentiometer

First, the stabilizing control system of the fuselage in gimbal system are considered, when the fuselage rolling and pitching angle are precisely measured by the potentiometers. The control system depicted in Fig.9 are used and rolling and pitching angles are controlled as to serve to the references. Total control system are depicted in Fig.10, where the anguler velocity of the main rotor, Ω is controlled constant by the motor input voltage, and yawing angle is controlled by the tailrotor pitch angle θ_r, by using single loop control systems respectively.

To control both the rolling and pitching angle by two inputs, θ_s , θ_c , a servo control system is designed. In this case, the flapping mode of (32) is furthere reduced bacause it is so fast compared with fuselage modes. The system is generally written as

$$\begin{cases} \dot{x} = A x + B u \\ y = C x + D u \end{cases} \qquad \text{----(33)}$$

and the augment system

$$\dot{\bar{x}} = \begin{bmatrix} A & 0 \\ C & 0 \end{bmatrix} \bar{x} + \begin{bmatrix} B \\ D \end{bmatrix} \bar{u} \qquad \text{----(34)}$$

are considered, then the feedback law to minimize the folloing performance index is calculated for a suitable Q and R pair.

$$J = \int_0^{\infty} (\bar{x}^T Q \bar{x} + \bar{u}^T R \bar{u})\, dt \qquad \text{----(35)}$$

The designed parameters and the control result are shown in Table 3 and Fig.11 respectively. In this case, the controlling time is 20 ms. and the four state are fed back by using output of potentiometers and their differences. The rolling and pitching angle are regulated well in the presence of disturbance.

Moreover, the model's output is simulated by minicomputer when this control input series is given into this model to investigate the reliability of the model. The result is depicted in Fig.12, where the model's output are compared with the potentiometer's output. Though there is a little time lag between them, the mode of motion and magnitude coincide nearly with the experimental one. So this mathematical model are considered to be valid.

5.2 Control by rate gyro and inclinometer

The gimbal system is also controlled by rate gyro and inclinometer as described before, to investigate their validity. The angular accelerations are measured directly by rate gyro, but the inclinometer outputs are combined with the angular acceleration of the fuselage as follows.

$$\theta_i = \theta_p + \delta \ddot{\theta}_p \qquad \text{----(36)}$$

$$\phi_i = \phi_p + \delta \ddot{\phi}_p \qquad \text{----(37)}$$

where δ is constant

So the observation equation (38) of the sensors is introduced.

$$\begin{bmatrix} \dot{\theta}_p \\ \dot{\phi}_p \\ \theta_i \\ \phi_i \end{bmatrix} = C^* \begin{bmatrix} \dot{\theta}_p \\ \dot{\phi}_p \\ \theta_p \\ \phi_p \end{bmatrix} + D^* \begin{bmatrix} \theta_s \\ \phi_c \end{bmatrix} \qquad \text{----(38)}$$

The fuselage rolling and pitching angle $[\theta_p, \phi_p]$ are obtained by solving this equation, and the control as before is executed by using them. The result are shown in Fig.13 and these sensors are found to be valid.

MATRIX A

$$\begin{bmatrix} -48.9 & 26.4 & 0.06 & 1.13 & 0.00 & 0.00 \\ -26.4 & -48.9 & -1.12 & 0.06 & 0.00 & 0.00 \\ 0.00 & 167.3 & 0.00 & 0.00 & 14.8 & 0.00 \\ -637.4 & 0.00 & 0.00 & 0.00 & 0.00 & 56.1 \\ 0.00 & 0.00 & 1.00 & 0.00 & 0.00 & 0.00 \\ 0.00 & 0.00 & 0.00 & 1.00 & 0.00 & 0.00 \end{bmatrix}$$

MATRIX B

$$\begin{bmatrix} -33.4 & -43.5 \\ 43.5 & -33.4 \\ 0.00 & 0.00 \\ 0.00 & 0.00 \\ 0.00 & 0.00 \\ 0.00 & 0.00 \end{bmatrix}$$

MATRIX C

$$\begin{bmatrix} 0.00 & 0.00 & 0.00 & 0.00 & 1.00 & 0.00 \\ 0.00 & 0.00 & 0.00 & 0.00 & 0.00 & 1.00 \end{bmatrix}$$

Table 1
Parameters ↑
and Eigenvalues →

	EIGENVALLUES
(1)	3.12 ±0.323
(2)	-6.45
(3)	-15.7
(4)	-41.4 ±31.4

Table 2 →
Eigenvalues of RC Helicopter

(1)	0.61 ±1.87
(2)	0.06 ±2.49
(3)	-5.20
(4)	-8.66
(5)	-39.9 ±34.6

MATRIX F 1

$$\begin{bmatrix} 0.052 & 0.003 & 0.385 & 0.117 \\ -0.008 & 0.018 & -0.129 & 0.300 \end{bmatrix}$$

MATRIX F 2

$$\begin{bmatrix} 0.012 & 0.005 \\ -0.005 & 0.012 \end{bmatrix}$$

MATRIX N

$$\begin{bmatrix} 0.297 & 0.103 \\ -0.115 & 0.212 \end{bmatrix}$$

Table 3 Designed Parameters

6. CONCLUSION

The mathematical model of an RC helicopter in hovering is derived and its reliability is investigated by using gimbal system with three axes of freedom. At the same time, the validity of the model rate gyro and the small inclinometer made by ourselves are also investigated. The mathematical model of flapping motion is derived by approximating the two blades as one rigid stick and the model of fuselage motion is also derived by using the flapping angle and average thrust of the main rotor. Since the vertical and yawing motions of the fuselage are decoupled from the other motions from the viewpoint of handler's experience and the derived mathematical model, the validity of the model is investigated by controlling the rolling and pitching motion of the model helicopter in gimbal system. The dynamics of the two kinds of sensors, model rate gyro and inclinometer, are also derived and it is found that their combined use is effective in spite of their time constants are not so fast.

In this experiment, the gimbal system is successfully controlled for safety and modelling error, however, in the case of the control of the real RC helicopter in the sky, the vertical motion may have to be controlled and the sensors to measure helicopter absolute position and velocity may have to be developed.

REFERENCES

(1) W. E. Hall, Jr. , Computational Models for the Synthesis of Rotary-Wing VTOL Aircraft Control System, Stanford Univ. (1971)

(2) W. Johnson, Helicopter Theory (Prinston University Press)

(3) A. R. S. Bramwell, Helicopter Dynamics (Edward Arnold Ltd. 1976)

(4) M. Okada and T. Oda, Self-Navigation System of Aircraft (Korona Ltd)

(5) Smith,H.W. and E.J.Davison, Design of industrial regurator, Proc. IEE,119 (1972) 1210-1216.

Appendix A

Notation

- a lift curve slope of main blade
- a_t lift curve slope of tail rotor blade
- B tip loss factor
- c main blade chord
- c_t tailrotor blade chord
- C_d main blade element profile drag coefficient
- g acceleration of gravity
- h height of rotor hub above vehicle center of mass
- h_t distance between tailrotor hub and vehicle center of mass
- I_B main blade flapping inertial moment
- $I_{\theta F}$ fuselage pitch inertial moment
- $I_{\phi F}$ fuselage roll inertial moment
- $I_{\psi F}$ fuselage yaw inertial moment
- K_B blade flapping tortional spring
- l diatance between gimbal center and fuselage mass center
- m_F mass of vehicle
- n rotation ratio of mainrotor and tailrotor
- R mainrotor radius
- R_t tailrotor radius
- Ω mainrotor angular velocity
- ρ air density
- θ main blade pitch angle
- θ_t tail blade pitch angle
- β flapping angle
- v mainrotor induced velocity
- v_t tailrotor induced velocity

$\frac{d}{dt}x = \dot{x}$

$\Omega_{\beta}^{\beta-I}$ mean angular velocity of β axes related to initial axes

subscript
- p : gimbal
- F : fuselage
- B : mainrotor blade
- t : tailrotor

double linear character like $\mathbb{L}$ means vector

$C_{LB} = \frac{1}{2}\rho ac$ $\qquad C_{Lt} = \frac{1}{2}\rho a_t c_t$

Appendix B

$$A_{11} = \frac{-C_{LB}B^4R^4\Omega}{\Delta I_B}\left(\Omega + \frac{K_B}{I_B}\right)$$

$$A_{12} = \frac{1}{\Delta}\left[\frac{-4K_B\Omega}{I_B} + \frac{(C_{LB}B^4R^4)^2\Omega^3}{4\ I_B^2}\right]$$

$$A_{13} = \frac{C_{LB}}{\Delta I_B}\left\{\frac{4}{3}B^3R^3\Omega^2\beta_0(1+h) - \frac{C_{LB}}{I_B}\left[\frac{B^8R^8}{2}\Omega^2\theta_0\beta_0 - \left(\frac{B^7R^7}{3}\Omega^2\theta_0 - \frac{B^6R^6}{2}\Omega v\right)(1+h) + \frac{B^7R^7}{3}\Omega v\beta_0\right]\right\}$$

$$A_{14} = \frac{1}{\Delta}\left\{\frac{C_{LB}}{I_B}\left[\left(\frac{8}{3}B^3R^3\Omega^2\theta - 2B^2R^2\Omega v\right)(h+1) + \frac{4}{3}B^3R^3v\beta_0 - 2B^4R^4\Omega^2\theta_0\beta_0\right] + 4\Omega^2 + \frac{C_{LB}^2}{I_B^2}\left[-\frac{1}{3}B^7R^7\Omega^2\beta_0(1+h) + \frac{1}{4}B^8R^8\Omega^2\right]\right\}$$

$$A_{32} = \frac{1}{I_{\theta P}}(F_{zo}(1+h) + 2K_B) \qquad A_{35} = \frac{1}{I_{\theta P}}m_F g l$$

$$A_{41} = \frac{1}{I_{\phi P}}(F_{zo}(1+h) + 2K_B) \qquad A_{46} = \frac{1}{I_{\phi P}}m_F g l$$

$$B_{11} = \frac{-C_{LB}^2}{4\Delta I_B^2}B^8R^8\Omega^3 \qquad B_{12} = \frac{-C_{LB}}{\Delta I_B}B^4R^4\Omega^3$$

$$\Delta = 4\Omega^2 + \frac{C_{LB}^2}{4I_B^2}B^8R^8\Omega^2$$

Simulation in Engineering Sciences
J. Burger and Y. Jarny (eds.)
Elsevier Science Publishers B.V. (North-Holland)

THE DESIGN OF AN ON-BOARD LOOK-AHEAD-SIMULATION FOR APPROACH

Dipl.-Ing. Jürgen Riepe

Institut für Flugführung
Technische Universität
D-3300 Braunschweig

Involved by the demand of better utilizing the terminal-capacity a new concept of flight guidance was developed in the "Institut für Flugführung" of the "Technische Universität" in Braunschweig. This on-board-system is used for determining an approach trajectory according to tower-demands and observing its exactly performance to enable a precise prediction of arrival at the runway.

It is based on a look-ahead-simulation that is much faster than real time:

1. THE FLIGHT GUIDANCE SYSTEM

A Model of the controlled aircraft is repetively sent along the planed trajectory, starting with the actual flight condition of the real aircraft that is characterized by position, course, altitude and wind-speed and passing some points predefined by ground control. The flight path that results in this simulated motion can be followed by the real aircraft without restrictions to the physical conditions. The time, that is computed to be arrival time, will become the valid future time for the real aircraft (Figure 1).

It is pilot's task, to perform the given demands by changing the path and the time. This is done by manipulating the intended speed, the decelleration point or - according to ground control - the position of the predefined points. After the next trajectory simulation cycle the predicted values are updated, the results of his manipulation are displayed and he can continue interactive planning. As flight tests have shown, the precision of the prediction lies within seconds and so they are much more exact than the estimation that is done usually.

The inaccuracy that ground control now has to accept, would decrease by operating the system essentially. As another advantage deviations from the preplanned path - for instance due to windspeed changing - are immediately indecated, an actualized precomputed trajectory is shown along with new recommended inputs for controlling future path. So corrections can be made without consequences to the actual flight dynamics.

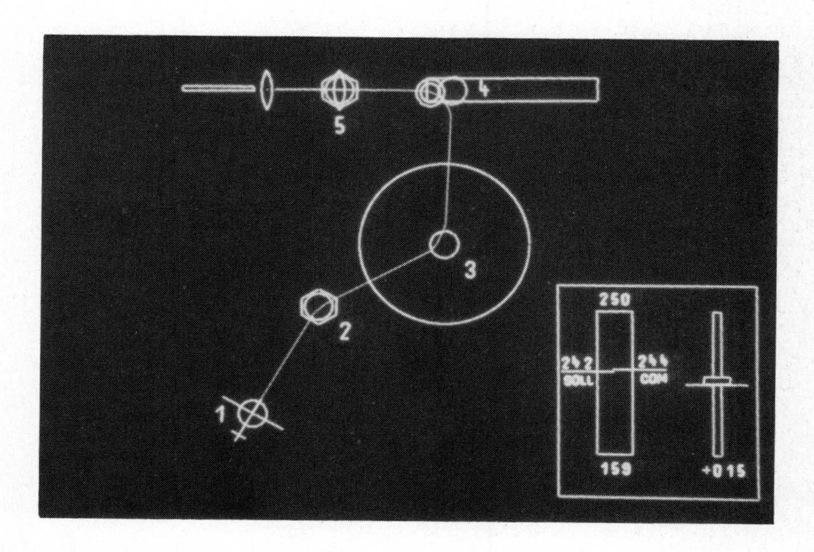

Figure 1: CRT-Display of Look-Ahead-Simulation

2. STRUCTURE OF THE PHYSICAL SYSTEM

The flight guidance system consists of

- simulation-computer-system
- cathod-ray-tube as a card display for the simulated trajectory and the geographical relations (runway, radio beacon, predefined points)
- control unit with keyboard, analog ministic (joystic) and an alphanumeric display as an input control
- alphanumeric command display indicating the actuell inputs the realtime aircraft needs to follow the precomputed path
- data link to flight equipment (air data computer, autopilot, navigation platform) for input of actual position, course and wind.

Aeronautical details have been published /1/; subject to this paper are the performance by processors and solutions of the problems resulting from high speed look-ahead computation.

3. TECHNICAL DEMANDS

The simulation as the essential part of the system could be done digital, analog or hybrid, but as it should not exceed size of other instrumental equipment when supported as standard only a digital is allowed. Although during development in the experimental plane a standard minicomputer is used for flexibility, a system can be realized meeting these requirements.

The main problems are demands concerning computational speed. Generally for an interactive man-machine-system it is necessary that input done by the man (pilot) is replied by the machine (Flight guidance system) so fast, that the answer seems to be immediate.

These desired short response times concerne to two different situations: The one is the direct echo to an input, for example the control display of a number of digits when pressing numeric keys, or the motion of a point on the CRT-Display when changing his position by help of the joystick. This is a closed loop including the man with a response time that must not be greater than about 100 ms, due to antropotechnical reasons. The other case is the time the system needs after manipulating the path to compute and show updated data like predicted time of arrival. This time lies within once and twice the time of a complete trajectory simulation. In this case, too, it is desired not to exceed the time of 0.1 s stated above. Indeed this cannot be more than a final goal as a detailed inspection of time conditions shows.

4. TIME CONDITIONS

Precomputation of the approach is convenient during the last 60 km to runway. Flying with a typical speed of 200 kts this lasts about 600 s. As shown in /1/ the trajectory of a controlled aircraft can be computed physically correct in intervalls of 1 s simulated time.

With 600 s as maximum time to simulate, the motion of the aircraft has to be computed for 600 path increments. To do this in the ideal time of 0.1, each increment computation may last 0.17 ms. For comparison: The corresponding time of the hybrid program for the real time simulator in the "Institut für Flugführung" is 33 ms.

A simplified version for test purposes when starting these researches needed 40 s for one simulation of a complete trajectory, that means 67 ms for each increment.

It clearly can be seen that the desired goal of 0.1 s as the time for computation of one path can hardly be obtained, but between 40 s and 0.1 s a solution has to be found that does not restrain operating to much. There is a typical critical situation when the pilot manipulates the future path by help of the joystick for example just to reduce a difference between desired and expected time. It may be necessary after wind changes. In this case, the desired input should be the time difference, but as this is a result of a complete trajectory simulation, feedback to the input has a delay of one trajectory cycle. So this input becomes iterativ, but mostly 3 ... 4 iteration steps are sufficient, because the data to manipulate are physical variables with a small variation range.

During this time the pilot reduces his attention to the real flight situation. According to experience it should not exceed 5s. Therefore simulation of a complete trajectory at least must not be greater than 1 ... 2 s. A comparison to the test version and the hybrid simulation shows, that even this demand cannot be performed by a standard type of computer with regular programming techniques. Systematic application of methods accelerating hard- and software are necessary.

5. INCREASING COMPUTATION EFFICIENCY

Getting more computing performance can be done by

- modifications to the computer structure, i.e. changing the hardware
- use of faster software with better programming techniques and algorithms.

Here it is necessary to do both.

5.1 Parallel computing

The flight guidance system mainly consists of two tasks

- repetitive computing the motion of the aircraft as part of the look ahead simulation
- interfacing to the real time periphery

These tasks are connected only by transfering some data. Program structure is absolutely different: The simulation is a permanently repetitive running computation whereas data transfer to the aircraft and dialog to the pilot are done only at certain times or when there are inputs by the pilot. In the latter case the program often runs idle, but it always has to be in standby for immediate dialog, that has highest priority and -when active- demands a remarkable part of capacity.

When tasks can be seperated so clearly distributed processing by parallel computers is an effective solution.

In this case the optimal structure consists of three digital processors (Figure 2):

1. the simulationprocessor for computing the trajectory
2. the displayprocessor for drawing the trajectory and showing some data
3. the dialogprocessor for I/O to pilot and aircraft instrumentation

- Because of compatibility to the digital part of the hybrid system used for the real time simulator noted above - a Pacer 100 -, the simulation processor is an EAI-Data-Pacer, with 32 k 16 bit words and about half the speed of a well known minicomputer, the PDP 11/34, when executing standard machine instructions.

- The display processor is a modified Plessey Miproc 16, that controls a x/y analog display by use of a dedicated developed graphic-controller. It includes information about structure and details of the image, and to display the trajectory and some special symbols like runway or radio beacon only coordinates and code to denote the type have to be given. Even alphanumeric symbols and bar indicators are generated directly from the binary value by the display processor.

- Dialog to the pilot and data I/O to the real time aircraft is done by a Z80 based microprocessor that is connected to

- keyboard
- analog ministick (joystick)
- input control display
- inertial navigation platform to get real time flight data
- command display to the pilot for recommended inputs
- command output to aircraft control or autopilot
- real time clock

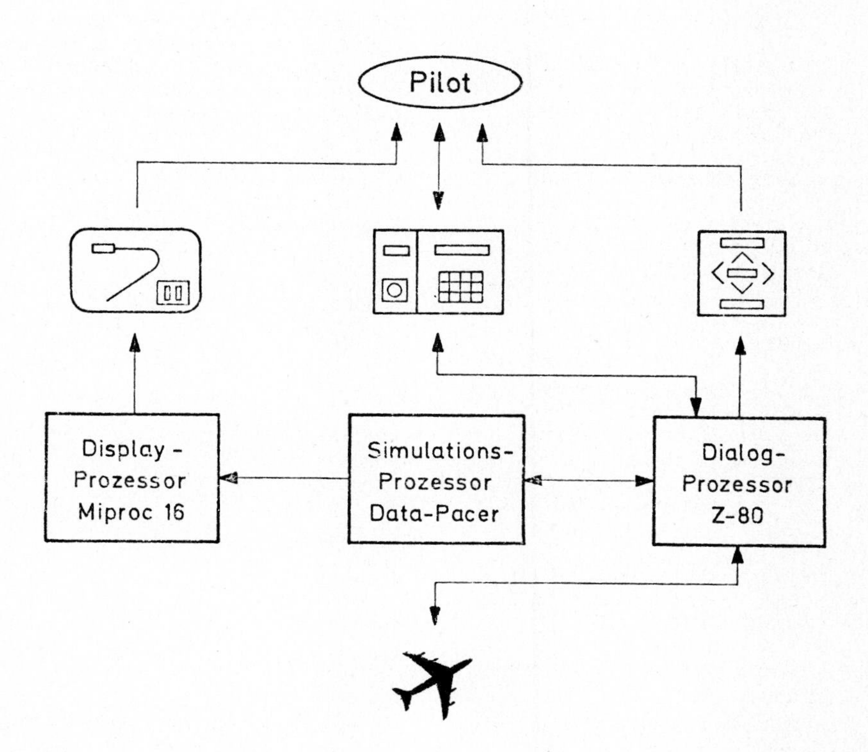

Figure 2: Processor- and Periphery-Configuration

Distributing of tasks saves most computing time, when each task becomes as small as possible, only few data transfers have to be done, and the transfers only demand minimum of processing time. Here the quantity of transfers is fixed by the necessary dialog between simulation and real time periphery, about 30 variables have to be transfered.

So also the amount of computing activity is fixed that has to be distributed, because all datas are transmitted in that code, they are used or determined by the simulation processor, either a binary integer number or as a 16 Bit fixed point number, scaled to a maximum value.

Changing of the code like the binary - BCD - conversion that is typical for I/O by help of keyboard or alphanumeric display have been given to the subprocessors, together with checks if an input has been legal (for example range violation) and interpreting joystick inputs.

The hardware interface between Simulationprocessor and both subprocessors has been built as a parallel channel for higher transfer rates with data buffers to avoid waitloops when transfering several data successively. Contrary to usual design these buffers are not shift registers but standard RAM (16 * 16 Bits or 32 * 8 Bits), that can be used as a 16 or 32 element parallel transfer register or, where needed, as a software controlled shift register.

Depending on the active program part both protocols are used by software control: one special data word determines type and size of a data block and its shake-hand conditions.

Another relief of the simulationprocessor is gained by adding a 16 Bit Hardware-Multiplier; that reduces execution time of the multiply instruction of the Data-Pacer from 70 µs - design depending rather slow - to 9 µs.

5.2 Time-effective Software

After application of these hardware enhancements, that cause the essential reduces of computation time, the speed of the most time-critical path has to be looked at next. As it is the look-ahead-simulation, further explanations will only concern to the Data-Pacer and its program.

Execution time of machine-instructions usually is not subject to change; so the software design has to enable high simulation speed.

- Except initialize part that is executed once only and that is written in Fortran IV, the complete program of the simulationprocessor is written in Assembler, coded absolutely time efficient. In the particular case it means, that not a standard, perhaps rather smart instruction sequence has been choosen, but another, that may seem to be more complicated, but indeed is the faster one. For example short loops with only a few passes have been opened and performed without repetition; so typical activities inside a loop like testing for the end and returning to the next pass are avoided.

- Floating point operations even when there is a fast floating point processor are slower than arithmetic operations that are done in the regular registers of the processor. Therefore only in the initialize part floating point variables are used for example to generate look-up tables or to scale later needed variables. Later on data only have 16 bit fixed point format, either as well known integer, in the range $-32768 \leqslant N < 32768$ or, when scaling to a maximum value is required, as so called "scaled fractions" in the range $-1.0 \leqslant x \leqslant 0.9999$. This format is similar to a floating point format with fixed exponent zero and 16 bit mantissa. Operations differ from integer operations in multiplying and dividing, as calculating with scaled fractions handles the more significant 16 bit of the 31 bit words, which are generated then.

- Optimizing computation speed has prior to be done with program parts that run most often; this part is the inner simulation loop for motion to the next path-increment. Instruction sequences that here run several times as library function or macro, for example trigonometric functions for course calculation, pythagoras function to determine distances and some functions that are known from analog computing for simulation of dynamic systems have been checked for fastest algorithms that do not reduce accuracy. These functions are:

- symmetrical and asymmetrical limiter
- integrator
- 1st order lag
- function look up by linear interpolating from a look up table

Integration is performed by the extrapolating trapezoidal rule

$$y_{n+1} = y_n + k \left(\frac{3}{2} f_{n+1} - \frac{1}{2} f_n \right) ,$$

where as an exception the integral value is not computed with 16 bit but with 32 bit accuracy. With this format the data is saved internally.

The 1st order lag results from the solution of the differential equation

$$\frac{dx}{dt} = \frac{1}{T} (y - x)$$

as

$$x_{i+1} = y_i + (x_i - y_i) \cdot k \quad \text{with} \quad k = e^{-\Delta t/T}$$

This is significantly more accurate especially for large step widths than the most often used formula

$$x_{i+1} = x_i + (y_i - x_i) \cdot F \quad \text{with} \quad F = \Delta t/T$$

that is derived from it by expansion on a power series and stopping after the first term.

Function look up from a table basis on linear interpolation with same distances between table-values. When the number of interpolation zones is an integer power of 2 computing in a binary system is much simpler, for a division can be substituted by a shift and a subtraction can be omitted.

In the present case the functions

$$y = SIN(x)$$

$$y = COS(x)$$

$$y = ATAN(x)$$

$$r = SQRT(x^2 + y^2)$$

are interpolated between 513 table values. The relative error that is caused by the linear interpolation is $1.2 \cdot 10^{-4}$ for the most critical function, the SIN and the COS with $0.3 \cdot 10^{-4}$ absolut accuracy for 16 bit numbers this error is tolerable. It even could be reduced to $0.9 \cdot 10^{-4}$ when using a 1025 sized table. The functions ATAN and - after some convertions - SQRT are generally more exact than SIN and COS because of their slow sweep /2/.

5.3 Enhancements to the machine instruction set

A very important acceleration of computation speed is gained by adding new instructions to the machine code set. In the Data-Pacer that was used as simulation processor during development the machine code is based on microprogram that is read from programmable-read-only-memories (PROMs). Modifying the microprogram is possible by exchanging the PROMs. As there have been unused address areas for microprogram-PROMs and also free bit combinations for machine instruction codes the functions descripted above have been added not as machine code routines but as one machine instruction with its respective microprogram part. So an integration for example is only one assembler instruction.

The considerable gain of time has two essential reasons:

- Each machine code instruction wastes time independently of the type of instruction itself that is needed for calculating the address of the instruction, read it from memory, decode it and find the corresponding microprogram-routine. Combining several instructions to one reduces this part.

- Microprogram-level enables a much more efficient handling of the computer-internal registers. The Data-Pacer contains 14 Registers with 16 bit that can be used then free, whereas on assembler level only 5 of these registers are accessible only for special purposes. Therefore during assembler programming it is often necessary to save intermediate results to memory by rather slow transfers. Besides on assembler level arithmetical and logical instructions can be performed only in one register, the accumulator. Normally this is a troublesome bottleneck, opposite to microprogramming, where each register has full facilities. So they can be used more economically to save time that had been used for intermediate storage.

Programming in microprogram-code can be seen as a third programming-level lower than FORTRAN and assembler. As an analogy to turning from FORTRAN to assembler, microprogramming enables decrease of computation time and more efficient utilization of hardware capabilities by more troublesome programming effort.

6. RESULTS

This flight guidance system has been performed by use of aspects noted above. Systems performance have been demonstrated connected to the flight simulator as well as incorporated in our test plane DO 28. Demands concerning flight technic have been fully satisfied. Hard- and software actions enabled 1.5 s as trajectory-simulation time.

Reduction of cycle time by hardware effort can hardly be measured, whereas the time saved by software techniques can be seen below.

As an example some typical computation times for microprogramm enhancements and their assembler-equivalents are shown:

Instruction	μ-Prog.	Assembler
Integrator	35	118
1st order lag	12.4	37.7
limiter	5	24
function look up	23.4	108.8

For reference: Function look up is derived from an EAI-supported software package. Indeed the original version can be handled more flexible, but it is functionally equivalent with an execution time of 699 μs instead of 23.4 μs.

It can be expected that further considerable reduction in computation time can be gained by excessive transfering assembler sequences to microprogram. Having sufficient addressable microprogram-space the complete inner loop, that computes the motion for the next path increment could be reduced to one machine instruction.

A further increase of simulation speed will be obtained for standard application, when the Data-Pacer as a somewhat slow minicomputer in respect to present technology would be replaced by a dedicated processor. This might be a processor built up in bipolar bit-slice-technology, for then a structure according to aspects noted above can be realized. Another reduction of trajectory cycle time to at least 30% can be expected, but in respect to actual innovation speed this apparently will not be the final solution.

References

/1/ Sundermeyer, P.: "Untersuchungen zur Verlagerung der Pilotentätigkeit auf eine höhere hierarchische Stufe der Flugführung", Dissertation, TU Braunschweig (1980).

/2/ Frieling, R.: "Increasing the computing speed of real time simulation by hardware supplements and selected software", 3rd Simulation Symposium, Interlaken, Switzerland (1980).

Simulation in Engineering Sciences
J. Burger and Y. Jarny (eds.)
Elsevier Science Publishers B.V. (North-Holland)

SIMULATION AND CONTROL OF A LONG FREIGHT TRAIN

Moustafa E. Ahmed and Mohamed M. Bayoumi

Department of Electrical Engineering
Queen's University
Kingston, Ontario
Canada

A Freight Train Simulator F.T.S. is developed to simulate the longitudinal dynamics and drawbar forces of trains consisting of multi-locomotives and freight cars over various terrain profiles. The F.T.S. has been used to design automatic control schemes and to assess their performance with respect to the coupler forces and speed variations as well as the energy consumption. This paper describes the main features of this simulator and some simulation results.

1. INTRODUCTION

Long freight trains consisting of multilocomotives and up to 150 large capacity cars have been used for hauling coal from mines in the Rocky Mountains to shipping facilities in the west of Canada. Problems associated with such heavy trains on mountainous terrains include breaking up or derailment of the train due to excessive coupler forces. A Freight Train Simulator F.T.S. is developed to simulate the longitudinal dynamics of the train over various terrain profiles in order to develop automatic control schemes [1-4] and to assess their performance with respect to the coupler forces and speed variations as well as the energy consumption. The following are some of the unique features of the F.T.S.

1 - The F.T.S. models and monitors the fast and slow modes along the train that are caused by the coupler dynamic characteristics. Such dynamics are not accurately predicted in other available simulation packages such as the: Association of American Railroads Train Operation Simulator AAR-TOS, or the Train Performance Calculator T.P.C.

2 - The braking system on each car is treated individually with a separate independent input to allow studying the electropneumatic brake systems in which the brake pipe, which transfers the brake signal pneumatically to the cars, is replaced by electric wires. In the meantime, the conventional air brake system can be introduced by an appropriate subroutine which simulates the brake pipe function. Such flexibility is not available in the other simulation programs.

3 - The operating commands are transferred to the F.T.S. through a user's subroutine. This feature is essential to implement complex train handling techniques in which the control decisions are taken on-line in accordance with the train status and the operating environments. The other simulators accept the control commands as input data, and hence all the control commands have to be determined apriori and can not be changed on line.

2. TRAIN MODEL

The train model presentation makes use of the studies of McLane et. al. [1], Gruber and Bayoumi [2-4], and the AAR train operation simulator documentations [5-7]. Each train member is modelled as shown in Figure 1.

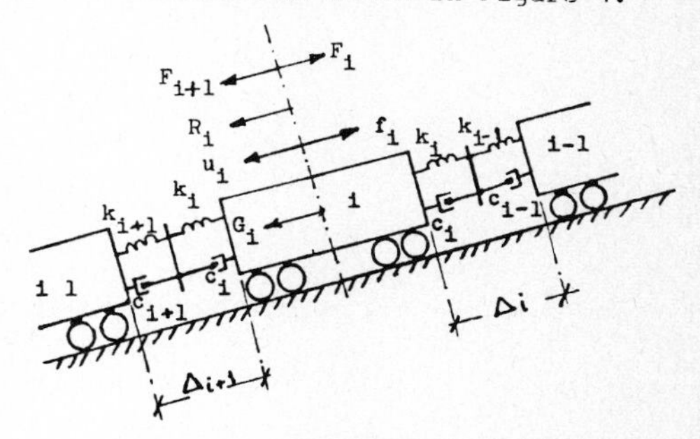

Figure 1. Model of the Train Member i.

The drawbar forces may be computed using the following formula.

$$\dot{v}_i = \frac{1}{m_i} [F_i(t)-F_{i+1}(t)-R_i(t)-G_i(t)+f_i(t)-u_i(t)]$$

$$\dot{\Delta}_i = v_{i-1}(t) - v_i(t) \qquad ..(1)$$

where

v_i = velocity of the ith train member

m_i = mass of the ith train member

Δ_i = the dynamic displacement between the train members i and i-1.

F_i, F_{i+1} = draft gear forces

R_i = total rolling and air dynamic resistance

G_i = the component of the gravitational force against the direction of motion

u_i = the total brake force

f_i = the throttle or dynamic brake force

Each coupler is represented by a spring with constant k_i and a damping coefficient c_i and a nonlinear dead zone DZ. Thus the draft gear force is given by

$$F_i = \delta_i k_i + c_i \dot{\delta}_i \qquad \text{if} \quad \Delta_i \geq DZ \qquad \ldots(2)$$
$$= 0.0 \qquad \text{if} \quad \Delta_i < DZ$$

where δ_i is the elongation (or compression) of the front coupler of the ith car. δ_i in turn is computed via the following non linear differential equation

$$\dot{\delta}_i = \frac{1}{c_i + c_{i+1}} [-(k_i + k_{i-1})\delta_i + d_1(\Delta_i, k_{i-1}, DZ) + d_2(\Delta_i, c_{i-1}, DZ, \dot{\Delta}_i)] \qquad \ldots(3)$$

where

$$d_1(x,y,z) = xy \quad \text{if} \quad x \geq z$$
$$= 0.0 \quad \text{if} \quad x < z$$

$$d_2(x,y,z,w) = xy \quad \text{if} \quad w \geq z$$
$$= 0.0 \quad \text{if} \quad w < z$$

The F.T.S. accepts cars and locomotives with different couplers characteristics; namely the spring constants k_i and the damping coefficients c_i. However the dead zone DZ is taken to be the same for all couplers.

The rolling resistance and the aerodynamic drag force are nonlinear functions of speed and are given by.

$$R_i = r_{oi} + r_{1i} v_i + r_{2i} v_i^2 \qquad \ldots(4)$$

where r_{oi}, r_{1i} and r_{2i} are constants which depend on the mass, the number of axles of the train member i, and its cross section area. They also depend on the location in the train if the train member i is a locomotive.

The simulation program computes the gravitational force G_i on each car individually. This feature is found to be necessary to study the effects of rapidly changing terrain profiles on the train dynamics.

The locomotives are assumed to be diesel-electric units in which the maximum allowable throttle/dynamic brake force obeys aproximately the following general curve

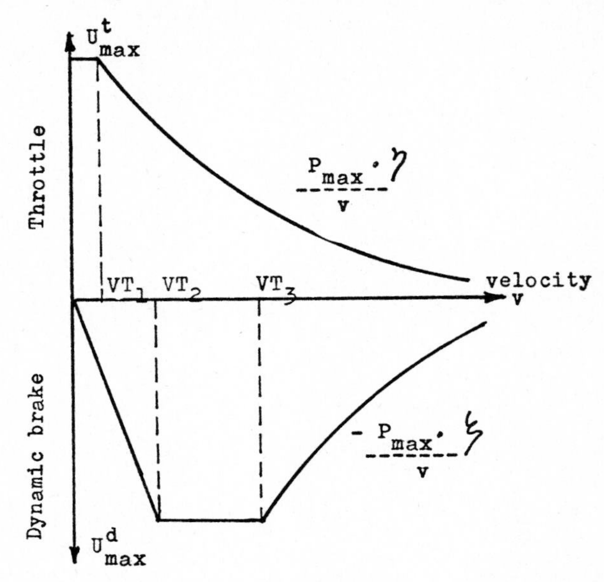

Figure 2. Typical Tractive Force Constraints of Diesel-Electric Locomotives.

where

P_{max} = the maximum power

= the transmission efficiency

= locomotive constant

VT_1, VT_2 and VT_3 = the transition velocities of the locomotive throttle/dynamic brake data.

In the dynamic brake state the traction motors act as generators which absorb the kinetic energy of the train and feed it to a dummy electric load. For a given velocity the throttle/dynamic brake range is quantized to the following approximate levels.

$$f = F_t = \frac{n}{8} \frac{P_{max}}{v(t)} \qquad \text{for throttle} \qquad \ldots(5)$$

$$f = F_{db} = -\frac{n}{8} \frac{P_{max}}{v(t)} \qquad \text{for dynamic brake} \qquad \ldots(6)$$

where

n = 0,1,2 ....., 8 is the power level.

In the analog Electropnemuatic brake system [8], the brake activation signal is carried by electric wires connected through the train. The electric signal activates an electro-pneumatic valve, see Figure 3, which passes air from a high pressure air reservoir to the brake cylinder in such a way that the brake cylinder pressure increases inversely proportional to the actuating current. The brake cylinder pressure P_c is related to the brake force u by the relation

$$u = \beta P_c \qquad \ldots(7)$$

where β is the train member brake force constant.

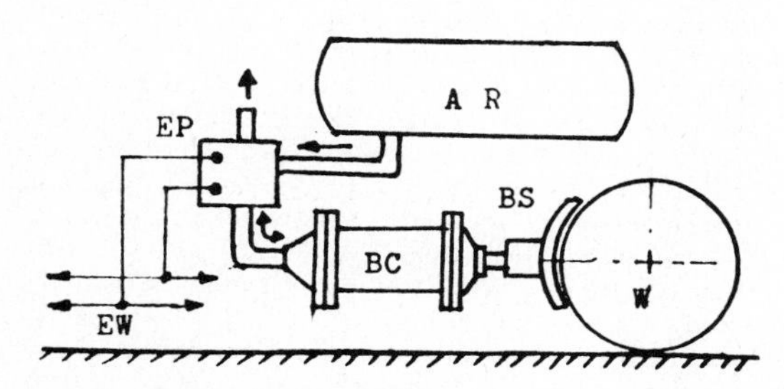

AR : Air Reservoir
BC : Brake Cylinder
BS : Brake Shoe
EP : Electro-Pneumatic valve
EW : Electric Wires
W : Wheel of a locomotive/car

Figure 3. A Car/Locomotive E.P. Brake System

The unsymmetric behaviour of the brake cylinder pressure during release and buildup is approximated by linear relation during increase and an exponential decay relation during release [6].

3. PROGRAM DESCRIPTION

In this section we provide a brief description of the various subroutines constituting the F.T.S. The F.T.S. is built in such a way to simplify the data setting phase as well as to produce the output results in the most convenient manner. Most of the relevant performance parameters are plotted directly using the calcomp plotter with minimum intervention by the user. A brief report of the simulation run is also printed. A trace back option prints a summary of the train status after each sampling period. The calling sequence of the various subroutines is summarized in the tree shown in Figure 4. The arrows go from the low level routines to the higher level ones. The letter M above an arrow indicates multicalls. The calling sequence coincides with the clock-wise direction.

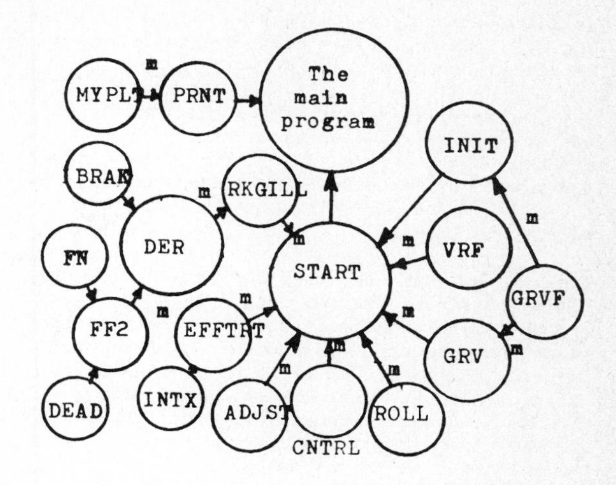

Figure 4. The Calling Sequence of the F.T.S. subroutines.

The main program reads data and performs conversion of units, if necessary, to the MKS system. It also prints out the title and a brief report on the simulation run. The START subroutine is responsible for the coordination of all technical computations. It initiates the train status by calling INIT which in turn calls GRVF to initialize the gravitational forces. Then START begins the main integration loop. Each integration step starts by calling VRF which updates the target velocity and the terrain section. GRV determines the slope of the ground under each car and finds the gravitational forces by calling GRVF. The ROLL subroutine computes the rolling resistance and the aerodynamic drag force on each car. CNTRL is a user's subroutine which returns at each call the control commands of the train. The ADJST subroutine is called only if manual control is used to translate the manual commands to throttle forces and brake forces. EFFTRT computes the effective throttle or dynamic brake according to the relation shown in Figure 2.

The INTX subroutine performs the quantization of the power levels as shown in equations 5 and 6. The integration is performed by RKGILL using the Runge-Kutta-Gill method. This integration method is chosen because it requires a minimum number of storage registers, and gives the highest attainable accuracy, i.e. controls the growth of round-off error and requires comparatively few instructions [9]. The subroutine DER computes the left handside of the nonlinear dynamic equation

$$\frac{dx}{dt} = f(x,t,u)$$

The unsymmetric behaviour of the brake cylinder pressure during release and build up is accommodated by subroutine BRAK. The subroutine FF2 and the Fortran functions DEAD and FN compute the coupler forces using the nonlinear relations 2 and 3 .

Due to obvious economical reasons the rolling resistance and the gravitational forces are updated about every 30 meters rather than every integration step. Updating the control commands can be controlled by the user control subroutine itself.

The ratio between the CPU time to the simulation time varies depending on many factors as the number of cars, the integration step, the number of measurement samples during the run, the number of terrain sections passed during the run, the computations involved in the User's control program ... etc. However, the CPU time may be estimated roughly using the following emperical formula.

$$t = \alpha_0 + \beta \left(\frac{1}{15} \; \frac{nc \times ts}{STEP} \right) \left(1 + .3 \; \frac{n_s - 100}{100} \right)$$

where

t : CPU time in seconds

α_0: The overhead time for loading, input/output and for the CALCOMP plotting

β : Constant depends on the computer power

t_s: The simulation time

STEP:The Integration Step

n_c : The number of cars + locomotives

n_s : The number of measurement samples during the run.

4. SOME SIMULATION RESULTS

In this section we provide two illustrative examples of the train handling techniques over a severe downhill. In the first example the train runs under manual operation in accordance with the current recommendations to avoid severe impacts between cars on the downhill. The air brakes are applied upon arriving to the downhill and the throttle is decreased gradually to idle. The dynamic brake plays here a minor role, only to assist the air brake. However, this technique has two serious drawbacks for long steep descent. In the first place it is a costly method since the application of brakes for long durations requires frequent replacement of the brake shoes. Second, as the brake shoes temperature increases the effectiveness of the brakes decreases, and hence the speed of the train starts to drift causing more heating and further deterioration of the brake. In the meantime the use of dynamic brakes on downhills requires high skill to estimate the proper amount of dynamic brake and the proper rate of application to avoid impacts between cars and possible derailment of the train.

A multivariable proportional Integral P.I. Controller together with an offline schedule has been developed to maintain the target speed and to minimize the coupler forces. Emphasis is given in this case to the use of dynamic brake to reduce the load on the air brake system. The P.I. Controller is chosen because it can achieve both velocity regulation and disturbance accommodation simultaneously. By this feature the controller provides the proper corrective action in case of any decrease in the effectiveness of one of the braking methods and eliminates the drift in the speed. The simulation tests for both cases are shown in Figures 5.a to 5.d and 6.a to 6.d respectively. The test example consists of 30 cars and two locomotives. The first locomotive is located at the front, the second is located after the first 20 cars. A summary of the train data and the test conditions are given in Tables 1 and 2.

	Locomotives	Cars	Units
Type	U25	GONDOLA	
Weight	139.	102.	Tons
Length	18.3	19.7	Meters
Max. Brake Force	$4.\ 10^5$	$1.47\ 10^5$	Newtons
Max. Power	2000.	--	H.P.
Locomotives transition velocities	8,13,26	--	Km/hr.
Couplers Spring Constant	70.10^5	70.10^5	Newton/meter
Couplers Damping Constant	$2.04\ 10^5$	$2.04\ 10^5$	N.Sec/meter

Table 1: Data of the 2-locomotive 30 car train

Terrain Section	1	2	3	Units
Length of the terrain section	724.	965.	5632.	Meters
Grade (% slope)	0.0	-2.	0.0	%
Target Velocity	64.36	64.36	64.36	km/hr
Total rolling and air resistance	$7.72\ 10^4$	$7.72\ 10^4$	$7.72\ 10^4$	Newton
Total Gravitational Forces	0.0	-6.4×10^5	0.0	Newton

Table 2: Data and Schedule Values for Simulation

Figure 5. Simulation Results For Manual Operation.

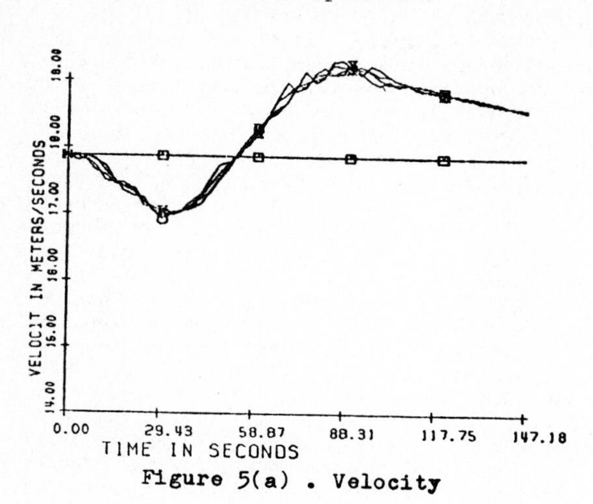

Figure 5(a) . Velocity

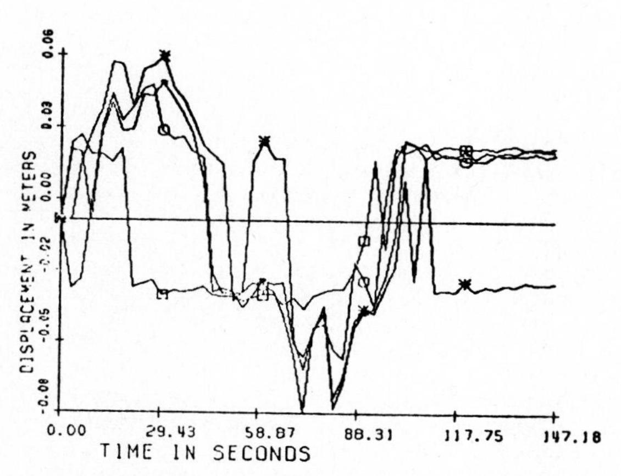

Figure 5(b) . Coupling displacement

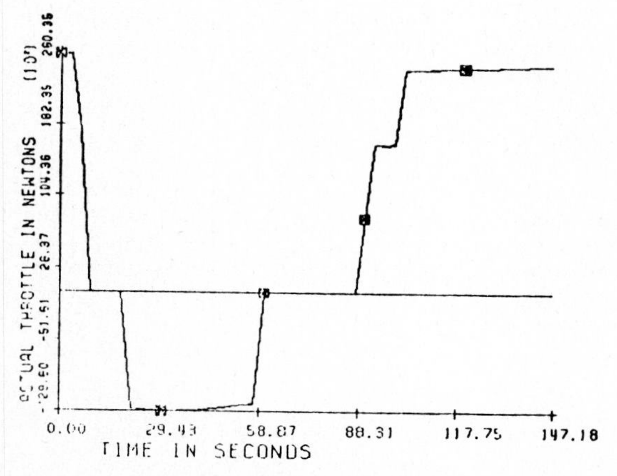

Figure 5(c) . Throttle

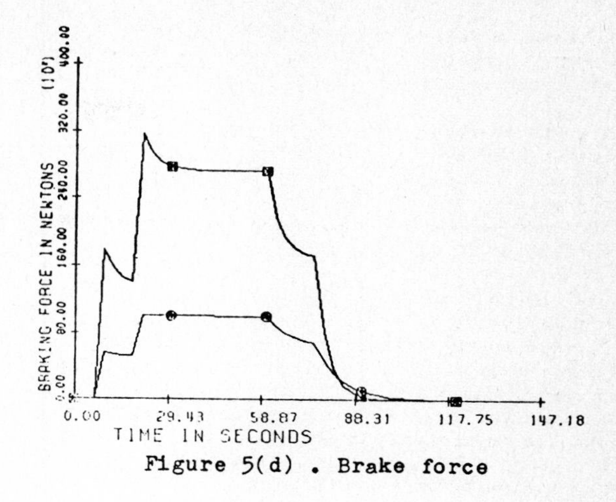

Figure 5(d) . Brake force

Figure 6. Simulation Results For Automatic Operation.

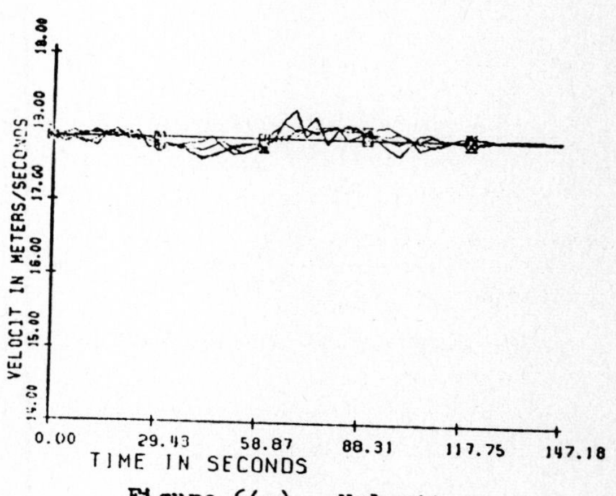

Figure 6(a) . Velocity

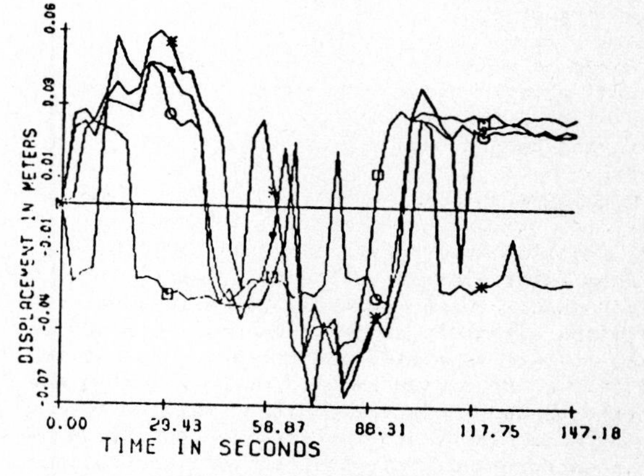

Figure 6(b) . Coupling displacement

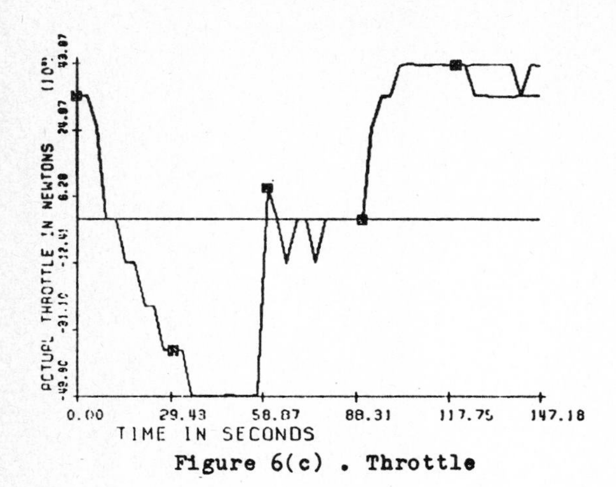

Figure 6(c) . Throttle

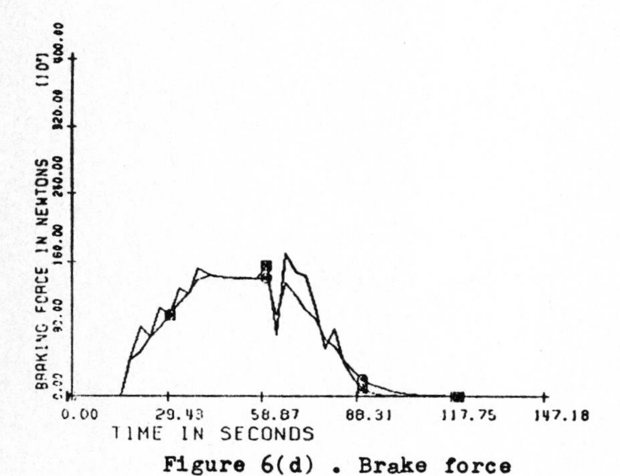

Figure 6(d) . Brake force

Fig.	Description	Graphic Symbol □	○	✱	#	⧗
(a)	Velocity	V_{ref}	V1	V10	V22	V32
(b)	Coupler displacement between train members	1,2	10,11	21,22	22,23	—
(c)	Throttle	1	1	22	—	—
(d)	Brake force	1	10	22	32	—

CONCLUSION

A simulation program for the longitudinal dynamics of freight trains has been built to improve the train handling techniques and to develop automatic control schemes. The F.T.S. can be put to a wide variety of applications, e.g., to study the impact of introducing the electropneumatic brake on the train handling and performance, to investigate the effect of allocation and distribution of locomotives along the train on the overall performance and the energy consumption.

REFERENCES

[1] McLane P.J.; Peppard, L.E., and Sundareswaran, K.K., Decentralized Feedback Control for Brakeless Operation of Multi-locomotive powered trains, IEEE Transaction on Automatic Control, Vol. AC-21 (1976) 358-363.

[2] Gruber, P. and Bayoumi, M.M., Train Braking Performance Studies Using Suboptimal Controllers, Part I, II, Report No. 79-11, 79-12, CIGGT (1979).

[3] Gruber, P. and Bayoumi, M.M., Suboptimal Control Strategies for Multilocomotive Powered Train, IEEE Transaction on Automatic Control, Vol. AC-27 (1982) 536-546.

[4] Gruber, P. and Bayoumi, M.M., State Estimation In Long Freight Trains, Report No. 79-21, CIGGT (1979).

[5] Low, E.M. and Garg, V.K., Train Operations Simulator Programmer's Manual, Report No. R-359, Association of American Railroads (1979).

[6] Low, E.M. and Garg, V.K., Train Operations Simulator Technical Documentations, Report No. R-269, Association of American Railroads (1978).

[7] Luttrell, N.W.; Gupta, R.K.; Low, E.M. and Martin, G.C., Train Operations Simulator - User's Manual, Report No. R. 198, Association of American Railroads (1977).

[8] Naughan Railway Consulting Services Inc., Electro-Pneumatic Train Brakes, Report No. TP1435, CIGGT (1977).

[9] Romanelli, M.J., Runge-Kutta Methods for Solution of Ordinary Differential Equations, in Ralston, A. and Wilf, H.S. (eds), Mathematical Methods For Digital Computers (John Wiley and Sons Inc., New York, 1960).

ACKNOWLEDGEMENT:

The authors would like to thank Messers G.W. English of the CIGGT, and W.S.C. McLaren and W.G. Rowan of the TDC for their cooperation and discussions. This work was done under a contract with Transport Develoment Centre, Canadian Department of Transport.

X. MECHANICAL SYSTEMS

Simulation in Engineering Sciences
J. Burger and Y. Jarny (eds.)
Elsevier Science Publishers B.V. (North-Holland)

A DYNAMIC SIMULATION OF WATER SUPPLY SYSTEMS AND ITS APPLICATION TO ADVANCED CONTROL SYSTEMS

S. Kobayashi, S. Saito, Y. Anbe and T. Arakawa

Heavy Apparatus Engineering Laboratory
Toshiba Corporation
Fuchu Works 1, Toshiba-cho, Fuchu-shi, Tokyo, Japan

This paper describes a new method for dynamic simulation of computer aided design of water supply systems in which the water supply networks are formulated as a general nonlinear composite dynamical system solved with the Backward Differentiation Formula. This simulator has already been used for the design of several actual plant control systems, the most recent being the one described in this paper.

1. INTRODUCTION

Recently the need has increased for simulation for analysis and synthesis of water supply control systems. Since the conventional algorithm in analysis of a typical pipe network gives only static characteristics, dynamic simulation must be programmed on a case-by-case basis. Given this requirement, there is an obvious need for an easily maintained dynamic simulator which can be used by non-specialists without requiring a great deal of time.

Among the many software packages for computer-aided design of control systems are CONPAK [1] for large, complex linear systems; DPS [2], a dynamic simulation language for chemical process engineering; a simplified dynamic modeling method for reservoir dynamics in large water distribution systems.[3]
This paper presents a new technique for simulation of water supply systems for computer-aided analysis and design of control systems. The simulator presented here has four specifications:

a) It can simulate a water supply system which can be described as a concentrated parameter system.
b) It can simulate a variety of processes which depend on the system construction and system element characteristics without additional programming. Simulation in this case is based solely on data input.
c) The control algorithm can be programmed with FORTRAN.
d) It can calculate dynamic characteristics of any water head and flow.

2. A MATHEMATICAL MODEL OF WATER SUPPLY SYSTEMS

In specification (b) of this simulator, the mathematical model of water supply systems must be very general. In this simulator the water supply systems are expressed by a directed graph. A reservoir, a branch point and an end of demand and supply are regarded as nodes. A water-way between interconnected nodes is regarded as a branch. The system elements such as pumps, valves and pipes are considered as loss or discharge head elements on branches. For the equation of flow continuity at each node, the balance of energy head for interconnected nodes and the flow head characteristics of system elements, the system model is formulated as:

$$\begin{cases} \mathbb{D} \cdot \mathbb{h} - \mathbb{L}\,(\,\mathbb{q},\ d\mathbb{q}/dt\,) = \mathbb{0} & (1) \\ \mathbb{D}_2 \cdot \mathbb{q} = \mathbb{0} & (2) \\ d\mathbb{h}_1/dt = \mathbb{S} \cdot \mathbb{D}_1 \cdot \mathbb{q} & (3) \end{cases}$$

where:
$\mathbb{D} \in R^{N \cdot b}$: incidence matrix (N*b matrix)
if $\mathbb{D}(i,j)=1$: water flows from node i into node j
if $\mathbb{D}(i,j)=-1$: water flows from node j into node i
if $\mathbb{D}(i,j)=0$: water doesn't flow from node i into node j or from node j into node i
$\mathbb{D}_1 \in R^{n_1 \cdot b}$: related to reservoirs
$\mathbb{D}_2 \in R^{n_2 \cdot b}$: related to branch points
$\mathbb{h}=(h_1,h_2,\ldots,h_N)^T \in R^N$; h_i : water head [mAq] at node i
$\mathbb{h}_1 \in R^{n_1}$: related to reservoirs
$\mathbb{q}=(q_1,q_2,\ldots,q_b)^T \in R^b$; q_i : water flow [m^3/sec] of branch i
$\mathbb{L}=(l_1,l_2,\ldots,l_b)^T \in R^b$; l_i : summation of loss and discharge head on branch i and is expressed by function of flow head characteristics (See Appendix)
$\mathbb{S}=(s_{ij}) \in R^{n_1 \cdot n_1}$; $s_{ij}=1/A_i \cdot \delta_{ij}$
A_i: bottom area of reservoir i
δ_{ij}: Kronecker's delta
N: number of nodes
b: number of branches
n_1: number of reservoirs
n_2: number of branch points
n_3: number of ends
and $N = n_1 + n_2 + n_3$ (4)

Now, let the number of equations be N ,

$$N = b + n_1 + n_2 \qquad (5)$$

The number of unknown variables m is :

$$m = b + N = b + (\,n_1 + n_2 + n_3\,) \qquad (6)$$

Therefore, the condition of ends (n_3) is given and the system equation can be solved.

<u>The Nonlinear Composite Dynamical System</u>

There are $d\mathbb{q}/dt$ terms implicitly in system equations (1) - (3). If there are no differential terms $d\mathbb{q}/dt$, and if the vector $\mathbb{q}$ can be expressed explicitly for the vector $\mathbb{h}$, in the form:

$$\mathbb{q} = F(\mathbb{h}) \quad (7)$$

where, F : a function

the vector $\mathbb{h}$ can be the only unknown variable in this system of equations. If the system equations are linear, it is not particularly difficult to formulate the system equations as in the form given in (7); or, if each flow head characteristic dh_e of system element can be expressed as a pipe friction loss head in the form:

$$dh_e = R_e \cdot q^v \quad (8)$$

where, v, R_e : constants

the system equation can be described explicitly with the vector $\mathbb{h}$ using the equation:

$$q = (dh_e / R_e)^{1/v} \quad (9)$$

But this simulator must calculate not only pipe characteristics but also various system elements. It is very difficult to change implicit system equations to explicit equations. That is to say, eliminating the vector $\mathbb{q}$ is not an effective approach to the problem. When the $d\mathbb{q}/dt$ term exists implicitly, the general Nonlinear Composite Dynamical System (NCDS) may be expressed as:

$$\begin{cases} f_i(x_1, x_2, \ldots, dx_1/dt, dx_2/dt, \ldots) = 0 & (10) \\ g_i(x_1, x_2, \ldots) = 0 & (11) \end{cases}$$

where, x_i: h_i or q_i, f_i, g_i: functions, $i \in \{1,\ldots,m\}$

must be changed to explicit equations, that is to say, simultaneous differential equations:

$$dx_i/dt = f'_i(x_1, x_2, \ldots) \quad (12)$$

where, f'_i : a function

Since this is generally difficult or impossible, the system model is treated as it is whether the term $d\mathbb{q}/dt$ exists or not.

3. THE ALGORITHM FOR A NUMERICAL ANALYSIS

3.1 Solving for the NCDS numerically

If the system is expressed only with static simultaneous equations, the Newton-Raphson method is commonly used; however, if the system is described only with explicit simultaneous differential equations, open algorithms such as the Runge-Kutta-Gill method may be used when the $d\mathbb{q}/dt$ term may be disregarded; but these methods will not suffice when the $d\mathbb{q}/dt$ term is necessary for simulation.
In the simulator presented in this paper, the Backward Differentiation Formula method is used.

The Gear-Nordsiek method and Backward Differentiation Formula (BDF)[4] are on the same base, and the BDF can solve NCDS directly. The BDF method is known to be the same as the Tabrou method[5] in which an approximate equivalent of dx_i/dt is given by Gear's algorithm. A principal advantage of BDF method is that the trial error E_k for the time step d is given numerically[6], and the time step can be automatically controlled using the Euclidian norm under the following criterion:

$$E_0 \geqq E_k(d) \quad (13)$$

where, E_0 : given

This is an important factor in reducing calculation time.

3.2 A simple decomposition algorithm

Even though the BDF method has the merit of automatic control of the time step, calculating time becomes too great for practical purposes when the system dimensions become relatively large. There are several graphical approaches to this problem, one being to decompose a total system into independent partial systems. Another approach is to decide the calculation order with block triangulation of the structual matrix[7]. Still another approach is to reduce dimensions by substitution and elimination of variables. This simulator employs the first method, decomposition. First, we solve for the vector $\mathbb{h}$ with the explicit algorithm under a given initial condition, and solve for the vector $\mathbb{q}$ with the BDF method. Then, the convergence criterion which is based on the error norm is checked. In many cases, the time constants of the flow dynamics ($\mathbb{q}$) and the reservoir dynamics ($\mathbb{h}$) are different and unacceptable error rarely occurs except at simulation starting time. This simple decomposition algorithm may be described in three steps.

step 1: devide the total branch set B into subsets B_i (i = 1,2,...,k) ;

$$B = B_1 \cup B_2 \cup \ldots B_k$$

where B: the set which consists of all branches, B_i: the set which consists of all branches that are connected to node i.

step 2: devide the total branch point set T into the subsets $\hat{T}_i$ with the division of the set B into the subsets $\hat{B}_i$ (i = 1,2,...,l) ;

$$T = \hat{T}_1 \cup \hat{T}_2 \cup \ldots \hat{T}_l$$

$$\hat{T}_i \cap \hat{T}_{i'} = \phi \quad (i, i' \in \{1,2,..,l\},\ i \neq i')$$

$$B = \hat{B}_1 \cup \hat{B}_2 \cup \ldots \hat{B}_l \cup \hat{B}_c$$

$$\hat{B}_i \cap \hat{B}_{i'} = \phi \quad (i, i \in \{1,2,..,l\},\ i \neq i')$$

where T: the set which consists of all branch points, $\hat{T}_i$: the set which consists of the connected branch points, and any branch point in $\hat{T}_i$ does not connect to any brnch point in $\hat{T}_{i'}$ (i,i' $\in$ {1,2,...,l}, i $\neq$ i'), $\hat{B}_i$: the set which consists of all branches connected to any branch point in $\hat{T}_i$, $\hat{B}_c$: the set which consists of all branches that are not included in any set $\hat{B}_i$ (i$\in${1,2,...,l})

step 3. devide the set $\hat{B}_c$ into subsets $\hat{B}_{ci}$ consisting of only one branch.

An exsample of the procedure is shown in Figure 1.

4. APPLICATION TO DESIGN OF CONTROL SYSTEMS

This simulator has been used for the design of several actual plant control systems. We will now describe the most recent application.

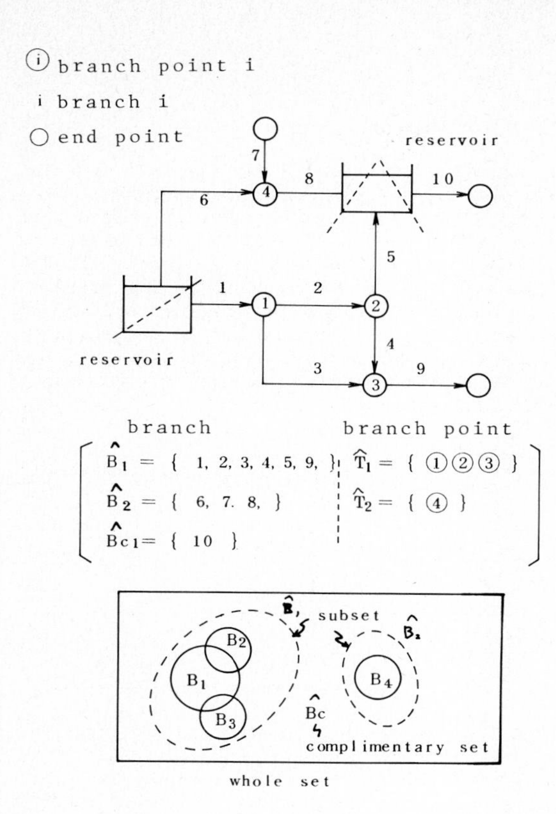

Figure 1 An example of the decomposition procedure

4.1 System Description

This simulator was used for the design of a plant control system of a booster pump station where the distance between the purification plant and the booster pump station was about 10 km. (See Figure 2) The booster pumps pump the water from the purification plant to a reservoir and directly to the distribution pipe network. For an ideal balance of energy consumption, the pumps should be used at night, while the potential energy of the reservoir is utilized by day; consequently, the water supply route is changed by pumps and valves, and the pressure on the input side of valve CV3 varies widely. To minimize leakage, the desired value of the water pressure is time varying.
Water distrubution pressure control is important because of the direct connection to consumers; incorrect pressure could cause a leak in a water pipe network. In conventional control systems, it is difficult to maintain the dynamical behavior because the characteristics of valves and hydraulic system vary with the volume of water flow. In response to this problem, an advanced controller,"The Real Time Tuning Controller " has been developed and applied.

4.2 Control System

Figure 3 is a schematic of the control system. This control system is based on a (PI-) control system with an automatic design calculation algorithm using real-time processing information as a feedforward compensation for the feedback control.For linear systems,there are auto-tuning

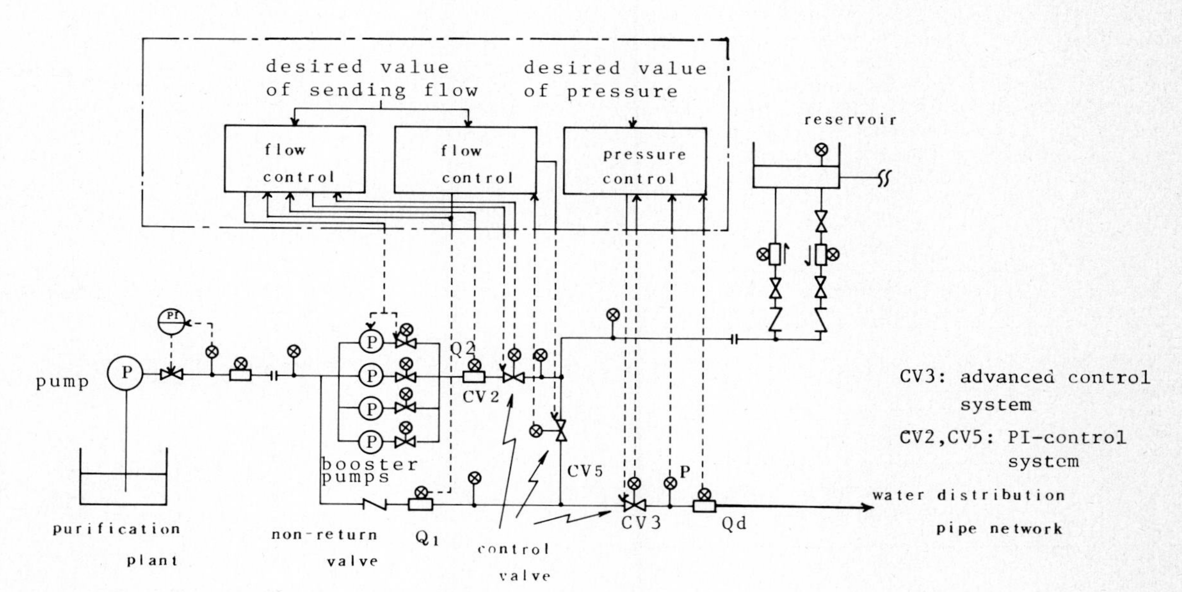

Figure 2 A water supply system

methods for PID- controller [8]. These methods include an identification algorithm of the process model, but in this case the controller must treat the process as nonlinear. If the identification algorithm is employed, for example the Recursive Least Square (RLS) method, the parameter values are automatically given, but the speed of following changes of processing nonlinear characteristics is very slow even if the forgetting factor is used. Thus, in this control system, the model for design calculation has to be provided in advance, expressed in the form of functions with process variables even though the identification algorithm is very charmfull. The model of this controller can treat the system described as follows:

$$c \cdot d(dx)/dt = - a \cdot dx + b(x,u,p) \cdot du + dv \qquad (14)$$

where x: controlled variable (pressure, p)
u: manipulated variable (opening of valve CV3)
v: disturbance (unmeasurable), p: process variable (measurable), "d" means a small change, a,b,c: parameter or non-linear function (a= 1, b=b (u,p), c=0, p: Qd)
The model can output the value of b on each sampling step. Control specifications for design calculations are:

(e) The rise time or the time to peak can be kept constant with a slight change in step response.

(f) The damping ratio can be kept constant when $c \neq 0$.

(g) The desired gain characteristics for the overall transfer function in the frequency domain can be maintained.

4.3 Simulation results and field data

Using the equations(1)-(3), the process of the system with the advanced controller was simulated. Figure 4(a) shows the step response. Figure 5 shows the behavior of the total system when the water supply route was changed. From these simulation results, the values of the rise time in the specification(e) were decided, and was applied to the actual water pressure control system. Figure 4(b) shows the field data.

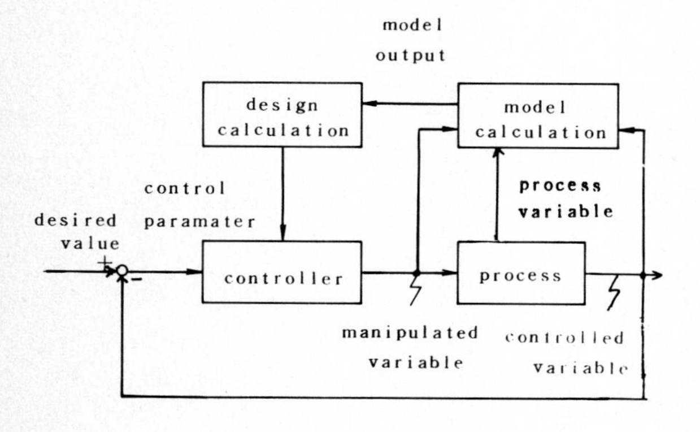

Figure 3 The configuration of the advanced controller

5. CONCLUSION

The dynamic simulation of water supply systems described in this paper is expressed as a general nonlinear composite dynamic system with graphic techniques for computer aided design of various systems. The NCDS is solved by the BDF method with a simple decomposition algorithm. The advanced controller presented in this paper is an application of the method to design, and the plant control system reported here is already in service and its characteristics have been confirmed.

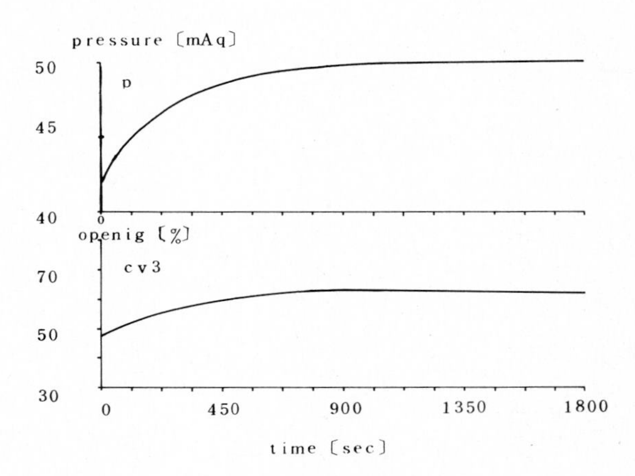

(a) Simulation results

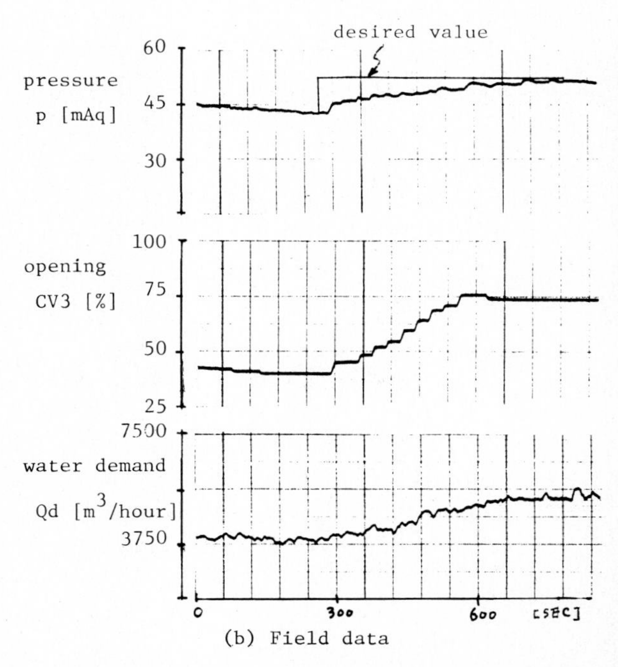

(b) Field data

Figure 4 Step response

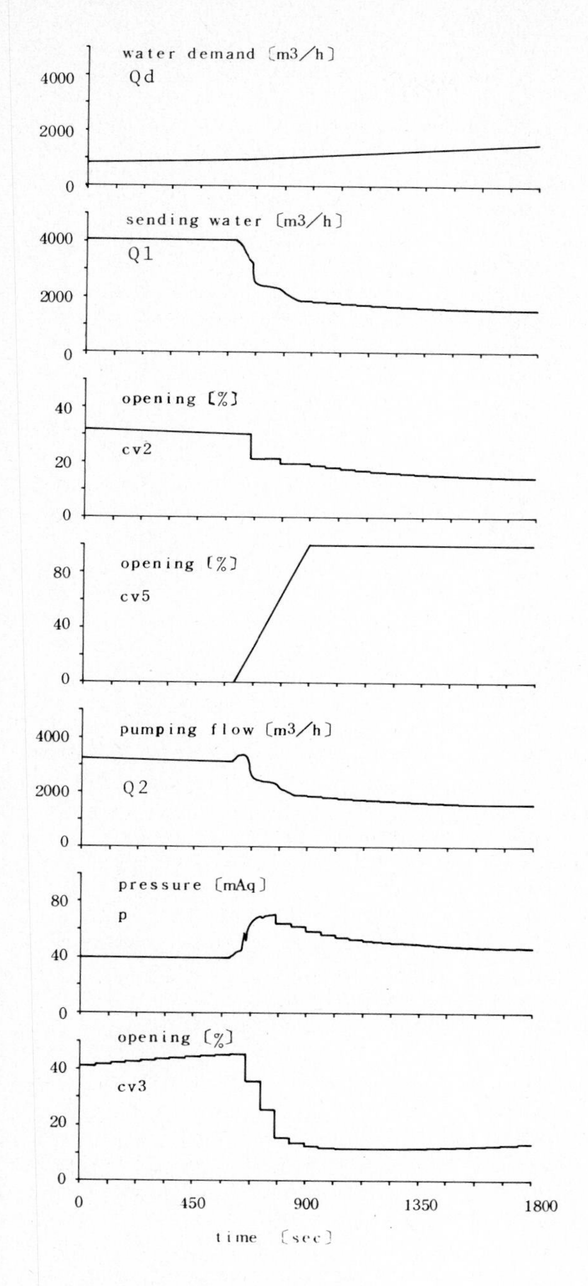

Figure 5 Simulation results
(changing of water-supply route)

REFERENCES

[1] K. Furuta, et al., Simulation in computer aided design for large complex systems, Simulation of control systems, I.Troch (ed.), IMACS, North-Holland, (1978) 63-65

[2] R.J.G. Sebastian et al., The structure and input language of DPS, Proc, 1982 Summer Computer Simulation Conference, SCS (1982) 531-536

[3] B. Coulbeck, Dynamic simulation of water distribution systems, IMACS trans. XXII (1980) 222-230

[4] R.Brayton, et al., A new efficient algorithm for solving differential-algebraic systems using implicit differentiation formula, proc.IEEE 60-1 (1972) 98-108

[5] G.D.Hachtel, et al., The sparce tabrau Approach to network analysis and design, IEEE CT 18-1 (1971) 101-113

[6] L.Chua, et al., Computer-aided analysis of electronic circuits, Englewood Cliffs, N.J. Rerentice-Hall (1975)

[7] D.V.Steward, Partitionizing and tearing systems of equations, SIAM J.Numer.Arnl.Ser.B2(1965) 345-365

[8] T.Sigemasa, et al., A closed loop auto-tuning method for digital PID controller, proc. IECI '81 IEEE (1981) 427-432

Appendix:

The functions which express the characteristics of the system elements (l_i in this paper) are explained in the reference [3], for example, except the pipe friction characteristics.

In this simulator, the relationship of the energy balance for the interconnected nodes with a pipe is formulated as:

$$(h_i - h_j) - (R_{ij} \cdot q_{ij} + R'_{ij} \cdot dq_{ij}/dt) = 0$$

where R_{ij}, R'_{ij}:parameters related to the pipe

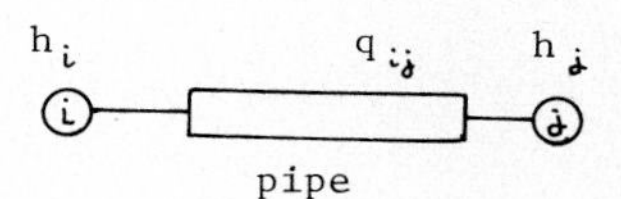

Simulation in Engineering Sciences
J. Burger and Y. Jarny (eds.)
Elsevier Science Publishers B.V. (North-Holland)

MATHEMATICAL MODEL FOR THE SOLIDIFICATION OF METAL CASTING IN A MOLD

Alfredo Bermúdez and José Durany

Departamento de Ecuaciones Funcionales
Universidad de Santiago
Santiago de Compostela, Spain

ABSTRACT.- A mathematical model to simulate heat transfer during solidification of a metal in a mold is considered.
It consist of a set of nonlinear partial differential equations which are solved by using a finite element method.
At each time step a nonlinear system has to be solved. For this an iterative algorithm is proposed.
Numerical results are given which show that the model can be used to predict formation of cavities in the metal due to shrinkage.

1. INTRODUCTION

In this paper a mathematical model of Stefan is considered, to simulate heat transfer during solidification of a metal in a mold.
Most metals and alloys contract on solidifying. Due to this shrinkage cavities can appear in the casting if liquid regions isolated from the riser are present at some stage of solidification (see for example, R.W. Ruddle [8], B. Chalmers [4], M. Flemings [6]). To achieve sound castings some appendages called hot tops are added and chills and insulators introduced in the mold.
Mathematical simulation of the solidification process should give information about optimal placement of these elements.
In order to consider boundary conditions on the surface of the mold, which would be easier in practice, the domain of the model has been taken to be that constitued by metal, mold, chills, insulators, etc.
Thus, an initial-boundary problem for a nonlinear partial differential equation has to be solved, for which a variational formulation is given.
Existence of a solution follows from a more general result given in A. Bermúdez, J. Durany and C. Saguez [1].
For numerical purposes another variational formulation is introduced by relaxing the requirement of continuity for test functions and introducing Lagrange multipliers on the boundaries between each two materials. It is a domain decomposition formulation which is similar to hybrid methods in elasticity.
In this way the global equation is equivalent to a set of heat transfer problems (one in each material) and, in particular, Kirchoff transformation can be done in order to solve the problem more easily.
Next a Galerkin discretisation is introduced by using implicit finite differences in time and triangular finite elements of degree one in space.
At each time step a nonlinear problem has to be solved. For this an iterative algorithm is proposed, which requires to compute the solution of several linear systems, one per material, in each iteration. Their coefficient matrices are constant so that assemblage and factorisation are done only once.

2. THE MODEL

Consider a material solidifying in a mold in which insulators and chills are introduced.
Let Ω_1 be the region occupied by this material and Ω_i $(i= 2,\dots,n)$ those occupied by mold, insulators, chills, etc.
Let $\Omega_{1L}(t)$ and $\Omega_{1S}(t)$ be the liquid and the solid regions at the time t, then we have

$$\rho_1(\theta_1)c_1(\theta_1)\frac{\partial\theta_1}{\partial t} - \nabla.(k_1(\theta_1)\nabla\theta_1) = 0 \tag{2.1}$$

in $\Omega_{1L}(t)$ and $\Omega_{1S}(t)$, a.e. $[0,T]$

$$\theta_1|_L = \theta_1|_S = \bar{\theta} \tag{2.2}$$

$$k_1(\bar{\theta})\nabla\theta_1|_L.\nu - k_1(\bar{\theta})\nabla\theta_1|_S.\nu = -\rho(\bar{\theta})Lv.\nu \tag{2.3}$$

(Stefan's condition)

on the solid-liquid interphase $\Sigma(t)$ (free boundary).

$$\rho_i(\theta_i)c_i(\theta_i)\frac{\partial\theta_i}{\partial t} - \nabla.(k_i(\theta_i)\nabla\theta_i)= 0 \qquad (2.4)$$

in $\Omega_i(t)$, a.e. $[0,T]$, $i= 2,\dots,n$

$$\theta_i= \theta_j \qquad (2.5)$$

$$k_i(\theta_i)\nabla\theta_i.\nu= k_j(\theta_j)\nabla\theta_j.\nu \qquad (2.6)$$

(trasmission conditions) on $\Lambda_{ij}= \bar{\Omega}_i \cap \bar{\Omega}_j$.

Boundary conditions, on the external surface of the mold, must be added to these equations. For example

$$\theta= g \ , \quad \text{on} \ \Gamma(t) \ , \quad \text{a.e.} \ [0,T] \qquad (2.7)$$

(where Γ denotes the boundary of $\Omega= \bigcup_{i=1}^{n} \Omega_i$).

Finally, we take the initial conditions

$$\theta_i(x,0)= \theta_{i0}(x) \quad \text{in} \quad \Omega_i \ , \quad i= 1,\dots,n \qquad (2.8)$$

The following notations have been used

θ: temperature
c: specific heat
L: latent heat
ν: unit normal vector
ρ: density
k: thermal conductivity
v: velocity of the liquid-solid interphase
$\bar{\theta}$: temperature of solidification

3. VARIATIONAL FORMULATION

The objetive is to get an equation which holds in the whole subdomain Ω_1 instead of in the liquid and solid separately.
From (2.1)-(2.3), by using a Green's formula, it can be obtained the so-called enthalpy formulation of the Stefan problem (see for example J.L. Lions [7], A. Bermúdez, C. Saguez [3]), i.e.:

$$\frac{\partial H_1(\theta_1)}{\partial t} -\nabla.(k_1(\theta_1)\nabla\theta_1)\ni 0 \qquad (3.1)$$

$$\text{in} \quad \Omega_1\times(0,T)$$

where $H_1(\theta)$ is the multivalued function relating temperature to local enthalpy per unit of volume given by

$$\int_0^\theta \rho_1(s)c_1(s)ds \qquad \text{if} \quad \theta<\bar{\theta}$$

$$\left[\int_0^{\bar{\theta}} \rho_1(s)c_1(s)ds, \int_0^{\bar{\theta}} \rho_1(s)c_1(s)ds+\rho_1(\bar{\theta})L\right] \qquad \text{if} \quad \theta= \bar{\theta}$$

$$\rho_1(\bar{\theta})L+ \int_0^\theta \rho_1(s)c_1(s)ds \qquad \text{if} \quad \theta>\bar{\theta}$$

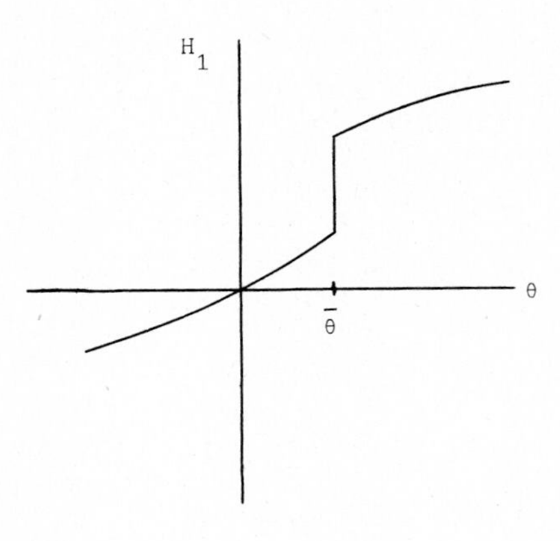

Fig. 1.- Enthalpy.

In a second step a variational formulation is obtained which holds in Ω. By multiplying (3.1) by a test function z in $H_0^1(\Omega)$, using a Green's formula and taking into account the transmission conditions (2.5)-(2.6), we get the following problem:
To find $\theta(x,t)$ definedin $\Omega\times(0,T)$ with $\theta(t)\in H^1(\Omega)$ such that

$$\int_\Omega \frac{\partial H(x,\theta)}{\partial t} zdx+ \int_\Omega k(x,\theta)\nabla\theta.\nabla zdx= 0 \qquad (3.2)$$

$$\text{for all } z \text{ in } H_0^1(\Omega)$$

$$\theta= g \qquad \text{on} \qquad \Gamma\times(0,T) \qquad (3.3)$$

$$\theta(x,0)=\theta_0(x) \qquad \text{in } \Omega \qquad (3.4)$$

where $H(x,\theta)$ and $k(x,\theta)$ are defined by

$$\left.\begin{array}{l} H(x,\theta)= H_i(\theta) \\ k(x,\theta)= k_i(\theta) \end{array}\right\} \quad \text{if } x\in\Omega_i \qquad (3.5)$$

and H_i is the enthalpy function in Ω_i.
Existence of a solution for (3.2)-(3.4) follows from a more general result given by A. Bermúdez, J. Durany, C. Saguez [1] in an abstract framework.

4. A DOMAIN DECOMPOSITION METHOD.

A variational formulation different from (3.2)-(3.4) is introduced by relaxing the continuity requirements for test functions. It involves Lagrange multipliers defined on the boundaries between two domains Ω_i and Ω_j, representing heat flux accross these boundaries.

Once Lagrange multipliers are known, the problem is equivalent to n nonlinear heat transfer equations, one for each domain Ω_i.

It can be shown (see J. Durany [5] for details) that (3.2)-(3.4) is equivalent to the following problem:

To find θ_i defined in $\Omega_i \times (0,T)$ with $\theta_i(t) \in H^1(\Omega_i)$ $(i= 1,\ldots,n)$ and λ_{ij} defined on $\Lambda_{ij} \times (0,T)$ $(i,j= 1,\ldots,n)$ such that

$$\sum_{i=1}^{n} \int_{\Omega_i} \frac{\partial H_i(\theta_i)}{\partial t} z_i \, dx + \sum_{i=1}^{n} \int_{\Omega_i} k_i(\theta_i) \nabla\theta_i \cdot \nabla z_i \, dx + \sum_{i,j=1}^{n} \int_{\Lambda_{ij}} \lambda_{ij}(z_i - z_j) \, d\Gamma = 0 \tag{4.1}$$

$$\forall z_i \in H^1(\Omega_i), \quad z_i = 0 \quad \text{on } \Gamma$$

$$i= 1,\ldots,n$$

$$\theta_i = \theta_j \quad \text{on} \quad \Lambda_{ij} \quad (i,j,= 1,\ldots,n) \tag{4.2}$$

$$\theta_i = g \quad \text{on} \quad \Gamma \quad (i=1,\ldots,n) \tag{4.3}$$

$$\theta_i(x,0) = \theta_0(x) \quad \text{in} \quad \Omega_i . \tag{4.4}$$

This formulation is very similar to the hybrid primal formulation considered in J.M. Thomas [10], to which we refer to for a more rigorous analysis.

Notice that the equation (4.1) is equivalent to the following set of equations $(i= 1,\ldots,n)$:

$$\int_{\Omega_i} \frac{\partial H_i(\theta_i)}{\partial t} z_i \, dx + \int_{\Omega_i} k_i(\theta_i) \nabla\theta_i \cdot \nabla z_i \, dx + \sum_{j=1}^{n} \int_{\Lambda_{ij}} (\lambda_{ij} - \lambda_{ji}) z_i \, d\Gamma = 0 \tag{4.5}$$

$$\forall z_i \in H^1(\Omega_i), \quad z_i = 0 \quad \text{on } \Gamma .$$

From (4.5) we deduce the equality

$$k_i(\theta_i) \frac{\partial \theta_i}{\partial \nu} = \sum_{j=1}^{n} (\lambda_{ji} - \lambda_{ij}) \quad \text{on} \quad \Lambda_{ij} \tag{4.6}$$

$$(i= 1,\ldots,n),$$

which gives a physical meaning to the Lagrange multipliers λ_{ij}.

Since in (4.5) k_i does not depend on x, the Kirchoff transformation can be done to avoid the nonlinearity of the diffusion term. Let y_i be - given by

$$y_i = \beta_i(\theta_i) \quad \text{with} \quad \beta_i(r) = \int_0^r k_i(s) ds \tag{4.7}$$

Then the problem (4.5) becomes

$$\int_{\Omega_i} \frac{\partial G_i(y_i)}{\partial t} z_i \, dx + \int_{\Omega_i} \nabla y_i \cdot \nabla z_i \, dx + \sum_{j=1}^{n} \int_{\Lambda_{ij}} (\lambda_{ij} - \lambda_{ji}) z_i \, d\Gamma = 0 \tag{4.8}$$

$$\forall z_i \in H^1(\Omega_i) \quad , \quad z_i = 0 \quad \text{on} \quad \Gamma$$

with $G_i = H_i \circ \beta_i^{-1}$.

5. DISCRETIZED PROBLEM

We introduced a finite element discretisation for the two-dimensional problem. Let τ_i^h be a triangulation covering Ω_i $(i= 1,\ldots,n)$, where h denotes the lenght of the longest triangle edge, and each triangle edge is either the edge of another triangle or has end points on the boundary of Ω_i. Moreover, assume that $\bigcup_{i=1}^{n} \tau_i^h$ is a triangulation of Ω.

Functions y_i and λ_{ij} are approximated, respectively, by elements of the following spaces:

$$V_i^h = \{ v \in C^0(\Omega_i) : \; v_{|K} \in P_1 \quad , \quad \forall k \in \tau_i^h \} \tag{5.1}$$

$$W_{ij}^h = \{ w \in C^0(\Lambda_{ij}) : \; w_{ij}{}_{|K \cap \Lambda_{ij}} \in P_0 \quad , \quad \forall k \in \tau_i^h \cup \tau_j^h \} \tag{5.2}$$

where P_r , $r= 0,1$ denotes the space of polynomials of degree less or equal than r.

On the other hand, an implicit finite difference scheme is considered for time discretisation, with time step $\Delta t = \frac{T}{N_T}$.

Thus, the discretized problem is the following: To find $y_i^{h\,m+1} \in V_i^h$ and $\lambda_{ij}^{h\,m+1} \in W_{ij}^h$ $(m=0,\dots,N_T-1)$ such that

$$\int_{\Omega_i} \frac{G_i(y_i^{h\,m+1}) - G_i(y_i^{h\,m})}{\Delta t} z_i^h \, dx + \int_{\Omega_i} \nabla y_i^{h\,m+1} \cdot \nabla z_i^h \, dx +$$

$$+ \sum_{j=1}^{n} \int_{\Lambda_{ij}} (\lambda_{ij}^{h\,m+1} - \lambda_{ji}^{h\,m+1}) z_i^h \, d\Gamma = 0 \qquad (5.3)$$

$$\forall z_i^h \in V_i^h \,, \quad z_i^h = 0 \quad \text{on } \Gamma .$$

$$\int_{\Lambda_{ij}} [\beta_i^{-1}(y_i^{h\,m+1}) - \beta_j^{-1}(y_j^{h\,m+1})] \mu_{ij}^h \, d\Gamma = 0 \qquad (5.4)$$

$$\forall \mu_{ij} \in W_{ij}^h$$

$$y_i^{h\,m+1} = \beta_i(g((m+1)\Delta t)) \qquad (5.5)$$

at points of Γ

$$y_i^{h\,0} \in V_i^h \,, \quad y_i^{h\,0} = \beta_i(\theta_0) \qquad (5.6)$$

at points of $\bar{\Omega}_i$.

To calculate $y_i^{h\,m+1}$ and $\lambda_{ij}^{h\,m+1}$ from $y_i^{h\,m}$ and $\lambda_{ij}^{h\,m}$ a nonlinear numerical system has to be solved. For this we use an iterative algorithm which is similar to duality methods given in A.Bermúdez, C.Moreno [2] (see also C.Saguez [9]). Let $q_i^{h\,m+1}$ be defined by $q_i^{h\,m+1} \in G_i(y_i^{h\,m+1}) - \omega_i \, y_i^{h\,m+1}$, where ω_i is an arbitrary positive real number. Then (5.3) can be written in the following equivalent form:

$$\frac{\omega_i}{\Delta t} \int_{\Omega_i} y_i^{h\,m+1} z_i^h \, dx + \int_{\Omega_i} \nabla y_i^{h\,m+1} \cdot \nabla z_i^h \, dx =$$

$$= - \sum_{j=1}^{n} \int_{\Lambda_{ij}} (\lambda_{ij}^{h\,m+1} - \lambda_{ji}^{h\,m+1}) z_i^h \, d\Gamma -$$

$$- \frac{1}{\Delta t} \int_{\Omega_i} q_i^{h\,m+1} z_i^h \, dx + \frac{1}{\Delta t} \int_{\Omega_i} G_i(y_i^{h\,m}) z_i^h \, dx \qquad (5.7)$$

$$\forall z_i^h \in V_i^h \,, \quad z_i^h = 0 \quad \text{on } \Gamma$$

$$q_i^{h\,m+1} = (G_i)_{\omega_i} (y_i^{h\,m+1} + \frac{1}{2\omega_i} q_i^{h\,m+1}) \qquad (5.8)$$

with

$$(G_i)_{\omega_i}(r) = 2\omega_i(r-s) \qquad (5.9)$$

s being the unique solution of the equation

$$\omega_i s + G_i(s) = 2\omega_i r \qquad (5.10)$$

(see [2] for a justification of (5.8)).

Equations (5.7), (5.8) and (5.4) suggest the following iterative algorithm:

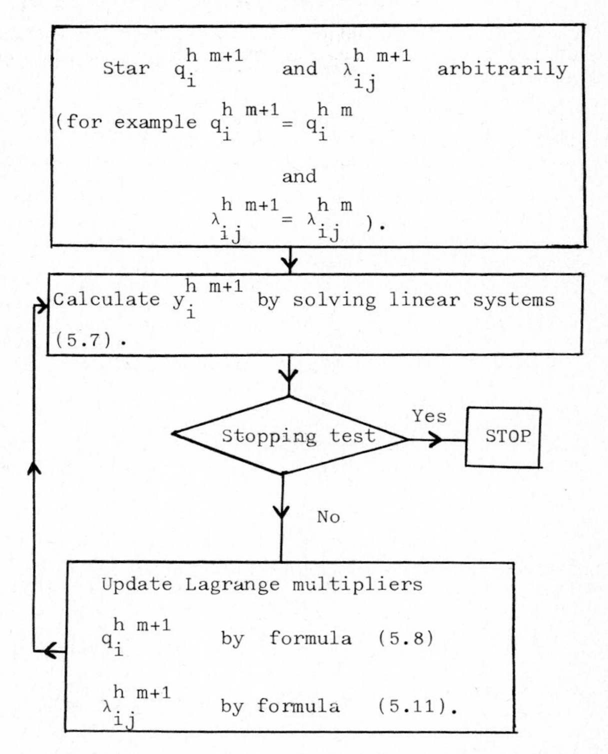

Updating of $\lambda_{ij}^{h\,m+1}$ is done by

$$\int_{\Lambda_{ij}} \lambda_{ij}^{h\,m+1} \mu_{ij}^h \, d\Gamma = \int_{\Lambda_{ij}} (\lambda_{ij}^{h\,m+1} + \alpha_{ij}(\beta_i^{-1}(y_i^{h\,m+1}) -$$

$$-\beta_j^{-1}(y_j^{h\,m+1})))\mu_{ij}^h \, d\Gamma \,, \quad \forall \mu_{ij}^h \in W_{ij}^h \qquad (5.11)$$

Notice that coefficient matrices of linear systems (5.7) do not depend on m. On the other hand, they are symmetric and positive definite so that they can be calculated and factorised only once. Thus, solving linear systems at each iteration only involves two triangular linear systems.

It can also be observed that computations in each of the two steps the proposed algorithm consists of can be performed in parallel.

6. NUMERICAL RESULTS

The method presented in this paper has been applied to simulate the solidification of a piece of steel in a sand mold. The same problem but introducing insulators and chill has also been considered.
We have taken the following physical data:

Ω_1: steel (piece)

$\rho_1 = 7 \frac{g}{cm^3}$ $\quad c_{1L} = 0.184 \frac{cal}{g\ ^\circ C}$

$L = 66.3 \frac{cal}{g}$ $\quad c_{1S} = 0.107 \frac{cal}{g\ ^\circ C}$

$k_{1L} = 0.065 \frac{cal}{cm\ ^\circ Cs}$

$k_{1S} = 0.175 \frac{cal}{cm\ ^\circ Cs}$

$\bar{\theta}_1 = 1502^\circ C$

Ω_2: sand (mold)

$\rho_2 = 1.61 \frac{g}{cm^3}$ $\quad c_2 = 0.267 \frac{cal}{g\ ^\circ C}$

$k_2 = 0.003 \frac{cal}{cm\ ^\circ Cs}$

$\Omega_3 = \Omega_4$: plaster (insulator)

$\rho_3 = \rho_4 = 1.1$ $\quad c_3 = c_4 = 0.2$

$k_3 = k_4 = 0.0004$

Ω_5: iron (chill)

$\rho_5 = 7.3$ $\quad c_5 = 0.12$

$k_5 = 0.15$

A net composed by 576 triangular finite elements has been considered, the degrees of freedom being 325.
Boundary conditions are the following:

i) On the external surface of the mold temperature is taken to be 25ºC.

ii) The external surface of the feeder is isolated, $\frac{\partial \theta}{\partial \nu} = 0$.

Initial temperature of the metal is supposed to be 1537ºC and that of the other elements (mold, insulators, chill) 25ºC.

Figures 2 to 5 give the position of the solid-liquid interphase after 30 and 60 seconds for each of the two cases considered.

Figures 6 and 7 show the last points solidifying (A,B,C) and, consequently, the places where shrinkage cavities could appear.

Finally, figure 8 gives evolution of temperature during solidification on line $\overline{PQ}$.

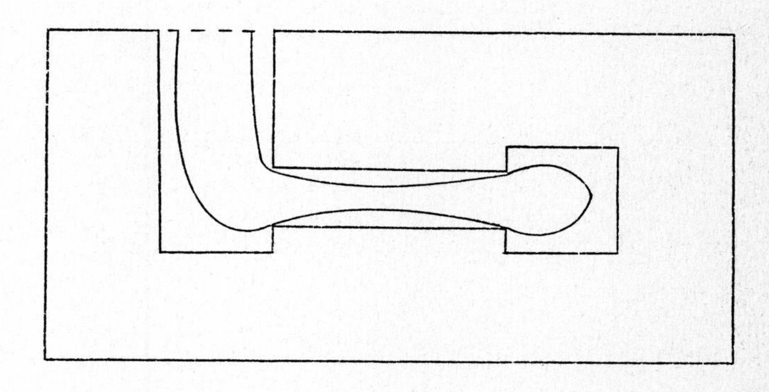

Figure 2.- Interphase after 30 seconds. (Metal-mold).

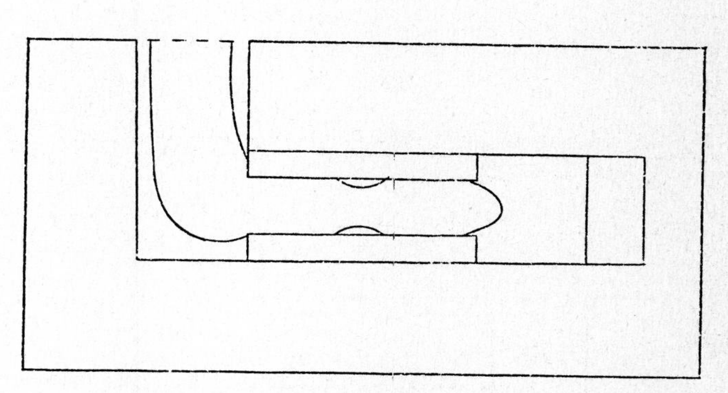

Figure 3.- Interphase after 30 seconds. (Metal-mold-insulators-chill).

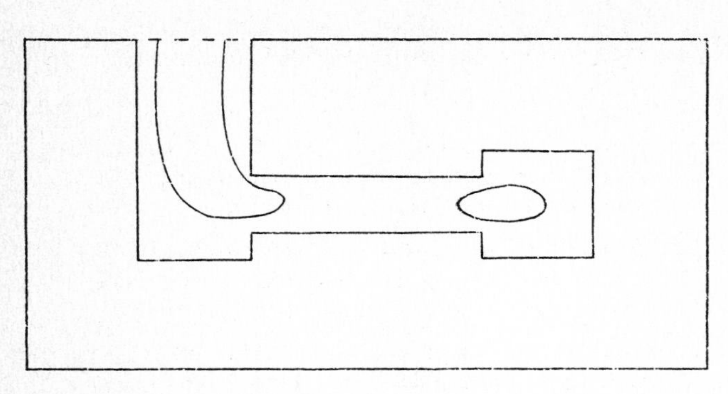

Figure 4.- Interphase after 60 seconds. (Metal-mold).

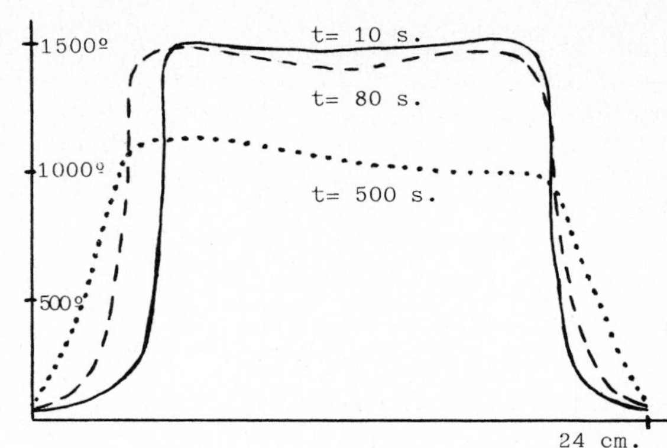

Figure 8.- Evolution of temperature on PQ.

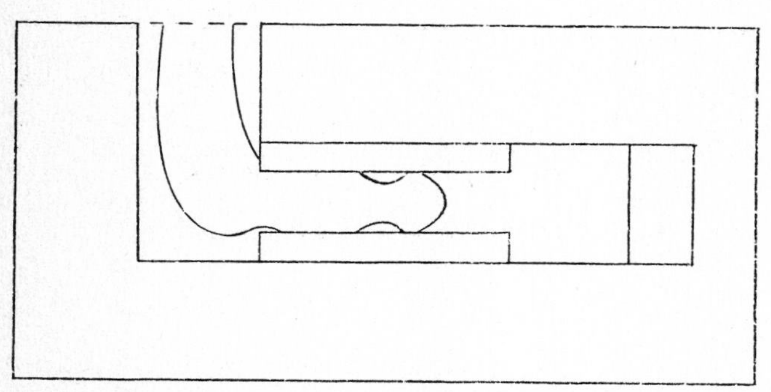

Figure 5.- Interphase after 60 seconds. (Metal-mold-insulators-chill).

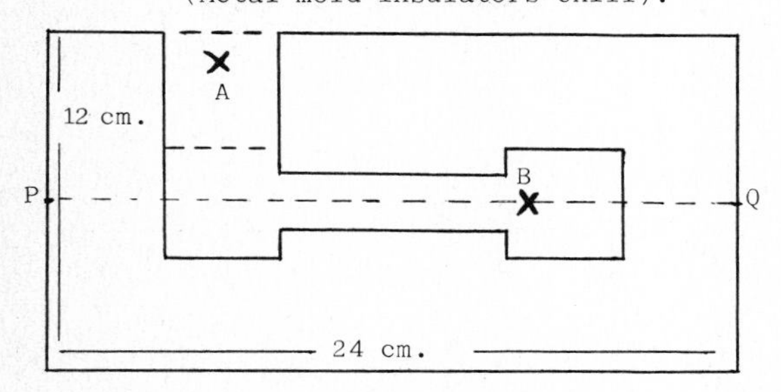

Figure 6.- Shrinkage cavities (metal-mold).

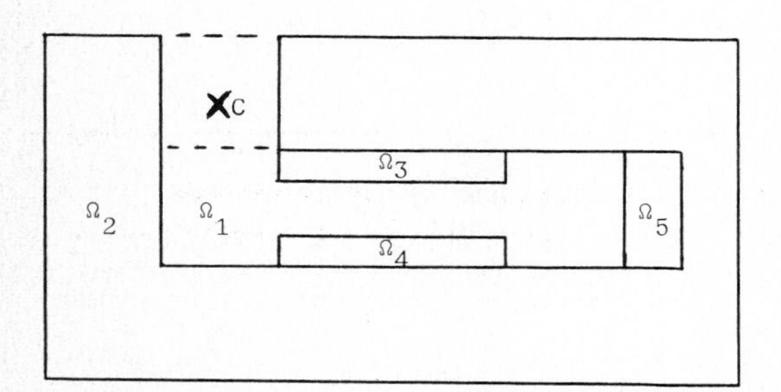

Figure 7.- Shrinkage cavity (metal-mold-insulators-chill).

REFERENCES

[1] Bermúdez, A., Durany, J. and Saguez, C. An existence theorem for an implicit nonlinear evolution equation. (à aparaître).

[2] Bermúdez, A. and Moreno, C. Duality methods for solving variational inequalities. Comp. and Math. Appl. 7 (1981) 43-58.

[3] Bermúdez, A., and Saguez, C. Etude numérique d'un problème de solidification d'un alliage Rapport de Recherche nº 104, INRIA, France (1981).

[4] Chalmers, B. Principles of solidification (John Wiley, New York, 1964).

[5] Durany, J. Contribución al estudio matemático del problema de Stefan en medios no homogéneos. Thèse, Dep. Ec. Func., Univ. Santiago (à paraître).

[6] Flemings, M. Solidification processing (Mc Graw Hill, New York, 1974).

[7] Lions, J.L. Quelques méthodes de résolution de problèmes aux limites non-linéaires (Dunod, Paris,1969).

[8] Ruddle, R.W. The solidification of castings (The Institute of Metals, London, 1957).

[9] Saguez, C. Contrôle optimal de systèmes à frontière libre. Thèse d'Etat, Univ. Tec. Compiègne (1980).

[10] Thomas, J.M. Sur l'analyse Numérique des méthodes d'éléments finis hybrides et mixtes. Thèse d'Etat. Univ. Paris 6 (1977).

Simulation in Engineering Sciences
J. Burger and Y. Jarny (eds.)
Elsevier Science Publishers B.V. (North-Holland)

COMPUTER-SIMULATION OF AN EXPERIMENTAL METHOD FOR NOTCH-SHAPE-OPTIMIZATION

Eckart Schnack

Institute for Technical Mechanics B
University of Karlsruhe, West-Germany

To reduce the weight of constructions and to save material for highly loaded machinery parts consisting of high-quality material, numerical optimization processes have been suggested by TVERGAARD [1] in 1973, by FRANCAVILLA, RAMAKRISHNAN and ZIENKIEWICZ [2] in 1975, and by KRISTENSEN and MADSEN [3] in 1976. In 1979, DURELLI and RAJAIAH [4] have developed an experimental method by means of the photoelasticity. As this method produces excellent results, on the other hand, however, proves to be a very time-wasting experimental method, requiring experimental knowledge, a numerical simulation of the experimental procedure is suggesting itself. It can be seen that the experimentators intuitively make use of wellknown laws of the notch-stress-theory. The generalization of the method leads to an optimization strategy, suggested by SCHNACK [5-9] in 1977. An example in the end will demonstrate the good agreement between the results of the experimental method and the numerical simulation.

1. INTRODUCTION

In 1979, DURELLI and RAJAIAH [4] optimized hole shapes for a big plate during uniaxial and biaxial in-plane loading (s. fig. 1).

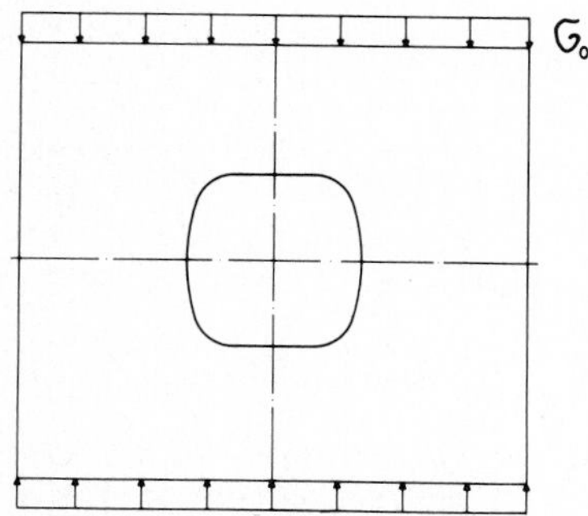

Fig. 1: Plate with optimal hole shape during uniaxial in-plane loading

For this, exterior geometrical limits for the hole were set, which fix the maximal extension of the cut. The plate is now manufactured from photoelastically active material with an estimated optimal profile and will then be submitted to stresses within an appropriate loading frame. A transillumination of the plate by monochromatical light produces the stress trajectories. By this, the tangential stress is known at the load-free surface. By handfiling at the modell, the maximal tangential stress is tried to be reduced to a minimum. For this, material, lying within only lightly loaded zones, is removed, so that these zones are more exposed to the flux of force and will releave, by this, the zone of the highest stresses.

This work is object to the simulation of the recently explained experimental method by an accurate numerical computing process. For this, an exact formulation of DURELLI's and RAJAIAH's [4] ideas is necessary at first.

2. COMPOSITION OF THE SIMULATION PROCEDURE

The geometrical limits of the hole shape inside the plate, determined by DURELLI and RAJAIAH [4], lead, if the terminology of the calculus of variations is used, to the definition of a variation domain, i.e. which means of a domain which may not be exceeded by the optimal hole shape, fig.2.

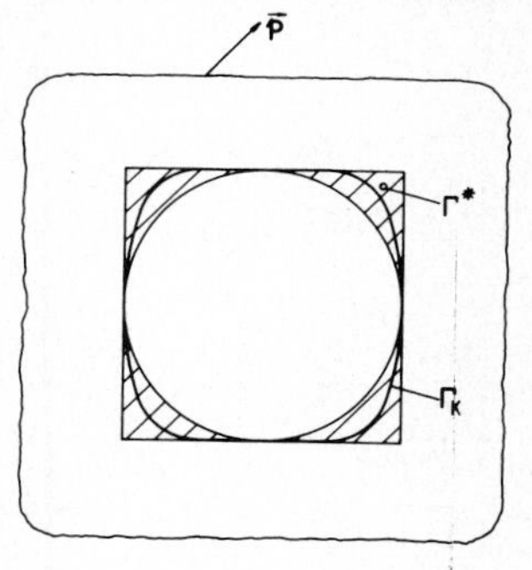

Fig. 2: Referring to the definition of the variation domain Γ^*

In their method for creating the optimal hole shape, DURELLI and RAJAIAH [4] make a serial arrangement of different radii along the hole shape, in a way that the radius is made smaller by removing material in lightly loaded zones. If this idea is analysed from the standpoint of the elasticity theory, analogies to the notch-stress-theory become evident:
The curvature of the notch is decisive for the stress peak. If the curvature is going up, the tangential stress is going up, too, if the curvature is falling down, the tangential stress is falling down, too.

This finding can be formulated as the following law: *The monotonization behavior of the tangential stress corresponds to the one of the curvature* $\chi = \frac{1}{\rho}$, *fig. 3.*

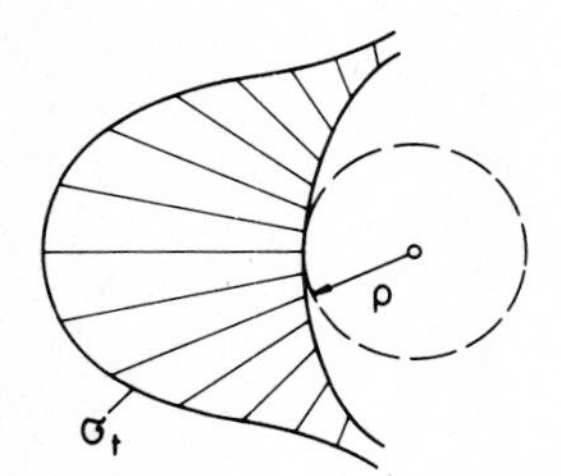

Fig. 3: Referring to the monotonization behavior of stress and curvature
σ_t tangential stress
ρ radius of curvature

DURELLI and RAJAIAH [4] furthermore used the method of the stress dislocation at the notch surface. By removal of material at lightly loaded zones, these will be done. This method can also be supported by the notch stress theory and is defined here as the reaction law of the notch effect and shall be formulated as another law: *The maximal tangential stress can be reduced by rising the minimal stress of the notch surface.* In the end, the numerical simulation of the photoelastical procedure has still to be executed for the determination of the stress field. There are two possibilities: either the boundary-element-method is used, or the finite-element-method is applied. Because the boundary-element-method is failing at sharp corners, the FEM is preferred.

In the next chapter the simulation method will be accurately formulated.

3. DESCRIPTION OF THE ALGORITHM

First, the material stress must be quantitatively fixed through the definition of the effective stress within the material. It is computed by the following rule consisting of the three components of the stress tensor for the present plane state of stress, the octahedral shear hypothesis presumed:

$$\bar{\sigma}_\mu^{\,2} = \bar{\sigma}_\mu^{\,2}(\{t\}) = (\sigma_x^2 + \sigma_y^2 - \sigma_x\sigma_y + 3\tau_{xy}^2), \quad (1)$$

$\{t\}$ being a design vector which describes the surface shape of the notch. The index μ characterizes the different load vectors, which cause the stresses in the plate, so that also the multiple loading problem can be treated, as shown by SCHNACK [10] in 1980. As it can be seen from fig. 4, the boundary stress vector $\{p_\mu\}$ is introduced at the surface ∂V_S, while the boundary displacement vector is specified at the surface ∂V_K. For the determination of the stress tensor, the equilibrium condition must be satisfied for the stress tensor τ_{km}:

$$\tau_{km,k} = 0 \quad \text{(without volume forces)}, \quad (2)$$

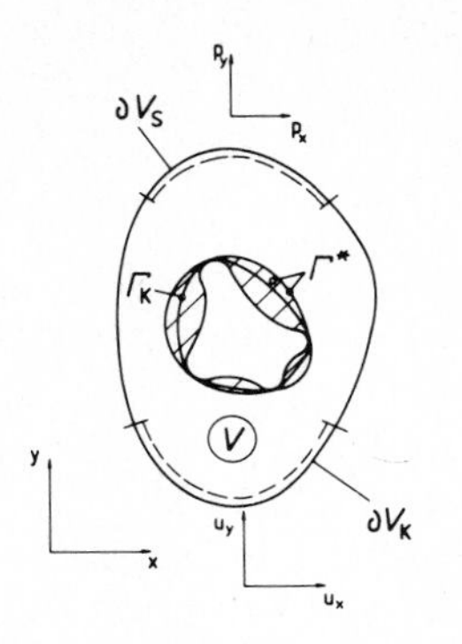

∂V_S boundary for boundary stress vector $\vec{p}_\mu$
∂V_K boundary for predetermined boundary-displacement vector $\vec{u}$
Γ_K notch boundary
Γ^* variation domain

Fig. 4: Geometrical scheme of the system symbols

the kinematical relation between the distortion tensor e_{km} and the displacement vector v_k must be considered:

$$e_{km} = \frac{1}{2}(v_{k,m} + v_{m,k}), \quad (3)$$

and the material law has to be considered, which describes the coherence between stress tensor and distortion tensor:

$$\tau_{km} = 2G\,(e_{km} + \frac{\nu}{1-2\nu}\,\delta_{km}\,e_{qq}), \quad (4)$$

and in the end, the compatibility condition has to be considered in form of the compatibility tensor i_{rs}:

$$i_{rs} = 0. \quad (5)$$

As fig. 4 shows, the treated optimization problem for realization of the computer-simulation can be described by means of the objective function:

$$\Phi = \max_{V} \bar{\sigma}_\mu = \max_{\Gamma_K} \bar{\sigma}_\mu = \Phi(\{t\}). \quad (6)$$

It is known from the notch stress theory, that the maximal effective stress occurs at the notch surface, so that the search for the maximum can be limited to the notch boundary. Now the minimum of Φ is searched by variation of $\{t\}$ in consideration of the secondary condition, i.e. which means that the notch contour may not leave the variation domain Γ^*, fig. 2 and 4:

$$\Gamma_K = \Gamma_K(\{t\}) \subset \Gamma^*. \quad (7)$$

The base for the computer-simulation of the experimental procedure of DURELLI and RAJAIAH [4], described as follows, is formed by the hypothesis formulated by R.V. BAUD [11] in 1934 and by NEUBER [12, 13] in 1971/72:

(I) The maximum stress situated at the surface can be minimized by the choice of a surface geometry with constant effective stress distribution $\bar{\sigma}_\mu$.

In 1977, this hypothesis was improved by SCHNACK [5] on:

(II) If no notch surface Γ_K with a constant effective stress distribution $\bar{\sigma}_\mu$ exists within a predetermined variation domain Γ^* between two fixed points, the occuring notch stress is minimal, provided that the bow length Λ of Γ_K, for which σ_μ is constant, will be maximal and not higher on $(\Gamma_K - \Lambda)$ than the constant tangential stress, fig. 5:

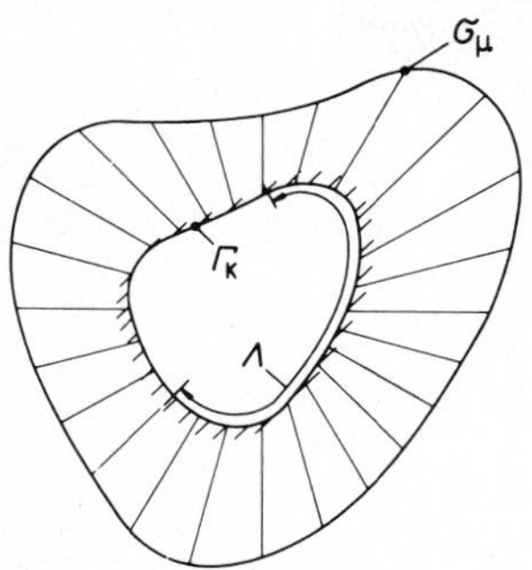

Fig. 5: Referring to the maximal bow length Λ with stress constancy

As already pointed out in chapter 2, an algorithm for the minimization of boundary stress maxima can be constructed by means of the laws of the notch stress theory, especially by the fade-away-law.

(III) The geometrical disturbance of the notch boundary Γ_K produces a quickly fading disturbance of the effective stress $\bar{\sigma}_\mu$ of the notch boundary (which is a consequence of the Saint-Venant-Principle).

(IV) The effective stress $\bar{\sigma}_\mu$ can be influenced by the change of the curvature χ of Γ_K, whilst the monotonization behavior of $\bar{\sigma}_\mu$ corresponds to the one of χ.

(V) Through the reaction law of the notch effect, the maximal effective stress $\bar{\sigma}_\mu$ of the notch boundary can be reduced; that means by rising of the minimal stress of the notch surface, the maximal stress is going down.

Consequently the following optimization rule results for the computer-simulation:

(VI) For attainment of a constant notch stress $\bar{\sigma}_\mu$ by forming of a constant notch stress distribution on Γ_K throughout a notch bow length as large as possible (s. (I) and (II)), the notch geometry is changed step by step, under consideration of (III), in a manner, that the momentary stress maximum is reduced by the local change of the curvature χ (s. (IV)), and at the same time the stress minimum is raised by a local change of the curvature χ (s. (V)).

Additionally a smoothing process for the notch contour Γ_K is inserted in a way that Γ_K will not only be changed at the local positions of the stress maximum or stress minimum, but even all boundary point coordinates by the principle of an arithmetic sequence will.

According to (IV) the boundary effective stress can be changed by the curvature. Because I use the FEM (s. chapter 2), I need only the discretized form of the structure, and therefore the curvature is defined by three coordinates. It can lead to the realization of the optimization principle by changing of the coordinates of the intermediate node j (fig. 6),

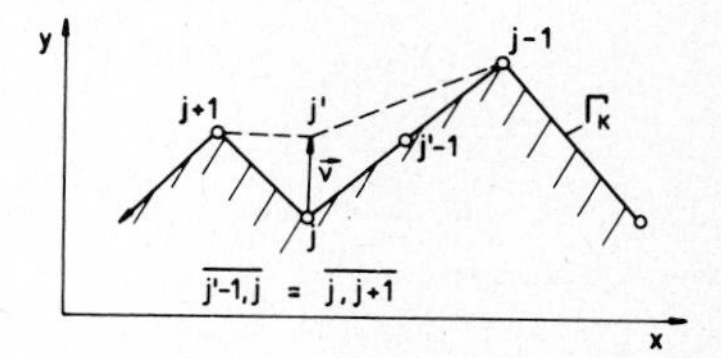

Fig. 6: Referring to the dislocation of node j by alteration of the local curvature
$\vec{V}$ displacement vector
Γ_K notch contour
x,y reference coordinates

or, if the position of node j is fixed by the definition of the variation domain, it can optionally be changed by alteration of the position of the nodes j-1 and/or j+1. The displacement of node j is fixed on the angle bisector of j-1, j, j+1. The displacement of node j on the angle bisector is effected with the dislocation of node j-1 to j'-1, so that the side bisector of j'-1, j, j+1 coincide.

After the direction of the displacement vector $\vec{V}$ is known, except the sign, by the situation of the angle bisector for the distortion of node j, value and sign of $\vec{V}$ must be determined. The value is depending on the average length $\tilde{L}$ of a particular boundary triangle side of Γ_K (e.g. the biggest one) and is regulated by the control parameters α and β.

$$|\vec{V}| = \alpha\beta\tilde{L}. \tag{8}$$

The size α is normally 0.1, so that during the first iteration step $|\vec{V}|$ is about 10% of $\tilde{L}$. As the optimal profile is approached further after each iteration step, the change of the contour by $|\vec{V}|$ must successively be diminished. For this purpose the parameter β is necessary. For the choice of α and β:

$$0 < \alpha < 1 \tag{9}$$
$$0 < \beta < 1. \tag{10}$$

The sign of $\vec{V}$ is controlled by means of parameter ξ ($\xi = -1$ or $+1$). During the optimization, the quantitative value of the curvature does not need to be known, but only the information is necessitated whether the curvature is locally convex, concave or zero in j. For this purpose the size ζ is necessary:

$$\zeta = -\frac{\det[w]}{|\det[w]|} \tag{11}$$

consisting of the auxiliary variable:

$$\det[w] = \begin{vmatrix} x_{j-1} & x_j & x_{j+1} \\ y_{j-1} & y_j & y_{j+1} \\ 1 & 1 & 1 \end{vmatrix}. \quad (12)$$

The new position of the node j, so the coordinates of j', are computed by:
if $\zeta \neq 0$, one obtains:

$$x_{j'} = x_j + \xi\ |\vec{V}|[(x_{j+1} + x_{j'-1})/2 - x_j]/F \quad (13)$$

$$y_{j'} = y_j + \xi\ |\vec{V}|[(y_{j+1} + y_{j'-1})/2 - y_j]/F \quad (14)$$

with

$$F = \sqrt{[(x_{j+1}+x_{j'-1})/2-x_j]^2 + [(y_{j+1}+y_{j'-1})/2-y_j]^2}. \quad (15)$$

If $\zeta = 0$, one obtains:

$$x_{j'} = x_j + \xi\ |\vec{V}|(y_{j+1} - y_j)/E \quad (16)$$

$$y_{j'} = y_j - \xi\ |\vec{V}|(x_{j+1} - x_j)/E \quad (17)$$

with

$$E = \sqrt{(y_{j+1} - y_j)^2 + (x_{j+1} - x_j)^2}. \quad (18)$$

The coordinates
$x_{j'-1}$ and $y_{j'-1}$ are computed as follows:

$$x_{j'-1} = x_j + \gamma(x_{j-1} - x_j) \quad (19)$$

$$x_{j'-1} = y_j + \gamma(y_{j-1} - y_j) \quad (20)$$

consisting of:

$$\gamma = \sqrt{\frac{(x_{j+1} - x_j)^2 + (y_{j+1} - y_j)^2}{(x_{j-1} - x_j)^2 + (y_{j-1} - y_j)^2}}. \quad (21)$$

For the stress analysis in computer-simulation, the photo-elasticity-method for determination of the stress field is replaced, as already mentioned, by the finite-element-method. Because the starting profile Γ_K^S is changed step by step by the described optimization procedure, a new stress computation becomes necessary after each optimization step, i.e. which means the displacement vector $\{r_i\}$ of the total structure in the i-th iteration step must be found, fig. 7:

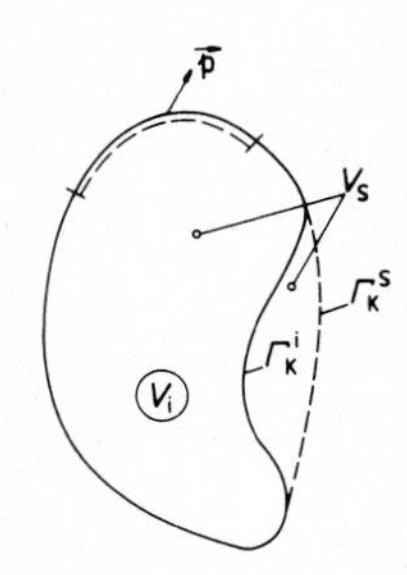

Fig. 7: Referring to the iterative computation of the displacement matrix $\{r_i\}$
Γ_K^S starting profile with volume V_S
Γ_K^i profile after i iteration steps with volume V_i

The equilibrium condition of the FE-structure provides:

$$[K_i]\{r_i\} = \{R_i\}. \quad (22)$$

By the Boole's matrix $[T_d]$, the displacement distribution can be filtered out for a single element, provided that the inverse of $[K_i]$ is known:

$$\{d\} = [T_d][K_i]^{-1}\{R_i\}. \quad (23)$$

The stress matrix $\{\sigma\}$ provides the Hooke's law from the distortion matrix $\{\varepsilon\}$, which is obtained from differentiation of the displacement vector $\{u\}$:

$$\{\varepsilon\} = [B]\{d\}. \quad (24)$$

The Hooke's law provides:

$$\{\sigma\} = [D]\{\varepsilon\}, \quad (25)$$

and so is valid:

$$\{\sigma\} = [D][B]\{d\}. \quad (26)$$

In order to avoid the necessity of solving the equation system (22) after each iteration step for the stress evaluation, the displacement matrix $\{r_i\}$ is determined from the inverse of the total stiffness matrix $[K_S]$ of the starting profile. Equ. (22) is extended by the starting stiffness matrix $[K_S]$.

$$\langle [K_S] - ([K_S] - [K_i])\rangle\ \{r_i\} - [R_i] = 0. \quad (27)$$

For the starting domain V_S as well as for the domain after i iteration steps V_i, the same kinematical boundary conditions are valid. So the notch shape Γ_K^i, as well as the notch shape of the starting profile Γ_K^S is load-free. Furthermore both domains show the same number of elements with the same displacement formulation group per element, with identic nodal parameters, so that Equ. (27) is existing.
For V_S is valid:

$$[K_S]\{r_S\} = [R_S], \quad (28)$$

because Γ_K is load-free, it can be written:

$$\{R_S\} = \{R_i\}. \quad (29)$$

From Equ. (27) follows:

$$\langle [K_S] - ([K_S] - [K_i])\rangle\ \{r_i\} - [K_S]\{r_S\} = 0. \quad (30)$$

For $\{r_i\}$, the following expression is formulated:

$$\{r_i\} = \{r_S\} + \{s\}, \quad (31)$$

so that from Equ. (30) can be derived:

$$[K_S]\{r_S\} + [K_S]\{s\} - ([K_S] - [K_i])(\{r_S\} + \{s\}) - [K_S]\{r_S\} = 0. \quad (32)$$

For $\{s\}$, this leads to:

$$\{s\} = [K_S]^{-1}([K_S] - [K_i])(\{r_S\} + \{s\}). \quad (33)$$

From this regulation, an iteration procedure for the computation of the correction $\{s\}$ can be derived:

$$\{s_k\} = [K_S]^{-1}([K_S] - [K_i])\{s_{k-1}\} + [K_S]^{-1} \cdot ([K_S] - [K_i])\{r_S\}. \quad (34)$$

The convergence of the iteration procedure can be saved, if for the spectral radius of the constant iteration matrix will be valid:

$$\rho\ ([K_S]^{-1}\langle [K_S] - [K_i]\rangle\ ) < 1. \qquad (35)$$

Examinations by SCHNACK [5, 7] in 1977/79 showed, that one iteration step is sufficiant to compute the norm of the new displacement vector $\{r_i\}$ up to a 1%-precision. The following formula is then valid:

$$\{r_i\} \cong \{r_S\} + [K_S]^{-1}\ ([K_S] - [K_i])\ \{r_{i-1}\}. \qquad (36)$$

4. CONCLUSIONS AND EXAMPLE COMPUTATIONS

The theoretical description of the computer-simulation leads, as shown in the previous chapter, to a non-linear optimization procedure, which must be appointed to the search strategy. So the question arises, if the computer-simulation of the experimental method of DURELLI and RAJAIAH [4] leads to an effective solution strategy, or, if perhaps another procedure from the group of the possible solution strategies will be more appropriate. Fig. 8 shows a plan.

A detailed discussion by SCHNACK and SPÖRL [14] in 1982 shows, that, with the random-search-strategy, a large number of stress computations must be made, due to the neglecting of informations about the iteration direction of the objective function. This does not admit higher valued finite elements for the FEM-computation, because otherwise, very high computing times, especially during the solution of Equ. (22) occur. The use of displacement formulations with only the P_1-polynome in the interpolation matrix is disturbing considerably the numerical results with reference to the stress values of the notch profiles. An essential disadvantage of the application of the gradient-strategy is caused by the numerical computation of the stress gradients. The application of the formulation of differences, as well as modifications of the FEM-algorithm provide only coarse approximations of the optimal profiles (s. QUEAU and TROMPETTE [15], 1980). In addition to this, during the DFP-procedure, the matrix $[H^k]$, as the approximating of the Hesse-matrix, becomes numerically unstable by influence of the rounding error; i.e. which means to lose its "positively definite"-character. Finally, the high computation effort of the penalty-function-technique is evident:

In every structural optimization problem, a series of non-linear optimization problems must be solved, and in each partial step, a one-dimensional minimization is effected anew.

In the contrary to this, a non-gradient procedure exists as a computer-simulation of an experimental procedure, which only necessitates the stress distribution along the notch profile to be optimized. Numerical problems, as a consequence of inaccurate gradient computation do not exist anymore. By the application of triangle elements with linearly varying strain, even sharply curved notches can be simulated, if the discretization is fine enough. Using statements out of the notch stress theory for the development of algorithms, the reference to the problem is much stronger in this procedure than in the above-mentioned ones and leads therefore to an extremely higher convergence speed.

The high value of the computer-simulation for obtaining the best notch shape possible, with smallest stress maximum, is shown by a practical problem. For a big pressure-loaded plate, the optimal hole shape is searched (s. fig. 1). Fig. 9 shows the FEM-structure with the optimal profile of SCHNACK and DURELLI/RAJAIAH. Fig. 10 shows the good correspondence to the experimental procedure of DURELLI and RAJAIAH [4]. The numerical procedure provides, however, some higher accuracy, but proves furthermore, that DURELLI and RAJAIAH [4] are right in the end with their statement, that RICHMOND [16] came to a totally wrong conclusion in 1939 about the optimal hole shape within the big plate during uniaxial load.

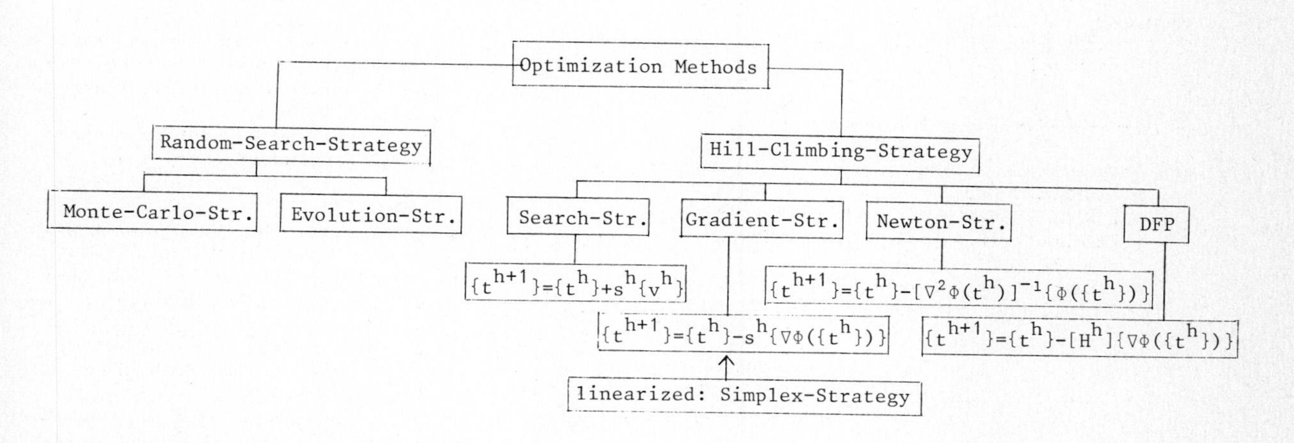

Fig. 8: Optimization strategies for the solution of non-linear optimization problems

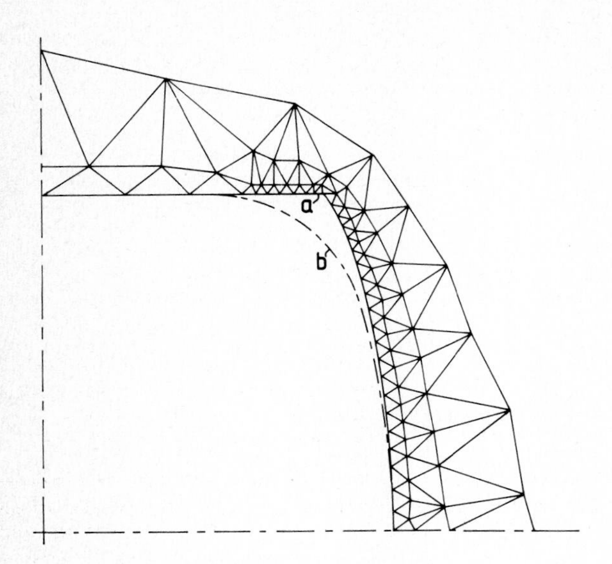

Fig. 9: Optimal notch profiles
a SCHNACK/SPÖRL
b DURELLI/RAJAIAH

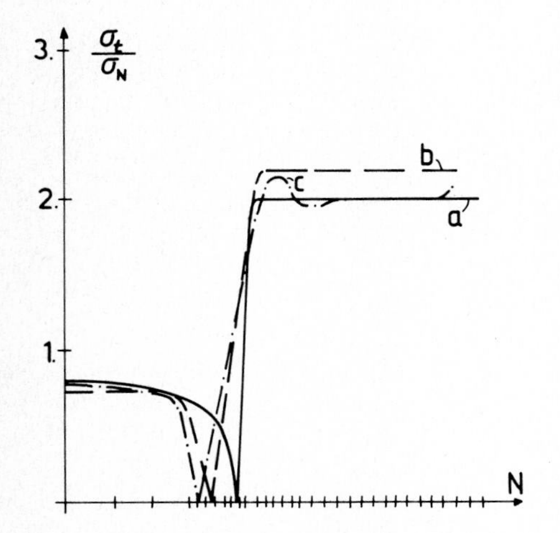

Fig. 10: Notch-stress curves
a SCHNACK/SPÖRL
b DURELLI/RAJAIAH with photo-elasticity
c FEM-analysis of notch profile from the DURELLI/RAJAIAH-solution
N nodes of the notch-profile

REFERENCES

[1] Tvergaard, V., On the optimum shape of a fillet in a flat bar with restrictions, Proc. IUTAM Symp. on Optimization in Struct. Design, Warsaw (1973), Springer Verlag, Berlin Heidelberg New York (1973).

[2] Francavilla, A., Ramakrishnan, C.V. and Zienkievicz, O.C., Optimization of shape to minimize stress concentrations, J. Strain Analysis Vol. 10 No. 2 (1975) 63-70.

[3] Kristensen, E.S. and Madsen, N.F., On the optimum shape of fillets in plates subjected to multiple in-plane loading cases, Int. J. Num. Meth. Engng. Vol. 10 No. 5 (1976) 1007-1019.

[4] Durelli, A.J. and Rajaiah, K., Quasi-square-hole with optimum shape in an infinite plate subjected to in-plane loading, (Oakland Univ. ONR-Report No. 49, 1979).

[5] Schnack, E., Ein Iterationsverfahren zur Optimierung von Spannungskonzentrationen, (Inaugural Dissertation, Univ. of Kaiserslautern, 1977).

[6] Schnack, E., Ein numerisches Verfahren zur Optimierung von Spannungskonzentrationen, ZAMM, Bd. 58, Heft 5 (1977) 122-123.

[7] Schnack, E., An optimization procedure for stress concentrations by the finite-element-technique, Int. J. Num. Meth. Engng., Vol. 14, No. 1 (1979) 115-124.

[8] Schnack, E., Proceedings of the Finite-Element Congress in Baden-Baden 13/14th Nov. 78: Optimization of Notched Structures using the FEM (ICOSS GmbH, Stuttgart, 1978).

[9] Schnack, E., Ein Iterationsverfahren zur Optimierung von Kerboberflächen (VDI-Verlag, VDI-Forschungsheft Nr. 589, Düsseldorf, 1978).

[10] Schnack, E., Optimierung von Spannungskonzentrationen bei Viellastbeanspruchung, ZAMM, Bd. 60 (1980) 151-152.

[11] Baud, R.V., Fillet profiles for constant stress, Product Engng. Vol. 5, No. 4 (1934) 133-134.

[12] Neuber, H., Zur Optimierung der Spannungskonzentration, Sonderdruck 9 aus den Sitzungsberichten 1971 der Bayerischen Akademie der Wissenschaften, Verlag der Bayerischen Akademie der Wissenschaften, München (1971) 127-136.

[13] Neuber, H., Zur Optimierung der Spannungskonzentration, Continuum Mechanics and Related Problems of Analysis, Moskau: Nauka (1972) 375-380.

[14] Schnack, E. and Spörl, U., Oberflächenoptimierung ebener und rotationssymmetrischer elastischer Kontinua mit der FEM (DFG-Bericht FEM-OPT-2D, Karlsruhe, 1982).

[15] Queau, J.P. and Trompette, Ph., Two-dimensional shape optimal design by the finite-element-method, Int. J. Num. Meth. Engng. Vol. 15 (1980) 1603-1612.

[16] Richmond, W.O., Discussion of R.D. Mindlin's paper on "Stress Distribution Around a Tunnel", Proc. ASCE, Vol. 65 (1939) 1465-1467.

XI. CHEMICAL SYSTEMS

Simulation in Engineering Sciences
J. Burger and Y. Jarny (eds.)
Elsevier Science Publishers B.V. (North-Holland)

NUMERICAL SIMULATION OF THE CONTROL SYSTEM OF A NITROGEN GASIFICATION PLANT

Mieczyslaw Metzger

Institute of Automatica, Silesian Technical University
Gliwice, Poland

The liquid nitrogen gasification plant supplies gas nitrogen to nuclear reactors in two regimes, in the rated duty and in the start-up. The paper presents the simplified mathematical model of the liquid nitrogen gasification process. The mathematical model of this process is a nonlinear differential system (15th order). Basing on the mathematical model of this process, the simulation study of the plant and its control system has been performed.

1. INTRODUCTION

A typical industrial plant of nuclear reactors requires a nitrogen feeding for each reactor in two regimes, the normal regime and the starting regime.

Figure 1 represents the nitrogen gasification plant which is studied in this paper. This is a typical industrial process, simplified in the most important elements.
The tank 1 contains 6200 kg of liquid nitrogen at pressure P_1 = 1600 kPa (ϑ_1 = 112 K). The luquid nitrogen is dispatched to the high pressure part of the plant were gas nitrogen is obtained at pressure P_{3H} = 6000 kPa and mass flow rate M_{SH} = 0.833 kg/sec for the reactor in starting regime, and to the low pressure part of the plant where gas nitrogen is obtained at pressure P_{3I} = 1000 kPa and mass flow rate M_{SI} = 0.00833 kg/sec for the reactor (or for reactors) in normal regime.

The high pressure part is composed of a piston pump 6 which drives the liquid nitrogen to the evaporator with steam heating 8 where nitrogen is gasified and heated at ambient temperature. The lower pressure part is composed of the atmospheric evaporator 4 where nitrogen is gasified and heated at ambient temperature by heat exchange with the atmosphere. In winter, the nitrogen must be heated in the steam heating process 5.

For the counteraction of a reduction of the pressure P_1 in the tank, a by-pass 2 is utilized with the steam heating evaporator 9.

In the plant, a very efficient regulation of pressures P_1, P_{31}, P_{3H} is required; the regulation of temperature ϑ_w in 8 is less critical. The regulation of pressure P_1 is performed with a P-regulator R_1; the regulations of P_{31}, P_{3H} and ϑ_w are performed with PI - regulators R_2, R_3, R_3 (Fig. 1).

The purpose of this paper is to show-how it has been possible to perform a control structure of this plant, with the help of numerical simulations, from a mathematical model of the nitrogen gasification plant.

2. MATHEMATICAL MODEL

The physical phenomena which occur in the different parts of the plant are very complex. For example, the evaporating process of nitrogen in the atmospheric evaporator 4 is similar to process in a monotubular boiler [1][2].

The main objective of the mathematical model formulation of a plant, being a study of interactions between regulators, the mathematical model must contain the whole description of the plant, hence, the description of the different constituents has to be simplified.

* Liquid nitrogen tank

In order to simplify the model the following hypotheses have been considered :

- the heat transfer between liquid and wall, between gas and wall and also between wall and ambient are negligible.
- in the tank, mixing in the liquid and in the gas is considered as perfect.
- the evaporating inertia of evaporator 9 is neglected. The evaporating process has been considered by increasing of nitrogen enthalpy ($i_L \rightarrow i_o$).

The model equations are :

- mass and thermodynamical balances

$$\frac{d\,(\varrho_L V_L)}{dt} = - M_x - M_{11} - M_{12} - M_v \qquad (1)$$

$$\frac{d\,(\varrho_1 V_1)}{dt} = M_x + M_v, \quad (V_1 + V_L = V_r = \text{const}) \qquad (2)$$

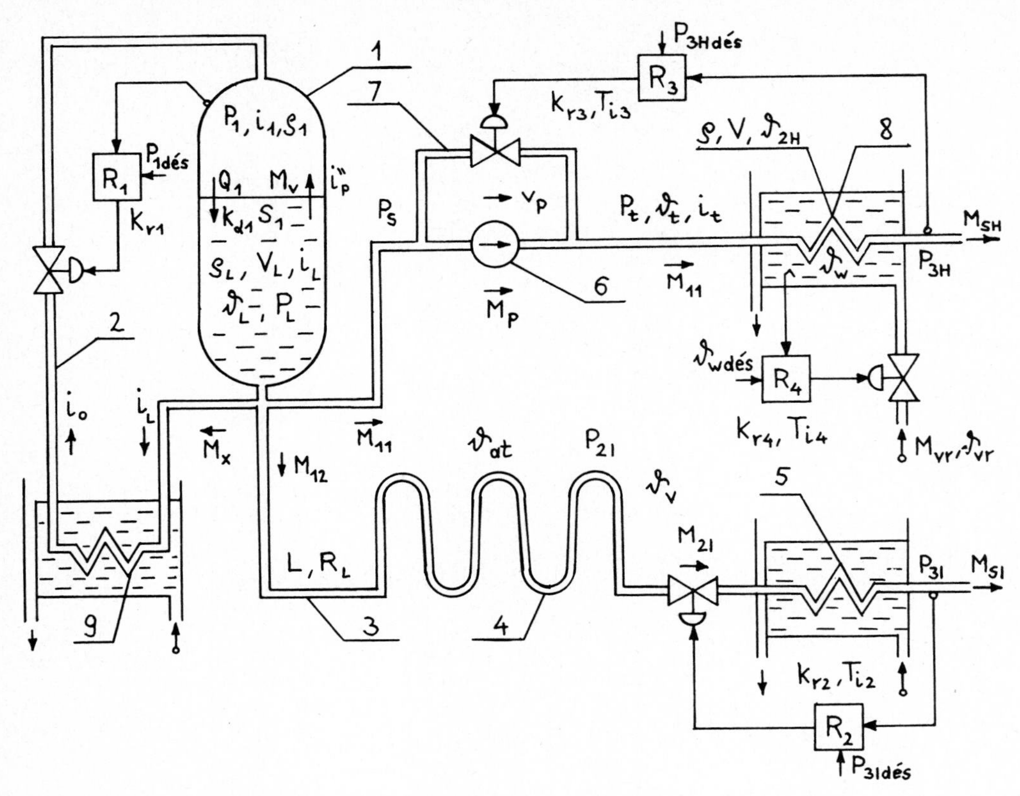

Fig. 1 : Installation and control system. ↕ - heating steam

$$\frac{d(\rho_L V_L i_L)}{dt} = - M_x i_L - M_{11} i_L - M_{12} i_L - M_v i''_p + Q_1 \quad (3)$$

$$\frac{d(\rho_1 V_1 i_1)}{dt} = M_x i_o + M_v i''_p - Q_1 \quad (4)$$

The state equation and the liquid steam equilibrium equation

$$P_1 = \rho_1 R \vartheta_1 \quad (5), \qquad P_L = P_L(\vartheta_L) \quad (6)$$

The flow rates of gas nitrogen and heat

$$M_v = k_p (P_L - P_1) \quad (7), \qquad Q_1 = k_{\alpha 1} S_1 (\vartheta_1 - \vartheta_L) \quad (8)$$

In [3] Luyben presents some suggestions with respect to the choice of the numerical value of k_p during numerical computation.

The thermodynamical relations

$$i_L = i_L(\vartheta_L) \quad (9), \qquad i_1 = i_1(\vartheta_1) \quad (10)$$

$$i''_p = i''_p(\vartheta_L) \quad (11)$$

By-pass and regulator

$$M_x = k_{x1} \sqrt{\frac{\rho_L V_L}{S_1}} \quad (12)$$

$$k_{x1} = k_{x1}(P_1, P_{1des}, k_{r1}) \quad (13)$$

✱ Tube 3 is considered as the lumped inertance L

$$\frac{dM_{12}}{dt} = \frac{1}{L}\left[R_L M_{12} + P_1 - P_{21} + \frac{g}{S_1}(\rho_L V_L)\right] \quad (14)$$

✱ Atmospheric evaporator

The atmospheric evaporator is considered as a lumped parameter system. The unitary flow rate of heat exchange between the ambient atmosphere and the liquid q (J/m^2 sec) is sufficient for the whole liquid nitrogen evaporation, near the evaporator input. Such a principle leads to the conclusion that the liquid nitrogen flows in the tube of the evaporator, like a 1-lenght piston, depending on flow rate M_{12} (Fig. 2). The remaining part of the evaporator is considered as a heated tank of gas nitrogen with variable thermal exchange surface.

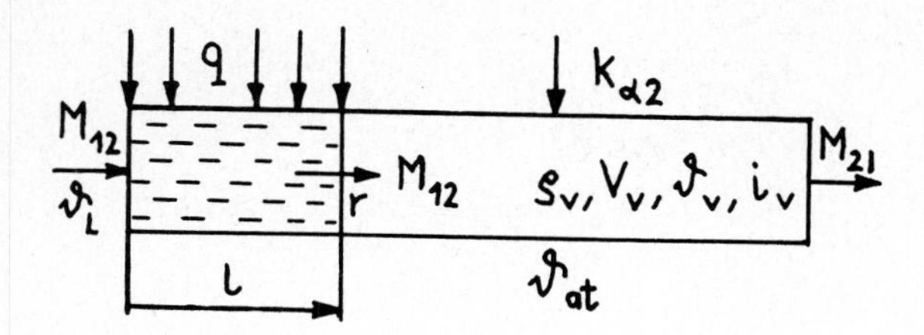

Fig. 2 : Model of the atmospheric evaporator.

The model equations are :

$$l = \frac{r}{q\,U_L} M_{12} \tag{15}$$

$$\frac{d\,(\varrho_v V_v)}{dt} = M_{12} - M_{2I} \tag{16}$$

$$\frac{d\,(\varrho_v V_v i_v)}{dt} = M_{12} i''_p - M_{2I}\, i_v + Q_2 \tag{17}$$

$$Q_2 = k_{\alpha 2}\,(F - \frac{r}{q} M_{12})\,(\vartheta_{at} - \vartheta_v) \tag{18}$$

$$P_{2I} = \varrho_v\, R \vartheta_v \tag{19}$$

$$i_v = i_v(\vartheta_v) \quad , \; i''_p = i''_p\,(\vartheta_L) \tag{20}$$

✱ The heated tube 5 is considered as a lumped capacity.

The model equations are :

$$\frac{dP_{3I}}{dt} = \frac{1}{C_{3I}}\,(M_{2I} - M_{SI}) \tag{21}$$

$$M_{2I} = k_{x2}\sqrt{P_{2I} - P_{3I}} \tag{22}$$

$$k_{x2} = k_{x2}\,(P_{3I},\, P_{3Ides},\, k_{r2},\, T_{i2}) \tag{23}$$

✱ The pump 6 and by-pass 7 are considered as a static system. Nevertheless, the mathematical description is very complex and this is the reason why a special algorithm is required for the simulation. In [4] Moszczynski established such an algorithm, enabling to calculate :

$$\{M_{11},\, M_p,\, \vartheta_t\} = f\,\{P_S, P_t,\, M_B,\, v_p,\, \vartheta_L\} \tag{24}$$

By-pass and regulator R_3 are described by equations :

$$M_B = k_{x3}\,(P_t - P_s) \tag{25}$$

$$k_{x3} = k_{x3}\,(P_{3H},\, P_{3Hdes},\, k_{r3},\, T_{i3}) \tag{26}$$

✱ The evaporator 8 is considered as a lumped parameter system (at a 6000 kPa level, it is difficult to make distinction between liquid phase and gaseous phase of nitrogen).

The model equations are :

$$\frac{d\,(\varrho V)}{dt} = M_{11} - M_{SH} \tag{27}$$

$$\frac{d\,(\varrho V i)}{dt} = M_{11} i_t - M_{SH}\, i + Q_3 \tag{28}$$

$$W\frac{d\vartheta_w}{dt} = M_{vr}\, r_1 + M_{vr}\, c_1\,(\vartheta_{vr} - \vartheta_w) - Q_3 \tag{29}$$

$$P_{3H} = P_{3H}\,(\varrho,\, \vartheta_{2H}) \tag{30}$$

$$Q_3 = k_w F_w\,(\vartheta_w - \vartheta_{2H}) \tag{31}$$

$$i = i\,(\vartheta_{2H}) \;,\; i_t = i_t\,(\vartheta_t) \tag{32}$$

$$M_{vr} = M_{vr}\,(\vartheta_w,\, \vartheta_{wdes},\, k_{r4},\, T_{i4}) \tag{33}$$

In the mathematical description of the regulators, equations (13) (23) (26) (33), the control variables were considered with saturation values. Equations (6) (11) (20) (30) (32) are linearised.

The model variables are defined as follows :

State variables

$x_1 = \varrho_L V_L,\; x_2 = \varrho_1 V_1,\; x_3 = \varrho_L V_L i_L,\; x_4 = \varrho_1 V_1 i_1$

$x_5 = M_{12},\; x_6 = \varrho_v V_v,\; x_7 = \varrho_v V_v i_v,\; x_8 = P_{3I},$

$x_9 = \varrho V,\; x_{10} = \varrho V i,\; x_{11} = \vartheta_w,\; x_{12} = P_{3H},$

$x_{13} = \frac{1}{T_{i2}} \int (P_{3Ides} - P_{3I})\, dt,$

$x_{14} = \frac{1}{T_{i3}} \int (P_{3Hdes} - P_{3H})\, dt,$

$x_{15} = \frac{1}{T_{i4}} \int (\vartheta_{wdes} - \vartheta_w)\, dt$

Input variables

$u_1 = M_{SI},\; u_2 = M_{SH},\; u_3 = v_p,\; u_4 = \vartheta_{at},$

$u_5 = \vartheta_{vr},\; u_6 = i_o,\; u_7 = k_{r1},\; u_8 = P_{1des},$

$u_9 = k_{r2},\; u_{10} = T_{i2},\; u_{11} = P_{3Ides},\; u_{12} = k_{r3},$

$u_{13} = T_{i3},\; u_{14} = P_{3Hdes},\; u_{15} = k_{r4},\; u_{16} = T_{i4},$

$u_{17} = \vartheta_{wdes}.$

Output variables

$y_1 = P_1,\; y_2 = P_{2I},\; y_3 = P_{3I},\; y_4 = \vartheta_1,\; y_5 = \vartheta_v,$

$y_6 = V_L$, $y_7 = 1$, $y_8 = M_v$, $y_9 = M_{12}$, $y_{10} = M_{2I}$,

$y_{11} = P_S$, $y_{12} = P_t$, $y_{13} = P_{3H}$, $y_{14} = \vartheta_{2H}$,

$y_{15} = \vartheta_w$, $y_{16} = M_{11}$, $y_{17} = M_p$

After calculation, the mathematical model can be written as follows :

$$\dot{\underline{x}} = f\,(\underline{x}, \underline{p}, \underline{u}) \tag{35}$$

$$\underline{y} = g\,(\underline{x}, \underline{p}, \underline{u}) \tag{36}$$

where - $\underline{p}$ is a vector of numerical values.

The non-linear differential system (35) was integrated by means of the 4th order Runge - Kutta method.

3. NUMERICAL SIMULATION

The numerical simulation of the mathematical model was performed in order to examine the structure of control of the plant and to introduce eventual alterations in the control structure or in the plant.
Simulations runs were defined by several possible control values $\underline{u}$, and allowed to observe output variables $\underline{y}$. The different regimes of the plant are obtained by choosing adequate initial conditions.
The simulation of the plant without regulators leads to a certain knowledge of its dynamics, and consequently of the values of coefficients : k_{r1}, k_{r2}, k_{r3}, k_{r4}, T_{i2}, T_{i3}, T_{i4}.

Results concerning several simulation runs were presented in [4] . Figure 3 illustrates one example for which initial conditions are such that the plant has not yet started at the initial time of the simulation - the nitrogen is not collected from the plant output ($M_{SI} = M_{SH} = 0$), the pump is uncoupled ($v_p = 0$), water in 8 is at ambient temperature.

The simulation started by applying a stepwise input $M_{SH} = 0.833$ (one reactor is in a start-up regime) and a stepwise input $M_{SI} = 0.0166$ (two reactors are in a normal regime).

The display of outputs enables the control system action to be examined, and the propagation of perturbations in the controlled plant, to be followed.
The discussion in [4] shows that the control system studied by simulation, was well choosen.

4. REFERENCES

[1] Profos P., Die Dynamik zwangsdurchströmter Verdampfer systeme. Regelungstechnik, No 12, (1962).

[2] Varcop L., Die Dynamik zwangsdurchströmter Verdampfersysteme unter Bertichsichtigung von Druck¨nderungen des Strömungsmediums. Regelungs-technik, No 9, (1967).

[3] Luyben W.L., Process Modeling, Simulation and Control for Chemical Engineers, Mc Graw Hill, (1973).

[4] Kuznik J., Krzyzanowski R., Metzger M., Moszczynski M., Modèle Mathématique de l'installation de gazéification de l'azote (en polonais), Rapport Interne de l'Institut d'Automatique, Ecole Polytechnique de Silésie, Novembre (1981).

5. NOTATION

ϱ_L, i_L - density and enthalpy of the liquid nitrogen in the tank in kg/m^3 and J/kg.

$\varrho_1, i_1, \varrho_v, i_v$ - density and enthalpy of the gas nitrogen in the tank and in the atmospheric evaporator in kg/m^3 and J/kg.

ϱ, i - density and enthalpy of the nitrogen in the evaporator 8 in kg/m^3 and J/kg.

i_o - enthalpy of the heated nitrogen in the bypass in J/kg.

i_p - enthalpy of the vaporising nitrogen in J/kg

V_L, V_1 - volume of the liquid nitrogen and gas nitrogen in m^3

V_r - complete volume of the tank in m^3

V_v - volume of the gas nitrogen in the atmospheric evaporator in m^3.

V - complete volume of the evaporator 8 in m^3

M_{11}, M_{12}, M_x, M_{2I}, M_{SI}, M_{SH}, M_B, M_p - mass flow rates of the gas nitrogen in the plant (fig. 1) in kg/sec.

M_v - mass flow rate of the vaporising nitrogen in kg/ sec

M_{vr} - mass flow rate of the heating vapour in kg/sec.

ϑ_1, ϑ_L - temperature of the gas and liquid nitrogen in the tank in K.

ϑ_v - temperature of the gas nitrogen in the atmospheric evaporator output in K.

ϑ_t - temperature of the nitrogenin the pump output in K.

ϑ_{2H} - temperature of the nitrogen in the output of the evaporator 8 in K.

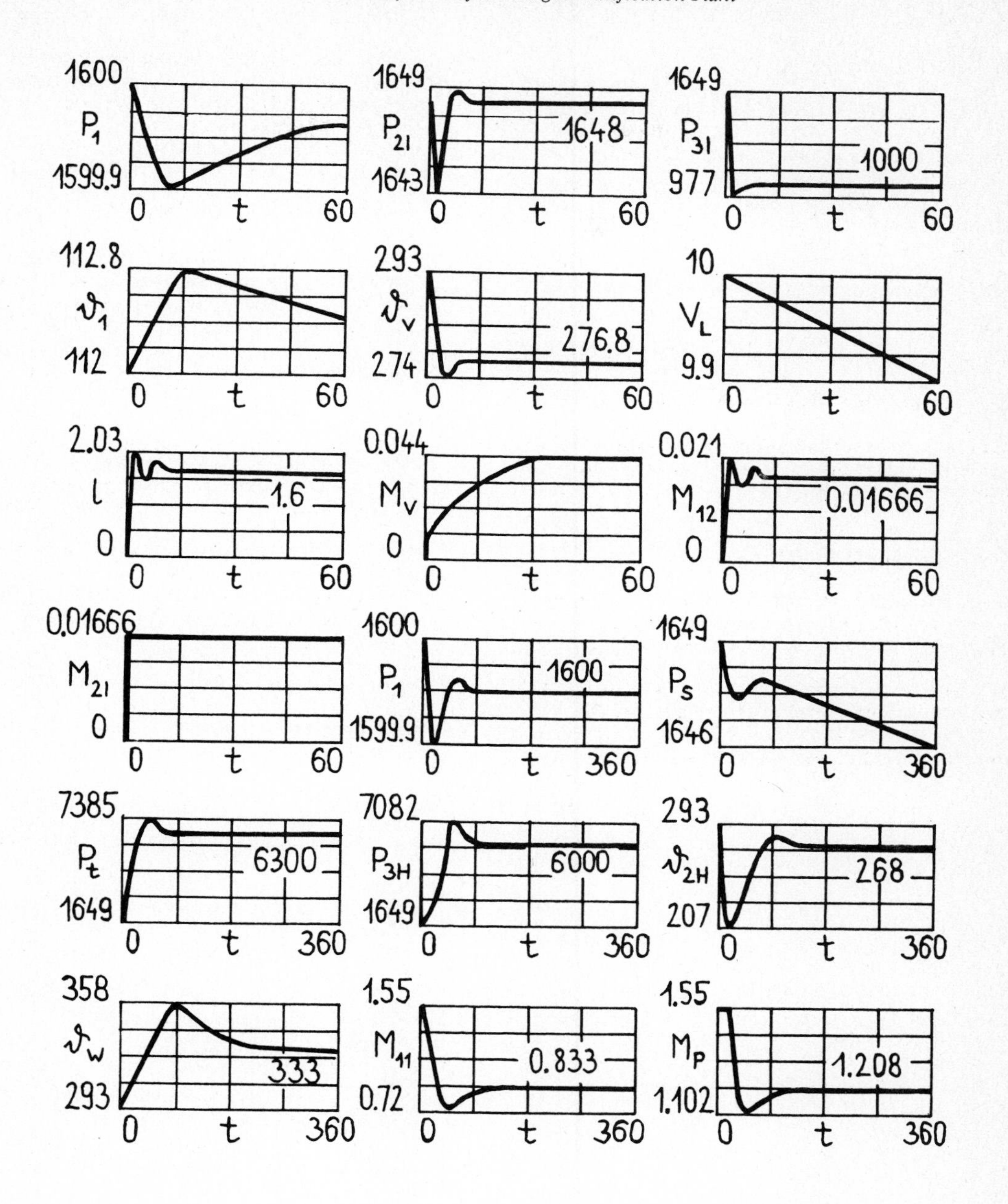

Fig. 3 : Example of the simulation - the output variable responses. Numerical values of the input variables are : M_{SI} = 0,00166, M_{SH} = 0,833, v_p = 0,0026, ϑ_{at} = 293, ϑ_{vr} = 433, i_o = 393578, k_{r1} = 0,001, P_{1des} = 1600, $k_{r2} = 10^{-6}$, T_{i2} = 5, P_{3Ides} = 1000, $k_{r3} = 10^{-8}$, T_{i3} = 20, P_{3Hdes} = 6000, k_{r4} = 0,1, T_{i4} = 20, ϑ_{wdes} = 333.

ϑ_{at} - ambient temperature in K.

ϑ_{vr} - temperature of the heating vapour in K.

ϑ_{w} - temperature of the water in the 8 in K

P_1, P_{2I}, P_{3I}, P_5, P_t, P_{3H} - pressure of the nitrogen in the different points of the plant (Fig. 1) in kPa.

P_L - partial pressure in the liquid nitrogen in kPa.

R - specific constant of the gas in J/kg K.

R_L - lumped resistance of the tube 3 in 1/m sec

L - lumped inertance of the tube 3 in 1/m.

S_1 - cross-surface of the tank in m^2

$k_{\alpha 1}$ - heat transfer coefficient between gas and liquid nitrogen in J/m^2 sec K.

K_p - coefficient of transfer molecular in 1/m sec

g - acceleration terrestrial in m/sec^2.

l - lenght of the piston of the liquid nitrogen in the atmospheric evaporator (Fig. 2) in m.

r - vaporising heat of the nitrogen in J/kg

q - unitary heat rate of the heat transfer between ambient atmosphera and liquid nitrogen in the atmospheric evaporator in J/m^2 sec.

U_L - perimeter of the atmospheric evaporator in m

F - lateral surface of the atmospheric evaporator in m^2.

$K_{\alpha 2}$ - overal heat transfer coefficient between the ambient atmosphere and gas nitrogen in the atmospheric evaporator in J/m^2 sec K.

C_{3I} - lumped volumetric capacity in m sec^2

v_p - volumetric rate of the pompe in m^3/sec.

W - thermal capacity of the water in the 8 in J/K

r_1 - vaporising heat of the water in J/kg.

c_1 - specific heat of the water in J/kg K.

k_w - overal heat transfer coefficient between the water and nitrogen in the evaporator 8 in J/m^2 sec K.

F_w - lateral surface of the evaporator 8 in m^2.

t - times in sec.

Simulation in Engineering Sciences
J. Burger and Y. Jarny (eds.)
Elsevier Science Publishers B.V. (North-Holland)

MODELLING AND SIMULATION OF A CATALYTIC REFORMER PLANT

M. Perroud , G. Bornard

Laboratoire d'Automatique de Grenoble
E.N.S.I.E.G. - I.N.P.G.
B.P. 46
38402 - Saint Martin d'Hères, France

1. INTRODUCTION

This model was introduced in order to be used for training refinery operators on a simulator.* This simulator already comprises various other models of distillation and conversion units, [1], [2], [3], which have been in operation since 1978.

In addition to the training of the operators reflexes the simulator has the pedagogical role of explaining the complex internal phenomena of the unit. The model must therefore be sufficiently representative, which demands a comprehensive understanding of the unit, but must also be capable of being run on a standard computer (program dimensions), and sufficiently fast (run-time) to be able to be used in real-time simulation.

2. THE MODEL

The catalytic reformation unit aims to produce car petrol from heavy fuels resulting from the first distillation.

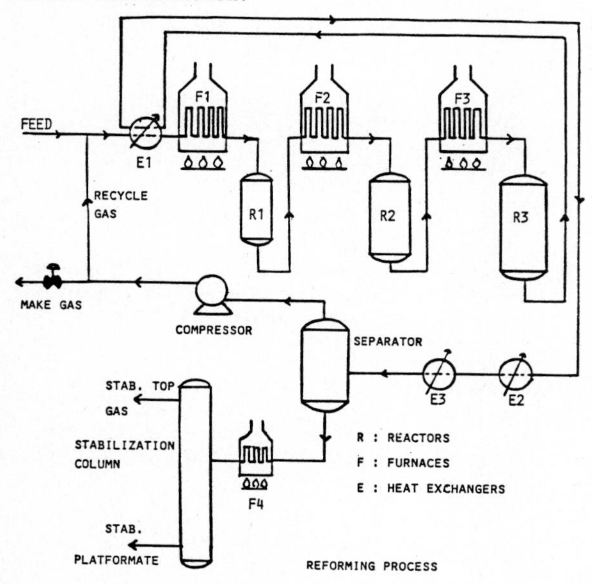

Figure 1

It is the chemical reaction which is the important part of the proceedings (Figure 1).

The complexity of the products treated and of the chemical reactions produced in the reactors requires a simplified model. Nevertheless the model must keep all its representativeness and its precise analysis of the physical-chemical phenomena in the unit to allow us to make a model of high pedagogical quality.

2.1. Description of process

The chemical reaction in the unit aims to increase the octane number rating (N.O.) of the feed in order to create fuel for cars (Normal & Super). Therefore one will seek to favour the chemical reactions producing compounds with a high octane number rating, by treating the feed at a high temperature (500°C) with the aid of a bifunctional zeolite catalyst.

2.2. Diagram of the reaction

We have taken into consideration the main reactions produced in the unit and adopted the following reactionary scheme (Figure 2).

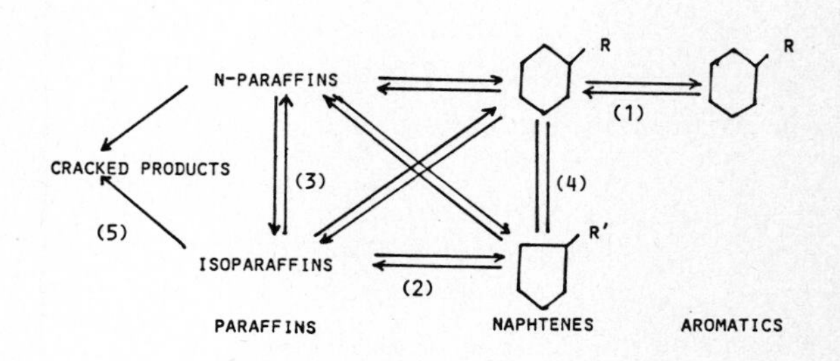

Figure 2

(1) Deshydrogenation of alkylcyclohexanes (aromatisation).

(2) Deshydrocyclisation of paraffins.

(3) Isomerisation of paraffins.

(4) Isomerisation of naphtenes.

(5) Hydrocracking of paraffins.

* *This study was sponsored by the SHELL-FRANCAISE Company.*

2.3. Simplifications

The conditions of operation and the quality of the reformate depend a lot upon the quality of the feed; In addition to the P.N.A. (Percentage of paraffins, naphtenes and aromatics) the length of the carbon chains is an important characteristic. Now, the mixture of feed + recycled gas consists of hydrocarbons whose length varies between 1 and 12 carbon atoms in general, depending on the origin of the naphta and the interval of distillation. We will only take into account the carbonic chains of 1 to 10 atoms of carbon (the amounts of C_{11} and C_{12} being in general negligible). The C_{10} will therefore include all the C_{10+} (C_{10} and higher).

For each class of carbonic chain there is a large quantity of types of molecules and isomeres. For example for the chain of 8 carbon atoms (C_8) there are 45 possible forms of molecules and isomeres. This large number of constituents with often very different characteristics makes it difficult to produce a model compatible with our needs. It is therefore necessary to simplify by regrouping the isomeres which have similar characteristics and properties. For each class of hydrocarbon of 6 to 10 atoms of carbon we have considered : one aromatic, two naphtenes (cyclohexano and cyclo-pentane), and three or four paraffins (one normal-paraffin and two (for C_6 and C_7) or three issoparaffins). We have also considered 6 light gasses : hydrogen, methane, ethane, propane, iso-butane and iso-pentane. All this allows us to reduce the number of constituents to 39 pseudo-constituents.

2.4. The Kinetics

We have used a kinetic analysis which takes into account the inhibitions of the catalyst due to the constituents not entering into the considered reaction. For example, for the reaction of the aromatisation of the naphtenes,

$$N \rightleftarrows A + 3\, H_2$$

and we have the following kinetic relation :

$$-\frac{dN}{dW} = \frac{k}{P^{\alpha}(H_2/HC)^{\beta}} \times \frac{P_N - \dfrac{PA\ PH_2^3}{K}}{(1 + \sum_i b_i P_i)^2}$$

Pi : partial pressure of constituent i

bi : the associated adsorption equilibrium constant

P : total pressure

α and β : constants connected with each type of reaction

K : equilibrium constant

K : kinetic constant.

2.5 Integration

The catalytic reforming reactor has a special kind of geometry : the feed travels radially inwards through the porous catalyst.

We will therefore integrate the model using the concentric ring ΔR (Figure 3).

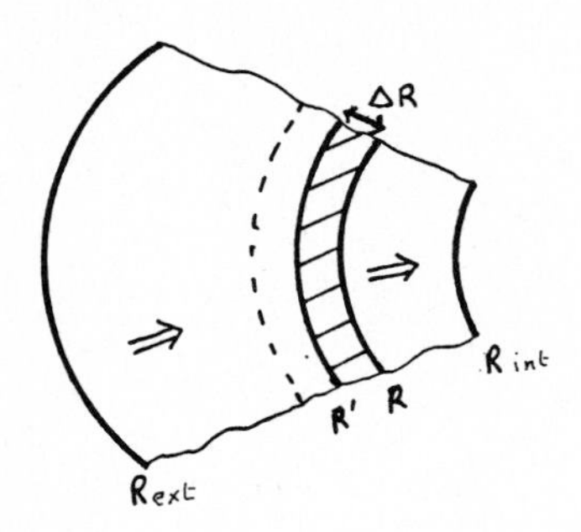

Figure 3

- Mass Balance :

$$\frac{X_{iR} - X_{iR'}}{\Delta R} = \Delta W \sum_{j=1}^{N.R} \alpha_j\, v_j$$

X_{iR} : Molar flow of constituent i in R

ΔW : Mass of catalyst in ΔR

α_j : Stochiometric coefficient of the reaction j

v_j : Kinetics of the reaction j.

- Thermal Balance :

$$\frac{T_R - T_{R'}}{\Delta R} = \Delta W \frac{\sum_{i=1}^{N.C} (\sum_{j=1} \alpha_j\, v_j \times H_i)}{\sum_{i=1}^{N.C} (X_i \times Cp_i)}$$

H_i : Enthalpie of constituent i

Cp_i : Heat capacity of constituent i.

Now, the simulation shows a drastic drop in temperature in the first part of the first reactor (excessive physical drop in the catalyst : more than 40°C over a few cms).

We have therefore taken into account the thermal exchanges being produced in the catalyst : the thermal conduction in the catalyst and the heat exchanges between the gas and the catalyst (Figures 4 and 5).

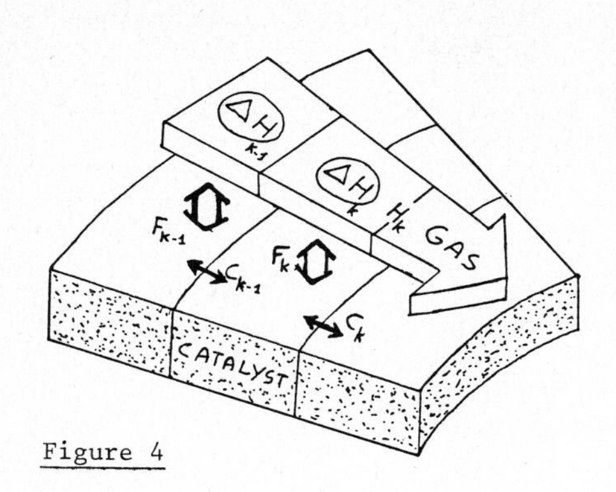

Figure 4

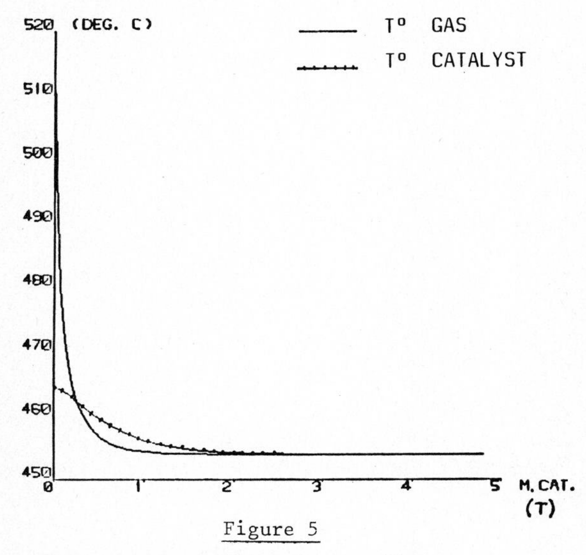

Figure 5

Temperature profiles in the first reactor.

H_o : Total enthalpy of gas at input to reactor.

H_k : Enthalpie of gas at output of narrow section K of catalyst.

C_k : Internal catalyst heat flow coming out of section k.

F_k : The heat exchanged between gas and catalyst in section k.

λ : The catalyst's thermal conduction coefficient.

γ : The thermal exchange coefficient (gas $\leftrightarrow$ catalyst).

S : The equivalent exchange surface area.

$$H_k + C_k = H_o$$

$$C_k = -\lambda S \frac{dT}{dR}$$

$$F_k : \gamma(T_{gas} - T_{catalyseur})_k$$

By linearizing the ΔH of the reaction, we obtain a linear system which allows us to calculate the temperature of the gas and of the catalyst for each narrow section k.

The non-linear system of first order differential method was originally integrated using an implicit method. Afterwards it was decided to use the more efficient Runge-Kutta method (Runge-Kutta-Fehlberg 4,5 with variable step size). And the sparse linear system obtained was solved using an adapted numerical method. [5]

$$\left[\begin{array}{l} \frac{\partial X_1}{\partial R} = f_1\ (X, T_G, R) \\ \vdots \\ \frac{\partial X_{NC}}{\partial R} = f_{NC}\ (X, T_G, R) \\ \begin{pmatrix} A \end{pmatrix} \begin{pmatrix} T_G \\ --- \\ T_M \end{pmatrix} = \begin{pmatrix} B \end{pmatrix} \end{array}\right.$$

3. INPUTS/OUTPUTS OF THE UNIT

Here we define the specific inputs/outputs for the correct running of the unit, as well as those presenting a pedagogical interest for the understanding of the complex phenomena which develop in each part of the unit.

3.1. Operational variables

The necessary variables for the running of the unit are :

- Feed flow
- Pressure
- Temperature (in fact the weighted mean of the temperatures throughout the 3 reactors: TMP).
- Recycling rate.

3.2. Inputs

- Feed flow
- Pressure
- Input temperature to the reactors
- Compressor speed (allows us to set the recycling rate).

3.3. Outputs

- Output temperature of the reactors (which are needed to find the TMP).
- Schilling density of the recycled gas.
- Output pressure at the compressor.
- Temperature of recycled gas.
  → Gives the purity of the recycled hydrogen gas.
- Flow of recycled gas.
  → This gives the recycling rate.
- The flows
  * of the gas deducted at the value.
  * of the stabilised platformate.
  * of the light gasses produced in the column.

3.4. Inputs/Outputs specific to the simulator

For pedagogical purposes it is possible to vary the type of feed (paraffinic, naphtenic), change the composition content ratio (C_6, C_7, C_8, C_9, C_{10}), and to modify the catalyst activity (to simulate the ageing of the catalyst due to parasitic coke formation).

As suplementary outputs there is the TMP and the recycling rate, as well as the product yields of hydrogen and C_5+, and of course the platformate quality : P.N.A. and N.O. Also available are the molecular weight and the liquid density of the feed and of the platformate. Other information available is the evolution of the composition of the feed flowing between each of the reactors, to show which reactions have taken place in which reactors (cracking, deshydrocyclisation, aromatisation...). We can also see the effect of injecting chlorine and water in the feed (regeneration of certain functions of the catalyst).

4. SIMULATION

The simulations that we have worked out have been based on the real unit. But there is no point in obtaining an exact concordance between the simulation and the real unit as we do not have access to all its variables anyway.

4.1. Simplified model

The distribution of the feed as a function of the length of the carbonic chains varies considerably depending on the origin of the crude and the conditions of distillation. For the sake of simplification wa had only included one class of constituents as being sufficiently representative of the feed. We therefore chose the C_8 constituents whose thermodynamic and kinetic characteristics are approximately equal to the net effect of the whole feed. Through the simulation of the model with such a feed had the advantage of being simple (10 constituents in total), the results obtained were too imprecise. We have therefore simulated the reforming of each class of constituents in order to observe their specific characteristics and their influence on the running of the unit.

Figures 6 and 7 show the influence of the choice of class of constituents on the temperature profiles and on the aromatic production of the three reactors. From these results we deduced that it was not possible to separate the specific characteristics of each class of constituents. (Great ΔT in R1 due to C_6, notable aromatisation of C_9 and C_{10} ...). We therefore decided to take the weighted mean of the characteristics of each class of constituent in order to obtain a feed similar to the real feed while keeping the model simple as possible. But the results obtained remained insufficient. The characteristic outputs of the model (product yield of platformate and hydrogen) were not close enough to those of the real unit under normal running conditions (bad equilibirum between the different types of reaction).

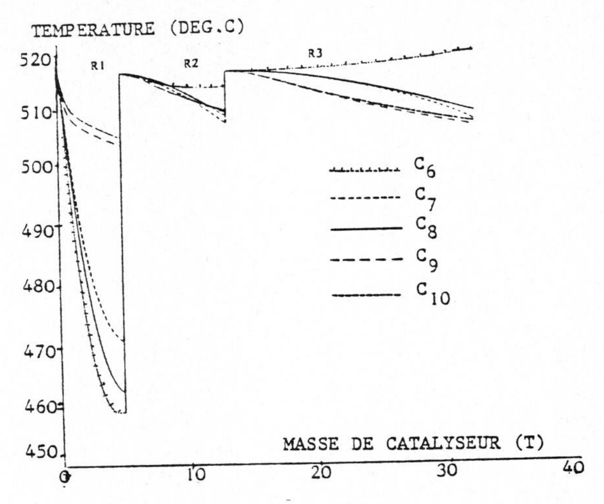

Figure 6. Temperature profiles.

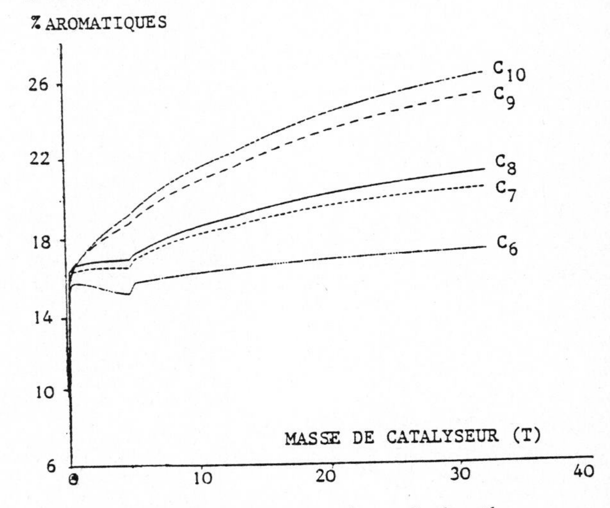

Figure 7. Aromatic production of the three reactors for each class of hydrocarbons.

4.2. Complete model

It was therefore necessary to keep a complete model of the feed (chains of 6 to 10 carbon atoms) this would allow us to verify the choice of kinetics, and to have a base from which simplifications of the model could be made.

The tests carried out with this model gave satisfactory results (Figure 8). The operators have at their disposal all the normal instruments of the unit on the control desk of the simulator. They also have the visualisation of the process on the graphic terminal, some curves having a particular pedagogical interst (profiles of temperature in the reactors for example). They also have at their disposition logging reports of tables 1, 2 and 3, where they can find all the internal variables to which they would not normally have access. This allows them to verify the unit is running correctly.

Table 3 shows clearly the different types of reactions produced in each reactor (aromatisation primarily in the first reactor, cracking and isomerisation of paraffins occuring mostly in the last reactor).

5. CONCLUSION

This static model of the catalytic reforming plant that we have produced gives satisfying results both in magnitude and tendency. It should be able to reproduce all the possible cases for the real unit and for all types of feed. Also the structure of the model allows us to have access to the kinetics of all the simulated reactions, which allows us to include important points in the running of the unit, like the influence of the ageing of the catalyst by coke formation and the injection of chlorine which regenerates certain functions of the catalyst.

6. REFERENCES

[1] Gauthier, J.P. and Panzarella, L., Simulateur d'entraînement en pétrochimie, Note Interne. Laboratoire d'Automatique de Grenoble. October 1979.

[2] Gauthier, J.P., Outils pour la Conception de Simulateurs d'Entraînement aux Procédés Industriels. Application à la Distillation. Thèse de Docteur-Ingénieur. INPG. January 1978.

[3] Lepage, C., Modélisation du Comportement Statique et Dynamique d'une Unité de Craquage Catalytique à Lits Fluidisés. Application à la Simulation Pédagogique. Thèse de Docteur-Ingénieur. INPG. Juin 1981.

[4] Fischer, B., Modélisation mathématique de la réformation catalytique. Thèse de Docteur-Ingénieur, Université Claude Bernard - Lyon. 1977.

[5] George, E., Forsythe, M., Malcolm, A.C.R. Moler, Computer Methods for Mathematical Computations (Prentice Hall). 1977.

[6] Henningsen, J., Bundgaard-Nielson, M., Catalytic reforming, British Chemical Engineering - Vol. 13, n° 11. 1970.

[7] Hevesi, J., Peter, I., Simulation and laboratory test of naphts reforming plants, Acata Chim. Acad. Sci. Hung (Vol. 99, n° 4). 1979.

[8] Krane, H.G., Groh, A.B., Schulman, B.L., Sinfelt, J.H., Reactions in catalytic reforming of naphtas, World Petroleum Congress Proceedings, 5th, New York (section III, paper 4). 1959.

[9] Lazenby, T.M., Waterfield, D.N., Bath, M. Rubie, P.E., Catalytic reforming model developpent. I. Chem. Eng. Symposium (series n° 35). Int. Chem. Eng. London. 1972.

[10] Smith, R.B., Kinetic analysis of naphta reforming with platinium catalyst, Chemical Engineering Progress (Vol. 55, n° 6). 1959

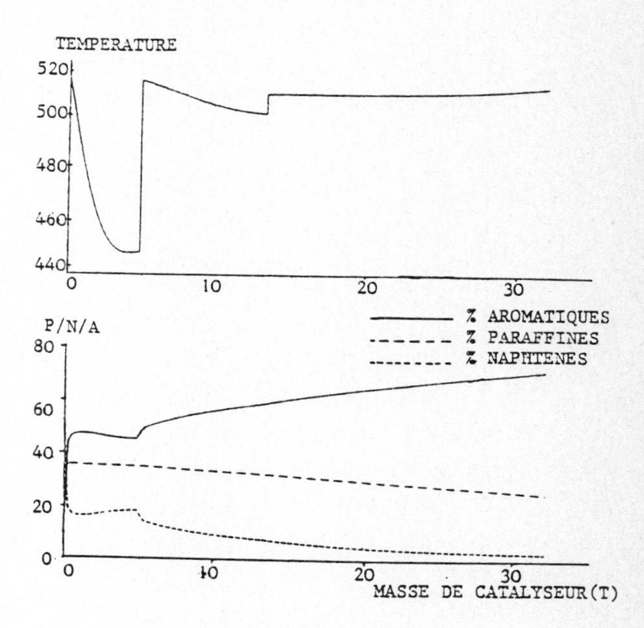

Figure 8

Paraffinic feed.

CONDITIONS DE MARCHE

DEBIT DE LA CHARGE	:	1811.	T/J
DEBIT DU GAZ RECYCLE	:	1505.	T/J
RAPPORT MOLAIRE H2/HC	:	7.16	
TEMPERATURE D'ENTREE DU REACTEUR 1	:	513.	DEG. C
TEMPERATURE D'ENTREE DU REACTEUR 2	:	513.	DEG. C
TEMPERATURE D'ENTREE DU REACTEUR 3	:	508.	DEG. C
PRESSION D'ENTREE DU REACTEUR 1	:	32.	BARS
PRESSION DE SORTIE DU REACTEUR 3	:	30.	BARS

BILAN THERMIQUE

REACTEUR	NUMERO 1	NUMERO 2	NUMERO 3
TEMP. D'ENTREE (DEG. C)	513.0	513.0	508.0
TEMP. DE SORTIE (DEG. C)	452.8	501.7	508.5
DELTA TEMP. (DEG. C)	-60.2	-11.3	0.5
DELTA TEMP. TOTAL (C)	-70.9	T.M.P.	504.2

Table 1

Logging reports of the process.

	CHARGE		PLATFORMAT STABILISE		GAZ DE RECYCLAGE		MAKE GAS + STAB. TOP LIQ. & GAS	
COMPOSITION	T/J	% VOL.	T/J	% VOL.	T/J	% VOL.	T/J	% VOL.
HYDROGENE	0.0	0.0	0.0	0.0	242.5	72.5	27.5	66.1
CHAINES EN C 1	0.0	0.0	0.0	0.0	345.9	13.0	32.4	9.8
CHAINES EN C 2	0.0	0.0	0.0	0.0	409.0	8.2	48.1	7.8
CHAINES EN C 3	0.0	0.0	0.0	0.0	329.2	4.5	80.4	8.8
CHAINES EN C 4	0.0	0.0	0.0	0.0	154.3	1.6	90.6	7.5
CHAINES EN C 5	0.0	0.0	82.2	7.5	23.9	0.2	0.0	0.0
CHAINES EN C 6	90.5	6.4	148.2	11.6	0.0	0.0	0.0	0.0
CHAINES EN C 7	724.4	43.7	561.0	38.6	0.0	0.0	0.0	0.0
CHAINES EN C 8	652.0	34.5	487.4	29.4	0.0	0.0	0.0	0.0
CHAINES EN C 9	181.1	8.5	136.3	7.3	0.0	0.0	0.0	0.0
CHAINES EN C 10	163.0	6.9	117.0	5.7	0.0	0.0	0.0	0.0
DEBIT TOTAL (T/J)	1811.		1532.		1505.		279.	
MASSE MOLAIRE (G)	107.8		100.7		9.1		13.5	
DENSITE LIQUIDE	0.743		0.779					
DENSITE SCHILLING					0.314		0.468	
% PARAFFINES { NP / IP }	52.9	{ 50.0 / 50.0 }	44.1	{ 33.8 / 66.2 }				
% NAPHTENES { 6N / 5N }	36.7	{ 70.0 / 30.0 }	1.7	{ 10.9 / 89.1 }				
% AROMATIQUES	10.4		54.3					
N.O ESTIME			96.5					
RENDEMENT C5 + (%)			84.6					
RENDEMENT H2 (%)			1.5					

Tables 2 and 3.

Logging reports of the process.

	ENTREE REACTEUR 1		SORTIE REACTEUR 1		SORTIE REACTEUR 2		SORTIE REACTEUR 3	
COMPOSITION	T/J	% VOL.	T/J	% VOL.	T/J	% VOL.	T/J	% VOL.
HYDROGENE	242.5	65.8	266.7	67.8	271.1	67.5	270.0	66.4
CHAINES EN C 1	345.9	11.8	348.9	11.1	359.2	11.2	378.3	11.7
CHAINES EN C 2	409.0	7.4	413.8	7.0	430.4	7.2	457.2	7.5
CHAINES EN C 3	329.2	4.1	336.5	3.9	363.2	4.1	409.6	4.6
CHAINES EN C 4	154.3	1.5	162.6	1.4	193.0	1.7	244.8	2.1
CHAINES EN C 5	23.9	0.2	34.1	0.2	66.6	0.5	106.1	0.7
CHAINES EN C 6	90.5	0.6	99.8	0.6	127.5	0.8	148.2	0.9
CHAINES EN C 7	724.4	4.0	704.5	3.7	653.4	3.4	561.0	2.9
CHAINES EN C 8	652.0	3.2	623.0	2.9	561.7	2.6	487.4	2.2
CHAINES EN C 9	181.1	0.8	172.5	0.7	155.2	0.6	136.3	0.6
CHAINES EN C 10	163.0	0.6	153.4	0.6	134.5	0.5	117.0	0.4
P.N.A. ESTIME SUR C5 +								
% PARAFFINES { NP / IP }	53.5	{ 48.8 / 51.2 }	53.9	{ 47.2 / 52.8 }	51.3	{ 41.2 / 58.8 }	44.9	{ 32.7 / 67.3 }
% NAPHTENES { 6N / 5N }	36.2	{ 70.0 / 30.0 }	11.3	{ 15.8 / 84.2 }	4.2	{ 6.1 / 93.9 }	1.7	{ 10.9 / 89.1 }
% AROMATIQUES	10.3		34.7		44.5		53.4	

Simulation in Engineering Sciences
J. Burger and Y. Jarny (eds.)
Elsevier Science Publishers B.V. (North-Holland)

IMPORTANCE OF SIMULATION IN DEEP BED FILTRATION

S. VIGNESWARAN

Environmental Engineering Division,
Asian Institute of Technology,
Bangkok, Thailand

In deep bed filtration, the removal of particles in the suspension is complex and depends on numerous filtration parameters. The mathematical models decribing deep bed filtration involve several filtration coefficients which are characteristic of the particular suspension to be filtered and the type of filter media used. If these coefficients are evaluated from a limited number of filter experiments, filtration performance can be simulated for any operating conditions which would be helpful in designing the filter in a rational manner. This paper presents the simulation of dual media and radial filters from a limited number of conventional (vertical down flow) filter experiments. The simulated and experimental results are found to be comparable which indicates the usefulness of simulation in deep bed filtration.

1. INTRODUCTION

Filtration, a solid-liquid separation process, has been used as final clarifying step in water treatment for more than a century. The removal mechanisms of particles in suspension is complex and depends on numerous filtration parameters like size, shape, density of filter grains and particles in suspension, viscosity of fluid, suspended solid concentration, porosity, filtration rate, chemical characteristics of particles, filter grain and fluid etc. in addition to filter depth and filtration time. In practice, the selection of design parameter is based on empirical values obtained from past experiences.

There are many mathematical models available to evaluate the filter performance. These mathematical models involve several filtration coefficients, which characterize the particular water to be filtered and the type of filter media used. These filter coefficients can be evaluated for a particular water and filter media from a limited number of laboratory scale experiments. Filtration performance for various conditions can then be simulated. This would facilitate in determining optimum design conditions without extensive experimental analysis.

This paper attempts to use the simulation method in designing dual media and radial filters which are two of the modified forms of conventional rapid filters.

2. FILTRATION SYSTEMS CONSIDERED

2.1 Dual Media Filtration

2.1.1 Mathematical Modelling of Dual Media Filtration

2.1.1.1 Equations Governing Deep Bed Filtration Removal Efficiency

Iwasaki (1937) proposed the first equation in deep bed filtration based on first order kinetics relating the change of concentration of suspended particles with filter depth and the local concentration

$$-\frac{\partial C}{\partial L} = \lambda C \qquad (1)$$

in which, C = local concentration of suspended particles
L = filter depth
λ = filter coefficient (which varies with dilter depth and time)

From the mass balance of suspended solids (the volume of particles removed from flowing suspension is equal to the particles accumulated in the pores), the following relationship can be established.

$$-v\frac{\partial C}{\partial L} = \frac{\partial \sigma}{\partial t} + \frac{\partial C}{\partial t}(f-\sigma) \qquad (2)$$

in which, f = porosity of filter
t = filtration time
v = approach velocity of filtration
σ = specific deposit

Since the change in concentration of particles in the pores with time is very small compared to specific deposit except in the beginning of filter run, the equation 2 can be simplified as follows:

$$-v\frac{\partial C}{\partial L} = \frac{\partial \sigma}{\partial t} \qquad (3)$$

To predict the local suspended solids concentration the relationship between λ and σ should also be known. There are many equations relating these two parameters. In the present study the following relationship is used (Ives, 1960).

$$\lambda = \lambda_o + c_1\sigma - \phi \frac{\sigma^2}{f_o - \sigma} \qquad (4)$$

in which, c_1 and ϕ are constants for a particular suspension and filter medium

Calculation of σ and λ

The specific deposit can be calculated from the following formula (Fox and Clasby, 1966) which is derived directly from its definition.

$$\sigma_{L_1}^{L_2} = \Sigma_{i=1}^{i} \frac{(C_1-C_2)_{i-1} + (C_1-C_2)_i}{2} (t_i - t_{i-1}) \cdot \left(\frac{v}{L_2 - L_1}\right) \qquad (5)$$

in which,

$\sigma_{L_1}^{L_2}$ = specific deposit between depth L_1 and L_2 at time t_i

$(C_1-C_2)_{i-1}$ = difference in concentration between depth L_1 and L_2 at time t_{i-1}

$(C_1-C_2)_i$ = difference in concentration between depth L_1 and L_2 at time t_i

The filter coefficient (λ) can be calculated at a particular time between two layers from the following formula which is a rearranged from of equation 1.

$$\lambda = \frac{\frac{-(C_1-C_2)_i}{L_1-L_2}}{(C_1)_i} \qquad (6)$$

Calculation of λ_o, C and ϕ:

Once the values of σ and λ are determined from the above equations using experimental local concentration values, the values of λ_o, c_1 and ϕ can be calculated from multiple linear regression method.

2.1.1.2 Equation Governing the Headloss

Mints (1966) developed the following relationship between headloss and concentration

$$H = H_o + k(C_o - C)t \qquad (7)$$

in which, H_o = clean bed headloss (which can be calculated from Kozney's equation)

k = constant

Calculation of k:

From the experimental values of H and C at different filtration times and depths, the value of k can be obtained which is a constant for a particular suspension and filter operating conditions.

2.1.2 Simulation of Dual media filter performance

2.1.2.1 Calculation of Single Medium Filter Mathematical Model Coefficients

The filter removal mechanism is so complex that the filter equations cannot be generalized. i.e a limited number of experiments have to be performed for the particular suspension and filter media to calculate the filter mathematical model coefficients such as λ_o, c_1, ϕ, k.

The values of λ_o, c_1, ϕ and k for sand and anthracite as filter media and Kaolin clay as suspension are summarized in Table 1.

Table 1 The filter mathematical model coefficients for different filter media (Filter depth = 40 cm; Filtration rate = 10 m/h)

Filter medium	λ_o	c_1	ϕ	k
Anthracite (1.2 to 1.6 mm)	0.064	24.04	6148.1	2413.9
Sand (0.60 to 0.79 mm)	0.089	37.80	4168.2	17472.9

From the single medium filter mathematical model coefficients, the dual media performance can be simulated with the assumption that there is no intermixing of filter media (anthracite and sand) at the interface.

2.1.2.2 Calculation of Theoretical Filter Run Lengths and Optimization of Dual Media Filters:

The theoretical filter run lengths based on concentration and headloss criteria are determined when the effluent concentration and headloss exceeded their specified limits. The upper limits of effluent concentration and headloss in the present study are fixed and are equal to 1 N T U and 70 cm respectively. An optimum filter run is achieved when the filter design and operation cause the filter to reach its headloss limit at the same time as filtrate quality deteriorates to an unacceptable value (Mints, 1966). The break-even optimization is adopted to find this optimum filter run (The break-even point is obtained when the graph of filter run length versus filter depth ratio for the given headloss limit intersects the similar graph for the given quality limit).

The simulated theoretical filter run values for different depth ratios of anthracite and sand are presented in Fig 1. The corresponding experimental values are also presented in this figure for verification purpose. The actual and the simulated results are comparable.

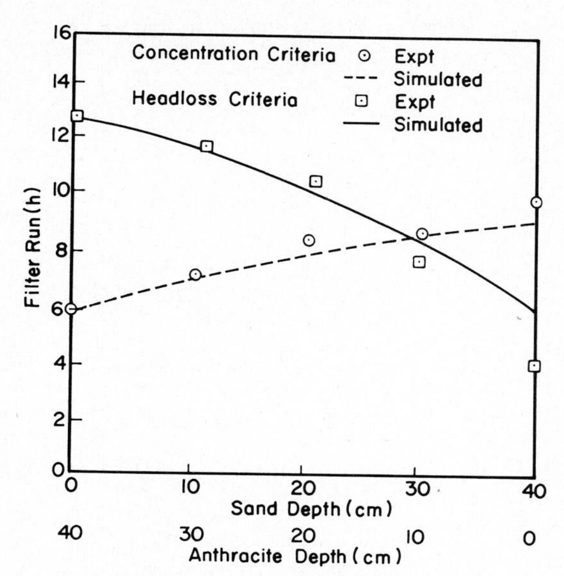

Fig. I - Simulated and Experimental Filter Runs for Different Filter Media Depth Ratios (Filtration Rate = $10\ m^3/m^2h$)

2.2 Radial Flow Filters

2.2.1 Mathematical Modelling of Radial Flow Filters

Governing equations of a radial flow filter are similar to that of a conventional (vertical down flow) filter. The only difference in the radial filter case is that the filtration rate varies along the radius in the flow direction (Eqs. 8 and 9)

$$\frac{\partial C}{\partial r} + \lambda^r C = 0 \qquad (8)$$

$$\frac{\partial C}{\partial r} + \frac{1}{v(r)} \frac{\partial \sigma}{\partial t} = 0 \qquad (9)$$

in which, λ^r = filter coefficient at radius r

$v(r)$ = filtration rate at radius r

As mentioned earlier, a relationship between λ^r and σ should be established to predict the local suspended solids concentration at different times. In this case the following simplified form of λ-σ relationship for conventional filter (Eq. 10) with the modification that takes into account the flow rate variation along the radius is used (Eq. 11)

$$\lambda = \lambda_o \left(1+\beta\frac{\sigma}{f_o}\right) \qquad (10)$$

$$\lambda^r = \lambda_o^{r_o} \left(\frac{r}{r_o}\right)^{\alpha} \left(1+\frac{\beta(r)}{f_o}\sigma\right) \qquad (11)$$

in which, r_o = radius of the inlet surface

$\beta(r)$ = radial filter mathematical model coefficient

$\lambda_o^{r_o}$ = initial filter coefficient at inlet surface

2.2.2 Simulation of Radial Filter Performance

As mentioned earlier, a laboratory scale experimental study is necessary to evaluate the mathematical model coefficients as they vary with the characteristics of suspended solids and filter medium. The values of mathematical model coefficients obtained, from conventional sand filtration of kaolin clay suspension are as summarized in Table 2.

Table 2 Values of λ_o and β at Different Filtration Rates (Influent Concentration = 100 mg/l; Filter depth = 2.5 cm; Media Size = 0.595 - 0.841 mm)

Filtration Rate (m/h)	λ_o	β
5	0.337	44.1
10	0.327	46.7
15	0.282	38.8

From these results following relationships can be established (Eqs. 12 and 13)

(i) The relationship between the initial filter coefficient (λ_o) and filtration rate (v) is:

$$\lambda_o = v^{-0.24} \qquad (12)$$

(ii) The model coefficient β does not vary significantly with the filtration rate and is assumed to be equal to a value of 44.

$$\beta(r) = 44 \qquad (13)$$

Hence equation 11 can be modified for the present experimental conditions as:

$$\lambda^r = \lambda_o^{r_o} \left(\frac{r}{r_o}\right)^{0.24} \left(1+44\frac{\sigma}{f_o}\right) \qquad (14)$$

Integrating equation 8 and combining with equation 9, one would obtain,

$$C(r+\Delta r, t+\Delta t) = \left(\frac{1-A}{1+A}\right) C\ (r, t+\Delta t) \qquad (15)$$

in which,

$$A = \lambda_o^{r_o}\left(\frac{r}{r_o}\right)^{0.24} \frac{\Delta r}{2} \left\{1+\frac{44}{f_o}\sigma(r,t+\Delta t)\right\}$$

Eq. 8 can be written in the finite difference form as follows:

$$\sigma(r,t+\Delta t) = \sigma(r,t)+\frac{v(r)\Delta t}{\Delta r}\{C(r,t)-C(r+\Delta r,t)\} \quad (16)$$

in which, $\sigma(r,t)$ and $\sigma(r,t+\Delta t)$ are the specific deposits at radius r and at times t and $t+\Delta t$ respectively
$C(r,t)$ and $C(r+\Delta r,t)$ are the suspended solids concentrations at time t and at radii r and $r+\Delta r$ respectively

The local concentrations at different times are simulated using eqs. 15 and 16 along with the boundary conditions listed below.

$$\sigma(r,0) = 0 \quad (17)$$

$$C(0,t) = C_o \quad (18)$$

$$C(r,0) = C_o.\exp\{-\frac{\lambda_o^{r_o}}{1.24 r_o^{0.24}} \cdot (r^{1.24} - r_o^{1.24})\} \quad (19)$$

The simulated effluent concentration curves of radial filter at different flow rates are presented in Fig. 2 together with the experimental curves. As in the earlier case, the simulated and experimental results are comparable.

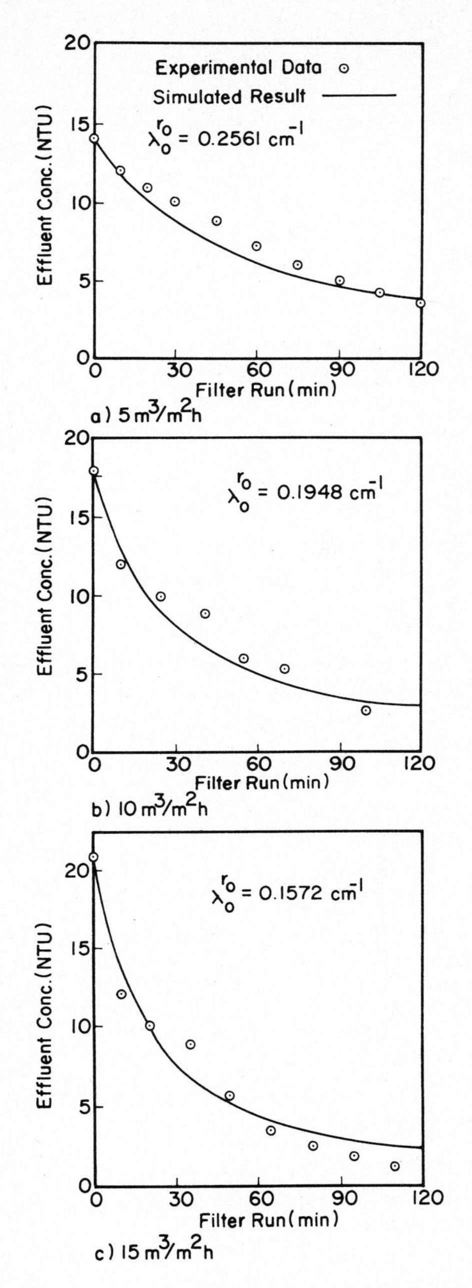

Fig. 2 – Simulated Effluent Concentrations for the Radial Filter

3. CONCLUSION

The dual media and radial filtration performances can be successfully simulated from a limited number of conventional filter experiments. Such a simulation study would not only facilitate the rational design of filters but also reduce substancially the number of experiments to be performed to set up the design criteria.

REFERENCES:

1 Fox, D.M. and Clasby, J.L., Experimental evaluation of sand filter theory, J. Amer. Soc. of Civil Eng. 92 (1966)

2 Ives, K.J., Rational design of filters, Proc. Inst. Civil Eng. 16 (1960)

3 Ives, K.J., Advances in deep bed filtration, Trans. Inst. Chem. Eng. 48 (1971)

4 Iwasaki, T., Some notes on sand filtration, J. Amer. Water. Works. Assoc. 29 (1937)

5 Vigneswaran, S., Comparative study of rapid filtration techniques for turbid water, M.Sc Tnesis No. 1407, AIT, Bangkok, Thailand (1978)

XII. MANIPULATORS – ROBOTS (1)

Simulation in Engineering Sciences
J. Burger and Y. Jarny (eds.)
Elsevier Science Publishers B.V. (North-Holland)

SIMULATION AND DYNAMICAL CONTROL OF A BUFFER STORAGE SYSTEM IN CHEMICAL INDUSTRY

J.M. CHARTRES, J.M. BARBEZ, D. MEIZEL

INSTITUT INDUSTRIEL DU NORD
Laboratoire d'Informatique Industrielle et d'Automatique
B.P. 48
59651 VILLENEUVE D'ASCQ CEDEX - FRANCE

The increasing complexity of manufacturing processes leads to a very perceptible parallel development of control and management tools.
From the example of an existing industriel system, we could implement an interesting structure, where the use of simulation enabled us to diversify the possibilities of regulation and control.
In a first chapter, we propose to introduce the existing system, to elaborate the aims of automation and to introduce the control system proposed.
The whole simulation unit together with its contribution in the control assembly are developed in a second chapter.

1. AUTOMATION OF THE STORAGE SYSTEM

1.1 Introducing the system

The A.P.C. factory (Nitrogen and Chemical Products) of Mazingarbe, from C.D.F. Chimie group manufactures nitric acid to make nitrates.

The set is composed with five acide making unities. These five production workshops (recorded AN2 to AN6) supply the storage system with variable flows (see Figure 1).

The latter which has been worked out time after time and according to the successive developments of the factory, consists in a net of eight tanks the capacities of which are varying between 100 and 500 m3. Each tank gets the supply of one or several production workshops and can ensure the acid production for the different consumer workshops (Nitramo I, Nitramo II and compounds) that have variable flows too.

The storage acts essentially as a buffer between working production and consumption workshops.

The whole storage capacity is 2200 m3. The maximum daily production can come up to 1700 m3 whereas the consumption varies up to 2200 m3 acid per day.

The present plant is equipped with a net of pipelines and pumps enabling numerous combinations of acid exchanges between production and consumption workshops together with internal transfers from a tank to another.

Then, it is possible to divert the acid flows with respect to input-output flows and the initial storage state in order to maintain the best mean acid storage.

1.2 System specification aim of the automation

The now-existing storage-system is entirely controlled by human operators, numerous interventions on the system are thus necessary and it requires at each moment an always ready staff of operators in order to modify the acid flows according to variations in the input-output flows or according to any working risks.

The goal of the automation of this storage unit is essentially financial by the fact that is should eliminate almost all human interventions whilst ensuring safety of operation and an easier and more versatile control.

The immediate financial goal of this automation is enforced by the fact that measuring devices replace human measurement of the tank-levels leading thus to a more efficient and easier management.

On an other hand, increasing safety of operation by automation leads to decrease the down time of production. In fact the most common accidents such as overflows are due to human interpretation errors while measuring (and writing) the tank-levels (such human errors cannot be avoided on long term).

With this aim in mind, the control of the storage unit must choose and establish the optimal network of pipe-lines in order to preserve the best mean storage capacity with respect to the state of the system storage (the level of the tank) and the measurement of the input-output flows.

We have defined a set of possible working strategy is characterized by one network of pipe-lines and thus by the state of the gates that "realizes" this network.

On the other hand, it had been decided to incorporate to the control-software a software-tool that performs simulation of the middle-term and long-term run of the unit. This software should be an aid to the decision in the case of human control (that always must be possible). The simulation of the nominal automatic run of the storage unit can also validate a decision concerning the production/consumption strategies on a long period such as, for example, a week-end.

The structure of the now-existing storage-unit has been obtained from the successive developments of the A.P.C. factory and, from an economic point of view the only changes that are allowed are constituted by the introduction of servo-gates in the pipe-network and measurement-devices for the tank-levels.

In short, the optimal automated storage-system obtained by the few previous possible modifications is as close as possible to the ideal system composed of one large tank whose volume is the sum of the volumes of all existing tanks (see Figure 2).

This large tank should naturally be connected to any production/consumption unit. The designed-system should then run as close as possible to the previous "optimal" system.

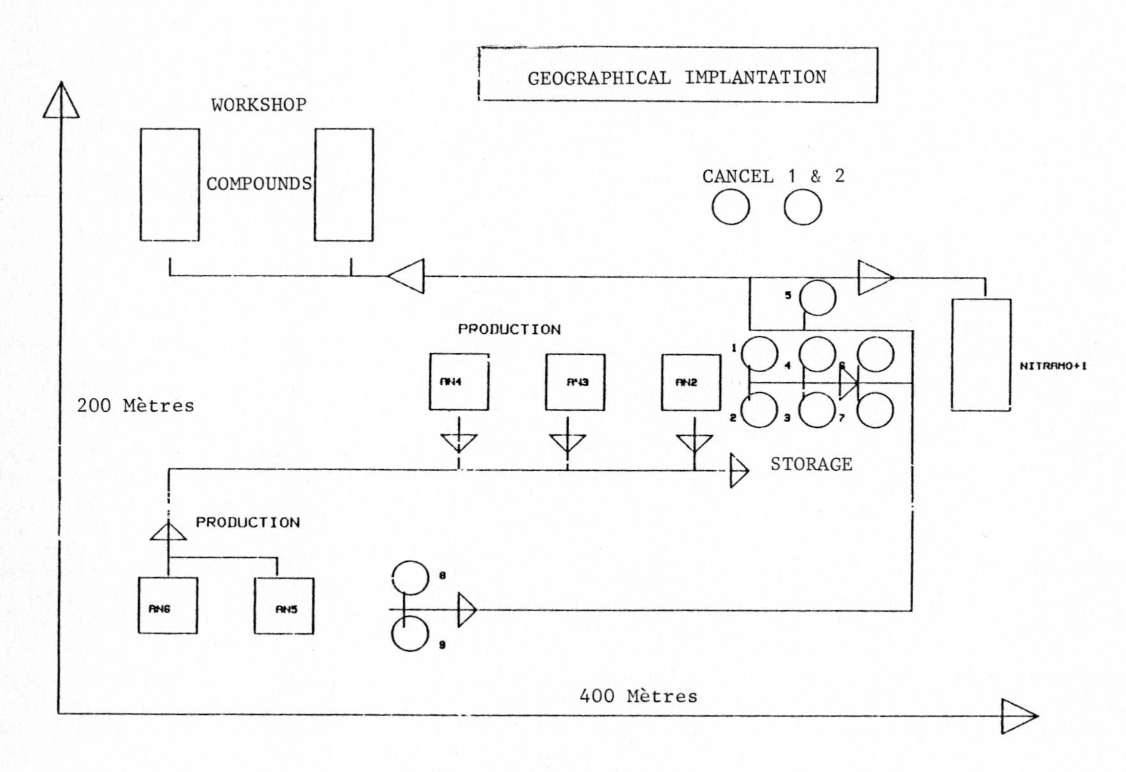

Figure 1

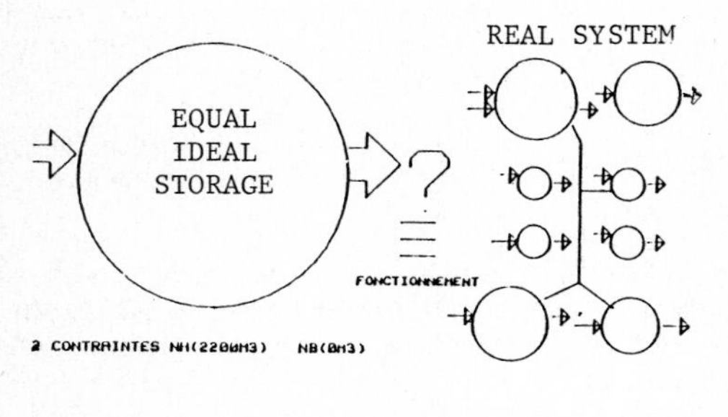

Figure 2

1.3 Structure of the controller

Analysing both the existing system and the desidered specification had led us to the control-structure drum up in Figure 3. We propose now to comment the different roles of the various blocks appearing on this scheme.

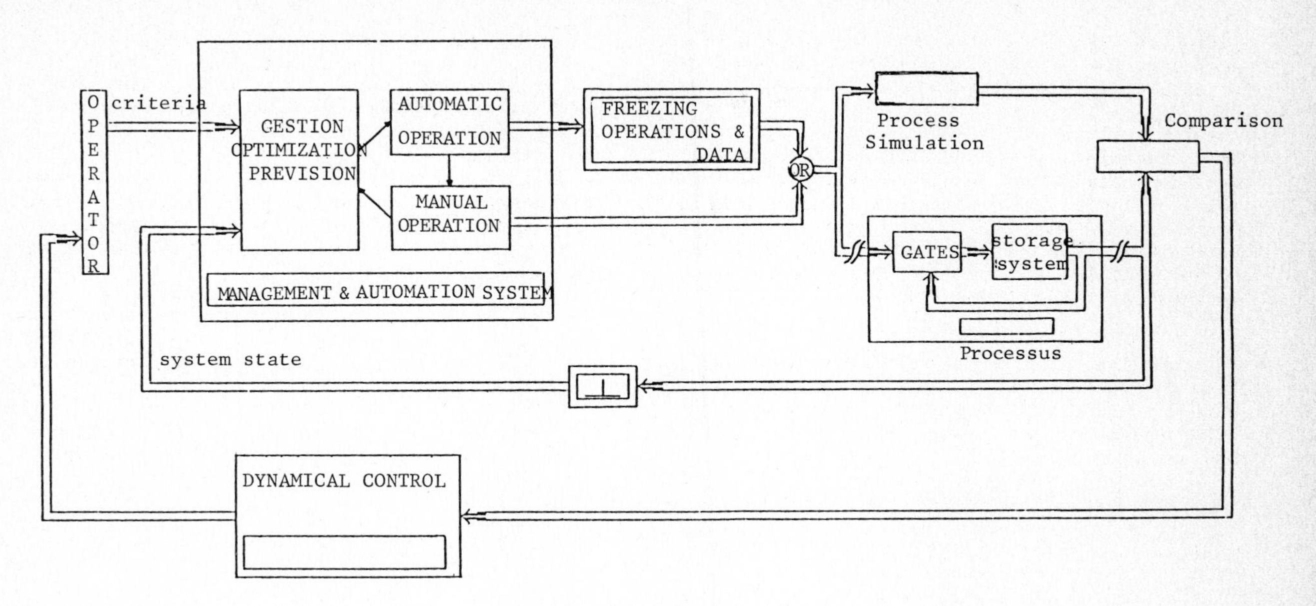

Figure 3

1.3.1 Storage management / automation system

It constitutes the discrete-time command of the process. It is materializes by a software that determines the control of the servo-gates within a sample-period of one hour with respect to the input/output specifications given by the (human) process-conductor and the "discretized" levels of the tanks.

The choice of the network of pipe-lines obtained by use of the servo-gates is performed by optimization of a criterion that tends to assign the acid-level in the various tanks.

The optimization is obtained by a simulation-test of the effect of the different admissible control available strategies memorized in a catalog.

The input-output links are thus established so that the storage-unit is the best "capacitor" as possible in order to take into account short-term perturbations without stopping the acid or fertilizers production. With such a control-policy, the only alarm is generated in the only cases where all the tanks are almost empty ou almost full.

1.3.2 Simulation block

It has a double role :

- The former is to compare the trajectories of the tanks-level with their nominal behaviour between two successive sample-time (dynamical observation). In such a case the simulation is performed in parallel to the process ans in the same time-scale. The redundancy introduced by this dynamical observation with respect to the one hour discrete-time control is usefull to diagnostic a defect occuring in the system (for instance in a servo-gate). Of course, it provides too quicker information about system's defect and it allows cure quickly this defect.

- The latter consists to use the simulation-block in a predictive way (in such a case, the simulation-block that gives the nominal future behaviour of the system is deconnected from the process and runs in an accelerated time). The storage management system (1.3.1) gives in this case, control-strategies with respect to the state obtained from the simulation. This utilisation answers to the demand for aid to the decision in giving the middle and long term consequences of a production/consumption policy.

2. SIMULATION

2.1 Simulation of the unit

The storage-model is composed of as many integrators as tanks. The time-constants that are due to the pipe-lines can be neglected with regards to the one hour control period.

Each control-strategy (pipe-line network) is depicted by a matrix whose elements are link between each pair of tanks.

By use of such a model with a small sampling periods (say 30 s), the continuous-time trajectories of the level of each tank is adequately approximated.

2.2 Simulation of the proposed automated unit

Off-line simulations performed on a DEC PDP 11/34 scientific computer leads to validate the previously defined control-structure and algorithms and to optimize the number of control-strategies and eventually a modification in the pipe-array.

In the simulation and the reality, the run of the process is possible as long as there is no alarm indicating that one tank is almost full or empty.

The results of this simulation have to be compared with the results obtained by the use of the "optimal" system composed of one "large" tank (see Fig. 2).

The comparison criteria taken into account are the mean run-time, the mean-upper and the lower storage volumes and the correspondent standard deviations that characterizes the process global-limits.

2.3 System global limits

In order to obtain the previously defined quantities, we propose to the simulation a catalog of 256 cases of input/output flows.

This catalog is representative of all possible situations that arise in reality. The initial state of the tanks is that they are half-full and the simulation of the automated system is performed until a blocking situation (one tank is either full or empty). At this step, we memorize the sum of the volume of tanks, the value of the stop-time and the number of times each control-strategy has been called. All infinite run-periods that arises for example when the sum of input floxs equate the sum of output flows are not taken into account in the results.

The final results are then summarized in terms of means values and standard deviations as below.

	Simulated case	Ideal case
Mean Upper-storage Volume	1950 m3	2200 m3
Standard deviation	99 m3	/
Mean Lower-storage Volume	157 m3	0 m3
Standard deviation	55 m3	/
Mean run-time	60 h	67 h
Number of control strategies	17	/

This table emphasizes the validity of the control structure and algorithms that lead to the desired specification (see Fig. 2).

2.4 Predictive oriented simulation

The previous simulation software can be run on the site based on the microprocessor control/management/simulation device.

It can be used, at the first sight, to predict sticking situations sufficiently early (10 hours in advance).

On the other hand, it enables to help the process conductor to obtain the best modifications in the working-points of the production/consumption units in order to go back into an asymptotically verifying checked situation.

In the case where a human operator takes back the control of the storage-unit, he can be informed about the controls that are to be implemented on the gates in the next ten hours.

For each run-hour, the execution time of the program is less than 2 mn and it provides the following results :

- time
- choosen control-strategy
- volume of each tank
- flows of each tank
- during the time before a sticking situation.

These results can all the same be obtained when the process runs with identified known defects.

This is the fact, for example, when a flood gate does not work. In such a case, control strategies that call a change of this flood gate are forbidden.

Such a modification is obviously introduced at the same time into the storage.management/control block.

3. CONCLUSION

The global specifications for the management and automation of the process led us to defined an adequate structure of a micro-processor based device.

A complete study of the performance of such a structure has been implemented in the Laboratory and validates the chosen structure.

The whole software has two levels in connection with the two levels of the control structure.

The former is constituted by a discrete time optimal control algorithm that produces a control-strategy during the one-hour sampling period. The optimized criterion leads to equate the acid-level in the various tanks. The latter is constituted by a dynamical observation of the state of the system continuously compared with its nominal evolution.

This scheme provides informations about a defect in the system and it helps the operators to know what defect it is.

The simulation block, informed about the identified defect can then be used in predictive mode to know if it is possible to run the process while reparation is performed.

The proposed structure, by its performances and its cost is economically optimal.

Simulation in Engineering Sciences
J. Burger and Y. Jarny (eds.)
Elsevier Science Publishers B.V. (North-Holland)

COMPUTER ASSISTED SYNTHESIS OF AN OPTIMAL DYNAMIC CONTROL FOR A ROBOT MANIPULATOR

J.Y. Grandidier, D. Girardeau, M. Monsion, C. Bouvet

Laboratoire d'Automatique, de Reconnaissance des Formes
et de Robotique Agricole
E.N.S.E.R.B. - 33405 TALENCE - FRANCE

We present a comparison by simulation of several control algorithms with given final position, for a robot manipulator powered by stepping motors. Traditional open loop control, using a table of frequencies has poor performance, due to changes of the manipulators dynamics during its movement. Optimum control in minimum time is achieved by a closed loop control. Costs caused us to seek an open loop approximation to the optimum, by considering simplified models of the dynamics of the manipulator.

1. INTRODUCTION

Control of an articulated system poses the following problem : what forces or torques must be applied to the various articulations in order to change the configuration of the system and bring the terminal organ orientation and position in space ? This problem may be solved by using the dynamic model of the system, from which the equations for the forces and couples may be derived (1).

The present work concerns the synthesis of a minimum time control for an articulated system powered by electric permanent-magnet step motors. The device is intended to automatically harvest fruit and vegetables.

The choice of motors was dictated by the useful properties of stepping motors, which are powered by pulses, and are thus perfectly adapted to controls from a digital calculator.

The motors function synchronously. There is a one to one correspondence between an electrical pulse and a rotation of the rotor, called the step. A constant speed of rotation corresponds to each combination of sequence and frequency of supply to the phases of the stator. Provided no steps are lost, the position and the speed of the rotor are thus controlled in an open loop, hence avoiding the addition of costly sensors.

The chances of loss of synchronism between the rotor and the power supply of the stator phases, increase with the work required of these motors, e.g. when we require optimal control in minimum time of a load of which the dynamics is not well controlled.

The synthesis of a least-time control for a given fixed load is a well solved problem (2). The optimum commutation frequencies of the supplies to the stator phases are calculated in advance and stored in memory.

The optimal command of loads with varying dynamics, such as articulated systems, requires the commutations to be calculated as functions of the position of the rotor. One is thus lead to a closed loop control with the necessary disadvantage of using position uncoders.

Cost and simplicity lead one to reduce the problem to one of an open loop control (3). The parameters of the model of the load are then given their worst possible values corresponding to the most difficult configurations of the manipulator. However, this produces a non-optimal control compared to a closed loop device.

In our simulations, we first produced a reference control, simulating closed loop control of the motors. Secondly we compared traditional open loop control using a frequency table, with the reference control. In order to limit costs, we chose to keep the open loop, but sought to approximate the optimal control by using simplified dynamical models of the load, which could be solved by the calculator in real time.

2. DESCRIPTION OF THE MANIPULATOR (Figure 1)

The manipulator studied here has three degrees of freedom :
- a horizontal translation
- a horizontal rotation
- a translation of the tool.

These 3 movements are controlled by 3 SLOSYN MO 93 stepping motors (200 steps/turn). The manipulator is mounted on a trolley (4) (5). The arm is thus light since it is mounted on the trolley and works in a hostile environment. The loads manipulated and the accuracy required are, however, small.

3. MODEL

3.1 Geometrical model

When a change of position is required of a robot, the user may ask that the tool or terminal organ should be brought to a final position specified by its coordinates X_F^i (e.g. cartesian coordinates of a part of the manipulator in R^3.

The present position of the manipulator may be determined from the generalized coordinates q^i by use of the transformation formulae :

$$X^i = F^i(q_1, \ldots, q_i, \ldots, q_n)$$

F constitutes the geometrical model of the manipulator.
The manipulator arm studied here has the equations :

$$X_1 = F^1(q_1,q_2,q_3) = q_1$$
$$X_2 = F^2(q_1,q_2,q_3) = l_2+l_3 \cos q_2+(n_4+q_3) \sin q_2$$
$$X_3 = F^3(q_1,q_2,q_3) = m_2+l_3 \sin q_2-(n_4+q_3) \cos q_2$$

Where the lenghts in metres are :

$l_2 = 0,125$ $m_2 = 0,180$ $l_3 = 0,1$ $n_4 = 0,6$.

3.2 Dynamical model

The dynamical model obtained from the Lagrangian equations can be expressed in matrix form as :

$$A \underline{\ddot{Q}} + B \underline{\dot{Q}\dot{Q}} + C \underline{\dot{Q}}^2 = \underline{C_m} \underline{G} - \underline{F} - K \underline{\dot{Q}} \quad (1)$$

where :
- $\underline{Q}$ is the generalized coordinate vector
- $\underline{\dot{Q}}$ is the generalized velocity vector
- $\underline{\ddot{Q}}$ is the generalized acceleration vector
- $\underline{\dot{Q}\dot{Q}}$ is the vector with components $\dot{q}_1\dot{q}_2$, $\dot{q}_1\dot{q}_3$ and $\dot{q}_2\dot{q}_3$.
- $\underline{\dot{Q}}^2$ is the vector of squared generalized velocities
- A is the 3x3 inertia matrix
- B is the 3x3 matrix of Coriolis forces
- C is the 3x3 matrix of centrifugal forces
- C_m is the vector of motor torques.

$$\underline{C_m} = \begin{pmatrix} -C_{m_1} \sin N_r (\theta_1 - n_1 \frac{2\Pi}{p}) \\ -C_{m_2} \sin N_r (\theta_2 - n_2 \frac{2\Pi}{p}) \\ -C_{m_3} \sin N_r (\theta_3 - n_3 \frac{2\Pi}{p}) \end{pmatrix}$$

where :

- . C_{mi} is the amplitude of the motor torque reflected to the output shaft.
- . θ_i is the angular postition of the rotor. If R is the step-down ratio, $\theta_i = R q_i$.
- . n_i is the number of commutations since the initial position
- . p is the number of steps of the motor (200 steps/turn)
- . $N_r = \theta_e/\theta_i$ where θ_e is the electric angle
- . $\underline{G}$ is the vector of gravitational forces
- . $\underline{F}$ is the vector of viscous frictional torques reflected to the secondary
- . K is the diagonal 3x3 matrix of coefficients of viscous frictional reflected to the secondary.

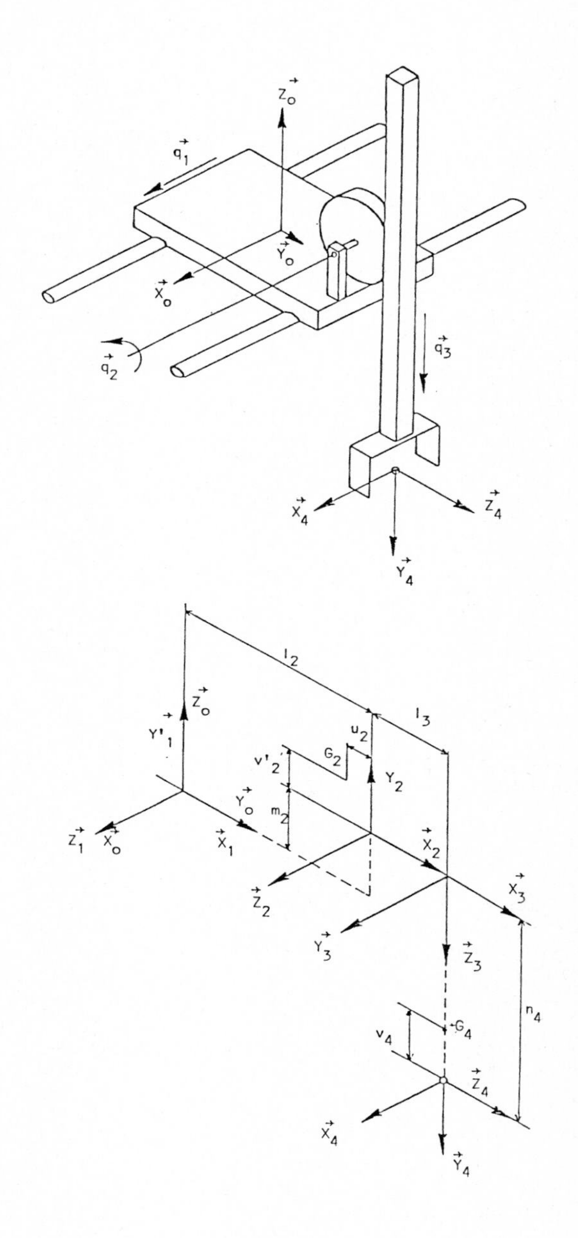

Figure 1 :
Diagram of the articulated arm for picking asparagus, with the origins relative to the different parts. The arm is shown according to RENAUD's convention (5) with all the coordinate values zero.

The arm studied here is defined by the values :

$$A = \begin{pmatrix} a_{11} & 0 & 0 \\ 0 & a_{22} & a_{23} \\ 0 & a_{32} & a_{33} \end{pmatrix}$$

with

$a_{11} = M_1 + M'_1$

$a_{22} = I_2 + M_3(\frac{l_4^2}{3} + l_3^2) + M_4(l_3^2 + (n_4 - v_4)^2) + (M_3 + M_4)q_3^2 + 2\,M_4(n_4 - v_4)q_3$

$a_{23} = a_{22} = -\,l_3(M_3 + M_4)$

$a_{33} = (M_3 + M_4) + M'_3$

where

M_i = mass of body i
M_1 = 16,0 kg
M'_1 = (inertia of motor 1 reflected to the secondary)
M'_1 = 22 kg
M_2 = 7,305 kg
M'_3 = (inertia of motor 3 reflected to the secondary)
M'_3 = 11 kg
M_4 = 2,1 kg
I_i = inertia of body i reflected to the secondary
I_2 = 0,88 m^2kg
V_i = ordinate of the centre of gravity G_i
$V_4 = -\,0,2$ m
$2l_4$ = length of the grip support, $l_4 = 0,45$ m

$$B = 2\begin{pmatrix} 0 & 0 & 0 \\ 0 & 0 & b \\ 0 & 0 & 0 \end{pmatrix}$$

where $b = (M_3 + M_4)q_3 + M_4(n_4 - V_4)$

$$C = \begin{pmatrix} 0 & 0 & 0 \\ 0 & 0 & 0 \\ 0 & c & 0 \end{pmatrix}$$

where $c = -\left((M_3 + M_4)q_3 + M_4(n_4 - V_4)\right)$

$$K = \begin{pmatrix} k_1 & 0 & 0 \\ 0 & k_2 & 0 \\ 0 & 0 & k_3 \end{pmatrix}$$

where K_1 = 204 N/m/s ; K_2 = 6,4 mN/rd/s ; K_3 = 56 N/m/s.

$$\underline{G} = \begin{pmatrix} 0 \\ g_2 \\ g_3 \end{pmatrix}$$

where $g_2 = -g\cos q_2\,\{n_2M_2 + m_3(M_3 + M_4)\} + g\sin q_2\,\{v_2M_2 - (n_4 - V_4)M_4 - (M_3 + M_4)q_3\}$

$g_3 = g(M_3 + M_4)\cos q_2$ where n_i is the abscissa of the centre of gravity G_i ; n_2 = 0,017 m
V_2 = 0,098 m

$$\underline{F} = \begin{pmatrix} f_1 \\ f_2 \\ f_3 \end{pmatrix}$$

where f_1 = 47 N ; f_2 = 9,6 mN ; f_3 = 20,5 N.

N.B. : Motor 1 is decoupled from the other two and its dynamics is independent of q_1, so we did not include it in our simulation. It can be time optimal controlled in open loop by a frequency table.

4. SIMULATION

We simulate the control of the arm between two points M_0 and M_1 (terminal position given) by numerically solving the differentical system (1).

It should be noted in what follows that the motor couple C_m is assumed to be constant. It is in fact now possible to obtain, with high-performance power supplies, nearly flat torque vs speed curves (7). This assumption allows us to explore the high speed domain where the dynamical model is most interesting.

4.1 Dynamic control in closed loop

Algorithm 1 consists in first accelerating as much as possible and then braking from some well chosen moment so that the motor stops dead at the right position.

The commutation during acceleration (respectively braking) is chosen when the phase between the rotor and the commuted phase is $\alpha_a = -\,45°$ (respectively α_d = 135°) (cf figure 2), so that maximum torque (or braking) is applied.

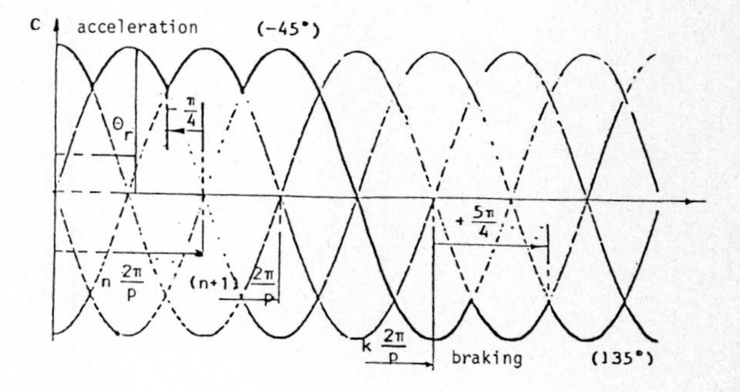

Figure 2

When the nth motor phase is powered, the maximum couple occurs for θ_r between $-\frac{5\pi}{4}$ and $-\frac{\pi}{4}$.
When θ_r reaches $\pi/4$ it is advantageous to commutate to the $(n+1)^{th}$ phase in order to maintain the torque.

The switch to slowing down is done when the kinetic energy of the manipulator reaches the brake energy over the number of steps remaining (8). The brake energy over one step is :

$W_{fi} = (C_i + K_i\dot{q}_i + f_i)\,\alpha p$

where C_i is the brake couple of the motor
k_i is the coefficient of viscous friction
f_i is the coefficient of dry friction
α_p is the angular step.

The brake energy, strictly equal to $N_i x W_{fi}$ is taken as $(N_i-0,5)W_{fi}$ to avoid avershooting the target. The constant 0,5 is an empirical choice.

LOAD THE NUMBER OF STEPS N_i OF THE FIRST COMMUTATION PERIOD OF EACH MOTOR

DYNAMICAL MODEL OF THE PROCESS CALCULATION OF $\ddot{q}$, $\dot{q}$, q.
INCREMENT THE TIME $t_i = t_i + \Delta t$

$50 \left[\theta_i - n_i \frac{2\pi}{\ } \right] > \alpha a$ ou αd — no

yes → commutation

$N_i = N_i - 1$

$N_i = 0$ — yes — STOP

Calculation of W_c and $W_F(N_i-0,5)$
$W_F (N_i - 0,5)$

acceleration ← yes — $W_c < W_F (N_i - 0,5)$ — no → braking

Algorithm 1 : Dynamic control in closed loop

4.2 Control by a frequency table

We simulated separately the movements of motors 2 and 3 taking account of the coupling introduced by the load. The load was assumed to be constant and to correspond to the worst possible situation, where the load is a maximum, and the resistive moment is maximum (starting) or minimum (stopping). This simulation determines the best commutation frequencies for a given load (8).

Algorithm 2 simulates the working of the motor, with a steady increase of speed, maximum speed and slowing down. The commutation frequencies are given in the corresponding tables.

The frequency tables in this paragraph were calculated off line from the following parameters (these parameters are reflected to the secondary).

MOTOR 2	MOTOR 3
C_{m_2} = 57,16 mN	C_{m_3} = 164 N
a_2 = 3,26 m^2kg	a_3 = 14,7 kg
k_2 = 12 mN/rd/s	k_3 = 56,5 N/m/s
f_3 = 19 mN	f_3 = 70 N

N.B. : Gravitational forces are included in the dry frictional terms, as is one Coriolis term, the other being included in k_2.

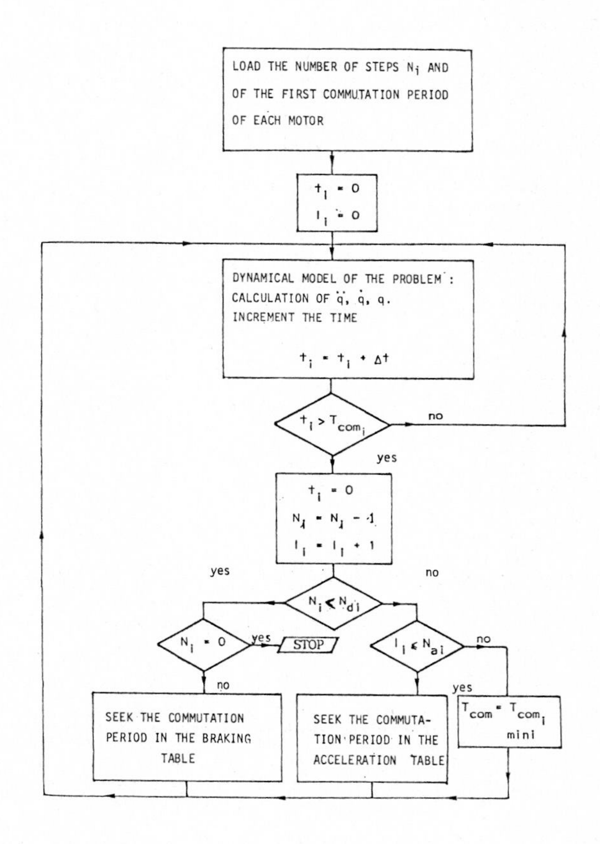

Algorithm 2 : Open loop control by a frequency table.

4.3 Open loop control with simplified dynamical models

We shall now try to approach the performance of closed loop controls by using an open loop control with commutations calculated from a simplified dynamical model of the loads on the motors.

Three simplified models were tested :
- inertia model accounting for variations of A with $\underline{Q}$.
- gravity model, taking into account the variations of G with $\underline{Q}$.
- Coriolis force model, taking into account the changes of B and C with $\underline{Q}$.

The simulation algorithm is realy the same as algorithm 2, except that the periods of commutation are determined otherwise. We describe below for the model of Coriolis and centrifugal forces, details of the determination of the commutation periods replacing the level corresponding to algorithm 2.

The acceleration of motor i for the coming step is given by the equation deduced from (1) :

$$\ddot{q}_i = (C_i - g_i - f_i - k_i \dot{g}_i - \sum_{j=1}^{3} \sum_{k=j+1}^{3} b_{ijk} \dot{q}_j \dot{q}_k - a_{ij} \ddot{q}_j) / a_{ii} \qquad (2)$$

where C_i is the mean torque over one step of the motor considered, $C_i = 0{,}9\ C_{mi}$.

$\dot{q}_i$, $\dot{q}_j$ and $\dot{q}_k$ are supposed to vary slowly and are estimated from

$$\dot{q}_l = \frac{\alpha_p}{T_{ol}} \qquad l = i,\ j,\ k$$

where T_{ol} is the commutation period of the previous step.

The new commutation period T_{com} is then obtained from the equation :

$$\frac{1}{2} \ddot{q}_i\ T_i^2 + \dot{q}_i\ T_i - \alpha_p = 0$$

where α_p is the geometrical step.

N.B. : The braking frequencies in the control by the simplified dynamical model are taken equal to the braking frequencies of the control by a frequency table.

Real-time calculation of the commutations was done with the numerical values :

1) Coriolis and centrifugal force model

MOTOR 2	MOTOR 3
C_2 = 51,4 mN	C_3 = 147,6 N
a_2 = 3,26 m^2kg	a_3 = 14,7 kg
k_2 = 6,4 mN/rd/s	k_3 = 56,5 N/m/s
f_2 = 19 mN	f_3 = 56,5 N

2) Gravitational model

MOTOR 2	MOTOR 3
C_2 = 51,4 mN	C_3 = 147,6 N
a_2 = 3,26 m^2kg	a_3 = 14,7 kg
k_2 = 12 mN/rd/s	k_3 = 56,5 N/m/s
f_2 = 10 mN	f_3 = 34 N

3) Inertial model

MOTOR 2	MOTOR 3
C_2 = 51,4 mN	C_3 = 147,6 N
k_2 = 12 mN/rd/s	k_3 = 56,5 N/m/s
f_2 = 19 mN	f_3 = 70 N

5. EXPERIMENTAL RESULTS

Plate 1 compares closed loop control and open loop control with a frequency table. The saving of time is clear.

Plate 2 shows the results of real time calculations for the simplified models.

In all three cases, we note that the arm does not arrive at the required final position. This is due to a loss of synchronism of the stepping motors. The rotor does not follow the turning field created in the stator by applying power successively to the phases.

The commutations are not controlled, as they would be in a closed loop command, so that the dephasing occurs at lower arm speeds than those reached in closed loop control.

Examination of the curves gives us the terms of the complete dynamical model which most influence a given motor. Thus motor 2 is very sensitive to the inertia term, whereas motor 3 is sensitive to the gravity term.

6. CONCLUSION

If, for reasons of cost, stepping motors must be controlled in an open loop, then their performance may be improved by seeking simplified dynamical models of their loads. These models should take into account those principal terms of the complete model, which contribute to the movement considered.

Control of articulated systems by stepping motors will be significantly improved only if there is some control of the rotor position. This will ensure synchronous functioning even in hostile agricultural environments, where the load parameters may vary widely.

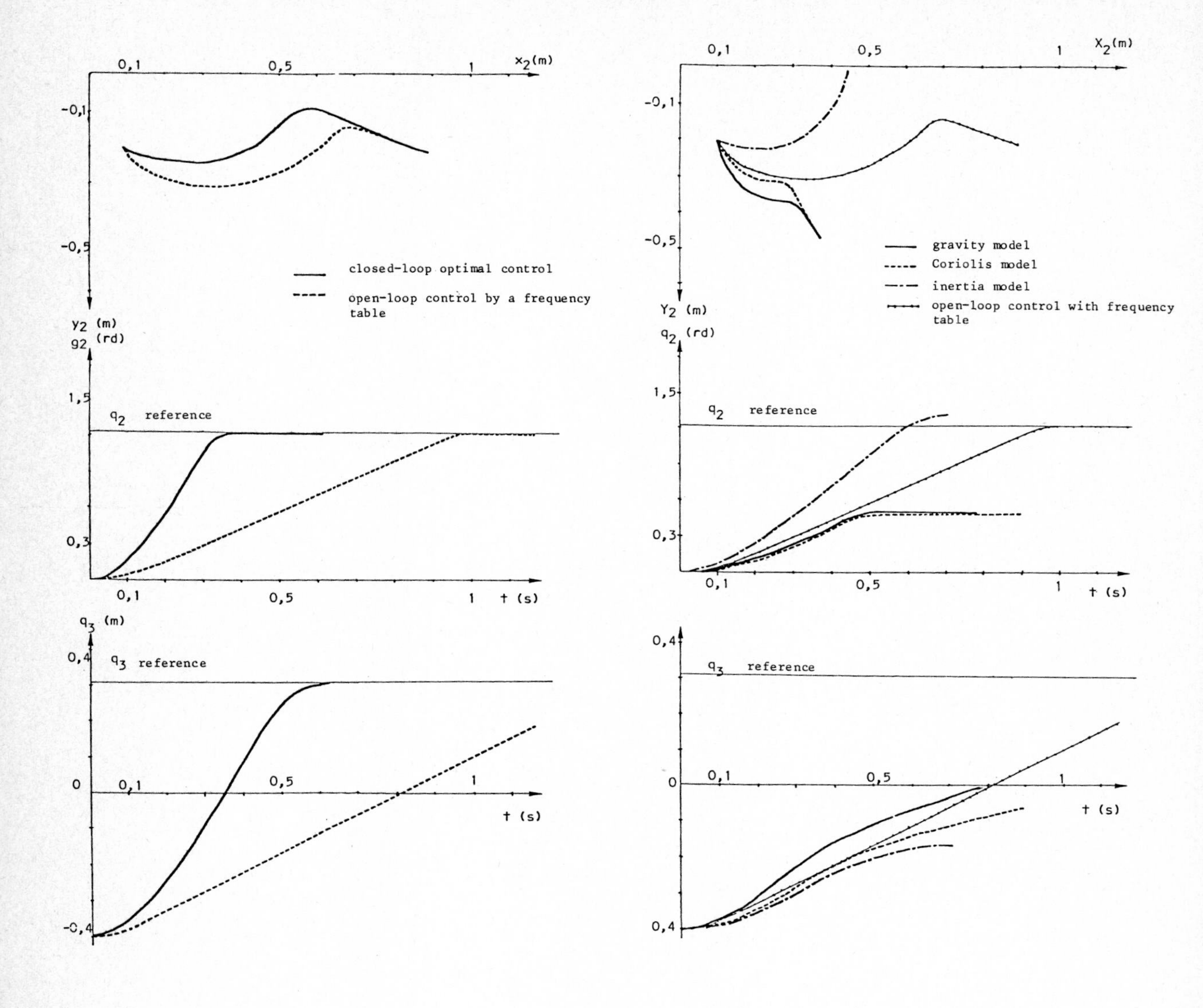

Plate 1

Plate 2

REFERENCES

(1) KHATIB,O., LLIBRE,M. and MAMPEY,R.
Fonction décision-commande d'un robot industriel, Rapport scientifique
CERT, TOULOUSE (juillet 1982)

(2) MIYAMOTO,H.
Modélisation et commande optimale d'un moteur pas à pas par microprocesseur
Thèse, INPI LAUSANNE (mai 1979)

(3) DUFAUT,M., GOEDEL,C., HUSSON,R. and PINCHON,D.
Génération de courbes d'accélérations adaptatives pour la commande d'un manipulateur équipé de moteurs pas à pas
2èmes journées d'études sur les moteurs pas à pas, EPF LAUSANNE (avril 1982)

(4) BAYLOU,P., MONSION,M., BOUVET,C. and BOUSSEAU,G.
La robotique agricole, Le Nouvel Automatisme (mars 1982)

(5) MONSION,M., BAYLOU,P., BOUVET,C. and BOUSSEAU,G.
Un robot mobile agricole, Etat de la robotique en France, Tome 1 (HERMES 1982)

(6) RENAUD,M.
Contribution à la modélisation et à la commande des robots manipulateurs
Thèse de doctorat d'Etat
Université P. Sabatier, TOULOUSE (1980)

(7) CARBON,C.
Application des moteurs pas à pas à la robotique
lères journées d'études sur les moteurs pas à pas
ENSEM NANCY (juin 1979)

(8) JUFER,M.
Transducteurs électromagnétiques
(Georgi 1979)

(9) GIRARDEAU,D., GRANDIDIER,J-Y., BOUVET,C. and MONSION,M.
Modélisation et simulation d'un moteur pas à pas en vue de sa commande optimale
Conférence internationale AMSE, PARIS (juillet 1982)

Simulation in Engineering Sciences
J. Burger and Y. Jarny (eds.)
Elsevier Science Publishers B.V. (North-Holland)

DESIGN OF A HIGH SPEED SEVEN-LINK-BIPED STABILIZED BY STATE FEEDBACK CONTROL

Tsutomu Mita

Department of Electrical Engineering, Chiba University
1-33, Yayoi cho, Chiba City, 260
Japan

The purpose of this paper is to propose a new control method of the biped locomotion composed of seven links and to show simulated and experimental data of the locomotive motion of the biped built in our laboratory. The control method depends mainly on the linear optimal state feedback regulator which stabilizes the stance supported by both legs as a commanded attitude. As the result the biped called CW-1 successfully walks any pre-determined steps such that one step takes only one second.

1. INTRODUCTION

There are many studies on simulation and realization of the biped locomotion (for example, [1]~[3]). Kato [1] made a biped called WL-9DR which is 3 dimensional model of the human lower body and walks arbitrary steps. However the motion is slow since one step takes 9 seconds in the existing circumstances. Miyazaki and Arimoto made a biped called "IDATEN" and realized a high speed locomotive motion [2]. Hemami et. al. [3] showed a simulated results of the attitude control of the biped.

In this paper we propose a control method of the biped using the linear optimal state regulator which stabilizes the stance supported by both legs as a commanded attitude, and show simulated and experimental data of the locomotive motion. As the result, the biped called CHIBA WALKER 1 (CW-1), which is a planar model of the human lower body, successfully walks any predetermined steps such that one step takes only one second. This shows that a usefulness of the typical modern control theory and also shows that the state regulator method should be used for further studies on realization of high speed biped of 3 dimensional model.

2. THE BIPED CW-1

The biped CW-1 is composed of seven links (one hip, two upper legs, two lower legs and two feet) as shown in Photo 1. Six DC servomotors attached at six joints control the relative angles q_i's of the joints.

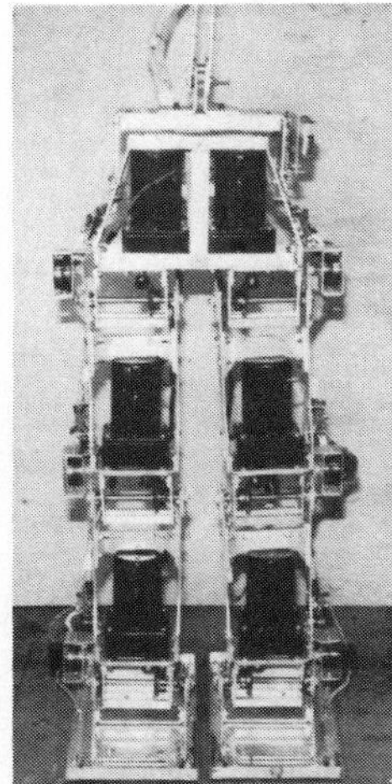

Photo 1
The biped CW-1

These angles are measured by six potentiometers. Four touch sensors are attached on the sole of each foot and they send an information whether the foot is on the ground or not. The microcomputor decides that the foot reaches the ground if one of the four sensors become "on" state. The hight of CW-1 is about 75 cm, and the length of the crotch is 60 cm which is about 3/4 multiple of Japanese adult human. The weight of CW-1 is about 15 kg. CW-1 is loosely constrained to the sagittal plane in the existing circumstances using two boards put on the hip and a pipe piercing through these two boards. The width of each board is about 5 cm and the diameter of the pipe is about 1.5 cm.

3. FOOT CONTROLLER AND MODELING

3.1 PARAMETERS AND FOOT CONTROLLER

The planar model of the seven-link-biped CW-1 can be illustrated by Figure 1. We define a_i, b_i and l_i are the length between the bottom and the center of mass of each link, the one between the top

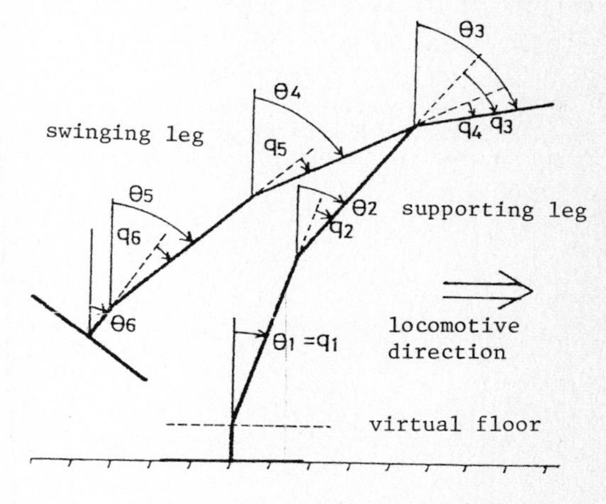

Figure 1 Seven link model of biped

Table 1 Parameters of CW-1

	M_i (Kg)	I_i (Kgm²)	l_i (m)	a_i (m)	b_i (m)
hip	4.20	0.014	0.200	0.085	0.115
lower leg, or upper leg	2.10	0.007	0.235	0.083	0.170
lower leg +foot	2.70	0.010	0.302	0.120	0.182

$\tilde{I}_i$ =0.366 (Kgm^2), D_i=11.42 (Nms), K_q=1.683(Nm/V)

and the center of mass, and the total length of each link, respectively. Also we define M_i, I_i, $\tilde{I}_i$ and D_i as the weight of each link, the moment of inertia around the center of mass of each link, the moment of inertia of each joint implying the actuater and the coefficient of viscous friction of each joint. Further Kq is defined as the torque/voltage conversion coefficient of the motor. The measured values of these parameters are shown in Table 1.

In the sequel, we assume that the robot CW-1 does not gallop, that is, one foot always touches the ground. Then the control of CW-1 is divided into two fundamental control strategies. The first is the foot control using a local feedback and the second is an attitude control using the state feedback control. In this section we explain the foot controller. The foot of the swinging leg is controlled so that it paralells the floor. This requirement comes from two reasons, (i) to prevent stumbling, it is better that the tip of the foot does not look down ward with respect to the floor, (ii) switching between the roles of the swinging leg and the supporting leg must be performed smoothely.

In Figure 1, θ_i is defined as the absolute angle of the i th link measured from the vertical. Then θ_3 can be written as $\theta_3=q_1+q_2+q_3$ using the relative angles of the supporting leg. On the other hand, it can be described by $\theta_3=q_4+q_5+q_6+\theta_6$ if the variables of the swinging leg is used where θ_6 is the absolute angle of the ankle of the swinging leg. These two equations yield

$$\theta_6=(q_1+q_2+q_3-q_4-q_5)-q_6 \triangleq r-q_6 \qquad (1)$$

which is considered to be a control deviation of the motor which controls q_6 as an output following the reference r. Therefore a local feedback $u_6=k\theta_6$ can regulate θ_6 to be zero since this control system is a type 1 servosystem provided k is chosen so that the closed loop system is stable. $\theta_6=0$ means that the foot of the swinging leg is parallel with the floor. Figure 2 shows the circuit of the transformation of the coordinate which is performed by means of operational amplifiers.

3.2 MODEL FOR THE ATTITUDE CONTROL

We assume that the above mentioned foot control is perfectly executed and the foot of the supporting leg is fixed to the floor until the swinging leg is transfered to the forward direction. Also we assume that the floor is given by the virtual floor shown in Figure 1 to simplify the problem and that the parameteres of the swinging leg is obtained assuming the lower leg implies the foot. Then the seven link model

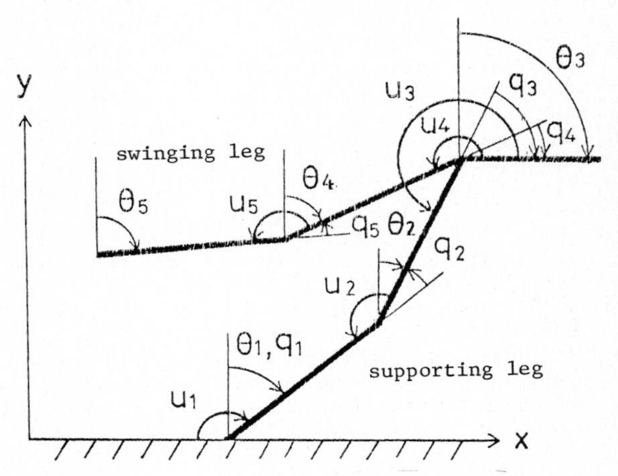

Figure 3 Five link model of the biped

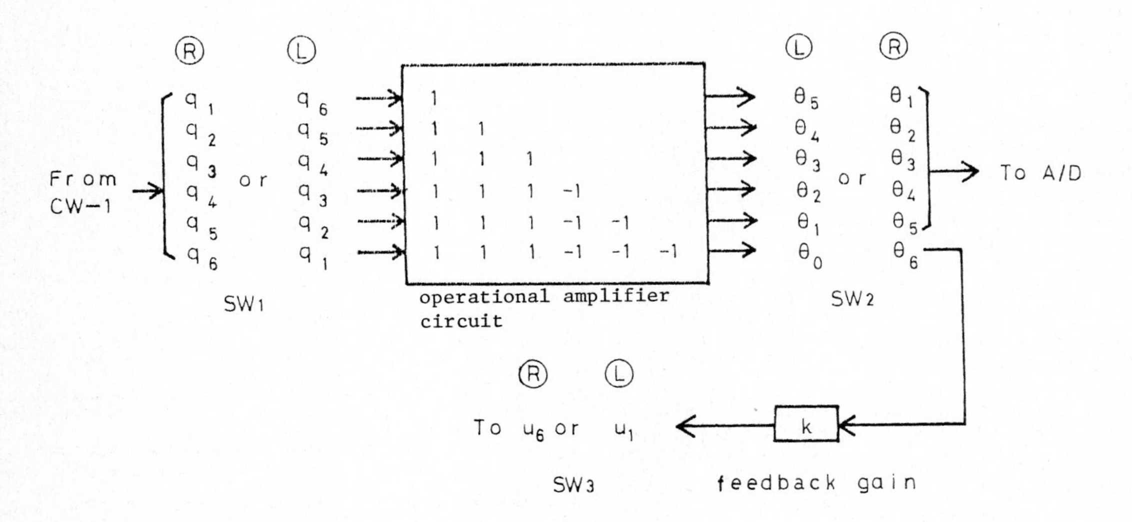

Figure 2 Coordinate transformer and analog switches

of CW-1 can be reduced to a five link model shown in Figure 3. If the origin of the sagittal plane is placed at the ankle of the supporting leg, the center of mass (x_i, y_i) of the i th link can be described using θ_i's. By use of these variables, we have the kinetic energy, the potential energy and the dissipative energy as

$$T=\frac{1}{2}\sum_{i=1}^{5}[I_i\dot{\theta}_i^2+M_i(\dot{x}_i^2+\dot{y}_i^2)+\tilde{I}_i\dot{q}_i^2] \quad (2)$$

$$U=\frac{1}{2}\sum_{i=1}^{5}M_i g y_i,\quad D=\frac{1}{2}\sum_{i=1}^{5}D_i\dot{q}_i^2 \quad (3)$$

respectively. The relation between θ_i's and q_i's is described by

$$q=K\theta,\qquad \theta=K^{-1}q \quad (4)$$

$$:\ q=(q_1,q_2,q_3,q_4,q_5)^T,\ \theta=(\theta_1,\theta_2,\theta_3,\theta_4,\theta_5)^T$$

$$K=\begin{bmatrix}1&0&&&\\-1&1&&&\\&-1&1&&\\&&1&-1&\\&&&1&-1\end{bmatrix},\quad K^{-1}=\begin{bmatrix}1&&&&\\1&1&&&\\1&1&1&&\\1&1&1&-1&\\1&1&1&-1&-1\end{bmatrix}$$

where the symbol T denotes the transposition.

If (2) and (3) are rewritten using θ, we have the following equation from Lagrange's equation of motion.

$$J(\theta)\ddot{\theta}+X(\theta)\dot{\theta}^2+Y\dot{\theta}+Z(\theta)=Eu \quad (5)$$

$$:\ J(\theta)=[L_{ij}\cos(\theta_i-\theta_j)]+K^T\mathrm{diag}(\tilde{I}_1,\ldots,\tilde{I}_5)K$$

$$X(\theta)=[L_{ij}\sin(\theta_i-\theta_j)],\ Y=K^T\mathrm{diag}(D_1,\ldots,D_5)K$$

$$Z(\theta)=-(G_1\sin\theta_1,\ldots,G_5\sin\theta_5)^T,\ E=KqK^T$$

$$\dot{\theta}^2\triangleq(\dot{\theta}_1^2,\ldots,\dot{\theta}_5^2)^T,\quad u=(u_1,\ldots,u_5)^T$$

where $[x_{ij}]$ denotes a matrix whose i,j th element is given by x_{ij} and u_i is defined as the control voltage of the i th motor. The parameters L_{ij} and G_i are determined from $M_i, I_i, \tilde{I}_i, a_i, b_i$ and l_i shown in the Table 1 (See the Appendix).

In the following we express (5) by

$$f[\ddot{\theta},\dot{\theta},\theta]=Eu \quad (6)$$

for the sake of simplicity.

4. ATTITUDE CONTROL AND GAIT CONTROL

4.1 ATTITUDE CONTROL USING STATE FEEDBACK

In this section, we derive a control method such that the robot takes a form shown in Figure 4 as a commanded attitude (equilibrium) $\bar{\theta}$. For this, the required set point $\bar{u}$ is given by

$$\bar{u}=E^{-1}f[0,0,\bar{\theta}] \quad (7)$$

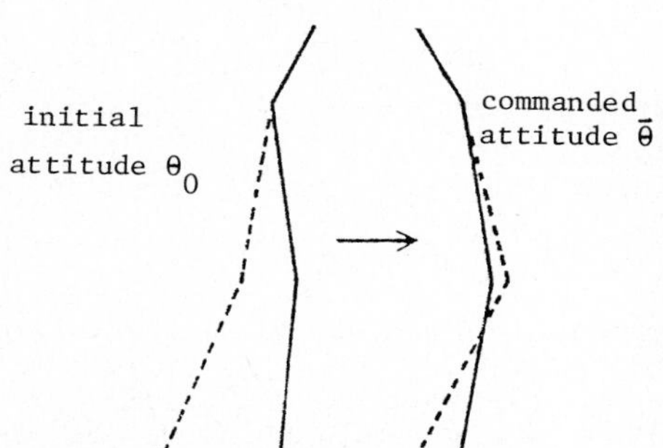

Figure 4 Attitude control

from (6). However such an attitude is unstable equilibrium, therefore we must give some stabilizing action to maintain the commanded attitude. We use state feedback control for this purpose. So we introduce small perturbed variables $d\theta$ and du and define

$$\theta=\bar{\theta}+d\theta,\quad u=\bar{u}+du \quad (8)$$

Then we have the following linearized system around $\bar{\theta}$ and $\bar{u}$ by substituting (8) into (6).

$$J\,d\ddot{\theta}+D\,d\dot{\theta}+H\,d\theta=E\,du \quad (9)$$

$$:\ J=\partial f/\partial\ddot{\theta}\big|_s,\quad D=\partial f/\partial\dot{\theta}\big|_s,\quad H=\partial f/\partial\theta\big|_s$$

where s denotes the data point $(0^T, 0^T, \bar{\theta}^T)^T$. This second order linearized system can be described by the state equation

$$\dot{x}=Ax+B\,du \quad (10)$$

$$:A=\begin{bmatrix}0&I_5\\-J^{-1}H,&-J^{-1}D\end{bmatrix},\quad B=\begin{bmatrix}0\\J^{-1}E\end{bmatrix},\quad x=\begin{bmatrix}d\theta\\d\dot{\theta}\end{bmatrix}$$

To determine the stabilizing input du, we use the linear optimal state feedback control which minimizes either

$$J=\int_0^\infty(d\theta^TW_1d\theta+du^TRdu)dt \quad (11)$$

or

$$J=\int_0^\infty(dq^TW_2dq+du^TRdu)dt \quad (12)$$

Then the optimal control du is given by

$$du=-Fx \quad (13a)$$

$$:\ F=R^{-1}B^TP \quad (13b)$$

$$PA+A^TP-PBR^{-1}B^TP+Q=0 \quad (13c)$$

where Q is given by either C^TW_1C or C^TW_2C corresponding to (11) and (12), respectively, provided C_1 and C_2 are definded by $(I_5, 0)$ and $(K,0)$.

The Riccati equation (13c) can be easily solved using eigenvectors of the Hamilton matrix and the EISPACK fortran program which calculates eigenvalues and eigenvectors of given matrix [4].

Then (8) and (13a) yield the real control law

$$u=-F(\theta^T,\dot{\theta}^T)^T+\bar{u}+F(\bar{\theta}^T,0^T)^T \quad (14)$$

In this expression $\dot{\theta}$ must be known. This can be estimated by observers, however, in this case the load of the microcomputer becomes heavy, therefore we use the three-point-differentiating formula to derive an approximated value of $\dot{\theta}$ from θ.

Since

$$\det\begin{vmatrix}sI-A&B\\C&0\end{vmatrix}=\text{constant}\ (\forall s) \quad (15)$$

holds for $C=C_1$ and $C=C_2$, the linearized system (A, B, C) has no invariant zeros [5], therefore

Table 2 Basic dtata of the attitude control

commanded attitude	$\bar{\theta}$ (deg.)	10, -10, -30, -15, 25
	$\bar{q}$ (deg.)	10, -20, -20, -15, -40
eigenvalues of A	2.82, 0.852, 0.135, -0.249, -3.30±j3.25 -7.36, -28.7, -29.0, -31.1	
weighting matrix W_1	diag(5000,5000, 10000, 10000, 10000)	
eigenvalues of A-BF	-3.53, -4.82, -6.43, -6.71 ± j2.51 -9.10 ± j9.06, -27.8, -28.5, -30.8	
initial attitude	θ_0(deg.)	5.6, -7.4, 27.7, 9.9, 23.4
	q_0(deg.)	5.6, -13.0, 35.1, 17.8, -13.5

all the 10 poles of the optimal closed loop system approach 5 sets of the second order Butterworth pattern [6] provided all the elements of W_1 and W_2 become large. This means that all the responses of θ_i's have good damping characteristics without large overshoot [5]. This shows the linear optimal feedback control is effective for the control of robots.

An example is shown in the following. When we select W_1 as shown in Table 2 and select $R=I_5$ (this is the same throughout all experiments of CW-1), the poles become the values shown in the same Table. The experimental response of CW-1 is shown in Figure 5(a). The simulated results, in which the equations (6) and (14) are solved by the Runge-Kutta method, is also shown in Figure 5(b). The initial state and the commanded state are already shown in Figure 4 and the numerical values are shown in Table 2.

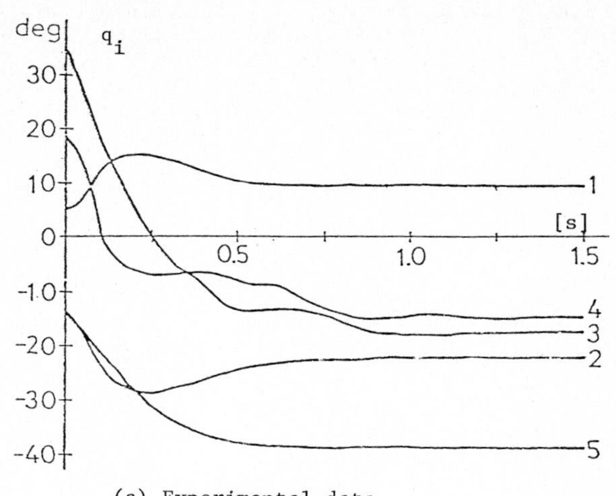

(a) Experimental data

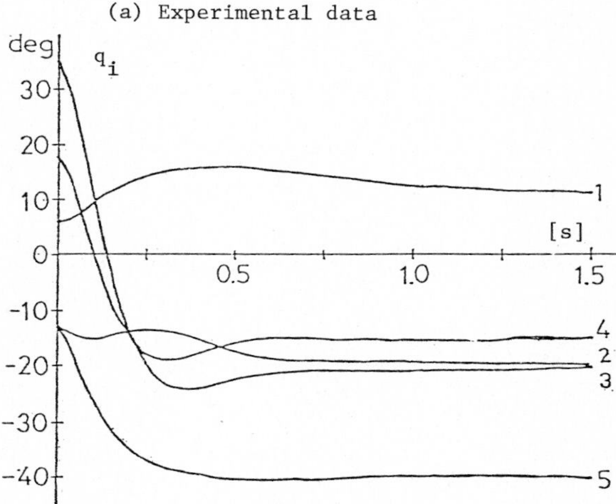

(b) Simulated data

Figure 5 Responses of q_i's

4.2 GAIT CONTROL

As illustlated in Figure 6, the gait control depends mainly on the attitude control in which the commanded attitude $\bar{\theta}$ is given by the stance supported by both legs. To maintain the locomotion , the variables of two legs are changed automatically when the foot of the swinging leg touches the ground. By this control the biped CW-1 successfully walks. The characteristics of this control is the following. (i) Any trajectory of each link is not required to be given and the control circuit becomes simple. (ii) Any small disturbance, for example stumbling, adding in the i th gait interval influences only the initial attitude of the i+1 th gait interval. Therefore the influence of the small disturbances is controlled after i+1 th step and the biped recovers the normal walking. (iii) The same action can be performed when the assumed constraint on the foot of the supporting leg fails and the robot falls down on the floor such that the stance supported by both legs differs from the desired commanded attitude.

An important and interesting problem is whether the foot of the swinging leg goes up or not in the gait interval provided only one commanded attitude is given ? The answer is affirmative if we select the weighting matrix W_1 or W_2 and tune the speed of the motion of each link. We show the experimental data in Figure 7. This is the case where CW-1 walks 6 steps such that the step length is 18cm and one step (gait interval) takes one second. The weighting matrix W_2 is chosen as W_2=diag(1000, 1000, 5000, 5000, 1000) in this experiment. The simulated data is omitted here but it is very near the experimental data except for $q_1$and q_6 .

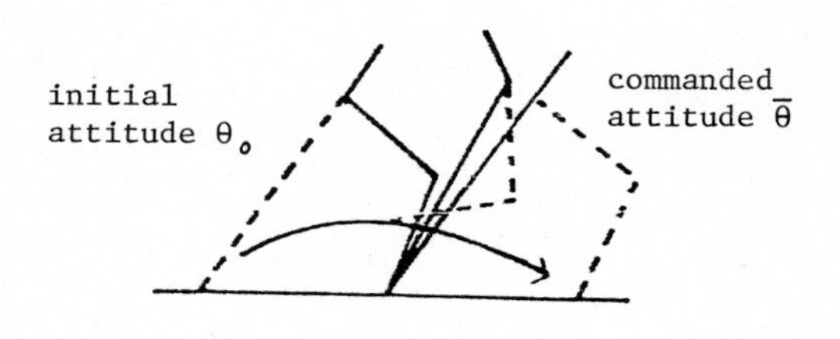

Figure 6 Gait control

5. EXPERIMENTAL SYSTEM

The experimental system is illustrated in Figure 8. The microcomputer (16 bit machine) calculates the approximated values of $\dot{\theta}$ and u in (14). It also decides the supporting leg using information sent from the touch sensors and sends variable selecting signal to the analog switches SW_1, SW_2, and SW_3 in the coordinate transformer already shown in Figure 2. The multiplications are performed using 32 bit fixed point method and upper 12 bit is transfered to D/A converter. The calculation time of u is about 4 milliseconds which becomes the sampling time of this system. With this sampling time, the optimal feedback

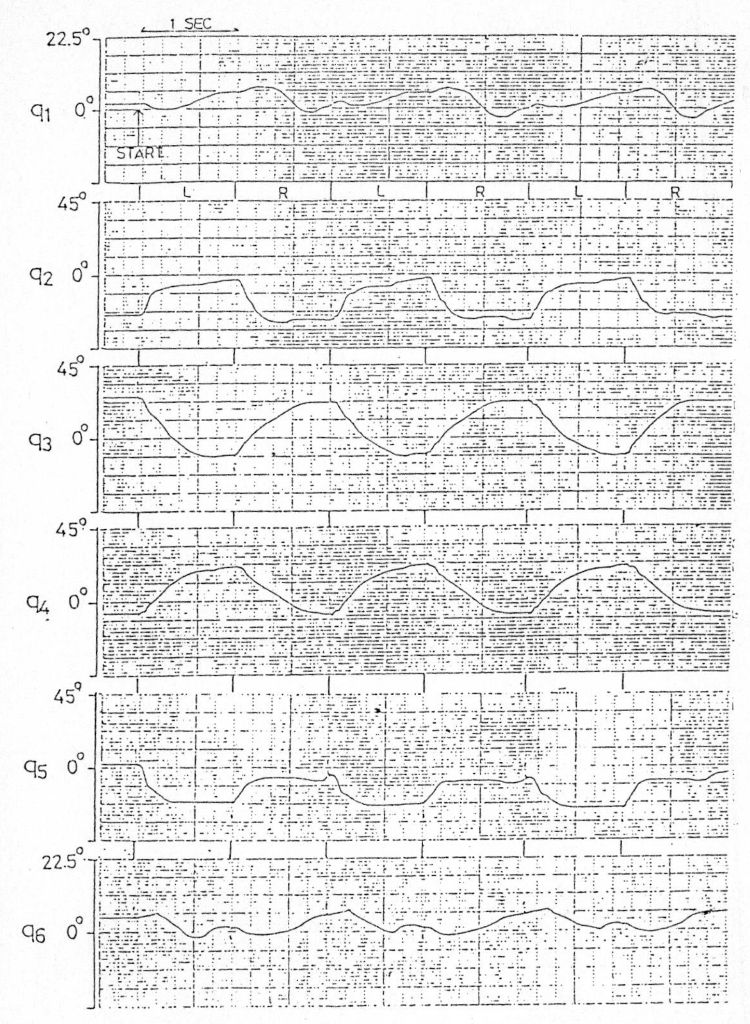

Figure 7 Experimental data of locomotion of CW-1

gain in (13b) is very close to the one obtained from the digital optimal feedback control provided Q and R are unchanged.

6. CONCLUDING REMARKS

We have shown a new and simple controlling method of a planar seven-link-biped using linear optimal state regulator. As the results, the biped CW-1 walks an arbitrary steps such that one step takes only one second when the step length is shorter than 20 cm. This shows the usefulness of the typical modern control theory for the further studies on the biped locomotion. However, when the step length becomes longer than 20 cm, the spin occurs and it disturbs the normal walking in the sagittal plane. In addition, the impact force, which arises when the foot of the swinging leg reaches the ground, becomes large and it gives the robot uncontrollable trembling. For these reasons, the maximal step length obtained in the existing circumstances is about 30 cm. However, in this case, it needs to take about one second to damp the trembling naturally. Photo 2 shown below is the locomotive motion of CW-1. In this experiment, the step length is 30 cm but one step takes 2 seconds since about one second waiting time is spent at the stance supported by two legs. Further W_2 is chosen as diag(1000, 5000, 1000, 2000, 1000).

We plan now the project CW-2 and the project CW-3. In CW-2 project, the kicking motion of the foot of swinging leg is introduced to suppress the spin of the robot. In the kicking motion, the center of mass of the robot is transfered to forward direction before beginning of the gait control [8]. In CW-3 project we want to realize 3 dimensional biped machine such that both rotational motion and balancing motion in the frontal plane can be controlled.

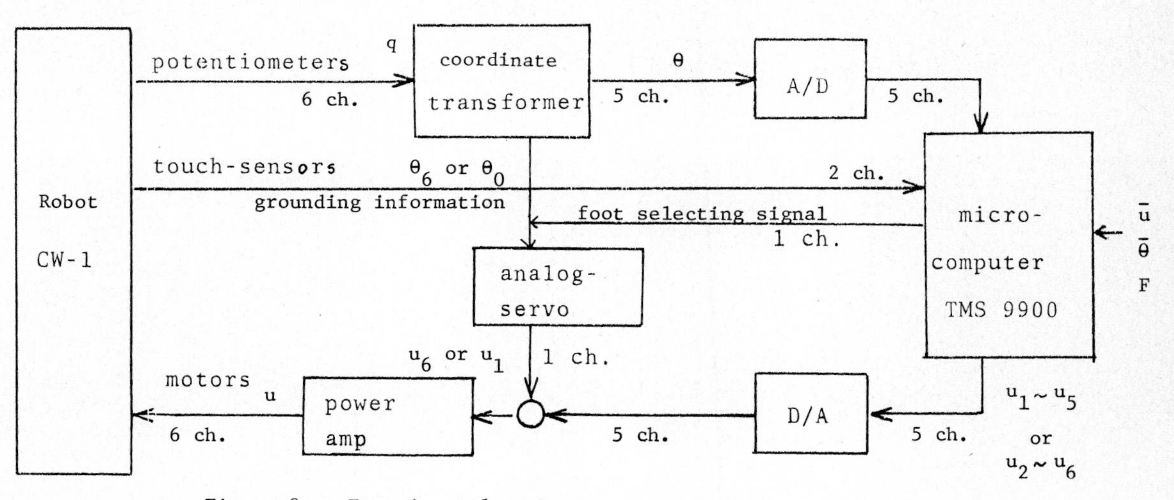

Figure 8 Experimental system

ACKNOWLEDGMENT

The author greatly appreciates the contributions of Toru Yamaguchi, Toshio Kashiwase, Atsuhiro Kawano, Youichiro Shingo, Yoshinori Sakanaka and many others of my laboratory to CW-1 project and CW-2 project. The author also greatly appreciates support and encouragement of these works by Prof. Taro Kawase of Chiba University and Prof. Katsuhisa Furuta of Tokyo Institute of Technology.

REFERENCES

[1] I. Kato, From Static Walking to Dynamic Walking (in Japanese), Preprint of 20 th SICE Conference (Sendai Japan, 1981), 291-292.

[2] F. Miyazaki and S. Arimoto, Implementation of a Hierarchical Control for Biped Locomotion, Preprint of 8 th IFAC world congress, XIV (Kyoto, Japan, 1981) 43-48.

[3] H. Hemami and R.L. Fransworth, Postual and Gait Stability of a planar Five Link Biped by Simulation, IEEE Trans. AC, 22-3 (1977), 452-458.

[4] B. Smith et. al., Matrix Eigensystem Routines -EISPACK Guide (Springer-Verlag, 1976).

[5] H. Kwakernaak and R. Sivan, Linear Optimal Control Systems (Wiely-Interscience, 1972).

[6] T. Takeda and T. Kitamori, On Some Properties of the Pole Configurations of Linear Multi-Input-Output Optimal Tracking Systems (in Japanese), Trans. SICE, 15-7 (1979), 858-865.

[7] T. Mita, Digital Control of High Speed Biped Locomotion (in Japanese), Computrol (Ohm Co. Ltd., 1983), to appear.

[8] T. Mita. T. Kashiwase and Y. Sakanaka, Feedback Control of Mechanical Systems under Linearized Constraints, Trans. IEEJ-C, 103-1 (1983), 17-24.

APPENDIX

(i) $L_{11}=I_1+M_1a_1^2+(M_2+M_3+M_4+M_5)l_1^2$

$L_{22}=I_2+M_2a_2^2+(M_3+M_4+M_5)l_2^2$, $L_{33}=I_3+M_3a_3^2$

$L_{44}=I_4+M_4b_4^2+M_5l_4^2$, $L_{55}=I_5+M_5b_5^2$

$L_{12}=M_2l_1a_2+(M_3+M_4+M_5)l_1l_2$, $L_{13}=M_3l_1a_3$

$L_{14}=-M_4l_1b_4-M_5l_1l_4$, $L_{15}=-M_5l_1b_5$

$L_{23}=M_3l_2a_3$, $L_{24}=-M_4l_2b_4-M_5l_2b_4$, $L_{25}=-M_5l_2b_5$

$L_{34}=L_{35}=0$, $L_{45}=M_5l_4b_5$ ($L_{ij}=L_{ji}$ for all i, j)

(ii)

$G_1=(M_1+M_2+M_3+M_4+M_5)a_1+(M_2+M_3+M_4+M_5)b_1$

$G_2=(M_2+M_3+M_4+M_5)a_2+(M_3+M_4+M_5)b_2$

$G_3=M_3a_3$

$G_4=-M_5a_4+(M_4+M_5)b_4$

$G_5=-M_5b_5$

(iii) We also experimented using the decoupling control [See: E. G. Gilbert, The Decoupling of Multivariable Systems by State Feedback, SIAM J. of Control, 7-1 (1969),50-63]and pole assigning method to determine F in (13a). However CW-1 did not walk when only one commanded attitude were given So we gave three commanded attitude in one gait interval.

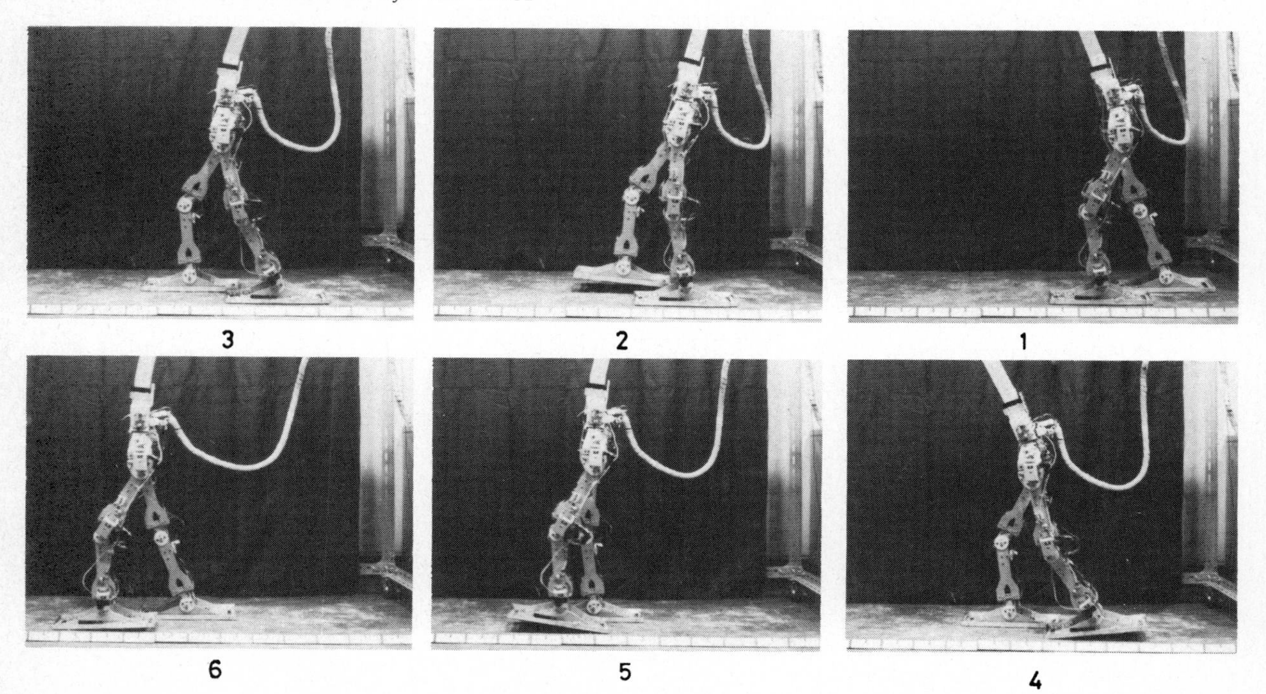

Photo 2 Walking motion of the CW-1

Simulation in Engineering Sciences
J. Burger and Y. Jarny (eds.)
Elsevier Science Publishers B.V. (North-Holland)

SIMULATION SUR UN MODELE DE ROBOT RIGIDE A DEUX DEGRES DE LIBERTE D'UNE COMMANDE NON LINEAIRE DECOUPLEE

F. Bournonville, J. Descusse

Laboratoire d'Automatique de l'ENSM,
Equipe de recherche associée au CNRS, Nantes, France

On présente dans cet article une technique de découplage non linéaire pour la classe des systèmes linéaires analytiques.
Cette technique est appliquée à un modèle de robot rigide à deux degrés de liberté en mouvement plan horizontal. Diverses simulations sont effectuées sur ce modèle, elles incluent notamment une commande en temps minimum.

1. INTRODUCTION

On présente dans cet article une technique de découplage pour la classe des systèmes linéaires analytiques. Son intérêt est de faire éclater un système à m entrées et m sorties en m sous systèmes indépendants, donc plus faciles à piloter. De plus, on a toujours la possibilité, lorsque le découplage est réalisable, de "linéariser" ces sous systèmes. Celle-ci augmente donc encore l'intérêt que l'on peut porter à une telle technique, dont les principaux aspects théoriques sont donnés au paragraphe 2.

Afin d'illustrer l'intérêt du découplage, nous avons pris comme exemple un modèle de robot rigide à deux degrés de liberté en mouvement plan horizontal. Son modèle mathématiques est donné au paragraphe 3. Le paragraphe 4, quant à lui, contient des résultats de simulations numériques réalisées sur le modèle précédent. Ces simulations incluent une commande en temps minimal, dans un problème de transfert d'un point à un autre du plan d'évolution.

2. LA RESOLUTION DU PROBLEME DE MORGAN DANS LE CAS LINEAIRE ANALYTIQUE

Dans ce paragraphe, nous allons énoncer un résultat qui généralise, au cas linéaire analytique, celui bien connu pour les systèmes linéaires et qui est dû à Falb et Wolovich [1].

Les considérations qui seront développées seront de nature locale, toutefois, pour alléger l'écriture, nous omettrons souvent de le préciser : ce sera sous entendu dans tout le texte.

On considère le système dit "linéaire analytique", $S_x(A,B,C)$ décrit par les équations

$$\begin{aligned} \mathring{x} &= A(x) + B(x)\,u \\ y &= C(x) \end{aligned} \qquad (2.1)$$

où $A(x)$, $B(x)$ et $C(x)$ sont des matrices de dimensions respectives nx1, nxm et mx1. Leurs coefficients sont des fonctions analytiques des composantes du vecteur d'état x qui est supposé appartenir à une variété analytique isomorphe à $\mathbb{R}^n$. Le vecteur de sortie y et le vecteur d'entrée u appartiennent, quant à eux, à des variétés analytiques isomporphes à $\mathbb{R}^m$ ($m\leq n$).

On considère la loi de commande $L_x(F,G)$ définie par

$$u = F(x) + G(x)\,v \qquad (2.2)$$

dans laquelle $F(x)$ et $G(x)$ sont des matrices de dimensions respectives mx1 et mxm. Nous supposerons que $G(x)$ est (localement) non singulière, autrement dit que son déterminant est non identiquement nul ; w est le nouveau vecteur d'entrée et appartient naturellement à une variété isomorphe à $\mathbb{R}^m$.

En appliquant la loi de commande (2.2) au système (2.1), on obtient le système en boucle fermée S_x (A,B,C,F,G), noté plus simplement S_x (F,G) et décrit par

$$\begin{aligned} \mathring{x} &= A(x) + B(x)\,F(x) + B(x)\,G(x)\,w \\ y &= C(x) \end{aligned} \qquad (2.3)$$

<u>Définition</u> : Le système S_x (A,B,C) est découplable par L_x (F,G) si, pour tout $i \in \underline{m}$, l'entrée w_i ($i^{è}$ composante de w) n'affecte que la sortie y_i ($i^{è}$ composante de y) pour le système S_x (F,G).

Pour énoncer une condition nécessaire et suffisante d'existence de telles loi de commande L_x (F,G) il nous faut, au préalable, introduire certains concepts.

On définit l'opérateur $\mathcal{L}_{i,k}(x)$ par

$$\mathcal{L}_{i,o}(x) = c_i(x) \quad i^e \text{ composante de } C(x)$$

$$\mathcal{L}_{i,k}(x) = \frac{\partial \mathcal{L}_{i,k-1}(x)}{\partial x} \cdot A(x) \quad k>0 \quad i \in \underline{m}$$

$$\text{où } \frac{\partial \mathcal{L}_{i,k-1}(x)}{\partial x} := \left[\frac{\partial \mathcal{L}_{i,k-1}(x)}{\partial x_1}, --, \frac{\partial\ i,k-1(x)}{\partial x_n}\right] \quad i \in \underline{m}$$

Pour tout $i \in \underline{m}$, on définit les "nombres caractéristiques" ρ_i par

$$\rho_i = \min\{k,\ \frac{\partial \mathcal{L}_{i,k-1}(x)}{\partial x} \cdot B(x) \not\equiv 0\ ,\ k \in \mathbb{N}^+\}\quad i \in \underline{m}$$

On supposera que ρ_i est fini et constant (au moins localement) pour $i \in \underline{m}$. Si ρ_i n'est pas défini celà implique que la sortie y_i n'est pas affecté par la commande u [2]. Nous écarterons ce cas que nous considèrerons comme mal posé.

Enfin nous définissons deux matrices $B^*(x)$ et $L^*(x)$, qui joueront un rôle capital dans la suite, de la manière suivante :

$$B^*(x) = \begin{bmatrix} \dfrac{\partial \mathcal{L}_{1,\rho_1-1}(x)}{\partial x} \\ \vdots \\ \dfrac{\partial \mathcal{L}_{m,\rho_m-1}(x)}{\partial x} \end{bmatrix} . B(x) \ , \quad L^*(x) = \begin{bmatrix} \mathcal{L}_{1,\rho_1} \\ \vdots \\ \mathcal{L}_{m,\rho_m} \end{bmatrix}$$

On peut maintenant énoncer la condition nécessaire et suffisante d'existence des lois $L_x(F,G)$. Elle est due initialement, comme condition suffisante, à Freund [3], puis, comme nécessaire et suffisante, à Sinha [4]. Dans ce dernier article la nécessité n'est toutefois pas réellement démontrée. La démonstration correcte apparaît pour la première fois dans [5], en utilisant la géométrie différentielle, et plus récemment dans [6], par une approche classique. Elle ne sera pas répètée ici.

<u>THEOREME</u> : Il existe une loi de commande $L_x(F,G)$ qui découple (localement) le système $S_x(A,B,C)$ si et seulement si $B^*(x)$ est non singulière. Sous cette condition, on peut choisir comme solution particulière

$$F^*(x) = -B^{*-1}(x)\, L^*(x)$$

$$G^*(x) = B^{*-1}(x)\, \Gamma$$

avec $\Gamma = \text{diag}\,(\lambda_i)$, $\lambda_i \neq 0$ $\quad i \in \underline{m}$

Le système $S_x(F^*,G^*)$ fait alors apparaître m sous sytèmes d'intégrateurs découplés décrits par

$$y_i^{(\rho_i)} = \gamma_i\, w_i \qquad i \in \underline{m}$$

Seul ce résultat sera d'importance dans la suite, aussi ne développerons nous pas plus longuement les aspects théoriques du découplage non linéaire.

3. APPLICATION A UN MODELE DE SYSTEME DE TIGES ARTICULEES EN MOUVEMENT PLAN HORIZONTAL

Le système considéré est représenté à la figure 1. Il est constitué de deux tiges articulées aux points O et P. Le point O est fixe, le point P est mobile. Deux couples u_1 et u_2 sont supposés agir sur les barres aux points O et P.

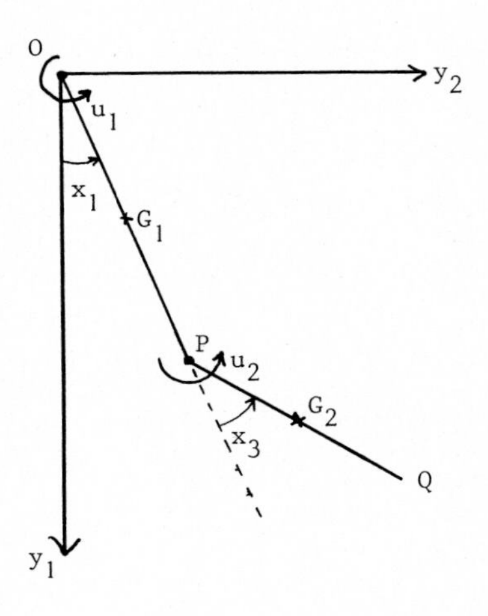

- Figure 1 -

Les barres sont supposées homogènes. Dans le but de simplifier les calculs ultérieurs, nous supposeront qu'il n'y a pas de frottements secs ni visqueux, et que le mouvement est horizontal pour éliminer les couples dûs à la pesanteur. Les barres sont de longueurs OP = a et PQ = b. Leurs masses respectives sont m_1 et m_2. Leurs centres de gravité respectifs sont G_1 et G_2. Nous avons $O\,G_1 = d_1$ et $P\,G_2 = d_2$. $I\,(A,G_1)$ et $I\,(B,G_2)$ désignent les moments d'inertie des barres en G_1 et G_2, respectivement, par rapport à un axe de rotation perpendiculaire au plan $(0,y_1,y_2)$.

La mise en équation est faite à l'aide des équations de Lagrange. Elle conduit au système suivant :

$$\begin{bmatrix} \mathring{x}_1 \\ \mathring{x}_2 \\ \mathring{x}_3 \\ \mathring{x}_4 \end{bmatrix} = \begin{bmatrix} x_2 \\ \dfrac{K_3(x_3)}{P(x_3)}\,[K_2(x_3)\,x_2^2 + N_2\,(2x_2x_4+x_4^2)] \\ x_4 \\ -\dfrac{K_3(x_3)}{P(x_3)}\,[K_1(x_3)\,x_2^2 + K_2(x_3)(2x_2x_4+x_4^2)] \end{bmatrix} + \frac{1}{P(x_3)} \begin{bmatrix} 0 & 0 \\ N_2 & -K_2(x_3) \\ 0 & 0 \\ -K_2(x_3) & K_1(x_3) \end{bmatrix} \begin{bmatrix} u_1 \\ u_2 \end{bmatrix}$$

avec

$$K_1(x_3) = I(A,G_1) + I(B,G_2) + m_1\ d_1^2 + m_2(a^2+d_2^2+2ad_2\cos x_3)$$

$$K_2(x_3) = I(B,G_2) + m_2\ (d_2^2+ad_2\cos x_3)$$

$$K_3(x_3) = m_2\ ad_2\sin x_3$$

$$N_2 = I(B,G_2) + m_2\ d_2^2$$

$$P(x_3) = [I(B,G_2) + m_2\ d_2^2][I(A,G_1) + m_1\ d_1^2] + m_2\ a^2[I(B,G_2) + m_2\ d_2^2\sin^2 x_3]$$

L'équation de sortie est donnée par

$$\begin{bmatrix} y_1 \\ y_2 \end{bmatrix} = \begin{bmatrix} a\cos x_1 + b\cos(x_1+x_3) \\ a\sin x_1 + b\sin(x_1+x_3) \end{bmatrix}$$

$$B^*(x) = \frac{1}{P(x_3)}\left[\begin{array}{c|c} b(K_2(x_3)-N_2)\sin(x_1+x_3)-a\ N_2\sin x_1 & a\ K_2(x_3)\sin x_1+b(K_2(x_3)-K_1(x_3))\sin(x_1+x_3) \\ \hline b(N-K_2(x_3))\cos(x_1+x_3)+a\ N_2\cos x_1 & b(K_1(x_3)-K_2(x_3))\cos(x_1+x_3)-a\ K_2(x_3)\cos x_1 \end{array}\right]$$

\- Tableau 1 -

l'état x et la commande u évoluent respectivement dans $\mathbb{R}^4$ et $\mathbb{R}^2$. La sortie y évolue, quant à elle, dans une variété non analytique, qui n'est autre que le domaine fermé constitué de la couronne centre 0 et de rayons $|b-a|$ et b+a. Pour pouvoir appliquer la théorie du §2, nous supposerons que y évolue dans $\mathbb{R}^2$, ce qui se traduira par des singularités lorsque le point Q arrivera sur les frontières de la couronne.

Compte tenu des équations du modèle, on obtient, tout calculs faits, (cf Tableau 1),

$$\rho_1 = \rho_2 = 2$$

$$\det\ B^*(x) = \frac{a\ b\ \mathrm{Sin}\ x_3}{P(x_3)} \qquad P(x_3) > 0 \quad \forall\ x_3\ ,$$

La matrice $B^*(x)$ est non singulière partout sauf pour $x_3 = k\Pi$, valeurs qui correspondent aux singularités mentionnées précédemment. Le découplage sera donc possible sauf lorsque les barres seront alignées.

Faute de place, nous n'indiquerons pas ici l'expression analytique de la commande u en fonction de x et de w.

Lorsque le découplage est possible le système éclate donc en deux doubles intégrateurs découplés

$$y_i^{(2)} = w_i \qquad i \in \underline{2}$$

Pour simplifier nous choisissons en effet $\gamma_i = 1$, $i \in \underline{2}$ (cf §2).

4. RESULTATS DE SIMULATION

Nous présenterons dans ce paragraphe des simulations destinées à illustrer les résultats théoriques précédents. Elles sont de trois ordres : simulations en boucle ouverte - simulation en boucle fermée c'est-à-dire avec découplage - simulation d'une commande en temps minimum sur le système découplé.

Pour fixer les idées, nous avons choisi pour les barres les caractéristiques suivantes :

$m_1 = m_2 = 6$ kg , a = b = 1 m , $d_1 = d_2 = 0{,}5$ m

$I(A,G_1) = I(B,G_2) = \frac{m\ a^2}{12} = 0{,}5\ \mathrm{Kg.m}^2$

avec ces valeurs numériques les coefficients des équations sont

$$K_1(x_3) = 2\ (5+3\cos x_3)$$

$$K_2(x_3) = 2 + 3\cos x_3$$

$$K_3(x_3) = 3\sin x_3$$

$$N = 2$$

$$P(x_3) = 16 - 9\cos^2 x_3$$

4.1 Simulations en boucle ouverte

L'objectif de ces simulations est d'appréhender la dynamique du système. Les courbes représentent la trajectoire décrite par l'extrémité Q du système de tiges articulées initialement au repos. La durée des simulations est de dix secondes.

La première simulation (Figure 2) est effectuée en appliquant un échelon de 5 m.N sur u_1, avec $u_2 = 0$

La seconde (Figure 3) est effectuée en appliquant un échelon de 2 m.N sur u_2 avec $u_1 = 0$

4.2 Simulations en boucle fermée

Le système est maintenant bouclé par la loi de commande obtenue grâce aux résultats du §2. Les simulations visent à mettre en évidence la propriété de découplage.

Pour le premier test de découplage, on applique à l'entrée un échelon unité sur w_1 tandis que w_2 est maintenu nul. L'évolution de l'extrémité Q au cours du temps est donnée à la figure 4. On constate à la figure 5 que la courbe qui représente l'évolution de y_1 au cours du temps est une parabole. Sur la courbe de la figure 5, l'évolution de y_2 en fonction du temps révèle une légère dérive, pouvant faire douter du découplage. Celle-ci résulte en fait de l'instabilité du double intégrateur. Elle peut être supprimée en transformant le double intégrateur en double constante de temps.

Le second test sur le découplage est effectué avec un échelon sur w_2 avec w_1 maintenu nul. Les conclusions sont identiques aux précédentes. Les courbes afférentes sont données aux figures 7, 8 et 9.

4.3 Simulation d'une commande en temps minimum sur le système découplé

A partir de conditions initiales données y(o), $\dot{y}(o) = 0$ on souhaite rejoindre le point $y(t_f)$ avec une vitesse finale $\dot{y}(t_f)$ nulle en minimisant le critère $\int_o^{t_f} dt.$

Les commandes w_i sont supposées prendre les valeurs ± 1, avec remise à zéro possible. Ce problème est, par exemple, celui du transfert d'une pièce d'un point à un autre du plan. L'avantage de le traiter sur le système découplé est d'avoir affaire à des systèmes linéaires indépendants. Le même problème envisagé sur le système en boucle ouverte est inextricable étant donné la complexité des équations du modèle. Il reste vrai cependant que le temps minimal calculé à partir de la commande w, ne sera pas le vrai temps minimal qui serait calculé à partir de la commande u.

La résolution du problème de commande optimale défini précédemment est classique et se trouve traitée dans divers ouvrages d'automatique, par exemple dans [7], elle ne sera pas répètée ici.

Pour la simulation, les conditions aux limites ont été fixées de la façon suivante

$y_1(o)=0,293m,\ y_2(o)=0,707; y_1(t_f)=1,8m, y_2(t_f)=0,5m$

Ce choix relève du souci d'éviter de passer, lors du transfert, par la position singulière $x_3 = k\Pi$.

La figure 10 donne la trajectoire d'écrite par l'extrémité Q. Le transfert est effectué en 2,44 secondes.

Les figures 11 et 12 représentent l'évolution au cours du temps de y_1 et y_2.

Les figures 13 et 14 donnent l'évolution dans le temps des couples u_1 et u_2. On constate que les amplitudes maximales sont tout à fait admissibles.

L'allure des commandes w_1 et w_2 est donnée aux figures 15 et 16.

BIBLIOGRAPHIE

[1] P.L. FALB, W.A. WOLOVICH, "Decoupling in the design and synthesis of multivariable control systems" IEEE Trans. Aut. Contr, AC 12, pp 651-659, 1967.

[2] D. CLAUDE, "Contribution à l'étude des systèmes non linéaires à l'aide des séries génératrices non commutatives" Thèse 3ème Cycle, Université de Paris 6, Spécialité mathématiques, Nov. 1981.

[3] E. FREUND, "The structure of decoupled non linear systems", Int. Jour. Contr., vol. 21, 3, pp '''-450, 1975.

[4] P.K. SINHA, "State feedback decoupling of non linear systems", IEEE Trans. Aut. Contr., AC 22, 3, pp 487-489, 1977.

[5] A. ISIDORI, A.J. KRENER, C. GORI-GIORGI, S. MONACO, "Nonlinear decoupling via feedback a differential geometric approach", IEEE Trans. Aut. Contr., AC 26, 2, pp 331-345, 1981.

[6] F. BOURNONVILLE, "Une introduction à la théorie du découplage pour les systèmes linéaires analytiques. Application à la commande d'un robot à deux degrés de liberté", DEA d'Automatique, ENSM, Nantes 1982.

[7] P. WASLIN, "Théorie de la commande et conduite optimale", Bibliothèque de l'Automaticien Vol. 33, Dunod, Paris 1969.

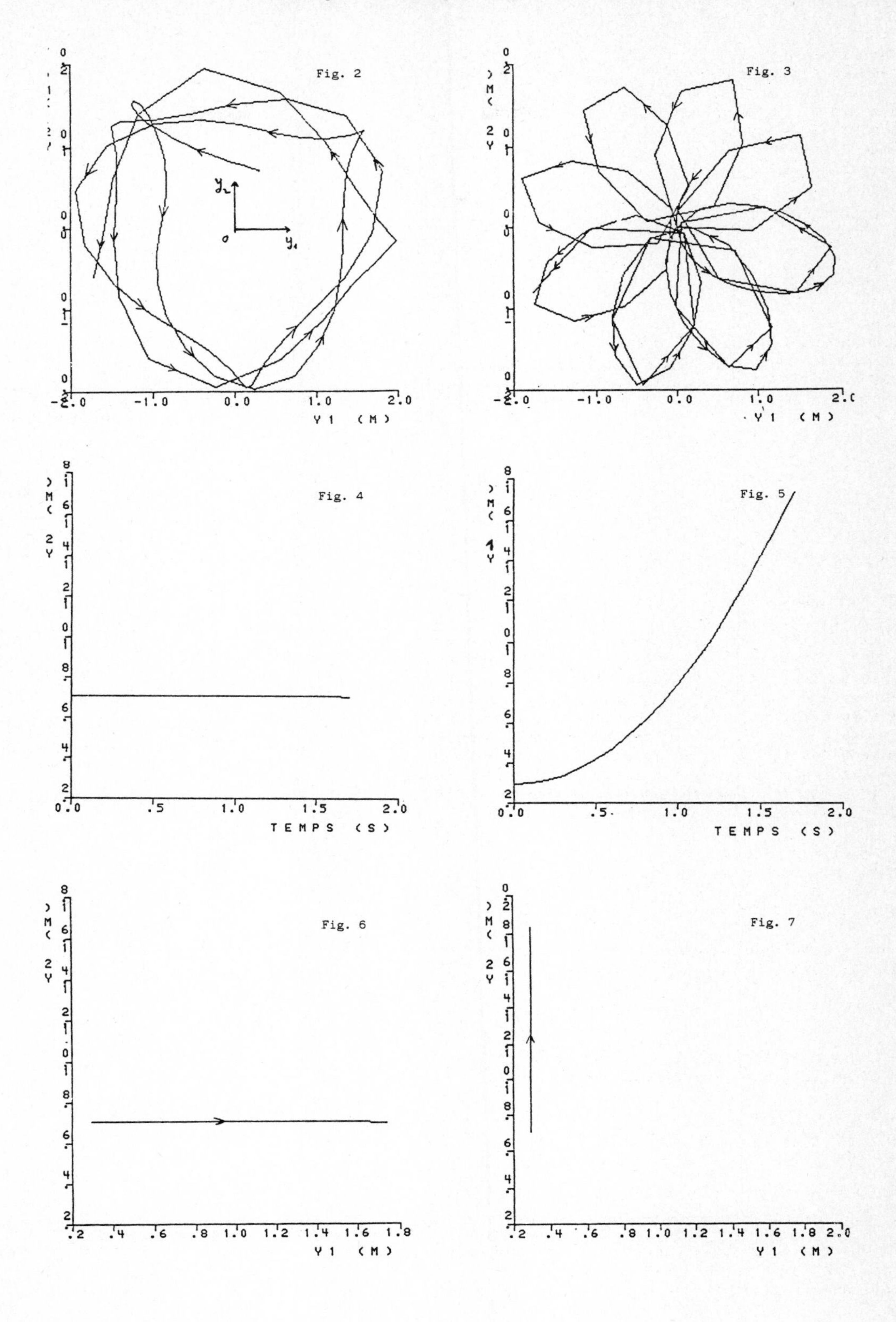
Fig. 2
Fig. 3
Fig. 4
Fig. 5
Fig. 6
Fig. 7
Y1 (M)
Y2 (M)
TEMPS (S)

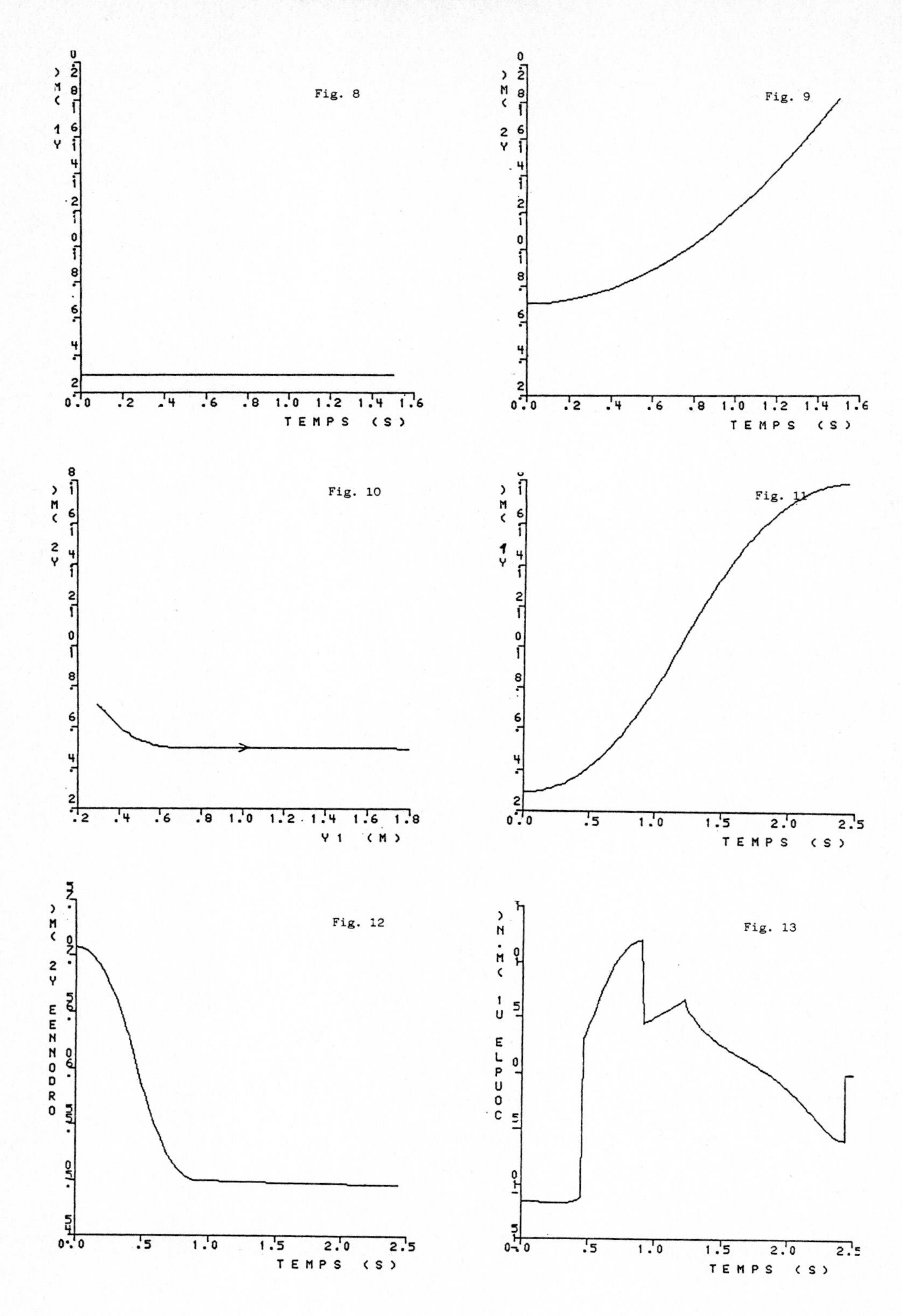
Fig. 8
0.0 .2 .4 .6 .8 1.0 1.2 1.4 1.6
TEMPS (S)
Fig. 9
0.0 .2 .4 .6 .8 1.0 1.2 1.4 1.6
TEMPS (S)
Fig. 10
.2 .4 .6 .8 1.0 1.2 1.4 1.6 1.8
Y1 (M)
Fig. 11
0.0 .5 1.0 1.5 2.0 2.5
TEMPS (S)
Fig. 12
0.0 .5 1.0 1.5 2.0 2.5
TEMPS (S)
Fig. 13
0.0 .5 1.0 1.5 2.0 2.5
TEMPS (S)

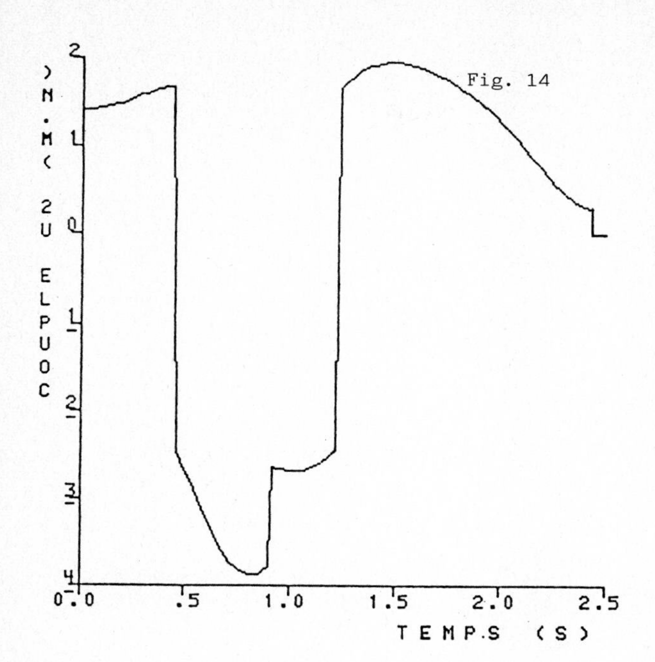

Fig. 14

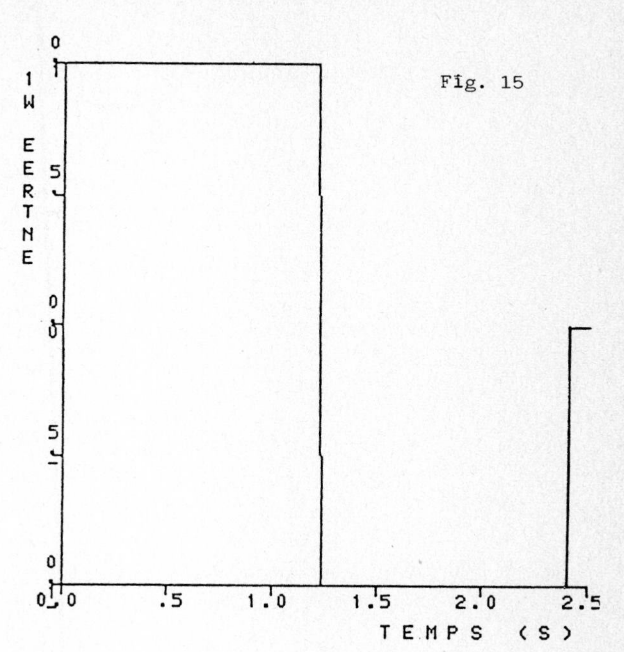

Fig. 15

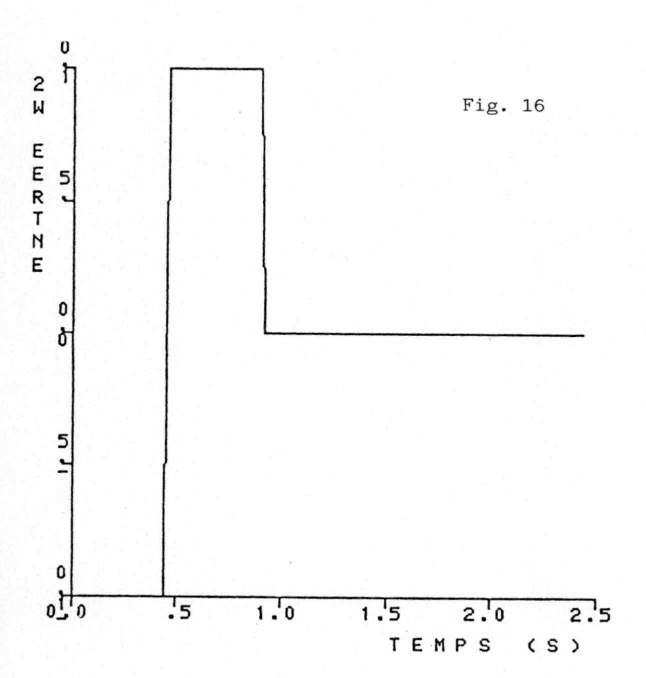

Fig. 16

XIII. THERMIC SYSTEMS

Simulation in Engineering Sciences
J. Burger and Y. Jarny (eds.)
Elsevier Science Publishers B.V. (North-Holland)

DYNAMIC ANALYSIS OF THE ROOM TEMPERATURE WITH NONSTATIONARY HEAT TRANSFER

Mladen Popović and Emir Humo

Electrotechnic Faculty, University of Sarajevo
Sarajevo, Yugoslavia

In this paper the simulation of the air temperature dynamics in the climatized room is discussed. In the first part of the paper a mathematical model for the simpler case with the stationary heat transfer is established. Dynamics is treated for two choosen types of room heating. In the second part the more sofisticate case with nonstationary heat transfer through the walls is considered. Dynamic behaviour this time is described by partial differential equations. For the both mathematical models the computer simulation and verification is realized . In the conclusion these two approaches are compared.

1. INTRODUCTION

The better understanding of the control processes dynamics is extremly important from the automatic control point of view. This problem is solved during the identification when according to the observings, experimental results and generalising the mathematical model of the real process is generated. The equations of the model describe the dynamics of the control variables, e.g. the process itself[1]. As for the mathematical models of control processes in heating, ventilation and air-conditioning (HVAC) systems they are established for the stationary (nominal) state mostly[2,3]. There are many advantages of such simplified models for this class of processes. But, the application of the modern control devices and computers demands more accurate dynamic models[4]. It gives much better regulation in all control loops. On the higher control levels the optimization and energy saving is easier too[5].

According to these reasons we discuss dynamic behaviour of the air temperature in the climatized rooms. It is the typical control process in HVAC. In the first part of the paper the mathematical for the simpler case with the stationary heat transfer is established. It is shown that these models are with concentrated parameters. Disscusion is made for two particular types of the room heating. In the second part the more sofisticate case with nonstationary heat transfer through the walls is considered. Dynamic behaviour of the temperature this time is described by partial differential equations.

2. MATHEMATICAL MODELS FOR THE ROOM TEMPERATURE

The seting of the mathematical models we shall disscused for two specific types of the room heating.

The first type has direct heating. Suppose it is realized by electric heating coil with power $p=RI^2$, as it is shown in Fig. 1 . As the body of

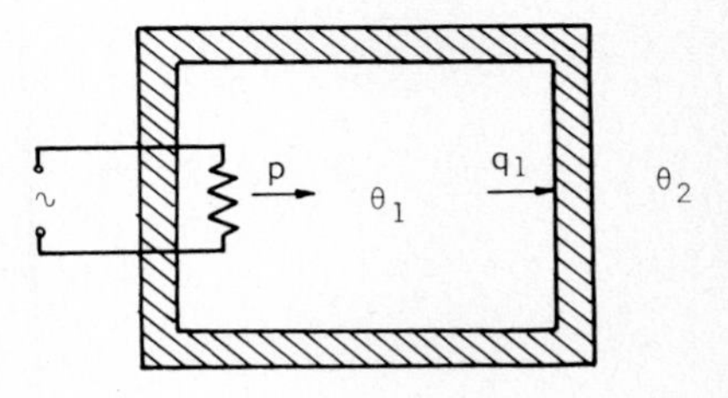

Figure 1. The room heated by electric heater.

the electric heater has the neglegable thermal capacitance, the generating of the of the heat is undependent of the air temperature in the room θ_1. Of course. temperature of the air near the heater is the greater then in the rest room space. It means that temperature θ_1 varies not only in the time but in the space too[5,6]. The good mixing of air is usually existed and so we can assume the equal temperature all over the room space. According to heat balance now we have

$$c_a \rho_a V_a \frac{d\theta_1(t)}{dt} + \sum q_{1i}(t) A_{wi} = p(t) \qquad (1)$$

where $q_{1i}(t)$ is the density of heat flow of losses $|W/m^2|$ through the i-th wall of area A_{wi}. One part of the heat loss remains in the walls and the rest goes outside.

If the walls are made of material with little thermal conductivity the heat loss will be very small, e.g. $\Sigma q_{1i} A_{wi} \approx 0$. In the reality some walls, specially the parts of walls as the windows and doors, have not so small thermal conductivity and respectively the heat loss is not neglegable.

The second type room has two air pipe: one for the outside fresh air θ_a in, and one for indoors air θ_1 out. The air mass flow is f(t). This is

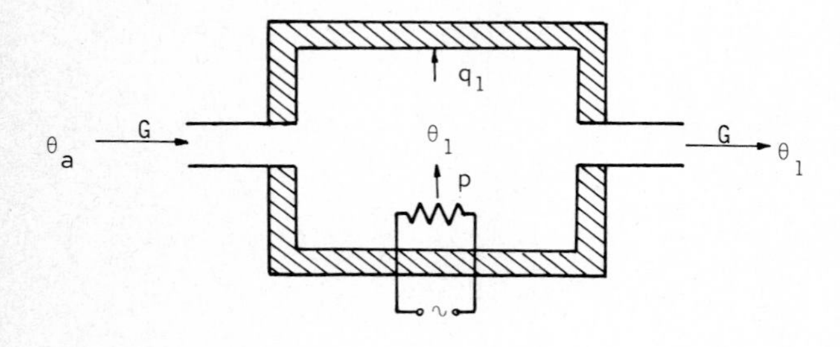

Figure 2. The room with simple ventilating system.

shown in Fig. 2. The equation of heat balance now is

$$c_a \rho_a V_a \frac{d\theta_1(t)}{dt} + \sum q_{1i}(t) A_{wi} + c_a f(t) \theta_1(t) = p(t) + c_a f(t) \theta_a(t) \qquad (2)$$

It is also possible to make more complex types of climatized room, but these simple two are considered here as basic. For example, the heater might be suited into input air-pipe. Such arangement is a variation of the second type room.

The equations (1) and (2) represent dynamic behaviour of the air temperature in the room. To solve them it is necessary to put the heat loss in functionally connection of indoors temperature $\theta_1(t)$, outside tempereature $\theta_2(t)$ and of airflow f(t). But, it is rather a complex problem. The heat loss depends on the conditions of the heat convection on both sides of all walls, and also depends on the heat conduction through the walls. Besides that, each one wall (including floor and ceiling) has different conditions of the heat transfer generally. It means that each wall has its own coefficient of thermal conductivity h, heat transfer area A, thickness δ and outside temperature. These problems will be disscused in the next section by solving just established mathematical models.

3. MODELING OF DYNAMICS FOR STATIONARY HEAT TRANSFER

Dynamic analysis of the heat transferthrough the wall is a complex problem. The basic mathematical difficulties arises because the wall temperature ia a distributed variable. The thermal capacity of the wall is the distributed variable consequently. Here we should disscus the solving of the equations (1) and (2) for the simpler case when the heat transfer is stationary.

3.1. The stationary heat transfer through wall

Take a look at Fig. 3. where the i-th wall of the air-conditioned room is represented. On the

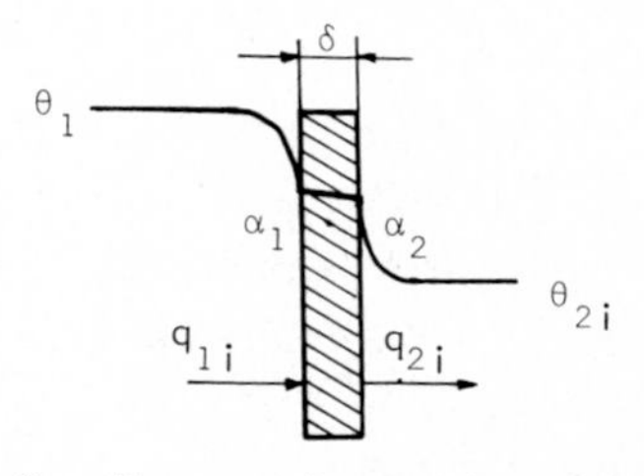

Figure 3. The stationary heat transfer

the air-conditioned room is represented. On the left side of the wall it is the heat convection from indoors air to the wall. On the right side it is convection from the wall to the outside air. Suppose the stationary the stationary heat transfer. Thus density of heat flow from the air to the wall q_{1i} is equal to density of the heat flow from the wall to the otside q_{2i}, and well known equation can be written

$$q_{1i} = h_i(\theta_1 - \theta_{2i}) \tag{3}$$

where $h_i=1/(\frac{1}{\alpha_{1i}} + \frac{\delta_i}{\lambda_i} + \frac{1}{\alpha_{2i}})$ is the overall heat transfer coefficient.

On the i-th wall can be j areas like doors and windows. Generally, each area has its own h_{ij}. In that case coresponding q_{1i} is computed by the equation $q_{1i}A_{wi}=\sum_j h_{ij}A_{ij}(\theta_1-\theta_{2i})$.

3.2. Temperature dynamic in the first type room

Put the equation for the heat loss (3) in the equation (1) which represents dynamic of indoors temperature for the first type of room heating. So it is

$$c_a\rho_aV_a\frac{d\theta_1(t)}{dt} + (\Sigma h_iA_{wi})\theta_1(t) = p(t) + \sum h_iA_{wi}\theta_{2i}(t) \tag{4}$$

As for the dynamic analysis it is more suitable to write the equation (4) in the next form

$$\frac{c_a\rho_aV_a}{\Sigma h_iA_{wi}}\frac{d\theta_1(t)}{dt} + \theta_1(t) = \frac{1}{\Sigma h_iA_{wi}}p(t) + \frac{1}{(\Sigma h_iA_{wi})}\sum h_iA_{wi}\Theta_{2i}(t) \tag{5}$$

Transform this equation into the Laplace domain. Thus Temperature $\Theta_1(s)$ can be explictly expressed as follows

$$\Theta_1(s) = \frac{1}{Ts+1}P(s) + \frac{K_i}{Ts+1}\Theta_{2i}(s) \tag{6}$$

These two transfer functions are the first order and aperiodic. It is ver y suitable for the computer simulation. They complitely describe dynamic changes of the indoors temperature $\Theta_1(s)$ in dependance of the electric heater power P(s) and of the all outside temperatures $\Theta_{2i}(s)$. Time constant is $T=c_a\rho_aV_a/\Sigma h_iA_{wi}$ and the gain factors are $K=1/\Sigma h_iA_{wi}$ and $K_i=h_iA_{wi}/\Sigma h_iA_{wi}$. All values necessary for calculation of these constants can be easily found.

3.3. Temperature dynamics in the second type room with stationary heat transfer

In the mathematical model (eq. 2) of the second type air-conditioned room there are two non-linear terms: $c_af(t)\theta_1(t)$ and $c_af(t)\theta_a$. By the linearization around the chosen aperating point equation (2) becomes

$$c_a\rho_aV_a\frac{d\theta_1(t)}{dt} + \sum h_iA_{wi}(\theta_1(t)-\theta_{2i}(t)) + c_ab_1\theta_1(t) +$$

$$+ c_ab_2f(t) = p(t) + c_ab_3\theta_a(t) + c_ab_4f(t) \tag{7}$$

where b_1 to b_4 are the coefficients of linearization.

In the Laplace domain temperature $\Theta_1(s)$ now is

$$\Theta_1(s) = \frac{K_1}{T_1s+1}P(s) + \frac{K_2}{T_1s+1}\Theta_a(s) + \frac{K_3}{T_1s+1}\Theta_{2i}(s) + \frac{K_4}{T_1s+1}F(s) \tag{8}$$

Transfer functions are the first order and aperiodic again. Time constant is $T_1=c_a\rho_aV_a/[\Sigma h_iA_{wi}+c_ab_1]$. Because of the influence of the term c_ab_1 in denominator time constant T_1 is less then for the first type room, e.g. $T_1<T$. According to this we conclude that existance of the airflow is very useful for indoors temperature dynamics. On the other hand the gain factor K_1 is less then K.

From the automatic point of view the input variables which influence on the indoors temperature are: the heat power of electric heater P(s), the temperature of the fresh air $\Theta_a(s)$, the temperature of the air on the front side of the walls $\Theta_{2i}(s)$ and the airflow F(s). The gain factors K_1, K_2, K_3 and K_4 define intensity of that influences respectively, e.g. the static characteristic between each input and indoors temperature.

Dynamic analysis of the indoors temperature in the case of stationary heat transfer of the heat loss is rather simple according to the transfer functions in the equations (6) and (8). The computer simulation is very efficient in this case. The computer modeling enables to find optimal responses. That is very important today for material and energy saving.

4. MODELING OF DYNAMICS FOR NONSTATIONARY HEAT TRANSFER

In the preceding section we supposed stationary heat transfer through the walls of the air-conditioned room, e.g. that thermal capacity of all walls is the concentrated parameter. But, in the case of nonstationary heat transfer thermal capacity of walls is distributed parameter. Temperature of wall θ_w behaves the same. It changes not only on time t but on the distance x from the indoors surface of the wall, and so is $\theta_w=\theta_w(t,x)$. This is true if there is not the temperature gradient on the wall surface, and if the wall is made of homogenous material.

Starting equation for dynamic analysis of the indoors temperature are (1) and (2) again. The first task now is to find out relation

$$Q_{1i}(s) = G_{1i}(s)\Theta_1(s) + G_{2i}(s)\Theta_{2i}(s) \qquad (9)$$

according to the nonstationary heat transfer through the wall. The equivalent equation for the stationary case was (3).

4.1. Nonstationary heat transfer through wall

Fig. 4 shows a wall when nonstationary heat transfer is dominant. Take a look at the inner layer of thickness dx. The heat balance for this layer is

$$q(t,x)-\left(q(t,x)+\frac{\partial q(t,x)}{\partial x}dx\right)=\frac{\partial}{\partial t}\left(c_w\rho_w\theta_w(t,x)dx\right) \qquad (10)$$

The specific heat c_w and density of wall material ρ_w do not vary with time, and so it yields

$$-\frac{\partial q(t,x)}{\partial x} = c_w\rho_w\frac{\partial\theta_w(t,x)}{\partial t} \qquad (11)$$

Besides that, according to the Fourier law of the heat conductivity through the observed layer it is

$$q(t,x) = -\lambda\frac{\partial\theta_w(t,x)}{\partial x} \qquad (12)$$

The partial differential equations (11) and (12) define nonstationary behaviour of the heat flow $q(t,x)$ through the wall and of wall temperature $\theta_w(t,x)$. To solve these equations as the first we shall make their Laplace transform on time variable t. It yields

$$-\frac{dQ(s,x)}{dx} = c_w\rho_w\Theta_w(s,x)$$
$$Q(s,x) = -\lambda\frac{d\Theta_w(s,x)}{dx} \qquad (13)$$

where $Q(s,x)$ and $\Theta_w(s,x)$ are the Laplace transforms of $q(t,x)$ and $\theta_w(t,x)$ respectively. System of differential equations (13) we can easy solve according to variable x. The solution is

$$Q(s,x)=-\lambda_i\left|C_1k\sqrt{s}\exp(k\sqrt{s}x)-C_2k\sqrt{s}\exp(-k\sqrt{s}x)\right|$$
$$\Theta_w(s,x)=C_1\exp(k\sqrt{s}x)+C_2\exp(k\sqrt{s}x) \qquad (14)$$

where $k=\sqrt{\frac{c_w\rho_w}{\lambda}}$, and $C_1=C_1(s)$ and $C_2=C_2(s)$ are the constants of the integration. These constants we shall define for the boundery conditions at x=0, e.g. $Q(s,0)=Q_1(s)$ and $\Theta_w(s,0)=\Theta_{w1}(s)$. According to this we have definitly

$$Q(s,x) = -\lambda k\sqrt{s}\,\mathrm{sh}\,k\sqrt{s}x\,\Theta_{w1}(s) + \mathrm{ch}\,k\sqrt{s}x\,Q_1(s)$$
$$\Theta_w(s,x) = \mathrm{ch}\,k\sqrt{s}x\,\Theta_{w1}(s) - \frac{1}{\lambda k\sqrt{s}}\mathrm{sh}\,k\sqrt{s}x\,Q_1(s) \qquad (15)$$

These equations complitely describe nonstationary

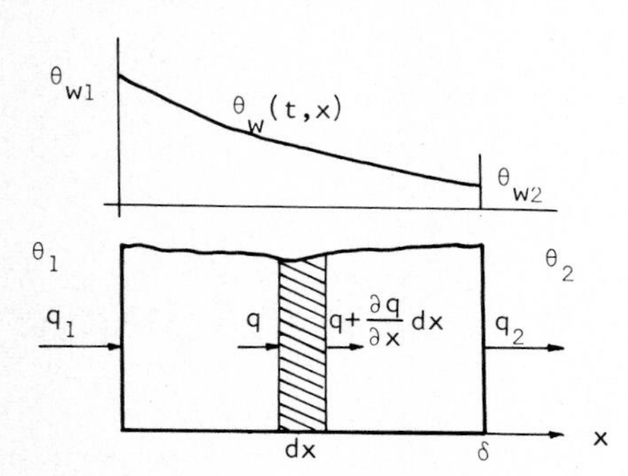

Figure 4. The nonstationary heat transfer through the wall

heat flow through the wall $Q(s,x)$ and nonstationary wall temperature $\Theta_w(s,x)$.

Now we have to use this result to find out extended shape of the equation (9). Tgerefore we shall first insert into (15) boundery conditions at $x=\delta$, e.g. $Q(s,\delta)=Q_2(s)$ and $\Theta_w(s,\delta)=\Theta_{w2}(s)$. It yields

$$Q_2(s) = -\lambda k\sqrt{s}\,\mathrm{sh}\,k\sqrt{s}\,\delta\,\Theta_{w1}(s) + \mathrm{ch}\,k\sqrt{s}\,\delta\,Q_1(s)$$
$$\Theta_{w2}(s) = \mathrm{ch}\,k\sqrt{s}\,\delta\,\Theta_{w1}(s) - \frac{1}{\lambda k\sqrt{s}}\mathrm{sh}\,k\sqrt{s}\,\delta\,Q_1(s) \qquad (16)$$

Secondly, to this system of equations (16) we have to add the equations for the heat convection on the both wall sides

$$Q_{1i}(s) = \lambda_1[\Theta_1(s) - \Theta_{w1}(s)]$$
$$Q_{2i}(s) = \lambda_2[\Theta_{w2}(s) - \Theta_{2i}(s)] \qquad (17)$$

It is simple to eliminate variables $Q_2(s)$ and $\Theta_{w1}(s)$ from system (16) and (17). So, at the end we get to extended form of (9) as it follows

$$Q_{1i}(s) = \frac{\mathrm{ch}\,k\sqrt{s}\,\delta + \alpha_2\,\mathrm{sh}\,k\sqrt{s}\,\delta}{D(s)}\Theta_1(s) - \frac{1}{D(s)}\Theta_{2i}(s) \qquad (18)$$

where denominator $D(s)=(\frac{1}{\alpha_1}\frac{1}{\alpha_2}\lambda k\sqrt{s}+\frac{1}{\lambda k\sqrt{s}})\,\mathrm{sh}\,k\sqrt{s}\,\delta + (\frac{1}{\alpha_1}+\frac{1}{\alpha_2})\,\mathrm{ch}\,k\sqrt{s}\,\delta$. Equation (18) can be for any wall. Only proper values have to be inserted. The term next to $\Theta_1(s)$ represents transfer function $G_{1i}(s)$ and next to $\Theta_{2i}(s)$ transfer function $G_{2i}(s)$.

4.2. Temperature dynamic in the first type room for nonstationary heat transfer

The transfer functions $G_{1i}(s)$ and $G_{2i}(s)$ are transcendental. Such transfer function are not suitable for practical dynamic analysis[7]. Because of that the approximation of the equation (18) is suggested. The approximation is made by linearization round the point s=0. If we neglect the high order terms this procedure gives simple transfer functions $G_{1i}(s)$ and $G_{2i}(s)$, namely

$$Q_{1i}(s) = \frac{K_{1i}}{T_i s + 1}\Theta_1(s) - \frac{1}{T_i s + 1}\Theta_{2i}(s) \qquad (19)$$

where gain factors $K_{1i}=\alpha_{1i}/(\alpha_{1i}+\alpha_{2i}+\frac{\alpha_{1i}\alpha_{2i}\delta_i}{\lambda_i})$ and time constant

$$T=c_{wi}\rho_{wi}\left(1+\frac{\alpha_{1i}\alpha_{2i}\delta_i}{2\lambda_i}+\frac{\alpha_{1i}\alpha_{2i}\delta_i^2}{6\lambda_i^2}\right)/\left(\alpha_{1i}+\alpha_{2i}+\frac{\alpha_{1i}\alpha_{2i}\delta_i}{\lambda_i}\right).$$

The differential equation (1) describes dynamics of the temperature in the first type room. If we make its Laplace transform and insert (19) which describes nonstationary heat loss we come to

$$\Theta_1(s) = \frac{P(s) + \sum\frac{K_{1i}A_{wi}}{T_i s+1}\Theta_{2i}(s)}{c_a\rho_a V_a s+\sum\frac{K_{1i}A_{wi}}{T_i s+1}} \qquad (20)$$

This is the final equation which describes dynamic behaviour of the temperature in the first type room with nonstationary heat transfer of loss. Its computer simulation is not difficult. Sometimes some difficulties can arise only because of the order of transfer function which strongly
depends on the number of the areas with the nonstationary heat transfer.

4.3. Temperature dynamic in the second type room with nonstationary heat transfer

On the same manner the final equation for dynamics of temperature in the secon type of the room heating is obtained. The equation (19) this time has to be inserted in the Laplace transform of the linearized equation (2). It yields

$$\Theta_1(s) = \frac{1}{N(s)}P(s)+\frac{1/(T_i s+1)}{N(s)}\Theta_{2i}(s)+\frac{c_a b_1}{N(s)}\Theta_a(s)+ + \frac{c_a(b_4-b_2)}{N(s)}F(s) \qquad (21)$$

where denominato $N(s)$ is $N(s)=c_a\rho_a V_a s+\sum\frac{K_{1i}A_{wi}}{T_{1i}s+1} + c_a b_1$.

As before the practical difficulties are possible only if the order of denominator N(s) is too high. As we see The order is equal to the number of the heat transfer areas plus one. Generally, the computer simulation of the temperature behaviour in the case of the non-stationary transfer of the heat loss through the walls according to this approach is very suitable.

5. CONCLUSIONS

An approach for setting up the mathematical models of indoors temperature in the air-conditioned room is discussed. The influence of the stationary and nonstationary transfer of the heat loss through the walls was apartly considered. The resulting models in both cases are made in the form of the simple transfer functions. The computer simulation according to this can be very useful for material and energy saving.

This approach can be easily applied to more complex types of room heating.

6. LIST OF SIMBOLS

$\theta(t)$ - temperature of the air, [^{0}C]
q(t) - density of heat flow, [W/m^2]
p(t) - power of the electric heater, [W]
f(t) - air mass flow, [kg/h]
$\Theta(s)$, Q(s), P(s), F(s) - the Laplace transforms of $\theta(t)$, q(t), p(t) and f(t) respectively
c - specific heat, [J/kg^{0}C]
ρ - density, [kg/m^3]
V - interior room volume, [m^3]
A - heat transfer area, [m^2]
δ - wall thickness, [m]
α-coefficient of heat transfer, [W/m^{2} ^{0}C]
λ - coefficient of conductivity, [W/m ^{0}C]
i - cardinal number index
a - index for variables concern the air
w - index for variables concern the wall
$_1$ - index for values on the interior side of wall
$_2$ - index for values on the out side of the wall

7. REFERENCES

[1] P.Eykhoff: System identification. Parameter and state estimation (John Wiley & Sons Ltd., New York, 1974).

[2] R.W.Haines: Control systems for heating, ventilating and air-conditioning (Van Nostrand Reinhold Comp., New York, 1977).

[3] S.J.Zrnić: Heating and climatization, on serbocroation (Naučna knjiga, Beograd, 1972).

[4] System description II, catalogue, (Sauter Comp., Bassel, 1980).

[5] E.Humo, M.Škrbić, M.Popović: Planning of microprocessor systems for energy saving, on serbocroation (S mposium Informatica '81, Ljubljana, 1981).

[6] A.G.Butkovskiy: Methods of control of distributed parameter systems, on russian (Science, Moscow, 1975).

[7] M.Popović: The dynamic of air temperature in the climatized room, on serbocroation (The fourth Symposium on Measurement, JUREMA Proceedings 27(1982) Part 2., Zagreb).

Simulation in Engineering Sciences
J. Burger and Y. Jarny (eds.)
Elsevier Science Publishers B.V. (North-Holland)

NUMERICAL SOLUTIONS OF THE THERMAL BEHAVIOUR OF AN EXTRUDER

BY

Yvon JARNY , Jacques BURGER

Laboratoire d'Automatique - ERA CNRS
1, rue de la Noë 44072 - NANTES CEDEX

An extruder designed to produce PVC profiles is modelled, from the thermal point of view, by a system of two coupled partial linear equations of the parabolic type and dimension 1 in space. This model is obtained from the thermal balance and each coefficient has physical significance. It is meant on the one hand to serve as a medium in technical studies of dimensioning on conceiving the machine, saving long and costly experiments, and, on the other hand, to allow numerical control of the extruder in continuous production, through minimization of a quadratic criterion. It is verified that the system is mathematically well posed, i.e. it allows a unique solution depending continuously on the data ; then the problem of the numerical solution is studied.
The first method studied consists in discretizing the two equations according to explicit schemes and solving them through checking the convergence conditions on the steps in space and time.
The second method studied consists in solving both equations globally by implicit discretizations. This leads to the solution, at each iteration, of a bi-tridiagonal linear system.
The last method studied uses implicit schemes : the two equations are solved one after the other, and the results of one carried into the other. The method leads, at each iteration and for each equation, to the solution of a tridiagonal linear system.
The survey ends with a comparison of those three methods, regarding the precision as well as the time needed for calculation.

1. INTRODUCTION

Consider the simulation of the thermal behaviour of an extruder designed for the production of PVC or polyethylene profiles.

The thermal model perfected must be able to serve in dimensioning surveys as well as in the calculation of an optimal control on operating the machine.

Two approaches are then available, depending on whether the extruder is considered as a system with localized or distributed parameters. A localized model |2| allows to simulate globally the thermal phenomena at work in a dynamic mode and to develop optimal control. This model requires the identification of about thirty coefficients and it is valid for only one extruder and only one location of the actioners and captors. Such a distributed parameter model as the one presented here consists in considering the extruder as a cylindrical exchanger. This model is worked out by assessing a thermal balance |1| and it includes two equations with coupled partial derivatives, whose parameters have immediate physical significance and are not numerous |4|.

The numerical approximation of the model's equations are presented after verifying that the system is mathematically well posed. Three solving methods are described and compared ; the first one uses explicit discretization schemes, the next two use implicit schemes.

2. THE MODEL'S EQUATIONS

The considered extruder consists of a cylinder heated at its wall by electrical resistors. Inside the cylinder, a motor-driven worm ensures the transformation of the polymere which is to be extruded.

Figure 1 : Diagram of an extruder

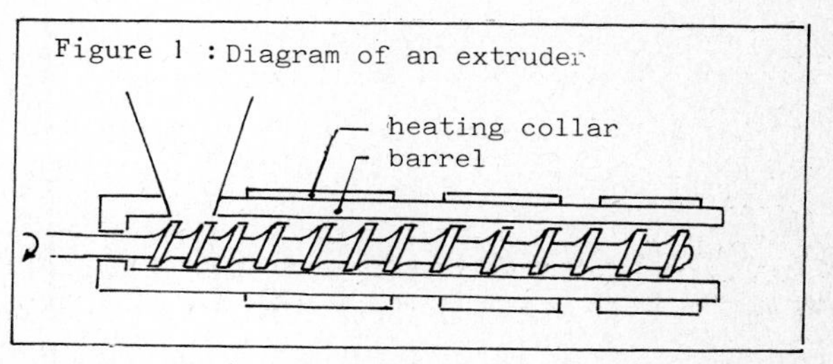

The thermal behaviour of the machine is modelled globally, considering it is equivalent to a cylinder inside which the matter proceeds with constant speed.

The matter and the barrel bearing the suffixes i = 1 and i = 2 respectively, we note :

C_i specific heat

θ_i temperature

ρ_i specific mass

λ_i coefficient of thermal diffusion

u_i velocity

q_i conduction heat flow

The heat equation writes as follows, when there is no source in the medium i |5|:

$$\rho_i \, C_i \left(\frac{\partial \theta_i}{\partial t} + \mathrm{div}(\theta_i u_i)\right) = - \mathrm{div}\, q_i \tag{1}$$

$$q_i = - \lambda_i \,\mathrm{grad}\, \theta_i . \tag{2}$$

When the medium changes, the Fourier transform is :

$$\lambda_1 \frac{\partial \theta_1}{\partial n} = \lambda_2 \frac{\partial \theta_2}{\partial n} \tag{3}$$

It is assumed that in the normal direction $\vec{n}$ at the boundary the flow q_n is such that :

$$q_n = h(s)\ (\theta_2-\theta_e) - q_u(s),\ s\varepsilon\Sigma \quad (4)$$

θ_e = temperature outside

q_n = flow brought by the collar at the wall of the cylinder.

The temperature of the matter when entering the machine being set, we assume unrestrictedly |3| that :

$$\theta_1(x = o) = \theta_e = \bar{o} \quad (5)$$

Writing that the temperatures depend only on the space variable x and time t, we get the following development equation on]O, 1[x]O, T[:

$$\left.\begin{array}{l} \dfrac{\partial\theta_i}{\partial t} + u_i \dfrac{\partial\theta_i}{\partial x} = D_i \dfrac{\partial^2\theta_i}{\partial x^2} + \sum_{j=1}^{2} K_{ij}\ \theta_i + F_i \\ \text{with } i = 1, 2\ ;\ u_2 = 0\ ;\ D_2 > D_1 > 0\ ; \\ \qquad K_{12} = - K_{11} > 0\ ; \\ \qquad - K_{22} > K_{21} > 0 \end{array}\right\} \quad (6)$$

We draw the boundary conditions of equations (4) and (5) :

$$\left.\begin{array}{l} D_i \dfrac{\partial\theta_i}{\partial x}(x = k) + g_{ik}\ \theta_i(x = k) = 0\ ; \\ \qquad k = 0,1\ ;\ i = 1, 2 \\ \qquad \theta_1(x=o) = o \\ \text{with } g_{11}, g_{21} > 0 \text{ et } g_{20} < 0. \end{array}\right\} \quad (7)$$

At initial time t = 0, we take without restriction :

$$\theta_i(t = 0) = 0\ ,\ \forall\ x\ \varepsilon(0,\ 1)\ ;\ i = 1,\ 2 \quad (8)$$

The forced terms F_i are written this way :

$$\left.\begin{array}{l} F_1\ (x,\ t) = \gamma(t)\ r(x) \\ F_2\ (x,\ t) = \sum_{j=1}^{p} X_j(x)\ W_j \end{array}\right\} \quad (9)$$

With :

- γ : motor power transmitted by the screw
- W_j : power supplied by the collar j to the barrel
- r, X_j : distribution functions.

The physical interpretation of the parameters in the model defined by equations (6) to (9) is the following :

u_1 : L/Ts = Length of the cylinder/average staying time for the matter.

$D_i = \lambda_i(\rho_i\ C_i)^{-1}$ characterizes the thermal diffusion in the medium i.

K_{22} characterizes the exchanges between the cylinder and outside

K_{12} and K_{21} characterizes the exchanges between media 1 and 2.

We have : $K_{12} \neq K_{21}$ if $\rho_1\ C_1 \neq \rho_2\ C_2$.

Putting $\theta = (\theta_1, \theta_2)$, it is suitable to write (6) in the form :

$$\frac{\partial\theta}{\partial t} + A\theta = F \quad (10)$$

with A defined by

$$A\theta = \begin{bmatrix} u_1 \dfrac{\partial}{\partial x} - D_1 \dfrac{\partial^2}{\partial x_2} - K_{11} & - K_{12} \\ - K_{21} & -D_2 \dfrac{\partial^2}{\partial x^2} - K_{22} \end{bmatrix} \begin{bmatrix} \theta_1 \\ \theta_2 \end{bmatrix}$$

3. EXISTENCE AND UNICITY OF THE SOLUTION

We are within the scope of the theory developed in |6|, and we take the proper notation.

Let H and V be two Hilbert spaces :

$$H = \{h = (h_1,\ h_2)\ \varepsilon\ [L^2(\Omega)]^2\}\quad \Omega =]0,1[$$

$$V = \{v = (v_1,\ v_2)\ \varepsilon\ [H^1(\Omega)]^2,\ v_1(0) = 0\}$$

H and V are provided with their usual norms.

The variational expression of (10) writes :

$$\int_\Omega \left(\frac{\partial\theta}{\partial t}\ v + (A\theta)\ v\right) dx = \int_\Omega F\ v\ dx,\ \forall\ v\ \varepsilon\ V.$$

We consider the family of forms $a(t;\phi,\Psi)$ defined on V x V such as :

$$a(t;\phi,\Psi) = \int_\Omega A\ \Phi\ \Psi\ dx\ ,\forall\ \Phi,\Psi\ \varepsilon\ V.$$

Developing and taking into account the conditions at the boundaries (7), we get :

$$a(t;\Phi,\Psi) = \sum_{i=1}^{2} \{ D_i \int_\Omega \frac{\partial\Phi_i}{\partial x}\frac{\partial\Phi_i}{\partial x}\ dx - \sum_j \int_\Omega K_{ij}\ \Phi_i\ \Psi_j\ dx\} + u_1 \int_\Omega \frac{\partial\Phi_i}{\partial x}\ \Psi_1\ dx + \sum_{i=1}^{2} g_{i1}\ \Phi_i(1)\Psi_i(1) - g_{20}\ \Phi_2(0)\Psi_2(0).$$

$a(t;.,.)$ is a bilinear form on V x V, and we verify |3| that it is continuous and coercive by applying the Schwartz inequality and a trace theorem.

Then we have the following result :

The problem - find $\theta\ \varepsilon\ L^2(0,T;V)$ so that

$$\frac{d\theta}{dt} + A\theta = F\ ,\ F \text{ given in } L^2(0,T;V')$$

$\theta(0) = \theta_o$, θ_o given in H

admits a unique solution, and the bilinear application $(F, \theta_o) \to \theta$ is continuous of $L^2(0,T;V') \times H$ in $L^2(0,T;V)$.

4. NUMERICAL APPROXIMATION

The system of equations (7) to (10) is solved by discretizing the spatial domain into N equal intervals of lengh h, according to a finite difference scheme. Three methods are considered for discretizing time t.

The interval (0,T) is divided into M equal intervals $]t_K, t_{K+1}[$ of length τ. We note :

$$\theta^k_{i,j} = \theta_i(x_j,t_k),\ j = 1,\dots, N+1;\ i = 1,2$$

$$\theta^k = \left[(\theta^k_{ij}, \theta^k_{2j})\right]_{j=1}^{N+1} \quad ; \theta = \left[\theta^k\right]_{k=0}^{M}$$

and we seek $\theta^k \in R^{N+1} \times R^{N+1}$, $k = 0,\dots, M$

4.1 Explicit scheme (S1). This scheme consists in taking the approximation :

$$\frac{d\theta^{k+1}}{dt} = (\theta^{k+1} - \theta^k)/\tau$$

and seeking θ so that :

$$\theta^{k+1} = (1-\tau\ A)\ \theta^k+\tau\ F^k\ ;\ \theta^{k+1} \varepsilon R^{N+1} \times R^{N+1}$$

θ^o given ; $K = 0,1,\dots,M$

The finite difference scheme leads to a matrix $(1 - \tau\ A)$ which is bitridiagonal.

The solution is fast but the scheme is conditionally stable.

4.2 Global implicit scheme (S2). We take the Crank-Nicolson approximation :

$$\frac{d\theta^{k+1/2}}{dt} = (\theta^{k+1}-\theta^k)/\tau$$

$$\theta^{k+1/2} = (\theta^{k+1}+\theta^k)/2$$

We seek θ so that :

$$(\tau^{-1}+A/2)\theta^{k+1} = (\tau^{-1}- A/2)\theta^k+F^{k+1/2}$$

θ^ogiven, $K = 0, 1,\dots, M$

The matrixes $(\tau^{-1} + A/2)$ and $(\tau^{-1} - A/2)$ are bitridiagonal. The solution takes this property into account and consists in calculating for $K = 0,\dots,M$:

1) $\omega^k = (\tau^{-1}-A/2)\theta^k+F^{K+1/2}$; $\omega^k \varepsilon R^{N+1} \times R^{N+1}$

2) $\theta^{k+1} = (\tau^{-1}+A/2)^{-1}\ \omega^k$; $\theta^{k+1} \varepsilon R^{N+1} \times R^{N+1}$

This scheme is unconditionally stable.

4.3 Shifted implicit scheme (S3). We take the previous approximation, but the two equations are written at different moments with shifts of $\tau/2$.

$$\frac{d\theta_1^{k+1/2}}{dt} + \sum_j A_{1j}\ \theta_j^{k+1/2} = F_1^{k+1/2}$$

$$\frac{d\theta_2^{k+1}}{dt} + \sum_j A_{2j}\ \theta_j^{k+1} = F_2^{k+1}$$

We then seek $\theta^k = (\theta_1^k, \theta_2^{k+1/2})$ so that

$$(\tau^{-1}+A_{11}/2)\theta_1^{k+1} = (\tau^{-1}-A_{11}/2)\theta_1^k$$

$$-A_{12}\theta_2^{k+1/2}+F_1^{k+1/2}$$

$$(\tau^{-1}+A_{22}/2)\theta_2^{k+3/2}= (\tau^{-1}-A_{22}/2)\theta_2^{k+1/2}$$

$$-A_{21}\theta_1^{k+1}+F_2^{k+1}$$

θ^o given $k = 0,1,\dots,M$

The solution consists in repeating the two stages of the previous scheme, but using boards sized (N + 1) instead of 2 x (N + 1), which reduces the number of operations.
This scheme is unconditionally stable. The matrixes are tridiagonal. We calculate successively :

1 $\omega_1^k = (\tau^{-1}-A_{11}/2)\theta_1^k-A_{12}\theta_2^{k+1/2}+F_1^{k+1/2}$

2 $\theta_1^{k+1} = (\tau^{-1}+A_{11}/2)^{-1}\omega_1^k$

3 $\omega_2^{k+1/2}= (\tau^{-1}-A_{22}/2)\ \theta_2^{k+1/2}-A_{21}\theta_1^{k+1}+F_2^{k+1}$

4 $\theta_2^{k+3/2} = (\tau^{-1}+A_{22}/2)^{-1}\omega_2^{k+1/2}$

for $k = 0,\dots,M$

5. COMPARISON OF THE METHODS

The simulation is performed on a mini-computer ; to compare the three diagrams presented we can consider :

- the number of arithmetical operations necessary to obtain θ^{K+1} knowing θ^K, therefore the calculating time for each time step.
- and the necessary memory space. This comparison must take stability into account : the stability of scheme (S1) is not assured unconditionally on the discretization steps of variables x and t. To study the stability of (S1) we take the following approximations :

$$\frac{\partial\theta_i^k}{\partial x}(x=x_j)=(\theta^k_{ij+1}-\theta^k_{ij-1})/2h$$

$$\frac{\partial^2\theta_i^k}{\partial x^2}(x=x_j) = (\theta^k_{ij+1}-2\theta^k_{ij}+\theta^k_{ij-1})/h^2$$

$h = 1/N$, $i = 1,\dots, N+1$

and the iterative scheme $\theta^{k+1} = A\theta^k$, $k = 0,\dots,M$ is numerically stable if the module of the eigen values of the A matrix are all less than one. We note :

$|\tilde{g}_i|=\sup_k|g_{ik}|$ et $H_i=(2\ D_i+|K_{ii}|h^2+|\tilde{g}_i|h)^{-1}$,

$i = 1, 2$

then, using the Gerschgorin Brauer theorem on the boundaries of the eigen values of a square matrix |7|, we get the following result : the scheme (S1) is stable if we take :

$$h < 2\, D_1/u_1 \text{ et } \tau/h^2 < \inf (H_1, H_2)$$

The number of arithmetical operations to compute θ^{k+1} knowing θ^k depends on N on the scheme used, we summarize the values on the following table :

Scheme	S1	S2	S3
Operations +	12N - 4	22N - 2	16N - 6
x	12N - 4	26N + 2	20N - 8

At each iteration, we see that for a given N, therefore for a given precision of the approximation by finite differences, the (S1) scheme is the best. But the stability condition leads, with the (S1) scheme, to very small values of τ, hence for a given horizon]0, T[very high values of M. So the (S3) scheme is the best for computing time.
Concerning the memory space, choosing N = 50, the three methods are implemented without difficulty on a 64 K bytes mini-computer.

6. EXAMPLES OF RESULTS

We try to simulate a machine with the following technical data :
- length of the barrel : 1 m
- diameter : 0,075 m
- number of heating collars : 3
- mass flow : 18 kg/hour.

The barrel temperature θ_2 is mesured at three different points and recorded during the following tests :
- heating of the barrel at constant power without matter, up to thermal balance
- introducing the matter, new balance
- step variations of the heating powers of each collar, constant flow.

The model is simulated according to the (S3) scheme. We perform again the tests described above with the following parameters, and we get the results of figure 3.

An example of parameters values :

$D_1 = 0{,}16\ m^2/hour$; $u_1 = 4 m/hour$; $- K_{11} = K_{12} = 50\ hour^{-1}$

$g_{11} = 2\ m/hour$; $D_2 = 0.1\ m^2/hour$; $- K_{22} = 11{,}2\ hour^{-1}$

$K_{21} = 10\ hour^{-1}$; $- g_{20} = 0{,}2\ m/hour$; $- g_{21} = 0.07\ m/hour$.

If we take N = 50, h = 0.02, the stability condition of the (S1) scheme leads to $\tau \leqslant 2\ 10^{-3}$ hour. With (S3) we can take $\tau = 25\ 10^{-3}$ hour = 90 seconds - this value corresponds to one tenth of the average time the matter stays in the machine. The distributing functions X_i (x), i = 1,2, 3 and r(x) described on figure 2

Figure 2 : distributing functions r, X_i

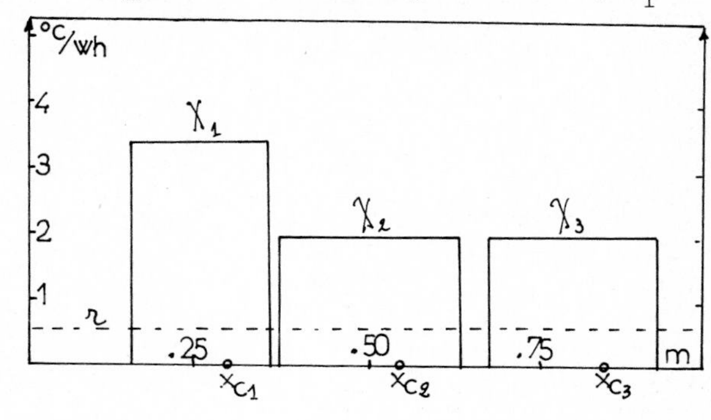

x_{c_i} , i = 1, 2, 3 : location of measure points.

During the simulation the digital computer is connected through A/D and D/A converters to an analog panel on which the desired power variations are simulated.

7. CONCLUSION

The description of the thermal behaviour of an extruder with two partial differential equations makes it possible to obtain a model involving a limited number of parameters.

The numerical solution of the equations is possible on a mini-computer, three methods have been described and compared. The one we have accepted is unconditionally stable on the integration steps and it makes it possible to solve the two equations in sequence, which means a certain freedom in choosing the time steps.

Comparison of simulation results to experiment results is satisfactory, thus it can be envisaged to use this model for purposes of dimensioning or functioning optimization.

Figure 3 : Simulation of the model : $\theta_2(x_1,t)$, $\theta_2(x_2,t)$, $u_1(t)$, $u_2(t)$, u_3 = 130 watts.

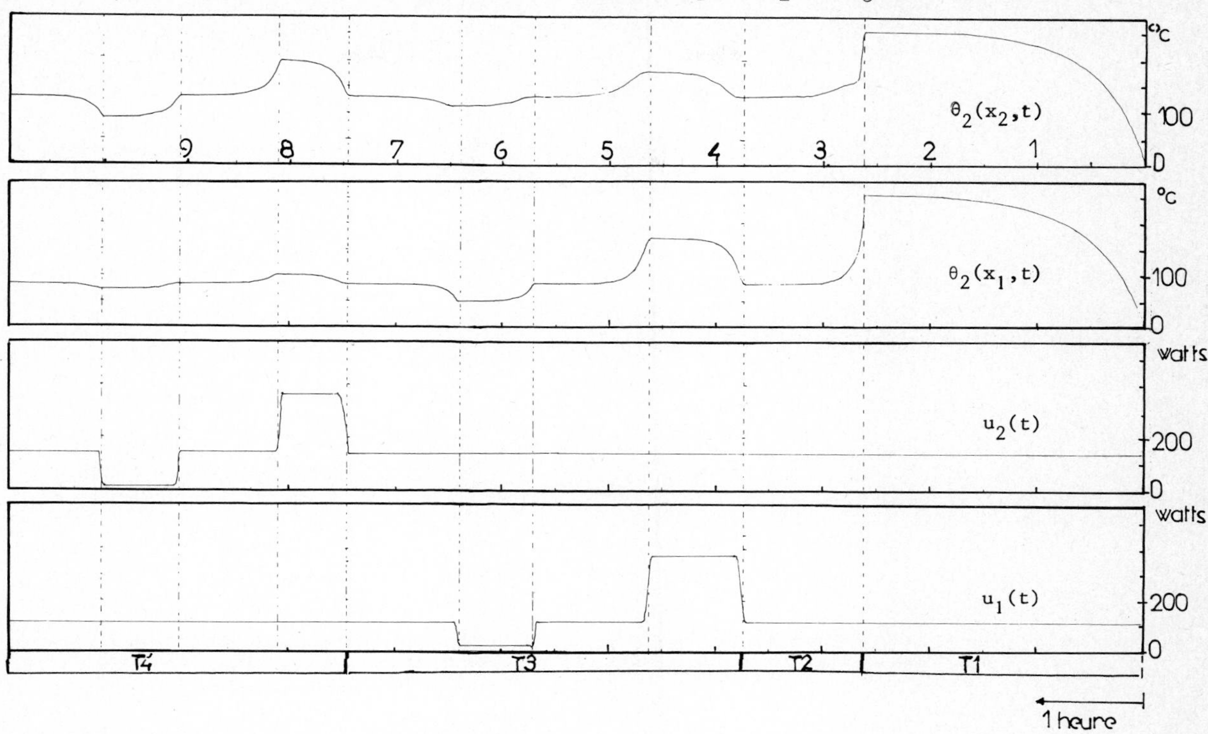

Period T_1 : heating of the barrel without matter.

Period T_2 : inserting the matter, PM = 760 watts.

Period T_3 : variations of u_1.

Period T_4 : variations of u_2.

REFERENCES

|1| Z. Tadmore, J. Klein : Engineering principles of plasticating extrusion" Van Nostrand Reinhold Company. 1970

|2| J.F. Lafay, Modélisation des pressions et de la température le long d'une extrudeuse. Proposition d'une commande. Thèse, Nantes, 1978.

|3| J. Burger, "Conduite automatique d'une extrudeuse décrite par un système d'équations aux dérivées partielles. Thèse, Nantes, 1980.

|4| Y. Jarny, Identification of spatially varying parameters for a distributed system. Application to a thermal process. Trans. of Imacs, vol. XXIII, n° 2, 1981.

|5| R. Bird, W. Steward, E. Lightfoot. Transport phenomena. J. Wiley, 1960.

|6| J.L.Lions, Contrôle optimal des systèmes gouvernés par des équations aux dérivées partielles. Dunod, Paris, 1968.

|7| E.D. Smith, Numerical Solution of Partial Differential Equations. Finite difference Methods. Oxford University-Press, 1978.

Simulation in Engineering Sciences
J. Burger and Y. Jarny (eds.)
Elsevier Science Publishers B.V. (North-Holland)

OPTIMISER-REGULATOR FOR LARGE BUILDINGS OR HOMES

V.J. BERWAERTS, hoogleraar;
J. BROEKX, M. SEGERS, J. STEVERLINCK, docenten;
G. BERWAERTS, E. CLAESEN , graduating students ind. ing.
KATHOLIEKE INDUSTRIELE HOGESCHOOL LIMBURG
Universitaire Campus DIEPENBEEK (Belgium)

The presented "optimiser-regulator" was developed in and by the Katholieke Industriële Hogeschool Limburg, Universitaire Campus Diepenbeek.
The parameters necessary for control and adjustment are measured and processed in a central computer (for large buildings) or in a thermostat (for homes).
Through application of the phaseplane method, the computer (or the thermostat for homes) can easily calculate the necessary set-ups such as:
-the time necessary for starting up the central unit;
-the dead time in the adjustment.
Whereas the starting-up of the central heating unit heppens with great power, the adjustment itself can either function with an on-off system with minimal deviations from the desired temperature, or with a PI adjustment algorithm by means of a modulating valve.
The optimiser (heating-up time) takes into account the momentary changes in room temperature requitements, with a retardation of about two hours.
The developed method, using an APPLE Computer, is being tested in the new KIHL building.
Compared with existing apparatus, this new system has, in our opinion, the following advantages :
-the exact set-up of heating-up is achieved in one day, and not after almost a week as with the existing systems.
-the cost price of the optimiser-thermostat is relatively low.

I. INTRODUCTION.

We assume that the heat dynamics of the building or the room can be represented in a simplified way as a first order process with retardation.

In the first instance, however, the retardation time will not be taken into account.

The basic formulae in our hypothesis are:

$$Q_{i(t)} - Q_{e(t)} = K_2 \frac{dT_{in(t)}}{dt} \quad (1)$$

$$Q_{e(t)} = K_1 \left[T_{in(t)} - T_{e(t)}\right] \quad (2)$$

In these formulae

$Q_{i(t)}$;heatflow from heating element to room

$Q_{e(t)}$:heatflow from room to environment outside of room

$T_{in(t)}$:average air temperature in room

$T_{e(t)}$:average environment temperature outside of room

K_2 :a constant;a.o. function of the magnitude of the exposed surfaces and of the heat transmission coefficient

Combination of formulae I and 2 results in:

$$Q_{i(t)} + K_{(t)} = K_2 \frac{dT_{in(t)}}{dt} + K_1 T_{in(t)}$$

With $K_{(t)} = K_1 T_{e(t)}$

and $A(t) : Q_{i(t)} + K_{(t)}$

we get $$T_{in(t)} = \frac{A}{K_1}(1 - e^{\frac{-t}{T_L}}) + T_{in(o)} e^{-\frac{t}{T_L}} \quad (3)$$

$$T_L = \frac{K_2}{K_1}$$

By means of an ON - OFF regulation, the magnitude $A_{(t)}$ can have two values:
in the On-position $A = Q_{i(t)} + K_{(t)}$
in the OFF-position $A = K_{(t)}$

Equation 3 is to be interpreted as follows: If the ON-position would last infinitely long, the average temperature in the room would be

$$T_{in(max)} = T_m + T_e$$

in which $T_m = \frac{Q_{i(t)}}{K_1}$

If the OFF-position would last infinitely long, the average temperature in the room would be

$$T_{IN(min)} = T_{e(t)}$$

Figure I represents these two positions on the abscissa or temperature axis.

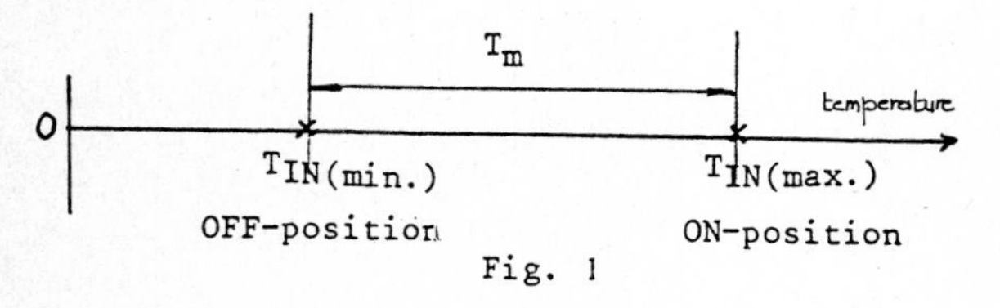

Fig. 1

2. Representation on the plane $\dot{T}_{in(t)}$, $T_{tn(t)}$ and interpretation.

By simple derivations we can represent the On- and OFF positions on the phaseplane by means of two straight lines. (Fig. 2.)

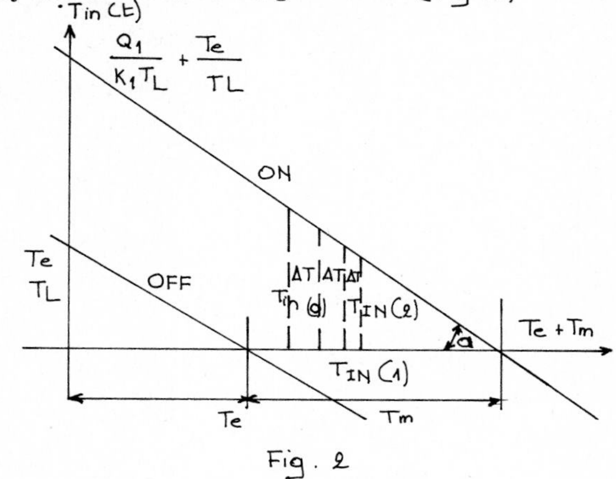

Fig. 2

The "ON-line" intersects the ordinate for

$$\dot{T}_{in(t)} = \frac{Q_{i(t)}}{K_1 T_L} + \frac{T_{e(t)}}{T_L}$$

whereas the 'OFF-line intersects the ordinate for

$$\dot{T}_{in(t)} = \frac{T_e}{T_L}$$

The intersections with the abscissa are respectively

for $T_{in(t)} = T_{e(t)} + T_{m(t)}$

and $T_{in(t)} = T_{e(t)}$

The interpretation of the heating system on the phase plane is as followx : we assume that at the moment of switching the hearing on, the average measured indoor temperature

$$T_{in(t)} = T_{IN(o)}$$

After a heating time of Δt time units (retardation time included) we measure an average room temperature $T_{in} = T_{IN(1)}$ °C.

The average indoor temperature during the time interval Δt is

$$\frac{T_{IN(1)} + T_{IN(2)}}{2}$$

The average corresponding $\dot{T}_{in}$ value is given by $\left[\frac{T_{IN(2)} - T_{IN(1)}}{\Delta T}\right] x60$ Δt in minutes

After a new prolonged heating during Δt minutes we measure the indoor temperature

$T_{IN(3)}$ ... etc.

The corresponding "rate" at which the temperature changes, can be found by reading the corresponding ordinate on the phase plane.

Not that the equal time interval Δt is not represented on the phase plane by means of equal differences :

$T_{IN(3)} - T_{IN(2)}$

$T_{IN(2)} - T_{IN(1)}$ ...

Through further derivations we get

$$T_{Li} = \frac{\Delta T \left[T_{IN(3)} - T_{IN(1)}\right]}{120 \left[2T_{IN(2)} - T_{IN(1)} - T_{IN(3)}\right]}$$

T_{Li} is the time constant of the building or the room, expressed in minutes, whereas $T_{IN(3)}, T_{IN(2)}$ .... etc. represents the indoor temperature T_{IN} measured every ΔT minutes.

$$T_m + T_e = \frac{\left[T_{IN(2)} - T_{IN(1)}\right]\left[T_{IN(3)} - T_{IN(2)}\right]}{2\left[2T_{IN(2)} - T_{IN(1)} - T_{IN(3)}\right]} + \frac{T_{IN(2)} + T_{IN(1)}}{2}$$

Considering that the ordinate gives the rate at which the room temperature changes- and taking into account that the starting room temperature is measured and the required temperature is set up, we can find after some calculation the necessary heating time to reach the required temperature.

$$T_{i+2} = \frac{T_{g(i+2)} - T_{IN(i+2)_o}}{\left[T_{e(i+2)} + T_{m(i+1)} - \frac{T_{g(i+2)} + T_{IN(i+2)_o}}{2}\right] tg_{\alpha i}}$$

with $$tg_{\alpha i} = \frac{120\left[2T_{IN(i+1)_2} - T_{IN(i+1)_1} - T_{IN(i+1)_3}\right]}{\Delta T\left[T_{IN(i+1)_3} - T_{IN(i+1)_1}\right]}$$

$$T_m(i+1) = \frac{\left[T_{IN(i+1)_2} - T_{IN(i+1)_1}\right]\left[T_{IN(i+1)_3} - T_{IN(i+1)_1}\right]}{2\left[2T_{IN(i+1)_2} - T_{IN(i+1)_1} - T_{IN(i+1)_3}\right]} + \frac{T_{IN(i+1)_2} + T_{IN(i+1)_1}}{2} - T_e(i+1)$$

In these formulea

$Tg_{(i+2)}$ represents the required for the day i+2 (e.g.tuesday)

$T_{IN(i+2)_o}$ the measured starting room temperature for the day i+2

$T_{e(i+2)}$ the measured outdoor temperature for the day(i+2)

$T_{m(i+1)}$ the calculated maximal temperature leap for the day i+1 (e.g. monday)

3. Practical execution.

Heating is started up the first day (i+1); the required heating time can't be calculated yet.

After a number of indoor temperature samlings T_{IN1}, T_{IN2} and T_{IN3} , as well as of the outdoor temperature $T_e(i+1)$, the parameters T_{mi} and $tg\alpha_i$ are calculated.

Not that for a valid measurement the average value of sixty samplings is taken.

When the required temperature $Tg_{(i+1)}$ is reached, the adjustement around the set up value can be executed either in the ON-OFF mode (taking the dead time into account) or by means of PI implementation.

At the required and set up point of time (stop time) of the day i+1 the heating unit will be switched off. The room temperature will drop gradually.

During the NOT-ON period (temperature drop) the indoor temperature $T_{IN(i+2)_o}$ as well as the outdoor temperature $T_{e(i+2)}$ will be measured.

Furthermore, the heating up time will be calculated at regular intervals; this heating ip time will be added to the real time.

Whem the sum of the heating time and the real time equals the set up point of time for the day i+2 , the heating unit will be switched on "softwarewise".

Adjustment around the set up temperature for the day i+2 will then again be executed either in ON-OFF mode or by means of PI implementation.

Not that for longer periods of non-occupation a frost-protection is built in.
The width of the hysteresis curve or, to be more concrete, the temperature variations around the set up temperature are decreased by switching on and off earlier (Fig. 4)

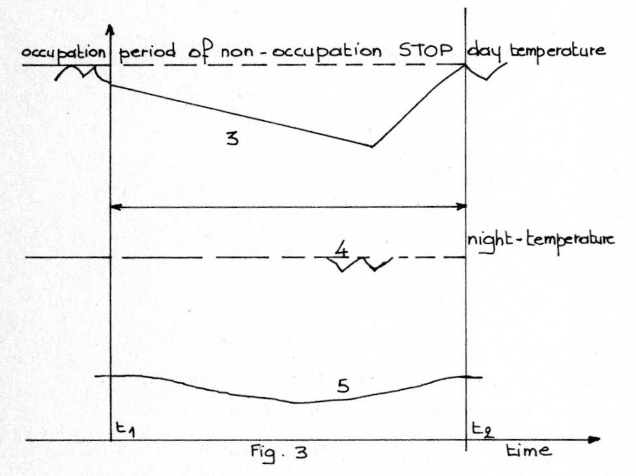

Fig. 3

To explain this, we'd like to refer to fig. 4. We assume that we are in ON-position in point A. (Heating unit is switched ON).

The required temperature is Tg. The heating system increases the room temperature with a steadily diminishing speed, according to theON-line. The real temperature can be read off the abscissa. Because switching off and on happens "softwarewise", the heating unit will be switched off as soon as the room temperature reaches the set up value Tg.

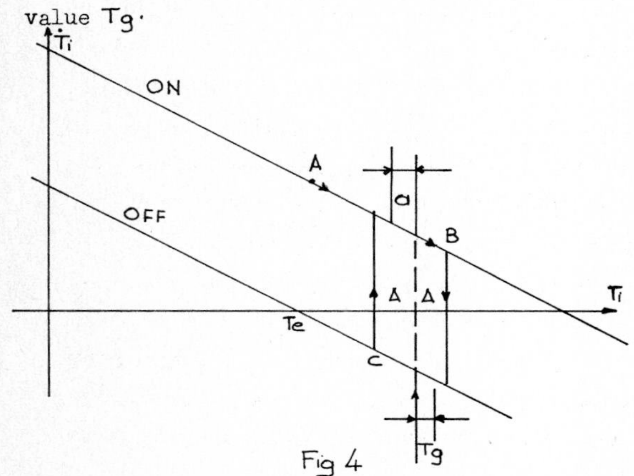

Fig 4

By the inertia of the process, the temperature will go on increasing until point B on the ON-line is reached. The room temperature is then Tg + Δ.

From this moment on, the evolution of the system is to be follewed on the OFF-line.

As soon as the room temperature will equal Tg , the heating system will be switched on again.

Nevertheless, the temperature will go on decreasing until point C is reached.

The maximal temperature variation is thus 2Δ. . It is clear that these variations will be smaller when switch off happens

at Tg − α

and switch on at Tg + α

In the realized thermostat-optimiser Δ is measured and α = f(Δ) is chosen.

4. Flow chart of control.

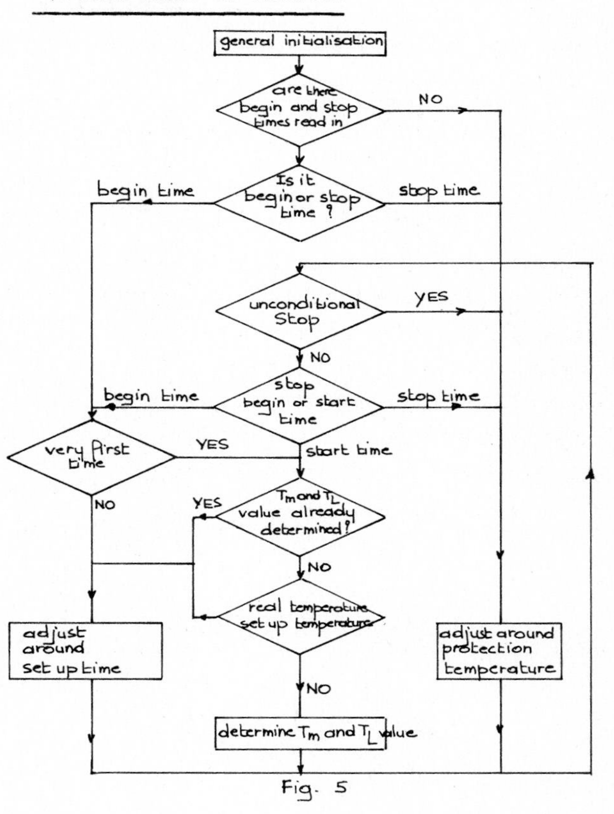

Fig. 5

5. General structure of the Optimiser-Thermostat.

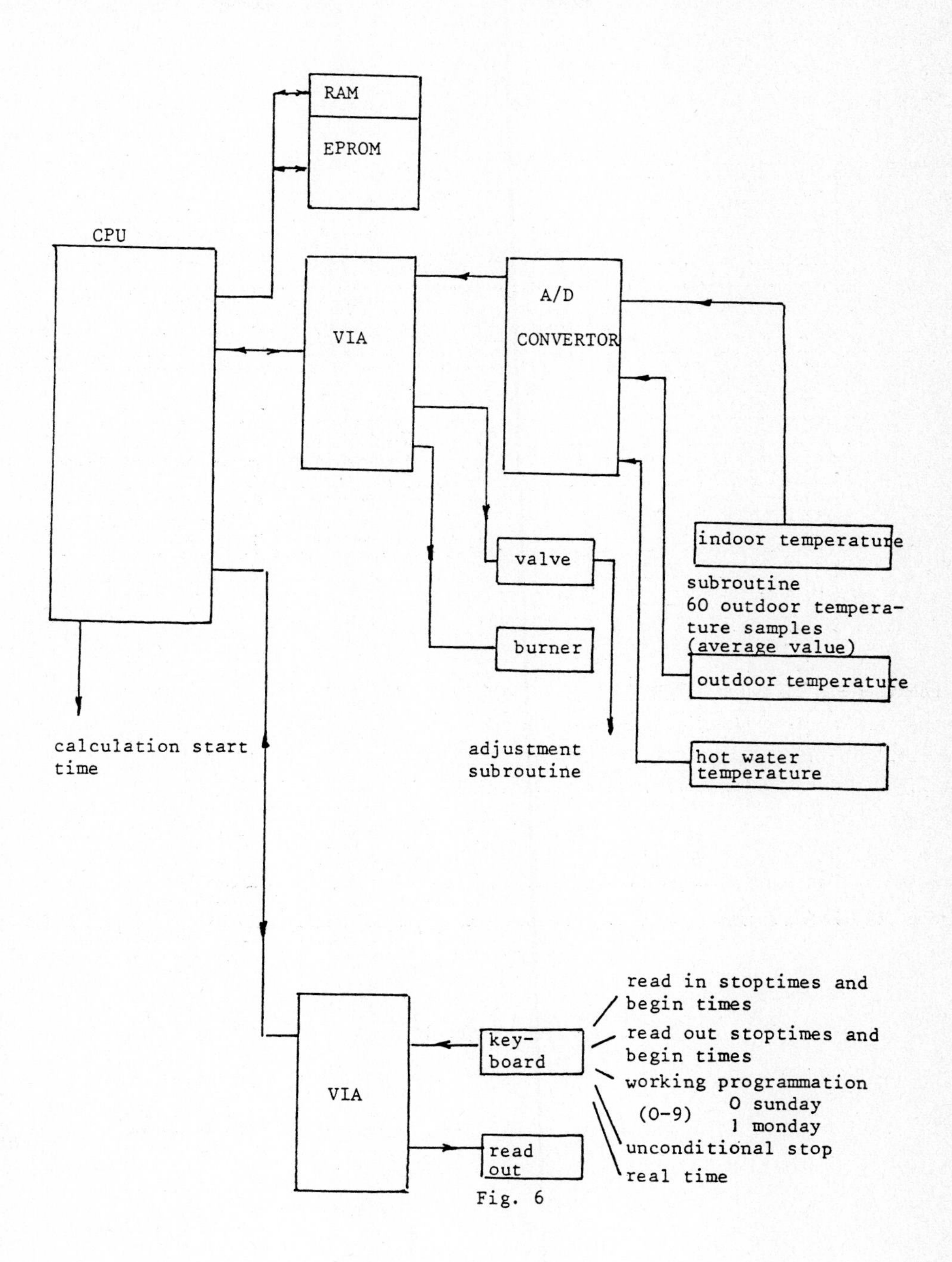

Fig. 6

6. Results of Simulation.

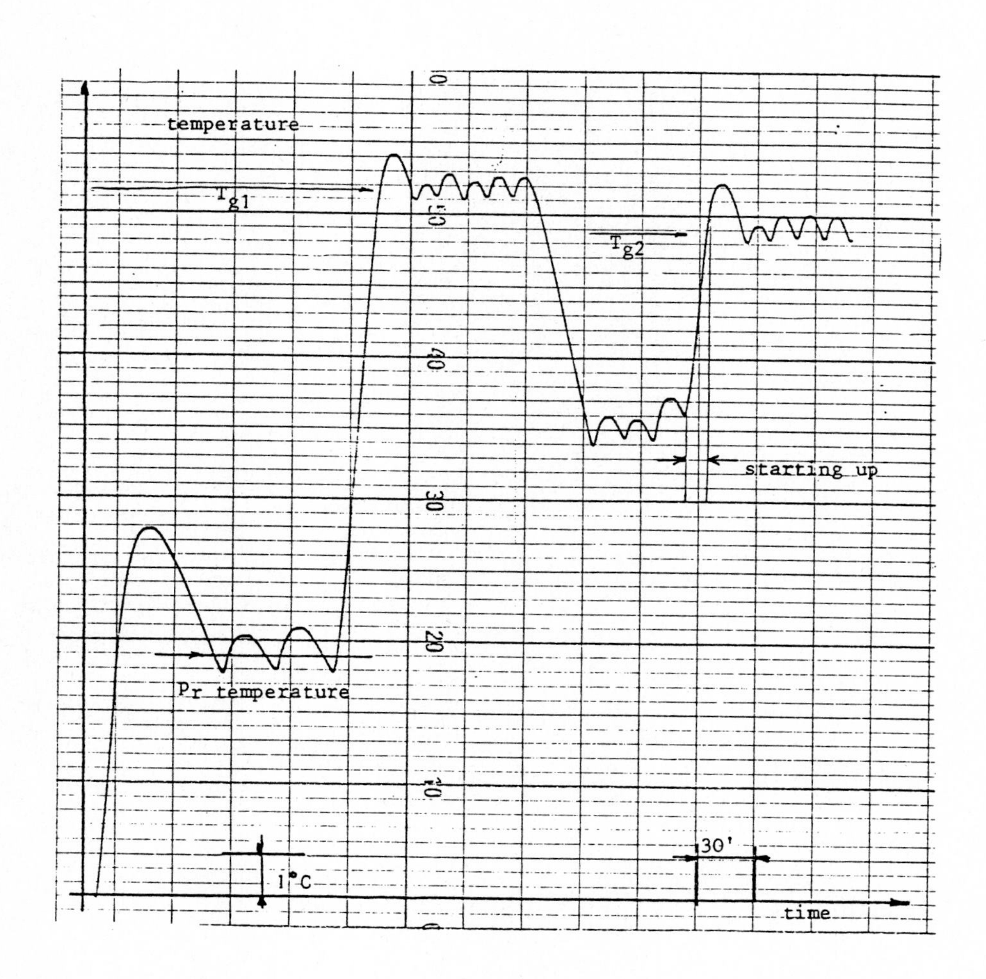

Fig. 7

XIV. VEHICLES – TRANSPORTATION (2)

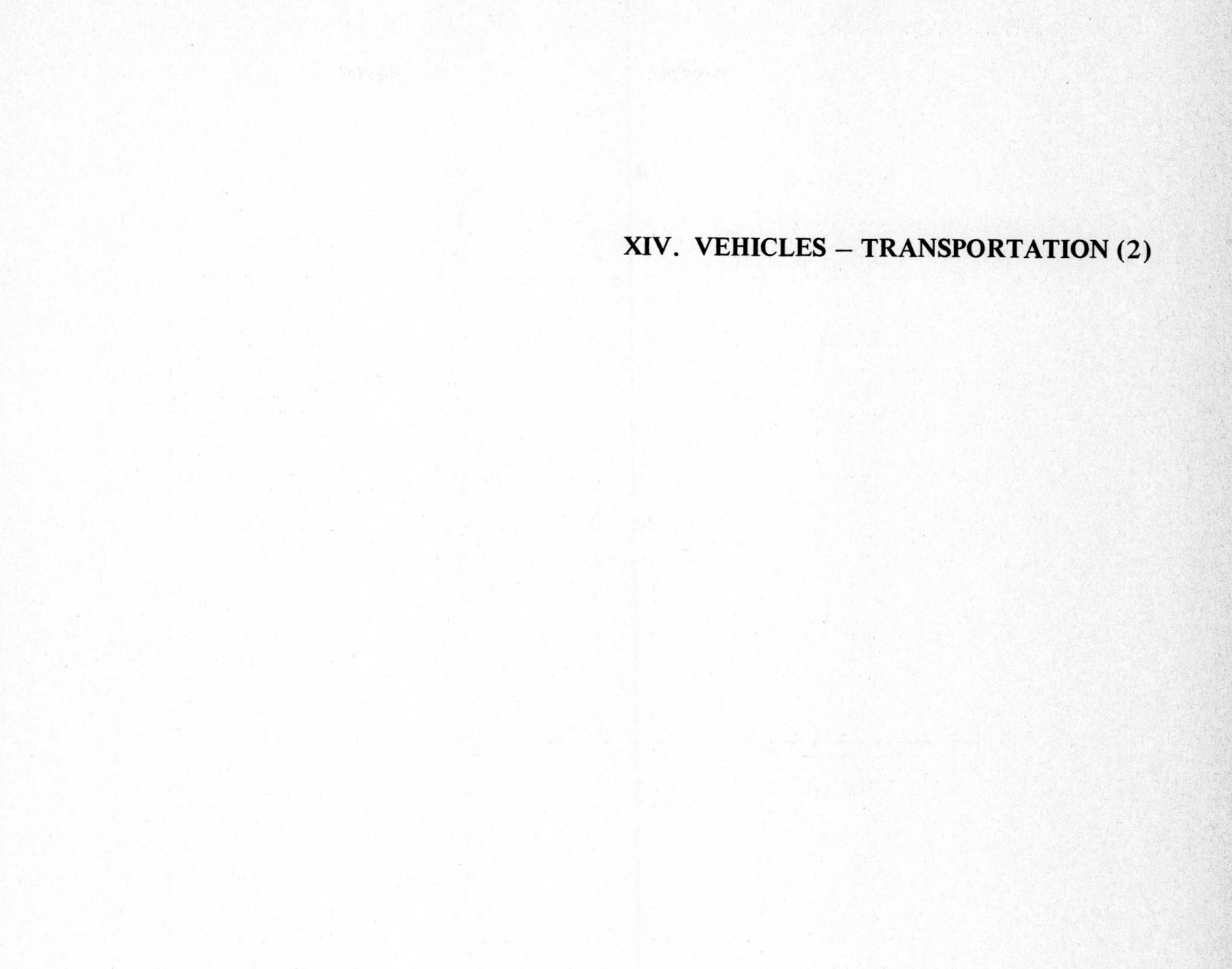

Simulation in Engineering Sciences
J. Burger and Y. Jarny (eds.)
Elsevier Science Publishers B.V. (North-Holland)

SIMULATION OF TRAIN OPERATION UNDER AUTOMATED CONTROL ON A METRO LINE

J. CASTET

INTERELEC (MATRA GROUP) (1)
53, rue du Commandant Rolland - 93350 Le Bourget - France

1. GENERAL

1.1 Notions of Automatic train Operation

The INTERELEC automatic pilot system carried on board trains, receives the data required for automatic train operation through a transmission line laid alongside the track.

The transmission line features the following form :

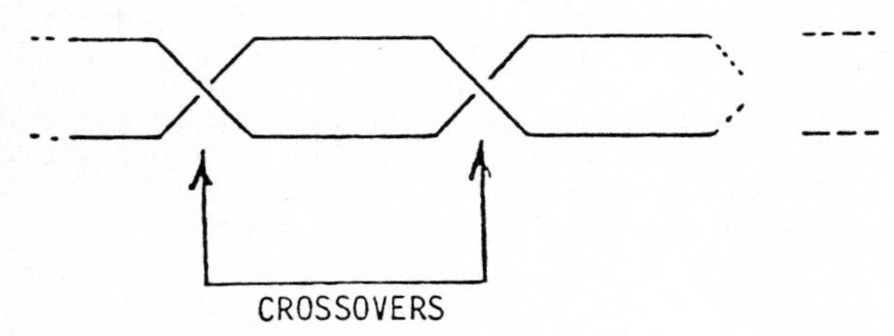

The command speed is represented by the length of the line section bounded by two successive crossovers. The greater the speed desired, the longer the sections will be. A digital simulation enables this transmission line to be worked out.

Operational safety is provided for by a speed-monitoring unit, (A.T.P. = Automatic Train Protection).

The unit consists of two modules :

- the first module verifies the travel time over a section; it triggers emergency braking when that time drops under a safety threshold,

- the second module verifies the track-to-train signal link; it triggers emergency braking when the received signal strength is too weak.

A second unit (A.T.O. or slaving), generates the traction/braking command signals as a function of the ground reference, model, and actual train speed. (A.T.O. = Automatic Train Operation).

The ground reference is computed in such a way that the automated running of the train is :

- comfortable (variation of acceleration limited : $\frac{\partial \gamma}{\partial t}$ < Jerk),

- efficient, train intervals maintained over an entire metro line,

- failsafe, insuring protection from obstacles; moving switchgear, bumper, train immediately down-line, etc.

1.2 Command Program

At least two transmission lines, called command programs, are associated with each stopping point. The first program maintains speed on a clear track, the length of the sections being generally constant. The second program is associated with reduction of train speed. The decrement of the distance between two consecutive sections is constant, resulting in a constant deceleration. The call-up of one or the other of the programs depends on the state of signalling.

Stopping Program

Stopping programs are designed for use :

- at stations,
- in shunting yards,
- between stations to respect the signals

In these cases, the automatic operating program sends the train a constant deceleration command from clear-track speed to a null speed. The stopping program determines the stopping distance during the braking phase. The train is slaved to a value of deceleration called "service deceleration", defined by the law of automatic operation as a function of the stopping program laid along the track. The stopping distance depends on the specific deceleration of the stopping program, the initial speed and on the longitudinal track-section profile.

1.3 Safety Distance

The safety distance represents the maximum distance into which no train can penetrate in the event of failure :

- of the automatic control system,
- of the rolling stock.

The safety distances depend on several parameters :

- performance of rolling stock
- minimum warranted value of emergency deceleration,
- pure delay equivalent to the application of emergency braking,
- maximum traction performance,
- performance of ATP safety circuits.

These distances also take into account :

- operating programs (command speed) embedded along the wayside,
- longitudinal track-section profile.

2. COMPUTATION OF WAYSIDE-INSTALLED PROGRAM SEQUENCES

2.1 Program Sequences

The program installed along the track consists of sequence lengths proportional to the command speed :

$$Ls = Vc \times TB$$

where :

Vc : command speed,
TB : base time set at 300 msec.

2.2 Computation of Stopping Program

2.2.1 Characteristics of Stopping-Program Deceleration

In order to maintain a constant force on the train during a deceleration command on a variable longitudinal track-section profile, a deceleration-command correction must be applied in accordance with the law :

$$\gamma p = \gamma o + k \times g \times p$$

where :

γp : deceleration program,
γo : deceleration program on the level (p = o),
k : gradient-effect corrector coefficient,
g : gravity-produced deceleration,
p : gradient value in thousandths.

2.2.2 Stopping Distance

$$DA = \frac{V^2}{2x\ (\gamma o + k \times g \times p)}$$

V represents the clear-track command speed from which deceleration is desired, down to the stopping of the train.

We can deduce that the change-over from a command speed V to an intermediate speed Vi is expressed :

$$DA' = \frac{V^2 - Vi^2}{2x\ (\gamma o + k \times g \times p)}$$

2.3 Case of a Variable Longitudinal Track-Section Profile

In this case, the stopping program consists of successive portions of parabola which are as a function of the longitudinal track-section profile.

3. SIMULATION OF TRAIN OPERATION ON WAYSIDE PROGRAM

3.1 Traction Model

At a given speed of Vi there is a corresponding acceleration of γi (see Figure 1) depending on the force commanded by the automatic control.

We can consider that the curves $\gamma T = f(V)$ are valid for a given masse.

The range falling between maximum and minimum acceleration is discretionalized at several points which correspond to various thresholds of a chronometric law. This means, that the automatic control sets a tractive control current proportionately as much higher as the actual train speed is separated from the command speed.

When changing over from a γi acceleration to an acceleration of $\gamma i + 1$, we take into account jerk limitation and transition time between the various modes and pure Rpt control delay.

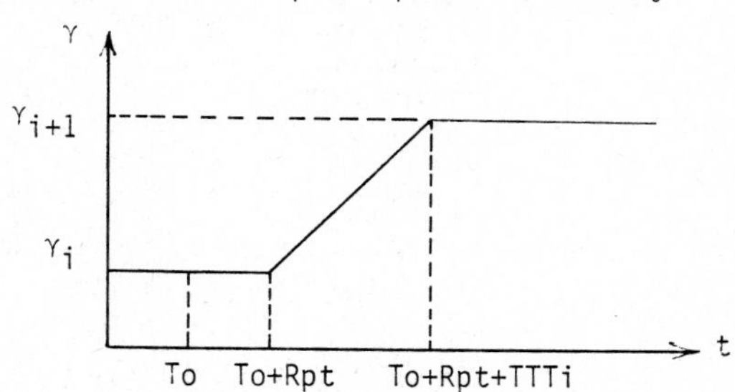

where :

Rpt : pure control delay,
TTTi : traction transition time between modes i and i + 1,
To : control initialization.

The transition time can be :

- either, a linear increase in the control,
- or, a combination of linear and exponential increase.

Acceleration γT thus found, is corrected by the gradient (m/S^2) and, depending on the case, by an equivalent deceleration due to curves.

3.2 Braking Model (Figure 2)

The automatic control law is also discretionalized. Let us consider a pure delay (Rpf) in the application of the controlled braking mode followed by transitional times between modes.

Response to small signals is tantamount to jerks.

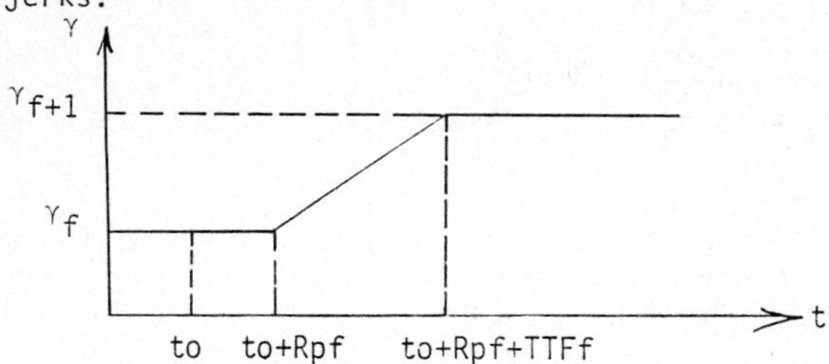

where : TTFf : braking transition time between modes f and f + 1,
Rpf : pure control delay,
to : represents control initialization

The transition time can be :

- either, a linear increase,
- or, a combination of linear and exponential increase.

3.3 Changeover to Coasting (neutral)

When a sequence travel time is about 300 msec. (base time), the automatic control calls for neither traction nor braking.

In this case, the train is subjected only to the effect of gravity, corrected by an equivalent deceleration due to curves. Also, resistance to straight-line forward motion is taken into account.

3.3.1 Resistance to Forward Motion

- In a straight line :

it is in the form of $\gamma R = AV^2 + BV + C$
where V = speed

- in a curve :

$\gamma RC = 500/R$ where R = radius of curve in meters.

3.4 "Normal" Operation

This is deduced from the accelerated operation previously computed by anamorphosis of the chronometric law in the tractive phase.

3.5 "Reduced" and "Slow" Operation

The chronometric law is anamorphosed in its tractive phase for the "reduced" operation.

For "slow" operation, reduced acceleration is applied, and traction is anamorphosed.

3.6 "Economy" Operation

Power consumption is simulated between two points leading to the determination of economical operation.

3.7 "Foul-weather" Operation

In elevated zones, in certain cases degraded operation is provided for during rainy spells.

4. COMPUTATION OF SAFETY DISTANCES

4.1 Failure Hypothesis Taken Into Account

4.1.1 Traction Blockage

We will consider that all motor units are active and that acceleration is maximal up to the triggering of emergency braking by the ATP circuits.

4.1.2 Inefficient Braking

We will consider that zero deceleration is applied until the triggering of emergency braking by the ATP. In this case, the train is only subjected to the effect of gravity due to the longitudinal track-section profile.

4.2 Performance of Circuits Forming the ATP

Two safety systems monitor the train operation.

4.2.1 "Cable Power-Monitoring Circuit" (S.L.)

This circuit initiates emergency braking upon loss of the signal transmitted by the program written into the wayside. This circuit operates principally when the train passes over the normal stopping point.

To ensure adequate availability, the circuit is amply time-delayed.

4.2.2 "Dynamic Safety" Circuit (S.D.)

The train speed is monitored at all points by this circuit and, as soon as the speed exceeds a threshold proportional to the command speed, the circuit generates a command applying emergency braking.

4.2.2.1 Overspeed Circuit (SD1)

4.2.2.1.1 Program Sequences

LS = Vc x TB

Thus, by comparison between the real travel time of a sequence and the base time TB, the automatic control system generates traction or braking commands.

Likewise, when the travel time of a sequence drops under a preset TSD value, the overspeed circuit (SD1) detects an overspeed condition.

4.2.2.1.2 When the train control equipment is in traction mode, the threshold limit is TSDT (Dynamic Safety Traction Time).

4.2.2.1.3 TSDF (Dynamic Safety Braking Time) is the threshold limit when the train control equipment is in braking mode.

4.2.2.2 Noise Immunity Circuit (SD2)

This circuit cuts-in in series after SD1 is triggered, and acts as a time-delay device. It filters out spurious noise or extraneous signals. At the expiration of the time-delayed period, the control circuits apply irreversible emergency braking.

4.2.3 Application of Emergency Braking

The computation of safety distances takes into account a pure delay equivalent to the application of emergency braking and the minimum warranted value of emergency deceleration.

4.3 Computation of Safety Distances

We simulate the preceding failures; we can, thereby, calculate the maximal overrunning of the normal stopping point in the event of a failure (see Figures 4 and 5).

4.4 Variable Profile Data Entry

The safety distances are obtained by means of EDP (Electronic Data Processing).

The various breaks in the profile are input at the level of the program determining the dual-level safety distances.

The stopping distance is computed as a function of the operating program imbedded in the wayside :

$$DA\ (Vi) = \sum_{i,j} \frac{(VI)^2 - (VJ)^2}{2 \times \gamma pij \times \frac{(TB)^2}{(TSD)}}$$

γpij represents the various command decelerations, and Vi, Vj represent the triggering speeds on alignement with the profile breaks.

The safety distance is computed by perfect integration of gravity deceleration over the entire length of the train.

5. SIMULATION PROGRAM

The general flowchart is shown in Figure 6. The characteristics and performance data of the system and automatic control unit are parametered.

Examples :

* Characteristics of Rolling Stock (traction and braking characteristics, train weight and consist, resistance to forward motion, etc.).

* Track Characteristics (gradients, curves, station locations, switchgear and bumpers, speed limitations, position of track circuits, etc.).

* Automatic Control Characteristics (noise immunity delay, safety thresholds, jerk value, nominal deceleration, etc.).

* System Performance (crew change and itinerary setup times, between-train intervals, set-out time, degraded operation performance, itinerary conditions, economy operation, various sheduled operating modes, etc.).

5.1 Simulation Method

An initial approach consists in making a parametrical study of certain characteristics. From the results, optimal values can be deduced (for example : service deceleration, station approach speed) for setting design performance ratings.

When the parameters are finalized, simulations (safety and slaving) enable definition of the programs which will be written into the wayside. The corresponding chart is shown in Figure 3.

5.2 Implementing Simulation Results

We can thus :

- compute the minimum theoretical intervals separating two consecutive trains,

- anticipate the spacing problems and define the signalling layout,

- compute the commercial line-speed,

- compute the line passenger-density,

- determine the number of trains required to maintain the desired intervals,

- define various modes of operation to insure traffic regulation,

- if necessary, estimate power consumption (possibility of obtaining economy mode operation).

5.3 Signalling System Layout

In the majority of cases, the signalling system design is optimized to match the performance of the automatic control system.

6. PERSPECTIVES

6.1 By integrating the results obtained previously on a complete metro line, it is possible to create traffic-regulation algorithms, with simulation of disruptive factors.

Extension to finalized adjustment of the electronic module : a response of the train to stimuli is simulated. Thus, certain electronic modules are developed in the laboratory, simulation partially substituting for on-site testing.

6.2 Future Generation of Automatic Control System

The new automatic control system is based on the computation of the train's commanded route by an on-board computer.

The intrinsic train characteristics are stored in the on-board computer's memory. The line characteristics are either stored in the computer's memory, or received via ground-to-train transmission. The signalling state is also transmitted by another track-to-train link. The computer, recognizing the position of the train from the data received, localizes the first obstacle immediately down-line and generates the safety limitation speed diagram which is not to be exceeded.

The safety speed Vs (x) is given by the following relation ships :

Let :

$$V^2_s(x) + (A \text{ sh } wz - B \text{ sh } w(z+T))\, Vs(x) + C = 0$$

Let :

$$V^2_s(x) + (A \sin wz - B \sin w(z+T))\, Vs(x) + C = 0$$

The parameters A, B, C, w depend on the characteristics of the train and the line; z is the detection time of train overspeed; T is the emergency-braking system response time.

The actual train speed at any point on the line must be less than the Vs speed for that point, and at the limit speed imposed by the track characteristics at that particular point.

The following tables give the parametric expressions of a, b, c :

1st case :

$$A = \frac{2\,\gamma_T}{w} \quad \text{and } w^2 = A > 0$$

$$B = -2\,\frac{\gamma_u}{w}$$

$$C = -c - a\,x^2 - 2\,x\,(b + \gamma_T\,(1-\text{ch } wz) + \gamma_u \text{ ch } w\,(z+T)) + \frac{2}{a}\,(\gamma_T{}^2\,(\text{ch } wz - 1) + b\,\gamma_T\,(\text{ch } wz - 1) + b\,\gamma_u\,(1 - \text{ch } w(z+T)) + \gamma_u + \gamma_T\,(\text{ch } wT - \text{ch } w(z+T))$$

2nd case :

$$A = 2\,\frac{\gamma_T}{w} \quad \text{and } w^2 = a < 0$$

$$B = -2\,\frac{\gamma_u}{w}$$

$$C = -c - a\,x^2 - 2\,x\,(b + \gamma_T\,(1 - \cos wz) + \gamma_u \cos w(z+T)) + \frac{2}{a}\,(\gamma_T{}^2\,(\cos wz - 1) + b\gamma_T\,(\cos wz - 1) + b\,\gamma_u\,(1 - \cos w\,(z+T)) + \gamma_u + \gamma_T\,(\cos wT - \cos w(z-T))$$

a = gradient of slope over the length of the train, b and c are constants dependent on the initial conditions.

γ_T = maximum train acceleration under traction,

γ_u = train deceleration.

7. CONCLUSION

Computerization, alone, enables us to integrate all the parameters having an interactive effect on automatic control.

The selected model is subsequently validated by full-scale tests on site. In general, an acceptable coincidence exists between the theoretical computations and the practical measured results.

-0-0-0-

(1)

Subsidiary of the MATRA GROUP, INTERELEC is a French Corporation specialized in fail-safe electronics and automation.

Among the prime projects developed, manufactured and installed by INTERELEC are the automated control systems for the Metros of Paris (12 out of 13 lines), Mexico City (7 lines), Rio de Janeiro (both lines), Santiago (both lines), and Caracas (both lines).

Additionally, INTERELEC is producing the fully-automated control system for VAL, Lille's new, driverless Metro. The Company has also been commissioned to supply the automated operating controls for the "C" line of the Lyon Metro.

INTERELEC will participate in the installations sheduled for Lagos and Hong Kong.

Based on the weight of its references, INTERELEC has become today's world leader in the field of automated operating controls. The Company's participation in the fitting out of transportation vehicles is being extended by furnishing operating aids to surface lines (busses, trolley-busses and streetcars) with its THESEE system which is being installed at Lille (on bus and streetcar lines) and at Clermont-Ferrand (on bus lines).

-0-0-0-

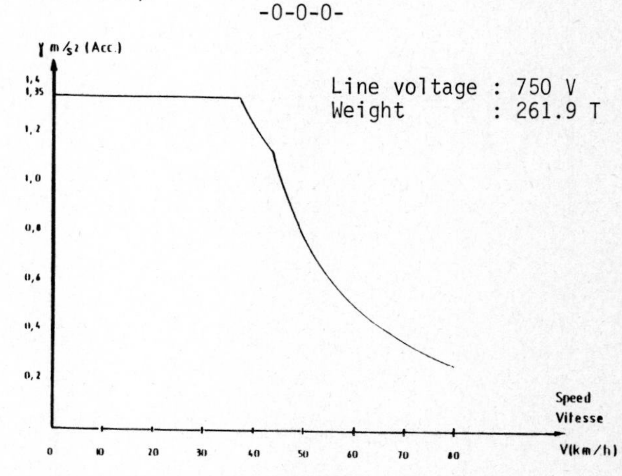

Figure 1 : traction performance, $\gamma_T = f(V)$

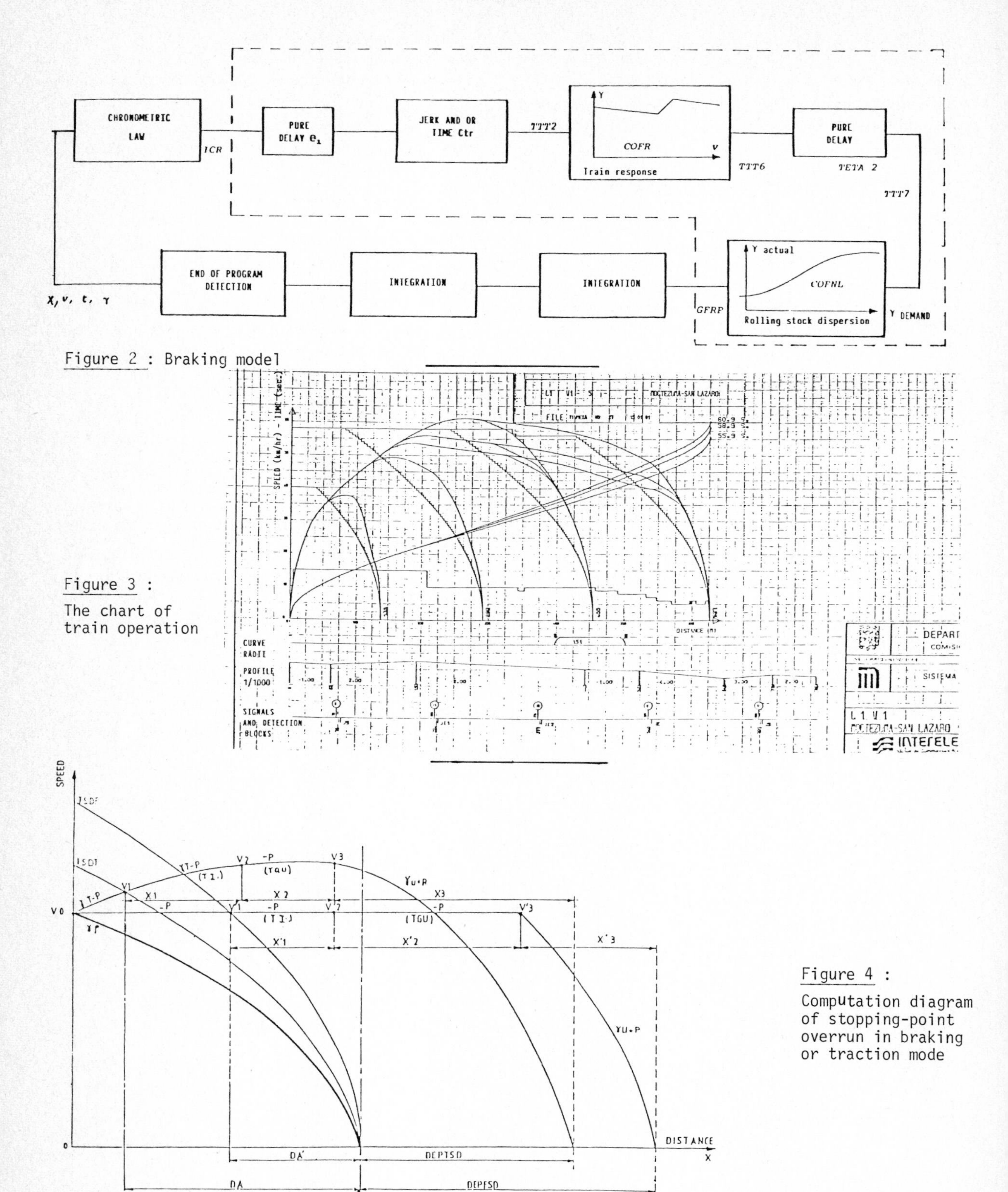

Figure 2 : Braking model

Figure 3 : The chart of train operation

Figure 4 : Computation diagram of stopping-point overrun in braking or traction mode

Theoretical stopping point

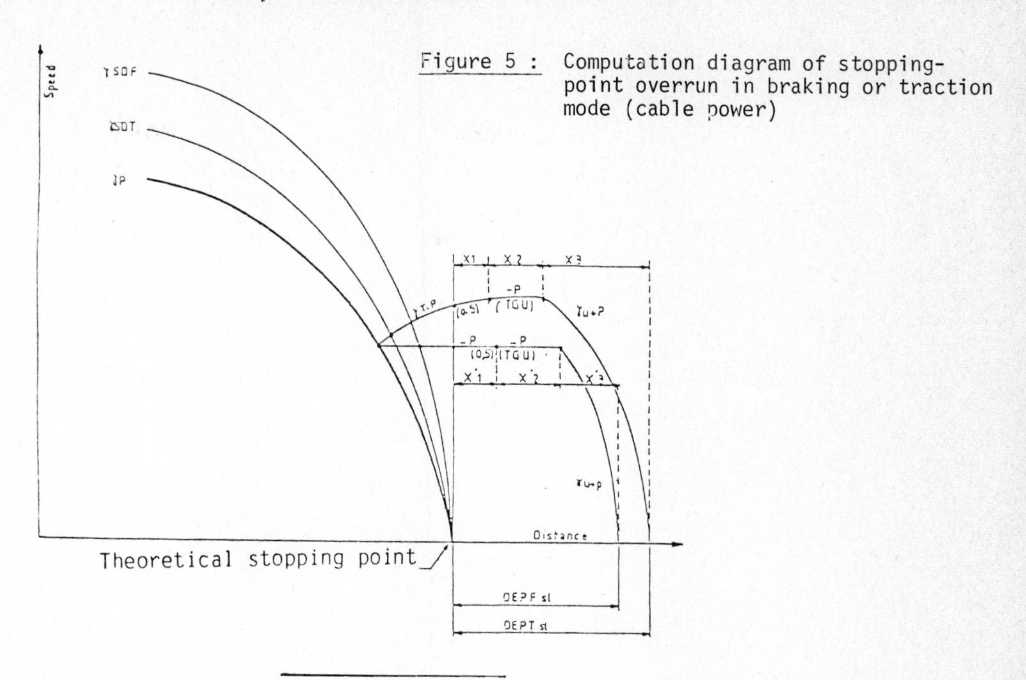

Figure 5 : Computation diagram of stopping-point overrun in braking or traction mode (cable power)

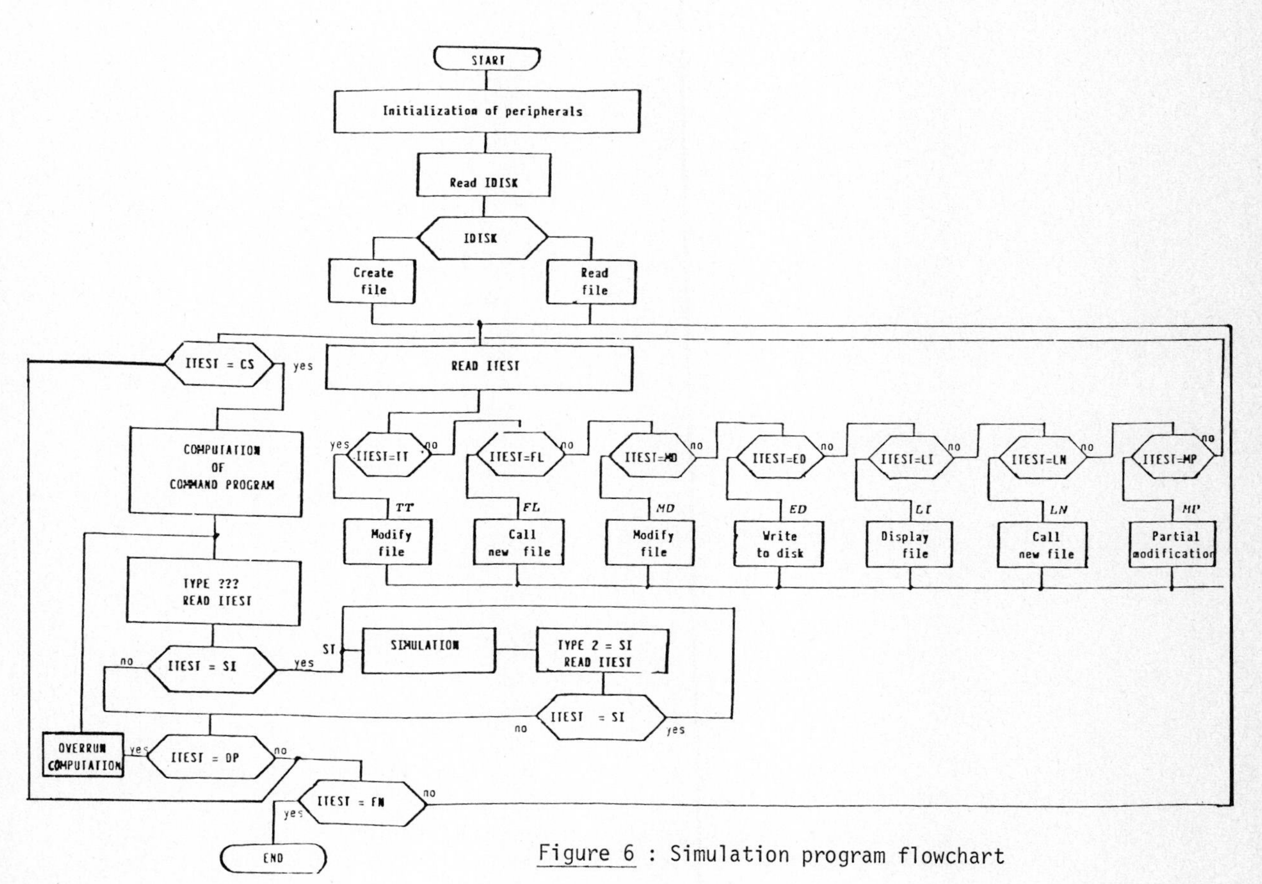

Figure 6 : Simulation program flowchart

Simulation in Engineering Sciences
J. Burger and Y. Jarny (eds.)
Elsevier Science Publishers B.V. (North-Holland)

SIMULATING THE ROAD BEHAVIOUR OF PRIVATE CARS

by M. DEGONDE

Engineer, Scientific Computation Department - PSA Research

Evolved mathematical models developed for the simulation of road behaviour have proven useful as an aid to designers in specifying and perfecting vehicles.

We describe a rather sophisticated model with 15 degrees of freedom which is in general use at the Direction des Etudes et Recherches of PSA and which represents a private car with four independent wheels.

INTRODUCTION

First developed some years ago to evaluate the impact of proposed safety regulations for private cars, analytical tools have been gradually reoriented to serve in the design and development of new vehicles.

Their usefulness as an aid to designers working with a certain number of problems relating to the dynamics of road behaviour is now established.

The methodology of these analyses is based essentially on simulation of vehicle movements in the course of typical driving situations such as in negotiating and leaving a curve, changing lanes, braking in a curve, deviation in a lateral wind and so on.

The Worthwhileness of simulating road behaviour and of quantifying the influence of the parameters determining this behaviour prior to their becoming irreversibly established in the design phase is obvious. Simulation of the phenomenon rests on the existence of sufficiently evolved and accurate mathematical models, used in an open loop, i.e. which take commands or inputs to the model such as rotation of the steering wheel, engine torque and the pressure of the braking system as given time functions.

To this end, we make use of a whole set of vehicle models, ranging in complexity from 3 degrees of freedom (simple plane-on-plane model) to 37 and 51 degrees of freedom, enabling a quite precise description of the vehicle.

The model most frequently used has 15 degrees of freedom and this is the one we will now describe, in its most recent version.

PART 1 : DESCRIPTION OF THE 15 DF MODEL

The vehicle being modelled is a 4-wheel independant suspension design, with either front or rear-wheel drive, moving over even, but not necessarily horizontal, ground.

1. Physical System and Parameterization

Consists of 6 solids, as follows (see Appendix 2, Fig. 1) :

- the body
- the 4 wheels
- the engine (reduced to its flywheel)

and of a material point representing the steering rack.

Each solid is represented in the conventional way by 6 positional parameters including 3 coordinates for the center of gravity and 3 Euler angles.

With each solid is associated a mass and the ellipsoid of inertia at the solid's center of gravity. The 36 positional parameters for the 6 solids are obviously not independant : in fact there are 22 equations inter-relating these parameters such that the "position" of the physical system depends on 15 independent parameters which, for convenience of interpretation, have been selected as follows :

X_G, Y_G, Z_G coordinates of body's center of gravity / fixed axis system

ψ, Θ, φ body / fixed axis system Euler angles

ζ_j all suspension / body axis system parameters (for example, the height of the wheel center)

$\varphi_{R,j}$ wheels' self-rotational angle / fixed axis system j = 1, 2, 3, 4

Z_{CR} steering rack lateral movement / body axis system

As concerns the front axle, the model takes into

account the most widely used suspension / steering technologies, i.e. the Mac-Pherson or parallelogram type, considered to be deformation-free. (Deformation is taken into account in the more complex models with 37 and 51 df).

This being the case, all the kinematic properties as well as the suspension and steering interactions are treated in great detail, as are any anti-dive devices that may be provided.

As concerns the rear axle, all independent suspension technologies - similarly deformation-free - can be represented in detail.

The gyroscopic effects stemming from the wheel's fast specific rotation are also taken into account.

One particularly important connection equation is that expressing, via the final drive and the differential gear, the relation between engine, drive wheels and suspended mechanical parts parameters which, specifically, makes it possible to represent the gyroscopic effects of the engine flywheel on the suspended hardware. The engine may be oriented either transversally or longitudinally.

The steering column is torsionally elastic.

2. Force Systems

Suspension Forces

The elastic forces can be any of the functions of wheel center elongation ; the model can take into account both metal springs and Citroën-type hydro-pneumatic suspensions. Forces due to the stops and sway bars are obviously included.

Damping forces can be any of the functions of ζ_j and ζ_j' and include a Coulomb friction term.

Aerodynamic Forces

The effect of a relative wind of any intensity and horizontal direction on the vehicle is represented by 3 forces and 3 aerodynamic moments (as recorded in a wind tunnel). This is a matter of approximation : the aerodynamic system is in fact computed as though it were stationary.

Forces Applied By the Ground to the Tires

The forces applied by ground to the tire are wholly defined (at least in the first approximation) as long as the state of the tire is given (on a certain ground). A precise definition of the state of a tire would lead beyond the scope of this presentation and involves notions of slip, of camber, etc.

For a series of tire steady states, we measure the 6 components (at wheel center) with a specific testing machine ; moreover, the state of a tire can also be calculated on the basis of wheel position parameters and their derivatives ; thus the 3 forces and 3 moments can be calculated at any time by interpolation.

Tire reactions are thus calculated as though they were a series of steady states, which is obviously only an approximation of a generally unsteady state phenomenon. (This also implies that we overlook the mass of the tire, which is a legitimate assumption).

Motive and Braking Forces

The torque produced by the power train (whether it be driving or resistive) can be either a given function of time or a function of the engine parameters.

With respect to braking, caliper or drum torques are calculated on the basis of main braking system pressure ; this being expressed as a function of time. We take into account the system limiting pressure on the rear. A variety of rear pressure limiting technologies are representable. The calipers and drums can be linked either to the suspended section or the nonsuspended section of the vehicle.

3. Model Inputs

These include :
- the rotation of the steering wheel
- the pressure in the braking system
- the torque of the power train, which data are a function of time ; inputs may also be any combination of the above.

4. Equations

How the mechanical system is represented in mathematical equations is summarized in Appendix 1, describing the Lagrange approach with multipliers and numerical solving.

PART 2 : MODEL UTILIZATION

A few of the different applications of the model, described below, illustrate the usefulness of such simulation.

It is known that the experimental measurement of a vehicle's actual trajectory is a very difficult problem.

Road tests are affected by random factors such as variations in adhesion, uneven ground and wind.

Moreover, uncertainty concerning the starting position and speed of a movement under study is such that the effects on resulting parameters such as trajectory, heading angle, etc. are often greater than that properly attributable to a modification of the vehicle which one wishes to evaluate.

These uncertainties are irreducible.

This being the case, it may be preferable to test the influence of such a design parameter via a simulation based on a sufficiently reliable model.

The Analytical Understanding of Certain Complex Motions

Thus, stepping off the accelerator, or braking in a curve. The movement involved is in fact a dynamic disturbance initiated by the transfer of load from the back to the front of the vehicle and the reversing of the force exercised by the ground on the drive wheels. This disturbance affects all the vehicle's degrees of freedom. The reactions (normal to the plane of the wheels) are modified, generally entailing a tangential and normal acceleration of the center of gravity, as well as an angular acceleration of the body.

The direction of these accelerations can be determined by simple reasoning over a few hundredths of a second at most from the commencement of the movement. Beyond this time, what happens cannot be described by logic and can only be described by simulation.

Second example : the influence of small initial speeds.

This involves the simulation of a movement closely approximating a simple change of lane with a reference vehicle.

The first simulation is performed with this vehicle having an initial speed directed down the center of the lane. Angular heading velocity is zero ; deviation after 3 seconds is 3.35 m.

We then reduce the stiffness of the front sway bar by 25 %. Under the same initial conditions as above, simulated deviation increases to 3.47 m after 3 seconds. When the movement of the reference vehicle is again simulated, this time with an angular heading speed of 0.02 rad/s, we obtain a deviation of 3.48 m.

In other words, the existence of this angular speed is responsible for an additional deviation of 13 cm after 3 seconds - a deviation differing only slightly in the final analysis from that provided by the modification of the sway bar.

In fact, this speed of 0.02 rad/s is rather minimal ; it is comparable to that resulting from minor adjustments made with the steering wheel to maintain the car as close as possible to the direction of entering the lane in the approach phase.

Simulation has thus made it possible to account for the observed difficulty of demontrating the effect of a modification, which is nonetheless significant, to an influencing design parameter during open loop tests. (Appendix 2). Figure 2 illustrates the deviation over time according to the parameter studied.

Dimensioning Computations

Simulation is used to determine the stresses at work in the ground-link components during various movements under extreme conditions : entering a curve, braking in a curve, etc. The calculated forces are applied to models of components in order to determine in terms of finite elements the strains and deformations.

Figure 3 shows the evolution of the forces and moments applied to the offside front wheel in a curve negotiated at 0.7 g.

Axles Optimization

Investigating axle kinematics providing an anti-roll effect.

Such an axle type, for which forces normal to the wheel contribute significantly to the motion of the suspension, can be simulated as it enters a curve and its effects on roll can be evaluated. Moreover, the influence of said axle on other dynamic particulars, in other kinds of movements, such as heading behaviour near a straight line, movement of the suspension proper, etc. can be simulated simultaneously.

The simulation provides a much more thorough solution than in the case of the simple study of "axle balance" due to the dynamic and transient nature of the movement being simulated and to the extent of vertical movement of the center of gravity, generally overlooked in the so-called "axle balance" reasoning. Figure 4 (Appendix 2) shows the roll angle and the movement of the center of gravity associated with different axles.

CONCLUSION

It has been possible, through various checks relating to complex movements to test the model's representiveness and such checks have shown an excellent agreement of parameters such as heading angle and normal and tangential accelerations, albeit a less satisfying agreement for others, such as the angle of pitch.

With respect to such movements, the model has shown an aptitude to predict with great accuracy the influence of the variation of this or that design parameter. This is an extremely important point for an investigative - and therefore a predictive -tool, which by itself justifies its use. All things considered, the main drawback resides in the basically summary way in which the tire model is established (overlooking the unsteady state properties). The repercussions for the simulation of highly unsteady state movements like those mentioned above - braking in a curve, alternating lane change, etc. - are difficult to evaluate. This specific point concerning the tire is currently under study.

This method of simulation, although it is only in its beginning stage of application and thus requires careful and continous comparison with acquired knowledge, is nonetheless already an operational tool.

Appendix 1

Guidelines Concerning Model Equations

Numerical Estimation Method - Computer Time.

The complexity of the motion equations and of the computations required to develop them is such that they would be difficult to include in this presentation. We will therefore limit ourselves to an outline of the procedure used to set up the equations, obtained by the Lagrange multiplier method.

Let $q = (q_i)_{i=1,\dots 37}$ stand for the set of 37 parameters of the 6 solids and of the rack.

Total kinetic energy for the system is expressed as :

$$2T = \sum_{s=1}^{6} 2T_s + 2T_{cr}$$

where $2T_s$ is the kinetic energy of solid s ; and $2T_{cr}$ is the kinetic energy of the rack.

For each solid s :

$$2T_s = \begin{pmatrix} X'_{cr} \\ Y'_{cr} \\ Z'_{cr} \end{pmatrix}^* M_s \begin{pmatrix} X'_s \\ Y'_s \\ Z'_s \end{pmatrix}$$

$$+ \begin{pmatrix} \Psi'_s \\ \Theta'_s \\ \varphi'_s \end{pmatrix}^* H^*(\Psi_s, \Theta_s, \varphi_s).\mathcal{J}_s.H(\Psi_s, \Theta_s, \varphi_s) \begin{pmatrix} \Psi'_s \\ \Theta'_s \\ \varphi'_s \end{pmatrix}$$

where : M_s stands for the mass of solid s

$\mathcal{J}_s$ the inertia matrix relative to the reference axis of solid s.

H is the matrix associated with the rotational vector in this reference axis.

$$2T_{cr} = \begin{pmatrix} X'_{cr} \\ Y'_{cr} \\ Z'_{cr} \end{pmatrix}^* M_{cr} \begin{pmatrix} X'_{cr} \\ Y'_{cr} \\ Z'_{cr} \end{pmatrix}$$

such that :

$$2T = (q')^*.\mathcal{H}(q).(q')$$

$\mathcal{H}(q)$ being a positive symmetric matrix whose coefficients are functions of the 37 parameters of position q_i.

The various relations are expressed in equations:

$$\mathcal{L}_k(q) = 0 \qquad k = 1,\dots 21$$

of which there are 21 in all ; plus a 22nd relation (concerning transmission) between the derivates of the parameters for the body, the engine and the wheels.

$$\sum_{l=1}^{37} \mathcal{V}_{22,l}(q)\, q'_l = 0$$

As such, the Lagrange equations are :

$$\frac{d}{dt}\left(\frac{\partial T}{\partial q'_i}\right) - \frac{\partial T}{\partial q_i} =$$

$$\sum_{\nu=1}^{22} L_\nu \mathcal{V}_{\nu,i}(q) + Q_i(q,q',t)$$

where $\mathcal{V}_{\nu,i}(q) = \frac{\partial \mathcal{L}_\nu}{\partial q_i}$ for $\nu = 1,\dots,22$ and $i = 1,\dots,37$

$Q_i(q,q',t)$ is the coefficient of the virtual work of the given forces (i.e. known in terms of the parameters, their derivatives and time) in the virtual displacement δ_{qi}.

L_ν is the Lagrange multiplier associated with connecting equation ν.

Elimination of the L_ν between the 37 Lagrange equations yields a differential system to the q''_i, $i = 1,\dots,37$ of the following form :

$$\mathcal{A}(q).q''_t = \mathcal{B}(q,q',t).$$

This is numerically solved using a classic, 4th order Runge-Kutta method.

The Fortran program requires :

0.10 seconds/step on a VAX 11/780

0.015 seconds/step on an IBM 3033

0.012 seconds/step on an AMDAHL V8

0.006 seconds/step on a CDC CYBER 176

A time step of 0.01 seconds is suitable in most cases. However, certain movements involving quick accelerations of the wheels require steps lasting less than 0.005 seconds.

APPENDIX 2

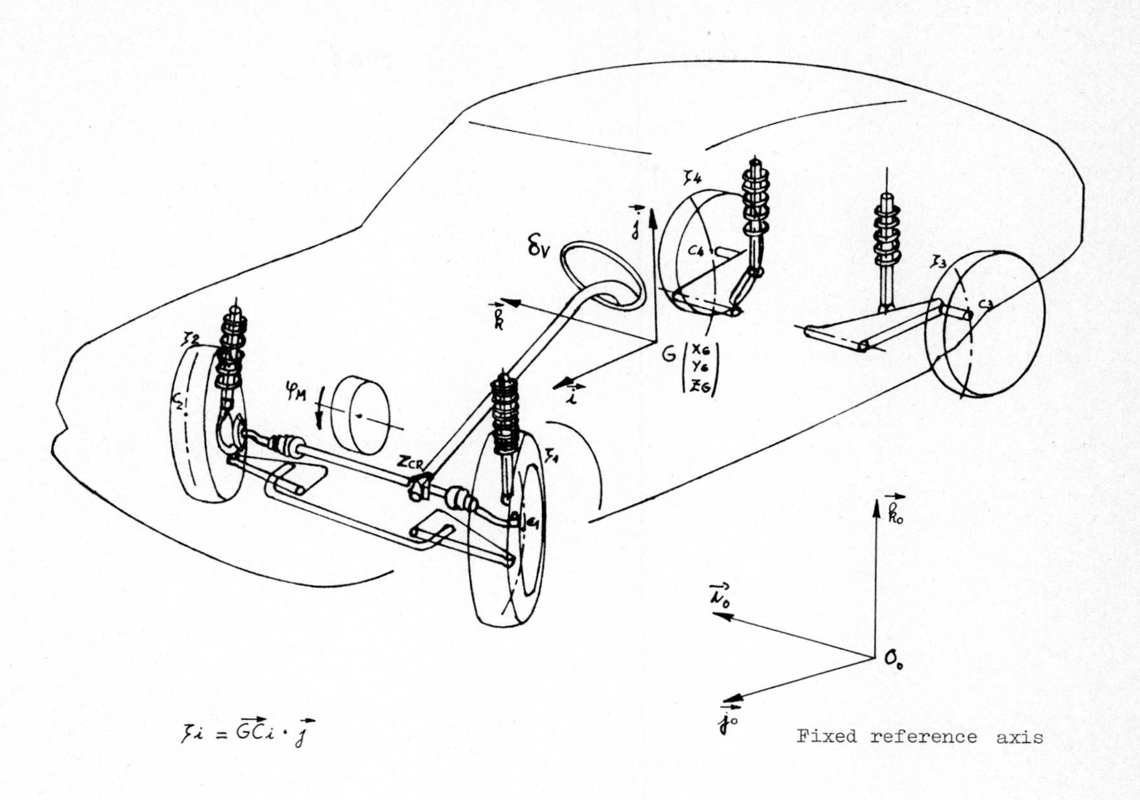

Euler angles defined

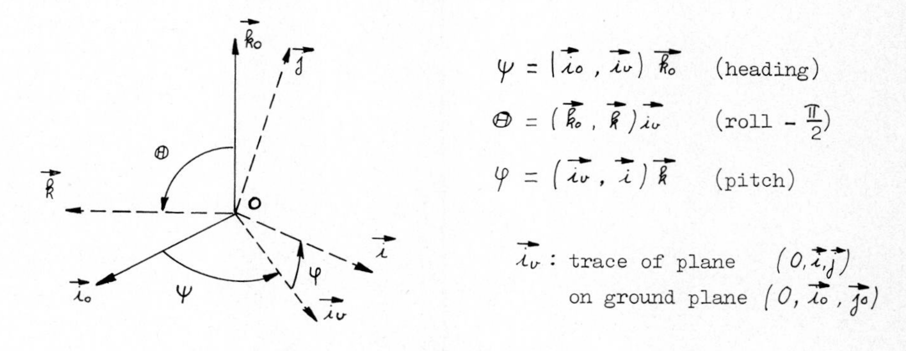

Figure 1 - Position Parameters

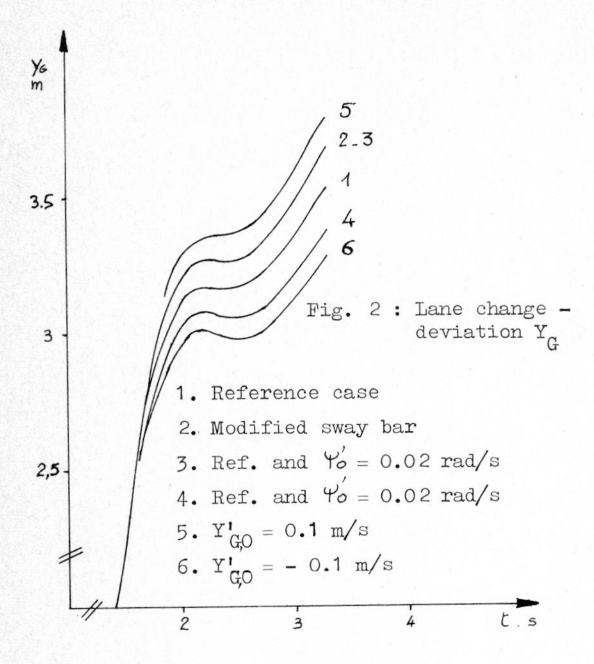

Fig. 2 : Lane change - deviation Y_G

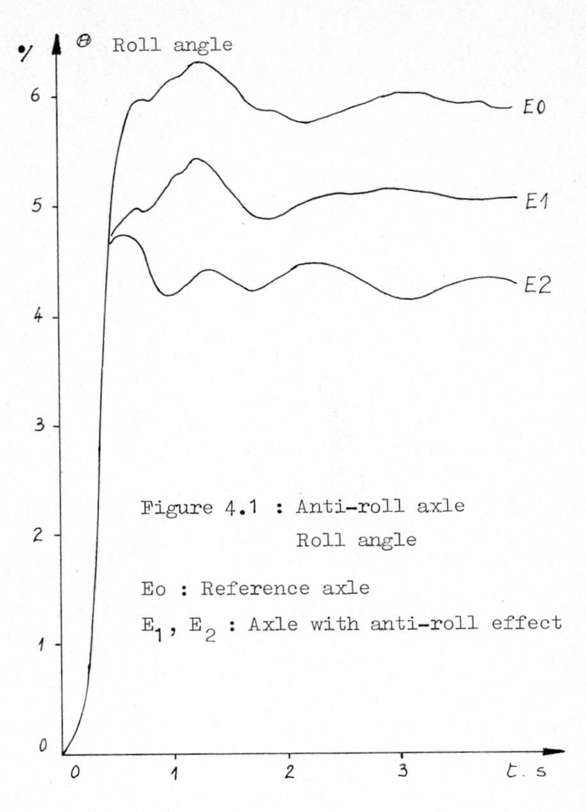

Figure 4.1 : Anti-roll axle
Roll angle

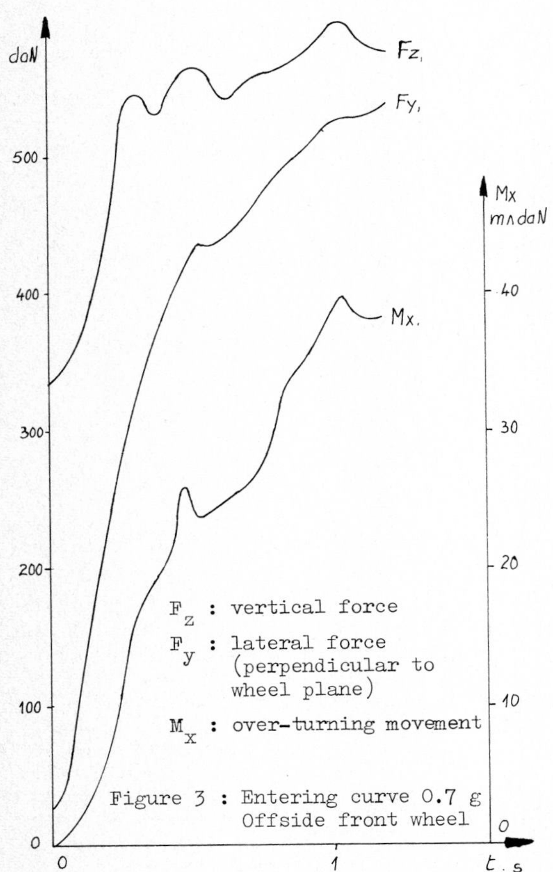

Figure 3 : Entering curve 0.7 g
Offside front wheel

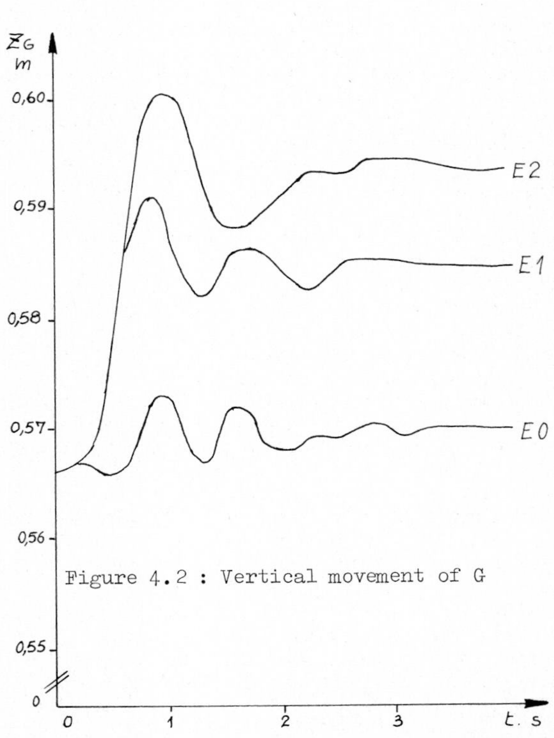

Figure 4.2 : Vertical movement of G

Simulation in Engineering Sciences
J. Burger and Y. Jarny (eds.)
Elsevier Science Publishers B.V. (North-Holland)

MODULAR PROGRAMMING STRUCTURE APPLIED TO THE SIMULATION OF NON-LINEAR AIRCRAFT MODELS

J.A. Hoogstraten and G.A.J. van de Moesdijk

Delft University of Technology
Department of Aerospace Engineering
Delft, The Netherlands

In developing a simulation, the actual system is usually modelled in terms of a mathematical formulation, which constitutes the basis for subsequent development of the simulation program model. This paper discusses the modular programming concept as applied to the implementation of a non-linear aircraft model on a digital computer. The modular simulation program structure provides a generally applicable program body, valid in principle for any type of aircraft. This is illustrated by the application of one program-package to the simulation of three different types of aircraft. Furthermore, the modular programming concept offers advantages in terms of portability in host/target application and easy access to subsystem modules.

INTRODUCTION

A popular statement about programming is: "Programming is an art, not a science". Although many aspects of translating a problem into machine-interpretable language tend to support this vision, the authors feel that, in particular where simulation programs are concerned, too often this is used as an excuse to justify program packages which are constructed in a fashion that is both haphazard and confusing to the outsider (meaning: Anyone except the programmer).

We think the excuse is not valid and is primarily meant to obscure the fact that, for reasons of expedience, construction of the program is usually begun before a clear-cut structure for the implementation of the mathematical model has been established. This is a common pitfall, and frequently the expedience gained in an early stage of the development process backfires upon the programmer, especially where large-scale complex systems are concerned. In further development or extension of the original program, ad hoc solutions and contributions from other programmers will eventually ensure total inaccessability.

The remedy seems to be obvious. Before initiation of program development, a basic structure to accommodate the mathematical model must be defined. During program development the programmer(s) must rigorously keep to the original scheme. Modifications are allowed only where these can be supported throughout the entire program.

Although basically simple, the scheme is easier stated than conformed to. Practice indicates that it is extremely tempting to rush those "last" statements that will yield a bug-free program into (the wrong) place. Besides, to maintain the desired versatility, the basic structure indicated above must meet severe demands.

First of all the basic structure must comprise the original numerical solutions to the problem. Also the structure should allow the inevitable extensions, redefinitions and more accurate versions of the original implementation to be accepted without undue programming effort.

The paper discusses a stepwise approach to the problem of constructing a basic structure. First, the differences will be emphasized between physical system and mathematical model, and, less obvious, between the mathematical model and its implementation. Focussing on the implementation, a structure for the mathematical model implementation of an aircraft will be treated. Subsequently the resulting mathematical model structure will be transformed into a transparant program structure. Finally, some examples of the concept will be given with respect to the simulation of aircraft and of some applications.

SIMULATION PROCESS IMPLEMENTATION

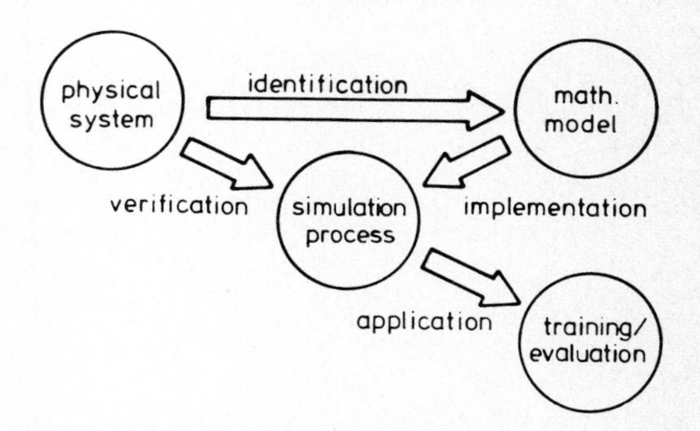

Fig. 1. Schematic survey of simulation development process.

In principle the applicability of simulation as an engineering tool, and the validity of simulation results, are limited only through the facilities available and the engineer's knowledge of the physical system. Consider fig. 1 which depicts the procedure of developing a simulation process. This figure illustrates that a mathematical model is formulated through mathematical/ physical and empirical analysis of the physical system, and through subsequent parameter identification. The process of implementation consists of consideration and reformulation of the mathematical model in terms of hardware and/or software. Verification of the resulting simulation process usually is based on qualitative and quantitative comparison with actual system behaviour.

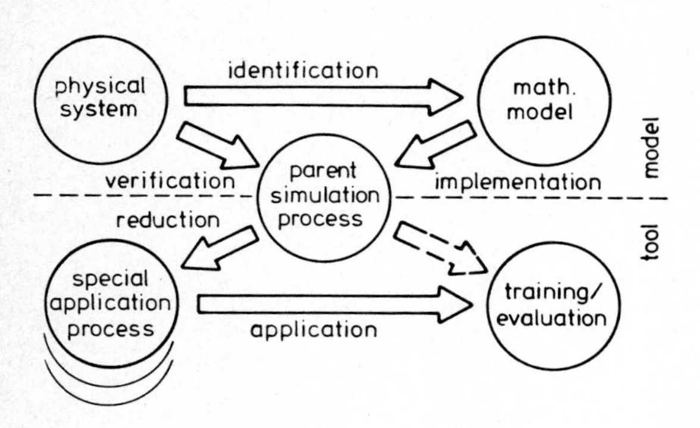

Fig. 2. Schematic survey of multiple-application simulation development.

When dealing with a variety of applications of essentially a single physical system, a natural extension is to insert an extra step in the development procedure (see fig. 2). This extra step entails the implementation of a single large-scale off-line simulation process, the parent simulation program, on suitable facilities (host environment). By virtue of its nature, the parent simulation process will both be complex and expansive. Large digital computers provide a suitable host environment to accommodate such processes and maintain flexibility. Special application processes, possibly using separate (target) facilities, may be derived from this large scale implementation and can easily be validated against the parent process.

Clearly the feasibility of the concept will stand or fall with the clarity of the parent simulation program structure. The modular programming concept applies a structure that is closely related to the physical system in terms of its constituent modules. Modular programming involves a strict definition of subsystem boundaries, and thus of program module input/output, prior to program implementation.

MATHEMATICAL MODEL STRUCTURE

Consider the aircraft (see fig. 3a) and in particular aircraft motion. The complexity of the mathematical model describing the aircraft motion will depend on the definition of the system boundaries (defining the scope of the simulation process, as well as the amount of detail included). A very restricted version is to consider the aircraft as a point of mass travelling through a conservative gravitational force field under the influence of a set of forces that are directly related to the aircraft's attitude and speed (see fig. 3b). Such a model is very well suited for computations with respect to aircraft performance. More sophisticated, mass distribution, forces' points of application and stabilizing moments may be introduced (see fig. 3c). A mathematical model covering the dynamic and kinematic equations of aircraft center of gravity motion is constructed relatively easy applying

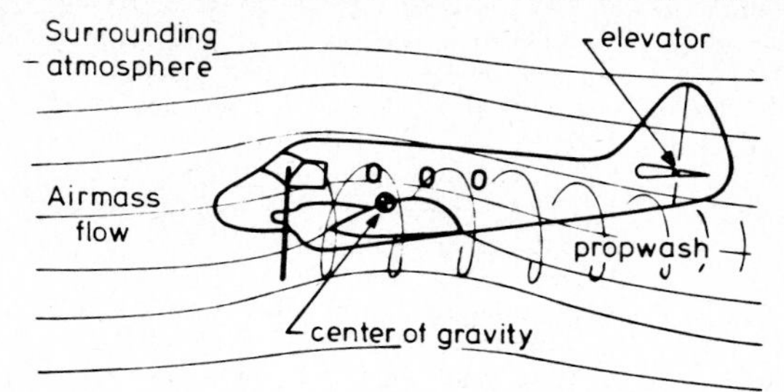

Fig. 3a. The aircraft in subsonic flight through the unperturbed atmosphere.

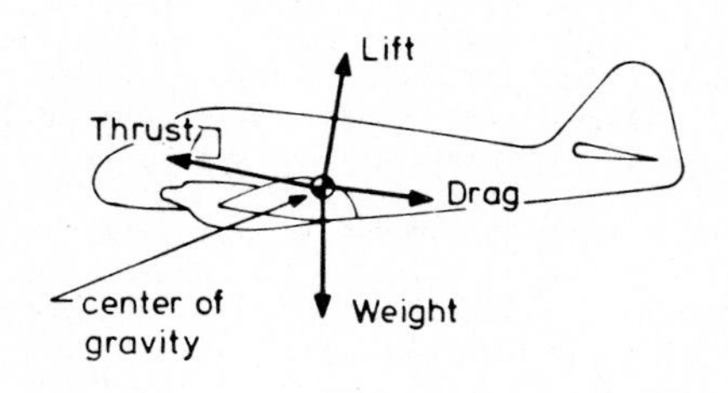

Fig. 3b. Basic aircraft performance model.

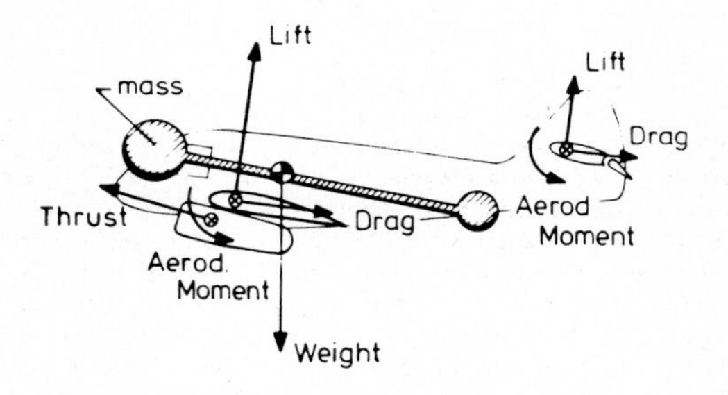

Fig. 3c. Basic aircraft motion model.

Newton's laws or Lagrange's theorem (see ref. 1, 2). Considering an n-dimensional set of variables x (and their time derivatives $\dot{x}$) related to the n translational and rotational degrees of freedom, a set of non-linear first-order differential equations may be specified:

$$\dot{x}(t) = f(\dot{x}(t), x(t), t) \qquad (1)$$

Equation (1) expresses the basic result that if all the coordinates x and velocities $\dot{x}$ are given at some instant in time, the accelerations $\ddot{x}$ at that instant and higher derivatives, are uniquely defined. Thus, equation (1) gives a closed mathematical formulation for the mechanical state and its evolution in time for any physical system.

Considering also a separate set of input signals, equation (1) may be rewritten as:

$$\dot{x}(t) = f(\dot{x}(t), x(t), u(t), t) \qquad (2)$$

Note that although for the physical system the variables $u(t)$ may take the form of control lever angles, pedal displacements, turbulence velocities etc., with respect to mechanics they always amount to a set of external forces and moments.

In representing the system through eq. (2), system boundaries have been defined so as to include motion of the aircraft center of gravity as a result of external inputs only. Although such a model may have its merits for some applications, further extension of system boundaries may be required. Considering for instance the generation of thrust, the mathematical model may further be extended to include the engine dynamics:

$$\dot{x}_e(t) = f_e\left(\dot{x}_e(t),\ x_e(t),\ u_e(t),\ t\right) \tag{3}$$

The inputs $u(t)$, viewed as forces and moments to the previously defined mathematical model (eq. (2)) result from contribution due to engine operation, and from external inputs $u_{ext}(t)$:

$$u(t) = g\left(\dot{x}_e(t),\ x_e(t),\ t\right) + u_{ext}(t) \tag{4}$$

Contributions due to engine operation may include magnitude and orientation of the thrust vector, gyroscopic moments etc. Conversely, engine operation will depend on parameters embodied in the model of eq. (2), and external inputs:

$$u_e(t) = g_e\left(\dot{x}(t),\ x(t),\ t\right) + u_{ext}(t) \tag{5}$$

Contribution to engine input due to aircraft motion may for instance result from flight altitude, effective air intake frontal area etc. External inputs include throttle position, surrounding atmosphere moisture etc.

Note that in eqs. (4) (5), "internal" inputs have been separated to result explicitly from engine operation resp. aircraft motion. Restriction to explicit expressions can usually be sustained in view of the previous selection of system boundaries (and physical manifestation of inputs as forces and momentes).

When considering the augmented system model in more detail, subsystem models may be seen to be interrelated through a set of algebraic relations:

$$0 = h\left(\dot{x}(t),\ x(t),\ \dot{x}_e(t),\ x_e(t),\ u_{ext}(t),\right) \tag{6}$$

(cross-coupling effects). Such relations may occur parallel to the previously defined inputs (see fig. 4b). As an alternative to fig. 4b, the structure of the augmented mathematical model may be arrayed as in fig. 4c. In fig. 4c, the explicit inter-relations (eqs. (4), (5)) are accommodated on off-diagonal elements, dynamic and algebraic relations (eqs. (2), (3) resp. eq. (6)) reside on diagonal elements.

Transformation of external input signals is depicted in a separate block to the right of the main structure. Interrelations between subdivisions of the main structure of fig. 4c occur as an anti-clockwise sequence, indicated by arrows (compare fig. 4b).

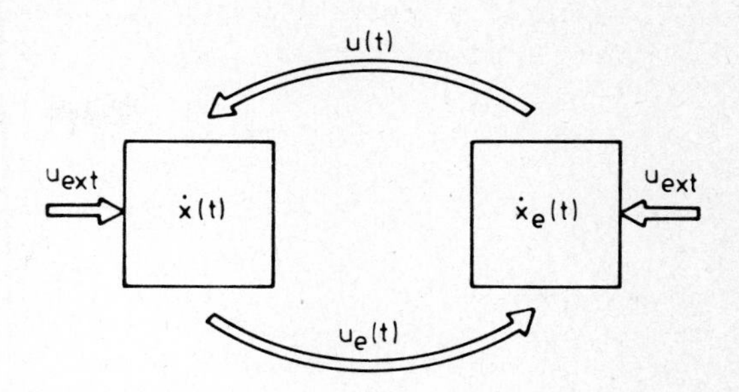

Fig. 4a. Dynamic system model augmentation.

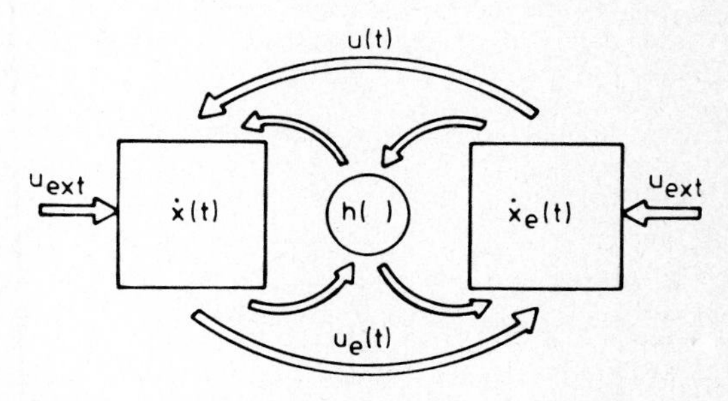

Fig. 4b. Dynamic system model interrelation.

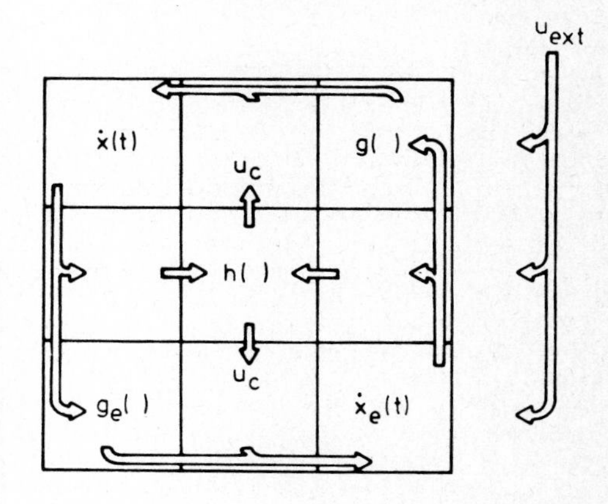

Fig. 4c. Schematic representation of system model interrelation.

Simulation of the augmented system model requires simultaneous solution of eqs. (2), (3):

$$x(t_1) = x(t_0) + \int_{t_0}^{t_1} \dot{x}(\tau)\ d\tau \tag{7}$$

$$x_e(t_1) = x_e(t_0) + \int_{t_0}^{t_1} \dot{x}_e(\tau)\ d\tau \tag{8}$$

and of equation (6):

$$u_c(t_0)=g_c\left(\dot{x}(t_0),x(t_0),\dot{x}_e(t_0),x_e(t_0),u_{ext}(t_0)\right) \tag{9}$$

(where contributions in eqs. (2), (3) due to cross-coupling have been indicated by u_c). Note that solution of the set of equations (2), (3), (4), (5), (6) through eqs. (7), (8), (9) for given $\dot{x}(t_k)$, $\dot{x}_e(t_k)$, $x(t_k)$, $x_e(t_k)$ and inputs $u_{ext}(t_k)$, does not yield $\dot{x}(t_{k+1})$. In implementing this solution on a digital computer a principal problem therefore arises in supplying a correct estimate of $\tilde{\dot{x}}(t_k,t_{k+1})$ for each computation cycle k (resulting in a solution $x(t_{k+1})$). This may be illustrated in fig. 4c by following the computational sequence indicated by the sequence of arrows, for given $\dot{x}(t_0)$, $\dot{x}_e(t_0)$, $x(t_0)$, $x_e(t_0)$ and external inputs $u_{ext}(t_i)$, i = 0, 1, 2, ... In fig. 4c. return to the starting point, which may in principle be any single subdivision, or to subdivisions already encountered, indicates closing a computational loop. Reversal in computational (arrow) sequence thus indicates creating an inner loop.

Inner loops may be avoided by providing a numerical estimate for $\tilde{\dot{x}}(t_k,t_{k+1})$, $\tilde{\dot{x}}_e(t_k,t_{k+1})$. Approximation of $\tilde{\dot{x}}(t_k,t_{k+1})$, $\dot{x}(t_k,t_{k+1})$ by $\dot{x}(t_k)$, $\dot{x}_e(t_k)$ introduces basic inaccuracies, though their effect may be alleviated by applying sufficiently high sample frequencies. Alternatively, $\tilde{\dot{x}}(t_k,t_{k+1})$, $\tilde{\dot{x}}_e(t_k,t_{k+1})$ may be acquired through iterative methods, involving recurrent solution of eqs. (7), (8), (9) for each cycle (see for instance ref. 3), thus creating an extra outer loop.

Numerical solutions as indicated above are necessitated by derivatives $\dot{x}$, $\dot{x}_e$ occurring implicitly in eqs. (2), (3), (4), (5), (6). The problem may also be tackled analytically. An explicit form for the expressions describing the dynamic characteristics of the (sub-) system (eqs. (2), (3)):

$$\dot{x}(t) = f(x(t),\ u(t),\ \dot{u}(t),\ t) \tag{10}$$

can be found through laborious substitution. Alternatively, proper choice of the state variables, or appropriate transformation (see ref. 2) may yield suitable expressions in the form of eq. (10). Ref. 4 describes a method to arrive at this result applying bond-graph techniques. Substitution of eq. (10) in eqs. (4), (5), (6) yields explicit relations containing as inputs only the state and input variables.

Note that this process of analytic substitution cancels the necessity for strictly adhering to the previously defined structure and computational sequence of fig. 4c. Its use is however advocated since it provides a very useful tool for detecting inadvertent inner loops. Furthermore, the structure of fig. 4c provides a uniform and transparent representation for further model augmentation, whether this involves implicit relations or not. This may further be illustrated with a model for the airflow around the aircraft, and its effect upon the previously defined models.

Direction and magnitude of forces and moments acting upon the aircraft may be related to aircraft altitude and motion through a more or less exhaustive model of the aerodynamics. This would yield a general dynamic model of the form:

$$\dot{x}_a(t) = f_a(\dot{x}_a(t),\ x_a(t),\ u_a(t),\ t) \tag{11}$$

where $u_a(t)$ specifies aircraft attitude and motion (specified through $x(t)$ and $\dot{x}(t)$), or concurrently, direction and magnitude of the unperturbed airflow. Also, the effects of engine operation, such as propeller wash or turbine exhaust flow may be included in $u_a(t)$.

Unfortunately, theory of aerodynamics does not allow construction of a comprehensive general model for the airflow around the aircraft. Simplified aerodynamic models are based on flight- or windtunnel-test results, and usually describe the airflow as a stationary or quasi-stationary process, so

$$u(t) = g(\dot{x}_e(t),\ x_e(t),\ u_a(t),\ t) + u_{ext}(t) \tag{12}$$

$$u_e(t) = g(\dot{x}\ (t),\ x(t),\ u_a(t),\ t) + u_{ext}(t) \tag{13}$$

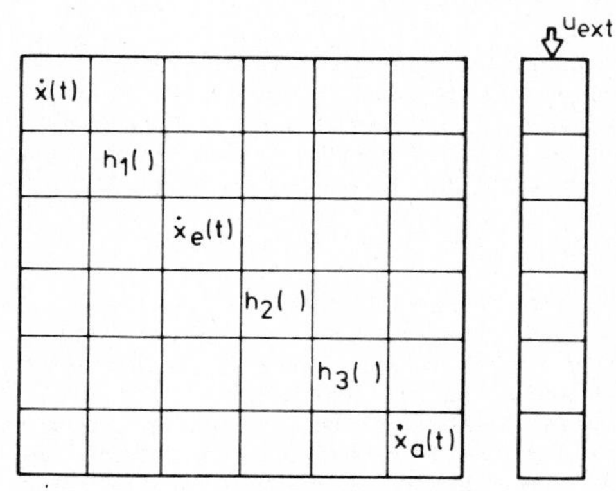

Fig. 5a. System containing three subsystems.

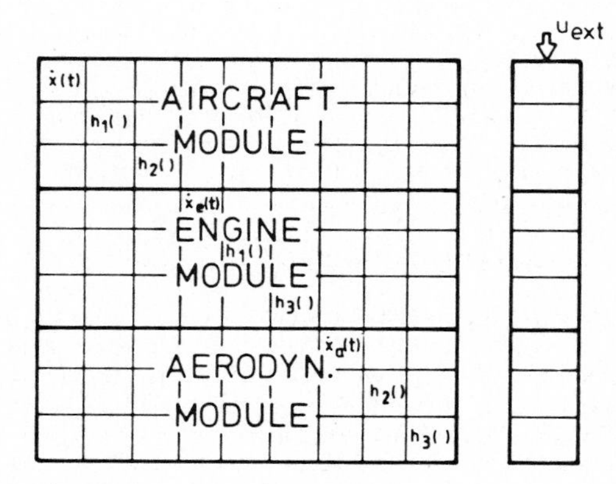

Fig. 5b. Modular structure.

Fig. 5a depicts, in a format analogous to fig. 4c, the structure of the resulting augmented mathematical model. External inputs have been augmented to contain inputs peculiar to the model of the aerodynamics. For a quasi-stationary or stationary model, the bottom diagonal element is a dummy subdivision.

Note that depending on the mathematical model used in extending the original model, dummy subdivisions may be created elsewhere also.

To reduce the number of subdivisions, the representation may be reorganized to yield fig. 5b. Note that, where coupling terms are involved that contribute to several models simultaneously, this will lead to identical computations occurring in multiple modules. It will be clear that system boundaries may be extended further, finally to enclose all relevant effects or subsystems occurring in the actual system. For the aircraft, the control surfaces together with their actuators, the landing gear, elasticity of the aircraft body, airmass flow perturbations, the condition of the surrounding atmosphere etc. may be modelled. Mathematical model structure may thus be represented in a table of growing dimensions. Essentially, the operation of extending the system boundaries shows a close correspondence to the process of augmenting the state-space representation in linear system theory. Note however that the "system matrix" in the modular programming concept contains modules for non-dynamical or interrelating (cross-coupling) subsystems, yielding a basically different representation. Apart from this, note that system augmentation may be carried out in scope as well as in depth, meaning that modules may be replaced in the overall structure and represented in more detail separately.

PROGRAMMING CONSIDERATIONS

This far, the construction of the mathematical model for the entire aircraft has been little more than a mental exercise. Model extensions have been indicated, but no exact mathematical description has been given or required. However, we have come across a systematic representation that is suitable to accept mathematical descriptions of varying depth (for each extension of the system boundaries) and widening scope (for further extensions). Each extension has been shown to satisfy the general relation:

$$\dot{x}(t) = f\big(\dot{x}(t),\ x(t),\ u(t),\ t\big) \qquad (2)$$

The simulation process consists of recurrent solution of this set of relations in the form (2) or, concurrently, of the augmented system:

$$\dot{x}'(t) = f\big(\dot{x}'(t),\ x'(t),\ u'(t),\ t\big) \qquad (14)$$

where x' and $\dot{x}'$ are vectors containing as subsets the state variables x and their time derivatives $\dot{x}$ of each subsystem. Furthermore u' is a vector that contains all inputs u that are not given through x' and $\dot{x}'$, and thus equals the external input signals u_{ext} denoted in the previous section.

The system may be simulated either by direct solution of eq. (14), involving creation of extra loops, or through solution of the explicit form:

$$\dot{x}'(t) = f'\big(x'(t),\ u'(t),\ \dot{u}'(t),\ t\big) \qquad (15)$$

which may be found through analytical substitution. The resulting simulation program, embodying either method, serves as a parent program for special application processes. Since applications in real-time simulation demand minimum processing time, the parent program should be tuned to this requirement. Consequently, explicit relations in the form (15) will be favoured for implementation.

For the solution of a set of non-linear explicit first-order differential equations various well-known numerical methods, based on integration, exist.

A general integration process may now be defined for the augmented state vector time derivative $\dot{x}'(t_k)$, yielding a solution $x'(t_{k+1})$ for given $x'(t_k)$, $u'(t)$. In the process the total input $u'(t_k)$ may be further distinguished as to consist of deterministic inputs i.e. pilot control inputs, constant wind speeds etc., and perturbations like gust velocities.

To compute $\dot{x}'(t_k)$ from $x'(t_k)$ and $u'(t_k)$ it will suffice merely to substitute the magnitudes for the latter vectors of variables in the mathematical model. Implementation of the previous section will yield several program modules, each with a distinct function. Of course the number of resulting modules may be reduced by grouping together, rowwise or columnwise, a set of modules. No definite preference for one or the other is clear, though practice suggests rowwise combination is easier to grasp since it leads to a module that combines a host of inputs into a restricted number of well-defined outputs. For dynamic systems, this output will consist of the subsystem state time derivative, for proportional systems it will consist of the input vector of (a) subsequent module(s). Also, the input to each module will consist of the state vector $x'(t_k)$ of the complete model, together with the input vector $u'(t_k)$, and, in some cases, the output of a proportional system (compare the engine model cross-coupling discussed in a previous section). Thus, and this is a main point, the only variables appearing in the argument list of the program modules are x', $\dot{x}'$, u' and some specific inputs u that can directly be retraced in the modular program structure table. All other variables appearing in this main data flow are programmatic shortcuts (to chaos).

There is one possible exception to this maxim. In the program some auxiliary variables may have specific physical meaning and may be of interest to the user. These are stored in the system output vector y:

$$y(t) = g\big(\dot{x}'(t),\ x'(t),\ u'(t),\ t\big) \qquad (16)$$

As already indicated above, y may also contain the higher order derivatives of the state $\ddot{x}'$, $\dddot{x}'$, etc. which may be found directly from $\dot{x}'(t)$, $x'(t)$, $u'(t)$ for any t, thus satisfying (16). Note that the vectors y as well as $\ddot{x}'$ can only appear in the output of a module, and may not take part in the dynamical simulation process. Whenever peculiar things are happening in your program you probably conveniently overlooked this when drafting the program source.

AIRCRAFT SIMULATION PROGRAM PACKAGE STRUCTURE

The resulting structure of a modular program package for the simulation of aircraft is depicted in fig. 6. The program contains five levels. The top, or main program level is a user-oriented structure containing user-specified information concerning the subroutines to be executed, background memory to be allocated etc. (job control) and the input data required for the simulation. Actually, a superior level is available to prepare and reorder this information in an interactive fashion (interactive job preparation program). The second level comprises the initial condition computation and the actual simulation process. Furthermore, two additional parts, at the start respectively at the end of the cycle, are included where data are read or stored, and where convenient conversions of data can be performed.

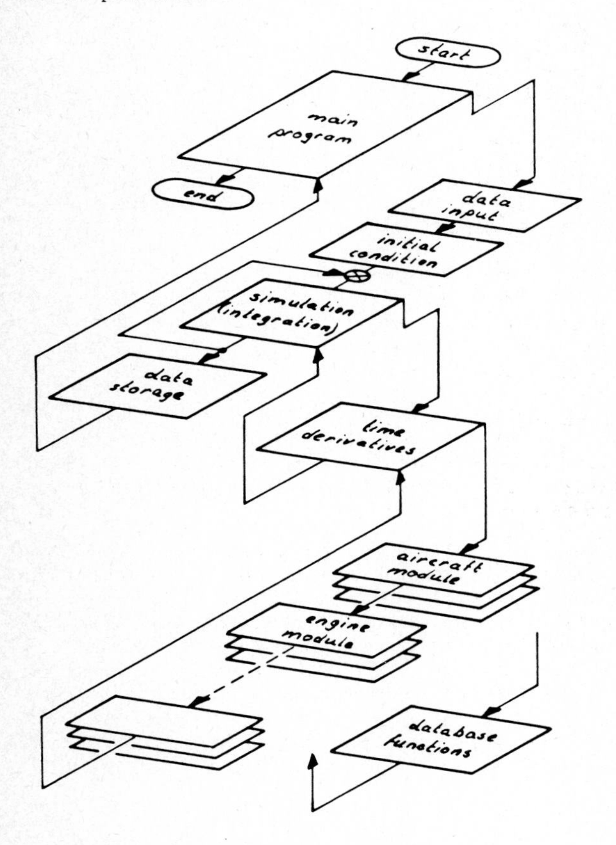

Fig. 6. Schematic view of modular aircraft simulation program package.

Proceeding downward, the third level contains the computation of the state vector time derivative and output vector of the complete augmented system. At this level, the total result of input signals $u'(t_k)$ and state $x(t_k)$ is computed from the various contributions computed separately in the lower level subroutines. These lower-level (level 4) program modules thus contain the actual (augmented) aircraft model, all other program parts mentioned before are fully aircraft-type independent. Level four program modules have access to a database of reference data containing relevant tabular functions or coefficients of analytical functions. To this end, a separate set of level 5 subroutines is available for data acquisition, table interpolation etc.

EXAMPLE I: IMPLEMENTATION OF AN AERODYNAMIC MODEL

The aircraft can be considered as a relatively small structure moving through the atmosphere under the influence of a set of external forces and moments. The mathematical formulation of the relation between aircraft state and external aerodynamic forces and moments is usually indicated as the aerodynamic model. Since the motion of coherent systems of particles can be expressed in generalized form, the mathematical models for different types of aircraft are primarily distinguished through their respective aerodynamic models.

In the non-linear simulation program, three different aerodynamic models (Boeing B747, De Havilland DHC-2 "Beaver", and Fokker F28 Mk 4000) are available.

The aerodynamic data used to construct the Fokker F28 aerodynamic model originally were given in the form of graphs. This necessitated conversion to programmable form, in this case a set of separately stored tabular representations. Intermediate points are found by numerical interpolation.

The model resulting from implementation in the parent simulation program was validated by comparison of actual and simulated time-histories. Also, the model was linearized for a set of reference conditions using a numerical linearization procedure (the mathematical concept of this procedure is treated elsewhere on this conference, ref 5). Besides yielding a cross-reference to the original graphical representation for an extensive set of reference conditions, the linear system representations allowed further simplification of the aerodynamic model through the elimination of minor effects.

Since the aerodynamic models are contained in separate program modules, these can be copied into target applications with relative ease, especially for target applications with a similar modular structure. Also, updates of the aerodynamic model may be put into effect by simply updating separately stored coefficients or by exchanging a single module.

EXAMPLE II: EVALUATION OF AN AUTOMATIC FLIGHT CONTROL SYSTEM

In the course of development of a digital flight control system, various control laws were designed using linear control systems theory.

Basically this involves the assumption of approximate linearity of the aircraft model for small deviations with respect to a chosen steady state of reference, and subsequent optimization of the controller for the resulting linear(-ized) system.

The non-linear and time-dependent effects which are ignored or approximated in the design process of the controller were further evaluated through non-linear simulation of the aircraft.

To stabilize the aircraft with respect to deviations from a desired flight altitude, a multivariable control law was designed for the DHC-2 Beaver experimental aircraft. Implementation of this "discrete-time optimal output feedback regulator" in the non-linear aircraft simulation program yielded a satisfactory improvement of the dynamical behaviour of the aircraft (see fig. 7). Results will eventually be verified in flight tests of the digitally controlled aircraft.

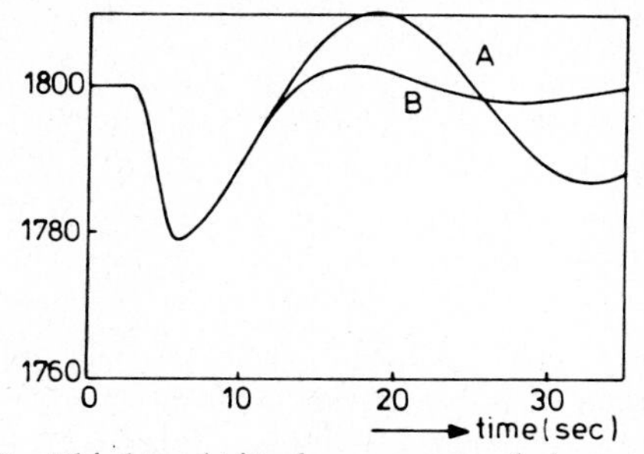

Fig. 7. Flight altitude response (m) on a "double-trapezoid" type deflection of the elevator.
A - uncontrolled aircraft
B - aircraft with altitude-hold controller.

Flight control laws will be implemented in an on-board micro-computer, controlling a set of hydraulic actuators linked to the aircraft control surfaces. Simulation of the micro-computer is contained in a separate program module, thus facilitating exchange of control laws. Control law implementation in the program module can be debugged separately before either simulation runs are performed, or micro-computer implementation is initiated.

EXAMPLE III: SOLUTION TO AN IMPLEMENTATION PROBLEM

On-line simulation of the Boeing B747 aircraft required implementation of a model for the aircraft landing gear. During simulation runs on the hybrid flight simulator equipment, peculiar effects were encountered when the aircraft was brought to a complete stop on the runway. The effect was most noticeable in an oscillating vertical motion of periodically changing amplitude. Taking the parent program landing gear module and submitting it to a series of linearizations for varying reference conditions (see ref. 5), the origin of the phenomenon could immediately be indicated.

Linearizing the landing gear strut model for varying strut compressions and compressing speeds yields the result given in fig. 8. From this figure it will be clear that in static load, the damping of the system approximates zero, yielding an indifferent system. Thus, small numerical inaccuracies or residuals will introduce a non-damped oscillatory motion, which is diminished again upon reaching larger amplitudes, where damping is reintroduced in the system (see fig. 8). The effect was eliminated by introducing a slightly non-zero gradient at static load in the strut simulation module.

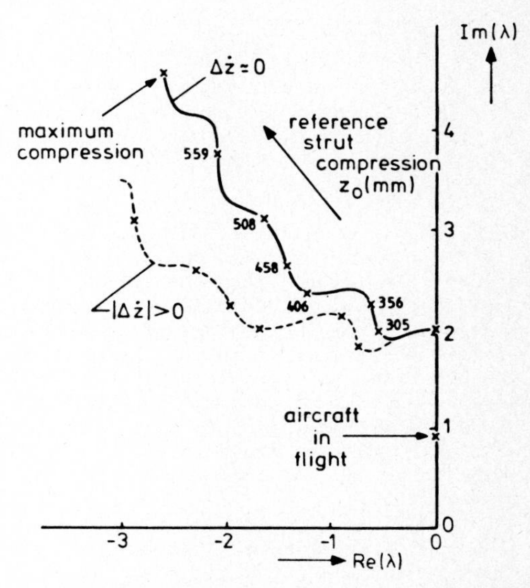

Fig. 8. Root-locus for linearized nose-gear model for varying steady state of reference z_o ($\dot{z}_o = 0$).

The example serves to illustrate both piecewise evaluation of subsystems and host/target application. The landing gear model could easily be evaluated since it could be isolated from the parent main program without any reprogramming. Also, evaluation was expedited by exploiting the host computer's facilities.

EXAMPLE IV: ON-LINE SIMULATION OF AIRCRAFT ELASTICITY

A very noticeable development in aviation is the increase in aircraft dimensions. As a consequence, dynamic aeroelasticity has become more and more interconnected with the stability of the aircraft. In most present simulation models quasi-static corrections to the aerodynamic data are inserted to account for these effects.

In on-line flight simulation this approximation tends to exhibit itself as a lack of realism, caused by the absence of specific forces and accelerations resulting from dynamic intercoupling between rigid body and flexible modes (see fig. 9, depicting vertical acceleration at the pilot's position as a result of pilot input).

To evaluate these effects, a mathematical model for the aeroelasticity of the Boeing B747 aircraft was developed and implemented in

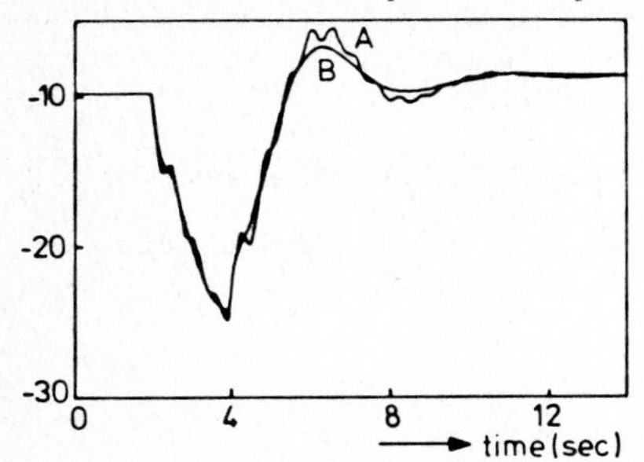

Fig. 9. Specific force at pilot's position (m/s^2) resulting from a "block"-type elevator deflection.
A - flexible aircraft.
B - rigid aircraft.

the parent simulation program, the Amdahl 470/V7 computer acting as the host. Subsequently, a model for the flexible modes, introducing eight degrees of freedom, was implemented in an on-line flight-simulation program running on a hybrid flight simulation installation. As a result of the computational requirements involved with the computation of these high-frequency modes, the program modules containing the aero-elastic model were implemented on the analog part of the target installation.

Although a development towards very flexible airframe structures with automatic control compensation can be envisaged (CCV, Control Configured Vehicles), simulation of aero-elasticity currently is of interest mainly for on-line applications. Nevertheless, construction of an off-line simulation module in the parent main progam greatly facilitated development of the mathematical concept and, finally, on a, rather restricted, on-line installation.

CONCLUSIONS

Aircraft simulation can range widely in the mathematical accuracy and/or the amount of side-effects that is required for particular applications. To reduce program development effort and cost for each application, an extensive main program, running on a large digital computer, was implemented. This parent main program can be used for further refinement of the overall mathematical model, both through more accurate modelling of subsystems, and by extending the number of subsystem models already included. Implementation of the refined mathematical model is facilitated by extensive host computer facilities and by a flexible structure of the parent simulation program.

The non-linear simulation program discussed was developed in a research and educational environment. Contributions were largely made by students with little or no previous experience in programming. Coordination of program development effort therefore proved to be a main point of concern and benefitted greatly from a clear definition of the program structure.

It should be remarked that the modular progamming concept results in slight increases in computation time and required storage, but transparency of the program structure and piecewise evaluation of system modules tend to reduce program developmend effort.

Versatility of the resulting program package was demonstrated at various occasions. Some examples have been given. Work on the program package is continuing.

REFERENCES

1. Etkin, B.: "Dynamics of atmospheric flight", John Wiley and Sons Inc., 1971.
2. Landau, L.D., Lifshitz, E.M.: "Mechanics", Pergamon Press, 1960.
3. Gear, C.W.: "Simultaneous numerical solution of differential-algebraic equations", Transactions on Circuit Theory, Vol. 18, 1971, 89-95.
4. Rosenberg, R.C.: "State-space formulation for bond-graph models of multiport systems", Journal of dynamic systems, measurements and control, Transactions of the ASME, Serie G, Vol. 93, 1971, 35-40.
5. Den Hollander, J.G., Hoogstraten, J.A., Van de Moesdijk, G.A.J.: "Linear approximation of non-linear systems based on Least Squares methods", IMACS symposium on Simulation in Engineering Sciences, May 9-11, 1983, Nantes, France.

XV. MANIPULATORS – ROBOTS (2)

Simulation in Engineering Sciences
J. Burger and Y. Jarny (eds.)
Elsevier Science Publishers B.V. (North-Holland)

VALIDATION OF AN ADAPTIVE ROBOT CONTROL STRUCTURE BY MEANS OF SIMULATION

R. Chaudet, Thomson CSF, Grenoble, France
J. O'Shea, Ecole Polytechnique, Montreal, Canada

A design objective being to control a three degree of freedom robot by means of microprocessors, for generating an optimal trajectory and for servoing each link, a control strategy was investigated. A study suggested an auto-adaptative structure with feedforward which naturally lends itself to a distributed system. However, the effects of the sensitivity to parameter variations had to be determined and the performances relative to other types of control systems suggested in the literature had to be compared. The great variability of the torques due to gravity, to the centrifugal and Coriolis forces, besides inertial complings, pointed out the simulation approach to be the best to reach such conclusions.

Simulation has been done twice in each case. The first time, assuming exact values for the parameters and, the second time, letting an error in the parameter estimation of $\sigma = 0.1$ in order to test the robustness of the control system being considered. Thus, it was found that the transient error is superior when using the system proposed by the authors but its response is as short and there is no steady state error. Simulation has therefore shown that the adaptive structure proposed by the authors, requiring only a small size memory and a limited real-time computational capability, constitutes an advantageous solution for robot control.

INTRODUCTION

Having as a goal to minimize the cost of a robot control system while keeping good performances, a search was done for control strategies which are not too demanding computer-wise.

To this end, many previously proposed solutions were considered. Thus, it readily appeared that the solutions which call for large size memories like the ones proposed by Raibert [1] and Khalil [2] as well as those imposing heavy real-time computational load to the CPU like the state feedback with quadratic optimal criterion method of Turner [3] or that of Gruver [4], had to be eleminated. Moreover, algorithms which do not allow for the robot to follow a time varying trajectory like the one proposed by Saridis [5] were not considered.

The preference was given to a distributed control structure with parallel processing using three and possibly four micro-computers without bulk memory. This permitted to adopt a control strategy [6] made up of a "guidance" unit which generates the optimal trajectory and of "pilot" units which perform the servoing of the linkages.

In order to test the validity of that strategy, the resulting control system was simulated as well as control systems based on the algorithm proposed by Khalil and the one proposed by Raibert. The comparison of the simulation results shows the merits of the control strategy proposed by the authors.

MODEL OF THE MANIPULATOR

For the purpose of simulation the "revolute" generic type of industrial manipulators, has been assumed: the six degreesof freedom consist of three jointed linkages, called the "arm", which execute the major displacements and which carry a "wrist" that has three more degrees of freedom. However, the amplitude and the velocity of the motions as well as the moment of inertia of the last three degree of freedom are comparatively small.

Therefore, as Raibert [1] did, one can consider the influence of the last three degrees of freedom on the first three linkages, during the displacement of an object along a trajectory, as negligeable. Thus, for the purpose of this paper, only the mass of the last three linkages was taken into consideration. It was lumped together with the mass of the load and their total mass was assumed to be ponctual and located at the end of the third linkage which is actually the moving point that traces the trajectory.

Hence, the manipulator being considered, is made-up of three jointed rigid linkages (fig. 1).

The first is $0_1 0_2$, which is fixed with respect to the cartesian axis but its rotation around the z-axis constitutes the first degree of freedom. The second and third degrees of freedom are the rotation around 0_2 and 0_3 respectively (these axes of rotation are parallel with each other and perpendicular to the plane containing 0_1, 0_2, 0_3). The length of each linkage is ℓ_1, ℓ_2, ℓ_3 while their mass is M_1, M_2, M_3. The angles θ_1, θ_2, θ_3 define the position of the manipulator while $\dot\theta_1$, $\dot\theta_2$, $\dot\theta_3$, is the rotational velocity of the joints. They form the state vector:

$$X = [\theta_1\ \theta_2\ \theta_3\ \dot\theta_1\ \dot\theta_2\ \dot\theta_3]^T$$

From the Euler-Lagrange equations, one obtains the model of the manipulator:

$$\Gamma = [A(\theta) + N^2 JM]\ddot\theta + G(\theta) + \dot\theta^T C_i\, \dot\theta + T_f(\dot\theta) \qquad (1)$$

where,

Γ = vector whose elements are the torques required to move the load

JM = diagonal matrix made up of the motors moment of inertia

N = diagonal matrix comprising the gear reducer ratios

$A(\theta)$ = matrix comprising the inertial coupling between the axes

$G(\theta)$ = torque vector due to gravitation

$\dot\theta^T C_i(\theta)\dot\theta$ = abuse of notation which stands for $(\dot\theta^T C_1\dot\theta, \dot\theta^T C_2\dot\theta, \dot\theta^T C_3\dot\theta)^T$

$C_i(\theta)$ = matrix of coefficients representing the Coriolis and centrifugal forces

$\ddot\theta$ = acceleration vector

T_f = friction vector

Since the moments of inertia matrix is nonsingular, it comes that:

$$\ddot\theta = [A(\theta) + N^2 JM]^{-1}\,[\Gamma - G(\theta) - \dot\theta^T C_i(\theta)\,\dot\theta - T_f]$$

which can be written in the form of a state equation as

$$\dot X = \begin{vmatrix} Z_3 & I_3 \\ Z_3 & Z_3 \end{vmatrix} X + \begin{vmatrix} Z_3 \\ [A(\theta) + N^2 JM]^{-1} \end{vmatrix} \Gamma' \qquad (2)$$

where

$$\Gamma' = \Gamma - G(\theta) - \dot\theta^T C_i(\theta)\,\dot\theta - T_f$$

Z_3 = 3x3 null matrix

I_3 = 3x3 identity matrix

Using feedforward, the control vector can be taken as:

$$u = [A(\theta) + N^2 JM]^{-1}\,\Gamma'$$

Then, one obtains a decoupled linear system that can be decomposed into three sub-systems:

$$\dot x_i = \begin{vmatrix} 0 & 1 \\ 0 & 0 \end{vmatrix} x_i + \begin{vmatrix} 0 \\ 1 \end{vmatrix} u_i \qquad (3)$$

where

$$x_i = \begin{vmatrix} \theta_i \\ \dot\theta_i \end{vmatrix} \quad \text{and } i = 1, 2, 3$$

This sums up the work done by many predecessors on modeling robots.

TYPES OF SIMULATED CONTROL SYSTEMS

As there are various means of implementing feedforward for decoupling and linearizing a robot control structure, some of the configurations have been categorized by the authors into types. Four of these have been compared by means of simulation:

Type I (proposed by Raibert [1])

Perfect decoupling and perfect compensation are assumed. It is implemented by means of pre-computed look-up tables: the decoupling matrix is obtained directly from the memory while the compensating vector is computed using the coefficients of $G(\theta)$ and $C_i(\theta)$ stored in memory besides necessitating thirteen multiplications and six additions.
For simulation purpose, a proportionnal compensator with tachometer feedback has been assumed (fig. 2).

Type II (Paul's approximation [7])

It is similar to the first type except that $\dot\theta^T C_i(\theta)\dot\theta$ and $T_f(\dot\theta)$ are made equalled to zero. The slower the motions, the better is that approximation. The quantity of data to be kept in memory is reduced by more than half and the values required for compensation are taken out of memory without requiring any calculation.

Type III (proposed by Khalil [3])

This configuration uses a predictive scheme (fig. 3) instead of relying on look-up tables. Decoupling gains and compensating signals are generated by computation assuming the trajectory to be known and using straight line interpolation for the determination of $A(\theta d) + N^2 JM$.

Type IV (proposed by the authors [6])

This configuration uses an auto-adaptive structure for decoupling and for compensating the nonlinear torques.

The forces acting on the linkages are compensated by means of an iterative algorithm that minimizes the acceleration error. This is why, later on, it is named the estimated compensation structure. Moreover, simulation shows that one can keep the decoupling matrix constant without altering the robot performances.

SIMULATION DETAILS

For each one of the simulated cases, CSMP was used for simulating the manipulator and its control system. Integrations were performed by the 4th order fixed step Runge-Kutta method. The TR version of CSMP has allowed to obtain curves directly from a CALCOMP plotter.

The simulation model was obtained from equation (1) taking into account the angular displacement limits of the linkages, the torque saturation of the motors (assuming current feedback) and the voltage saturation of the amplifiers. In order to determine how sensitive the measured performances of the simulated system are, relative to the model discrepancies, simulation was repeated in each case using erroneous parameters for computing the decoupling matrix and the compensating vector. Different values were given to those errors for different computer runs.

OPTIMAL TRAJECTORY

In order to minimize the torques required, the cross-section of the linkages, the amplitude of the vibration and in order to obtain a control law which can be relatively easy to implement, the cost function that has been chosen minimizes the acceleration along the trajectory from the origin to the destination of the object moved by the robot. From the coordinates of those two points, the guidance unit in the actual system will thus generate a continuous trajectory in the form of six reference signals $\theta_i^d(t)$ and $\dot{\theta}_i^d(t)$. As a matter of fact, the minimum of the cost function

$$J_i = 1/2 \int_o^{tf} u_i^2 \, dt$$

for the following initial and final conditions:

$$\theta_i(0) = \alpha \qquad \theta_i(t_f) = \delta$$

$$\dot{\theta}_i(0) = 0 \qquad \dot{\theta}_i(t_f) = 0$$

yields [8]:

$$\theta_i^d(t) = \frac{2(\alpha-\delta)}{t_f^3} t^3 - \frac{3(\alpha-\delta)}{t_f^2} t^2 + \alpha$$

$$\dot{\theta}_i^d(t) = \frac{6(\alpha-\delta)}{t_f^3} t^2 - \frac{6(\alpha-\delta)}{t_f^2} t \tag{4}$$

The determination of t_f is a difficult problem since one must take into account all possible saturations. It has been more thoroughly investigated in [9] but simulation permits to easily find t_f by successive trials. Therefore equation (4) has been used for generating the trajectories during the simulation. When simulating type IV, the divergence of the algorithm due to saturation was prevented by keeping T_c constant as long as saturation existed. Thus it has been possible to test the system for small value of t_f.

COMPARISON OF THE RESULTS

The four types of control systems described above have been simulated as well as some modified versions. Hence, the comparison bears on height different cases. In some of them, the decoupling matrix has been made constant in taking for $A(\theta)$ the value obtained for $\theta_2 = 0$ and $\theta_3 = \pi/2$. In other cases, noisy values (σ=0.1) have been assumed for the parameters used in $G(\theta)$, $C_i(\theta)$. However, for each type, the proportional plus derivative controller has been used with $K_p = 72$ and $K_v = 12$.

The optimal trajectory shown on fig. 5 has served as input in each one of the eight following cases:

1. Control with constant decoupling but no compensating torque [Type 2]
2. Control with constant decoupling and perfect compensation
3. Control with constant decoupling and estimated compensation [Type 4 modified]
4. Control with perfect decoupling and estimated compensation [Type 4]
5. Control with perfect decoupling and perfect compensation [Type 1]
6. Control with perfect decoupling and simplified compensation
7. Predictive control [Type 3]
8. Simplified predictive control

The criteria adopted for comparison purposes are:

1. Time required for each degree of freedom to reach the reference trajectory within 10^{-3} rad.: T_1, T_2, T_3 and $T = \max(T_1, T_2, T_3)$.

2. The integral of the absolute values of the errors: E_1, E_2, E_3 and $E = E_1 + E_2 + E_3$.

3. The absolute values of the maximum errors: EM_1, EM_2, EM_3. Tables 1 and 2 sum up the results that were obtained.

CONCLUSIONS

Simulation allows one to conclude that:

i) for zero steady state error, the compensation of gravitational torques is necessary;

ii) the compensation of torques due to Coriolis and centripetal forces decreases the overshoot and the integral of the absolute values of the errors;

iii) performances of the predictive control compared to the type I control worsened as the speed increased;

iv) in case 3 the performances are roughly the same as those of more complex systems as seen in cases 5 and 7.

In general, the simulation results indicate that type IV system (proposed by the authors), incorporating an auto-adaptive compensation loop, will give the robot the same robustness and the same performances as the other types that were simulated while requiring more modest computational capabilities.

REFERENCES

[1] Raibert, M.H., Horn B.K.P., Manipulator control using the configuration space method, The Industrial Robots, vol. 5, No.2, June 1978, pp. 69-73.

[2] Khalil, W., Liégeois, A., Fournier, A., Commande dynamique des robots, RAIRO Automatique vol. 13, No. 2, 1979, pp. 189-201.

[3] Turner, T.L., Parameter estimation and path control for a three-joint robot manipulator, Master Thesis, North Carolina State University, 1979.

[4] Gruver, W.A., Hedges, J.C., Snyder, W.E., Hierarchical control of large scale linear systems with an application to robotics, Proc. of the IFAC Workshop on Control Applications of Nonlinear Programming, 1979, pp. 1979, pp. 93-98.

[5] Saridis, G.N., Lee, C.S.G., An approximation theory of optimal control for trainable manipulators, IEEE Trans. on Systems, Man, Cybernetics, vol. SMC-9, No. 3, March 1979, pp. 152-159.

[6] Chaudet, R., O'Shea, J., Une structure de commande décentralisée pour la robotique, Proc. of the Canadian Conference on Industrial Computer Systems, May 1982, pp. 801-806.

[7] Paul, R., Modeling trajectory calculation and servoing of a computer controlled arm, Ph.D. Thesis, Stanford University, 1972.

[8] Athans, M. et Falb, L.B., Optimal control, McGraw Hill, 1966, pp. 684-693.

[9] Chaudet, R., Etude comparée de systèmes de commande d'un robot par simulation, Thèse de maîtrise, Ecole Polytechnique de Montréal, septembre 1981.

Figure 1

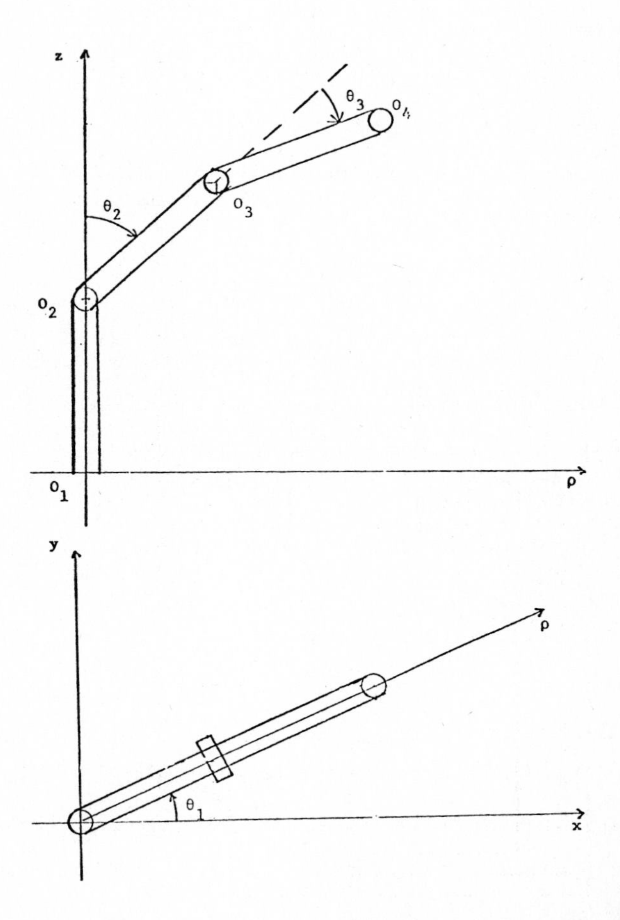

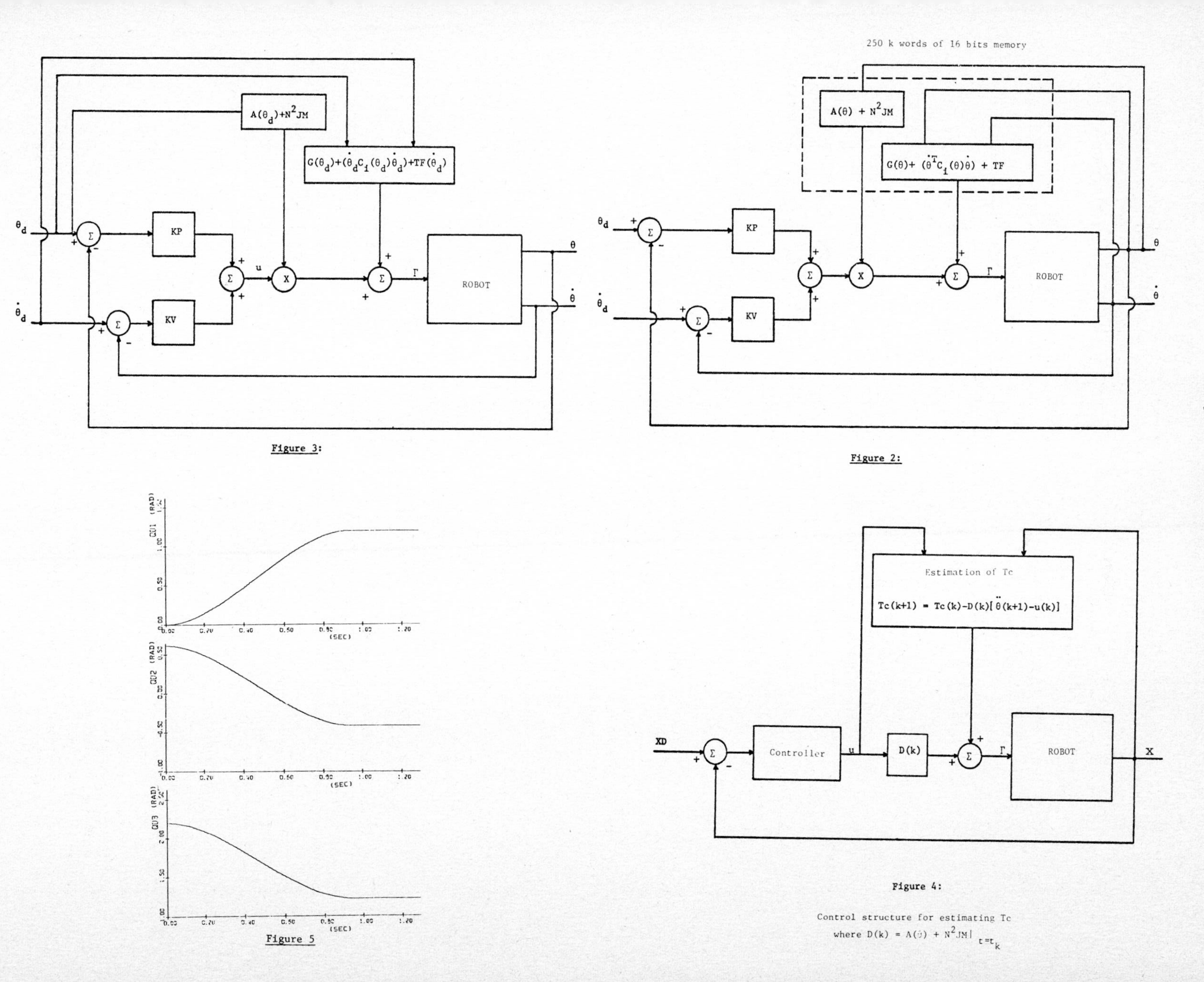

Figure 3:

Figure 2:

Figure 5

Figure 4:

Control structure for estimating Tc where $D(k) = A(\theta) + N^2JM|_{t=t_k}$

TABLE 1

(σ = 0)

ICOMM	1	2	3	4	5	6	7	8
T1	1.35	1.33	1.70	1.59	1.59	1.59	1.58	1.58
T2	1.65	1.65	1.53	1.58	1.58	1.58	1.62	1.62
T3	∞	1.61	1.56	1.57	1.57	1.57	1.60	1.60
T	∞	1.65	1.70	1.59	1.59	1.59	1.62	1.62
E1	.0566	.0438	.0558	.0553	.0546	.0692	.0556	.0678
E2	.0766	.0488	.0506	.0501	.0504	.0512	.0491	.0529
E3	∞	.0457	.0459	.0456	.0454	.0514	.0478	.0531
E	∞	.1383	.1522	.1510	.1504	.1718	.1525	.1737
EM1	.0756	.0701	.0778	.0809	.0785	.1018	.0[illegible]16	.1000
EM2	.0982	.0882	.0804	.0759	.0760	.0689	.0768	.0674
EM3	.0716	.0704	.0690	.0677	.0674	.0766	.0734	.0795
EM	.2454	.2287	.2272	.2245	.2219	.2473	.2318	.2469

TABLE 2

(σ = 0.1)

ICOMM	1	2	3	4	5	6	7	8
T1	1.35	1.33	1.70	1.61	1.54	1.55	1.53	1.54
T2	1.65	1.68	1.53	1.53	1.68	1.69	1.76	1.77
T3	∞	1.62	1.56	1.59	1.54	1.54	1.55	1.55
T	∞	1.68	1.70	1.61	1.68	1.69	1.76	1.77
E1	.0566	.0448	.0558	.0553	.0523	.0652	.0536	.0642
E2	.0766	.0496	.0505	.0517	.0634	.0642	.0626	.0667
E3	∞	.0463	.0459	.0456	.0432	.0496	.0462	.0515
E	∞	.1406	.1522	.1516	.1589	.1791	.1625	.1824
EM1	.0756	.0703	.0778	.0804	.0749	.0950	.0781	.0940
EM2	.0982	.0867	.0804	.0796	.0898	.0837	.0904	.0803
EM3	.0739	.0706	.0690	.0669	.0636	.0733	.0700	.0762
EM	.2454	.2276	.2272	.2269	.2283	.2520	.2385	.2505

Simulation in Engineering Sciences
J. Burger and Y. Jarny (eds.)
Elsevier Science Publishers B.V. (North-Holland)

SIMULATION OF AN ELECTROPNEUMATIC DRIVE FOR INDUSTRIAL ROBOTS

K. Desoyer, P. Kopacek and I. Troch

Department of Mechanical Engineering and Department of Technical Mathematics
Technical University of Vienna
Vienna, Austria

The paper deals with the development of simple models for the pneumatic drive of an industrial handling device or robot consisting of a solenoid valve and a pneumatic cylinder. Starting from a nearly exact model various simplifications are discussed and a simple input-output model is tested and compared by simulations on a hybrid computer. It can be shown that the simple model seems to be sufficiently accurate for many practical applications.

1. INTRODUCTION

Handling devices and industrial robots will be of great interest in the very near future. Such devices are driven electrically, pneumatically or hydraulically and are controlled electrically in most cases. Their application for assembling operations requires a high accuracy of the position control.

Today various control algorithms are available for these purposes but they are very complicated and consequently too expensive for most practical applications. The development of simpler and sufficient accurate positioning control algorithms requires mathematical models of the dynamic behaviour of the handling device or the robot. These models should be on the one side simple to applicate in practice and on the other side sufficient accurate enough for approximate description of the dynamic behaviour.

As a first step in this direction a simple input-output model of a pneumatic drive used for handling devices will be tested by means of a hybrid computer and the results are compared with those of a complicated nonlinear but more exact model.

2. MODEL OF A PNEUMATIC DRIVE

A scheme of a pneumatic drive for various axes of an industrial handling device is shown in Fig.1. An electric input signal u_{el} actuates a pneumatic solenoid valve. Its output pressures $\hat{p}_1$ and $\hat{p}_2$ drive the pneumatic cylinder.

The model equations of a similar electrohydraulic servosystem are derived in [1] . These can be adapted for the pneumatic system [2] . Under the assumptions that

- the system is rigid
- the system is leakproof
- the air flow characteristic of the valve is linear
- the supply pressure p_o and the return pressure p_R are constant.

the following equations hold for the dynamic behaviour of the various parts of the drive with the symbols given in Fig.1.

Solenoid valve

$$\ddot{\varphi} + 2D\omega_o\dot{\varphi} + \omega_o^2\varphi = \omega_o^2 \frac{V}{\phi_{max}} u_{el} \qquad (1)$$

This is a second order lag element with the time constant $1/\omega_o$, the damping D and the gain V/ϕ_{max}.

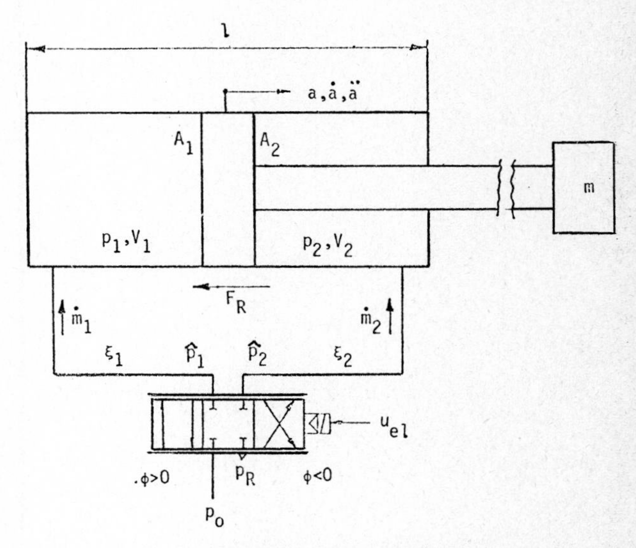

Fig. 1: Scheme of the electropneumatic servosystem

Pneumatic transmission lines

If the state changes in the lines are regarded isothermal and in the cylinder chambers isentropic, the mass flows $\dot{m}_1$ and $\dot{m}_2$ depending on the pressures in the cylinder chambers (p_1,p_2) will be given by

$$\dot{m}_1 = A_1\dot{a}\rho_o\left(\frac{p_1}{p_o}\right)^{\frac{1}{\varkappa}} + V_1\left(\frac{p_o}{p_1}\right)^{\frac{\varkappa-1}{\varkappa}} \cdot \frac{1}{\varkappa.R.T_o}\dot{p}_1 \qquad (2a)$$

$$\dot{m}_2 = -A_2\dot{a}\rho_o\left(\frac{p_2}{p_o}\right)^{\frac{1}{\varkappa}} + V_2\left(\frac{p_o}{p_2}\right)^{\frac{\varkappa-1}{\varkappa}} \cdot \frac{1}{\varkappa.R.T_o}\dot{p}_2 \qquad (2b)$$

with the values of the normal state (ρ_o, p_o, T_o) and the gas constant R.

Mass flows in the solenoid valve

With respect to the pressure losses in the lines characterized by the equivalent resistance number and the direction of motion of the valve considered by the signum function the mass flow in the valve is given by

$$\dot{m}_1 = \dot{m}_o\varphi\sqrt{2}\left[\Big|0,5.(1+\mathrm{sgn}\varphi)+0,5.\frac{p_R}{p_o}.(1-\mathrm{sgn}\varphi) - \frac{p_1}{p_o} - \eta_1\dot{m}_1^2\,\mathrm{sgn}\,\varphi\Big|\right]^{1/2} \qquad (3a)$$

$$\dot{m}_2 = -\dot{m}_o\varphi\sqrt{2}\left[\Big|0,5.(1-\mathrm{sgn}\varphi)+0,5.\frac{p_R}{p_o}.(1+\mathrm{sgn}\varphi) - \frac{p_2}{p_o} + \eta_2\dot{m}_2^2\,\mathrm{sgn}\,\varphi\Big|\right]^{1/2} \qquad (3b)$$

with the abbreviations

$$\eta_1 = \frac{\xi_1}{2p_o\rho_oA_1^2} \quad \text{and} \quad \eta_2 = \frac{\xi_2}{2p_o\rho_oA_2^2}$$

Friction force

Usually the friction force F_R in the cylinder is considered as a sum of a speed proportional part $f.\dot{a}$, the Coulomb friction F_o and the static friction $F_H.\exp(-c/\dot{a})$

$$F_R = \begin{cases} f\dot{a} + \left[F_o + F_H\exp(-c/\dot{a})\right]\mathrm{sgn}\,\dot{a} & \text{for } \mathrm{sgn}\,\dot{a} = \mathrm{sgn}\,\ddot{a} \\ f\dot{a} + F_o\,\mathrm{sgn}\,\dot{a} & \text{for } \mathrm{sgn}\,\dot{a} = -\mathrm{sgn}\,\ddot{a} \end{cases} \qquad (4)$$

Motion of the cylinder

The dynamic behaviour of the cylinder is given by

$$m\ddot{a} = A_1p_1 - A_2p_2 - F_R \qquad (5)$$

From the equations (1) to (5) follows with the state variables

$$x_1 = \varphi \qquad x_4 = \frac{p_2}{p_o}$$
$$x_2 = \dot{x}_1 \qquad \dot{x}_5 = \frac{a}{l} \qquad (6)$$
$$x_3 = \frac{p_1}{p_o} \qquad x_6 = \dot{x}_5$$

and the constants

$$\alpha_1 = \sqrt{2}\frac{\dot{m}_o}{A_1 l \rho_o} \qquad \delta = \frac{f}{m}$$
$$\alpha_2 = \alpha_1\frac{A_1}{A_2} \qquad \varepsilon = \frac{F_o}{ml} \qquad (7)$$
$$\beta = \frac{A_1p_o}{ml} \qquad \mu = \frac{F_H}{ml}$$
$$\gamma = \frac{A_2p_o}{ml} \qquad \nu = cl$$

a system of 6, mainly nonlinear, differential equations of first order describing the dynamic behaviour of the pneumatic drive

$$\dot{x}_1 = x_2$$
$$\dot{x}_2 = -2D\omega_ox_2 - \omega_o^2x_1 + \omega_o^2\frac{V}{\phi_{max}}u_{el}$$
$$\dot{x}_3 = \frac{\varkappa x_3}{0,5+x_5}\left\{\frac{\alpha_1x_1}{x_3^{1/\varkappa}}\left[\Big|0,5(1+\mathrm{sgn}x_1) + 0,5\frac{p_R}{p_o}(1-\mathrm{sgn}x_1) - \eta_1\dot{m}_1^2\,\mathrm{sgn}\varphi - x_3\Big|\right]^{1/2} - x_6\right\} \qquad (8)$$
$$\dot{x}_4 = \frac{\varkappa x_4}{0,5-x_5}\left\{\frac{-\alpha_2x_1}{x_4^{1/\varkappa}}\left[\Big|0,5(1-\mathrm{sgn}x_1) + 0,5\frac{p_R}{p_o}(1+\mathrm{sgn}x_1) + \eta_2\dot{m}_2^2\,\mathrm{sgn}\varphi - x_4\Big|\right]^{1/2} + x_6\right\}$$
$$\dot{x}_5 = x_6$$
$$\dot{x}_6 = \beta x_3 - \gamma x_4 - \delta x_6 - \left[\varepsilon + 0,5\,\mathrm{sgn}x_6 + \mu.\exp(-\nu|x_6|).\mathrm{sgn}\dot{x}_6\right]\mathrm{sgn}x_6$$

Boundary conditions of the state variables are

$$|x_1| < 1 \qquad |x_5| \leqq K$$
$$x_3 \leqq 1 \qquad x_6 = 0 \quad \text{for} \quad x_5 = K$$
$$x_4 \leqq 1$$

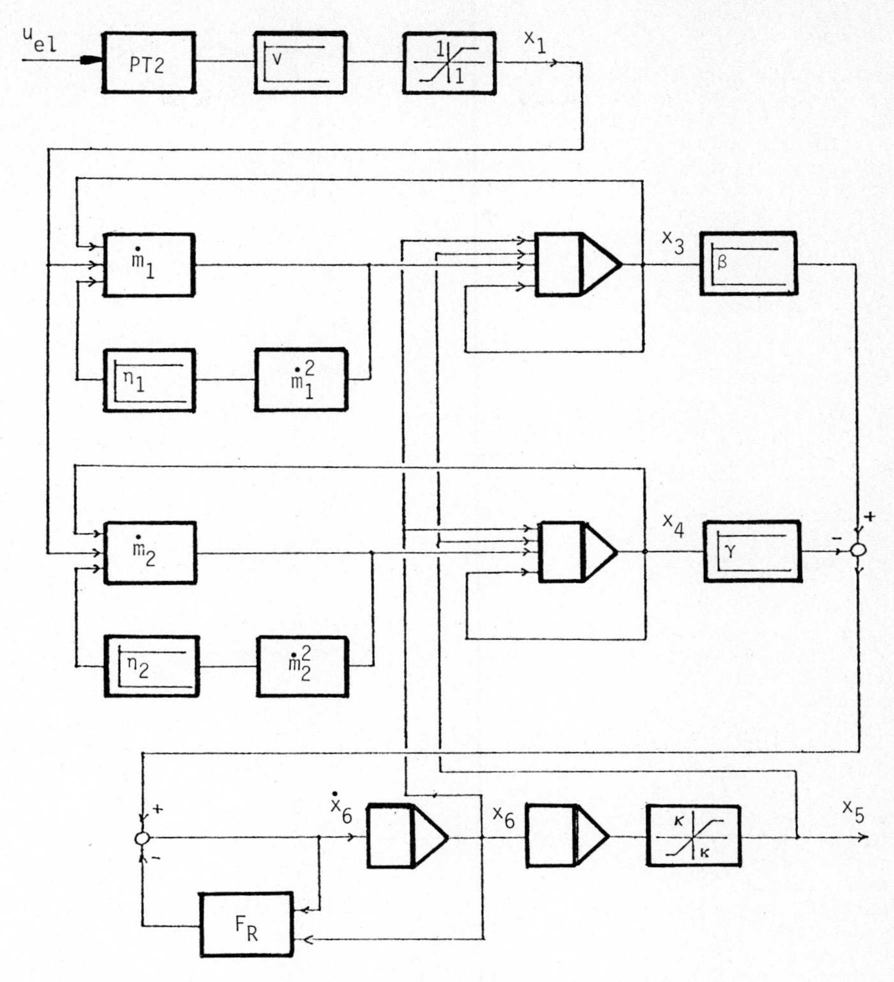

Fig.2: Block diagram of the model (equations 8)

The corresponding block diagramm to these state equations is given in Fig.2.

This model is too complicated for some practical applications. Therefore simplifications are necessary in most cases. These are usually

- because of the small time constants T, the model of the servo valve- a second order lag element - may be replaced by a simple proportional element.
- the line pressure losses may be neglected. Then the output pressures of the valve are the same as in the chambers of the pneumatic cylinder
- the friction force is only proportional to the speed of the piston.

All these simplifications together yield a model of fourth order in which some terms of the remaining equations are zero.

Another method of simplifying the model equations (8) starts from tests on real handling devices. As pointed out in [5] and also shown in Fig.4 from the step response (distance a as a function of time for a stepwise change of the input signal u_{el}) the dynamic behaviour may be approximately described by an integral element with first order lag and dead time. Their well known transfer function is

$$W(s) = \frac{K_I}{s} \cdot \frac{1}{1+T_1 s} e^{-T_t s} \qquad (9)$$

Herein K_I is the integral action factor which is corresponding to the final speed of the piston, T_1 the time constant and T_t the dead time. The latter two depend on the load and the pressures. In contrast to the state space equations (8) only the distance a and by differentiation the speed $\dot{a}$ can be calculated from this input-output description.

3. SIMULATION

The simulation of the state equations (8) was carried out on the Hybrid Computer system EAI Pacer 600 at the Technical University of Vienna. This computer system is provided with an AUTO-PATCH-system. It allows, in connection with the hybrid processor HYBSYS, the automatic development of switching connections from a terminal into a completely machine independent syntax. The main advantages for the practically oriented user are the simple programming of complicated model equations. The necessary switching connections are found and patched automatically within a short time, scaling and synchronisation are done by the processor, aids for documentation of the program as well as of the results (ploting routines) are available. Consequently, the user can concentrate on the problem and its solution and need not worry about programming details. Once a problem has been treated by the processor it can be stored and is available within a short time with all parameters etc. having the same values as in the moment of storing.

The main part of the AUTOPATCH equipment is an analogue switching matrix with 5120 integrated MOSFET switches. This matrix consists of two submodules, each with 64 inputs and 128 outputs, and is connected with the analogue parallel-processor EAI 680 by means of trunklines containing preprogrammed analogue computing elements (MAKROS). Furthermore the analogue computer is coupled by control and datalines with the standard interface 693. By means of a special controller at the standard I/O channel the switching matrix can be programmed first and then the matrix inputs and outputs can be read out. A standard channel serves to activate the 64 operation controllines (FOCL). A separate logic processor (Pipelined Boolean Processor) developped at Hybrid Computing Center Vienna is connected with preprogrammed Macros in the parallel logic of the analogue computer. Detailed informations about this concept may be found in [3] and [4] .

These facilities have been used to program the equations (8) on the hybrid computer. Fig. 3 shows the declaration part for all elements and parameters as well as the description of the input-output relations for the first three equations. The description of a program consists in

1) Reformulation of the equations as a series of differential and algebraic equations so that standard MACROS (Integrator,Multiplicator,...) can be used.
2) Declaration of the parameters to be used and assignment of their numerical values (the latter can be done also at computation time).
3) Description of the input-output relation for each MACRO. With this in mind, Fig.3 in connection with equations (8) and the block-diagram of Fig.2 is rather self-explainatory. It should be noted that most of step 1 can be omitted in case the equations are written directly for the block diagram of Fig.2 without describing several blocks by a single equation. Therefore, the HYBSYS processor can be very easily handled by engineers.

```
DECLAR PAR:K01=35,K02=625,K04=1.4,K05=.5
DECLAR PAR:K06=1.6835,K07=.9286,K08=1.87
DECLAR PAR:K09=.143,K010=.1,K011=163
DECLAR PAR:K012=147,K013=1,K014=9.8
DECLAR PAR:K015=0,K016=.48,K017=.143
DECLAR INT:X2,X3,X4,KX1,KX2,KX3
DECLAR SUM:ALL1,ALL2,ALL3,ALL4,X1,X5,X6
DECLAR SUM:NAH1,NAH2,WU1,WU2,PEM,MOZ
DECLAR MULT:MA1,MA2,MA3,MA4
DECLAR ABS:WUA,WUB
DECLAR SQRT:WUR,WURZ
DECLAR DIV:DUR1,DUR2,DUR3,DUR4
DECLAR COMP:LV,Z
DECLAR DSWI:HUB,GESCH
DECLAR SWI:SCH


DECLAR KX3=X2
DECLAR X1=KX3


DECLAR ALL1=-K01*X2,-K02*X1,K02
DECLAR X2=ALL1


DECLAR WU1=K07,-X3
DECLAR WUA=WU1
DECLAR WUR=WUA
```

Fig. 3: Part of the HYBSIS program

The results were plotted (Fig.5,6) and lists of vaiables as well as of the patching connections were printed.

In the same way the simple model given by equation (9) was programmed. Clearly this is very short. Its parameters K_I,T and T_t were determined approximately from the step response of the state veriable x_5 as shown in Fig.4.
The difference of the distances (BETR1) and the speeds (BETR2) as well as their integrals over the time serves for comparison of the two models. These curves are alos plotted by the computer.

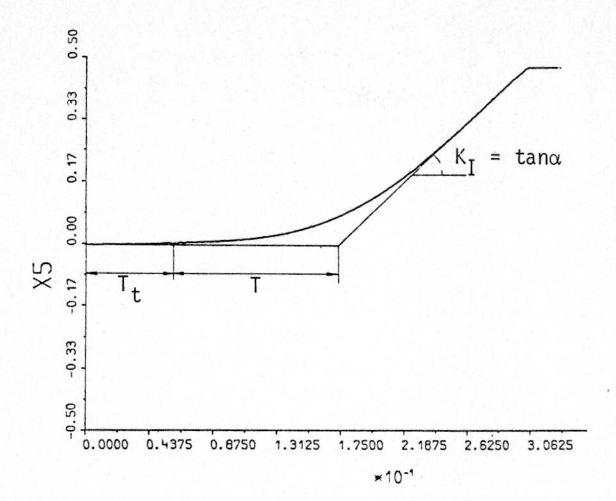

Fig. 4: Step response and approximation

The tests were carried out for a pneumatic servo-system with the following construction parameters:

$$m = 1 \text{ kg}; \quad l = 30 \text{ cm}; \quad A_1 = 7 \text{ cm}^2 \; ; \; A_2 = 6{,}3 \text{ cm}^2$$

The supply air pressure was p_0 =7.0 bar and the return pressure p_R=1.0 bar. The servo valve has a time constant T=0,04 sec and a damping D=0.7.

The curves for the distance and the speed of the piston over the time resulting from the exact model (8) and the simplified model (9) are collected in Fig.5 for a stepwise change of the input signal u_{el} of the servo valve. While curves for the distance (Fig.5a) are nearly identical considerable differences occur for the speeds. The main reason for this fact lies in a inflection point near the maximum of distance in Fig.4. This may also be shown in Fig.6 where the differences of the distances BETR1 (Fig.6a) and the speeds BETR2 (Fig.6b) of the exact and the simplified model

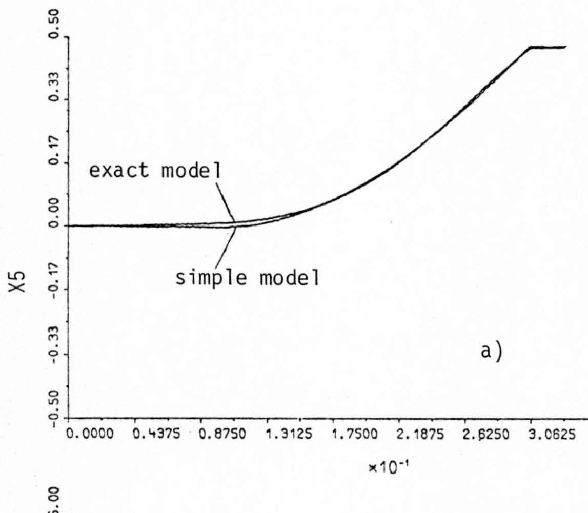

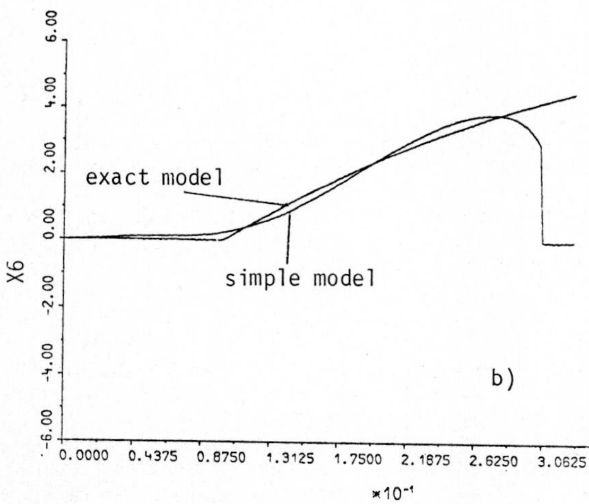

Fig. 5: Distances and speeds of both models

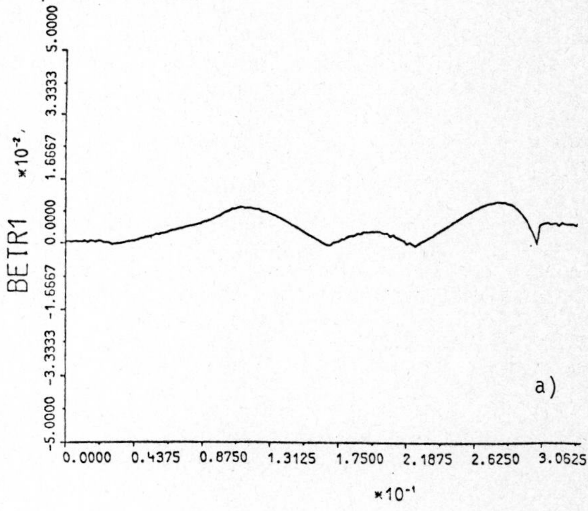

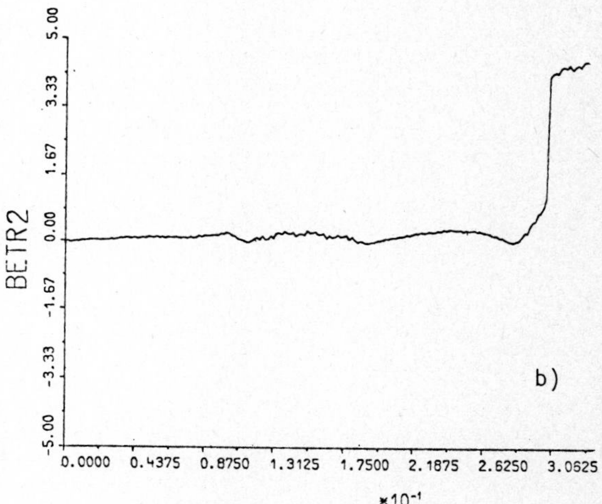

Fig. 6: Comparison of the two models

are plotted. It must be observed that the scale of Fig.6a which shows the difference of distances differs by a factor 10^2 from that of Fig.5a.

If, for practical purposes, only the distance and the speed are necessary as model outputs, the simplified model may be used with sufficient accuracy as shown by the tests just described.

SUMMARY

The performance of a very simple model for the dynamic behaviour of an electropneumatic drive for an industrial robot or handling device was tested by means of a hybrid computer system equipped with an AUTOPATCH system. An accurate, mainly nonlinear state space model, well known for electrohydraulic drives, was simulated and the results compared with the simple input-output model. The simulation results show that the simple model can be applicated for most practical purposes under the assumptiom that only the distance and the speed of the piston is required. For theoretical applications (e.g. design of state space controllers and/or observers) the derived state space model must be taken. For comparison of both models serve the step responses of distances and speeds and as performance index for their valididation the differences over time. Tests on various drives of handling devices with different sets of construction parameters confirm the test results in this paper.

REFERENCES

[1] Quetting,P.: Design of an observer for electrohydraulical servosystems. Report 6/79, Inst.für Meß-,Steuer-und Regelungstechnik, GH Duisburg, 1979 (in German).

[2] Müllner,P.: Dynamic behaviour of electropneumatical servo systems. Diplomarbeit, TU-Wien, 1982 (in German).

[3] Kleinert,W.: The new Autopatch-Hardware. Interface 1980, H.15/16m p.11-22. (in German).

[4] Troch,I.: Introduction in the use of a analogue computer with Autopatch-system by means of HYBSYS, Inst.f.Techn.Math., TU-Wien, 1981.(IN GERMAN)

[5] Kopacek,P.: Microcomputer control of manipulators and assembling machines. Preprints of the 7th IFAC Congress, Vol.XIV, p.49-53, Kyoto, 1981.

XVI. ELECTRICAL AND ELECTROMECHANICAL SYSTEMS

Simulation in Engineering Sciences
J. Burger and Y. Jarny (eds.)
Elsevier Science Publishers B.V. (North-Holland)

SIMULATION OF PIECEWISE LINEAR STRUCTURED ELECTRIC CIRCUITS
PETRI-NET REPRESENTATION OF SEMI-CONDUCTORS FUNCTIONING IN COMMUTATION

F. BORDRY - H. FOCH - M. METZ

Laboratiore d'Electrotechnique et d'Electronique Industrielle
E.R.A. au C.N.R.S. n° 536
ENSEEIHT
2 rue Camichel 31071 TOULOUSE Cédex

In the study of switching (and especially semi-conductor) devices simulation methods play an important role in both the design of systems and the analysis of their functioning. A numerical simulation of semi-conductor electric circuits is proposed in this study. The method which the present authors describe as global, does not require a priori knowledge of the state sequences of the semi-conductor nor of the commutation instants. Particular attention has been paid to the modelling of the semi-conductors. A generalized switch (on/off), whose output is a function of the state of the switch is defined. State changes are defined by logical functions which take into account the nature of the switch (i.e. diode, thyristor, transistor or composite semi-conductor). This generalized switch will be modelled by a Petri-Net. Using this representation an easier maintenance (service and software improvement) as well as functioning dependability problems of the semi-conductor devices can be ensured. The simulation of an electric energy static convertor, with the inclusion of a functioning fault, is presented and used as an application.

1. INTRODUCTION

The main difficulty involved in taking into account non-linear elements, which the semi-conductors are, lies in determining the existence and the linking of the elementary sequences defined by the conduction or non-conduction of these components. Using analytical methods based on functioning decomposition in a succession of known sequences ("sequence simulation"), their employment can only with difficulty be envisaged in the design an analysis of new systems. The search for a simulation method which only uses knowledge of the circuit's structure, of its parameters and the control order of the switching control ("global simulation") is therefore a major concern. In a global simulation the semi-conductors are modelled individually ; the linking of functioning sequences is done automatically due to the functioning properties of the models.

This study starts with a description of the determination of the models. The method and architecture of the simulation program is then presented. An example demonstrates the application possibilities of the program.

2. SIMULATION METHOD

2.1. Semi-conductor modelling

2.1.1. Introduction

The choice of the semi-conductor electric model is an important point in the design of a simulation method. Both the accuracy and complexity of the simulation depend on its degree of sensitivity.

A number of electrically equivalent diagrams which represent different semi-conductors exist.

There are considered by the present authors as functioning in commutation (the notion of the switch in opposition with a linear component). This choice involves the representation of a semi-conductor by a mono-output logical system (switch On or Off).

This binary representation of a semi-conductor is conducive to a number of models :

- the semi-conductors are modelled in the form of perfect switches. In the off-state, the branches on which they are situated disappear from the graph. The topology is thus variable and a particular system of equations corresponds to each sequence (1), (2).

- the semi-conductors are modelled by a second order circuit (inductance series and parallel RC circuit). The semi-conductor itself is considered as a perfect switch ; when the semi-conductor is off the variable state "current" associated to it is forced to be nil. The topology is fixed and the device can thus be described by a single system of equations but of high order (3).

- the semi-conductors are modelled by controlled voltage (or current) sources. The voltage at the bounds of a voltage source corresponding to a semi-conductor in the on-state is nil, and the one corresponding to a semi-conductor in the off-state is determined so as to have a nil current in the concerned voltage source (dual proposition in the representation by a current source). The topology is fixed, there is a single system of equations, but at each instant the value of the controlled sources corresponding to the off semi-conductors (on) has to be calculated (4).

- the semi-conductors are modelled in the form of a binary resistance, varying on a large scale according to whether they are off (high resistance : R off) or conductors (low resistance : R on). There is a single topology and a single state vector ; the state equation coefficients depend on the value taken by the binary resistances (5), (6).

This latter modelling was chosen for the present study. It is certainly the most interesting because of its very physical aspect which lends itself especially to considerable simplicity of implementation. In this representation the input vector is constituted of the voltage at the bounds of the power electrodes (a variable which is also representative of the current) (Vak) and of the control signal (G).

2.1.1. State notion of a semi-conductor

A semi-conductor cannot be represented by a combinatory system : output does not only depend on the value of the inputs at a given instant, but also on the history of the semi-conductor. For example, a thyristor with a positive anode-cathode voltage and to which no control signal is applied can be on or off.

The sequential aspect of the representation of a semi-conductor is conducive to defining the latter's state which the present authors consider to be a minimum item of information which must be known at a given moment to foresee the evolution of the system from knowledge of the inputs. The state of a semi-conductor does, of course, depend on the sensitivity of the chosen model (inclusion of reverse voltage bias time, recovered charges, etc.). The switch output (on or off) only identifies with the state of the switch for simplified models ; it is generally a state vector component.

2.1.3. Generalized switch

In existing simulation programs the semi-conductors are specified (i.e. diode, thyristor, transistor, etc.) and are processed separately in the course of the programm. So as to have a single model for all semi-conductors the present study defined the notion of a generalized switch. It rests on the introduction of a characteristic variable of the switch in its state vector. The advantages of this definition, apart from simplying the program, are numerous :

- easy maintenance of the simulation program. The introduction of new switches does not pose any problems (dual thyristor, self-turn-on transistor, etc.).

- taking into account composite semi-conductors as switches (anti-parallel thyristor-diode, series transistors with a diode, etc.).

- study of semi-conductor degradation (the anti-parallel thyristor-diode switch can degenerate into a thyristor or a diode according to the type of fault ; furthermore, the thyristor or diode can become an open or short-circuit).

2.1.4. Representation of switches by Petri nets

Having introduced the notion of the generalized switch, the problem of the choice of a functional representation tool is posed. For this study the Petri net representation, which includes a good compromise between generality and simplicity of implementation, was chosen (7), (8).

Defining the elementary Petri net representing a generalized switch, depends on the nature of the semi-conductors under consideration. Figure 1 shows the network comprising the three classical semi-conductors. The places characterize the state of the switch and the transitions are functions of the commutation conditions. The definition of an initial marking is, of course, necessary.

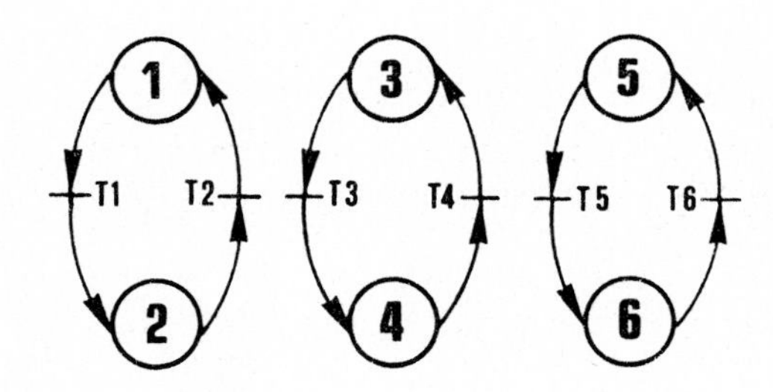

Places	Transitions
1. diode off	T1 : Vak $>$ 0
2. diode on	T2 : Vak $\leqslant$ 0 ($\sim$ Iak $\leqslant$ 0)
3. thyristor off	T3 : Vak $>$ 0 and G = 1
4. thyristor on	T4 : Vak $\leqslant$ 0 ($\sim$ Iak $\leqslant$ 0)
5. transistor off	T5 : G = 1
6. transistor on	T6 : G = 0

Figure 1

This representation, which is linked to the notion of the generalized switch, gives a high evolutionary capacity to the software. Some examples can be cited :

1) Taking into account new switches by modification of the logical functions attached to the transitions (receptivities) or by increasing the number of places in the network :

1.a - dual thyristor : the receptivities associated with transitions t3 and t4 are exchanged.

1.b - anti-parallel diode-thyristor : the network is extended as shown in figure 2.

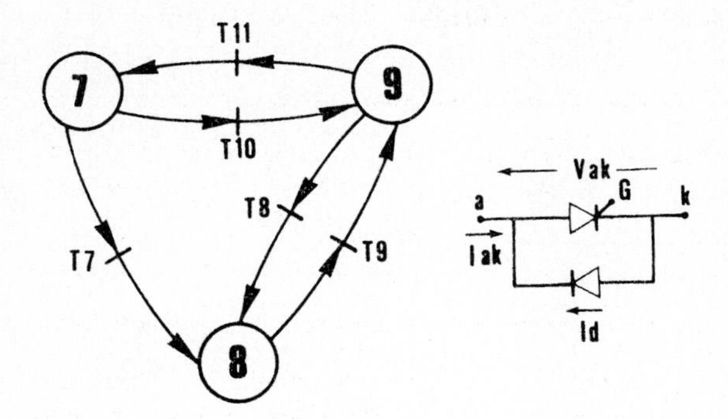

Places

7. thyristor off, diode off
8. thyristor on, diode off
9. thyristor off, diode on

Transitions

T7 : G = 1
T8 : Vak > 0 and G = 1
T9 : Vak ⩽ 0 (∿ Iak ⩽ 0)
T10 : Vak < 0
T11 : Vak ⩾ 0 (∿ Iak ⩾ 0)
(∿ Id ⩽ 0)

Figure 2

In a study of the degradation of this switch is required, it suffices to define the arcs and transitions linking places 7, 8 and 9 to places 1, 2, 3 and 4.

2) Taking into account reverse voltage bias time (Trev) at the bounds of a thyristor can be effected either by considering three states for the thyristor : thyristor off and Trev > Toff (Toff : turn-off time), thyristor off and Trev < Toff, thyristor on, or by modifying the receptivity associated with t3 : Vak > 0 and (G = 1 or Trev < Toff) (reduction of previous network).

By defining the logical states 1 and 0 for a conducting and off switch (switch output) an output matrix S (in the case of figure 2, S = (010101)) is associated to the Petri net. Output S of a switch will be determined from the marking (M) at that instant and of So : S = So.M. This logical signal S determines the binary resistance value (Rscr) modelling the semi-conductor (Rscr = S.R on + $\bar{S}$.Roff).

2.2. Expressing the device in equations

The automatic expression of the device in equations is one of the determining elements of the generality of the program. Its aim is to define the equations of the circuit in state form (10):

$$dX/dt = A.X + B.U + E.dU/dt$$
$$Y = C.X + D.U$$

If the graph associated with the circuit is considered, each component is a branch linking two nodes of the graph. A tree is determined (a subset of the graph joining all the nodes without forming a loop) which is single if the numbering order of the branches is fixed (i.e. voltage sources, capacitors, resistances, inductances, current sources). State vector X is constitued of voltages at the bounds of the capacitors belonging to the tree and the currents in the inductances which do not belong to it. The input vector U contains voltage and current sources. Output vector Y represents the voltages at the bounds of the resistive elements. Matrixes A, B, C, D and E are calculated from the topology and the value of the device's elements (including representative resistances of the switches) ; they are therefore constant between two changes of state of a semi-conductor.

2.3. Resolution of state equations

With the knowledge of the state equations of the system between two commutations, the following solution in the form of a recurrent equation is obtained :

$$X(t + T) = \exp(A.T)X(t) + A.^{-1}(\exp(A.T)-I).B.U(t) + A.^{-1}(\exp(A.T)-I)E.dU/dt(t)$$

The calculation of matix exp(A.T) must be accurate and quick (6). Calculation step T is an observation step and is thus not linded to the smallest time constants of the circuits. Vector U must be constant for the duration of the calculation step. Variable sources (sinusoidal in particular) are represented by a piecewise constant function of equal length in the calculation step. The standard step, comprised between 1/100th and 1/200th of the functioning period, thus of the supply period if the sources are variable, gives a good approximation while ensuring a good definition of the different results.

2.4. Determination of the changes of states instants of the semi-conductors

A change of state of a semi-conductor takes place if a sensitized transition is fired, that is to say if the event corresponding to this transition occurs (firable transition). It is important to know the exact instant when the transition becomes firable. Two types of cases are possible :

- the instant is known (control signal) ; a repositioning is then made to the instant using a reduced step ;

- the instant is not known (a passage by zero of an anode-cathode voltage, anode-cathode voltage becoming higher than the avalanche voltage, etc.) as soon as an event of this type is detected, the step is reduced from the previous instant so that the variable causing the event is included in a prefixed threshold.

3. SIMULATION PROGRAM

It was decided to elaborate a structured program so as to include high adaptation flexibility (i.e. maintenance an diffusion).

The simulation programm comprises three main parts :

- determination of the circuit structure ;
- circuit simulation over a given interval ;
- results output (tables, layout).

Figures 3, 4 and 5 describe the general architecture of these parts.

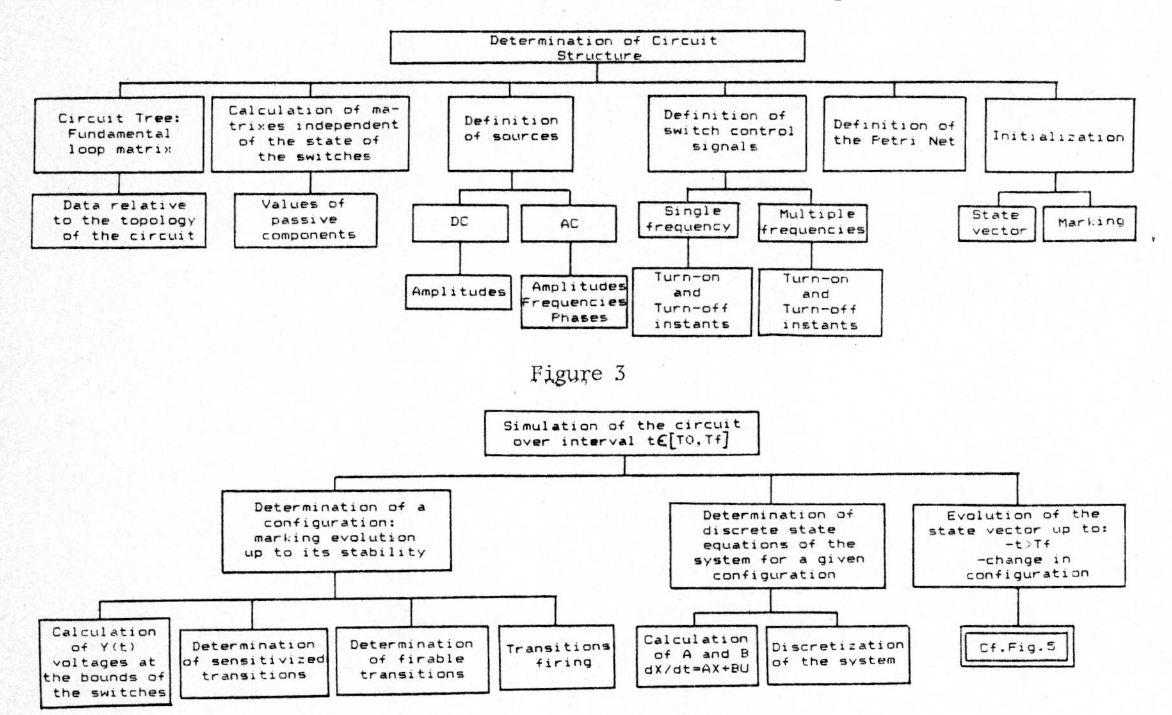

Figure 3

Figure 4

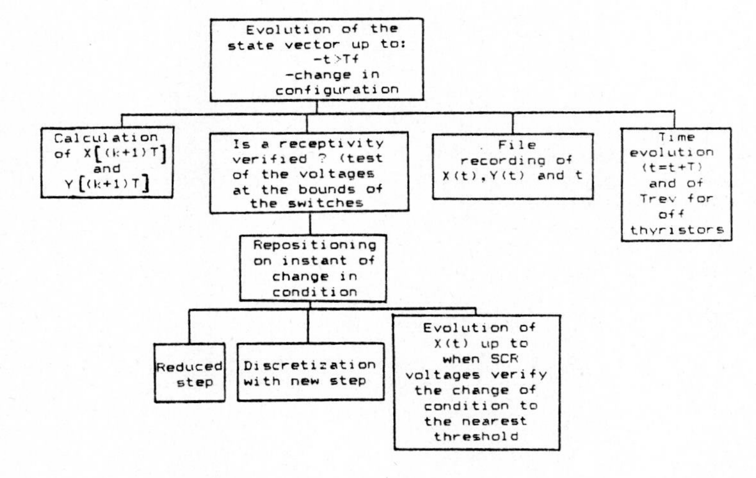

Figure 5

4. APPLICATION EXAMPLE

A natural self-adaptative chopper is presented as application (figure 6).

Figures 7 and 8 illustrate a transitory voltage regime showing a loss of control after the third functioning cycle. This fault is due to the fact that the reverse voltage bias time at the bounds of the main thyristor (Tp) becomes lower than its recovery time.

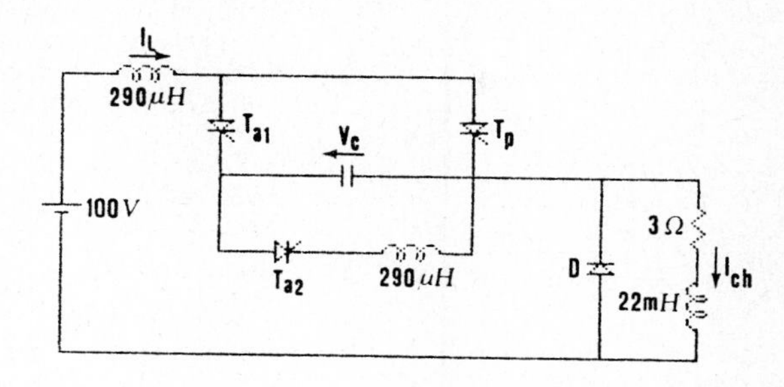

Figure 6

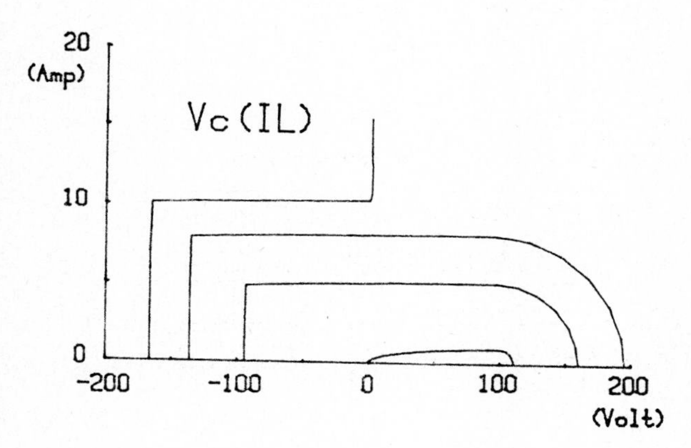

Figure 7

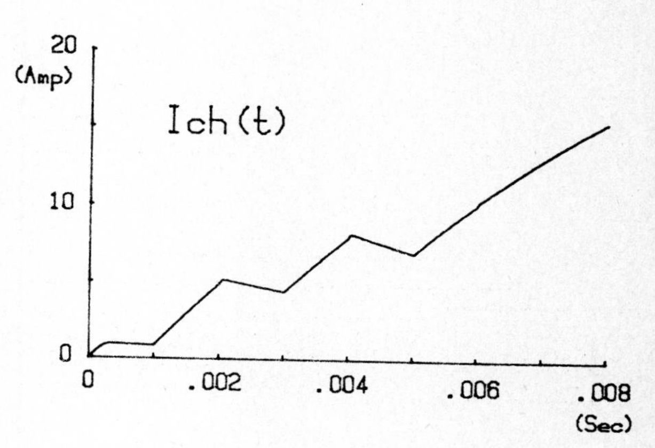

Figure 8

5. CONCLUSION

An example has demonstrated the aid which a global program simulation can bring to the study of a semi-conductor electric circuit. In addition, the use of a "generalized switch" within the framework of global simulations allows the study of all types of transitories and especially those conducive to a functioning fault (cataleptic and degradation faults).

This simulation method therefore seems to be a powerful tool in both the design and implementation of semi-conductor electric circuits functioning in commutation, as well as in the study of their dependability

6. REFERENCES

(1) REVANKAR G.N.
"Digital computation of SCR chopper circuits", IEEE Trans. Ind. Electron. Contr. Instrum., Vol. IECI-20, pp 20-23, Feb 1973

(2) REVANKAR G.N.
"Topological approach to thyristor-circuit analysis", Proceedings IEE, Vol. 120, n° 11, Nov. 1973

(3) RAJAGOPALAN V., SANKARA K.
"User's manual on topological analysis of power electronic converter systems : ATOSEC Version 1", Université du Québec à Trois Rivières, Canada

(4) LAKATOS L.
"A new method for simulating power semi-

conductor circuits", IEEE Trans. Ind. Electron. Contr. Instrum., Vol. IECI-26, n° 1, Feb. 79

(5) KUTMAN T.
"A method of digital computation for SCR circuits", IEEE Trans. Ind. Electron. Contr. Instrum., Vol. IECI-21, n° 2, pp 80-83, May 1974

(6) SCHONEK J.
"Simulation numérique des convertisseurs statiques. Elaboration d'un programme général. Application à la conception et l'optimisation des convertisseurs", Thèse de Docteur Ingénieur, INP Toulouse, 1977

(7) VALETTE R., COURVOISIER M.
"Systèmes de commandes en temps réel", Editions SCM, 1980

(8) THELLIEZ S.
"Pratique séquentielle et réseaux de Pétri" Editions Eyrolles, 1978

(9) CHERON Y.
"Application des règles de la dualité à la conception de nouveaux convertisseurs à transistors de puissance. Synthèse du thyristor dual. Domaine d'application", Thèse de Docteur Ingénieur, INP Toulouse, 1982

(10) BALABANIAN N., BICKART T.
"Electrical network theory", Wiley, New York, 1979

(11) TRANNOY B., FOCH H., METZ M.
"Autoadaptative forced commutation circuits for thyristor static converters", P.C.I. 81, Munich (Proceedings, pp 267-277)

Simulation in Engineering Sciences
J. Burger and Y. Jarny (eds.)
Elsevier Science Publishers B.V. (North-Holland)

NON LINEAR IDENTIFICATION, APPLICATION TO AN ELECTROMECHANICAL SYSTEM.

M. GAUTIER

IUT, 12 Rue de la Fonderie - 71200 LE CREUSOT - FRANCE.

This paper deals with a method of non linear identification withan application to the simulation of the input-ouput behaviour of a motor-generator rotating group. The process is characterized by a discrete time state affine system which is a polynomial extension of bilinear systems. Parameters of this models are determined with several linear models identified around each working point. The global non linear model can replace all the linear models and follow the dynamic and static evolution of the process over a wide range of operating conditions.

I - INTRODUCTION.

New techniques of non linear identification are applied to the modelization of the speed behaviour of a motor generator rotating group.
The steps of the method are as following :

- Identification of several linear representation models for several working points. Any linear identification method can be used.

- Characterization of the non linear system by a discrete time non linear state model.

- Identification of the parameters of the non linear model by a least squares method which uses the results of the first step.

II - DESCRIPTION OF THE PROCESS.

We are interested in the following system

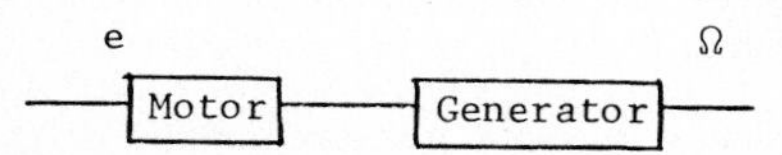

The generator supplies an autonomous load. $e(t)$ is the commandable input, $\Omega(t)$ is the rotor speed.
The particular configuration is studied : $y(t)$ is the output.

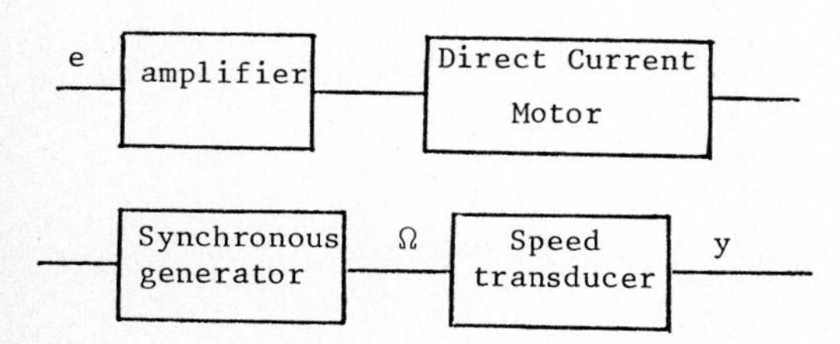

The power amplifier is a thyristor bridge which supplies the armature of a DC motor.
For small variations of Ω around Ω_o one can define an average power P_{ao} supplied by the generator and fixed by the impedance of the plant.
For one value of Pa_o, one can identify the step response of the process with that of a second order system.

$$G(s) = \frac{G_o \, \omega_o^2}{\omega_o^2 + 2m\,\omega_o s + s^2} = \frac{y(s)}{e(s)}$$

Yet, each parameter G_o, m, ω_o varies with P_{ao} and with the value $e = E$ of the step.
A simple discrete time model of this non linear model is required.

III - NON LINEAR DISCRETE-TIME REPRESENTATION MODEL.

III.1. Linear model about an operating point
The system controlled by a microprocessor can be represented by the block diagram.

e(t) / e_n → $B_o(s)$ → $G(s)$ → y(t) / y_n

T = sampling period, $e_n = e(nT)$, $y_n = y(nT)$

$B_o(s) = \frac{1}{s}(1 - e^{-Ts})$: zero order hold

$G_1 = B_o \times G$, is realized by the two order linear model

$$\begin{vmatrix} q_1 \\ q_2 \end{vmatrix}_{n+1} = \begin{vmatrix} \phi_{11} & \phi_{12} \\ \phi_{21} & \phi_{22} \end{vmatrix} \begin{vmatrix} q_1 \\ q_2 \end{vmatrix}_n + \begin{vmatrix} \Gamma_1 \\ \Gamma_2 \end{vmatrix} e_n$$

Or $\underline{q}_{n+1} = \phi\, \underline{q}_n + \Gamma\, e_n$

And $y_n = \begin{vmatrix} H_{11} & H_{12} \end{vmatrix} \underline{q}_n$ or $y_n = H\, \underline{q}_n$

$q_n = \begin{vmatrix} q_1 \\ q_2 \end{vmatrix}_n$ is the state space vector.

Matrices ϕ and Γ depend on e and $P_{ao} = \xi$.

Their coefficients are identified from step responses of the system : $e_n = E$, $n > o$.

A test number k is characterized by $\{ \xi_k, E_k \}$ and leads to

$\underline{q}_{n+1} = \phi_k\, \underline{q}_n + \Gamma_k E_k$, $y_n = H\, \underline{q}_n$

Matrices ϕ_k and Γ_k can be identified by any linear identification method.

In the present one input-one output case two simple ways are possible.

Using the z transform of $G_{1k}(s) = B_o(s) \times G_k(s)$, the recurrent equation giving y(nT) is calculated.

$$G_{1k}(z) = \frac{y(z)}{e(z)} = \frac{b_1^k z^{-1} + b_2^k z^{-2}}{1 + a_1^k z^{-1} + a_2^k z^{-2}}$$

and $y_n = b_1^k e_{n-1} + b_2^k e_{n-2} - a_1^k y_{n-1} - a_2^k y_{n-2}$

$e_n = E_k$, $n > o$, $e_n = o$, $n < o$.

Writing this equation for N values of n and with $N > 4$, one obtains a linear system of N equations, with 4 unknowns, which is resolved by a least squares method.

So, the parameters of G_{1k} are identified directly from the step sampled responses, without using $G_k(s)$.

Sometimes, $G_k(s)$ has been first identified, and parameters $\{ G_{ok}, \omega_{ok}, m_k \}$ are known.

The calculation of $G_{1k}(z)$ should be explicited and the parameters b_1, b_2, a_1, a_2 expressed by $\{ G_o, \omega_o, m \}$

So, with $T_r = \omega_o T$ and with $m > 1$:

$a_1 = -(e^{P_1 T_r} + e^{P_2 T_r})$, $P_1 = -(m + \sqrt{m^2 - 1})$

$a_2 = (e^{P_1 T_r} \times e^{P_2 T_r})$, $P_2 = -(m - \sqrt{m^2 - 1})$

The interest of this method, compared with that of direct identification of the discrete time model, should be examinated later (3.2).

Let $q_{1n} = y_n$ and $q_{2n} = a_2\, q_{1n-1} + b_2\, e_{n-1}$

The recurrent equation :

$y_n = b_1^k e_{n-1} + b_2^k e_{n-2} - a_1^k y_{n-1} - a_2^k y_{n-2}$

should be written :

$$\begin{vmatrix} q_1 \\ q_2 \end{vmatrix}_{n+1} = \begin{vmatrix} -a_1^k & 1 \\ -a_2^k & 0 \end{vmatrix} \begin{vmatrix} q_1 \\ q_2 \end{vmatrix}_n + \begin{vmatrix} b_1^k \\ b_2^k \end{vmatrix} e_n$$

$$y_n = \begin{vmatrix} 1 & 0 \end{vmatrix} \begin{vmatrix} q_1 \\ q_2 \end{vmatrix}_n$$

Then matrices can be expressed :

$$\phi_k = \begin{vmatrix} -a_1^k & 1 \\ -a_2^k & 0 \end{vmatrix} \quad \Gamma_k = \begin{vmatrix} b_1^k \\ b_2^k \end{vmatrix} \quad H = \begin{vmatrix} 1 & 0 \end{vmatrix}$$

A second possible way consists in choosing state variables :

$q_{1n} = y(nT)$, $q_{2n} = \mathring{y}(nT)$

Then ϕ has the expression

$$\phi = \frac{1}{p_2 - p_1} \begin{vmatrix} p_2 e^{p_1 T_r} - p_1 e^{p_2 T_r} , & \dfrac{e^{p_2 T_r} - e^{p_1 T_r}}{\omega_o} \\ \omega_o (e^{p_1 T_r} - e^{p_2 T_r}) , & p_2 e^{p_2 T_r} - p_1 e^{p_1 T_r} \end{vmatrix}$$

with p_1, p_2, T_r défined former ($m > 1$)

ϕ and Γ from the discrete time model should be calculated after the identification of the continuous model G(s), i-e $\{ G_o, \omega_o, m \}$.

III.2. Non linear representation model.

- Characterization.

The former linear model should be written as an équivalent bilinear formulation.

$$\begin{vmatrix} q_1 \\ q_2 \\ 1 \end{vmatrix}_{n+1} = \begin{vmatrix} \phi_{11} & \phi_{12} & 0 \\ \phi_{21} & \phi_{22} & 0 \\ 0 & 0 & 1 \end{vmatrix} + \begin{vmatrix} 0 & 0 & \Gamma_1 \\ 0 & 0 & \Gamma_2 \\ 0 & 0 & 0 \end{vmatrix} e_n \begin{vmatrix} q_1 \\ q_2 \\ 1 \end{vmatrix}_n$$

or $\underline{q}'_{n+1} = \left| A'_o + A'_1 e_n \right| \underline{q}'_n$ and $y_n = |1\ 0\ 0| \underline{q}'_n = H \underline{q}'_n$

The bilinear model becomes :

$$\begin{vmatrix} q_1 \\ q_2 \\ 1 \end{vmatrix}_{n+1} = \begin{vmatrix} \phi_{11} & \phi & \Gamma_1 e_n \\ \phi_{21} & \phi_{22} & \Gamma_1 e_n \\ 0 & 0 & 1 \end{vmatrix} \begin{vmatrix} q_1 \\ q_2 \\ 1 \end{vmatrix}_n$$

or $\underline{q}'_{n+1} = D_k \underline{q}'_n$

For test number k defined by the values of the step $e_n = E_k$ and the average power $p_{ao} = \xi_k$, a matrix is obtained :

$$D_k = \begin{vmatrix} \phi_{11}^k & \phi_{12}^k & \Gamma_1^k & Ek \\ \phi_{21}^k & \phi_{22}^k & \Gamma_2^k & Ek \\ 0 & 0 & & 1 \end{vmatrix}$$

This writing is quite formal, but is interesting because the varying coefficients of the matrix D_k can be approximated by polynomials fonctions in the u_r which are monomials in $\{ \xi, e_n \}$.

Example :

$$\underline{u} = \left[1, \xi, \xi^2, e_n, e_n\xi, e_n\xi^2\right] = |u_o, u_1, \ldots, u_5|$$

It becomes :

$$\phi_{11}^k(\xi_k, E_k) = a_{11}^1 \xi_k + a_{11}^2 \xi_k^2 + a_{11}^3 E_k + a_{11}^4 \xi_k E_k + a_{11}^5 E_k \xi_k^2 + a_{11}^o$$

$$\phi_{i,j}^k(\xi_k\ E_k) = a_{ij}^o + a_{ij}^1 \xi_k + \ldots a_{ij}^5 E_k \xi_k^2$$

$$\Gamma_1^k E_k(\xi_k, E_k) = a_{13}^o + a_{13}^1 \xi_k + \ldots + a_{13}^5 E_k \xi_k^2$$

Let Ar be the matrix defined by the coefficients a_{ij}^r

with i = (1, 2, 3) ; j = (1, 2, 3) ; r = (0 à 5)

$$A_o = \begin{vmatrix} a_{11}^o & a_{12}^o & a_{13}^o \\ a_{21}^o & a_{22}^o & a_{23}^o \\ 0 & 0 & 1 \end{vmatrix} \quad Ar = \begin{vmatrix} a_{11}^r & a_{12}^r & a_{13}^r \\ a_{21}^r & a_{22}^r & a_{23}^r \\ 0 & 0 & 0 \end{vmatrix} r = (1 à 5)$$

$$\underline{q}'_n + 1 = \left| A_o + \sum_{r=1}^{5} Ar\, u_r \right| \underline{q}'_n, \quad y_n = |1\ 0\ 0|\ \underline{q}'_n$$

As can be seen, a generalization of bilinear systems is obtained with an affine state form and new inputs u_r which are monomials in $\{ \xi, e_n \}$.

The main problem of this identification method is the determination of the products of inputs u_r.

The knowledge of $\{ G_{ok}, \omega_{ok}, m_k \}$ can greatly help us because the coefficients of the matrices ϕ_k and Γ_k depend on them. By using limited developments one can deduce the inputs u_r.

Example :

$$a_2^k = e^{-2mk\ \omega_{ok} T} \simeq 1 - 2\, mk\, \omega_{ok} T + 2(m_k\, \omega_{ok}\, T)^2$$

$$m_k = m_{11} + m_{12}\, \xi_k + m_{13}\, E_k, \quad \omega_{ok} = \text{constant} = \omega_o$$

$$a_2^k = 1 - 2m_{11}\, \omega_o T - 2m_{12}\, \omega_o\, T\, \xi_k - 2m_{13}\, \omega_o\, T\, E_k$$
$$+ 2\,\omega_o^2\, T^2 \times (m_{11}^2 + m_{12}^2 \xi_k^2 + m_{13}^2\, E_k^2 + 2m_{11}\, m_{12}$$
$$\xi_k + 2m_{12}\, m_{13}\, \xi_k\, E_k + \ldots) = a_{10} + a_{11}\ \xi_k$$
$$+ a_{12} \xi_k^2 + a_{13} E_k + a_{14} E_k \xi_k\ a_{15} E_k \xi_k^2$$

Then one can choose $\underline{u} = \left| 1, \xi, \xi^2, e_n, e_n\xi, e_n\xi^2 \right|$

If G(s) is unknown or if limited developments are too complicated, $\underline{u}$ must be choosen a priori. The efficiency of the choice will be tested by simulation.

- identification of matrices Ar.

For each test k, $\{ \xi_k, E_k \}$, linear and non linar models must be identical.

$$\underline{q}'_n + 1 = D_k\, \underline{q}'_n = \left(\sum_{r=o}^{5} Ar\ u_r \right) \underline{q}'_n$$

Six linear equations in a_{ij} are obtained

$$d_j^k = \sum_{r=o}^{5} a_{ij}\ u_r^k \qquad i = (1,2)\ ,\ j(1,2,3).$$

D_k results of the linear identification (3-1)

Ne tests are worked out, with $N_e > 6$.

For each (i,j), a_{ij}^r, r = o to 5, is the solution of a linear least squares problem

$$\begin{vmatrix} 1, \xi_1, \xi_1^2, E_1, E_1\xi_1^2 \\ \\ 1, \xi_{Ne} \cdots\cdots E_{Ne}\, \xi_{Ne}^2 \end{vmatrix} \begin{vmatrix} a_{ij}^o \\ \\ a_{ij}^5 \end{vmatrix} = \begin{vmatrix} d_{ij} \\ \\ d_{ij}^{N_e} \end{vmatrix}$$

Six linear systems have to be solved giving the a_{ij}.

Remark :

The non linear obtained model generalizes the linear model with variable parameters. Indeed, in this case ϕ_k and Γ_k only depend on ξ_k. For each working point ξ_k, the process is linear in the control E_k. In our formulation, the matrices Ar would take the following particular form :

$$A_o = \begin{vmatrix} a_{11}^o & a_{12}^o & 0 \\ a_{21}^o & a_{22}^o & 0 \\ 0 & 0 & 1 \end{vmatrix}, Ar = \begin{vmatrix} a_{11}^r & a_{12}^r & 0 \\ a_{21}^r & a_{22}^r & 0 \\ 0 & 0 & 0 \end{vmatrix}, \quad r = (1,2)$$

$$A_r = \begin{vmatrix} 0 & 0 & a_{13}^r \\ 0 & 0 & a_{23}^r \\ 0 & 0 & 0 \end{vmatrix}, \quad r = (3, 4, 5)$$

$$\text{with } B_r = \begin{vmatrix} a_{11}^r & a_{12}^r \\ a_{21}^r & a_{22}^r \end{vmatrix} \quad r = (0,1,2)$$

$$\text{and } B_r = \begin{vmatrix} a_{13}^r \\ a_{23}^r \end{vmatrix}, \quad r = (3,4,5)$$

So, the initial model would become

$$\begin{vmatrix} q_1 \\ q_2 \end{vmatrix}_{n+1} = \left| B_o + B_1 \xi_k + B_2 \xi_k^2 \right| \begin{vmatrix} q_1 \\ q_2 \end{vmatrix}_n$$

$$+ \left| B_3 + B_4 \xi_k + B_5 \xi_k^2 \right| e_n$$

That is a linear model with varying parameters.

$$\underline{q}_{n+1} = \phi(\xi_k)\, \underline{q}_n + \Gamma(\xi_k)\, e_n , \; y_n = |10|\, q_n$$

IV. RESULTS AND SIMULATION.

The 2 models presented in (3.1) have been identified in order to test the influence of the linear model structure upon the validity of the final non linear model.
We have pointed out 8 tests of step responses on a 4 KVA generator for ξ_k = 0%, 30%, 60%, 100% of the nominal power rating and, for each value of ξ_k, two amplitude step E_k = 1V and 2 V corresponding to a step speed variation of 10% around the nominal speed Ω_o = 1500 rpm when ξ_k = 60%.
For every test, $\{ G_{ok}, \omega_{ok}, m_k \}$ are identified on a second order transfert function Gk(s).

The matrices ϕ_k and Γ_k have been calculated for k = 1 to 8, and then matrices

Ar have been calculated for r = 0 to 5. Step responses of the process and model n° 2 have been drawn on the figures.

The model issued from z-transform(model n°1) and that from the output and its derivation (model n°2) give excellent results. In these two cases the gain variation G_o from 6 to 4.35 is simulated with a maximum difference of 10%. For the dynamic characteristics and even for a large variation of time responses from 0,5 s to 18 s the maximum relative error does n't exceed 10% at all.
The Numerical results have been pointed out for matrices Ar.

V - CONCLUSION.

The non linear model obtained can simulate the large variations of the static and dynamic characteristics (rate from 1 to 40) of the process.
This method suits very well for modelization of non linear processes, linearizable around an operating point.
The possibility to use simple transfert function (1st, 2nd order), which are common for engineers, is a major advantage for this method.
Finally, the characterization with discrete time state model permits a fast and simple simulation on microprocessor.

BIBLIOGRAPHY

[1] S.A.Billings-Identification of nonlinear systems, a survey, IEE Proc.,127,(1980) 272-285.

[2] H.Dang Van Mien and D. Normand-Cyrot - A nonlinear identification by using a polynomial generalization of regular systems, in "Advances in Cont."(D. G.Lainiotis and N.S. Tzannes eds.),300-307,Reidel,Dordrecht,1980.

[3] H. Dang Van Mien and D. Normand-Cyrot - Non linear state-affine identification methods, applications to electrical power plants.Proc. IFAC Symp. Aut. Control, Power Generation, Distribution and Protection Pretoria (1980), 449-462.

[4] D. Normand-Cyrot - Utilisation de certaines familles algébriques de systèmes non linéaires à quelques problèmes de filtrage et d'identification. Thèse de 3e cycle, Université Paris VII, 1978.

[5] D. Normand-Cyrot -Identification par systèmes à état-affine et application aux centrales électriques, "Outils et Modèles Mathématiques pour l'Automatique, l'analyse de systèmes et le Traitement du Signal" (I.D. Landau ed.)409-417,CNRS, Paris, 1981.

[6] E.D. Sontag-Realization theory of discrete-time nonlinear systems. Part I : The bounded case, IEEE Trans. Circuits and Syst. 26, (1979), 342-356.

[7] M.Fliess and D. Normand-Cyrot -La propriété d'approximation des systèmes réguliers (ou bilinéaires) "Outils et Modèles Mathématiques pour l'Automatique, l'Analyse des Systèmes et le Traitement du Signal", (I.D. Landau ed.), 379-384,CNRS, Paris, 1981.

[8] R.R.Mohler -Bilinear control processes with applications to engineering, ecology and medicine, Academic Press, New-York, 1973.

[9] H. Espana and I.D. Landau : Reduced order bilinear models for distillation columns, Automatica, 14, 345-355, 1978.

[10] S. Beghelli and R. Guidorzi : Bilinear systems identification from input-output sequences. IV IFAC Symp. On Identification and System Parameter Estimation, Tbilisi, Sept. 1976.

[11] A. Bellini and G. Figalli -A bilinear observer of the state of the induction machine, Ricerche di Automatica,Vol.9,n°1, June 1978.

MATRICES A_r

Sampling Period : T = 0,05s

Model n° 1

$$A_o = \begin{vmatrix} 1 & 1 & 0,468.10^{-2} \\ -0,015 & 0 & 0,784.10^{-2} \\ 0 & 0 & 1 \end{vmatrix}$$

$$A_1 = \begin{vmatrix} -0,146 & 0 & -0,0644 \\ 0,147 & 0 & -0,046 \\ 0 & 0 & 0 \end{vmatrix}$$

$$A_2 = \begin{vmatrix} 0,070 & 0 & 0,0172 \\ -0,0871 & 0 & 0,495.10^{-2} \\ 0 & 0 & 0 \end{vmatrix}$$

$$A_3 = \begin{vmatrix} -0,190.10^{-2} & 0 & 0,0393 \\ 0,311.10^{-2} & 0 & -0,738.10^{-2} \\ 0 & 0 & 0 \end{vmatrix}$$

$$A_4 = \begin{vmatrix} -0,0116 & 0 & 0,170 \\ 0,219.10^{-2} & 0 & 0,0199 \\ 0 & 0 & 0 \end{vmatrix}$$

$$A_5 = \begin{vmatrix} 0,0158 & 0 & -0,755.10^{-2} \\ -0,0149 & 0 & 0,0426 \\ 0 & 0 & 0 \end{vmatrix}$$

Modèl n° 2

$$A_o = \begin{bmatrix} 0{,}992 & 0{,}169.10^{-2} & 0{,}467.10^{-2} \\ -0{,}169 & 0{,}0139 & 0{,}243 \\ 0 & 0 & 1 \end{bmatrix} \qquad A_3 = \begin{bmatrix} 0{,}489.10^{-3} & -0{,}234.10^{-3} & 0{,}0393 \\ 0{,}0234 & -0{,}24.10^{-2} & 0{,}602 \\ 0 & 0 & 0 \end{bmatrix}$$

$$A_1 = \begin{bmatrix} -0{,}0113 & -0{,}169.10^{-3} & -0{,}0644 \\ 0{,}0169 & -0{,}135 & -2{,}16 \\ 0 & 0 & 0 \end{bmatrix} \qquad A_4 = \begin{bmatrix} -0{,}594.10^{-2} & 0{,}184.10^{-2} & 0{,}170 \\ -0{,}184 & -0{,}569.10^{-2} & 3{,}78 \\ 0 & 0 & 0 \end{bmatrix}$$

$$A_2 = \begin{bmatrix} -0{,}563.10^{-2} & 0{,}328.10^{-2} & 0{,}0172 \\ -0{,}328 & 0{,}0759 & 0{,}433 \\ 0 & 0 & 0 \end{bmatrix} \qquad A_5 = \begin{bmatrix} 0{,}112.10^{-2} & -0{,}163.10^{-3} & -0{,}755.10^{-2} \\ 0{,}0163 & 0{,}0147 & 0{,}669 \\ 0 & 0 & 0 \end{bmatrix}$$

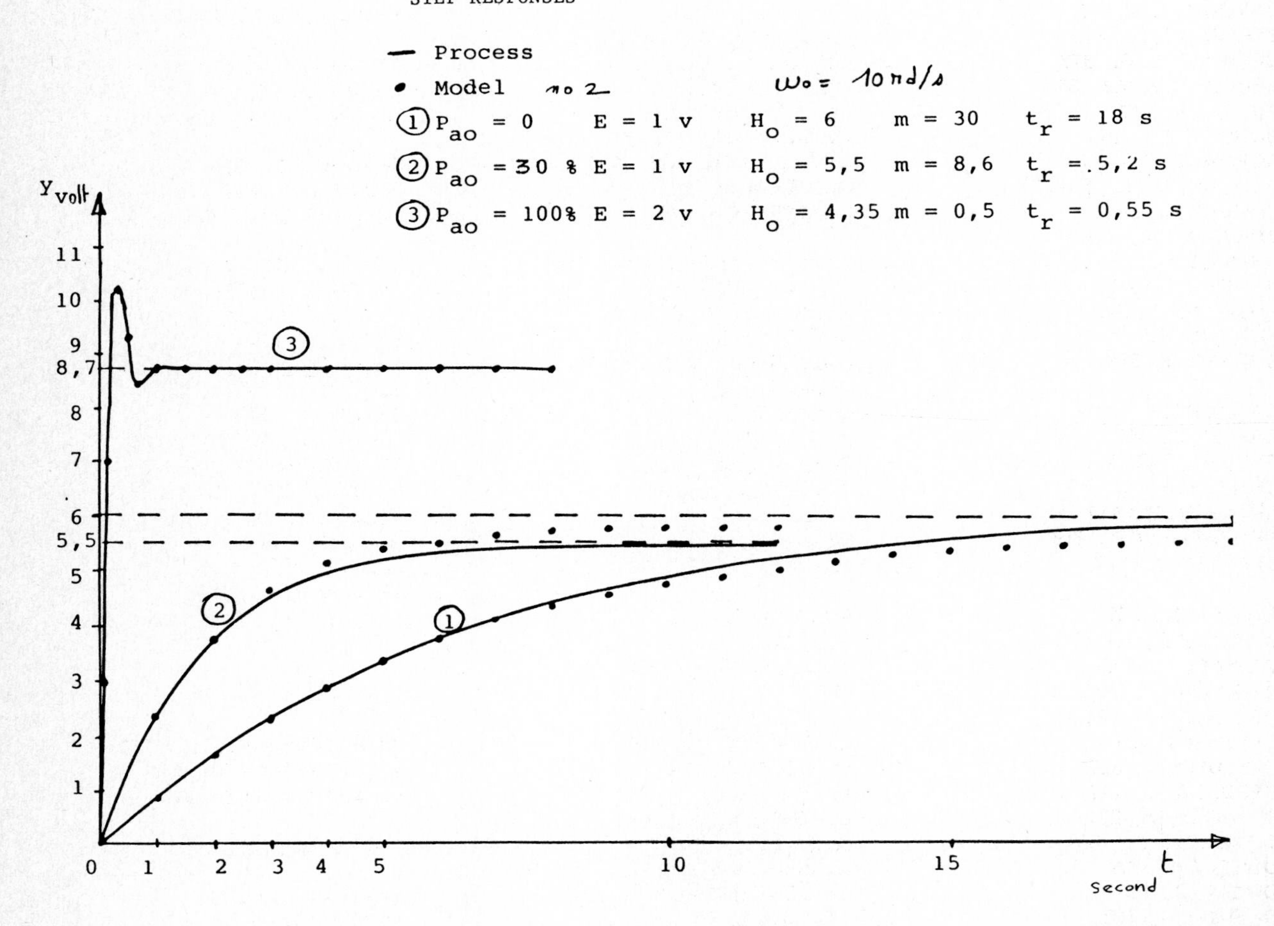

AUTHOR INDEX

AHMED, M.E., 291
AMBRÓZY, D., 207
ANBE, Y., 299
ANDERSSON, N., 249
ARAKAWA, T., 299

BABARY, J.P., 213
BARBARINO, A., 233
BARBEZ, J.M., 337
BARTHELMES, M., 73
BAYOUMI, M.M., 291
BERMÚDEZ, A., 305
BERWAERTS, G., 377
BERWAERTS, V.J., 377
BIELEC, P., 147
BORDRY, F., 425
BORNARD, G., 325
BOURNONVILLE, F., 357
BOUVET, C., 343
BREBBIA, C.A., 109
BRESSLER, P., 73
BROEKX, J., 377
BUBENKO, J., 249
BÜNZ, D., 73
BURGER, J., 373

CAMACHO, E.F., 55
CASTET, J., 387
CELLIER, F.E., 3
CHARTRES, J.M., 337
CHAUDET, R., 411
CHUBAROV, E.P., 219
CLAESEN, E., 377

DANILA, N., 239
DE BUYSER, D., 153
DE WAEL, L., 153
DEGONDE, M., 395
DEGRAEVE, F., 263
DEKKER, L., 21
DESCUSSE, J., 357
DESOYER, K., 417
DI PAOLA, M., 271
DONCARLI, C., 185
DOST, M.H., 139
DUQUÉ, C., 227
DURANY, J., 305

ELARABY, M.E., 165
EL KADER, A.A., 165

FOCH, H., 425
FORTUNA, L., 87, 233
FRANQUELO, L.G., 55
FURUTA, K., 279

GALLO, A., 87, 233
GARCIN, G., 263
GAUTIER, M., 431
GIRARDEAU, D., 343
GITTON, D., 185
GOREZ, R., 227
GOUYON, J.P., 213
GRANDIDIER, J.Y., 343
GRABOWIECKI, K.A., 133
GUESDON, M., 263
GÜTSCHOW, K., 73

HABIBOLLAHZADEH, H., 249
HAJ NASSAR, R., 147
HEEGER, J., 73
den HOLLANDER, J.A., 103
HOOGSTRATEN, J.A., 103,401
HUMO, E., 367
HULME, A.J., 257

INOU, N., 179
IONESCU, D.C., 239

JARNY, Y., 373
JOOSTEN, S., 127

KOBAYAHSI, S., 299
KOPACEK, P., 417
KUBYSHKIN, V.A., 219

LA CAVA, M., 87, 233
LARSSON, G., 249
LE GOFF, P., 33
LE LETTY, L., 93
LEMKE, H.-J., 73
LENNGREN, M., 121
LOPEZ-CORONADA, J., 93
LOZANO, J., 55

MARCHENKO, V.M., 193
MARCOCCI, L., 61
MASLOWSKI, A., 81
MEIZEL, D., 147, 337
MERABET, A.A., 67
METZ, M., 425
METZGER, M., 319
MITA, T., 343
van de MOESDIJK, G.A.J., 103, 401
MONSION, M., 343
MUSCOLINO, G., 271

OHYAMA, Y., 279
O'SHEA, J., 411
OUSTALOUP, A., 199

PERROUD, M., 325
POPOVIĆ, M., 367
POTERASU, V.P., 271

RIEPE, J., 285
ROODA, J.E., 127

SAITO, M., 299
SCHNACK, E., 311
SEGERS, M., 377
SIBONY, M., 173
SINHA, P.K., 257
SJELVGREN, D., 249
SOUZA LEAO, J., 213
SPELTA, S., 61
STAN, N., 239
STEVERLINCK, J., 377
SWIERNIAK, A., 159

TROCH, I., 417

UMETANI, Y., 179

VANSTEENKISTE, G.C., 153
VICHNEVETSKY, R., 41
VIGNESWARAN, S., 331

WHITE, R.C., 67

YAMAMOTO, Y., 121
YAMONO, O., 279